VOLUME ONE HUNDRED AND FORTY TWO

CURRENT TOPICS IN DEVELOPMENTAL BIOLOGY

Molecular Mechanisms of Neural Development and Insights into Disease

CURRENT TOPICS IN DEVELOPMENTAL BIOLOGY

"A meeting-ground for critical review and discussion of developmental processes"
A.A. Moscona and Alberto Monroy (Volume 1, 1966)

VOLUME ONE HUNDRED AND FORTY TWO

CURRENT TOPICS IN DEVELOPMENTAL BIOLOGY

Molecular Mechanisms of Neural Development and Insights into Disease

Edited by

GREG J. BASHAW
Department of Neuroscience, Perelman School of Medicine, University of Pennsylvania, Philadelphia, PA, United States

ELSEVIER

ACADEMIC PRESS
An imprint of Elsevier

Academic Press is an imprint of Elsevier
50 Hampshire Street, 5th Floor, Cambridge, MA 02139, United States
525 B Street, Suite 1650, San Diego, CA 92101, United States
The Boulevard, Langford Lane, Kidlington, Oxford OX5 1GB, United Kingdom
125 London Wall, London, EC2Y 5AS, United Kingdom

First edition 2021

Notices
Knowledge and best practice in this field are constantly changing. As new research and experience broaden our understanding, changes in research methods, professional practices, or medical treatment may become necessary.

Practitioners and researchers must always rely on their own experience and knowledge in evaluating and using any information, methods, compounds, or experiments described herein. In using such information or methods they should be mindful of their own safety and the safety of others, including parties for whom they have a professional responsibility.

To the fullest extent of the law, neither the Publisher nor the authors, contributors, or editors, assume any liability for any injury and/or damage to persons or property as a matter of products liability, negligence or otherwise, or from any use or operation of any methods, products, instructions, or ideas contained in the material herein.

ISBN: 978-0-12-815281-2
ISSN: 0070-2153

For information on all Academic Press publications
visit our website at https://www.elsevier.com/books-and-journals

Publisher: Zoe Kruze
Editorial Project Manager: Shellie Bryant
Production Project Manager: Denny Mansingh
Cover Designer: Greg Harris

Typeset by SPi Global, India

Contents

Contributors

Sandy Alvarez
Department of Neurobiology; Molecular Biology Interdepartmental Doctoral Program, University of California, Los Angeles, CA, United States

E.S. Anton
UNC Neuroscience Center and the Department of Cell and Molecular Physiology, University of North Carolina School of Medicine, Chapel Hill, NC, United States

Greg J. Bashaw
Department of Neuroscience, Perelman School of Medicine, University of Pennsylvania, Philadelphia, PA, United States

Thomas Biederer
Department of Neurology, Yale School of Medicine, New Haven, CT, United States

Allison M. Bond
Department of Neuroscience and Mahoney Institute for Neurosciences, Perelman School of Medicine, University of Pennsylvania, Philadelphia, PA, United States

Víctor Borrell
Instituto de Neurociencias, Consejo Superior de Investigaciones Científicas & Universidad Miguel Hernández, Sant Joan d'Alacant, Spain

Caley J. Burrus Lane
Department of Cell Biology; Department of Neurobiology, Duke University Medical Center, Durham, NC, United States

Samantha J. Butler
Department of Neurobiology; Eli and Edythe Broad Center of Regenerative Medicine and Stem Cell Research, University of California, Los Angeles, CA, United States

Karina Chaudhari
Department of Neuroscience, Perelman School of Medicine, University of Pennsylvania, Philadelphia, PA, United States

Dhrubajyoti Chowdhury
Department of Neurology, Yale School of Medicine, New Haven, CT, United States

Suna L. Cranfill
Department of Neuroscience, Perelman School of Medicine, University of Pennsylvania, Philadelphia, PA, United States

Cagla Eroglu
Department of Cell Biology; Department of Neurobiology, Duke University Medical Center; Duke Institute for Brain Sciences; Regeneration Next Initiative, Duke University, Durham, NC, United States

Magdalena Götz
Institute for Stem Cell Research, Helmholtz Zentrum München & Biomedical Center, Ludwig-Maximilians Universitaet, Planegg-Martinsried, Germany

Jiami Guo
Hotchkiss Brain Institute, Alberta Children's Hospital Research Institute, and the Department of Cell Biology and Anatomy, University of Calgary, Calgary, AB, Canada

Ellie S. Heckscher
Department of Molecular Genetics and Cell Biology; Program in Cell and Molecular Biology; Committee on Development, Regeneration, and Stem Cell Biology; Grossman Institute for Neuroscience, University of Chicago, Chicago, IL, United States

Fadi Jacob
Department of Neuroscience and Mahoney Institute for Neurosciences, Perelman School of Medicine, University of Pennsylvania, Philadelphia, PA; The Solomon H. Snyder Department of Neuroscience; Medical Scientist Training Program, Johns Hopkins University School of Medicine, Baltimore, MD, United States

Kristy O. Johnson
Center for Neuroscience Research, Children's National Research Institute; Institute for Biomedical Sciences, The George Washington University School of Medicine, Washington, DC, United States

Julie L. Lefebvre
Program for Neurosciences and Mental Health, Hospital for Sick Children; Department of Molecular Genetics, University of Toronto, Toronto, ON, Canada

Siling Liu
UNC Neuroscience Center and the Department of Cell and Molecular Physiology, University of North Carolina School of Medicine, Chapel Hill, NC, United States

Wenqin Luo
Department of Neuroscience, Perelman School of Medicine, University of Pennsylvania, Philadelphia, PA, United States

Julia L. Meng
Department of Molecular Genetics and Cell Biology; Program in Cell and Molecular Biology, University of Chicago, Chicago, IL, United States

Guo-li Ming
Department of Neuroscience and Mahoney Institute for Neurosciences; Department of Cell and Developmental Biology; Institute for Regenerative Medicine; Department of Psychiatry, Perelman School of Medicine, University of Pennsylvania, Philadelphia, PA, United States

Jordan G. Schnoll
Department of Neuroscience and Mahoney Institute for Neurosciences, Perelman School of Medicine, University of Pennsylvania, Philadelphia, PA, United States

Jeremy Simon
UNC Neuroscience Center and the Department of Cell and Molecular Physiology, University of North Carolina School of Medicine, Chapel Hill, NC, United States

Hongjun Song
Department of Neuroscience and Mahoney Institute for Neurosciences; Department of Cell and Developmental Biology; Institute for Regenerative Medicine; The Epigenetics Institute, Perelman School of Medicine, University of Pennsylvania, Philadelphia, PA, United States

Christabel X. Tan
Department of Cell Biology, Duke University Medical Center, Durham, NC, United States

Jason W. Triplett
Center for Neuroscience Research, Children's National Research Institute; Department of Pediatrics, The George Washington University School of Medicine, Washington, DC, United States

Mia X. Trupiano
UNC Neuroscience Center and the Department of Cell and Molecular Physiology, University of North Carolina School of Medicine, Chapel Hill, NC, United States

Supraja G. Varadarajan
Department of Neurobiology, Stanford University, Palo Alto, CA, United States

Ana Villalba
Instituto de Neurociencias, Consejo Superior de Investigaciones Científicas & Universidad Miguel Hernández, Sant Joan d'Alacant, Spain

Katherine Watters
Department of Neurology, Yale School of Medicine, New Haven, CT; Neuroscience Graduate Program, Tufts University School of Medicine, Boston, MA, United States

Yixin Zang
Department of Neuroscience, Perelman School of Medicine, University of Pennsylvania, Philadelphia, PA, United States

Preface

"The brain ... is a complicated, intricately woven tissue, like nothing else we know of in the universe, but it is composed of cells, as any tissue is. They are, to be sure, highly specialized cells, but they function according to the laws that govern any other cells. Their electrical and chemical signals can be detected, recorded and interpreted and their chemicals can be identified; the connections that constitute the brain's woven feltwork can be mapped."

David H. Hubel

The development of the nervous system is a fantastically complex process that depends on the precise orchestration of a series of developmental events, beginning with the specification of the right kinds and numbers of neurons and glia, and culminating with their organization and refinement into interconnected networks to form functional neural circuits. The coordinated activity of these circuits underlies our ability to perceive, move, think, feel, learn, and remember.

The process of neural development can be divided into a number of discrete, but overlapping, steps including neurogenesis, the specification of distinct types of neurons and glia, neuronal migration, axon guidance and dendrite morphogenesis, synapse formation, and activity-dependent refinement of synaptic connections. Studies using diverse experimental models and approaches have provided major insights into these developmental events; however, the mechanistic understanding of any one of these processes at the molecular level is still far from complete. Not surprisingly, disruption of these developmental events can lead to a myriad of neural developmental disorders, and knowledge gained by studying disease mechanisms can also enhance our knowledge of normal development. Indeed, important insights into the pathogenesis of neural developmental disorders originate from the study of normal development in a range of model organisms. More recently, induced pluripotent stem cells and sophisticated tissue culture techniques, complemented by the study of human mutations in animal models, have extended our understanding of these processes in humans.

12 chapters in this volume of *Current Topics in Developmental Biology* highlight the most recent advances in unraveling the molecular mechanisms of nervous system development, and how this work has enriched our understanding of developmental disorders. To cover different areas of developmental neurobiology, I recruited a diverse group of emerging or

established leaders to provide snapshots of rapidly developing areas of investigation and the essential unanswered questions in their respective fields. I have given free rein to the authors to focus their chapters however they chose, and for this reason, not all of the chapters are equally weighted to the coverage of both normal development and developmental disorders. I also encouraged investigators to include their current graduate students and postdoctoral fellows as authors on their chapters and am pleased that so many took this opportunity.

The volume opens with three chapters that focus on the development of the mammalian cerebral cortex, with a particular emphasis on neurogenesis and the diversification of cell types, and how disruptions in different aspects of normal cortical development contribute to disease. In the first chapter, Villalba, Götz, and Borrell review the current state of the art in the regulation of cortical neurogenesis. Their chapter provides a historical perspective to the field and discusses how genetic regulatory mechanisms define cell types and lineage relationships in the cortex, and how in turn deciphering these mechanisms may illuminate the understanding of the evolution of cortical neurogenesis. In the second chapter, Bond, Ming, and Song explore the origins of adult neural stem cells (NSCs) in the mammalian brain. They discuss how defining the factors and mechanisms that shape the adult NSC pool contribute to the understanding of the remarkable plasticity of the adult brain and form the foundation for the capacity for life-long neurogenesis. In the third chapter, Liu, Trupiano, Simon, Guo, and Anton provide an overview of the diverse and essential roles of primary cilia in cortical development and disorders. Their chapter reviews the intricate subcellular machinery that bestows on the primary cilium the ability to act as a critical signaling hub that integrates developmental and homeostatic signals to contribute to the normal development and function of the brain.

After the right numbers of and types of neurons and glia are specified, they must form an intricate set of connections to establish circuits. The second group of four chapters includes reviews of cellular and molecular mechanisms of axon guidance and dendrite morphogenesis, and how in turn these events allow for the precise assembly of neural circuits. In the fourth chapter, Zang, Chaudhari, and I, review current insights into how axon guidance pathways are spatially and temporally regulated and how guidance receptors signal to regulate attraction and repulsion in response to extracellular signals. In Chapter 5, Alvarez, Varadarajan, and Butler review the current understanding of the cellular and molecular mechanisms that control axon guidance of commissural neurons in the developing spinal cord.

This chapter highlights the fact that even in arguably the most extensively studied axon guidance event, fundamental new discoveries are still being made. In Chapter 6, Lefebvre turns our attention to the mechanisms that mediate dendrite morphogenesis, drawing examples from work in multiple experimental systems that highlight key concepts in how proper shaping and elaboration of dendritic territories underlies information flow in neural circuits. In Chapter 7, Johnson and Triplett highlight the mechanisms of axon guidance and targeting in the context of building and aligning sensory maps in the visual system. Their review describes both activity-independent and -dependent mechanisms that allow the retina, superior colliculus, and cortex to construct aligned representations of visual space.

Once axons and dendrites form, navigate to the correct regions, and acquire their complex arborization patterns, they must form synapses with the correct partner neurons. The precise recognition and choice of synaptic partners are made even more challenging to understand, because of the fact that the "correct" partners are intermingled with a large number of incorrect targets in neurite dense neuropil structures. The next two chapters in the volume focus on the mechanisms of synapse formation and maturation, with an emphasis on the critical roles of diverse families of conserved synaptic adhesion molecules and the essential functions that are mediated by astrocyte–neuronal interactions to regulate these events. In Chapter 8, Chowdhury, Watters, and Biederer catalog the large repertoire of trans-synaptic adhesion molecules, and highlight the mechanisms that allow these molecules to control initial wiring decisions, to reject incorrect partner choices, to promote maturation, and to contribute to synaptic pruning. In Chapter 9, Tan, Lane, and Eroglu describe the growing body of literature that has identified key roles for astrocytes in providing instructive signals to promote synapse formation and maturation. This chapter illustrates the importance of bidirectional adhesion and secretion-based interactions between neurons and glia that underpin normal synapse formation and function. Both chapters also discuss the implications of perturbing these developmental mechanisms for neural disease.

In order to think more holistically about how discrete steps of neuronal circuit development instruct specific behavioral outputs at the organismal level, it is useful to focus on the development of distinct neuronal systems that are dedicated to specific functions. The next two chapters review recent work that is directed at understanding the development of motor circuits and somatosensory circuits, respectively. In Chapter 10, Meng and Heckscher discuss how stem cells generate specific cohorts of neurons that function together in mature motor circuits, with a particular focus on the Drosophila larval motor system. This chapter highlights how precisely timed expression of temporal transcription factors contributes to motor circuit assembly by regulating the fate, maturation and synaptic partner choice of discrete groups of neurons. In Chapter 11, Cranfill and Luo review the mechanisms underlying the development of somatosensory circuits, with a particular emphasis on how extrinsic and intrinsic programs are integrated to control the diversification of somatosensory neurons. In addition, this chapter illustrates how understanding the organization of somatosensory circuitry can offer insights into normal processing of pain and itch sensation, as well as disease processes underlying congenital pain-related disorders.

While studies using a diverse array of animal models have provided and will continue to provide important conceptual advances that have fueled the understanding of normal development and neural developmental disorders, these models are inherently limited because they do not adequately recapitulate human-specific features that are relevant to disease. Hence, there is a clear need to develop tools and methods to more directly study neural development and the pathogenesis of disease in humans. In 12th and final chapter of this volume, Jacob, Schnoll, Song, and Ming, describe the exciting development of new technologies to generate region-specific human brain organoids to study development and disease. As some of the pioneers of these techniques, Song, Ming, and their colleagues describe recent advances that are allowing investigators to better mimic the in vivo developing human brain, as well as describing obstacles that need to be overcome to maximize the potential of these technologies for future studies.

Acknowledgments

I would like to thank the editorial staff at CTDB, including Shellie Bryant, Zoe Kruze, Naiza Ermin Mendoza, and Denny Mansingh for all of their help putting this volume together, and especially for their generous flexibility with the deadlines for receiving chapter submissions. I would also like to acknowledge all of the graduate students, postdoctoral fellows and principal investigators who contributed to the 12 excellent chapters in this volume. Finally, I am very appreciative of the many anonymous peer-reviewers who volunteered their time to provide valuable and constructive feedback to the authors on their review chapters.

Greg J. Bashaw
Professor of Neuroscience
Perelman School of Medicine
University of Pennsylvania

CHAPTER ONE

The regulation of cortical neurogenesis

Ana Villalba[a], Magdalena Götz[b], and Víctor Borrell[a,*]

[a]Instituto de Neurociencias, Consejo Superior de Investigaciones Científicas & Universidad Miguel Hernández, Sant Joan d'Alacant, Spain

[b]Institute for Stem Cell Research, Helmholtz Zentrum München & Biomedical Center, Ludwig-Maximilians Universitaet, Planegg-Martinsried, Germany

*Corresponding author: e-mail address: vborrell@umh.es

Contents

Abstract

The mammalian cerebral cortex is the pinnacle of brain evolution, reaching its maximum complexity in terms of neuron number, diversity and functional circuitry. The emergence of this outstanding complexity begins during embryonic development, when a limited number of neural stem and progenitor cells manage to generate myriads of neurons in the appropriate numbers, types and proportions, in a process called neurogenesis. Here we review the current knowledge on the regulation of cortical

Current Topics in Developmental Biology, Volume 142
ISSN 0070-2153
https://doi.org/10.1016/bs.ctdb.2020.10.003

neurogenesis, beginning with a description of the types of progenitor cells and their lineage relationships. This is followed by a review of the determinants of neuron fate, the molecular and genetic regulatory mechanisms, and considerations on the evolution of cortical neurogenesis in vertebrates leading to humans. We finish with an overview on how dysregulation of neurogenesis is a leading cause of human brain malformations and functional disabilities.

1. Introduction

The cerebral cortex is the seat of our higher cognitive abilities. Composed by trillions of neurons and glial cells of a still incompletely charted diversity, which are interconnected in outstandingly complex and exquisitely precise, yet remarkably plastic patterns, the mammalian cerebral cortex is by far the most complex biological structure. The mature cerebral cortex is the result of a developmental process that ends with the maturation of neural connections and the refinement of functional circuit assemblies. It all begins during embryogenesis, when a relatively limited number of neural stem cells (NSCs) generates the vast numbers and diversity of cortical neurons, a process called neurogenesis, followed by the generation of a rich diversity of glial cells. In this chapter, we will review the current knowledge on the mechanisms that regulate neurogenesis during cortical development. We will begin with a ground description of the process of cortex development and its main cellular players and mechanisms. This will be followed by a more detailed description of the regulation of neurogenesis in different neural stem and progenitor cells, and then of mechanisms regulating the eventual fate of cortical neurons. Next, we will review the current knowledge on the main molecular mechanisms controlling cortical neurogenesis, from classic signaling cascades to the more recent discoveries on transcriptomic and epigenomic regulation. Finally, a summary of mechanisms and hypotheses about the evolution of cortical neurogenesis, and how its dysregulation leads to human brain disease, will bring us to the closing remarks.

2. Cerebral cortex development

Development of the cerebral cortex begins with the formation of the telencephalic vesicles and the specification of the prospective territories, when the cortical primordium is solely composed of a monolayer of neuroepithelial stem cells (NECs; Fig. 1A, B) (Bishop, Goudreau, & O'Leary, 2000;

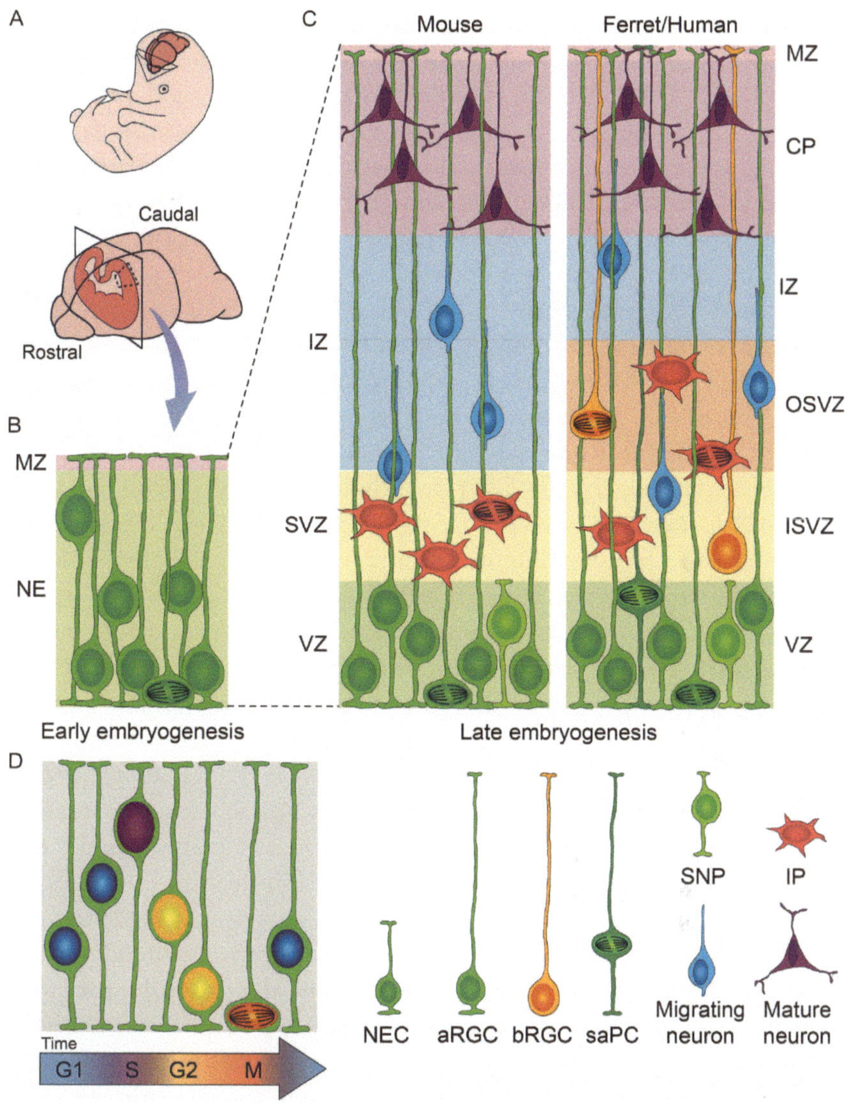

Fig. 1 Mammalian cortical development and cellular components. (A) Schema of a mammalian embryo and its forebrain, with a schematic drawing of the telencephalon in coronal section (dark pink). (B, C) Structure and cell types of the cortical primordium at early (B) and late (C) stages of development. The pseudostratified neuroepithelium (NE) is formed by proiferative neuroepithelial cells (NECs). At late stages, multiple germinal (VZ, SVZ) and post-mitotic layers (IZ, CP, MZ) are distinguishable, which are more specialized in the folded ferret and human cortex than in the smooth mouse cortex. Whereas the main cellular components are very similar, the abundance of basal progenitors is highly increased in ferret and human. (D) Schema of Interkinetic Nuclear Migration occurring in parallel to the cell cycle phases. aRGC, apical Radial Glia Cell; bRGC, basal Radial Glia Cell; saPC, subapical Progenitor Cell; SNP, Short Neural Precursor; IP, Intermediate Progenitor; VZ, Ventricular Zone; SVZ (Outer and Inner), Subventricular Zone; IZ, Intermediate Zone; CP, Cortical Plate; MZ, Marginal Zone.

Sur & Rubenstein, 2005). These are highly polarized progenitor cells that extend two thin processes away from the soma in the radial direction, one contacting the apical surface of the neuroepithelium (apical process) and the other contacting the basal lamina (basal process) (Sidman & Rakic, 1973). Adjacent NECs are attached to each other by the apical process at the level of the apical surface, via adherent and tight junctions (Taverna, Gotz, & Huttner, 2014). NECs proliferate in a self-amplificative mode, producing two new NECs from each division. During each cell cycle, their nucleus moves along the entire apical-basal extent in a highly stereotyped fashion, called Interkinetic Nuclear Migration (INM) (Takahashi, Nowakowski, & Caviness, 1993). Mitosis occurs at the apical surface, during G1 the nucleus moves toward the basal side, where it remains during DNA synthesis (S phase), and at G2 it moves back from the basal to the apical side, where it undergoes mitosis again (Fig. 1D) (Hinds & Ruffett, 1971; Sidman & Rakic, 1973; Takahashi, Nowakowski, & Caviness, 1995a, 1995b). NECs perform INM in an asynchronous manner, conferring the neuroepithelium its typical false multilayered appearance (pseudostratified). INM is evolutionarily conserved, although the migration flow, cell density and VZ thickness differ between species (Miyata, Okamoto, Shinoda, & Kawaguchi, 2015).

After a few rounds of cell division and amplification, NECs lose tight junctions while retaining apical adherent junctions, and become apical Radial Glial Cells (aRGCs) (Taverna et al., 2014). aRGCs are the primary type of cortical progenitor cell, divide at the apical surface, express the Paired-box transcription factor 6 (Pax6) and their lineage gives rise to all cortical excitatory neurons (Malatesta et al., 2003). In fact, the emergence of aRGCs marks the initiation of cortical neurogenesis (Fig. 1C). Very similar to NECs, aRGCs have an apical and a basal process extended radially and anchored to the apical and basal sides of the cortical primordium (Rakic, 1972; Sidman & Rakic, 1973). aRGCs also undergo INM, but becomes restricted to the apical domain, next to the ventricular surface (Attardo, Calegari, Haubensak, Wilsch-Brauninger, & Huttner, 2008; Haubensak, Attardo, Denk, & Huttner, 2004; Noctor, Flint, Weissman, Dammerman, & Kriegstein, 2001; Takahashi et al., 1993). This defines the primary germinal layer of the cortical primordium: the Ventricular Zone (VZ) (Boulder_Committee, 1970). The VZ contains the cell body of aRGCs and Short Neural Precursors (SNPs), a second and less frequent type of apical progenitor cell with either no basal process or a short one restricted to the VZ (Stancik, Navarro-Quiroga, Sellke, & Haydar, 2010). In the developing murine cerebral cortex SNPs are largely neurogenic, with limited self-renewal capacity, while in other brain regions like the ventral telencephalon they divide several times, serving as amplifying progenitors (Pilz et al., 2013).

At the onset of neurogenesis (circa embryonic day (E) 11 in mouse), most aRGC divisions produce additional aRGCs (self-amplification) and, in a small proportion, neurons. As development proceeds, the rate of aRGC self-amplification diminishes concomitant with an increase in asymmetric divisions to generate secondary progenitor cells and neurons (Miyata et al., 2004; Takahashi et al., 1995b; Takahashi, Nowakowski, & Caviness, 1996). These new secondary progenitors migrate from the apical surface, where they are born, to the basal border of the VZ, where they divide, hence called Basal Progenitors (BPs) (Miyata et al., 2004; Noctor, Martinez-Cerdeno, Ivic, & Kriegstein, 2004; Noctor, Martínez-Cerdeño, & Kriegstein, 2008; Taverna et al., 2014). BPs accumulate forming a secondary germinal zone: the Subventricular Zone (SVZ) (Boulder_Committee, 1970; Noctor et al., 2004, 2008; Takahashi et al., 1995b). In mouse, the majority of basal progenitors are Intermediate Progenitor Cells (IPCs): small cells with no obvious polarity, multiple short processes and marked by the expression of the T-box transcription factor 2 (Tbr2) (Englund et al., 2005). IPCs produce most of the excitatory neurons to all layers of the murine cerebral cortex (Haubensak et al., 2004; Kowalczyk et al., 2009; Noctor et al., 2004; Vasistha et al., 2015). At later stages, aRGCs produce two additional types of progenitor cells that divide away from the apical surface. Subapical Progenitor Cells (saPCs) have an apical process similar to aRGCs, but divide midway within the VZ (Pilz et al., 2013) and are very scarce in cortex development of species with a small and smooth cortex, like mouse and marmoset monkey, but abundant in those with a large and folded cortex, like ferret and sheep (Pilz et al., 2013), pointing to a possible role in cortex expansion and folding (Borrell & Götz, 2014).

The second type of basal progenitor is called basal Radial Glia Cell (bRGC; also named outer radial glia cell, oRG (LaMonica, Lui, Hansen, & Kriegstein, 2013)). bRGCs share many similarities with aRGCs, including Pax6 expression and the presence of a basal process extended radially and anchored to the basement membrane (Fietz et al., 2010; Hansen, Lui, Parker, & Kriegstein, 2010; Reillo, De Juan Romero, García-Cabezas, & Borrell, 2011; Wang, Tsai, Lamonica, & Kriegstein, 2011). Contrary to aRGCs, however, bRGCs do not have a process extended to the apical surface nor divide apically, but instead divide basally within the SVZ. bRGCs generally do not undergo INM, but they undergo mitotic somal translocation (rapid basal movement of the soma immediately before mitosis) in the human cortex and, to a limited extent, in non-human primates (Betizeau et al., 2013; Hansen et al., 2010; LaMonica et al., 2013). In the developing small and smooth mouse cortex bRGCs are scarce, mostly not self-renewing but neurogenic, and express Pax6 but not Tbr2 (Shitamukai, Konno, & Matsuzaki, 2011; Wang et al., 2011). In contrast, in the developing cortex of species with large and folded brains such as primates, carnivores and ungulates, bRGCs are very abundant, with high self-amplificative capacity, and a large proportion co-express Pax6 and Tbr2 (Betizeau et al., 2013; Martínez-Martínez et al., 2016; Reillo et al., 2011).

The generation of neurons directly by aRGCs in the VZ is called Direct Neurogenesis, as opposed to the production of neurons indirectly from aRGCs through BPs, called Indirect Neurogenesis (Cárdenas et al., 2018; Laguesse et al., 2015). Because of the amplificative nature of BPs, indirect neurogenesis increases neuron production several-fold compared to direct neurogenesis (Hevner, Hodge, Daza, & Englund, 2006; Kriegstein, Noctor, & Martinez-Cerdeno, 2006). Indirect neurogenesis is the predominant mode of cortical neuron production in mammals, in contrast to reptiles, and it is greatly amplified in primates and other species with a prominently large cortex (Cárdenas et al., 2018; Dehay & Kennedy, 2007; Haubensak et al., 2004; Reillo et al., 2011). Accordingly, a fundamental difference between species with a small and smooth cortex versus a large and folded cortex is the thickness and complexity of the SVZ (Fig. 1C), with 10–50% of mitoses (the rest occurring in the VZ) in the former and up to 90% in the latter (Betizeau et al., 2013; Fietz et al., 2010; Hansen et al., 2010; Kelava et al., 2012; Lukaszewicz et al., 2005; Pilz et al., 2013; Reillo et al., 2011; Reillo & Borrell, 2012; Smart, Dehay, Giroud, Berland, & Kennedy, 2002). In addition, this thick SVZ is subdivided in Inner and Outer SVZ (ISVZ, OSVZ; Fig. 1) (Dehay & Kennedy, 2007; Kawasaki, Toda, & Tanno, 2013;Lukaszewicz et al., 2005 ; Reillo et al., 2017 ; Smart et al., 2002). The OSVZ is particularly thick and prominent in macaque and human embryos, containing up to 65% of cortical mitoses at the peak stages of neurogenesis (Lukaszewicz et al., 2005; Reillo et al., 2017; Smart et al., 2002). The OSVZ and its constituent progenitor cells play key roles in promoting the formation of cortical folds (Borrell & Reillo, 2012; Fernandez et al., 2016; Llinares-Benadero & Borrell, 2019; Reillo et al., 2011). Loss of function experiments in ferrets that specifically reduce bRGC proliferation in OSVZ led to a reduction in the size of cortex folds (Matsumoto, Shinmyo, Ichikawa, & Kawasaki, 2017; Reillo et al., 2011), and even their complete loss (Poluch & Juliano, 2015).

Gain-of-function experiments promoting the over-proliferation of OSVZ progenitor cells in ferrets and marmosets increases cortical surface area and folding (Heide et al., 2020; Masuda et al., 2015; Nonaka-Kinoshita et al., 2013). Similar genetic experiments in mouse, where the abundance of bRGCs was increased several-fold, caused folding of the otherwise smooth cortex (Stahl et al., 2013; Wang, Hou, & Han, 2016).

Both ISVZ and OSVZ are densely populated by a complex mixture of IPCs and bRGCs, which in primates and carnivores come in different flavors depending on their cell morphology, lineage potential and transcription factor expression (Fig. 1C) (Betizeau et al., 2013; Borrell & Götz, 2014; Reillo et al., 2011; Reillo & Borrell, 2012). In primates, IPCs and multiple morphotypes of bRGCs are highly proliferative, producing excitatory neurons and additional basal progenitors. In fact, progenitor cells in the primate OSVZ have an unmatched plasticity of lineage potential, each progenitor type generating virtually any other type as well as neurons, albeit at different frequencies (Betizeau et al., 2013). The regulation and developmental importance of these transitions remains to be elucidated, but their being so conspicuous suggests significant relevance.

Neurogenesis in the cerebral cortex takes place during a protracted period of time, spanning 30–50% of gestation depending on the species (Rakic, 1974, 1995; Takahashi, Goto, Miyama, Nowakowski, & Caviness, 1999). During this period, the rate of neurogenesis by cortical progenitors changes gradually but dramatically (Takahashi et al., 1996; Takahashi, Nowakowski, & Caviness, 1994). In mouse and rat, at the onset of neurogenesis progenitor cells mostly undergo symmetric proliferative divisions producing additional progenitors and few neurons. Then, during most of the neurogenic period they divide in an asymmetric mode, generating one aRGC and either a BP (indirect neurogenesis) or a neuron (direct neurogenesis) (Attardo et al., 2008; Haubensak et al., 2004; Noctor et al., 2001; Noctor et al., 2004). At the end of neurogenesis, most aRGCs produce differentiating progenitors or ependymal cells, thereby depleting the NSC pool and terminally ending neurogenesis. The gradual increase in neurogenesis is caused by the progressive lengthening of the cell cycle (Arai et al., 2011; Calegari, Haubensak, Haffner, & Huttner, 2005; Dehay & Kennedy, 2007; Lange, Huttner, & Calegari, 2009; Takahashi et al., 1995a). In species with a large and folded cortex, like primates, the cell lineage dynamics of cortical progenitors is more complex than in rodents. In macaque monkey embryos, an early period of steadily increasing neurogenesis to generate deep layer neurons is followed by an interim period when progenitor cells preferentially self-expand (low neurogenesis). This expansion is essential so that neurogenesis can later proceed at very high rates to produce the vast amounts of superficial layer neurons (Betizeau et al., 2013). Nevertheless, changes in cell cycle exit are always coupled to variations in cell cycle length (Betizeau et al., 2013), as in rodents.

Once excitatory neurons are born in the cortical primordium, they initiate radial migration to their final destination in the cortical surface. It has been suggested that the earliest born neurons emerging from aRGCs lose their apical anchoring and migrate to the cortical plate by contracting/shrinking their basal process, a mechanism called somal translocation (Nadarajah, Brunstrom, Grutzendler, Wong, & Pearlman, 2001; Nadarajah & Parnavelas, 2002). At mid-stages of neurogenesis, radial migration often begins with neurons translocating to the SVZ (unless they are already born there by indirect neurogenesis), where they acquire a multipolar configuration extending multiple short processes. Neurons spend a few hours in this morphology moving for very short distances tangentially within the SVZ (Tabata & Nakajima, 2003). Multipolar migration seems to be important for the intermixing of cortical neurons with different clonal origin, influenced e.g. by neurite extension and under the control of Eph receptor tyrosine kinases (Dimidschstein et al., 2013; Torii, Hashimoto-Torii, Levitt, & Rakic, 2009). Eventually, neurons re-acquire polarity when leaving the SVZ and extend a single leading process oriented radially toward the pial surface, and a short trailing process in the opposite direction. This re-polarization of migrating neurons to exit the SVZ seems to be a highly vulnerable event with many pathways essential for it (Cooper, 2014), including the centrosomal microtubule-organizing activity, which differs profoundly between the migration in and out of the SVZ (Camargo Ortega et al., 2019). Once re-polarized, neurons establish intimate contact with the basal process of RGCs (either aRGCs or bRGCs), which serves as substrate and guide for their migration to the cortical surface (Anton, Kreidberg, & Rakic, 1999; Anton, Marchionni, Lee, & Rakic, 1997; Marin & Rubenstein, 2003; Misson, Austin, Takahashi, Cepko, & Caviness, 1991; Rakic, 1972; Sidman & Rakic, 1973). The single leading process of migrating cortical neurons frequently forms lateral branches (Martínez-Martínez et al., 2019), which are likely used to sense the extracellular environment for molecular guidance cues, as in migrating inhibitory interneurons (Martini et al., 2009). These branches are more frequent, elaborate and form at wider angles in ferret than mouse, and they are a means for neurons to shift laterally between radial glia fibers, as they progress in outward migration (Martínez-Martínez et al., 2019). This has been proposed to be a central mechanism contributing to cortical surface area expansion and folding

(Borrell & Götz, 2014; Borrell & Reillo, 2012; Fernandez et al., 2016; Lui, Hansen, & Kriegstein, 2011).

Neurons migrate through the Intermediate Zone (IZ) and the cortical plate (CP), overpassing previous cohorts of neurons, until contacting the marginal zone (MZ), where radial migration finishes and neurons detach from the radial fiber (Marin & Rubenstein, 2003). Neurogenesis in the cerebral cortex occurs in a strictly orderly fashion, where neurons destined to form the deep cortical layers are produced first, followed by neurons forming the immediately overlying layer, and so on, until neurons forming the most superficial layer are generated last (Rakic, 1974; Takahashi et al., 1999). This temporal sequence is known as inside-out gradient of neurogenesis (Rakic, 1974). During radial migration, neurons already start growing their primary axon from the trailing process (Auladell, Martinez, Alcantara, Super, & Soriano, 1995; Noctor et al., 2004). Once they finish radial migration, neurons begin differentiating their dendritic arbors and further elaborating their axonal arbor (Borrell & Callaway, 2002; Callaway, 1998; Callaway & Borrell, 2011; Callaway & Katz, 1992; 1990; Callaway & Lieber, 1996; Sidman & Rakic, 1973; Takahashi, Misson, & Caviness, 1990). Notably, cells with morphological resemblance to bRGCs after neurogenesis serve as glial progenitors (Marín-padilla, 1995; Voigt, 1989), later re-named "astrocyte intermediate progenitor cells" (Kriegstein & Alvarez-Buylla, 2009). If these cells are different from the neurogenic bRGCs, or the same generating first neurons and later astrocytes, remains to be determined.

Cortical inhibitory interneurons (INs) follow a developmental program quite different from that of excitatory neurons, presented above (Lim, Mi, Llorca, & Marin, 2018; Wamsley & Fishell, 2017). Owing to the limited space available in this review, the remaining sections will deal exclusively with aspects related to neurogenesis within the cerebral cortex primordium, producing excitatory neurons.

3. Regulation of neurogenesis in apical progenitors

As mentioned in the previous section, immediately prior to the onset of neurogenesis NECs transition to aRGCs, thus becoming more committed to a neural fate. aRGCs are the primary progenitor cell type of the cerebral cortex and give rise to all subsequent progenitor cell types and excitatory neurons (Borrell & Reillo, 2012; De Juan Romero, Bruder, Tomasello, Sanz-Anquela, & Borrell, 2015; Taverna et al., 2014). Given the high polarity of aRGCs, their intracellular distribution of mRNA, proteins and organelles is also asymmetric. The normal progression of neurogenesis requires a precise distribution of these components within the subcellular compartments, which often varies depending on progenitor cell type and even phase of the cell cycle (Arai & Taverna, 2017; Pilaz & Silver, 2017; Taverna et al., 2016).

3.1 Apical process

The apical process of aRGCs, spanning from the cell body to the ventricular surface, represents only 2% of their plasma membrane (Fig. 2A, A″) (Kosodo et al., 2004; Veeraval, O'Leary, & Cooper, 2020). Although it is a small part of aRGCs, it plays key roles in their biology, dynamics and fate.

3.1.1 Adherens junctions

aRGCs remain attached to each other via the most apical domain of the apical process, establishing calcium-dependent cell adhesion via adherent junctions (AJs; Fig. 2A″) (Taverna et al., 2014). AJs are composed by a complex mixture of cell-cell adhesion proteins (including N-cadherin, αE-catenin, β-catenin, Afadin and Plekha7) (Stocker & Chenn, 2015; Tavano et al., 2018), cell polarity proteins (such as Par3, Par6, aPKC, Cdc42, Pals, Lgl and Dlg5) (Bultje et al., 2009; Izumi et al., 1998; Lin et al., 2000; Nechiporuk & Raible, 2008; Veeraval et al., 2020) and actin regulators (RhoA, Cdc42 or Neogenin) (Cappello et al., 2006; Veeraval et al., 2020). AJs establish the limit between the apical and lateral membranes of the apical process, thus preserving the apical-basal polarity of aRGCs. The combined set of AJs from all aRGCs establishes an apical belt, which is key to maintain the structural integrity of the VZ and to endure mechanical forces exerted from the CSF (Cerebro-spinal fluid) and neighboring cells. Defects in the apical AJ belt lead to the appearance of severe cortical malformations, including cell heterotopias and hydrocephaly (Veeraval et al., 2020). Accordingly, overexpression of AJ protein repressors, or ablation of AJ proteins, drive the delamination of aRGCs to basal positions, where they undergo mitosis. For example, loss of AJ may cause the retraction of the basal process of aRGCs, thus leading to radial migration defects and formation of heterotopias (Cappello et al., 2012). At late stages of neurogenesis the complex Par3/Pax6/aPKC is normally downregulated, and hence the early loss of Par3 from aRGCs causes their premature exit from the cell cycle (Costa, Wen, Lepier, Schroeder, & Gotz, 2008). Disruptions in the normal trafficking or turnover of AJ proteins, such as FlnA, Arfgef2 or Numb, mimic AJ failure and lead to aRGC polarity

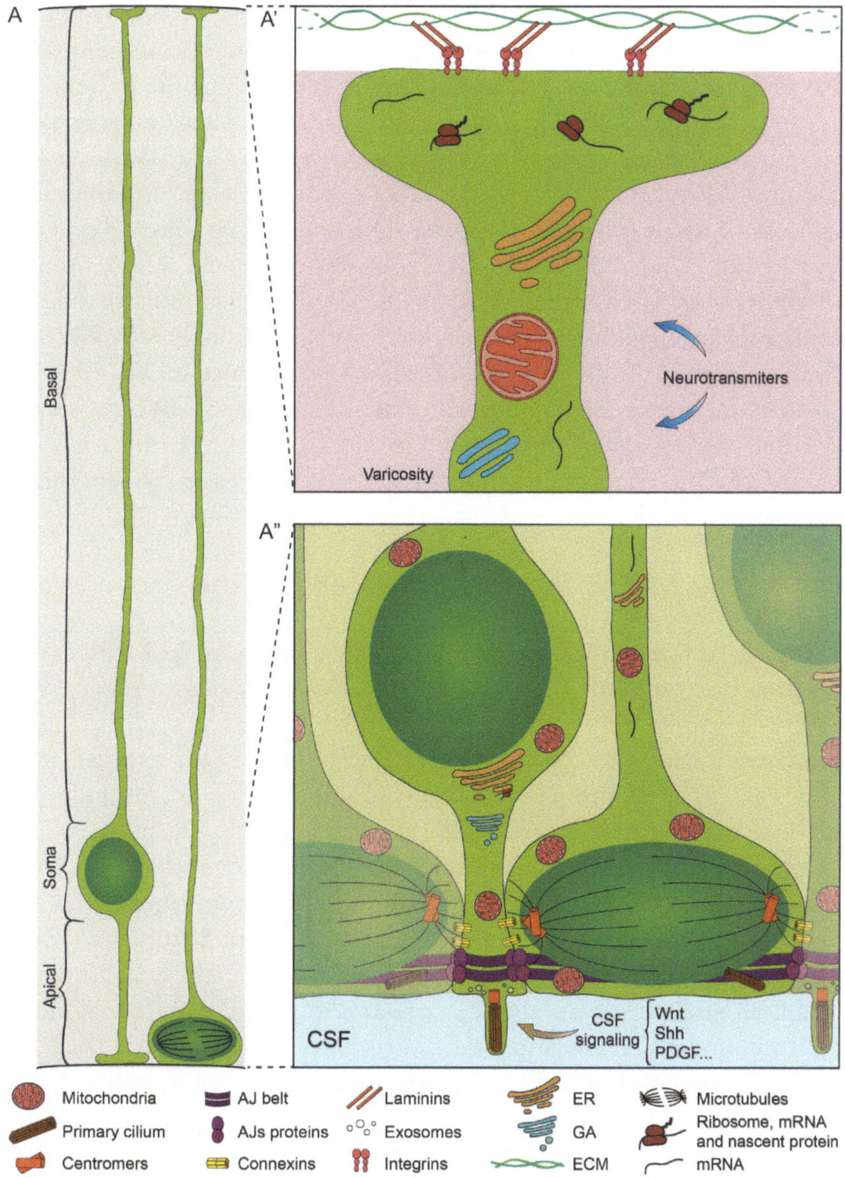

Fig. 2 Subcellular components of apical Radial Glia Cells and their distribution. (A) Schematic of two aRGCs, one in interphase (left) and one in mitosis (right), illustrating similarities and differences in subcellular compartments. (A′) Schema of the basal end-foot showing integrin-mediated attachment to the extracellular matrix (ECM) at the basal lamina. mRNA molecules transported along the basal process are locally translated at the basal end-foot, and trans Golgi apparatus (GA) is preferentially found in varicosities. (A″) Schematic of the apical domain of the VZ highlighting two aRGCs, one in interphase (left) and one in mitosis (right). The apical adherens junction (AJ) belt maintains aRGCs attached to each other and confers structural cohesion to the VZ.

disturbance (Bultje et al., 2009; Rasin et al., 2007). Accordingly, over-expression of AJ components blocks NPC delamination in some instances (Tavano et al., 2018), but it is not sufficient to overcome the delamination promoted by the centrosomal protein Akna, which mediates centrosomal microtubule organizing activity (Camargo Ortega et al., 2019).

Apical AJs retain aRGCs within the VZ, which also ensures their interaction with extrinsic signals from the CSF and neighboring cells, such as Wnt ligands (Chenn & Walsh, 2002; Stocker & Chenn, 2015). Some AJ components such as β-catenin participate in signaling pathways that directly regulate the proliferation and fate of progenitor cells (Chenn & Walsh, 2002; Hirabayashi et al., 2004). Given the central importance of AJs for aRGC function, their expression and asymmetric inheritance in aRGCs is tightly regulated during development (Bultje et al., 2009; Martínez-Martínez et al., 2016), and once proneural genes start to be expressed their production is gradually silenced (Itoh et al., 2013).

3.1.2 Primary cilium

The apical surface of NECs and aRGCs is specifically marked with the surface protein Prominin1 (Prom1) and decorated with a single primary cilium (Gotz & Huttner, 2005). This is a small organelle that protrudes from the apical membrane into the ventricular cavity, where it is exposed to the CSF (Fig. 2A″). The primary cilium is structured by a microtubule-based axoneme attached to a basal body, the mother centriole. The primary cilium is key for the transduction of extracellular signaling cues, including the Sonic hedgehog (Shh), Wnt and PDGF pathways, well known regulators of neurogenesis and progenitor cell fate (Goetz & Anderson, 2010; Park, Jang, & Lee, 2019). The ciliary protein Arl13b plays key roles in the establishment of the apical-basal polarity of aRGCs, which is impaired upon loss of the primary cilium early in development. Loss of apical-basal polarity in aRGCs leads to the inversion of polarity in the entire cerebral cortex and disruption of its exquisite laminar organization (Higginbotham et al., 2013). The precise mechanism of action of Arl13b implicated in these defects is not well understood, but it may be due to disruptions on Shh signaling at the cilium (Gigante, Taylor, Ivanova, Kahn, & Caspary, 2020; Higginbotham et al., 2013). The release of exosomes at the primary cilium and apical surface further supports their key relevance in signal transduction (Dubreuil, Marzesco, Corbeil, Huttner, & Wilsch-Brauninger, 2007; Wood & Rosenbaum, 2015).

Following mitosis, the primary cilium and mother centriole of aRGCs are endocytosed and asymmetrically inherited by one of the two daughter cells. Inhibition of this reabsorption, or blockade of asymmetric inheritance, drives cell cycle exit and neuronal differentiation (Gabriel et al., 2016; Wang et al., 2009). At early stages of neurogenesis, asymmetric inheritance of the centrioles and primary cilium are key in defining progenitor cell fate, favoring the self-renewal of progenitor cells: the daughter cell retaining the mother centriole and primary cilium remains as aRGC (Paridaen, Wilsch-Brauninger, & Huttner, 2013; Wang et al., 2009). The rapid restoration of the primary cilium in the apical surface renders the cell immediately ready to continue receiving pro-proliferative signals from the CSF, thus potentiating the aRGC fate. Genes essential for primary cilium structure and function, such as *Arl13b*, *Kif3a* and *Ift88*, have been removed in a variety of conditional mutant mice (Nestin-Cre, Emx1-Cre and hGFAP-Cre) with no overt defects in aRGC organization or neurogenesis (Han et al., 2008; Higginbotham et al., 2013; Insolera, Bazzi, Shao, Anderson, & Shi, 2014; Snedeker et al., 2017). However, these genetic mutations are often produced after aRGCs and their polarity have already been established, so the precise role of the primary cilium in aRCGs and cortical neurogenesis remains to be defined. Retention and reconstruction of the primary cilium are influenced by the levels of N-Cadherin (Das & Storey, 2014). Ablation of Ninein, a protein specific to the mature centriole, disrupts the asymmetric segregation of centrosomes, depleting the progenitor pool size (Wang et al., 2009). Similarly, the interphase centrosome protein Akna, localized at the subdistal appendages of the mother centriole, organizes centrosomal microtubules and promotes the delamination of progenitor aRGCs from the VZ to the SVZ (Camargo Ortega et al., 2019). The detachment and delamination of BPs or differentiated neurons is also preceded by the regrowth of the primary cilium basolaterally, instead of apically (Wilsch-Bräuninger, Peters, Paridaen, & Huttner, 2012). At late stages of neurogenesis, the strong association between the mother centriole and the ciliary membrane weakens, favoring neuron differentiation (Paridaen et al., 2013; Wang et al., 2009; Yamashita et al., 2007).

3.1.3 Cerebrospinal fluid

The apical surface of aRGCs is exposed to the CSF of the telencephalic lateral ventricles. The primitive CSF in the early embryo is a combination of amniotic fluid trapped after neural tube closure and secretions from the choroid plexus epithelium (Lehtinen et al., 2011; Lehtinen & Walsh, 2011).

Mass spectrometry analyses of embryonic CSF reveal a complex mixture of proteins that provide growth and survival signals to aRGCs, including extracellular matrix components, osmotic pressure regulators, ion carriers, hormone-binding proteins, lipid metabolism regulators, enzymes and their regulators. This composition seems to be conserved across species (Bueno & Garcia-Fernàndez, 2016; Cavanagh et al., 1983; Gato et al., 2004; Parada, Gato, & Bueno, 2005; Zappaterra et al., 2007). Because of the liquid nature of the CSF, its constituent signaling factors reach distant targets within the telencephalic ventricle throughout the entire rostro-caudal and latero-medial extent of the cortical VZ. CSF proteins may also have region-specific distribution, establishing local signaling gradients (Cavanagh et al., 1983; Zappaterra et al., 2007). While the detailed composition of the CSF remains to be fully characterized, some of its identified components are known to regulate neurogenesis. These include Insulin-like growth factor, Fibroblast growth factor-2, Slit proteins, Sonic hedgehog, Retinoic acid and Wnt signaling, among others (Borrell et al., 2012; Huang et al., 2010; Lehtinen et al., 2011; Lehtinen & Walsh, 2011; Martín et al., 2006; Raballo et al., 2000). Therefore, the CSF plays multiple and critical roles on the complex regulation of cortical neurogenesis (Johansson et al., 2013).

3.1.4 Endoplasmatic reticulum and Golgi apparatus

The endoplasmatic reticulum (ER) and Golgi apparatus (GA) are essential organelles for the proper asymmetric production and distribution of proteins prior to, and following, cell division. Prior to mitosis, aRGCs disassemble their GA, releasing proteins such as Numb and Cadherin that directly affect AJ protein composition and corticogenesis (Arai & Taverna, 2017; Taverna et al., 2016; Zhou et al., 2007). The subcellular localization of the GA varies depending on the progenitor cell type: while in aRGCs it is confined to the apical process, it becomes re-organized and pericentrosomal in delaminating BPs (Arai & Taverna, 2017; Taverna et al., 2016). On the other hand, secretory compartments related to the GA are distributed along the varicosities of the basal process of aRGCs (Fig. 2A′, A″) (Coquand et al., 2020). In contrast, the ER is placed in both apical and basal processes of aRGCs (Taverna et al., 2016), and plays a role in regulating the mode of neurogenesis, with impaired protein translation favoring direct neurogenesis (Laguesse et al., 2015).

3.2 Soma

The soma of aRGCs marks the limit between the apical and the basal process. Its exact apical-basal position within the VZ varies along the cell cycle due to INM. Interestingly, uncoupling INM from the cell cycle does not prevent cell division, but defects in the cell cycle may disrupt INM (Taverna & Huttner, 2010). Multiple factors contribute to the normal occurrence of INM. Human mutations in the cytoskeletal protein LIS1 alter INM and lead to the severe loss of cortical folds (lissencephaly) (Gambello et al., 2003; Taverna et al., 2014). The extracellular matrix protein family Laminin is also important in the regulation of INM, which is disrupted as a result of mutations in Laminin γ1, driving mitoses to basal positions (Tsuda et al., 2010). The regular INM movements require a physical barrier at the basal border of the VZ (Watanabe, Kawaue, & Miyata, 2018).

During cell division, the mitotic spindle separates homologous chromosomes and sets up the plane of cell division. In highly polarized tissues like mammalian epithelia, and in neuroblasts of *Drosophila melanogaster* embryos, the orientation of the mitotic cleavage plane is a critical determinant of daughter cell fate. Following this principle, cortical aRGC divisions with vertical cleavage plane (perpendicular to the ventricular surface), which divide the apical domain and AJs, should give rise to identical daughter aRGCs. In contrast, horizontal or oblique cleavage plane divisions should have an asymmetric outcome, producing one aRGC and a different cell (Chenn & McConnell, 1995; Taverna et al., 2014). In support of this model is the vertical cleavage plane division of aRGCs during self-expansion early and switch to horizontal plane divisions generating BPs (LaMonica et al., 2013). Also experimental randomization of mitotic cleavage planes (increasing non-vertical planes) augments asymmetric inheritance of the apical process from the mother aRGC (Konno et al., 2008). However, randomization of mitotic cleavage planes fails to modify neuronal production (Konno et al., 2008; Lancaster & Knoblich, 2012). Also at later stages, when functionally asymmetric divisions occur, most aRGC divisions have a vertical cleavage plane (Fietz et al., 2010; Konno et al., 2008; Kosodo et al., 2004; Noctor et al., 2008). An alternative model suggests that cleavage plane orientation is correlated with progenitor fate rather than cellular output (Noctor et al., 2008). Then again, there is no correlation between spindle orientation and the symmetry of distribution of Par3, an essential component of AJs in aRGCs (Bultje et al., 2009). In summary, the relationship between mitotic cleavage plane and cortical cell fate remains far from clear in light of the existing evidence (Lancaster & Knoblich, 2012; Mora-Bermúdez & Huttner, 2015).

The location of the mitotic plate within the cell determines the symmetric or asymmetric inheritance of organelles, cellular components and molecular factors that exert a key influence on the fate of daughter cells, particularly those related with the apical surface (Bultje et al., 2009; Chenn & McConnell, 1995; Lancaster & Knoblich, 2012; Mora-Bermúdez & Huttner, 2015; Rasin et al., 2007). The opposite sides of the mitotic spindle are not perfectly identical in size, and the degree of this size asymmetry is related to the probability of progenitor cell self-renewal, higher for the daughter cell with the smaller spindle half (Delaunay, Cortay, Patti, Knoblauch, & Dehay, 2014). One organelle key in cortical progenitor cell fate and differentiation is the mitochondrion. Mitochondria are classically viewed only as a cell's power source and a redox regulator system, but mitochondrial dysfunction also affects progenitor cell cycle exit and neuronal differentiation (Homem et al., 2014; Khacho et al., 2017). NSCs have elongated mitochondrial networks, like astrocytes, both of which are biased to glycolytic metabolism, while neurons have shorter mitochondria and predominantly use oxidative phosphorylation, linked to high energy demands (Homem et al., 2014; Khacho et al., 2016; Khacho, Harris, & Slack, 2019; Mils et al., 2015). Mice mutant for proteins mediating mitochondria fission show soft and smaller cerebellum and forebrain (Khacho et al., 2016; Wakabayashi et al., 2009). In the cerebral cortex, a recent study demonstrates that mitochondria fission promotes neuronal fate, whereas mitochondria fusion after mitosis redirects daughter cells toward self-renewal (Iwata, Casimir, & Vanderhaeghen, 2020). Changes in the generation of reactive oxygen species (ROS) activate redox regulators, which directly mediate transcriptional modifications driving neuronal differentiation (Khacho et al., 2016). Of particular relevance is the case of the human-specific gene *ARHGAP11B*, a component of the mitochondrial inner membrane that promotes glutaminolysis (Namba et al., 2020). *ARHGAP11B* enhances the generation and amplification of cortical BPs, thereby driving surface area expansion and folding of the cerebral cortex (Florio et al., 2015; Namba et al., 2020).

3.3 Basal process

The basal process of aRGCs is a very thin membranous extension of the cell that spans radially from the soma (in the VZ) to the pial surface. This nearly covers the entire cortical thickness, which may be up to millimeters long in primates (Rakic, 1972; Sidman & Rakic, 1973). The basal process displays swellings or varicosities throughout its length. At the marginal zone, immediately beneath the pia, it forms multiple branches, and each individual branch contacts the basal lamina with an end-foot (Fig. 2A′) (Coquand et al., 2020; Fietz & Huttner, 2011; Rakic, 1972; Sidman & Rakic, 1973; Takahashi et al., 1990; Weissman, Noctor, Clinton, Honig, & Kriegstein, 2003). The basal lamina is rich in integrins and extracellular matrix proteins, which signal to RGCs via G-protein coupled receptors in their basal process (Sidman & Rakic, 1973; Takahashi et al., 1990; Taverna et al., 2014). The attachment and interaction between the basal process and the basal lamina is critical in the regulation of cerebral cortex development (Hartfuss et al., 2003; Super, Del Rio, Martinez, Perez-Sust, & Soriano, 2000). For example, Integrin β1 is the most abundant at basal end feet and regulates their regeneration after mitosis (Fujita et al., 2020; Long & Huttner, 2019; Radakovits, Barros, Belvindrah, Patton, & Müller, 2009). Defects in integrin signaling lead to alterations in proliferation, neuronal migration and differentiation, basal process detachment and cell death, suggesting that integrin signaling is important not only at the basement membrane (Campos et al., 2004; Drago, Nurcombe, & Bartlett, 1991; Fietz et al., 2010; Leone et al., 2005; Long & Huttner, 2019; Radakovits et al., 2009; Schmid & Anton, 2003). Other mutations in basal membrane proteins that interfere with basal process contact seem to dramatically affect neuronal migration and accommodation in layers, but not cell proliferation or neurogenesis (Haubst, Georges-Labouesse, De Arcangelis, Mayer, & Götz, 2006).

At very early stages of cortical development, the basal processes of NECs and aRGCs may occasionally split longitudinally during mitosis and be inherited by both daughter cells (Kosodo et al., 2008). However, in most cases when one RGC self-amplifies to produce two RGCs, one daughter cell inherits the maternal basal process and the other grows a new one (Arai & Taverna, 2017; Betizeau et al., 2013; Hansen et al., 2010; Martínez-Martínez et al., 2016). In this case, given the uniqueness of the basal process, its asymmetric inheritance upon cell division may be yet one more signal influencing cell fate (Konno et al., 2008; LaMonica et al., 2013). In addition to anchoring RGCs to the basal membrane, the long basal process of aRGCs contains multiple mRNA species that are transported within, and translated locally at the basal end feet (Pilaz, Lennox, Rouanet, & Silver, 2016; Pilaz & Silver, 2017; Tsunekawa et al., 2012). These include proteins regulating the cell cycle, like Ccnd2, and that lengthen G1, S, and M phases of the cell cycle (Arai et al., 2011; Lange et al., 2009; Pilaz et al., 2016). Ccnd2 along with other signals promotes proliferation, suggesting that the daughter cell inheriting the basal process may be prone to proliferate and self-renew, while the other may undergo differentiation (Shitamukai et al., 2011; Tsunekawa et al., 2012). Thus, the asymmetric inheritance of the basal process represents an important source of bias in the content of these molecules by daughter cells, and yet another source of asymmetry in their fate. Finally, the basal process may also contribute to determine the position of the cytokinesis machinery, and hence the division plane, which may in turn affect the daughter cell fate, as mentioned above. In this regard, RhoA, Anillin and/or F-actin are located at the base of the basal process and regulate the orientation of the division plane (Kosodo et al., 2008). In spite of these many recent advances on individual aspects of basal process inheritance, it remains unclear precisely how these interact and are coordinated to determine cell fate (Kosodo & Huttner, 2009; Miyata, Kawaguchi, Okano, & Ogawa, 2001; Tsunekawa et al., 2012).

In addition to the long basal process of aRGCs and bRGCs, BP cells (bRGCs and IPCs) extend multiple short processes. The abundance of these short processes is regulated, at least in part, by the palmitoylation of PALMD, and is directly linked to the proliferative activity of these BPs, suggesting that those processes may be an entry site for pro-proliferative signals (Kalebic et al., 2019).

4. Regulation of neurogenesis in BPs

As mentioned above, cortical neurons are produced directly by aRGCs (direct neurogenesis), but mostly indirectly via BPs (indirect neurogenesis). Hence, BPs are fundamental for the massive production of neurons during the formation of large and folded cortices (Llinares-Benadero & Borrell, 2019). Due to their basal position, BPs are far away from apical sources of signaling molecules that regulate cell proliferation, such as the CSF. Alternative sources of extracellular signals that regulate BP proliferation include migrating INs. As mentioned, inhibitory neurons migrate tangentially from their birthplace at the medial and caudal ganglionic eminences to the neocortex, mostly along the BP-filled SVZ (Marin & Rubenstein, 2003). The normal pace of tangentially-migrating INs is important for IPC proliferation, likely via the release of paracrine signals, which eventually impacts IP numbers and the abundance of upper layer cortical neurons

(Silva et al., 2018). This regulation is critical to fine-tune the balanced abundance of excitatory neurons and INs, essential for the correct functioning of cortical circuits.

Crosstalk regulation is not limited to interactions between cells within the neural lineage, but also with other cell types. Microglia are the resident cells of the immune system in the cerebral cortex, and are heavily involved in brain cell signaling. During corticogenesis, a variety of microglia subtypes come into direct contact with cortical progenitor cells, and likely regulate their numbers by phagocytosis (Cunningham, Martinez-Cerdeno, & Noctor, 2013; Noctor et al., 2019). Signals from the vasculature also influence cortical progenitor self-renewal, neurogenesis and neural differentiation, via regulation of tissue oxygen levels and release of diffusible growth factors (Javaherian & Kriegstein, 2009; Lange et al., 2016; Moreno et al., 2015).

Thus, the size of the BP population is regulated during its generation by the number of cells delaminating from the VZ and turning into BPs, the duration of this process and their amplification (self-renewal) within the SVZ. The generation of BPs is anticipated within the VZ as subsets of aRGCs start to up-regulate BP fate determinants, such as the transcription factors (TFs) AP2γ (Tfap2c), Cux2, Neurogenin1 and 2, Sox4 and Tbr2. This is the case at mRNA and protein level, and the aRGCs positive for these BP fate determinants are then referred to as differentiating NSCs or nascent BPs (Kawaguchi et al., 2008; Pinto et al., 2008). Accordingly, deletion of these fate determinants leads to reduced numbers of BPs, affecting their generation (Arnold et al., 2008; Chen, Lee, Pourmorady, Sock, & Donoghue, 2015; Pinto et al., 2009; Sessa, Mao, Hadjantonakis, Klein, & Broccoli, 2008) and amplification within the SVZ (Cubelos et al., 2008). Neurogenins have multiple roles in neurogenesis, as discussed below, including driving delamination from the VZ and BP generation (Hand et al., 2005; Li et al., 2012; Miyata et al., 2004) often accompanied by terminal symmetric divisions. Interestingly, these TFs form a cross-regulatory network (Pinto et al., 2009; Sessa et al., 2017) thereby stabilizing the fate of SVZ cells, and mediating resilience in case of mutation of one of these factors. Moreover, many of these TFs that are involved in formation and expansion of the BP pool must be turned off for cells leaving the SVZ. Thus, this TF network regulates both seeding and retention of cells within the SVZ, like the centrosomal protein Akna regulated by Sox4 (Camargo Ortega et al., 2019), and thereby determines the size of this important zone in ontogeny and phylogeny (see below).

5. Regulation of cortical neuron fate

A key aspect in the regulation of cortical neurogenesis is the specification of neuronal fate, the definition of which types of neurons are generated and when. As mentioned in the first section, neurons forming the different cortical layers are generated sequentially following an inside-out gradient, with those destined to deep layers born first, and those destined to superficial layers born last (Rakic, 1974). The basis for our mechanistic understanding of this process lies on pioneering studies by Susan McConnell using cell transplantation strategies in developing ferrets (McConnell, 1985). By taking neural progenitor cells from the cortical primordium of one embryo, labeling them and transplanting into an embryo of a different developmental stage, she demonstrated that at each age cortical progenitor cells are imprinted with an intrinsic program to heavily bias neuron production to only one layer (McConnell, 1988; McConnell, 1989; McConnell & Kaznowski, 1991). Intriguingly, progenitor cells transplanted while in S-phase of the cell cycle change the fate of their daughter neurons to that of those being born in the host embryo. This demonstrates that these progenitors respond to extracellular cues, which seem to define the fate of daughter neurons (McConnell, 1995; McConnell & Kaznowski, 1991). However, the capacity or competence of cortical progenitor cells to respond to extrinsic cues, and change the fate of their daughter neurons, becomes gradually limited as development progresses: early progenitors have the potential to produce neurons for all layers, but late progenitors can only produce late, upper layer neurons (Desai & McConnell, 2000; Frantz & McConnell, 1996). This led to the concept of progressive fate restriction of cortical neurogenesis (Desai & McConnell, 2000).

The temporal regulation and fate restriction of cortical neurogenesis involves a number of transcription factors expressed sequentially by progenitor cells, and thereon by their daughter neurons (Kriegstein et al., 2006; Molyneaux, Arlotta, Menezes, & Macklis, 2007). This classical view was directly challenged by analyses of mice where a reporter gene is expressed in the lineage of cells expressing Cux2, one of the above transcription factors (Franco et al., 2012). This proposed the existence of two main classes of cortical progenitors, one producing neurons to upper layers and another to deep layers. However, this model has been largely disregarded in the field because of the results from subsequent studies. The distinction of aRGC subtypes based on Cux2 expression has been refuted by other labs using similar strategies (Eckler et al., 2015; Guo et al., 2013). The study of neuronal clones identified with mosaic analysis with double markers (MADM), and also with retroviral vectors, refutes the existence of distinct neurogenic progenitor populations. Instead, it reveals that cortical neurons destined to deep and superficial layers are produced essentially by a rather homogeneous population of aRGCs (Gao et al., 2014; Llorca et al., 2019).

The concept of progressive fate restriction of cortical neurogenesis has been revisited recently using high-throughput single-cell RNA sequencing (scRNAseq), which allows enquiring the genetic and molecular mechanisms that regulate this process. In agreement with the concept proposed by McConnell in the 90's, analyses in mice suggest that the transcriptional states of apical progenitors (essentially aRGCs) progress during developmental time, and at each stage they interact with conserved differentiation programs to generate neuronal diversity (Telley et al., 2019). Based on their transcriptomic signatures, the transcriptional states of aRGCs are proposed to shift along development from internally regulated early on, to regulated by external signals later. This is however inconsistent with the temporal fate restriction model, and the classic experimental findings by cell transplantation in ferret, which instead demonstrated that the fate potential of cortical progenitors is highly malleable by environment at early but not late stages (Desai & McConnell, 2000; Frantz & McConnell, 1996).

A key limitation of McConnell's transplantation studies and interpretation of results is that the nature and diversity of cortical progenitor cell types was a decade into the future to be elucidated. Again, scRNAseq analyses have become an unmatched tool with unprecedented resolution to address this issue. Indeed, a recent scRNAseq analysis suggests a key difference between aRGCs and IPCs in the temporal plasticity of their neuron fate potential (Oberst et al., 2019). On the one hand, aRGCs seem to be temporally plastic and able to re-enter past molecular and neurogenic states when exposed to an earlier-stage environment. In contrast, IPCs are proposed to be committed progenitors that lack such retrograde fate plasticity (Oberst et al., 2019). While this is a fascinating new model compatible with earlier work, this study also bears some inconsistencies with previous analyses that will need to be sorted out, primarily the capacity of late aRGCs to be completely re-specified to produce early neurogenic fates (Frantz & McConnell, 1996).

6. Intrinsic factors—Key transcriptional mechanisms

The coordinated unfolding of the many complex events that are involved in the formation of the cerebral cortex requires an exquisite orchestration of multiple molecular pathways, which acquire different relevance at different steps. Several transcription factors (TFs) have been involved in this regulation, with Pax6, Neurogenins and Fox transcription factors being best understood at the molecular level. These will be discussed first before we move on to the signaling cascades acting on progenitor cell proliferation and neuronal differentiation.

6.1 Transcription factors promoting aRGC self-renewal

Given the multitude of cell biological hallmarks retaining aRGCs in the VZ versus preparing their delamination, a series of key transcriptional regulators have been identified that promote self-renewal and retention of aRGCs in the VZ, such as Foxg1, Pax6, Sox9 and Trnp1 (see detailed description below and (Manuel et al., 2011; Scott et al., 2010; Stahl et al., 2013; Xuan et al., 1995)). These factors often antagonize the expression of key transcriptional regulators for BP fate as described above. Thus, subsets of aRGCs can be discriminated at each point in time that are biased to differentiate (the nascent BPs described above) co-expressing Tbr2, Sox4, Ngn2 and AP2γ, versus those that self-renew and remain in the VZ, do not express the BP fate determinants, but e.g. higher levels of Trnp1 (Esgleas et al., 2020; Stahl et al., 2013). Several of these TFs are subject to extrinsic regulation, e.g., by phosphorylation mediated by growth factor-derived signals that e.g. profoundly alter the function of Ngn2 at mid-neurogenesis from proneural to sub-type-specification (Li et al., 2012). So how are these cascades of transcription factors regulated and coordinated given the importance of SVZ formation and expansion by extrinsic signals?

6.1.1 Pax6

This paired-box transcription factor exerts multiple actions during cortical development via regulating gene expression in progenitor cells, where its precise effects are heavily context-dependent (Manuel, Mi, Mason, & Price, 2015; Mi et al., 2013). For example, it regulates cell cycle length and exit in a cell-autonomous manner, but in such a way that loss of Pax6 induces shortening of the cell cycle at early stages of cortical development, while it has the opposite effect at late stages. Moreover, the magnitude of these regulatory effects is linked to its expression levels, where cortical areas with higher *Pax6* expression levels are more profoundly affected (Mi et al., 2013). Pax6 also regulates progenitor proliferation and cell cycle re-entry in a dose- and context-dependent manner. Premature cell-cycle exit is caused by high Pax6 expression at early stages of corticogenesis, but then also by loss of Pax6 at late stages (Haubst et al., 2004; Manuel et al., 2015; Wong et al., 2015). Pax6 also acts as neural fate determinant in the developing cortex. The number of neurons produced by RGCs is directly related to the dose of Pax6, where its loss leads to a 50% reduction in neurons (Schmahl, Knoedlseder, Favor, & Davidson, 1993), and its overexpression increases their number (Heins et al., 2002). In particular, Pax6 is critical for the generation and specification of late born cortical neurons, where its conditional ablation results in a dramatic reduction of superficial layer neurons (Georgala, Manuel, & Price, 2011; Tuoc et al., 2009). Pax6 also participates in the acquisition of the glutamatergic phenotype throughout cortical development, but with a particularly important influence at late stages, when the absence of Pax6 reduces the levels of cortical and glutamatergic markers (Schuurmans et al., 2004). The transcriptional regulation of *Pax6* itself is also complex, with four different transcription start sites, several distant regulatory elements, and several splice variants that potentially exert distinct functions (Haubst et al., 2004; Manuel et al., 2015; Walcher et al., 2013). Interestingly, Pax6 utilizes its modular nature of 3 DNA-binding domains to exert distinct functions. A constitutive subdomain of the paired domain is involved in its neurogenic fate and promoting cell cycle exit, while its splice variant promotes proliferation and self-renewal along with the homeobox-domain mediating survival (Ninkovic et al., 2010; Walcher et al., 2013). Pax6 is also a direct regulator of centrosomal proteins and some of the cell biological hallmarks discussed above (Tylkowski, Yang, Hoyer-Fender, & Stoykova, 2015). In summary, Pax6 regulates cortical neurogenesis in a dose- and context-dependent manner by influencing proliferation, differentiation, neurogenesis and cell fate specification (Manuel et al., 2015). Not surprisingly, loss-of-function mutations in human *PAX6* cause severe neurological and psychiatric dysfunctions, and important structural defects (Bamiou et al., 2007; Davis et al., 2008; Maekawa et al., 2009; Malandrini et al., 2001; Manuel et al., 2015).

6.1.2 Basic-helix-loop-helix transcription factors

Basic-helix-loop-helix (bHLH) TFs are downstream effectors of Notch signaling that regulate neural fate choice. Hes1 is a transcriptional repressor expressed in neural progenitor cells that shows oscillatory expression at the mRNA level, where disruption of these oscillations alters the maintenance of progenitor cell proliferation (Imayoshi et al., 2013; Imayoshi, Ishidate, & Kageyama, 2015; Ochi, Imaizumi, Shimojo, Miyachi, & Kageyama, 2020). For example, sustained high expression of *Hes1* in progenitor cells represses the expression of cell cycle regulators, leading to cell cycle arrest (Baek, Hatakeyama, Sakamoto, Ohtsuka, & Kageyama, 2006). *Hes1* oscillations may result from intronic-dependent expression delays, as changes in their length lead to unstable oscillations (Ochi et al., 2020). In addition, negative feedback through *miR-9* may further promote and provide robustness to *Hes1* oscillations (Bonev, Stanley, & Papalopulu, 2012; Goodfellow, Phillips, Manning, Galla, & Papalopulu, 2014). Mathematical modeling shows that the intrinsic biological noise of the *Hes1/miR-9* oscillatory system favors the progenitor fate, while in contrast *Hes5* oscillations become periodic as differentiation progresses (Manning et al., 2019; Phillips et al., 2016). Other mechanisms regulating *Hes1* expression and cortical progenitor cell fate include Robo signaling. In the parietal mouse cerebral cortex, signaling via the transmembrane receptors Robo1 and Robo2 enhances *Hes1* transcription, and this restrains the generation of IPCs in favor of aRGC self-renewal (Borrell et al., 2012). Oscillations of *Hes1* expression levels repress the expression of other bHLH proneural genes like *Ascl1*, *Neurog2* and *Dll1*, thus driving their oscillatory expression (Imayoshi et al., 2015, 2013; Shimojo et al., 2016; Shimojo, Ohtsuka, & Kageyama, 2008). Similar to *Hes1*, the oscillatory expression of *Ascl1* induces progenitor cell proliferation, whereas sustained expression promotes neuronal differentiation (Imayoshi et al., 2013; Ochi et al., 2020). *Hes* genes also negatively regulate the expression of other genes involved in progenitor cell fate decisions, including *Insm1* (Farkas et al., 2008).

Other proneural bHLH transcription factors are Neurogenins (Ngn). Ngn1 and Ngn2 are crucial factors for neurogenesis in the cerebral cortex (Hirabayashi et al., 2004; Schuurmans & Guillemot, 2002). On the one hand, Ngn1 and Ngn2 dimerize with ubiquitous bHLH proteins and bind to specific DNA sequences, activating tissue-specific gene expression that promotes neuronal differentiation (Cau, Gradwohl, Fode, & Guillemot, 1997). On the other hand, Ngn1 blocks alternative fates by, for example, inhibiting astrocyte differentiation via sequestering an essential complex that activates the transcription of astrocyte differentiation genes, and suppressing JAK/STAT signaling pathways (Sun et al., 2001). Importantly, the proneural role of Ngns is subject to regulation by signaling and finishes by mid-neurogenesis in the developing cortex. This is mediated by phosphorylation switching the target regulation of Ngns (Li et al., 2012).

7. Signaling pathways regulating progenitor cell lineage

A number of signaling cascades regulate the lineage and fate of aRGCs. Robo signaling, via modulating the Notch pathway, plays a central role in regulating the balance between direct and indirect neurogenesis in the developing cerebral cortex. As explained in a later section, this mechanism is evolutionarily conserved, and the modulation of *Robo-Dll1* expression levels during vertebrate evolution seems to have driven changes in the primary mode of neurogenesis leading to cortical expansion (Cárdenas et al., 2018). With regard to the generation and lineage of BPs (IPCs and bRGCs), one key regulator is Insulinoma-associated 1 (Insm1). This transcription factor prevents aRGC expansion and drives their delamination to basal positions, thus promoting the generation of IPCs. This is achieved by repressing the expression of Plekha7, a component of the AJ belt (Farkas et al., 2008; Tavano et al., 2018). The absence of *Insm1* causes a decrease in cortical neurons of both upper and lower layers, likely due to a reduction in IPCs. Concomitantly, overexpression of *Insm1* increases IPC production and self-amplification, at the expense of aRGCs. Interestingly, Robo and Insm1 signaling seem to be part of a robust regulatory system promoting aRGC fate, as *Robo2* expression is reduced in *Insm1* mutant mouse embryos (Farkas et al., 2008).

The expansion of BPs seems to require the existence of an appropriate niche in the SVZ, which may come at least in part from BPs themselves. For example, the SVZ is rich in extracellular matrix components that promote BP proliferation, and that are expressed by BPs themselves. Expression of these ECM components is promoted by Sox9, which itself enhances BP proliferation (Güven et al., 2020). Hence, Sox9 promotes BP proliferation directly as well as indirectly, via the extracellular matrix. Accordingly, Sox9 is expressed in highly proliferative progenitor cells promoting their proliferation and self-renewal, such as aRGCs in mouse, ferret and human, as well as BPs in ferret and human, but not in progenitors that are mostly neurogenic, such as BPs in mouse (Güven et al., 2020; Scott et al., 2010). A similar role is played by the transcriptional activator YAP, downstream effector of the Hippo signaling pathway (Kostic et al., 2019). Loss of YAP expression in ferret and human cortex reduces BP proliferation, while it is promoted in mouse upon expression of constitutively active YAP (Kostic et al., 2019). In mice, YAP is also active in aRGCs and promotes their proliferation (Cappello et al., 2013), consistent with both aRGCs and bRGCs sharing many regulators and both being NSCs.

The Wnt pathway has a very profound influence on cell fate decisions. An extracellular Wnt ligand binds to a receptor of the Frizzled family and activates intracellular signaling. Wnt signaling is in fact a complex set of signal transduction pathways, three of which are most prevalent: the canonical, the non-canonical planar cell polarity, and the non-canonical Wnt/calcium signaling. Activation of the canonical Wnt/β-catenin signaling at early stages of cortical development promotes progenitor cell proliferation and represses neuronal differentiation (Chenn & Walsh, 2002; Stocker & Chenn, 2015; Woodhead, Mutch, Olson, & Chenn, 2006). Importantly, Wnt/β-catenin signaling must be gradually downregulated during development to transition from early aRGC self-amplification to late neurogenesis. Otherwise, constitutive Wnt signaling leads to generating an over-grown neuroepithelium largely devoid of neurons (Chenn & Walsh, 2002). Similar to other pathways, the role of Wnt on neural progenitor cells seems to be stage- and ligand-dependent, as loss of Wnt7a signaling at late stages disrupts neuronal differentiation, for example (Hirabayashi et al., 2004; Stocker & Chenn, 2015).

Another complex and evolutionarily conserved signaling pathway with fundamental roles in development of the central nervous system is the Sonic Hedgehog (Shh) signaling pathway. Shh signaling is very important in cell lineage regulation and may influence neurogenesis at several levels. Shh is a ligand molecule that binds to the Patched transmembrane receptor (Ptch) and its co-receptors (Cdon, Boc and Gas1) (Allen et al., 2011; Marigo, Davey, Zuo, Cunningham, & Tabin, 1996; Stone et al., 1996). Under baseline conditions, Ptch is bound to Smoothened (Smo), another transmembrane protein. Binding of Shh to Ptch releases Smo, which then initiates an intracellular signaling cascade ultimately activating the zinc-finger transcription factor Gli (Yabut & Pleasure, 2018). Loss of Shh or Smo at early developmental stages causes lengthening of the cell cycle, thus impairing progenitor cell proliferation, reducing genesis of BPs and causing microcephaly (Komada et al., 2008). Conversely, loss of Ptch1 at early stages (all Smo is unbound) increases symmetric aRGCs divisions (Shikata et al., 2011). However, these effects are context-dependent, as loss of Ptch1 at late stages abolishes aRGC self-renewal, while this is increased by overexpression of constitutively active Smo at mid-corticogenesis (Wang et al., 2016).

Shh signaling is also critical for the specification and expansion of a specific subtype of bRGCs (expressing *Hopx*), leading to cortical folding of the otherwise smooth mouse cortex (Matsumoto, Tanaka, Horiike, & Shinmyo, 2020). Shh signaling is critically dependent on the primary cilium, without which Shh signals cannot be transmitted in vertebrates, an evolutionary invention profoundly discriminating this signaling in vertebrates and invertebrates. In the developing murine cerebral cortex, Ptch receptors bind Shh on the cilia of aRGCs (Stocker & Chenn, 2015). Shh is also involved in other non-neurogenic processes such as dorso-ventral neural patterning, cortical gliogenesis and specification of neuronal types (Yabut & Pleasure, 2018).

8. Extrinsic signals

Beyond the intrinsic components of cortical progenitor cells that influence their behavior and fate, multiple extracellular signals also contribute to this regulation and thus, to cortical neurogenesis. Transcriptomic analyses reveal that at early developmental stages, apical cortical progenitors follow genetic programs that are largely intrinsic, but then at late stages of neurogenesis they gradually shift to be more receptive to extracellular signals (Telley et al., 2019). These signals may come from different sources, including the progenitor cells themselves, mature neurons, endothelial cells and CSF (Lehtinen et al., 2011; LoTurco, Owens, Heath, Davis, & Kriegstein, 1995; Mayer et al., 2019; Reillo et al., 2017; Weissman, Riquelme, Ivic, Flint, & Kriegstein, 2004). For example, IPCs have been shown to interact with aRGCs via Notch signaling, thus influencing aRGC behavior and fate (Nelson, Hodge, Bedogni, & Hevner, 2013). A dynamic and complex network of long-range and filopodia-like cell-cell contacts establish Notch signaling communication between IPCs and the other classes of apical and BP cells, which possibly is important to maintain Notch activity throughout development (Nelson et al., 2013). Similarly, axonal projections from extracortical regions innervate the cerebral cortex at early and late embryogenesis, influencing its development (Antón-Bolaños et al., 2019; Dehay & Kennedy, 2007; Dehay, Savatier, Cortay, & Kennedy, 2001; Reillo et al., 2017). This influence may come from the release of neurotransmitters, known to modulate progenitor cell proliferation, neuronal migration and differentiation, likely via inducing calcium transients (LoTurco et al., 1995; Mayer et al., 2019; Weissman et al., 2004). Calcium waves diffuse via connexin hemichannels between cortical progenitor cells and promote their proliferation (Weissman et al., 2004). The resting membrane potential of aRGCs gradually decreases along the neurogenetic period, which seems to increase their neurogenic potential (Vitali et al., 2018). Intriguingly, a very low resting membrane potential in astrocytes does not make them neurogenic, which seems to indicate that if this feature is relevant for neurogenesis it may be so restricted to aRGCs. This suggests an interface of regulatory mechanisms with the membrane potential that will be intriguing to unravel. Radial communication between postmitotic areas and progenitor cells may take place by bidirectional calcium propagation through the basal process of RGCs (Reillo et al., 2017), a mechanism of long distance communication induced by Notch signaling and growth factors (Rash, Ackman, & Rakic, 2016).

9. Genome-wide transcriptomic and epigenomic regulation

Understanding the genetic mechanisms that regulate brain development, and in particular development of the cerebral cortex, has been addressed for decades at the level of individual genes. Mutant mouse lines knock-out for a single gene, or a handful of related genes, have been used to shed light on their necessity for specific aspects of cortical development. The next generation of mutant mouse strains, where the gene or genes of interest are specifically and solely eliminated from a particular tissue or cell type of interest (conditional), and/or during a time window of interest (inducible), allow a significant refinement of this understanding. However, this also led eventually to the realization that all events in cortical development are regulated by more than just one gene, but by a true network of genes and signaling cascades that are functionally interrelated. Thus, understanding the regulation of cortical development and neurogenesis necessarily requires a holistic approach, which begins with a characterization and understanding at the genomic level.

9.1 Transcriptomes of germinal zones

Understanding the genomic regulation of cerebral cortex development was first addressed by characterizing the transcriptomic landscape of its germinal zones (Fig. 3A). High throughput transcriptomic characterization of individual germinal layers was first done using microarray technology. In the embryonic mouse cortex this was pioneered by Rakic and colleagues, who found significant differences between germinal zones and non-germinal zones

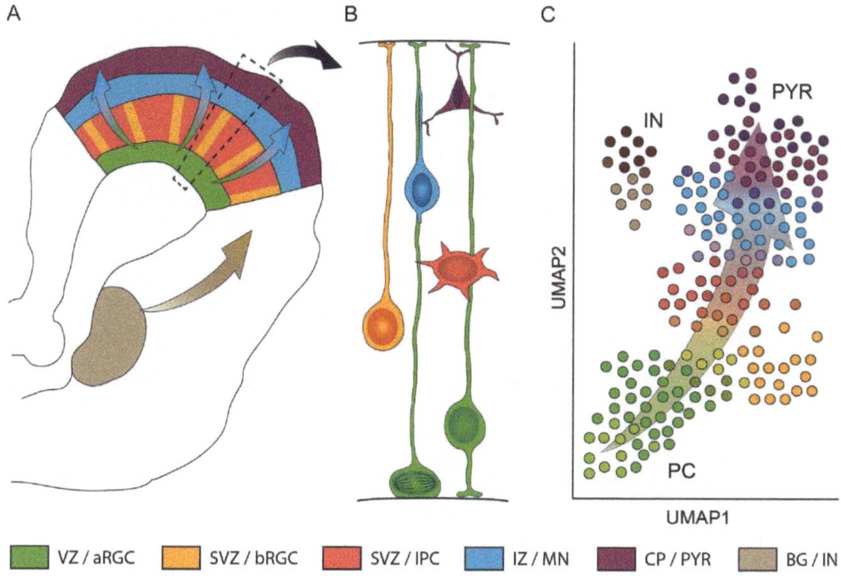

Fig. 3 Transcriptomic profiling of the developing cerebral cortex. (A) Schematic of a human telencephalic hemisphere in coronal section showing the main layers, with distinct transcriptomic profiles: VZ (green), SVZ (pink/orange), IZ (blue) and CP (purple). VZ and SVZ generate excitatory neurons that migrate radially to the CP (blue arrows). The medial ganglionic eminence (beige) generates inhibitory interneurons (IN) that migrate tangentially to the cortex. (B) Different cortical cell populations exhibit distinct transcriptomic profiles: aRGCs (green), bRGCs (orange), IPCs (pink), migrating neurons (blue) and mature neurons (purple). (C) Analyses at single cell resolution identify unique transcriptomic profiles, identifying the continuum of cell types and their lineage trajectories, from progenitor cells (PC) to mature pyramidal (PYR) and inhibitory (IN) neurons.

and, most importantly, between VZ and SVZ (Ayoub et al., 2011). This landmark study identified transcriptional programs containing largely uncharacterized genes in addition to transcripts necessary for stem cell maintenance, neurogenesis, migration, and differentiation. Similar analyses soon followed using next-generation sequencing (RNAseq). By combining laser capture microdissection and RNAseq, the individual cortical germinal zones of mouse embryos were analyzed, confirming that the transcriptome of the VZ is quite distinct from the SVZ, which remarkably displays a greater resemblance with the CP (non-germinal) than with the VZ (Fietz et al., 2012). With an interest in understanding human-specific features of cortical development, this and other studies extended the transcriptomic analysis to germinal zones of the human embryo cerebral cortex. In contrast to mouse, the transcriptomes of the human ISVZ and OSVZ were found to be similar to each other, and much more similar to the VZ than to the CP (Fietz et al., 2012; Miller et al., 2014). In more general terms, a high proportion of the human transcriptome has been found to be expressed in the developing cerebral cortex, with nearly half being differentially regulated between developmental stages or across brain regions following gradients of expression (Johnson et al., 2009; Kang et al., 2011; Miller et al., 2014). Differentially expressed genes seem to be more frequently associated with human-specific evolution of putative cis-regulatory elements, and distinct cortical regions and neurodevelopmental processes are bound to a large number of patterns of specific gene expression and alternative splicing (Johnson et al., 2009). At the functional level, Huttner and colleagues (Fietz et al., 2012) were the first to pinpoint a candidate molecular mechanism, by highlighting differential expression of extracellular matrix-associated genes that include distinct sets of collagens, laminins, proteoglycans and integrins. This suggested that cell adhesion and cell–extracellular matrix interactions influence the proliferation and self-renewal of neural progenitors in the developing human neocortex, as later shown experimentally in mouse and ferret (Fietz et al., 2010; Güven et al., 2020; Long & Huttner, 2019; Stenzel, Wilsch-Brauninger, Wong, Heuer, & Huttner, 2014).

Transcriptomic analyses of individual germinal zones have also been fundamental to investigate more specific questions of cortical development and evolution. A seminal study in macaque embryos used microRNA sequencing to investigate genetic mechanisms regulating the definition of area-specific features (Arcila et al., 2014). A large number of primate-specific miRNAs were identified, including many that target cell-cycle proteins, again highlighting an evolutionary pressure for the regulation of primate-specific features of cortical development (Arcila et al., 2014). Seminal insights into the genetic regulation of cerebral cortex folding were also obtained via transcriptomics and the exquisite microdissection of cortical regions and germinal layers (De Juan Romero et al., 2015; Martínez-Martínez et al., 2016). This unraveled the existence of discrete domains of gene expression defining the eventual formation (mapping) of cortical folds and fissures (De Juan Romero et al., 2015; de Juan Romero & Borrell, 2017; Fernandez et al., 2016). A similar strategy was followed to investigate changes in the transcriptional landscape of the VZ along development, identifying prominent variations under tight temporal control that underlie the formation of the OSVZ (Martínez-Martínez et al., 2016). Proof-of-concept for the functional involvement of some of those mapping genes in folding of the cerebral cortex has been experimentally demonstrated in mouse (del Toro et al., 2017; Stahl et al., 2013) and ferret

(Martínez-Martínez et al., 2016; Masuda et al., 2015; Toda, Shinmyo, Dinh Duong, Masuda, & Kawasaki, 2016).

A fundamental limitation of bulk tissue transcriptomic analyses is the absence of detailed spatial information, the subtleties of which are missed even using tissue microdissection. To this end, high-throughput analyses of gene expression in the embryonic brain of mouse and macaque monkey have been performed by In Situ Hybridization, highlighting the many similarities, but also dramatic differences, between these species (Bakken et al., 2016; Diez-Roux et al., 2011).

9.2 Transcriptomes of germinal cells

As presented in the first sections, the developing mammalian cerebral cortex contains a remarkable cellular heterogeneity, both between germinal zones and within them. Consequently, there is also significant genetic and transcriptional heterogeneity within each germinal zone. Insights into the transcriptomic basis of observed cellular heterogeneity were first obtained by bulk transcriptomic analyses (microarray or RNAseq) of distinct cell populations isolated by Fluorescence Activated Cell Sorting (FACS; Fig. 3B). In a seminal study by Götz and colleagues, a combination of FACS and microarray analysis was used to characterize the transcriptional profile of two populations of aRGCs in the embryonic mouse cortex, distinguished by their cell amplification and neurogenic potential (Pinto et al., 2008). This screen unraveled major new regulators of aRGC self-renewal (Martínez-Martínez et al., 2016; Stahl et al., 2013) and BP cell formation (Camargo Ortega et al., 2019; Pinto et al., 2009). A similar strategy was used to investigate the transcriptomic basis of cortical progenitor cell fate commitment during neurogenesis, by comparing progenitors fated for self-amplification versus neuron production (Aprea et al., 2013; Artegiani et al., 2015). The results from these screens provided not only convincing evidence of remarkable transcriptional heterogeneity among aRGCs and neurogenic progenitor cells, but also of its striking temporal dynamics during development, and revealed specific sets of genes with key functions in cerebral cortex development (Artegiani et al., 2015; Ramesh et al., 2016).

The success of high-throughput transcriptomic studies in mouse prompted the use of similar strategies for the identification of genetic programs involved in the human-specific development and expansion of the cerebral cortex. Profiting from known anatomical and molecular features that distinguish progenitor cell populations in the human cortex, including aRGCs, bRGCs and IPCs, these were sorted and their bulk transcriptome was profiled (Fig. 3B) (Florio et al., 2015; Johnson et al., 2015). This showed that, at the transcriptional level, the diversity of cortical progenitor cells is greater in human and ferret than in mouse, confirming previous anatomical discoveries (Fietz et al., 2010; Hansen et al., 2010; Reillo et al., 2011; Reillo & Borrell, 2012). These findings further allowed suggesting a greater neuronal lineage commitment and differentiation of self-renewing progenitors in human than in other species (Johnson et al., 2015). More importantly, the transcriptomic analysis of sorted cells led to the identification of *ARHGAP11B*. This gene is selectively and highly expressed in human RGCs (apical and basal), but not in mouse or marmoset, and has a key and prominent role in the generation and amplification of BPs, thus becoming the first human-specific gene discovered with a likely role in the evolutionary expansion of the human cortex (Florio et al., 2015; Florio, Borrell, & Huttner, 2017; Heide et al., 2020).

Transcriptomic profiles are profoundly shaped by epigenetic regulation. High-throughput sequencing technology has also enabled characterizing the epigenetic landscape of embryonic cortical layers and cell types, and understanding the epigenetic regulation of cortical development and neurogenesis (Albert et al., 2017; Pataskar et al., 2016; Thakurela, Sahu, Garding, & Tiwari, 2015; Ypsilanti & Rubenstein, 2016). This has revealed, for example, the highly dynamic nature of distal gene regulation during neurogenesis, which gets progressively restricted to distinct genomic regions as neurons acquire a post-mitotic, terminally differentiated state. These distal accessible and active regions serve as target sites for distinct transcription factors that function in a stage-specific manner to contribute to the transcriptional program underlying neuronal commitment and maturation (Thakurela et al., 2015). One of such transcription factors essential in neurogenesis and cortical development is NeuroD1. This has been shown to directly bind regulatory elements of neuronal genes that are developmentally silenced by epigenetic mechanisms. This targeting is sufficient to reprogram the transcription factor landscape, convert heterochromatin to euchromatin, and increase chromatin accessibility (Pataskar et al., 2016). Such transcriptional induction of neuronal fate genes is maintained via epigenetic memory. At the level of cortical progenitor cell types, genome-wide maps of histone methylation have been described and compared for distinct cell populations in the developing mouse neocortex (Albert et al., 2017; Elsen et al., 2018). Extensive differences in H3K27me3 modifications exist between neural progenitor cell subtypes coinciding with major developmental and cell biological transitions, with dynamic changes occurring on promoters of several crucial transcription factors. This is the case for Tbr2, the regulator of neurogenic IPCs. H3K27me3 methylation at the Tbr2 locus results in its reduced expression and lower IPC abundance (Albert et al., 2017).

A crucial epigenetic regulation is mediated by chromatin looping and the 3D organization bringing enhancer and promoter elements close to each other. The first comprehensive 3D chromatin analysis in ES-derived neural stem and progenitors, and cortex-derived RGCs and BPs, revealed profound 3D reorganization accompanying the transcriptional response to major neurogenic factors (Bonev et al., 2017). Hence, cell fate specification and neurogenesis during embryonic cortical development involves the concerted action of transcription factors and epigenetic mechanisms.

Epitranscriptomics, or the epigenetic modification of the transcriptome, has recently been identified as an additional layer of genetic regulation of cortical development, particularly neurogenesis (Yoon et al., 2017; Yoon, Vissers, Ming, & Song, 2018). Methylation of mRNAs in adenosine residues drives their decay, which is critical for proper transcriptional pre-patterning in cortical neurogenesis. Accordingly, loss of Mettl factors, the prevalent methyltransferases, dramatically alters the transcriptional landscape promoting progenitor cell amplification in detriment of neurogenesis (Yoon et al., 2017).

9.3 Developmental transcriptomics at the single-cell level

Discoveries on new cortical progenitor cell types, the specialized roles of each type, and their distinct relevance in different stages and aspects of cortical development, have evidenced the fundamental importance of studying this complex process at single cell resolution. A pioneer study by Matsuzaki and colleagues reported a single-cell transcriptomic analysis performed semimanually on only few hundreds of cells of the developing brain (Kawaguchi et al., 2008). This seminal work revealed the important concept of transcriptional priming of the BP cell lineage already in aRGCs. With the advent of microfluidics-based single cell technologies (scRNAseq) hundreds of cells could be analyzed systematically and in parallel (Frazer et al., 2017; Pollen et al., 2015, 2014). Subsequent major technological breakthroughs on the massively parallel analysis of individual cells have enabled augmenting the depth of analysis and the number of cells analyzed by up to three orders of magnitude (Han et al., 2020; Hodge et al., 2019; Tasic et al., 2018). Recent developments now allow analyzing chromatin accessibility on single cells (scATACseq), alone and in combination with scRNAseq (Chen, Lake, & Zhang, 2019; Zhu et al., 2019), offering the opportunity to understand the regulation of cortical neurogenesis with unprecedented resolution (Fig. 3C).

Comparative single cell studies of the developing cerebral cortex in human and non-human primates have revealed global intraspecies (ontogenetic) and interspecies (phylogenetic) transcriptomic differences between brain regions (Zhu et al., 2018), as expected from previous bulk transcriptomic analyses (Bakken et al., 2016; Miller et al., 2014). Transcriptomic differences exhibit hourglass-shaped patterns, with prenatal patterns displaying transient topographic gradients, followed by perinatal convergence, and then postnatal patterns that reflect functional hierarchy (Zhu et al., 2018). Importantly, genes exhibiting heterotopic and heterochronic divergence include those linked to autism spectrum disorder and schizophrenia, suggesting the existence of human specializations in gene expression programs that, while favoring brain complexification, may also favor the appearance of pathological alterations of social interaction. This notion is further substantiated by transcriptomic and epigenomic analyses of the human brain during prenatal and postnatal development, across brain regions and/or cell types (Li et al., 2018), strongly supporting a potential neurodevelopmental origin and genomic basis of neuropsychiatric risks.

Single cell studies have also successfully identified discrete cell types as well as temporally and spatially restricted trajectories of RGC maturation and neurogenesis in the developing human telencephalon (Fig. 3C) (Nowakowski et al., 2017). These lineage-specific trajectories reveal significant temporal variations in gene expression among progenitor cell types, where the expression of neurogenic transcription factors dominate in early RGCs. Intriguingly, only modest transcriptional differences have been found among RGCs across cortical areas, in spite of the existence of robust typological distinctions among maturing neurons. This suggests a mixed model where topographical, typological and temporal hierarchies govern cell-type diversity and lineage in the developing human telencephalon (Nowakowski et al., 2017). Regarding bRGCs, scRNAseq analyses show that bRGCs preferentially express genes related to extracellular matrix formation, migration and progenitor cell stemness (Pollen et al., 2015). These molecular features are related to distinctive behaviors of bRGCs, including their extensive proliferative potential, hence suggesting that bRGCs directly support the SVZ niche at the molecular level, promoting the developmental and evolutionary expansion of the human cerebral cortex (Pollen et al., 2015).

Insights into the mechanisms of cortical neurogenesis and its variation along development have been obtained also with single-cell transcriptomic studies in mouse, using a high-temporal resolution strategy (Telley et al., 2019, 2016). These studies have identified a set of temporally patterned genes that drive the state of aRGCs from internally driven to more extrinsically driven. This temporal progression of aRGC states is epigenetically regulated by the Polycomb repressor complex 2 (PRC2). The age-dependent molecular states of aRGCs are transmitted to their progeny as successive ground states, onto which early neuronal differentiation programs are applied (Telley et al., 2019). These programs are then directed by a series of transcriptional waves whose proper sequence is critical for the normal progression through development. Waves provide discrete time windows during which specific transcriptional complexes are simultaneously present and can thus interact (Telley et al., 2016). At later developmental stages, these waves are complemented by environment-dependent signals (Telley et al., 2019).

10. Evolution of cortical neurogenesis

One of the most fascinating questions about cortical development in general, and neurogenesis in particular, is how it unfolded during evolution. Unfortunately, our knowledge on this is highly fragmentary and incomplete, because most of our understanding of cortical development is based on studying the mammalian neocortex. The mammalian neocortex derives during development from an embryonic subdivision of the telencephalic vesicles called dorsal pallium, which also exists in the embryonic brain of birds and reptiles. The more broadly accepted view is that the avian hyperpallium and the reptilian dorsal cortex are homologous to the mammalian neocortex (Luzzati, 2015; Puelles, 2001; Puelles, 2017). Although some controversy remains, this view has been further supported recently by single-cell transcriptomic analyses (Desfilis, Abellan, Sentandreu, & Medina, 2018; Tosches et al., 2018).

During the evolution of mammals, neocortical size and complexity increased disproportionately relative to other brain regions (Finlay & Darlington, 1995; Florio & Huttner, 2014). Cortical expansion during the evolution of amniotes resulted from the combination of increased neuronal production and changes in cell packing (Geschwind & Rakic, 2013; Herculano-Houzel, Collins, Wong, Kaas, & Lent, 2008). For example, some parrots and songbirds have much smaller brains than some primates but twice as many neurons, hence those primates have a much lower neuronal density (Olkowicz et al., 2016). Differences in neurogenesis result from fundamental variations in neural progenitor cells during development: their types, abundance and lineage dynamics (Cárdenas & Borrell, 2020; Kriegstein et al., 2006; Llinares-Benadero & Borrell, 2019). Whereas aRGCs are highly conserved across phylogeny, the existence of BPs and an SVZ, their diversity and neurogenic potential, are much more divergent (Cárdenas et al., 2018; Cheung et al., 2010; De Juan Romero & Borrell, 2015; Martinez-Cerdeno, Noctor, & Kriegstein, 2006; Molnar, 2011).

As mentioned in the first section, indirect neurogenesis is the predominant mode of cortical neuron production in mammals, greatly amplified in primates and other species with a prominently large cortex (Cárdenas et al., 2018; Dehay & Kennedy, 2007; Haubensak et al., 2004; Reillo et al., 2011). In contrast to mammals, the existence of BPs and SVZ in other, more basal clades remains under debate. In the embryonic dorsal pallium of cartilaginous fish, at the base of vertebrate phylogeny, Tbr2 positive cells have been identified forming a distinct SVZ and undergoing basal mitosis, which suggests that basic features of the SVZ and mechanisms for its development may have already existed in the ancestor common to all jawed vertebrates (Docampo-Seara et al., 2018). In amphibians, however, non-apical mitoses have been observed at extremely low frequency in the developing dorsal pallium, and these are negative for Tbr2 (Moreno & Gonzalez, 2017; Nomura et al., 2016). Similarly, there is no evidence of BPs nor SVZ in lepidosaur reptiles (snakes, lizards and geckos), but only the presence of non-proliferative Tbr2 positive cells (Cárdenas et al., 2018; Martinez-Cerdeno et al., 2015; Nomura, Gotoh, & Ono, 2013). Only a subset of archosaurs (turtles and birds, but not crocodiles) display a small number of Tbr2 positive non-apical mitoses in the dorsal cortex, but these are generally not organized in a distinct SVZ (Charvet, Owerkowicz, & Striedter, 2009; Cheung, Pollen, Tavare, Deproto, & Molnár, 2007; Clinton, Cunningham, Kriegstein, Noctor, & Martinez-Cerdeno, 2014; Martinez-Cerdeno et al., 2015; Nomura et al., 2016). In orders of birds with an ancient evolutionary origin, including chicken and dove, a small number of basal mitotic cells positive for Tbr2 align in the hyperpallium, defining a vestigial SVZ (Martinez-Cerdeno et al., 2015). In contrast, the presence of an SVZ is more obvious in orders of birds that emerged more recently, like *Passeriformes* (parakeets) and *Psittaciformes* (songbirds), potentially related to the greater complexity of their brains (Charvet & Striedter, 2008). Hence, it seems that a distinct SVZ, and prominent indirect neurogenesis, emerged or expanded during the transition sauropsids-mammals. The gradual transition from a

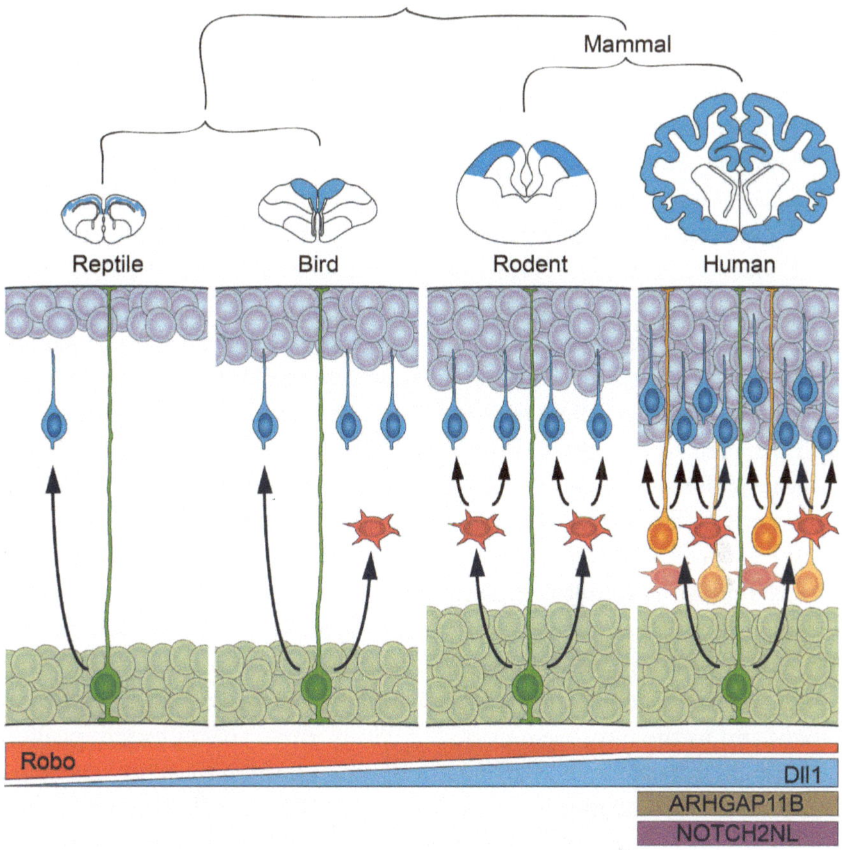

Fig. 4 Evolution of cortical neurogenesis. In reptiles and birds most cortical neurons are generated by direct neurogenesis from aRGCs (green), whereas in mammals there is a massive prevalence of indirect neurogenesis, via IPCs (red) and bRGCs (orange). This evolutionary switch in the mode of neurogenesis is driven by changes in expression levels of Robo and Dll1. The appearance of new genes in the recent human lineage (ARHGAP11B, NOTCH2NL) contributed to increase the generation of basal progenitors and a further expansion of cerebral cortex size. Increased in neuron numbers were accompanied by a biased expansion in cortical surface area over thickness (bottom, blue areas in brains sections). *Adapted from Cárdenas, A., Villalba, A., de Juan Romero, C., Picó, E., Kyrousi, C., & Tzika, A.C., et al. (2018). Evolution of cortical neurogenesis in amniotes controlled by robo signaling levels. Cell, 174(3), 590–606.e21. https://doi.org/10.1016/j.cell.2018.06.007.*

prevalent direct neurogenesis, as found in amphibians, reptiles and some birds, to a prevalent indirect neurogenesis, typical of mammals, was critical in leading to the remarkable expansion of the cerebral cortex during vertebrate evolution (Fig. 4) (Cheung et al., 2010; Kriegstein et al., 2006).

At the genetic and molecular level, the mechanisms that emerged during evolution driving the emergence of indirect neurogenesis, and hence expansion of the cerebral cortex, are just beginning to be unraveled. Several proteins and signaling pathways play highly conserved key roles in cortical development across phylogeny. For example, the secreted protein Sonic hedgehog (Shh) promotes amplificative divisions of aRGCs in the chick spinal cord and the mouse neocortex (Komada, 2012; Komada et al., 2008; Saade et al., 2013). Similarly, the transcription factor Pax6 is critical in regulating the proliferation and fate of aRGCs across amniotes, where a fine-tuning of its expression in a spatial-temporal fashion and dosage-dependent manner modulates these functions differently across amniotes (Asami et al., 2011; Manuel et al., 2015; Mi et al., 2013; Quinn et al., 2007; Sansom et al., 2009; Wong et al., 2015; Yamashita et al., 2018).

One of the most fundamental and highly conserved signaling pathways regulating cell fate across vertebrate phylogeny is Notch (Gaiano & Fishell, 2002; Pierfelice, Alberi, & Gaiano, 2011; Yoon & Gaiano, 2005). The levels of Notch signaling activity determine the rate of neural differentiation across amniotes, providing an ideal entry point to regulate the rate of neurogenesis in the dorsal pallium during amniote evolution, from reptiles to mammals. Notch activity depends on the expression levels of its ligands (*Delta-like* and *Jagged*), and in the developing dorsal pallium, these are regulated by Robo signaling (Cárdenas et al., 2018). Although best known for their central role in axon guidance during brain development (Blockus & Chedotal, 2014; Chedotal, 2019; Dickson & Gilestro, 2006), Robo signaling also regulates neurogenesis in mouse and flies (Andrews et al., 2008; Borrell et al., 2012; Mehta & Bhat, 2001) in a dose-dependent manner. Specifically in the mouse telencephalon, high levels of Robo1/2 signaling in aRGCs promotes direct neurogenesis, low levels favor aRGC self-renewal, whereas absence of expression promotes indirect neurogenesis via generation of IPCs (Borrell et al., 2012; Cárdenas et al., 2018). This modulation of the aRGC lineage and mode of neurogenesis, depending on the levels of Robo receptor signaling, is conserved across amniotes from reptiles to birds and mammals, and is mediated through the interaction with the Notch pathway (Cárdenas et al., 2018). In the dorsal cortex of snake embryos, the medial hyperpallium of chick and the mouse olfactory bulb, where direct neurogenesis predominates, expression levels of *Robo* are high and *Dll1* is low. Reciprocally, in the lateral hyperpallium and the mouse neocortex, where indirect neurogenesis is more abundant or predominates, *Robo* levels are low and *Dll1* is high (Cárdenas et al., 2018). Experimental manipulations in the developing dorsal cortex of snake, chick and mouse embryos in vivo, and in human cerebral organoids in vitro, demonstrate that these reciprocal expression levels of *Robo* and *Dll1* are necessary and sufficient to define the rates of direct and indirect neurogenesis. At the mechanistic level, high Robo expression reduces *Dll1* and increases *Jag1/2* expression, the alternative Notch ligands, which induces direct neurogenesis (Cárdenas et al., 2018). Hence, the mode of neurogenesis is defined by the specific ligand that activates Notch signaling, as it also occurs in other cellular systems (Cárdenas et al., 2018; Nandagopal et al., 2018). In the wider perspective of cortical evolution, the expression level of *Robo* receptors in aRGCs of the embryonic cortex became attenuated during amniote phylogeny, resulting in decreased *Jag1/2* and increased *Dll1* expression levels, altogether changing the balance from direct to indirect neurogenesis, which dramatically expanded neuron number and cortex complexity (Cárdenas et al., 2018).

The evolution of cortical neurogenesis did not stop at the ancestor of mammals, but it continued in the subsequent phylogenic lineages. Remarkably, genetic evolution was particularly fast in the human lineage, especially in the last few million years and in genes related to brain development (Dorus et al., 2004; Sousa, Meyer, Santpere, Gulden, & Sestan, 2017). Many of these genetic modifications were found in so-called Human Accelerated Regions (HARs), which affect gene regulatory elements essential in brain development, including progenitor cell proliferation and amplification (Pollard et al., 2006; Silver, 2016), demonstrating their relevance for the recent evolution of the human neocortex (Boyd et al., 2015). In addition to changes in regulatory elements, mounting evidence demonstrates the importance of newly-emerged human-specific genes, in driving human brain evolution and expansion (Dennis & Eichler, 2016; Florio et al., 2017). Gene duplications, creating paralogs from ancestral genes, was one key mechanism (Long, VanKuren, Chen, & Vibranovski, 2013). This is the case for *ARHGAP11B*, which arose from partial duplication of *ARHGAP11A* after the divergence between the lineages leading to chimpanzee and human (Florio et al., 2015; Florio, Namba, Paabo, Hiller, & Huttner, 2016). This new gene acquired a function completely new and unrelated to the ancestral gene, strongly promoting the formation of self-renewing BPs, and thus further increasing indirect neurogenesis and cortical expansion (Florio et al., 2015; Kalebic et al., 2018; Namba et al., 2020). Similarly, the *NOTCH2* locus was duplicated partially, generating three paralogs that are expressed in cortical progenitor cells: *NOTCH2NLA*, *B* and *C* (Fiddes et al., 2018; Florio et al., 2018; Suzuki et al., 2018). Testing the function of these paralogs in human embryonic stem cells and cerebral organoids demonstrated their role in promoting progenitor cell amplification and delayed neurogenesis, which ultimately translates in greater neuron production and cortical expansion (Fiddes et al., 2018; Suzuki et al., 2018). These examples demonstrate the fundamental importance of human-specific genetic modifications, both in regulatory regions and coding sequences, in the evolutionary expansion of the human neocortex.

Beyond individual genes, a key driver in human evolution appear to be large intra-chromosomal duplications organized in an interspersed fashion reaching an excess in the common ancestors of humans and African great apes (Marques-Bonet & Eichler, 2009; She et al., 2004). Much of the interspersed chromosomal duplications are focused around common sequence elements, the so-called 'core duplicons'. This genomic architecture promotes not only increased genetic variation, but also large scale structural diversity mediated by non-allelic homologous recombination. This generates a non-linear acceleration of large scale genomic changes fueling new genes in an unprecedented number, hence serving as playground for adaptations in evolution.

Recent assessments of structural genome variations among modern humans show a strong bias for these to occur at the tips of chromosomes (Audano et al., 2019), which is consistent with increased SNV divergence between chimpanzee and humans in the vicinity of telomeres (Mikkelsen et al., 2005). Large structural variants have a great likelihood to affect genes as well as gene expression, by adding, eliminating or altering regulatory sequences (Eichler, 2019). Importantly, this highly frequent occurrence of structural variation in the human genome, which contributes to human evolution and divergence, also contributes very significantly to disease and disease susceptibility (Eichler, 2019).

11. Human disease

Alterations in mechanisms of neurogenesis are responsible for severe neurological human diseases [for a comprehensive review of human cortical malformations and their genetic basis, please see (Barkovich, Guerrini, Kuzniecky, Jackson, & Dobyns, 2012; Fernandez et al., 2016)]. Among the most striking developmental malformations are those related with brain size, either loss (microcephaly) or gain (megalencephaly), and cortex folding, either loss (lissencephaly) or excess (polymicrogiria). All these conditions lead to intellectual disability, epileptic seizures and, potentially, autism (Barkovich et al., 2012). Microcephaly is caused by a variety of deleterious disruptions in different cellular processes that affect the amount of neurons produced, including premature depletion and/or death of progenitor cells. Microcephaly has been mostly linked to mutations in genes coding for centrosomal-associated proteins, or microcephalins (MCPH) (Bizzotto & Francis, 2015). For example, mutations in abnormal spindle–like microcephaly associated protein (ASPM) are the most frequent cause of human microcephaly, and *ASPM* encodes for a protein important to maintain the orientation of the mitotic cleavage plane (Bond et al., 2003; Fish, Kosodo, Enard, Pääbo, & Huttner, 2006). Human microcephaly is also linked to mutations in genes related to AJ assembly and apical-basal RGC polarity, like *FLNA* (Sheen, 2014) and *ARFGEF2* (Sheen & Walsh, 2003), and to activation of the unfolded protein response, which promotes direct neurogenesis and thus reduced neuron production (Frank, Ge, Xie, Zhou, & Tsai, 2010; Laguesse et al., 2015). Many of these cellular mechanisms are also altered in microcephaly occurring upon Zika virus infection, in addition to increased progenitor cell apoptosis (Alfano, Gladwyn-Ng, Couderc, Lecuit, & Nguyen, 2019; Gladwyn-Ng et al., 2018; Laguesse et al., 2015). In contrast to microcephaly, megalencephaly is a much less well understood malformation that seems to result from an overproduction of neural cells and/or a decrease in programmed cell death. The genetic basis of megalencephaly remains poorly explored (Barkovich et al., 2012).

Alterations in cortex folding have been mostly studied with regard to the loss of folds (partial, pachygyria; or complete, lissencephaly), which has been largely linked to defects in neuronal migration. These are usually caused by loss-of-function mutations in cytoskeletal proteins required for the movement of migrating neurons, such as DCX and LIS, or signaling proteins regulating this highly stereotyped movement, like RELN (Barkovich et al., 2012; des Portes et al., 1998; Francis et al., 1999; Hong et al., 2000; Jaglin & Chelly, 2009; Pilz et al., 1998). But defects in neuronal migration also arise from deficiencies in the migratory scaffold, the radial glia fiber system composed by the basal processes of RGCs (Sidman & Rakic, 1973). The discovery of bRGCs shed new light into this mechanism, with the new concept that bRGCs force the divergence of the radial fiber scaffold. This causes the lateral dispersion of radially-migrating neurons, hence the local expansion of cortical surface area, eventually the formation of folds (Borrell & Götz, 2014; Borrell & Reillo, 2012; Lui et al., 2011; Reillo et al., 2011). Accordingly, some of the cytoskeletal protein mutations causative of lissencephaly are also implicated in the proliferation and fate of RGCs (Bershteyn et al., 2017; Mi Moon, Hippenmeyer, Luo, & Wynshaw-Boris,

2020; Yingling et al., 2008). Cortical defects related to an excess of minifolds (polymicrogyria) are much less well understood mechanistically, but causative mutations again affect genes important in the proliferation and fate of cortical RGCs, including *PAX6, EOMES, CCND2, NR2F1, GPR56* and *PIK3R2* (Baala et al., 2007; Bertacchi et al., 2020; Mirzaa et al., 2015, 2014; Mitchell et al., 2003; Piao et al., 2004). Deficient or impaired neuronal migration may also cause other cortical alterations with a milder functional impact, such as periventricular nodular heterotopia (Buchsbaum et al., 2020; Cappello et al., 2013; Klaus et al., 2019).

Aside from malformations of brain size and cortex folding, severe neurological defects also emerge from ciliopathies, a set of human disorders caused by mutations in genes necessary for the formation and maintenance of cilia, including the primary cilium of aRGCs. Defective primary cilia in the central nervous system lead to neurological defects, intellectual disability and ataxia. As mentioned above, mutations in genes important for primary cilia (ciliopathy genes), like *ARL13B*, cause profound disruptions in neurogenesis (Cantagrel et al., 2008; Park et al., 2019).

12. Concluding remarks

The regulation of neurogenesis in the cerebral cortex is a highly complex process, involving multiple types of neural stem and progenitor cells organized in specialized germinal layers. The behavior and fate decisions of these cells are subject to multiple layers of genetic and epigenetic regulation, determining the activity level of multiple and intersecting signaling cascades. We are just beginning to understand how these regulatory mechanisms emerged and changed during vertebrate evolution, a knowledge recently accelerated thanks to the advent of transformative technologies (i.e. single cell transcriptomics, genome editing) that allow genetic analysis and experimental access to an ever-growing spectrum of model animal species. It is by taking full advantage of these and other emerging technologies and analytical methods that will allow exploring in the future the evolutionary mechanisms that steered the size and complexity of the cerebral cortex leading to its wonderful diversity across mammals.

Acknowledgments

We thank members of our lab for insightful discussions. We apologize to all researchers whose work could not be cited owing to space limitation. Work in our lab is supported by grants to V.B. from the ERC Consolidator Grant CORTEXFOLDING, the Spanish State Research

Agency (PGC2018-102172-B-I00) and through the "Severo Ochoa" Programme for Centers of Excellence in R&D (ref. SEV-2017-0723), and to M.G. from the ERC Advanced Grant ChroNeuroRepair.

References

Albert, M., Kalebic, N., Florio, M., Lakshmanaperumal, N., Haffner, C., Brandl, H., et al. (2017). Epigenome profiling and editing of neocortical progenitor cells during development. *The EMBO Journal, 36*, 2642–2658. https://doi.org/10.15252/embj.201796764.

Alfano, C., Gladwyn-Ng, I., Couderc, T., Lecuit, M., & Nguyen, L. (2019). The unfolded protein response: A key player in Zika virus-associated congenital microcephaly. *Frontiers in Cellular Neuroscience, 13*, 94. https://doi.org/10.3389/fncel.2019.00094.

Allen, B. L., Song, J. Y., Izzi, L., Althaus, I. W., Kang, J. S., Charron, F., et al. (2011). Overlapping roles and collective requirement for the coreceptors GAS1, CDO, and BOC in SHH pathway function. *Developmental Cell, 20*, 775–787. https://doi.org/10.1016/j.devcel.2011.04.018.

Andrews, W., Barber, M., Hernadez-Miranda, L. R., Xian, J., Rakic, S., Sundaresan, V., et al. (2008). The role of Slit-Robo signaling in the generation, migration and morphological differentiation of cortical interneurons. *Developmental Biology, 313*, 648–658. https://doi.org/10.1016/j.ydbio.2007.10.052.

Anton, E. S., Kreidberg, J. A., & Rakic, P. (1999). Distinct functions of alpha3 and alpha(v) integrin receptors in neuronal migration and laminar organization of the cerebral cortex. *Neuron, 22*, 277–289. https://doi.org/S0896-6273(00)81089-2 [pii].

Anton, E. S., Marchionni, M. A., Lee, K. F., & Rakic, P. (1997). Role of GGF/neuregulin signaling in interactions between migrating neurons and radial glia in the developing cerebral cortex. *Development, 124*, 3501–3510.

Antón-Bolaños, N., Sempere-Ferràndez, A., Guillamón-Vivancos, T., Martini, F. J., Pérez-Saiz, L., Gezelius, H., et al. (2019). Prenatal activity from thalamic neurons governs the emergence of functional cortical maps in mice. *Science, 364*, 987–990. https://doi.org/10.1126/science.aav7617.

Aprea, J., Prenninger, S., Dori, M., Ghosh, T., Monasor, L. S., Wessendorf, E., et al. (2013). Transcriptome sequencing during mouse brain development identifies long non-coding RNAs functionally involved in neurogenic commitment. *The EMBO Journal, 32*, 3145–3160. https://doi.org/10.1038/emboj.2013.245.

Arai, Y., Pulvers, J. N., Haffner, C., Schilling, B., Nusslein, I., Calegari, F., et al. (2011). Neural stem and progenitor cells shorten S-phase on commitment to neuron production. *Nature Communications, 2*, 154. https://doi.org/ncomms1155 [pii]10.1038/ncomms1155.

Arai, Y., & Taverna, E. (2017). Neural progenitor cell polarity and cortical development. *Frontiers in Cellular Neuroscience, 11*, 384. https://doi.org/10.3389/fncel.2017.00384.

Arcila, M. L., Betizeau, M., Cambronne, X. A., Guzman, E., Doerflinger, N., Bouhallier, F., et al. (2014). Novel primate miRNAs coevolved with ancient target genes in germinal zone-specific expression patterns. *Neuron, 81*, 1255–1262. https://doi.org/10.1016/j.neuron.2014.01.017.

Arnold, S. J., Huang, G. J., Cheung, A. F. P., Era, T., Nishikawa, S. I., Bikoff, E. K., et al. (2008). The T-box transcription factor Eomes/Tbr2 regulates neurogenesis in the cortical subventricular zone. *Genes & Development, 22*, 2479–2484. https://doi.org/10.1101/gad.475408.

Artegiani, B., Jesus Domingues, A. M., Bragado Alonso, S., Brandl, E., Massalini, S., Dahl, A., et al. (2015). Tox: A multifunctional transcription factor and novel regulator of mammalian corticogenesis. *The EMBO Journal, 34*, 896–910. https://doi.org/10.15252/embj.201490061.

Asami, M., Pilz, G. A., Ninkovic, J., Godinho, L., Schroeder, T., Huttner, W. B., et al. (2011). The role of Pax6 in regulating the orientation and mode of cell division of progenitors in the mouse cerebral cortex. *Development*, *138*, 5067–5078. https://doi.org/10.1242/dev.074591.

Attardo, A., Calegari, F., Haubensak, W., Wilsch-Brauninger, M., & Huttner, W. B. (2008). Live imaging at the onset of cortical neurogenesis reveals differential appearance of the neuronal phenotype in apical versus basal progenitor progeny. *PLoS One*, *3*, e2388. https://doi.org/10.1371/journal.pone.0002388.

Audano, P. A., Sulovari, A., Graves-Lindsay, T. A., Cantsilieris, S., Sorensen, M., Welch, A. M. E., et al. (2019). Characterizing the major structural variant alleles of the human genome. *Cell*, *176*, 663–675.e19. https://doi.org/10.1016/j.cell.2018.12.019.

Auladell, C., Martinez, A., Alcantara, S., Super, H., & Soriano, E. (1995). Migrating neurons in the developing cerebral cortex of the mouse send callosal axons. *Neuroscience*, *64*, 1091–1103.

Ayoub, A. E., Oh, S., Xie, Y., Leng, J., Cotney, J., Dominguez, M. H., et al. (2011). Transcriptional programs in transient embryonic zones of the cerebral cortex defined by high-resolution mRNA sequencing. *Proceedings of the National Academy of Sciences of the United States of America*, *108*, 14950–14955. https://doi.org/1112213108 [pii] 10.1073/pnas.1112213108.

Baala, L., Briault, S., Etchevers, H. C., Laumonnier, F., Natiq, A., Amiel, J., et al. (2007). Homozygous silencing of T-box transcription factor EOMES leads to microcephaly with polymicrogyria and corpus callosum agenesis. *Nature Genetics*, *39*, 454–456. https://doi.org/ng1993 [pii] 10.1038/ng1993.

Baek, J. H., Hatakeyama, J., Sakamoto, S., Ohtsuka, T., & Kageyama, R. (2006). Persistent and high levels of Hes1 expression regulate boundary formation in the developing central nervous system. *Development*, *133*, 2467–2476. https://doi.org/10.1242/dev.02403.

Bakken, T. E., Miller, J. A., Ding, S. L., Sunkin, S. M., Smith, K. A., Ng, L., et al. (2016). A comprehensive transcriptional map of primate brain development. *Nature*, *535*, 367–375. https://doi.org/10.1038/nature18637.

Bamiou, D. E., Campbell, N. G., Musiek, F. E., Taylor, R., Chong, W. K., Moore, A., et al. (2007). Auditory and verbal working memory deficits in a child with congenital aniridia due to a PAX6 mutation. *International Journal of Audiology*, *46*, 196–202. https://doi.org/10.1080/14992020601175952.

Barkovich, A. J., Guerrini, R., Kuzniecky, R. I., Jackson, G. D., & Dobyns, W. B. (2012). A developmental and genetic classification for malformations of cortical development: Update 2012. *Brain*, *135*, 1348–1369. https://doi.org/10.1093/brain/aws019aws019 [pii].

Bershteyn, M., Nowakowski, T. J., Pollen, A. A., Di Lullo, E., Nene, A., Wynshaw-Boris, A., et al. (2017). Human iPSC-derived cerebral organoids model cellular features of lissencephaly and reveal prolonged mitosis of outer radial glia. *Cell Stem Cell*, *20*, 435–449 e4. https://doi.org/10.1016/j.stem.2016.12.007.

Bertacchi, M., Romano, A. L., Loubat, A., Tran Mau-Them, F., Willems, M., Faivre, L., et al. (2020). NR2F1 regulates regional progenitor dynamics in the mouse neocortex and cortical gyrification in BBSOAS patients. *The EMBO Journal*, *39*, e104163. https://doi.org/10.15252/embj.2019104163.

Betizeau, M., Cortay, V., Patti, D., Pfister, S., Gautier, E., Bellemin-Menard, A., et al. (2013). Precursor diversity and complexity of lineage relationships in the outer subventricular zone of the primate. *Neuron*, *80*, 442–457. https://doi.org/10.1016/j.neuron.2013.09.032. S0896-6273(13)00863-5 [pii].

Bishop, K. M., Goudreau, G., & O'Leary, D. D. (2000). Regulation of area identity in the mammalian neocortex by Emx2 and Pax6. *Science*, *288*, 344–349.

Bizzotto, S., & Francis, F. (2015). Morphological and functional aspects of progenitors perturbed in cortical malformations. *Frontiers in Cellular Neuroscience*, *9*, 30. https://doi.org/10.3389/fncel.2015.00030.

Blockus, H., & Chedotal, A. (2014). The multifaceted roles of Slits and Robos in cortical circuits: From proliferation to axon guidance and neurological diseases. *Current Opinion in Neurobiology*, *27*, 82–88. https://doi.org/10.1016/j.conb.2014.03.003.

Bond, J., Scott, S., Hampshire, D. J., Springell, K., Corry, P., Abramowicz, M. J., et al. (2003). Protein-truncating mutations in ASPM cause variable reduction in brain size. *American Journal of Human Genetics*, *73*, 1170–1177. https://doi.org/S0002-9297(07)61979-X [pii]10.1086/379085.

Bonev, B., Mendelson Cohen, N., Szabo, Q., Fritsch, L., Papadopoulos, G. L., Lubling, Y., et al. (2017). Multiscale 3D genome rewiring during mouse neural development. *Cell*, *171*, 557–572.e24. https://doi.org/10.1016/j.cell.2017.09.043.

Bonev, B., Stanley, P., & Papalopulu, N. (2012). MicroRNA-9 Modulates Hes1 ultradian oscillations by forming a double-negative feedback loop. *Cell Reports*, *2*, 10–18. https://doi.org/10.1016/j.celrep.2012.05.017.

Borrell, V., & Callaway, E. M. (2002). Reorganization of exuberant axonal arbors contributes to the development of laminar specificity in ferret visual cortex. *The Journal of Neuroscience*, *22*, 6682–6695. https://doi.org/10.1523/jneurosci.22-15-06682.2002.

Borrell, V., Cárdenas, A., Ciceri, G., Galcerán, J., Flames, N., Pla, R., et al. (2012). Slit/Robo signaling modulates the proliferation of central nervous system progenitors. *Neuron*, *76*, 338–352. https://doi.org/10.1016/j.neuron.2012.08.003.

Borrell, V., & Götz, M. (2014). Role of radial glial cells in cerebral cortex folding. *Current Opinion in Neurobiology*, *27*, 39–46. https://doi.org/10.1016/j.conb.2014.02.007.

Borrell, V., & Reillo, I. (2012). Emerging roles of neural stem cells in cerebral cortex development and evolution. *Developmental Neurobiology*, *72*, 955–971. https://doi.org/10.1002/dneu.22013.

Boulder_Committee. (1970). Embryonic vertebrate central nervous system: Revised terminology. *The Anatomical Record*, *166*, 257–261.

Boyd, J. L., Skove, S. L., Rouanet, J. P., Pilaz, L. J., Bepler, T., Gordân, R., et al. (2015). Human-chimpanzee differences in a FZD8 enhancer alter cell-cycle dynamics in the developing neocortex. *Current Biology*, *25*, 772–779. https://doi.org/10.1016/j.cub.2015.01.041.

Buchsbaum, I. Y., Kielkowski, P., Giorgio, G., O'Neill, A. C., Di Giaimo, R., Kyrousi, C., et al. (2020). ECE 2 regulates neurogenesis and neuronal migration during human cortical development. *EMBO Reports*, *21*, 1–24. https://doi.org/10.15252/embr.201948204.

Bueno, D., & Garcia-Fernàndez, J. (2016). Evolutionary development of embryonic cerebrospinal fluid composition and regulation: An open research field with implications for brain development and function. *Fluids Barriers CNS*, *13*, 1–12. https://doi.org/10.1186/s12987-016-0029-y.

Bultje, R. S., Castaneda-Castellanos, D. R., Jan, L. Y., Jan, Y. N., Kriegstein, A. R., & Shi, S. H. (2009). Mammalian Par3 regulates progenitor cell asymmetric division via notch signaling in the developing neocortex. *Neuron*, *63*, 189–202. https://doi.org/10.1016/j.neuron.2009.07.004.

Calegari, F., Haubensak, W., Haffner, C., & Huttner, W. B. (2005). Selective lengthening of the cell cycle in the neurogenic subpopulation of neural progenitor cells during mouse brain development. *The Journal of Neuroscience*, *25*, 6533–6538. https://doi.org/10.1523/JNEUROSCI.0778-05.2005.

Callaway, E. M. (1998). Local circuits in primary visual cortex of the macaque monkey. *Annual Review of Neuroscience*, *21*, 47–74.

Callaway, E. M., & Borrell, V. (2011). Developmental sculpting of dendritic morphology of layer 4 neurons in visual cortex: Influence of retinal input. *The Journal of Neuroscience*, *31*, 7456–7470. https://doi.org/10.1523/JNEUROSCI.5222-10.2011.

Callaway, E. M., & Katz, L. C. (1990). Emergence and refinement of clustered horizontal connections in cat striate cortex. *The Journal of Neuroscience*, *10*, 1134–1153.

Callaway, E. M., & Katz, L. C. (1992). Development of axonal arbors of layer 4 spiny neurons in cat striate cortex. *The Journal of Neuroscience*, *12*, 570–582.

Callaway, E. M., & Lieber, J. L. (1996). Development of axonal arbors of layer 6 pyramidal neurons in ferret primary visual cortex. *The Journal of Comparative Neurology*, *376*, 295–305.

Camargo Ortega, G., Falk, S., Johansson, P. A., Peyre, E., Broix, L., Sahu, S. K., et al. (2019). The centrosome protein AKNA regulates neurogenesis via microtubule organization. *Nature*, *567*, 113–117. https://doi.org/10.1038/s41586-019-0962-4.

Campos, L. S., Leone, D. P., Relvas, J. B., Brakebusch, C., Fässler, R., Suter, U., et al. (2004). β1 integrins activate a MAPK signalling pathway in neural stem cells that contributes to their maintenance. *Development*, *131*, 3433–3444. https://doi.org/10.1242/dev.01199.

Cantagrel, V., Silhavy, J. L., Bielas, S. L., Swistun, D., Marsh, S. E., Bertrand, J. Y., et al. (2008). Mutations in the cilia gene ARL13B lead to the classical form of Joubert syndrome. *American Journal of Human Genetics*, *83*, 170–179. https://doi.org/S0002-9297(08)00406-0 [pii]10.1016/j.ajhg.2008.06.023.

Cappello, S., Attardo, A., Wu, X., Iwasato, T., Itohara, S., Wilsch-Brauninger, M., et al. (2006). The Rho-GTPase cdc42 regulates neural progenitor fate at the apical surface. *Nature Neuroscience*, *9*, 1099–1107. https://doi.org/nn1744 [pii]10.1038/nn1744.

Cappello, S., Bohringer, C. R., Bergami, M., Conzelmann, K. K., Ghanem, A., Tomassy, G. S. (2012). A radial glia-specific role of RhoA in double cortex formation. Neuron 73, 911–924. https://doi.org/10.1016/j.neuron.2011.12.030S0896-6273(12)00080-3 [pii].

Cappello, S., Gray, M. J., Badouel, C., Lange, S., Einsiedler, M., Srour, M. (2013). Mutations in genes encoding the cadherin receptor-ligand pair DCHS1 and FAT4 disrupt cerebral cortical development. Nature Genetics 45, 1300–1308. https://doi.org/10.1038/ng.2765ng.2765 [pii].

Cárdenas, A., & Borrell, V. (2020). Molecular and cellular evolution of corticogenesis in amniotes. *Cellular and Molecular Life Sciences*, 77, 1435–1460. https://doi.org/10.1007/s00018-019-03315-x.

Cárdenas, A., Villalba, A., de Juan Romero, C., Picó, E., Kyrousi, C., Tzika, A. C., et al. (2018). Evolution of cortical neurogenesis in amniotes controlled by robo signaling levels. *Cell*, *174*(3), 590–606.e21. https://doi.org/10.1016/j.cell.2018.06.007.

Cau, E., Gradwohl, G., Fode, C., & Guillemot, F. (1997). Mash1 activates a cascade of bHLH regulators in olfactory neuron progenitors. *Development*, *124*, 1611–1621.

Cavanagh, M. E., Cornelis, M. E., Dziegielewska, K. M., Evans, C. A., Lorscheider, F. L., Mollgard, K., et al. (1983). Comparison of proteins in CSF of lateral and IVth ventricles during early development of fetal sheep. *Brain Research*, *313*, 159–167. https://doi.org/10.1016/0165-3806(83)90213-4.

Charvet, C. J., Owerkowicz, T., & Striedter, G. F. (2009). Phylogeny of the telencephalic subventricular zone in sauropsids: Evidence for the sequential evolution of pallial and subpallial subventricular zones. *Brain, Behavior and Evolution*, *73*, 285–294. https://doi.org/10.1159/000230673.

Charvet, C. J., & Striedter, G. F. (2008). Developmental species differences in brain cell cycle rates between northern bobwhite quail (Colinus virginianus) and parakeets (Melopsittacus undulatus): Implications for mosaic brain evolution. *Brain, Behavior and Evolution*, *72*, 295–306. https://doi.org/10.1159/000184744.

Chedotal, A. (2019). Roles of axon guidance molecules in neuronal wiring in the developing spinal cord. *Nature Reviews. Neuroscience*, *20*, 380–396. https://doi.org/10.1038/s41583-019-0168-7.

Chen, S., Lake, B. B., & Zhang, K. (2019). High-throughput sequencing of the transcriptome and chromatin accessibility in the same cell. *Nature Biotechnology*, *37*, 1452–1457. https://doi.org/10.1038/s41587-019-0290-0.

Chen, C., Lee, G. A., Pourmorady, A., Sock, E., & Donoghue, M. J. (2015). Orchestration of neuronal differentiation and progenitor pool expansion in the developing cortex by SoxC genes. *The Journal of Neuroscience*, *35*, 10629–10642. https://doi.org/10.1523/JNEUROSCI.1663-15.2015.

Chenn, A., & McConnell, S. K. (1995). Cleavage orientation and the asymmetric inheritance of Notch1 immunoreactivity in mammalian neurogenesis. *Cell*, *82*, 631–641.

Chenn, A., & Walsh, C. A. (2002). Regulation of cerebral cortical size by control of cell cycle exit in neural precursors. *Science*, *297*, 365–369.

Cheung, A. F., Kondo, S., Abdel-Mannan, O., Chodroff, R. A., Sirey, T. M., Bluy, L. E., et al. (2010). The subventricular zone is the developmental milestone of a 6-layered neocortex: Comparisons in metatherian and eutherian mammals. *Cerebral Cortex*, *20*, 1071–1081. https://doi.org/bhp168 [pii] 10.1093/cercor/bhp168.

Cheung, A. F. P., Pollen, A. A., Tavare, A., Deproto, J., & Molnár, Z. (2007). Comparative aspects of cortical neurogenesis in vertebrates. *Journal of Anatomy*, *211*, 164–176. https://doi.org/JOA769 [pii] 10.1111/j.1469-7580.2007.00769.x.

Clinton, B. K., Cunningham, C. L., Kriegstein, A. R., Noctor, S. C., & Martinez-Cerdeno, V. (2014). Radial glia in the proliferative ventricular zone of the embryonic and adult turtle, Trachemys scripta elegans. *Neurogenes*, *1*, e970905. https://doi.org/10.4161/23262125.2014.970905.

Cooper, J. A. (2014). Molecules and mechanisms that regulate multipolar migration in the intermediate zone. *Frontiers in Cellular Neuroscience*, *8*, 386. https://doi.org/10.3389/fncel.2014.00386.

Coquand, L., Victoria, G. S., Tata, A., Brault, J. B., Guimiot, F., Fraisier, V., et al. (2020). A dendritic-like microtubule network is organized from swellings of the basal fiber in neural progenitors. *bioRxiv*, 2020.03.16.993295. https://doi.org/10.1101/2020.03.16.993295.

Costa, M. R., Wen, G., Lepier, A., Schroeder, T., & Gotz, M. (2008). Par-complex proteins promote proliferative progenitor divisions in the developing mouse cerebral cortex. *Development*, *135*, 11–22. https://doi.org/10.1242/dev.009951.

Cubelos, B., Sebastian-Serrano, A., Kim, S., Moreno-Ortiz, C., Redondo, J. M., Walsh, C. A., et al. (2008). Cux-2 controls the proliferation of neuronal intermediate precursors of the cortical subventricular zone. *Cerebral Cortex*, *18*, 1758–1770. https://doi.org/10.1093/cercor/bhm199.

Cunningham, C. L., Martinez-Cerdeno, V., & Noctor, S. C. (2013). Microglia regulate the number of neural precursor cells in the developing cerebral cortex. *The Journal of Neuroscience*, *33*, 4216–4233. https://doi.org/10.1523/JNEUROSCI.3441-12.2013.

Das, R. M., & Storey, K. G. (2014). Apical abscission alters cell polarity and dismantles the primary cilium during neurogenesis. *Science*, *343*, 200–204. https://doi.org/10.1126/science.1247521.

Davis, L. K., Meyer, K. J., Rudd, D. S., Librant, A. L., Epping, E. A., Sheffield, V. C., et al. (2008). Pax6 3′ deletion results in aniridia, autism and mental retardation. *Human Genetics*, *123*, 371–378. https://doi.org/10.1007/s00439-008-0484-x.

De Juan Romero, C., & Borrell, V. (2015). Coevolution of radial glial cells and the cerebral cortex. *Glia*, *63*, 1303–1319. https://doi.org/10.1002/glia.22827.

de Juan Romero, C., & Borrell, V. (2017). Genetic maps and patterns of cerebral cortex folding. *Current Opinion in Cell Biology*, *49*, 31–37. https://doi.org/10.1016/j.ceb.2017.11.009.

De Juan Romero, C., Bruder, C., Tomasello, U., Sanz-Anquela, J. M., & Borrell, V. (2015). Discrete domains of gene expression in germinal layers distinguish the development of gyrencephaly. *The EMBO Journal*, *34*, 1859–1874. https://doi.org/10.15252/embj.201591176.

Dehay, C., & Kennedy, H. (2007). Cell-cycle control and cortical development. *Nature Reviews. Neuroscience*, *8*, 438–450.

Dehay, C., Savatier, P., Cortay, V., & Kennedy, H. (2001). Cell-cycle kinetics of neocortical precursors are influenced by embryonic thalamic axons. *The Journal of Neuroscience*, *21*, 201–214.

del Toro, D., Ruff, T., Cederfjäll, E., Villalba, A., Seyit-Bremer, G., Borrell, V., et al. (2017). Regulation of cerebral cortex folding by controlling neuronal migration via FLRT adhesion molecules. *Cell*, *169*, 621–635.e16. https://doi.org/10.1016/j.cell.2017.04.012.

Delaunay, D., Cortay, V., Patti, D., Knoblauch, K., & Dehay, C. (2014). Mitotic spindle asymmetry: A Wnt/PCP-regulated mechanism generating asymmetrical division in cortical precursors. *Cell Reports*, *6*, 400–414. https://doi.org/10.1016/j.celrep.2013.12.026.

Dennis, M. Y., & Eichler, E. E. (2016). Human adaptation and evolution by segmental duplication. *Current Opinion in Genetics & Development*, *41*, 44–52. https://doi.org/10.1016/j.gde.2016.08.001.

des Portes, V., Pinard, J. M., Billuart, P., Vinet, M. C., Koulakoff, A., Carrie, A., et al. (1998). A novel CNS gene required for neuronal migration and involved in X-linked subcortical laminar heterotopia and lissencephaly syndrome. *Cell*, *92*, 51–61. https://doi.org/S0092-8674(00)80898-3 [pii].

Desai, A. R., & McConnell, S. K. (2000). Progressive restriction in fate potential by neural progenitors during cerebral cortical development. *Development*, *127*, 2863–2872.

Desfilis, E., Abellan, A., Sentandreu, V., & Medina, L. (2018). Expression of regulatory genes in the embryonic brain of a lizard and implications for understanding pallial organization and evolution. *The Journal of Comparative Neurology*, *526*, 166–202. https://doi.org/10.1002/cne.24329.

Dickson, B. J., & Gilestro, G. F. (2006). Regulation of commissural axon pathfinding by slit and its Robo receptors. *Annual Review of Cell and Developmental Biology*, *22*, 651–675. https://doi.org/10.1146/annurev.cellbio.21.090704.151234.

Diez-Roux, G., Banfi, S., Sultan, M., Geffers, L., Anand, S., Rozado, D., et al. (2011). A high-resolution anatomical atlas of the transcriptome in the mouse embryo. *PLoS Biology*, *9*, e1000582. https://doi.org/10.1371/journal.pbio.1000582.

Dimidschstein, J., Passante, L., Dufour, A., vandenAmeele, J., Tiberi, L., Hrechdakian, T., et al. (2013). Ephrin-B1 controls the columnar distribution of cortical pyramidal neurons by restricting their tangential migration. *Neuron*, *79*, 1123–1135. https://doi.org/10.1016/j.neuron.2013.07.015.

Docampo-Seara, A., Lagadec, R., Mazan, S., Rodríguez, M. A., Quintana-Urzainqui, I., Candal, E., et al. (2018). Study of pallial neurogenesis in shark embryos and the evolutionary origin of the subventricular zone. *Brain Structure & Function*, *223*, 3593–3612. https://doi.org/10.1007/s00429-018-1705-2.

Dorus, S., Vallender, E. J., Evans, P. D., Anderson, J. R., Gilbert, S. L., Mahowald, M., et al. (2004). Accelerated evolution of nervous system genes in the origin of Homo sapiens. *Cell*, *119*, 1027–1040.

Drago, J., Nurcombe, V., & Bartlett, P. F. (1991). Laminin through its long arm E8 fragment promotes the proliferation and differentiation of murine neuroepithelial cells in vitro. *Experimental Cell Research*, *192*, 256–265. https://doi.org/10.1016/0014-4827(91)90184-v.

Dubreuil, V., Marzesco, A. M., Corbeil, D., Huttner, W. B., & Wilsch-Brauninger, M. (2007). Midbody and primary cilium of neural progenitors release extracellular membrane particles enriched in the stem cell marker prominin-1. *The Journal of Cell Biology*, *176*, 483–495. https://doi.org/10.1083/jcb.200608137.

Eckler, M. J., Nguyen, T. D., McKenna, W. L., Fastow, B. L., Guo, C., Rubenstein, J. L. R. R., et al. (2015). Cux2-positive radial glial cells generate diverse subtypes of neocortical projection neurons and macroglia. *Neuron*, *86*, 1100–1108. https://doi.org/10.1016/j.neuron.2015.04.020.

Eichler, E. E. (2019). Genetic variation, comparative genomics, and the diagnosis of disease. *The New England Journal of Medicine*, *381*, 64–74. https://doi.org/10.1056/NEJMra1809315.

Elsen, G. E., Bedogni, F., Hodge, R. D., Bammler, T. K., MacDonald, J. W., Lindtner, S., et al. (2018). The epigenetic factor landscape of developing neocortex is regulated by transcription factors Pax6- -> Tbr2- -> Tbr1. *Frontiers in Neuroscience*, *12*, 571. https://doi.org/10.3389/fnins.2018.00571.

Englund, C., Fink, A., Lau, C., Pham, D., Daza, R. A., Bulfone, A., et al. (2005). Pax6, Tbr2, and Tbr1 are expressed sequentially by radial glia, intermediate progenitor cells, and postmitotic neurons in developing neocortex. *The Journal of Neuroscience*, *25*, 247–251. https://doi.org/25/1/247 [pii]10.1523/JNEUROSCI.2899-04.2005.

Esgleas, M., Falk, S., Forné, I., Thiry, M., Najas, S., Zhang, S., et al. (2020). Trnp1 organizes diverse nuclear membrane-less compartments in neural stem cells. *The EMBO Journal*, *39*(16), e103373. https://doi.org/10.15252/embj.2019103373.

Farkas, L. M., Haffner, C., Giger, T., Khaitovich, P., Nowick, K., Birchmeier, C., et al. (2008). Insulinoma-associated 1 has a panneurogenic role and promotes the generation and expansion of basal progenitors in the developing mouse neocortex. *Neuron*, *60*, 40–55. https://doi.org/10.1016/j.neuron.2008.09.020.

Fernandez, V., Llinares-Benadero, C., Borrell, V., Fernández, V., Llinares-Benadero, C., & Borrell, V. (2016). Cerebral cortex expansion and folding: What have we learned? *The EMBO Journal*, *35*, 1021–1044. https://doi.org/10.15252/embj.201593701.

Fiddes, I. T., Lodewijk, G. A., Mooring, M., Bosworth, C. M., Ewing, A. D., Mantalas, G. L., et al. (2018). Human-specific NOTCH2NL genes affect notch signaling and cortical neurogenesis. *Cell*, *173*, 1356–1369 e22. https://doi.org/10.1016/j.cell.2018.03.051.

Fietz, S. A., & Huttner, W. B. (2011). Cortical progenitor expansion, self-renewal and neurogenesis-a polarized perspective. *Current Opinion in Neurobiology*, *21*, 23–35. https://doi.org/S0959-4388(10)00187-X [pii]10.1016/j.conb.2010.10.002.

Fietz, S. A., Kelava, I., Vogt, J., Wilsch-Brauninger, M., Stenzel, D., Fish, J. L., et al. (2010). OSVZ progenitors of human and ferret neocortex are epithelial-like and expand by integrin signaling. *Nature Neuroscience*, *13*, 690–699. https://doi.org/nn.2553 [pii] 10.1038/nn.2553.

Fietz, S. A., Lachmann, R., Brandl, H., Kircher, M., Samusik, N., Schroder, R., et al. (2012). Transcriptomes of germinal zones of human and mouse fetal neocortex suggest a role of extracellular matrix in progenitor self-renewal. *Proceedings of the National Academy of Sciences of the United States of America*, *109*, 11836–11841. https://doi.org/10.1073/pnas.1209647109.

Finlay, B. L., & Darlington, R. B. (1995). Linked regularities in the development and evolution of mammalian brains. *Science*, *268*, 1578–1584.

Fish, J. L., Kosodo, Y., Enard, W., Pääbo, S., & Huttner, W. B. (2006). Aspm specifically maintains symmetric proliferative divisions of neuroepithelial cells. *Proceedings of the National Academy of Sciences of the United States of America*, *103*, 10438–10443. https://doi.org/0604066103 [pii] 10.1073/pnas.0604066103.

Florio, M., Albert, M., Taverna, E., Namba, T., Brandl, H., Lewitus, E., et al. (2015). Human-specific gene ARHGAP11B promotes basal progenitor amplification and neocortex expansion. *Science*, *347*, 1465–1470. https://doi.org/10.1126/science.aaa1975.

Florio, M., Borrell, V., & Huttner, W. B. (2017). Human-specific genomic signatures of neocortical expansion. *Current Opinion in Neurobiology*, *42*, 33–44. https://doi.org/10.1016/j.conb.2016.11.004.

Florio, M., Heide, M., Pinson, A., Brandl, H., Albert, M., Winkler, S., et al. (2018). Evolution and cell-type specificity of human-specific genes preferentially expressed in progenitors of fetal neocortex. *Elife*, *7*. https://doi.org/10.7554/eLife.32332.
Florio, M., & Huttner, W. B. (2014). Neural progenitors, neurogenesis and the evolution of the neocortex. *Development*, *141*, 2182–2194. https://doi.org/10.1242/dev.090571.
Florio, M., Namba, T., Paabo, S., Hiller, M., & Huttner, W. B. (2016). A single splice site mutation in human-specific ARHGAP11B causes basal progenitor amplification. *Science Advances*, *2*, e1601941. https://doi.org/10.1126/sciadv.1601941.
Francis, F., Koulakoff, A., Boucher, D., Chafey, P., Schaar, B., Vinet, M. C., et al. (1999). Doublecortin is a developmentally regulated, microtubule-associated protein expressed in migrating and differentiating neurons. *Neuron*, *23*, 247–256.
Franco, S. J., Gil-Sanz, C., Martinez-Garay, I., Espinosa, A., Harkins-Perry, S. R., Ramos, C., et al. (2012). Fate-restricted neural progenitors in the mammalian cerebral cortex. *Science*, *337*, 746–749. https://doi.org/10.1126/science.1223616. 337/6095/746 [pii].
Frank, C. L., Ge, X., Xie, Z., Zhou, Y., & Tsai, L. H. (2010). Control of activating transcription factor 4 (ATF4) persistence by multisite phosphorylation impacts cell cycle progression and neurogenesis. *The Journal of Biological Chemistry*, *285*, 33324–33337. https://doi.org/10.1074/jbc.M110.140699.
Frantz, G. D., & McConnell, S. K. (1996). Restriction of late cerebral cortical progenitors to an upper-layer fate. *Neuron*, *17*, 55–61. https://doi.org/10.1016/S0896-6273(00)80280-9.
Frazer, S., Prados, J., Niquille, M., Cadilhac, C., Markopoulos, F., Gomez, L., et al. (2017). Transcriptomic and anatomic parcellation of 5-HT3A R expressing cortical interneuron subtypes revealed by single-cell RNA sequencing. *Nature Communications*, *8*, 14219. https://doi.org/10.1038/ncomms14219.
Fujita, I., Shitamukai, A., Kusumoto, F., Mase, S., Suetsugu, T., Omori, A., et al. (2020). Endfoot regeneration restricts radial glial state and prevents translocation into the outer subventricular zone in early mammalian brain development. *Nature Cell Biology*, *22*, 26–37. https://doi.org/10.1038/s41556-019-0436-9.
Gabriel, E., Wason, A., Ramani, A., Gooi, L. M., Keller, P., Pozniakovsky, A., et al. (2016). CPAP promotes timely cilium disassembly to maintain neural progenitor pool. *The EMBO Journal*, *35*, 803–819. https://doi.org/10.15252/embj.201593679.
Gaiano, N., & Fishell, G. (2002). The role of notch in promoting glial and neural stem cell fates. *Annual Review of Neuroscience*, *25*, 471–490. https://doi.org/10.1146/annurev.neuro.25.030702.130823.
Gambello, M. J., Darling, D. L., Yingling, J., Tanaka, T., Gleeson, J. G., & Wynshaw-Boris, A. (2003). Multiple dose-dependent effects of Lis1 on cerebral cortical development. *The Journal of Neuroscience*, *23*, 1719–1729.
Gao, P., Postiglione, M. P., Krieger, T. G., Hernandez, L., Wang, C., Han, Z., et al. (2014). Deterministic progenitor behavior and unitary production of neurons in the neocortex. *Cell*, *159*, 775–788. https://doi.org/10.1016/j.cell.2014.10.027.
Gato, A., Martin, P., Alonso, M. I., Martin, C., Pulgar, M. A., & Moro, J. A. (2004). Analysis of cerebro-spinal fluid protein composition in early developmental stages in chick embryos. *Journal of Experimental Zoology. Part A, Comparative Experimental Biology*, *301*, 280–289. https://doi.org/10.1002/jez.a.20035.
Georgala, P. A., Manuel, M., & Price, D. J. (2011). The generation of superficial cortical layers is regulated by levels of the transcription factor Pax6. *Cerebral Cortex*, *21*, 81–94. https://doi.org/10.1093/cercor/bhq061.
Geschwind, D. H., & Rakic, P. (2013). Cortical evolution: Judge the brain by its cover. *Neuron*, *80*, 633–647. https://doi.org/10.1016/j.neuron.2013.10.045S0896-6273(13)00997-5 [pii].

Gigante, E. D., Taylor, M. R., Ivanova, A. A., Kahn, R. A., & Caspary, T. (2020). ARL13B regulates Sonic hedgehog signaling from outside primary cilia. *eLife*, *9*. https://doi.org/10.7554/eLife.50434.

Gladwyn-Ng, I., Cordon-Barris, L., Alfano, C., Creppe, C., Couderc, T., Morelli, G., et al. (2018). Stress-induced unfolded protein response contributes to Zika virus-associated microcephaly. *Nature Neuroscience*, *21*, 63–71. https://doi.org/10.1038/s41593-017-0038-4.

Goetz, S. C., & Anderson, K. V. (2010). The primary cilium: A signalling centre during vertebrate development. *Nature Reviews. Genetics*, *11*, 331–344. https://doi.org/10.1038/nrg2774.

Goodfellow, M., Phillips, N. E., Manning, C., Galla, T., & Papalopulu, N. (2014). microRNA input into a neural ultradian oscillator controls emergence and timing of alternative cell states. *Nature Communications*, *5*, 3399. https://doi.org/10.1038/ncomms4399.

Gotz, M., & Huttner, W. B. (2005). The cell biology of neurogenesis. *Nature Reviews. Molecular Cell Biology*, *6*, 777–788.

Guo, C., Eckler, M. J., McKenna, W. L., McKinsey, G. L., Rubenstein, J. L. R., & Chen, B. (2013). Fezf2 expression identifies a multipotent progenitor for neocortical projection neurons, astrocytes, and oligodendrocytes. *Neuron*, *80*, 1167–1174. https://doi.org/10.1016/j.neuron.2013.09.037.

Güven, A., Kalebic, N., Long, K. R., Florio, M., Vaid, S., Brandl, H., et al. (2020). Extracellular matrix-inducing Sox9 promotes both basal progenitor proliferation and gliogenesis in developing neocortex. *eLife*, *9*. https://doi.org/10.7554/eLife.49808.

Han, Y. G., Spassky, N., Romaguera-Ros, M., Garcia-Verdugo, J. M., Aguilar, A., Schneider-Maunoury, S., et al. (2008). Hedgehog signaling and primary cilia are required for the formation of adult neural stem cells. *Nature Neuroscience*, *11*, 277–284. https://doi.org/10.1038/nn2059.

Han, X., Zhou, Z., Fei, L., Sun, H., Wang, R., Chen, Y., et al. (2020). Construction of a human cell landscape at single-cell level. *Nature*, *581*, 303–309. Springer US https://doi.org/10.1038/s41586-020-2157-4.

Hand, R., Bortone, D., Mattar, P., Nguyen, L., Heng, J. I. T., Guerrier, S., et al. (2005). Phosphorylation of Neurogenin2 specifies the migration properties and the dendritic morphology of pyramidal neurons in the neocortex. *Neuron*, *48*, 45–62. https://doi.org/S0896-6273(05)00736-1 [pii] 10.1016/j.neuron.2005.08.032.

Hansen, D. V., Lui, J. H., Parker, P. R., & Kriegstein, A. R. (2010). Neurogenic radial glia in the outer subventricular zone of human neocortex. *Nature*, *464*, 554–561. https://doi.org/nature08845 [pii]10.1038/nature08845.

Hartfuss, E., Forster, E., Bock, H. H., Hack, M. A., Leprince, P., Luque, J. M., et al. (2003). Reelin signaling directly affects radial glia morphology and biochemical maturation. *Development*, *130*, 4597–4609.

Haubensak, W., Attardo, A., Denk, W., & Huttner, W. B. (2004). Neurons arise in the basal neuroepithelium of the early mammalian telencephalon: A major site of neurogenesis. *Proceedings of the National Academy of Sciences of the United States of America*, *101*, 3196–3201.

Haubst, N., Berger, J., Radjendirane, V., Graw, J., Favor, J., Saunders, G. F., et al. (2004). Molecular dissection of Pax6 function: The specific roles of the paired domain and homeodomain in brain development. *Development*, *131*, 6131–6140. https://doi.org/10.1242/dev.01524.

Haubst, N., Georges-Labouesse, E., De Arcangelis, A., Mayer, U., & Götz, M. (2006). Basement membrane attachment is dispensable for radial glial cell fate and for proliferation, but affects positioning of neuronal subtypes. *Development*, *133*, 3245–3254. https://doi.org/10.1242/dev.02486.

Heide, M., Haffner, C., Murayama, A., Kurotaki, Y., Shinohara, H., Okano, H., et al. (2020). Human-specific ARHGAP11B increases size and folding of primate neocortex in the fetal marmoset. *Science*, *369*, 546–550. https://doi.org/10.1126/science.abb2401.
Heins, N., Malatesta, P., Cecconi, F., Nakafuku, M., Tucker, K. L., Hack, M. A., et al. (2002). Glial cells generate neurons: The role of the transcription factor Pax6. *Nature Neuroscience*, *5*, 308–315. https://doi.org/10.1038/nn828.
Herculano-Houzel, S., Collins, C. E., Wong, P., Kaas, J. H., & Lent, R. (2008). The basic nonuniformity of the cerebral cortex. *Proceedings of the National Academy of Sciences of the United States of America*, *105*, 12593–12598. https://doi.org/10.1073/pnas.0805417105.
Hevner, R. F., Hodge, R. D., Daza, R. A., & Englund, C. (2006). Transcription factors in glutamatergic neurogenesis: Conserved programs in neocortex, cerebellum, and adult hippocampus. *Neuroscience Research*, *55*, 223–233. https://doi.org/S0168-0102(06)00067-8 [pii]10.1016/j.neures.2006.03.004.
Higginbotham, H., Guo, J., Yokota, Y., Umberger, N. L., Su, C. Y., Li, J., et al. (2013). Arl13b-regulated cilia activities are essential for polarized radial glial scaffold formation. *Nature Neuroscience*, *16*, 1000–1007. https://doi.org/10.1038/nn.3451.
Hinds, J. W., & Ruffett, T. L. (1971). Cell proliferation in the neural tube: An electron microscopic and golgi analysis in the mouse cerebral vesicle. *Zeitschrift für Zellforschung und Mikroskopische Anatomie*, *115*, 226–264. https://doi.org/10.1007/bf00391127.
Hirabayashi, Y., Itoh, Y., Tabata, H., Nakajima, K., Akiyama, T., Masuyama, N., et al. (2004). The Wnt/beta-catenin pathway directs neuronal differentiation of cortical neural precursor cells. *Development*, *131*, 2791–2801. https://doi.org/10.1242/dev.01165.
Hodge, R. D., Bakken, T. E., Miller, J. A., Smith, K. A., Barkan, E. R., Graybuck, L. T., et al. (2019). Conserved cell types with divergent features in human versus mouse cortex. *Nature*, *573*, 61–68. https://doi.org/10.1038/s41586-019-1506-7.
Homem, C. C. F., Steinmann, V., Burkard, T. R., Jais, A., Esterbauer, H., & Knoblich, J. A. (2014). Ecdysone and mediator change energy metabolism to terminate proliferation in Drosophila neural stem cells. *Cell*, *158*, 874–888. https://doi.org/10.1016/j.cell.2014.06.024.
Hong, S. E., Shugart, Y. Y., Huang, D. T., Shahwan, S. A., Grant, P. E., Hourihane, J. O., et al. (2000). Autosomal recessive lissencephaly with cerebellar hypoplasia is associated with human RELN mutations. *Nature Genetics*, *26*, 93–96.
Huang, X., Liu, J., Ketova, T., Fleming, J. T., Grover, V. K., Cooper, M. K., et al. (2010). Transventricular delivery of sonic hedgehog is essential to cerebellar ventricular zone development. *Proceedings of the National Academy of Sciences of the United States of America*, *107*, 8422–8427. https://doi.org/10.1073/pnas.0911838107.
Imayoshi, I., Ishidate, F., & Kageyama, R. (2015). Real-time imaging of bHLH transcription factors reveals their dynamic control in the multipotency and fate choice of neural stem cells. *Frontiers in Cellular Neuroscience*, *9*, 1–6. https://doi.org/10.3389/fncel.2015.00288.
Imayoshi, I., Isomura, A., Harima, Y., Kawaguchi, K., Kori, H., Miyachi, H., et al. (2013). Oscillatory control of factors determining multipotency and fate in mouse neural progenitors. *Science*, *342*, 1203–1208. https://doi.org/10.1126/science.1242366.
Insolera, R., Bazzi, H., Shao, W., Anderson, K. V., & Shi, S. H. (2014). Cortical neurogenesis in the absence of centrioles. *Nature Neuroscience*, *17*, 1528–1535. https://doi.org/10.1038/nn.3831.
Itoh, Y., Moriyama, Y., Hasegawa, T., Endo, T. A., Toyoda, T., & Gotoh, Y. (2013). Scratch regulates neuronal migration onset via an epithelial-mesenchymal transition-like mechanism. *Nature Neuroscience*, *16*, 416–425. https://doi.org/10.1038/nn.3336nn.3336 [pii].
Iwata, R., Casimir, P., & Vanderhaeghen, P. (2020). Mitochondrial dynamics in postmitotic cells regulate neurogenesis. *Science*, *369*, 858–862. https://doi.org/10.1126/science.aba9760.

Izumi, Y., Hirose, T., Tamai, Y., Hirai, S. I., Nagashima, Y., Fujimoto, T., et al. (1998). An atypical PKC directly associates and colocalizes at the epithelial tight junction with ASIP, a mammalian homologue of Caenorhabditis elegans polarity protein PAR-3. *The Journal of Cell Biology*, *143*, 95–106. https://doi.org/10.1083/jcb.143.1.95.

Jaglin, X. H., & Chelly, J. (2009). Tubulin-related cortical dysgeneses: Microtubule dysfunction underlying neuronal migration defects. *Trends in Genetics*, *25*, 555–566. https://doi.org/S0168-9525(09)00189-9 [pii]10.1016/j.tig.2009.10.003.

Javaherian, A., & Kriegstein, A. (2009). A stem cell niche for intermediate progenitor cells of the embryonic cortex. *Cerebral Cortex*, *19*(Suppl. 1), i70–i77. https://doi.org/10.1093/cercor/bhp029bhp029 [pii].

Johansson, P. A., Irmler, M., Acampora, D., Beckers, J., Simeone, A., & Gotz, M. (2013). The transcription factor Otx2 regulates choroid plexus development and function. *Development*, *140*, 1055–1066. https://doi.org/10.1242/dev.090860.

Johnson, M. B., Kawasawa, Y. I., Mason, C. E., Krsnik, Ž., Coppola, G., Bogdanović, D., et al. (2009). Functional and evolutionary insights into human brain development through global transcriptome analysis. *Neuron*, *62*, 494–509. https://doi.org/S0896-6273(09)00286-4 [pii] 10.1016/j.neuron.2009.03.027.

Johnson, M. B., Wang, P. P., Atabay, K. D., Murphy, E. A., Doan, R. N., Hecht, J. L., et al. (2015). Single-cell analysis reveals transcriptional heterogeneity of neural progenitors in human cortex. *Nature Neuroscience*, *18*, 637–646. https://doi.org/10.1038/nn.3980.

Kalebic, N., Gilardi, C., Albert, M., Namba, T., Long, K. R., Kostic, M., et al. (2018). Human-specific ARHGAP11B induces hallmarks of neocortical expansion in developing ferret neocortex. *eLife*, 7. https://doi.org/10.7554/eLife.41241.

Kalebic, N., Gilardi, C., Stepien, B., Wilsch-Brauninger, M., Long, K. R., Namba, T., et al. (2019). Neocortical expansion due to increased proliferation of basal progenitors is linked to changes in their morphology. *Cell Stem Cell*, *24*, 535–550 e9. https://doi.org/10.1016/j.stem.2019.02.017.

Kang, H. J., Kawasawa, Y. I., Cheng, F., Zhu, Y., Xu, X., Li, M., et al. (2011). Spatio-temporal transcriptome of the human brain. *Nature*, *478*, 483–489. https://doi.org/10.1038/nature10523nature10523 [pii].

Kawaguchi, A., Ikawa, T., Kasukawa, T., Ueda, H. R., Kurimoto, K., Saitou, M., et al. (2008). Single-cell gene profiling defines differential progenitor subclasses in mammalian neurogenesis. *Development*, *135*, 3113–3124. https://doi.org/10.1242/dev.022616.

Kawasaki, H., Toda, T., & Tanno, K. (2013). In vivo genetic manipulation of cortical progenitors in gyrencephalic carnivores using in utero electroporation. *Biology Open*, *2*, 95–100. https://doi.org/10.1242/bio.20123160.

Kelava, I., Reillo, I., Murayama, A. Y., Kalinka, A. T., Stenzel, D., Tomancak, P., et al. (2012). Abundant occurrence of basal radial glia in the subventricular zone of embryonic neocortex of a lissencephalic primate, the common marmoset callithrix jacchus. *Cerebral Cortex*, *22*, 469–481. https://doi.org/10.1093/cercor/bhr301.

Khacho, M., Clark, A., Svoboda, D. S., Azzi, J., MacLaurin, J. G., Meghaizel, C., et al. (2016). Mitochondrial dynamics impacts stem cell identity and fate decisions by regulating a nuclear transcriptional program. *Cell Stem Cell*, *19*, 232–247. https://doi.org/10.1016/j.stem.2016.04.015.

Khacho, M., Clark, A., Svoboda, D. S., MacLaurin, J. G., Lagace, D. C., Park, D. S., et al. (2017). Mitochondrial dysfunction underlies cognitive defects as a result of neural stem cell depletion and impaired neurogenesis. *Human Molecular Genetics*, *26*, 3327–3341. https://doi.org/10.1093/hmg/ddx217.

Khacho, M., Harris, R., & Slack, R. S. (2019). Mitochondria as central regulators of neural stem cell fate and cognitive function. *Nature Reviews. Neuroscience*, *20*, 34–48. https://doi.org/10.1038/s41583-018-0091-3.

Klaus, J., Kanton, S., Kyrousi, C., Ayo-Martin, A. C., Di Giaimo, R., Riesenberg, S., et al. (2019). Altered neuronal migratory trajectories in human cerebral organoids derived from individuals with neuronal heterotopia. *Nature Medicine*, *25*, 561–568. https://doi.org/10.1038/s41591-019-0371-0.

Komada, M. (2012). Sonic hedgehog signaling coordinates the proliferation and differentiation of neural stem/progenitor cells by regulating cell cycle kinetics during development of the neocortex. *Congenital Anomalies*, *52*, 72–77. https://doi.org/10.1111/j.1741-4520.2012.00368.x.

Komada, M., Saitsu, H., Kinboshi, M., Miura, T., Shiota, K., & Ishibashi, M. (2008). Hedgehog signaling is involved in development of the neocortex. *Development*, *135*, 2717–2727. https://doi.org/10.1242/dev.015891.

Konno, D., Shioi, G., Shitamukai, A., Mori, A., Kiyonari, H., Miyata, T., et al. (2008). Neuroepithelial progenitors undergo LGN-dependent planar divisions to maintain self-renewability during mammalian neurogenesis. *Nature Cell Biology*, *10*, 93–101. https://doi.org/ncb1673 [pii]10.1038/ncb1673.

Kosodo, Y., & Huttner, W. B. (2009). Basal process and cell divisions of neural progenitors in the developing brain. *Development, Growth & Differentiation*, *51*, 251–261. https://doi.org/10.1111/j.1440-169X.2009.01101.x.

Kosodo, Y., Roper, K., Haubensak, W., Marzesco, A. M., Corbeil, D., & Huttner, W. B. (2004). Asymmetric distribution of the apical plasma membrane during neurogenic divisions of mammalian neuroepithelial cells. *The EMBO Journal*, *23*, 2314–2324. https://doi.org/10.1038/sj.emboj.76002237600223 [pii].

Kosodo, Y., Toida, K., Dubreuil, V., Schenk, J., Kiyokage, E., Attardo, A., et al. (2008). *Cytokinesis of neuroepithelial cells can divide their basal process before anaphase* (pp. 3151–3163). https://doi.org/10.1038/emboj.2008.227.

Kostic, M., Paridaen, J., Long, K. R., Kalebic, N., Langen, B., Grubling, N., et al. (2019). YAP activity is necessary and sufficient for basal progenitor abundance and proliferation in the developing neocortex. *Cell Reports*, *27*, 1103–1118 e6. https://doi.org/10.1016/j.celrep.2019.03.091.

Kowalczyk, T., Pontious, A., Englund, C., Daza, R. A., Bedogni, F., Hodge, R., et al. (2009). Intermediate neuronal progenitors (basal progenitors) produce pyramidal-projection neurons for all layers of cerebral cortex. *Cerebral Cortex*, *19*, 2439–2450. https://doi.org/bhn260 [pii]10.1093/cercor/bhn260.

Kriegstein, A., & Alvarez-Buylla, A. (2009). The glial nature of embryonic and adult neural stem cells. *Annual Review of Neuroscience*, *32*, 149–184. https://doi.org/10.1146/annurev.neuro.051508.135600.

Kriegstein, A., Noctor, S., & Martinez-Cerdeno, V. (2006). Patterns of neural stem and progenitor cell division may underlie evolutionary cortical expansion. *Nature Reviews. Neuroscience*, 7, 883–890.

Laguesse, S., Creppe, C., Nedialkova, D. D., Prévot, P. P., Borgs, L., Huysseune, S., et al. (2015). A dynamic unfolded protein response contributes to the control of cortical neurogenesis. *Developmental Cell*, *35*, 553–567. https://doi.org/10.1016/j.devcel.2015.11.005.

LaMonica, B. E., Lui, J. H., Hansen, D. V., & Kriegstein, A. R. (2013). Mitotic spindle orientation predicts outer radial glial cell generation in human neocortex. *Nature Communications*, *4*, 1665. https://doi.org/10.1038/ncomms2647ncomms2647 [pii].

Lancaster, M. A., & Knoblich, J. A. (2012). Spindle orientation in mammalian cerebral cortical development. *Current Opinion in Neurobiology*, *22*, 737–746. https://doi.org/10.1016/j.conb.2012.04.003.

Lange, C., Huttner, W. B., & Calegari, F. (2009). Cdk4/cyclinD1 overexpression in neural stem cells shortens G1, delays neurogenesis, and promotes the generation and expansion of basal progenitors. *Cell Stem Cell*, *5*, 320–331. https://doi.org/S1934-5909(09)00284-7 [pii]10.1016/j.stem.2009.05.026.

Lange, C., Turrero Garcia, M., Decimo, I., Bifari, F., Eelen, G., Quaegebeur, A., et al. (2016). Relief of hypoxia by angiogenesis promotes neural stem cell differentiation by targeting glycolysis. *The EMBO Journal, 35*, 924–941. https://doi.org/10.15252/embj.201592372.

Lehtinen, M. K., & Walsh, C. A. (2011). Neurogenesis at the brain-cerebrospinal fluid interface. *Annual Review of Cell and Developmental Biology, 27*, 653–679. https://doi.org/10.1146/annurev-cellbio-092910-154026.

Lehtinen, M. K., Zappaterra, M. W., Chen, X., Yang, Y. J., Hill, A. D., Lun, M., et al. (2011). The cerebrospinal fluid provides a proliferative niche for neural progenitor cells. *Neuron, 69*, 893–905. https://doi.org/10.1016/j.neuron.2011.01.023.

Leone, D. P., Relvas, J. B., Campos, L. S., Hemmi, S., Brakebusch, C., Fässler, R., et al. (2005). Regulation of neural progenitor proliferation and survival by β1 integrins. *Journal of Cell Science, 118*, 2589–2599. https://doi.org/10.1242/jcs.02396.

Li, S., Mattar, P., Zinyk, D., Singh, K., Chaturvedi, C. P., Kovach, C., et al. (2012). GSK3 temporally regulates neurogenin 2 proneural activity in the neocortex. *The Journal of Neuroscience, 32*, 7791–7805. https://doi.org/10.1523/JNEUROSCI.1309-12.2012.

Li, M., Santpere, G., Kawasawa, Y. I., Evgrafov, O. V., Gulden, F. O., Pochareddy, S., et al. (2018). Integrative functional genomic analysis of human brain development and neuropsychiatric risks. *Science, 362*(6420), eaat7615. https://doi.org/10.1126/science.aat7615.

Lim, L., Mi, D., Llorca, A., & Marin, O. (2018). Development and functional diversification of cortical interneurons. *Neuron, 100*, 294–313. https://doi.org/10.1016/j.neuron.2018.10.009.

Lin, D., Edwards, A. S., Fawcett, J. P., Mbamalu, G., Scott, J. D., & Pawson, T. (2000). A mammalian PAR-3-PAR-6 complex implicated in Cdc42/Rac1 and aPKC signalling and cell polarity. *Nature Cell Biology, 2*, 540–547. https://doi.org/10.1038/35019582.

Llinares-Benadero, C., & Borrell, V. (2019). Deconstructing cortical folding: Genetic, cellular and mechanical determinants. *Nature Reviews. Neuroscience, 20*, 161–176. https://doi.org/10.1038/s41583-018-0112-2.

Llorca, A., Ciceri, G., Beattie, R., Wong, F. K., Diana, G., Serafeimidou-Pouliou, E., et al. (2019). A stochastic framework of neurogenesis underlies the assembly of neocortical cytoarchitecture. *eLife, 8*. https://doi.org/10.7554/eLife.51381.

Long, K. R., & Huttner, W. B. (2019). How the extracellular matrix shapes neural development. *Open Biology, 9*, 180216. https://doi.org/10.1098/rsob.180216.

Long, M., VanKuren, N. W., Chen, S., & Vibranovski, M. D. (2013). New gene evolution: Little did we know. *Annual Review of Genetics, 47*, 307–333. https://doi.org/10.1146/annurev-genet-111212-133301.

LoTurco, J. J., Owens, D. F., Heath, M. J., Davis, M. B., & Kriegstein, A. R. (1995). GABA and glutamate depolarize cortical progenitor cells and inhibit DNA synthesis. *Neuron, 15*, 1287–1298. https://doi.org/0896-6273(95)90008-X [pii].

Lui, J. H., Hansen, D. V., & Kriegstein, A. R. (2011). Development and evolution of the human neocortex. *Cell, 146*, 18–36. https://doi.org/S0092-8674(11)00705-7 [pii] 10.1016/j.cell.2011.06.030.

Lukaszewicz, A., Savatier, P., Cortay, V., Giroud, P., Huissoud, C., Berland, M., et al. (2005). G1 phase regulation, area-specific cell cycle control, and cytoarchitectonics in the primate cortex. *Neuron, 47*, 353–364.

Luzzati, F. (2015). A hypothesis for the evolution of the upper layers of the neocortex through co-option of the olfactory cortex developmental program. *Frontiers in Neuroscience, 9*, 162. https://doi.org/10.3389/fnins.2015.00162.

Maekawa, M., Iwayama, Y., Nakamura, K., Sato, M., Toyota, T., Ohnishi, T., et al. (2009). A novel missense mutation (Leu46Val) of PAX6 found in an autistic patient. *Neuroscience Letters, 462*, 267–271. https://doi.org/10.1016/j.neulet.2009.07.021.

Malandrini, A., Mari, F., Palmeri, S., Gambelli, S., Berti, G., Bruttini, M., et al. (2001). PAX6 mutation in a family with aniridia, congenital ptosis, and mental retardation. *Clinical Genetics*, *60*, 151–154. https://doi.org/10.1034/j.1399-0004.2001.600210.x.

Malatesta, P., Hack, M. A., Hartfuss, E., Kettenmann, H., Klinkert, W., Kirchhoff, F., et al. (2003). Neuronal or glial progeny: Regional differences in radial glia fate. *Neuron*, *37*, 751–764.

Manning, C. S., Biga, V., Boyd, J., Kursawe, J., Ymisson, B., Spiller, D. G., et al. (2019). Quantitative single-cell live imaging links HES5 dynamics with cell-state and fate in murine neurogenesis. *Nature Communications*, *10*(1), 2835. https://doi.org/10.1038/s41467-019-10734-8.

Manuel, M. N., Martynoga, B., Molinek, M. D., Quinn, J. C., Kroemmer, C., Mason, J. O., et al. (2011). The transcription factor Foxg1 regulates telencephalic progenitor proliferation cell autonomously, in part by controlling Pax6 expression levels. *Neural Development*, *6*, 9. https://doi.org/10.1186/1749-8104-6-9.

Manuel, M. N., Mi, D., Mason, J. O., & Price, D. J. (2015). Regulation of cerebral cortical neurogenesis by the Pax6 transcription factor. *Frontiers in Cellular Neuroscience*, *9*, 70. https://doi.org/10.3389/fncel.2015.00070.

Marigo, V., Davey, R. A., Zuo, Y., Cunningham, J. M., & Tabin, C. J. (1996). Biochemical evidence that patched is the Hedgehog receptor. *Nature*, *384*, 176–179. https://doi.org/10.1038/384176a0.

Marin, O., & Rubenstein, J. L. (2003). Cell migration in the forebrain. *Annual Review of Neuroscience*, *26*, 441–483.

Marín-padilla, M. (1995). Prenatal development of fibrous (white matter), protoplasmic (gray matter), and layer I astrocytes in the human cerebral cortex: A Golgi study. *The Journal of Comparative Neurology*, *357*, 554–572. https://doi.org/10.1002/cne.903570407.

Marques-Bonet, T., & Eichler, E. E. (2009). The evolution of human segmental duplications and the core duplicon hypothesis. *Cold Spring Harbor Symposia on Quantitative Biology*, *74*, 355–362. https://doi.org/10.1101/sqb.2009.74.011.

Martín, C., Bueno, D., Alonso, M. I., Moro, J. A., Callejo, S., Parada, C., et al. (2006). FGF2 plays a key role in embryonic cerebrospinal fluid trophic properties over chick embryo neuroepithelial stem cells. *Developmental Biology*, *297*, 402–416. https://doi.org/10.1016/j.ydbio.2006.05.010.

Martinez-Cerdeno, V., Cunningham, C. L., Camacho, J., Keiter, J. A., Ariza, J., Lovern, M., et al. (2015). Evolutionary origin of Tbr2-expressing precursor cells and the subventricular zone in the developing cortex. *The Journal of Comparative Neurology*, *523*, 433–447. https://doi.org/10.1002/cne.23879.

Martinez-Cerdeno, V., Noctor, S. C., & Kriegstein, A. R. (2006). The role of intermediate progenitor cells in the evolutionary expansion of the cerebral cortex. *Cerebral Cortex*, *16*(Suppl. 1), i152–i161.

Martínez-Martínez, M. Á., Ciceri, G., Espinós, A., Fernández, V., Marín, O., & Borrell, V. (2019). Extensive branching of radially-migrating neurons in the mammalian cerebral cortex. *The Journal of Comparative Neurology*, *527*, 1558–1576. https://doi.org/10.1002/cne.24597.

Martínez-Martínez, M. Á., De Juan Romero, C., Fernández, V., Cárdenas, A., Götz, M., & Borrell, V. (2016). A restricted period for formation of outer subventricular zone defined by Cdh1 and Trnp1 levels. *Nature Communications*, 7, 11812. https://doi.org/10.1038/ncomms11812.

Martini, F. J., Valiente, M., Lopez Bendito, G., Szabo, G., Moya, F., Valdeolmillos, M., et al. (2009). Biased selection of leading process branches mediates chemotaxis during tangential neuronal migration. *Development*, *136*, 41–50. https://doi.org/10.1242/dev.025502136/1/41 [pii].

Masuda, K., Toda, T., Shinmyo, Y., Ebisu, H., Hoshiba, Y., Wakimoto, M., et al. (2015). Pathophysiological analyses of cortical malformation using gyrencephalic mammals. *Scientific Reports*, *5*, 15370. https://doi.org/10.1038/srep15370.

Matsumoto, N., Shinmyo, Y., Ichikawa, Y., & Kawasaki, H. (2017). Gyrification of the cerebral cortex requires FGF signaling in the mammalian brain. *eLife*, *6*. https://doi.org/10.7554/eLife.29285.

Matsumoto, N., Tanaka, S., Horiike, T., & Shinmyo, Y. (2020). A discrete subtype of neural progenitor crucial for cortical folding in the gyrencephalic mammalian brain. *eLife*, *9*, 1–26.

Mayer, S., Chen, J., Velmeshev, D., Mayer, A., Eze, U. C., Bhaduri, A., et al. (2019). Multimodal single-cell analysis reveals physiological maturation in the developing human neocortex. *Neuron*, *102*, 143–158 e7. https://doi.org/10.1016/j.neuron.2019.01.027.

McConnell, S. K. (1985). Migration and differentiation of cerebral cortical neurons after transplantation into the brains of ferrets. *Science*, *229*, 1268–1271.

McConnell, S. K. (1988). Fates of visual cortical neurons in the ferret after isochronic and heterochronic transplantation. *The Journal of Neuroscience*, *8*, 945–974.

McConnell, S. K. (1989). The determination of neuronal fate in the cerebral cortex. *Trends in Neurosciences*, *12*, 342–349.

McConnell, S. K. (1995). Constructing the cerebral cortex: Neurogenesis and fate determination. *Neuron*, *15*, 761–768.

McConnell, S. K., & Kaznowski, C. E. (1991). Cell cycle dependence of laminar determination in developing neocortex. *Science*, *254*, 282–285.

Mehta, B., & Bhat, K. M. (2001). Slit signaling promotes the terminal asymmetric division of neural precursor cells in the Drosophila CNS. *Development*, *128*, 3161–3168.

Mi, D., Carr, C. B., Georgala, P. A., Huang, Y. T., Manuel, M. N., Jeanes, E., et al. (2013). Pax6 exerts regional control of cortical progenitor proliferation via direct repression of Cdk6 and hypophosphorylation of pRb. *Neuron*, *78*, 269–284. https://doi.org/10.1016/j.neuron.2013.02.012.

Mi Moon, H., Hippenmeyer, S., Luo, L., & Wynshaw-Boris, A. (2020). LIS1 determines cleavage plane positioning by regulating actomyosin-mediated cell membrane contractility. *Elife*, *9*. https://doi.org/10.7554/eLife.51512.

Mikkelsen, T. S., Hillier, L. W., Eichler, E. E., Zody, M. C., Jaffe, D. B., Yang, S. P., et al. (2005). Initial sequence of the chimpanzee genome and comparison with the human genome. *Nature*, *437*, 69–87. https://doi.org/10.1038/nature04072.

Miller, J. A., Ding, S. L., Sunkin, S. M., Smith, K. A., Ng, L., Szafer, A., et al. (2014). Transcriptional landscape of the prenatal human brain. *Nature*, *508*, 199–206. https://doi.org/10.1038/nature13185nature13185 [pii].

Mils, V., Bosch, S., Roy, J., Bel-Vialar, S., Belenguer, P., Pituello, F., et al. (2015). Mitochondrial reshaping accompanies neural differentiation in the developing spinal cord. *PLoS One*, *10*, 1–15. https://doi.org/10.1371/journal.pone.0128130.

Mirzaa, G. M., Conti, V., Timms, A. E., Smyser, C. D., Ahmed, S., Carter, M., et al. (2015). Characterisation of mutations of the phosphoinositide-3-kinase regulatory subunit, PIK3R2, in perisylvian polymicrogyria: A next-generation sequencing study. *Lancet Neurology*, *14*, 1182–1195. https://doi.org/10.1016/S1474-4422(15)00278-1.

Mirzaa, G. M., Parry, D. A., Fry, A. E., Giamanco, K. A., Schwartzentruber, J., Vanstone, M., et al. (2014). De novo CCND2 mutations leading to stabilization of cyclin D2 cause megalencephaly-polymicrogyria-polydactyly-hydrocephalus syndrome. *Nature Genetics*, *46*, 510–515. https://doi.org/10.1038/ng.2948.

Misson, J. P., Austin, C. P., Takahashi, T., Cepko, C. L., & Caviness, V. S., Jr. (1991). The alignment of migrating neural cells in relation to the murine neopallial radial glial fiber system. *Cerebral Cortex*, *1*, 221–229.

Mitchell, T. N., Free, S. L., Williamson, K. A., Stevens, J. M., Churchill, A. J., Hanson, I. M., et al. (2003). Polymicrogyria and absence of pineal gland due to PAX6 mutation. *Annals of Neurology*, *53*, 658–663. https://doi.org/10.1002/ana.10576.
Miyata, T., Kawaguchi, A., Okano, H., & Ogawa, M. (2001). Asymmetric inheritance of radial glial fibers by cortical neurons. *Neuron*, *31*, 727–741.
Miyata, T., Kawaguchi, A., Saito, K., Kawano, M., Muto, T., & Ogawa, M. (2004). Asymmetric production of surface-dividing and non-surface-dividing cortical progenitor cells. *Development*, *131*, 3133–3145.
Miyata, T., Okamoto, M., Shinoda, T., & Kawaguchi, A. (2015). Interkinetic nuclear migration generates and opposes ventricular-zone crowding: Insight into tissue mechanics. *Frontiers in Cellular Neuroscience*, *8*, 1–11. https://doi.org/10.3389/fncel.2014.00473.
Molnar, Z. (2011). Evolution of cerebral cortical development. *Brain, Behavior and Evolution*, *78*, 94–107. https://doi.org/10.1159/000327325.
Molyneaux, B. J., Arlotta, P., Menezes, J. R., & Macklis, J. D. (2007). Neuronal subtype specification in the cerebral cortex. *Nature Reviews. Neuroscience*, *8*, 427–437.
Mora-Bermúdez, F., & Huttner, W. B. (2015). Novel insights into mammalian embryonic neural stem cell division: Focus on microtubules. *Molecular Biology of the Cell*, *26*, 4302–4306. https://doi.org/10.1091/mbc.E15-03-0152.
Moreno, M., Fernández, V., Monllau, J. M., Borrell, V., Lerin, C., & De La Iglesia, N. (2015). Transcriptional profiling of hypoxic neural stem cells identifies calcineurin-NFATc4 signaling as a major regulator of neural stem cell biology. *Stem Cell Reports*, *5*, 157–165. https://doi.org/10.1016/j.stemcr.2015.06.008.
Moreno, N., & Gonzalez, A. (2017). Pattern of neurogenesis and identification of neuronal progenitor subtypes during pallial development in Xenopus laevis. *Frontiers in Neuroanatomy*, *11*, 24. https://doi.org/10.3389/fnana.2017.00024.
Nadarajah, B., Brunstrom, J. E., Grutzendler, J., Wong, R. O., & Pearlman, A. L. (2001). Two modes of radial migration in early development of the cerebral cortex. *Nature Neuroscience*, *4*, 143–150.
Nadarajah, B., & Parnavelas, J. G. (2002). Modes of neuronal migration in the developing cerebral cortex. *Nature Reviews. Neuroscience*, *3*, 423–432. https://doi.org/10.1038/nrn845.
Namba, T., Dóczi, J., Pinson, A., Xing, L., Kalebic, N., Wilsch-Bräuninger, M., et al. (2020). Human-specific ARHGAP11B acts in mitochondria to expand neocortical progenitors by glutaminolysis. *Neuron*, *105*, 867–881.e9. https://doi.org/10.1016/j.neuron.2019.11.027.
Nandagopal, N., Santat, L. A., LeBon, L., Sprinzak, D., Bronner, M. E., & Elowitz, M. B. (2018). Dynamic ligand discrimination in the notch signaling pathway. *Cell*, *172*, 869–880 e19. https://doi.org/10.1016/j.cell.2018.01.002.
Nechiporuk, A., & Raible, D. W. (2008). FGF-dependent mechanosensory organ patterning in zebrafish. *Science*, *320*, 1774–1777. https://doi.org/10.1126/science.1156547.
Nelson, B. R., Hodge, R. D., Bedogni, F., & Hevner, R. F. (2013). Dynamic interactions between intermediate neurogenic progenitors and radial glia in embryonic mouse neocortex: Potential role in Dll1-Notch signaling. *The Journal of Neuroscience*, *33*, 9122–9139. https://doi.org/10.1523/JNEUROSCI.0791-13.2013.33/21/9122 [pii].
Ninkovic, J., Pinto, L., Petricca, S., Lepier, A., Sun, J., Rieger, M. A., et al. (2010). The transcription factor Pax6 regulates survival of dopaminergic olfactory bulb neurons via crystallin αA. *Neuron*, *68*, 682–694. https://doi.org/10.1016/j.neuron.2010.09.030.
Noctor, S. C., Flint, A. C., Weissman, T. A., Dammerman, R. S., & Kriegstein, A. R. (2001). Neurons derived from radial glial cells establish radial units in neocortex. *Nature*, *409*, 714–720.
Noctor, S. C., Martinez-Cerdeno, V., Ivic, L., & Kriegstein, A. R. (2004). Cortical neurons arise in symmetric and asymmetric division zones and migrate through specific phases. *Nature Neuroscience*, 7, 136–144.

Noctor, S. C., Martínez-Cerdeño, V., & Kriegstein, A. R. (2008). Distinct behaviors of neural stem and progenitor cells underlie cortical neurogenesis. *The Journal of Comparative Neurology, 508*, 28–44. https://doi.org/10.1002/cne.21669.

Noctor, S. C., Penna, E., Shepherd, H., Chelson, C., Barger, N., Martinez-Cerdeno, V., et al. (2019). Periventricular microglial cells interact with dividing precursor cells in the nonhuman primate and rodent prenatal cerebral cortex. *The Journal of Comparative Neurology, 527*, 1598–1609. https://doi.org/10.1002/cne.24604.

Nomura, T., Gotoh, H., & Ono, K. (2013). Changes in the regulation of cortical neurogenesis contribute to encephalization during amniote brain evolution. *Nature Communications, 4*, 2206. https://doi.org/10.1038/ncomms3206.

Nomura, T., Ohtaka-Maruyama, C., Yamashita, W., Wakamatsu, Y., Murakami, Y., Calegari, F., et al. (2016). The evolution of basal progenitors in the developing non-mammalian brain. *Development, 143*, 66–74. https://doi.org/10.1242/dev.127100.

Nonaka-Kinoshita, M., Reillo, I., Artegiani, B., Martínez-Martínez, M. A., Nelson, M., Borrell, V., et al. (2013). Regulation of cerebral cortex size and folding by expansion of basal progenitors. *The EMBO Journal, 32*, 1817–1828. https://doi.org/10.1038/emboj.2013.96.

Nowakowski, T. J., Bhaduri, A., Pollen, A. A., Alvarado, B., Mostajo-Radji, M. A., Di Lullo, E., et al. (2017). Spatiotemporal gene expression trajectories reveal developmental hierarchies of the human cortex. *Science, 358*, 1318–1323. https://doi.org/10.1126/science.aap8809.

Oberst, P., Fièvre, S., Baumann, N., Concetti, C., Bartolini, G., & Jabaudon, D. (2019). Temporal plasticity of apical progenitors in the developing mouse neocortex. *Nature, 573*, 370–374. https://doi.org/10.1038/s41586-019-1515-6.

Ochi, S., Imaizumi, Y., Shimojo, H., Miyachi, H., & Kageyama, R. (2020). Oscillatory expression of Hes1 regulates cell proliferation and neuronal differentiation in the embryonic brain. *Development, 147*. https://doi.org/10.1242/dev.182204.

Olkowicz, S., Kocourek, M., Lucan, R. K., Portes, M., Fitch, W. T., Herculano-Houzel, S., et al. (2016). Birds have primate-like numbers of neurons in the forebrain. *Proceedings of the National Academy of Sciences of the United States of America, 113*, 7255–7260. https://doi.org/10.1073/pnas.1517131113.

Parada, C., Gato, A., & Bueno, D. (2005). Mammalian embryonic cerebrospinal fluid proteome has greater apolipoprotein and enzyme pattern complexity than the avian proteome. *Journal of Proteome Research, 4*, 2420–2428. https://doi.org/10.1021/pr050213t.

Paridaen, J. T., Wilsch-Brauninger, M., & Huttner, W. B. (2013). Asymmetric inheritance of centrosome-associated primary cilium membrane directs ciliogenesis after cell division. *Cell, 155*, 333–344. https://doi.org/10.1016/j.cell.2013.08.060S0092-8674(13)01091-X [pii].

Park, S. M., Jang, H. J., & Lee, J. H. (2019). Roles of primary cilia in the developing brain. *Frontiers in Cellular Neuroscience, 13*, 1–10. https://doi.org/10.3389/fncel.2019.00218.

Pataskar, A., Jung, J., Smialowski, P., Noack, F., Calegari, F., Straub, T., et al. (2016). NeuroD1 reprograms chromatin and transcription factor landscapes to induce the neuronal program. *The EMBO Journal, 35*, 24–45. https://doi.org/10.15252/embj.201591206.

Phillips, N. E., Manning, C. S., Pettini, T., Biga, V., Marinopoulou, E., Stanley, P., et al. (2016). Stochasticity in the miR-9/Hes1 oscillatory network can account for clonal heterogeneity in the timing of differentiation. *eLife, 5*, 1–33. https://doi.org/10.7554/eLife.16118.

Piao, X., Hill, S. S., Bodell, A., Chang, B. S., Basel-Vanagaite, L., Straussberg, R., et al. (2004). G protein-coupled receptor-dependent development of human frontal cortex. *Science, 303*, 2033–2036. https://doi.org/10.1126/science.1092780.

Pierfelice, T., Alberi, L., & Gaiano, N. (2011). Notch in the vertebrate nervous system: An old dog with new tricks. *Neuron, 69*, 840–855. https://doi.org/10.1016/j.neuron.2011.02.031.

Pilaz, L. J., Lennox, A. L., Rouanet, J. P., & Silver, D. L. (2016). Dynamic mRNA transport and local translation in radial glial progenitors of the developing brain. *Current Biology*, *26*, 3383–3392. https://doi.org/10.1016/j.cub.2016.10.040.

Pilaz, L. J., McMahon, J. J., Miller, E. E., Lennox, A. L., Suzuki, A., Salmon, E., et al. (2016). Prolonged mitosis of neural progenitors alters cell fate in the developing brain. *Neuron*, *89*, 83–99. https://doi.org/10.1016/j.neuron.2015.12.007.

Pilaz, L. J., & Silver, D. L. (2017). Moving messages in the developing brain-emerging roles for mRNA transport and local translation in neural stem cells. *FEBS Letters*, *591*, 1526–1539. https://doi.org/10.1002/1873-3468.12626.

Pilz, D. T., Matsumoto, N., Minnerath, S., Mills, P., Gleeson, J. G., Allen, K. M., et al. (1998). LIS1 and XLIS (DCX) mutations cause most classical lissencephaly, but different patterns of malformation. *Human Molecular Genetics*, *7*, 2029–2037.

Pilz, G. A., Shitamukai, A., Reillo, I., Pacary, E., Schwausch, J., Stahl, R., et al. (2013). Amplification of progenitors in the mammalian telencephalon includes a new radial glial cell type. *Nature Communications*, *4*, 3125. https://doi.org/10.1038/ncomms3125.

Pinto, L., Drechsel, D., Schmid, M. T., Ninkovic, J., Irmler, M., Brill, M. S., et al. (2009). AP2γ regulates basal progenitor fate in a region- and layer-specific manner in the developing cortex. *Nature Neuroscience*, *12*, 1229–1237. https://doi.org/10.1038/nn.2399.

Pinto, L., Mader, M. T., Irmler, M., Gentilini, M., Santoni, F., Drechsel, D., et al. (2008). Prospective isolation of functionally distinct radial glial subtypes—Lineage and transcriptome analysis. *Molecular and Cellular Neurosciences*, *38*, 15–42. https://doi.org/10.1016/j.mcn.2008.01.012S1044-7431(08)00026-2 [pii].

Pollard, K. S., Salama, S. R., Lambert, N., Lambot, M. A., Coppens, S., Pedersen, J. S., et al. (2006). An RNA gene expressed during cortical development evolved rapidly in humans. *Nature*, *443*, 167–172. https://doi.org/10.1038/nature05113.

Pollen, A. A., Nowakowski, T. J., Chen, J., Retallack, H., Sandoval-Espinosa, C., Nicholas, C. R., et al. (2015). Molecular identity of human outer radial glia during cortical development. *Cell*, *163*, 55–67. https://doi.org/10.1016/j.cell.2015.09.004.

Pollen, A. A., Nowakowski, T. J., Shuga, J., Wang, X., Leyrat, A. A., Lui, J. H., et al. (2014). Low-coverage single-cell mRNA sequencing reveals cellular heterogeneity and activated signaling pathways in developing cerebral cortex. *Nature Biotechnology*, *32*, 1053–1058. https://doi.org/10.1038/nbt.2967.

Poluch, S., & Juliano, S. L. (2015). Fine-tuning of neurogenesis is essential for the evolutionary expansion of the cerebral cortex. *Cerebral Cortex*, *25*, 346–364. https://doi.org/10.1093/cercor/bht232.

Puelles, L. (2001). Thoughts on the development, structure and evolution of the mammalian and avian telencephalic pallium. *Philosophical Transactions of the Royal Society of London. Series B, Biological Sciences*, *356*, 1583–1598. https://doi.org/10.1098/rstb.2001.0973.

Puelles, L. (2017). Comments on the updated tetrapartite pallium model in the mouse and chick, featuring a homologous claustro-insular complex. *Brain, Behavior and Evolution*, *90*, 171–189. https://doi.org/10.1159/000479782.

Quinn, J. C., Molinek, M., Martynoga, B. S., Zaki, P. A., Faedo, A., Bulfone, A., et al. (2007). Pax6 controls cerebral cortical cell number by regulating exit from the cell cycle and specifies cortical cell identity by a cell autonomous mechanism. *Developmental Biology*, *302*, 50–65. https://doi.org/S0012-1606(06)01105-5 [pii]10.1016/j.ydbio.2006.08.035.

Raballo, R., Rhee, J., Lyn-Cook, R., Leckman, J. F., Schwartz, M. L., & Vaccarino, F. M. (2000). Basic fibroblast growth factor (Fgf2) is necessary for cell proliferation and neurogenesis in the developing cerebral cortex. *The Journal of Neuroscience*, *20*, 5012–5023. https://doi.org/20/13/5012 [pii].

Radakovits, R., Barros, C. S., Belvindrah, R., Patton, B., & Müller, U. (2009). Regulation of radial glial survival by signals from the meninges. *The Journal of Neuroscience*, *29*, 7694–7705. https://doi.org/10.1523/JNEUROSCI.5537-08.2009.

Rakic, P. (1972). Mode of cell migration to the superficial layers of fetal monkey neocortex. *The Journal of Comparative Neurology*, *145*, 61–83.
Rakic, P. (1974). Neurons in rhesus monkey visual cortex: Systematic relation between time of origin and eventual disposition. *Science*, *183*, 425–427.
Rakic, P. (1995). A small step for the cell, a giant leap for mankind: A hypothesis of neocortical expansion during evolution. *Trends in Neurosciences*, *18*, 383–388.
Ramesh, V., Bayam, E., Cernilogar, F. M., Bonapace, I. M., Schulze, M., Riemenschneider, M. J., et al. (2016). Loss of Uhrf1 in neural stem cells leads to activation of retroviral elements and delayed neurodegeneration. *Genes & Development*, *30*, 2199–2212. https://doi.org/10.1101/gad.284992.116.
Rash, B. G., Ackman, J. B., & Rakic, P. (2016). Bidirectional radial Ca(2+) activity regulates neurogenesis and migration during early cortical column formation. *Science Advances*, *2*, e1501733. https://doi.org/10.1126/sciadv.1501733.
Rasin, M. R., Gazula, V. R., Breunig, J. J., Kwan, K. Y., Johnson, M. B., Liu-Chen, S., et al. (2007). Numb and Numbl are required for maintenance of cadherin-based adhesion and polarity of neural progenitors. *Nature Neuroscience*, *10*, 819–827. https://doi.org/nn1924 [pii]10.1038/nn1924.
Reillo, I., & Borrell, V. (2012). Germinal zones in the developing cerebral cortex of ferret: Ontogeny, cell cycle kinetics, and diversity of progenitors. *Cerebral Cortex*, *22*, 2039–2054. https://doi.org/10.1093/cercor/bhr284.
Reillo, I., De Juan Romero, C., Cárdenas, A., Clascá, F., Martínez-Martinez, M. A., & Borrell, V. (2017). A complex code of extrinsic influences on cortical progenitor cells of higher mammals. *Cerebral Cortex*, *27*, 4586–4606. https://doi.org/10.1093/cercor/bhx171.
Reillo, I., De Juan Romero, C., García-Cabezas, M. Á., & Borrell, V. (2011). A role for intermediate radial glia in the tangential expansion of the mammalian cerebral cortex. *Cerebral Cortex*, *21*, 1674–1694. https://doi.org/10.1093/cercor/bhq238.
Saade, M., Gutiérrez-Vallejo, I., LeDréau, G., Rabadán, M. A., Miguez, D. G., Buceta, J., et al. (2013). Sonic hedgehog signaling switches the mode of division in the developing nervous system. *Cell Reports*, *4*, 492–503. https://doi.org/10.1016/j.celrep.2013.06.038.
Sansom, S. N., Griffiths, D. S., Faedo, A., Kleinjan, D. J., Ruan, Y., Smith, J., et al. (2009). The level of the transcription factor Pax6 is essential for controlling the balance between neural stem cell self-renewal and neurogenesis. *PLoS Genetics*, *5*, e1000511. https://doi.org/10.1371/journal.pgen.1000511.
Schmahl, W., Knoedlseder, M., Favor, J., & Davidson, D. (1993). Defects of neuronal migration and the pathogenesis of cortical malformations are associated with Small eye (Sey) in the mouse, a point mutation at the Pax-6-locus. *Acta Neuropathologica*, *86*, 126–135. https://doi.org/10.1007/BF00334879.
Schmid, R. S., & Anton, E. S. (2003). Role of integrins in the development of the cerebral cortex. *Cerebral Cortex*, *13*, 219–224. https://doi.org/10.1093/cercor/13.3.219.
Schuurmans, C., Armant, O., Nieto, M., Stenman, J. M., Britz, O., Klenin, N., et al. (2004). Sequential phases of cortical specification involve neurogenin-dependent and -independent pathways. *The EMBO Journal*, *23*, 2892–2902. https://doi.org/10.1038/sj.emboj.7600278.
Schuurmans, C., & Guillemot, F. (2002). Molecular mechanisms underlying cell fate specification in the developing telencephalon. *Current Opinion in Neurobiology*, *12*, 26–34.
Scott, C. E., Wynn, S. L., Sesay, A., Cruz, C., Cheung, M., Gaviro, M. V. G., et al. (2010). SOX9 induces and maintains neural stem cells. *Nature Neuroscience*, *13*, 1181–1189. https://doi.org/10.1038/nn.2646.
Sessa, A., Ciabatti, E., Drechsel, D., Massimino, L., Colasante, G., Giannelli, S., et al. (2017). The Tbr2 molecular network controls cortical neuronal differentiation through complementary genetic and epigenetic pathways. *Cerebral Cortex*, *27*, 3378–3396. https://doi.org/10.1093/cercor/bhw270.

Sessa, A., Mao, C. A., Hadjantonakis, A. K., Klein, W. H., & Broccoli, V. (2008). Tbr2 directs conversion of radial glia into basal precursors and guides neuronal amplification by indirect neurogenesis in the developing neocortex. *Neuron*, *60*, 56–69. https://doi.org/S0896-6273(08)00807-6 [pii]10.1016/j.neuron.2008.09.028.
She, X., Jiang, Z., Clark, R. A., Liu, G., Cheng, Z., Tuzun, E., et al. (2004). Shotgun sequence assembly and recent segmental duplications within the human genome. *Nature*, *431*, 927–930. https://doi.org/10.1038/nature03062.
Sheen, V. L. (2014). Filamin A mediated Big2 dependent endocytosis: From apical abscission to periventricular heterotopia. *Tissue Barriers*, *2*, 1–6. https://doi.org/10.4161/tisb.29431.
Sheen, V. L., & Walsh, C. A. (2003). Developmental genetic malformations of the cerebral cortex. *Current Neurology and Neuroscience Reports*, *3*, 433–441.
Shikata, Y., Okada, T., Hashimoto, M., Ellis, T., Matsumaru, D., Shiroishi, T., et al. (2011). Ptch1-mediated dosage-dependent action of Shh signaling regulates neural progenitor development at late gestational stages. *Developmental Biology*, *349*, 147–159. https://doi.org/10.1016/j.ydbio.2010.10.014.
Shimojo, H., Isomura, A., Ohtsuka, T., Kori, H., Miyachi, H., & Kageyama, R. (2016). Oscillatory control of Delta-like1 in cell interactions regulates dynamic gene expression and tissue morphogenesis. *Genes & Development*, *30*, 102–116. https://doi.org/10.1101/gad.270785.115.
Shimojo, H., Ohtsuka, T., & Kageyama, R. (2008). Oscillations in notch signaling regulate maintenance of neural progenitors. *Neuron*, *58*, 52–64. https://doi.org/10.1016/j.neuron.2008.02.014.
Shitamukai, A., Konno, D., & Matsuzaki, F. (2011). Oblique radial glial divisions in the developing mouse neocortex induce self-renewing progenitors outside the germinal zone that resemble primate outer subventricular zone progenitors. *Journal of Neuroscience*, *31*, 3683–3695. https://doi.org/31/10/3683 [pii]10.1523/JNEUROSCI.4773-10.2011.
Sidman, R. L., & Rakic, P. (1973). Neuronal migration, with special reference to developing human brain: A review. *Brain Research*, *62*, 1–35.
Silva, C. G., Peyre, E., Adhikari, M. H., Tielens, S., Tanco, S., Van Damme, P., et al. (2018). Cell-intrinsic control of interneuron migration drives cortical morphogenesis. *Cell*, *172*, 1063–1078.e19. https://doi.org/10.1016/j.cell.2018.01.031.
Silver, D. L. (2016). Genomic divergence and brain evolution: How regulatory DNA influences development of the cerebral cortex. *BioEssays*, *38*, 162–171. https://doi.org/10.1002/bies.201500108.
Smart, I. H., Dehay, C., Giroud, P., Berland, M., & Kennedy, H. (2002). Unique morphological features of the proliferative zones and postmitotic compartments of the neural epithelium giving rise to striate and extrastriate cortex in the monkey. *Cerebral Cortex*, *12*, 37–53.
Snedeker, J., Schock, E. N., Struve, J. N., Chang, C. F., Cionni, M., Tran, P. V., et al. (2017). Unique spatiotemporal requirements for intraflagellar transport genes during forebrain development. *PLoS One*, *12*(3), e0173258. https://doi.org/10.1371/journal.pone.0173258.
Sousa, A. M. M., Meyer, K. A., Santpere, G., Gulden, F. O., & Sestan, N. (2017). Evolution of the human nervous system function, structure, and development. *Cell*, *170*, 226–247. https://doi.org/10.1016/j.cell.2017.06.036.
Stahl, R., Walcher, T., De Juan Romero, C., Pilz, G. A., Cappello, S., Irmler, M., et al. (2013). Trnp1 regulates expansion and folding of the mammalian cerebral cortex by control of radial glial fate. *Cell*, *153*, 535–549. https://doi.org/10.1016/j.cell.2013.03.027.
Stancik, E. K., Navarro-Quiroga, I., Sellke, R., & Haydar, T. F. (2010). Heterogeneity in ventricular zone neural precursors contributes to neuronal fate diversity in the postnatal neocortex. *The Journal of Neuroscience*, *30*, 7028–7036. https://doi.org/10.1523/JNEUROSCI.6131-09.2010.

Stenzel, D., Wilsch-Brauninger, M., Wong, F. K., Heuer, H., & Huttner, W. B. (2014). Integrin alphavbeta3 and thyroid hormones promote expansion of progenitors in embryonic neocortex. *Development*, *141*, 795–806. https://doi.org/10.1242/dev.101907141/4/795 [pii].

Stocker, A. M., & Chenn, A. (2015). The role of adherens junctions in the developing neocortex. *Cell Adhesion & Migration*, *9*, 167–174. https://doi.org/10.1080/19336918.2015.1027478.

Stone, D. M., Hynes, M., Armanini, M., Swanson, T. A., Gu, Q., Johnson, R. L., et al. (1996). The tumour-suppressor gene patched encodes a candidate receptor for Sonic hedgehog. *Nature*, *384*, 129–134. https://doi.org/10.1038/384129a0.

Sun, Y., Nadal-Vicens, M., Misono, S., Lin, M. Z., Zubiaga, A., Hua, X., et al. (2001). Neurogenin promotes neurogenesis and inhibits glial differentiation by independent mechanisms. *Cell*, *104*, 365–376. https://doi.org/10.1016/S0092-8674(01)00224-0.

Super, H., Del Rio, J. A., Martinez, A., Perez-Sust, P., & Soriano, E. (2000). Disruption of neuronal migration and radial glia in the developing cerebral cortex following ablation of Cajal-Retzius cells. *Cerebral Cortex*, *10*, 602–613.

Sur, M., & Rubenstein, J. L. (2005). Patterning and plasticity of the cerebral cortex. *Science*, *310*, 805–810. https://doi.org/310/5749/805 [pii]10.1126/science.1112070.

Suzuki, I. K., Gacquer, D., Van Heurck, R., Kumar, D., Wojno, M., Bilheu, A., et al. (2018). Human-specific NOTCH2NL genes expand cortical neurogenesis through delta/notch regulation. *Cell*, *173*, 1370–1384 e16. https://doi.org/10.1016/j.cell.2018.03.067.

Tabata, H., & Nakajima, K. (2003). Multipolar migration: The third mode of radial neuronal migration in the developing cerebral cortex. *The Journal of Neuroscience*, *23*, 9996–10001.

Takahashi, T., Goto, T., Miyama, S., Nowakowski, R. S., & Caviness Jr., V. S. (1999). Sequence of neuron origin and neocortical laminar fate: Relation to cell cycle of origin in the developing murine cerebral wall. *The Journal of Neuroscience*, *19*, 10357–10371.

Takahashi, T., Misson, J. P., & Caviness Jr., V. S. (1990). Glial process elongation and branching in the developing murine neocortex: A qualitative and quantitative immunohistochemical analysis. *The Journal of Comparative Neurology*, *302*, 15–28.

Takahashi, T., Nowakowski, R. S., & Caviness Jr., V. S. (1993). Cell cycle parameters and patterns of nuclear movement in the neocortical proliferative zone of the fetal mouse. *The Journal of Neuroscience*, *13*, 820–833.

Takahashi, T., Nowakowski, R. S., & Caviness Jr., V. S. (1994). Mode of cell proliferation in the developing mouse neocortex. *Proceedings of the National Academy of Sciences of the United States of America*, *91*, 375–379.

Takahashi, T., Nowakowski, R. S., & Caviness Jr., V. S. (1995a). The cell cycle of the pseudostratified ventricular epithelium of the embryonic murine cerebral wall. *The Journal of Neuroscience*, *15*, 6046–6057.

Takahashi, T., Nowakowski, R. S., & Caviness Jr., V. S. (1995b). Early ontogeny of the secondary proliferative population of the embryonic murine cerebral wall. *The Journal of Neuroscience*, *15*, 6058–6068.

Takahashi, T., Nowakowski, R. S., & Caviness Jr., V. S. (1996). The leaving or Q fraction of the murine cerebral proliferative epithelium: A general model of neocortical neuronogenesis. *The Journal of Neuroscience*, *16*, 6183–6196.

Tasic, B., Yao, Z., Graybuck, L. T., Smith, K. A., Nguyen, T. N., Bertagnolli, D., et al. (2018). Shared and distinct transcriptomic cell types across neocortical areas. *Nature*, *563*, 72–78. https://doi.org/10.1038/s41586-018-0654-5.

Tavano, S., Taverna, E., Kalebic, N., Haffner, C., Namba, T., Dahl, A., et al. (2018). Insm1 induces neural progenitor delamination in developing neocortex via downregulation of the adherens junction belt-specific protein Plekha7. *Neuron*, *97*, 1299–1314.e8. https://doi.org/10.1016/j.neuron.2018.01.052.

Taverna, E., Gotz, M., & Huttner, W. B. (2014). The cell biology of neurogenesis: Toward an understanding of the development and evolution of the neocortex. *Annual Review of Cell and Developmental Biology*, *30*, 465–502. https://doi.org/10.1146/annurev-cellbio-101011-155801.

Taverna, E., & Huttner, W. B. (2010). Neural progenitor nuclei IN motion. *Neuron*, *67*, 906–914. https://doi.org/10.1016/j.neuron.2010.08.027.

Taverna, E., Mora-Bermudez, F., Strzyz, P. J., Florio, M., Icha, J., Haffner, C., et al. (2016). Non-canonical features of the Golgi apparatus in bipolar epithelial neural stem cells. *Scientific Reports*, *6*, 21206. https://doi.org/10.1038/srep21206.

Telley, L., Agirman, G., Prados, J., Amberg, N., Fièvre, S., Oberst, P., et al. (2019). Temporal patterning of apical progenitors and their daughter neurons in the developing neocortex. *Science*, *364*(6440), eaav2522. https://doi.org/10.1126/science.aav2522.

Telley, L., Govindan, S., Prados, J., Stevant, I., Nef, S., Dermitzakis, E., et al. (2016). Sequential transcriptional waves direct the differentiation of newborn neurons in the mouse neocortex. *Science*, *351*, 1443–1446. https://doi.org/10.1126/science.aad8361.

Thakurela, S., Sahu, S. K., Garding, A., & Tiwari, V. K. (2015). Dynamics and function of distal regulatory elements during neurogenesis and neuroplasticity. *Genome Research*, *25*, 1309–1324. https://doi.org/10.1101/gr.190926.115.

Toda, T., Shinmyo, Y., Dinh Duong, T. A., Masuda, K., & Kawasaki, H. (2016). An essential role of SVZ progenitors in cortical folding in gyrencephalic mammals. *Scientific Reports*, *6*, 29578. https://doi.org/10.1038/srep29578.

Torii, M., Hashimoto-Torii, K., Levitt, P., & Rakic, P. (2009). Integration of neuronal clones in the radial cortical columns by EphA and ephrin-A signalling. *Nature*, *461*, 524–528. https://doi.org/nature08362 [pii]10.1038/nature08362.

Tosches, M. A., Yamawaki, T. M., Naumann, R. K., Jacobi, A. A., Tushev, G., & Laurent, G. (2018). Evolution of pallium, hippocampus, and cortical cell types revealed by single-cell transcriptomics in reptiles. *Science*, *360*, 881–888. https://doi.org/10.1126/science.aar4237.

Tsuda, S., Kitagawa, T., Takashima, S., Asakawa, S., Shimizu, N., Mitani, H., et al. (2010). FAK-mediated extracellular signals are essential for interkinetic nuclear migration and planar divisions in the neuroepithelium. *Journal of Cell Science*, *123*, 484–496. https://doi.org/10.1242/jcs.057851.

Tsunekawa, Y., Britto, J. M., Takahashi, M., Polleux, F., Tan, S. S., & Osumi, N. (2012). Cyclin D2 in the basal process of neural progenitors is linked to non-equivalent cell fates. *The EMBO Journal*, *31*, 1879–1892. https://doi.org/10.1038/emboj.2012.43.

Tuoc, T. C., Radyushkin, K., Tonchev, A. B., Piñon, M. C., Ashery-Padan, R., Molnár, Z., et al. (2009). Selective cortical layering abnormalities and behavioral deficits in cortex-specific Pax6 knock-out mice. *The Journal of Neuroscience*, *29*, 8335–8349. https://doi.org/10.1523/JNEUROSCI.5669-08.2009.

Tylkowski, M. A., Yang, K., Hoyer-Fender, S., & Stoykova, A. (2015). Pax6 controls centriole maturation in cortical progenitors through Odf2. *Cellular and Molecular Life Sciences*, *72*, 1795–1809. https://doi.org/10.1007/s00018-014-1766-1.

Vasistha, N. A., Garcia-Moreno, F., Arora, S., Cheung, A. F., Arnold, S. J., Robertson, E. J., et al. (2015). Cortical and clonal contribution of Tbr2 expressing progenitors in the developing mouse brain. *Cerebral Cortex*, *25*, 3290–3302. https://doi.org/10.1093/cercor/bhu125.

Veeraval, L., O'Leary, C. J., & Cooper, H. M. (2020). Adherens junctions: Guardians of cortical development. *Frontiers in Cell and Development Biology*, *8*, 1–17. https://doi.org/10.3389/fcell.2020.00006.

Vitali, I., Fievre, S., Telley, L., Oberst, P., Bariselli, S., Frangeul, L., et al. (2018). Progenitor hyperpolarization regulates the sequential generation of neuronal subtypes in the developing neocortex. *Cell*, *174*, 1264–1276 e15. https://doi.org/10.1016/j.cell.2018.06.036.

Voigt, T. (1989). Development of glial cells in the cerebral wall of ferrets: Direct tracing of their transformation from radial glia into astrocytes. *The Journal of Comparative Neurology*, *289*, 74–88.

Wakabayashi, J., Zhang, Z., Wakabayashi, N., Tamura, Y., Fukaya, M., Kensler, T. W., et al. (2009). The dynamin-related GTPase Drp1 is required for embryonic and brain development in mice. *The Journal of Cell Biology*, *186*, 805–816. https://doi.org/10.1083/jcb.200903065.

Walcher, T., Xie, Q., Sun, J., Irmler, M., Beckers, J., Öztürk, T., et al. (2013). Functional dissection of the paired domain of Pax6 reveals molecular mechanisms of coordinating neurogenesis and proliferation. *Development*, *140*, 1123–1136. https://doi.org/10.1242/dev.082875.

Wamsley, B., & Fishell, G. (2017). Genetic and activity-dependent mechanisms underlying interneuron diversity. *Nature Reviews. Neuroscience*, *18*, 299–309. https://doi.org/10.1038/nrn.2017.30.

Wang, L., Hou, S., & Han, Y. G. (2016). Hedgehog signaling promotes basal progenitor expansion and the growth and folding of the neocortex. *Nature Neuroscience*, *19*, 888–896. https://doi.org/10.1038/nn.4307.

Wang, X., Tsai, J. W., Imai, J. H., Lian, W. N., Vallee, R. B., & Shi, S. H. (2009). Asymmetric centrosome inheritance maintains neural progenitors in the neocortex. *Nature*, *461*, 947–955. https://doi.org/10.1038/nature08435.

Wang, X., Tsai, J. W., Lamonica, B., & Kriegstein, A. R. (2011). A new subtype of progenitor cell in the mouse embryonic neocortex. *Nature Neuroscience*, *14*, 555–561. https://doi.org/nn.2807 [pii]10.1038/nn.2807.

Watanabe, Y., Kawaue, T., & Miyata, T. (2018). Differentiating cells mechanically limit the interkinetic nuclear migration of progenitor cells to secure apical cytogenesis. *Development*, *145*, 1–7. https://doi.org/10.1242/dev.162883.

Weissman, T., Noctor, S. C., Clinton, B. K., Honig, L. S., & Kriegstein, A. R. (2003). Neurogenic radial glial cells in reptile, rodent and human: From mitosis to migration. *Cerebral Cortex*, *13*, 550–559.

Weissman, T. A., Riquelme, P. A., Ivic, L., Flint, A. C., & Kriegstein, A. R. (2004). Calcium waves propagate through radial glial cells and modulate proliferation in the developing neocortex. *Neuron*, *43*, 647–661.

Wilsch-Bräuninger, M., Peters, J., Paridaen, J. T. M. L., & Huttner, W. B. (2012). Basolateral rather than apical primary cilia on neuroepithelial cells committed to delamination. *Development*, *139*, 95–105. https://doi.org/10.1242/dev.069294. dev.069294 [pii].

Wong, F. K., Fei, J. F., Mora-Bermudez, F., Taverna, E., Haffner, C., Fu, J., et al. (2015). Sustained Pax6 expression generates primate-like basal radial glia in developing mouse neocortex. *PLoS Biology*, *13*, e1002217. https://doi.org/10.1371/journal.pbio.1002217.

Wood, C. R., & Rosenbaum, J. L. (2015). Ciliary ectosomes: Transmissions from the cell's antenna. *Trends in Cell Biology*, *25*, 276–285. https://doi.org/10.1016/j.tcb.2014.12.008.

Woodhead, G. J., Mutch, C. A., Olson, E. C., & Chenn, A. (2006). Cell-autonomous beta-catenin signaling regulates cortical precursor proliferation. *The Journal of Neuroscience*, *26*, 12620–12630.

Xuan, S., Baptista, C. A., Balas, G., Tao, W., Soares, V. C., & Lai, E. (1995). Winged helix transcription factor BF-1 is essential for the development of the cerebral hemispheres. *Neuron*, *14*, 1141–1152. https://doi.org/10.1016/0896-6273(95)90262-7.

Yabut, O. R., & Pleasure, S. J. (2018). Sonic hedgehog signaling rises to the surface: Emerging roles in neocortical development. *Brain Plasticity*, *3*, 119–128. https://doi.org/10.3233/bpl-180064.

Yamashita, N., Morita, A., Uchida, Y., Nakamura, F., Usui, H., Ohshima, T., et al. (2007). Regulation of spine development by semaphorin3A through cyclin-dependent kinase 5 phosphorylation of collapsin response mediator protein 1. *Journal of Neuroscience*, *27*, 12546–12554. https://doi.org/27/46/12546 [pii]10.1523/JNEUROSCI.3463-07.2007.

Yamashita, W., Takahashi, M., Kikkawa, T., Gotoh, H., Osumi, N., Ono, K., et al. (2018). Conserved and divergent functions of Pax6 underlie species-specific neurogenic patterns in the developing amniote brain. *Development*, *145*(8), dev159764. https://doi.org/10.1242/dev.159764.

Yingling, J., Youn, Y. H., Darling, D., Toyo-Oka, K., Pramparo, T., Hirotsune, S., et al. (2008). Neuroepithelial stem cell proliferation requires LIS1 for precise spindle orientation and symmetric division. *Cell*, *132*, 474–486. https://doi.org/S0092-8674(08)00126-8 [pii] 10.1016/j.cell.2008.01.026.

Yoon, K., & Gaiano, N. (2005). Notch signaling in the mammalian central nervous system: Insights from mouse mutants. *Nature Neuroscience*, *8*, 709–715. https://doi.org/nn1475 [pii]10.1038/nn1475.

Yoon, K. J., Ringeling, F. R., Vissers, C., Jacob, F., Pokrass, M., Jimenez-Cyrus, D., et al. (2017). Temporal control of mammalian cortical neurogenesis by m6A methylation. *Cell*, *171*, 877–889. https://doi.org/10.1016/j.cell.2017.09.003.

Yoon, K. J., Vissers, C., Ming, G. L., & Song, H. (2018). Epigenetics and epitranscriptomics in temporal patterning of cortical neural progenitor competence. *The Journal of Cell Biology*, *217*, 1901–1914. https://doi.org/10.1083/jcb.201802117.

Ypsilanti, A. R., & Rubenstein, J. L. R. (2016). Transcriptional and epigenetic mechanisms of early cortical development: An examination of how Pax6 coordinates cortical development. *The Journal of Comparative Neurology*, *524*, 609–629. https://doi.org/10.1002/cne.23866.

Zappaterra, M. D., Lisgo, S. N., Lindsay, S., Gygi, S. P., Walsh, C. A., & Ballif, B. A. (2007). A comparative proteomic analysis of human and rat embryonic cerebrospinal fluid. *Journal of Proteome Research*, *6*, 3537–3548. https://doi.org/10.1021/pr070247w.

Zhou, Y., Atkins, J. B., Rompani, S. B., Bancescu, D. L., Petersen, P. H., Tang, H., et al. (2007). The mammalian golgi regulates numb signaling in asymmetric cell division by releasing ACBD3 during mitosis. *Cell*, *129*, 163–178. https://doi.org/10.1016/j.cell.2007.02.037.

Zhu, Y., Sousa, A. M. M., Gao, T., Skarica, M., Li, M., Santpere, G., et al. (2018). Spatiotemporal transcriptomic divergence across human and macaque brain development. *Science*, *362*(6420), eaat8077. https://doi.org/10.1126/science.aat8077.

Zhu, C., Yu, M., Huang, H., Juric, I., Abnousi, A., Hu, R., et al. (2019). An ultra high-throughput method for single-cell joint analysis of open chromatin and transcriptome. *Nature Structural & Molecular Biology*, *26*, 1063–1070. https://doi.org/10.1038/s41594-019-0323-x.

CHAPTER TWO

Ontogeny of adult neural stem cells in the mammalian brain

Allison M. Bond[a], Guo-li Ming[a,b,c,d], and Hongjun Song[a,b,c,e,*]
[a]Department of Neuroscience and Mahoney Institute for Neurosciences, Perelman School of Medicine, University of Pennsylvania, Philadelphia, PA, United States
[b]Department of Cell and Developmental Biology, Perelman School of Medicine, University of Pennsylvania, Philadelphia, PA, United States
[c]Institute for Regenerative Medicine, University of Pennsylvania, Philadelphia, PA, United States
[d]Department of Psychiatry, Perelman School of Medicine, University of Pennsylvania, Philadelphia, PA, United States
[e]The Epigenetics Institute, Perelman School of Medicine, University of Pennsylvania, Philadelphia, PA, United States
*Corresponding author: e-mail address: shongjun@pennmedicine.upenn.edu

Contents

Abstract

Neural stem cells (NSCs) persist into adulthood in the subgranular zone (SGZ) of the dentate gyrus in the hippocampus and in the ventricular-subventricular zone (V-SVZ) of the lateral ventricles, where they generate new neurons and glia cells that contribute to neural plasticity. A better understanding of the developmental process that enables NSCs to persist beyond development will provide insight into factors that determine the size and properties of the adult NSC pool and thus the capacity for life-long neurogenesis in the adult mammalian brain. We review current knowledge regarding the developmental origins of adult NSCs and the developmental process by which embryonic NSCs transition into their adult form. We also discuss potential mechanisms that might regulate proper establishment of the adult NSC pool, and propose future directions of research that will be key to unraveling how NSCs transform to establish the adult NSC pool in the mammalian brain.

Current Topics in Developmental Biology, Volume 142
ISSN 0070-2153
https://doi.org/10.1016/bs.ctdb.2020.11.002

1. Introduction

Neural stem cells (NSCs) are resident stem cells of the nervous system and are responsible for populating the brain with neurons and glia during development. Though their numbers decline as development is completed, some NSCs are retained where they continue to generate neurons and glia throughout adulthood. The role of NSCs in the adult brain is very different than during development—NSCs generate progeny during development to establish neural circuits, whereas NSCs generate progeny in adulthood to modify circuits through neural plasticity. In particular, neurogenesis, or the generation of new neurons from NSCs, provides a source of new neurons which can be added to neural circuits to induce plasticity (Ma, Bonaguidi, Ming, & Song, 2009; Ming & Song, 2011). Active neurogenesis does not occur in all regions of the adult mammalian brain under physiological conditions, but instead is limited to two discrete regions: NSCs in the subgranular zone (SGZ) of the hippocampal dentate gyrus (DG) region generate dentate granule neurons (DGNs) and astrocytes, and NSCs in the ventricular-subventricular zone (V-SVZ) of the lateral ventricles generate olfactory bulb interneurons and oligodendrocytes (Bond, Ming, & Song, 2015; Ming & Song, 2005). A fundamental question in NSC biology is how NSCs are maintained in the SGZ and V-SVZ beyond development. To get at this question, we need to understand the developmental ontogeny of adult NSCs.

The earliest neural precursors in the embryonic brain are neuroepithelial cells, which divide symmetrically to expand the precursor pool. Neuroepithelial cells transform into radial glia (RG) by acquiring features associated with glial cells, such as astroglial marker expression and initiation of specialized contact with endothelial cells in the developing vasculature (Kriegstein & Alvarez-Buylla, 2009). RG first divide asymmetrically to generate neurons and neuronal progenitors, called intermediate progenitor cells (IPCs), and later divide to generate glia and glial progenitors (Kriegstein & Alvarez-Buylla, 2009; Namba & Huttner, 2017). While the majority of embryonic RG are not maintained beyond development, a small proportion of RG transform into adult NSCs and are maintained throughout life. More specifically, adult NSCs in the SGZ and V-SVZ are called RGLs (radial glia-like NSCs) (Bonaguidi et al., 2011) and B cells (Doetsch, Garcia-Verdugo, & Alvarez-Buylla, 1997), respectively. Though the behavior and function of adult NSCs has been extensively studied, the developmental ontogeny of adult NSCs, particularly the transformation of RG into adult NSCs, has only recently become a focus of adult NSC research (Berg, Bond, Ming, & Song, 2018; Obernier & Alvarez-Buylla, 2019). One defining feature of adult NSCs is their quiescent state, a reversible state in which a cell has exited cell cycle but retains the capacity to divide (Urban, Blomfield, & Guillemot, 2019). The transition from a proliferative state to quiescence is thought to be a critical step in the transformation from a developmental RG state to an adult NSC state and has been used as a developmental milestone marking the transition to an adult NSC phenotype.

In this chapter, we present an overview of our current understanding of the ontogeny of adult NSCs in the mammalian brain, focusing on studies in rodent models. We start by discussing embryonic origins of adult NSCs, including the developmental timing of RG to adult NSC transformation, the lineage relationship between RG of the embryonic brain and adult NSCs, and differences in the embryonic origins of adult NSCs in the SGZ and V-SVZ. Next, we review how intrinsic cellular and molecular properties of NSCs change as they transform from RG into adult NSCs, and the developmental establishment of adult NSC niches, which are specialized microenvironments that support NSC maintenance and continuous neurogenesis. Finally, we summarize potential cellular and molecular mechanisms that regulate transformation of RG to adult NSCs. We conclude with remaining unanswered questions and future directions. We hope that by taking a developmental perspective to understand how adult NSCs emerge from development, we will gain insight into mechanisms that maintain long-term NSC potential, which can be applied to enhance plasticity and regenerative capacity of the adult mammalian brain.

2. Embryonic origins of adult neural stem cells

Adult NSCs in the SGZ and V-SVZ have different embryonic origins and arise through distinct developmental processes (Fig. 1). The major differences between origins of adult SGZ and V-SVZ NSCs are (1) the timing of entry into quiescence and (2) the continuity of progeny fate. Within each of the two adult NSC populations, there also appears to be some variety in their embryonic origin, which has been linked to adult NSC heterogeneity. Thus, understanding developmental origins of adult NSCs may provide insight into adult NSC function. In this section, we discuss the developmental timing of adult NSC emergence and lineage relationships between adult NSCs, embryonic RG and differentiated cell types in the brain.

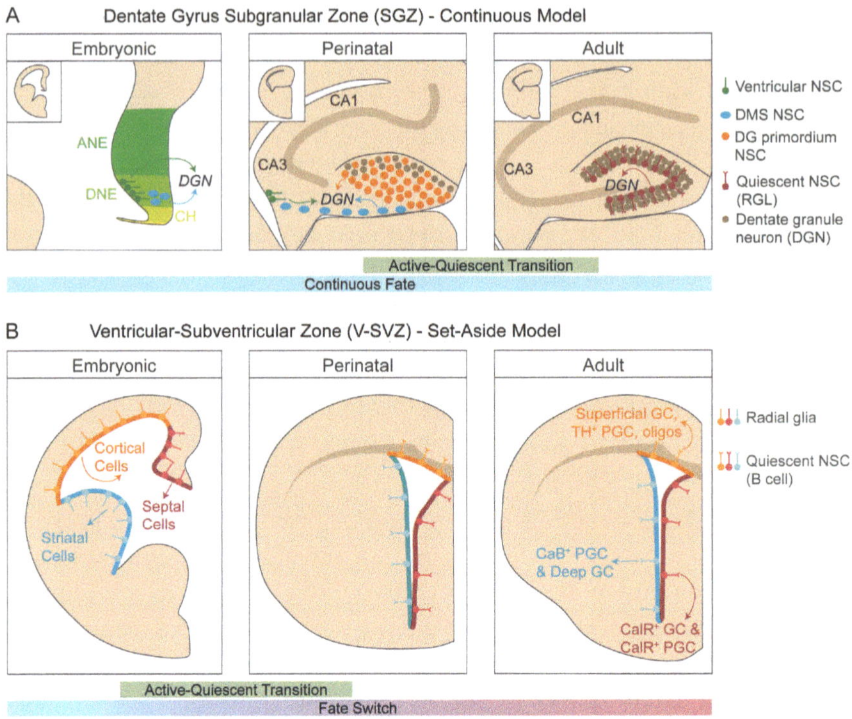

Fig. 1 Two models for the origin of adult neural stem cells in the subgranular zone and the ventricular-subventricular zone. (A) Radial glia (RG) in the embryonic dentate neuroepithelium (DNE) divide and begin to detach from the ventricle to enter the dentate migratory stream (DMS). Dentate gyrus (DG) neural stem cells (NSCs) from the DMS continuously arrive in the dentate primordium perinatally, where they divide and contribute to peak DG cytogenesis. During early postnatal development, DG NSCs transition to a quiescent state in the SGZ until reactivated in adulthood. At each developmental stage, DG NSCs give rise only to dentate granule neurons (DGNs) and astrocytes, but not CA1 or CA3 neurons, which originate from the ammonic neuroepithelium (ANE). CH, cortical hem. (B) RG in the embryonic ventricular zone generate cells for different regions of the brain depending on their location along the ventricle—RG in the medial pallium, dorsal pallium, and subpallium give rise to cells in the septum, cortex, and striatum, respectively. During mid-embryonic development, a subpopulation of RG transition to a quiescent state and remain quiescent until reactivated in adulthood. In adulthood, quiescent NSCs (B cells) give rise to different types of olfactory bulb interneurons and oligodendrocytes depending on their location along the ventricle, which corresponds to their embryonic origin—dorsal B cells give rise to superficial granule cells (GCs), tyrosine hydroxylase (TH)-expressing periglomerular cells (PGCs) and oligodendrocytes, medial B cells give rise to calretinin (CalR)-expressing GCs and PGCs, and ventral B cells give rise to calbindin (CaB)-expressing PGCs and deep GCs. Therefore, NSCs in the V-SVZ lineage undergo a fate switch between embryonic development and adulthood.

2.1 Origins of subgranular zone adult neural stem cells

The emergence of adult RGLs in the SGZ is inextricably linked to DG development. To understand origins of adult RGLs, we must first understand DG development (Fig. 1A). The earliest DG precursors in mice emerge around embryonic day 10.5 (E10.5) in the dentate neuroepithelium (DNE). The DNE is a small region of the medial pallium flanked dorsally by the ammonic neuroepithelium (ANE), which gives rise to CA regions of the hippocampus, and ventrally by the cortical hem, a source of Wingless-related (Wnt) and bone morphogenetic protein (BMP) signaling that patterns the hippocampus and neocortex (Altman & Bayer, 1990; Bayer & Altman, 1974; Caronia-Brown, Yoshida, Gulden, Assimacopoulos, & Grove, 2014). RG in the ANE and the DNE maintain distinct progeny fate specification, only generating neurons for their respective regions of the hippocampus throughout development (Berg et al., 2019). During early embryonic brain development, precursors in the DNE divide to expand the precursor population and then begin to migrate medially to forge the dentate migratory stream (DMS) around E15.5. Unlike most RG which remain at the ventricular surface, embryonic RG in the DG detach from the ventricle and migrate in the DMS to the primitive DG where the majority of cytogenesis occurs during perinatal development (Bayer & Altman, 1974; Berg et al., 2019). The proliferative capacity of the DG NSC population changes drastically during early postnatal development (Fig. 1A). Co-localization of the proliferative marker Mcm2 with NSC markers Hopx and Nestin revealed that DG NSCs are highly proliferative through postnatal day 3 (P3), and become less proliferative starting at P7, with a rapid decline in NSC proliferation between P3 and P14 (Berg et al., 2019). Additionally, retroviral labeling in the neonatal DG showed that dividing precursors give rise to adult RGLs, in addition to DG neurons and astrocytes (Namba et al., 2005). Thus, the first postnatal week is a time of peak cytogenesis in the DG and a turning point in the proliferative capacity of DG NSCs. Multiple studies have used thymidine analog label-retaining experiments to show that most DG NSCs enter long-term quiescence during the first postnatal week, with a peak at P3 (Berg et al., 2019; Noguchi, Castillo, Nakashima, & Pleasure, 2019; Ortega-Martínez & Trejo, 2015). Though the majority of developmental cytogenesis occurs during the first postnatal week, morphogenesis and organization of DG sub-structures, such as the hilus, granule cell layer, molecular layer and SGZ, is not complete until after P14 (Bayer & Altman, 1974; Nicola, Fabel, & Kempermann, 2015).

Separate lineage-tracing studies have identified two potentially different embryonic sources of adult SGZ RGLs. The first study used a *Gli1-CreER*^{T2}

mouse line to fate-map perinatal Hedgehog (Hh)-responsive precursors located in the ventricular zone near the ventral hippocampus at E17.5. These Hh-responsive precursors migrate dorsally over the course of neonatal development to populate the entire septo-temporal axis of the DG with long-lived adult RGLs (Li, Fang, Fernandez, & Pleasure, 2013). The second study used the *Hopx-CreER*^{T2} mouse line to clonally lineage trace embryonic precursors located in the DNE at E11.5 (Berg et al., 2019). Hopx-expressing embryonic precursors divide, migrate along the DMS and arrive in the primitive DG by E18.5. Postnatally (at P8 and P30), each Hopx^{+} precursor-derived clone contained RGLs, in addition to DGNs and astrocytes, and cells within a single clone could spread along up to ~900 μm of the septotemporal axis, which is ~30% of the entire adult DG (Berg et al., 2019). Throughout embryonic and early postnatal development, RG in the dentate ventricular zone constantly divide and migrate along the DMS to the primitive DG (Nelson et al., 2020). The above two studies identify embryonic RG populations that migrate away from the dentate VZ at different times: the Hopx^{+} precursors are located at the ventricle in the dentate neuroepithelium at E11.5 but have largely re-located in the primitive DG by late embryonic development, while the Hh-responsive precursors are located at the ventricle during late embryonic development, specifically in the ventral hippocampus. It remains unclear whether adult RGLs generated by these two different populations of embryonic NSCs are functionally different. These studies highlight temporal and spatial differences in embryonic NSCs within the dentate germinal zone, which may instruct adult RGL heterogeneity. For example, the dorsal-ventral location along the ventricle or timing of migration away from the ventricle could define different subpopulations of adult RGLs. Future studies will need to examine how differences in developmental origin may impact adult RGL function, potentially generating heterogeneity within the RGL population in the adult hippocampus (Bonaguidi et al., 2016; Gebara et al., 2016).

2.2 Origins of ventricular-subventricular zone adult neural stem cells

Quiescent adult V-SVZ B cells are generated through a distinct process that occurs well before the cessation of development (Fig. 1B). Two separate studies used different label-retaining assays (BrdU or H2B-GFP) to determine that the majority of adult V-SVZ B cells enter a slow-cycling, quiescent state between E13.5 and E15.5 in mice and remain largely quiescent until at least young adulthood (P21–P28) (Fuentealba et al., 2015;

Furutachi et al., 2015). Ependymal cells, which line the walls of brain ventricles in the adult and play a critical role in cerebral spinal fluid (CSF) flow through coordinated beating of their cilia, are also generated between E13.5 and E15.5 from embryonic RG (Del Bigio, 2010; Spassky et al., 2005). Two recent studies using multiple methods of clonal lineage tracing revealed that ependymal cells and B cells share a common embryonic precursor (Ortiz-Álvarez et al., 2019; Redmond et al., 2019). A subpopulation of RG divides between E13.5 and E15.5 to give rise to differentiated ependymal cells and quiescent B cells; most B cells are generated through E-B divisions and most ependymal cells are generated through E-E divisions, though a small number of B–B cell divisions also occur (Ortiz-Álvarez et al., 2019). Ependymal cells and B cells then undergo a lengthy maturation process from E17.5 to P21, during which coordinated expansion of the ependymal cell apical domain and reduction of the B cell apical domain ultimately results in the formation of the characteristic pinwheel ultrastructure in the adult V-SVZ (Mirzadeh, Merkle, Soriano-Navarro, Garcia-Verdugo, & Alvarez-Buylla, 2008; Redmond et al., 2019). Thus, B cells originate from a distinct subpopulation of RG that become quiescent during embryogenesis and can generate ependymal cells, a cell type that is fundamental to the physical structure of the adult V-SVZ niche.

Adult B cell heterogeneity is well-defined and has been directly linked to B cell location and embryonic origin (Fig. 1B). As a population, adult B cells can generate many different subtypes of olfactory bulb interneurons, as well as oligodendrocytes, but the fate of individual B cells is restricted to certain cell types. First, neurons and oligodendrocytes are generated from different B cell populations. Live imaging of single adult V-SVZ NSCs and their progeny in vitro revealed that individual B cells generate either neurons or oligodendrocytes, but never both (Ortega et al., 2013). In addition, most B cell-derived oligodendrogenesis occurs in the dorsal V-SVZ, with only scarce oligodendrogenesis occurring in the ventral V-SVZ, suggesting that B cell fate is spatially organized (Ortega et al., 2013). Similarly, the regional location along the ventricle dictates which interneuron subtype is generated by neurogenic B cells (Merkle et al., 2014; Merkle, Mirzadeh, & Alvarez-Buylla, 2007; Ortega et al., 2013; Ventura & Goldman, 2007). For example, B cells located in the dorsal V-SVZ generate mostly superficial granule cells (GCs) and dopaminergic (TH^+) periglomerular cells (PGCs), and a small number of glutamatergic juxtaglomerular neurons, B cells in the ventral V-SVZ produce deep GCs and calbindin ($CalB^+$) PGCs, and B cells in the medial V-SVZ generate calretinin ($CalR^+$) GCs and $CalR^+$ PGCs

(Brill et al., 2009; Delgado & Lim, 2017; Lledo, Merkle, & Alvarez-Buylla, 2008). The positional identity and subsequent fate potential appear to be an intrinsic property of the B cells because region-specific V-SVZ NSC cultures can maintain their positional identity after multiple serial passages in vitro and also when heterotopically grafted (Delgado, Lu, & Lim, 2016; Merkle et al., 2007). Though it was initially thought that the adult V-SVZ arose from the lateral ganglionic eminence (LGE) region of the embryonic brain, it is now clear that other regions of the developing telencephalon contribute to the adult V-SVZ as well (Fig. 1B). Clonal lineage tracing using genetically barcoded retrovirus showed that adult B cells share a common embryonic precursor with cells from the cortex, striatum, or septum, but lose this relationship around E15.5 when RG generate quiescent B cells (Fuentealba et al., 2015). The number of clones analyzed in this study was low and future independent approaches will be needed to validate these results. Regionalization of fate specification in the embryonic brain occurs when morphogen gradients induce discrete territories of transcription factor expression within the VZ (Alvarez-Buylla, Kohwi, Nguyen, & Merkle, 2008). Some of these regions are retained in the adult V-SVZ, suggesting that adult B cells maintain a memory of their embryonic positional identity through consistent transcription factor expression, which results in NSC fate potential heterogeneity in the adult brain (Brill et al., 2008; Fuentealba et al., 2015; Merkle et al., 2014; Young, Fogarty, Kessaris, & Richardson, 2007). Recent work showed that sonic hedgehog (Shh) signaling induces a ventral identity in embryonic NSCs, which is then maintained by mixed-lineage leukemia 1 (Mll1)-dependent epigenetic mechanisms throughout development and into adulthood (Delgado et al., 2020). Thus, adult V-SVZ B cell fate is predetermined by regional patterning of embryonic precursors and is directly linked to their embryonic origin.

In summary, developmental origins of adult NSCs in the SGZ and V-SVZ are distinct. Precursors to SGZ RGLs generate DGNs and glia throughout DG development and become quiescent and transform into RGLs during the first postnatal week (Fig. 1A). RGLs continue to generate DGNs when re-activated during adulthood. SGZ RGLs are therefore generated through a process that is integrated into DG development, and DG NSCs maintain continuous fate specification from development through adulthood. In contrast, V-SVZ B cells are generated through a separate process in which a subpopulation of RG are set aside in a quiescent state embryonically, while most RG continue proliferating and contributing to brain development (Fig. 1B). Precursors to V-SVZ B cells first generate neurons and glia for the cortex, striatum or septum before entering quiescence embryonically. Then quiescent V-SVZ B cells generate olfactory bulb interneurons and oligodendrocytes when they are reactivated in adulthood, exhibiting a different fate than their embryonic counterparts. Adult NSC heterogeneity has been directly linked to the embryonic origin of NSCs in the V-SVZ, but a potential link between origin and heterogeneity in the SGZ has not been explored. Further investigation into the developmental origin(s) of adult SGZ NSCs may reveal new insights into NSC heterogeneity and the relationship between origin and function.

3. Transformation of neural stem cell properties from development to adulthood

Transformation of developmental RG into adult NSCs is a multi-step process that begins with entry into a quiescent state. Quiescence is a common characteristic of many adult somatic stem cell populations and is critical to their maintenance and function (Urban et al., 2019; van Velthoven & Rando, 2019). In addition to entrance into quiescence, there are many other cellular properties, such as morphology, progeny fate and cell division mode, that change as NSCs transition from a developmental to adult state. We discuss how properties of NSCs in the SGZ and V-SVZ change during the transition from development to adulthood and associated molecular changes.

Morphological features, such as cell polarity, influence NSC function. Embryonic RG in the developing brain have an apical endfoot contacting the ventricle and extend a basal process that contacts the pial surface (Kriegstein & Alvarez-Buylla, 2009; Taverna, Gotz, & Huttner, 2014). The apical endfoot plays a role in signaling from the CSF, intercellular communication, and cell–cell adhesion, while the basal endfoot is critical for conveying extracellular matrix signals from the basal lamina (Taverna et al., 2014). Adult NSCs in the V-SVZ maintain contact with the ventricle through their apical processes, which cluster together in the center of ependymal cell pinwheel structures where they can access the CSF (Mirzadeh et al., 2008). However, DG NSCs lose ventricular contact early on during development when they enter the DMS (starting at ~E15.5 in mice), well before they transition to quiescence and an adult RGL state (Fig. 2) (Berg et al., 2019). The mechanisms that direct DG NSCs to detach from the ventricle and enter the DMS remain largely unexplored, but the process is reminiscent of outer radial glia (oRG) observed in higher order mammals, such as in the human cortex (Florio & Huttner, 2014; Molnár et al., 2019). Recent work in the mouse

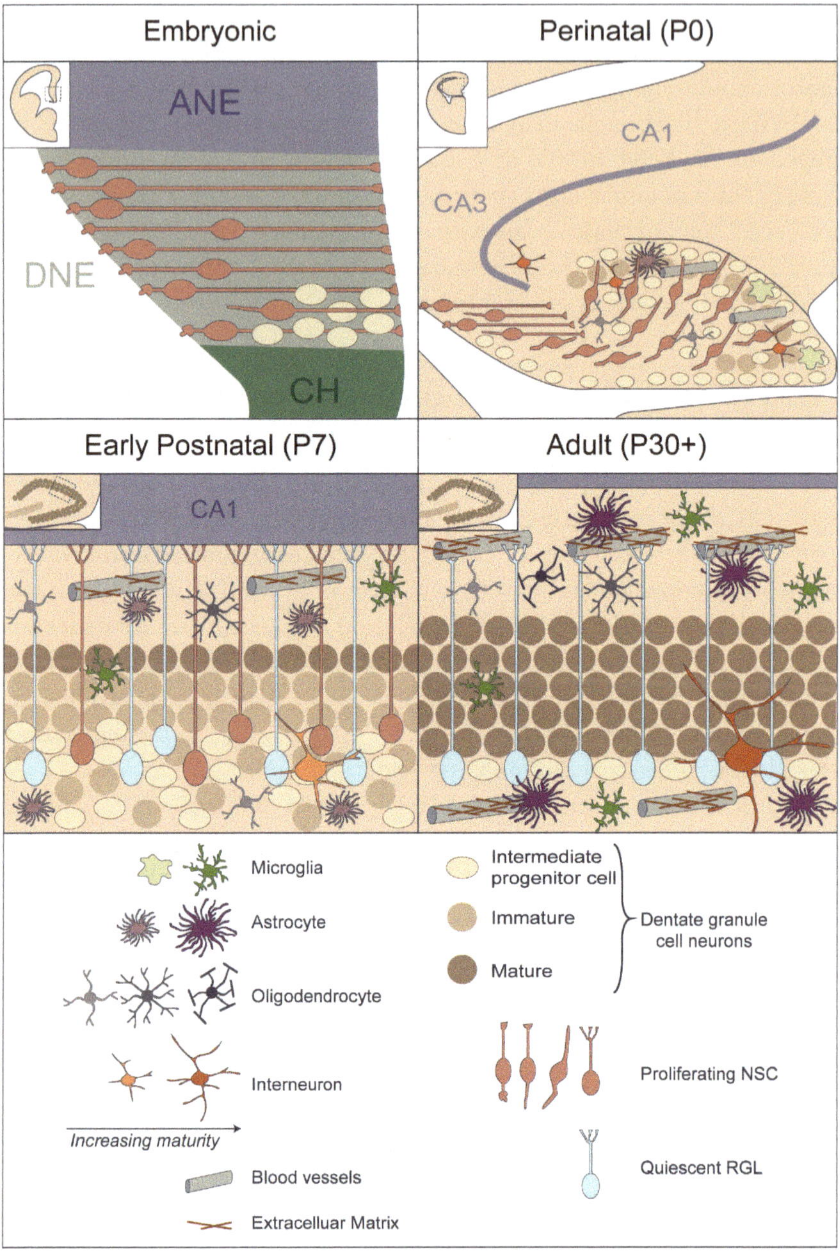

Fig. 2 Establishment of the adult neural stem cell pool and mature neurogenic niche in the dentate gyrus of the hippocampus. Embryonic dentate gyrus (DG) radial glia (RG) in the dentate neuroepithelium (DNE) divide to expand the RG population and generate intermediate progenitor cells, which serve as scaffolding for migrating RG that have

suggests that DG NSCs follow pioneering $Tbr2^+$ IPCs through the DMS and into the primitive DG and that this interaction is mediated by Notch signaling (Nelson et al., 2020). The characteristic radial morphology of adult RGLs and cell body localization in the SGZ begins to emerge between P3 and P7 and is fully present by P14, correlating with the transition into quiescence (Fig. 2) (Berg et al., 2019; Brunne et al., 2010; Nicola et al., 2015; Seki et al., 2014). In addition, while the basal processes of developmental RG reach the pial surface, the basal processes of adult NSCs in both the SGZ and V-SVZ are more likely to contact the vasculature, where they receive important contact-mediated signaling (Mirzadeh et al., 2008; Palmer, Willhoite, & Gage, 2000; Tavazoie et al., 2008).

NSC fate is guided by temporal and spatial patterning during development, creating neuronal subtype diversity in the brain. The V-SVZ B cell lineage follows this developmental principle and undergoes a fate switch from development to adulthood, initially generating neurons for the septum, cortex or striatum prior to entering quiescence and then generating olfactory bulb interneurons when adult NSCs are reactivated (Fig. 1B) (Fuentealba et al., 2015). Though olfactory bulb interneurons begin to be generated from RG around E14.5, the majority of olfactory bulb interneurons are generated perinatally (Batista-Brito, Close, Machold, & Fishell, 2008; Bayer, 1983). However, embryonic precursors to adult B cells enter quiescence between E13.5 and E15.5, suggesting that developmentally-born and adult-born olfactory bulb interneurons are generated by distinct NSC populations. The mechanism that induces a change in fate potential in the V-SVZ B cell lineage remains unknown. A recent study showed that Shh triggers the fate switch of cortical RG from generating glutamatergic pyramidal neurons to generating astrocytes, oligodendrocytes and GABAergic olfactory bulb interneurons during late embryonic development (Zhang et al., 2020). Shh ligand is

---

detached from the ventricle and entered the dentate migratory stream (DMS). The dentate primordium emerges during perinatal development, a time of peak cytogenesis and cell migration. NSCs continuously migrate into the dentate primordium from the DMS, peak neurogenesis occurs, astrocytes are generated, angiogenesis occurs, and microglia, oligodendrocyte precursor cells and immature interneurons migrate into the DG. The gross morphology of the DG begins to take shape during early postnatal development, as the DG NSCs transition to a quiescent state and transform into radial glia-like NSCs (RGLs). Cytogenesis declines and many cell types, including neurons and glia, begin a maturation process that extends across the first 3–4 postnatal weeks. The DG gains its adult form around P30 when most RGLs are quiescent and located in the SGZ, and other cell types have reached full maturation.

secreted from ventral sources during embryonic development, but during perinatal development migrating cortical interneurons and cells in the choroid plexus provide a dorsal source of Shh, which induces the fate switch of cortical RG (Winkler et al., 2018). Thus, it is possible that B cell olfactory bulb interneuron fate is acquired in a similar manner during late embryonic development and after B cells have entered quiescence. This is in stark contrast to DG NSCs, which exhibit continuity of fate specification, participating in developmental DGN generation embryonically before entering quiescence and continuing to generate DGNs throughout adulthood when reactivated (Fig. 1A) (Berg et al., 2019). It remains unknown whether the DGN fate is permanently induced in DG RG by morphogens during early embryonic development or if it is continuously reinforced by environmental cues throughout development. In addition, DGNs born at different times during development/adulthood have distinct morphological features, physiology, and rates of survival, suggesting that DG NSCs may generate different subtypes of DGNs at different times (Ciric, Cahill, & Snyder, 2019; Cole et al., 2020; Kerloch, Clavreul, Goron, Abrous, & Pacary, 2019; Mathews et al., 2010; Save, Baude, & Cossart, 2019; Snyder, 2019).

In addition to fate potential, other functional properties change in NSCs across time, including self-renewal capacity and cell division mode. Analysis of in vitro neurosphere cultures of DG progenitors suggested that P7 progenitors are intrinsically more proliferative and more capable of forming neurospheres than P28 progenitors (Gilley, Yang, & Kernie, 2011). Time-lapse imaging of glial fibrillary acid protein (GFAP)-expressing DG NSCs between P4 and P6 revealed largely symmetric cell division to generate either 2 NSCs or 2 IPCs (Namba et al., 2011), while clonal lineage-tracing and live in vivo imaging have revealed largely asymmetric cell division of adult RGLs (Bonaguidi et al., 2011; Pilz et al., 2018). In contrast, the cell division of adult B cell precursors is largely asymmetric to generate IPCs and then one ependymal cell and one quiescent B cell (Kriegstein & Alvarez-Buylla, 2009; Ortiz-Álvarez et al., 2019; Redmond et al., 2019), but adult V-SVZ B cells divide symmetrically, with mostly consuming divisions and only a small proportion of self-renewing divisions (Obernier et al., 2018). There are undoubtedly additional NSC property differences between developmental and adult NSCs, and understanding how these properties change over time will provide insight into mechanisms driving the transition from developmental to adult NSCs.

Many molecular changes accompany cellular property changes during the transition from developmental RGs into adult NSCs. For example, astroglial marker gene expression increases with the transformation into adult NSCs. DG NSCs do not express the astroglial marker brain lipid binding protein (Blbp) until around P3, when it is expressed very weakly, and then expression of Blbp in NSCs rapidly increases around P7 as NSCs extend radial processes and become localized to the SGZ (Matsue et al., 2018; Nicola et al., 2015). V-SVZ NSCs concomitantly downregulate expression of RC2 (Nestin), an RG marker, and increase GFAP expression around P6, the time when ependymal cells and B cells mature (Merkle, Tramontin, García-Verdugo, & Alvarez-Buylla, 2004). Other studies have used more global methods to investigate developmental changes in NSC gene expression. An early study used microarrays to compare the molecular profiles of NSCs from P7 to P28 DG in vivo and found differential expression of genes involved in cell adhesion, lipid metabolism, glutamate transport and gap junctions (Gilley et al., 2011). More recently, RNA-sequencing (RNA-seq) studies of NSCs across developmental time points have identified a distinct molecular shift in NSCs that corresponds to the timing of entry into quiescence (Berg et al., 2019; Hochgerner, Zeisel, Lönnerberg, & Linnarsson, 2018; Yuzwa et al., 2017). For example, single-cell RNA-seq of the DG across a spectrum of ages identified a molecular shift in NSCs during early postnatal development (Hochgerner et al., 2018), and single-cell RNA-seq of cortical NSCs across embryonic development identified a molecular shift in NSCs between E11.5 and E13.5 (Yuzwa et al., 2017). Further bulk RNA-seq characterization of the molecular evolution of Hopx-expressing DG NSCs at embryonic, postnatal and adult stages suggests that changes in cell signaling and metabolism correspond to the transition from developmental NSCs to adult RGLs (Berg et al., 2019). RNA-sequencing studies have also shown that developmental RG share a core molecular identity with their adult counterparts in both the SGZ and V-SVZ, which is conserved across all stages of development and represents a defined NSC signature, including expression of genes involved in NSC maintenance and cell cycle, as well as less well-characterized genes within the neural precursor context (Berg et al., 2019; Yuzwa et al., 2017). Though molecular changes likely underlie shifts in NSC properties as developmental RG transform into adult NSCs, few studies have linked changes in the molecular landscape to changes in NSC behavior across development. Future studies will need to determine causal links between molecular mechanisms discovered by single-cell sequencing studies and NSC behavior in vivo to identify mechanisms that instruct the transition from developmental RG to adult NSC. More in depth studies that break down the transformation of developmental RG into adult NSCs into simpler sequential steps will be required to understand the mechanisms that regulate this complex process (Shin et al., 2015).

4. Establishment of adult neurogenic niches

The adult neurogenic niches are composed of both neural and non-neural cell types. These cells transduce local and long-range signals that support NSC function and modulate NSC behavior, such as the balance between quiescence and activation. Each cell type has its own developmental timeline of generation, migration and maturation, such that the adult neurogenic niche is gradually assembled over the course of development (Bjornsson, Apostolopoulou, Tian, & Temple, 2015). Here we review how the SGZ (Fig. 2) and V-SVZ (Fig. 3) neurogenic niches evolve during development and how the timing of niche development might affect NSC function and behavior.

Multiple non-neural cell types, such as choroid plexus, vascular cells and microglia, contribute to the adult neurogenic niches. The choroid plexus is a secretory tissue located in each ventricle. The choroid plexus secretes growth factors and morphogens into the CSF, which regulate the growth, proliferation and maintenance of embryonic RG and adult V-SVZ B cells that have direct contact with the CSF (Lehtinen et al., 2013). The cocktail of secreted factors changes across development and age, resulting in dynamic regulation of NSC behavior (Lehtinen et al., 2011; Silva-Vargas, Maldonado-Soto, Mizrak, Codega, & Doetsch, 2016). For example, the concentration of CSF-borne insulin-like growth factor 1 (Igf1) peaks during late embryonic development and promotes progenitor proliferation (Lehtinen et al., 2011). In addition to secreted factors, the CSF supplied by the choroid plexus provides mechanical pressure which is required for normal brain enlargement and development (Desmond & Jacobson, 1977). Vascular cells, including endothelial cells, pericytes and vascular smooth muscle cells, play a multifaceted role in the adult NSC niches, providing oxygen through blood flow and providing secreted and contact-derived neurogenic signals (Palmer et al., 2000; Shen et al., 2004). The composition of blood-borne circulatory factors changes across development due to the developmental timeline of source cell types, and the effects of blood-borne factors on the niche are mediated by the level of vascular infiltration. Angiogenesis in the brain begins around E9.5–E10, and peak endothelial cell proliferation occurs at the end of the first postnatal week (Marin-Padilla, 1985; Robertson, Du Bois, Bowman, & Goldstein, 1985). Throughout development, the neural progenitor population promotes angiogenesis and blood vessel maturation, while the vasculature promotes NSC self-renewal, proliferation and neurogenesis, creating a symbiotic relationship

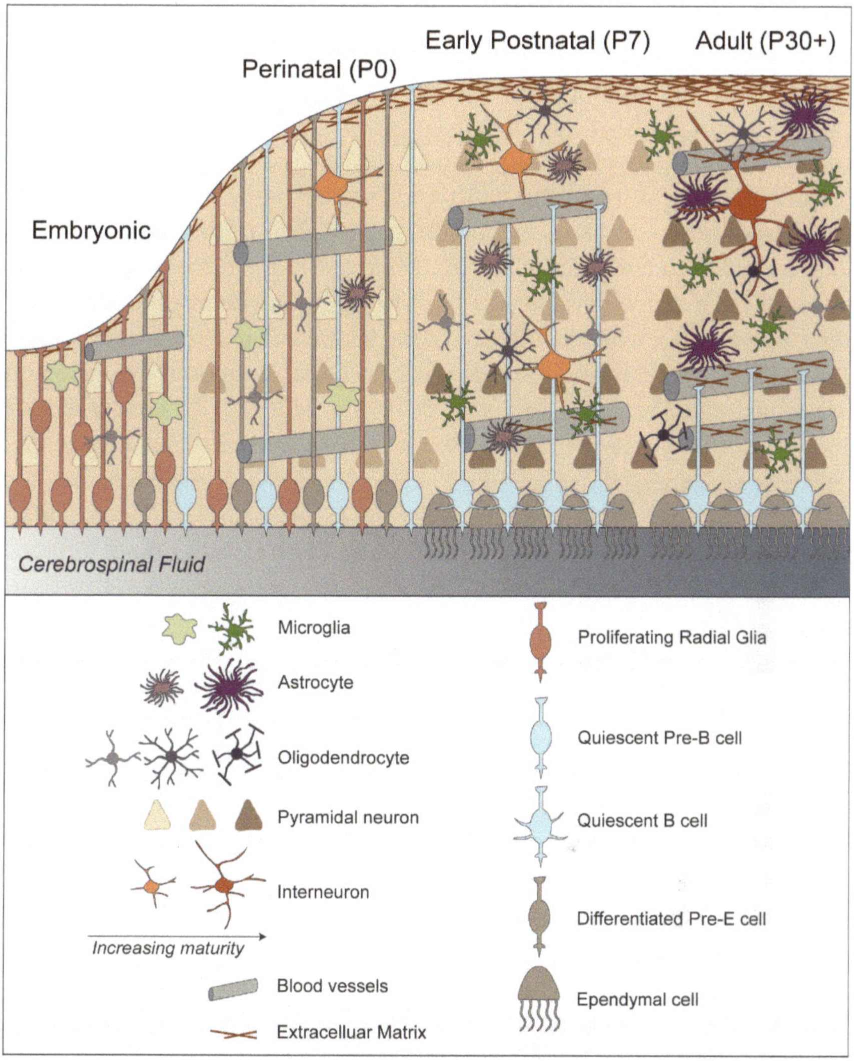

Fig. 3 Establishment of the adult neural stem cell pool and mature neurogenic niche in the ventricular-subventricular zone. Embryonic radial glia (RG) divide to expand the RG population and then to generate neurons. During mid-embryonic development, some RG divide to generate immature ependymal cells and quiescent RG. Infiltration of oligodendrocyte precursor cells and microglia and angiogenesis begin in the embryonic ventricular zone and continue throughout perinatal development. In addition, astrocyte generation and interneuron migration occur perinatally, as embryonically born neurons begin to mature. During early postnatal development, ependymal cells become multiciliated and become distinguishable from quiescent B cells, gliogenesis declines, and cellular maturation of many cell types continues. The V-SVZ gains its adult form around P30 when ependymal cells have expanded to their full size, niche cell types have completed their maturation and quiescent B cells can become reactivated to generate olfactory bulb interneurons. In addition to these stepwise changes, there are gradual changes in the composition of the cerebral spinal fluid and the accumulation of extracellular matrix over the course of development.

which lasts into adulthood (Gerhardt et al., 2004; Ma & Huang, 2015; Palmer et al., 2000). Microglia are resident macrophages which mediate immune responses in the central nervous system (Tay, Savage, Hui, Bisht, & Tremblay, 2017). Microglia begin colonizing the brain around E9.5 and undergo rapid proliferation to expand between E10.5 and P0 (Kierdorf et al., 2013). The microglial population continues to increase during the first 2 postnatal weeks, decreases by 50% from week 3 to 6, and then stabilizes to adult levels (Nikodemova et al., 2015). Microglia in the developing brain have an amoeboid morphology and play a role in supporting neurogenesis and clearing apoptotic cells, while microglia in the adult brain have a ramified morphology and play a larger role in synaptic regulation, but continue to clear apoptotic cells in neurogenic zones (Sierra et al., 2010; Tay et al., 2017). Thus, the role of non-neural cell types in regulating NSCs changes across development as these cell types are generated, migrate and mature in the NSC environment.

Neuronal activity contributes to the adult neurogenic niche and affects NSC behavior directly through neurotransmitter signaling and indirectly through activity-induced release of neurogenic factors from neurons (Berg, Belnoue, Song, & Simon, 2013; Jang & Song, 2008; Song, Sun, Olsen, Ming, & Song, 2015; Vicidomini, Guo, & Sahay, 2020). Adult NSCs express receptors for many neurotransmitters, including both glutamate (excitatory) and GABA (inhibitory) (Berg et al., 2013; Shin et al., 2015). Activation of $GABA_A$ receptors maintains quiescence in both SGZ RGLs and V-SVZ B cells (Giachino et al., 2014; Liu, Wang, Haydar, & Bordey, 2005; Song et al., 2012). Both adult neurogenic niches are innervated by neural inputs from local and long-range projections. For example, local parvalbumin interneurons directly regulate DG RGLs by tonic GABA activation (Song et al., 2012), and hypothalamic proopiomelanocortin neurons send projections that innervate the anterior ventral V-SVZ and promote proliferation of B cells (Paul, Chaker, & Doetsch, 2017). DGN activity induces expression of pro-neurogenic factors like brain-derived neurotrophic factor (BDNF) and fibroblast growth factor (FGF) and downregulates expression of anti-neurogenic factors like secreted frizzled-related protein 3 (sFRP3), a Wnt inhibitor (Jang et al., 2013; Ma et al., 2009; Sun et al., 2015). Neuronal regulation of adult NSCs has primarily been studied in the adult when neurons are fully mature, so it remains unknown when adult-like neuronal regulation of NSCs is established during development. Most neurogenesis occurs embryonically, as does axonal targeting, but synapse development occurs during the first 3 postnatal weeks (Reemst, Noctor, Lucassen, & Hol, 2016). Specifically, DG neurogenesis is a protracted process that peaks even later during perinatal development (Bayer & Altman, 1974). Thus, neuronal regulation of NSCs is likely dynamic until neuronal maturation is complete during the postnatal period.

Glia also play a critical role in regulating adult NSCs through paracrine signaling (Falk & Götz, 2017; Ma, Ming, & Song, 2005; Song, Stevens, & Gage, 2002). Astrocytes provide many different secreted and contact-derived signals that regulate adult NSCs in the SGZ and V-SVZ. For example, astrocyte derived interleukins 1β and 6 (IL-1β and IL-6) and ephrin-B signaling promote neuronal differentiation in the SGZ (Ashton et al., 2012; Barkho et al., 2006), and astrocyte derived ligands of the BMP, Wnt and Notch signaling pathways regulate self-renewal and neuronal differentiation in the V-SVZ (Ferrón et al., 2011; Moreno-Estellés et al., 2012; Peretto et al., 2004). Astrogliogenesis in the brain begins in late embryonic stages (E16–E18) and occurs throughout the first 2weeks of postnatal life (Bandeira, Lent, & Herculano-Houzel, 2009). Once born, astrocytes gradually acquire mature astrocyte functionality over the first postnatal month (Akdemir, Huang, & Deneen, 2020; Bushong, Martone, & Ellisman, 2004). Oligodendrocytes are the myelinating cells of the CNS that form a myelin sheath to facilitate rapid transmission of axon potentials and provide metabolic support to the axons. Demyelination through administration of cuprizone and rapamycin inhibits adult SGZ RGL proliferation and reduces the dendritic spine densities of adult-born DGNs through unknown molecular mechanisms (Zhang et al., 2020). To date, the role of the oligodendrocyte lineage in adult neurogenic niches appears to be indirect through modulation of neuronal circuitry, but it is also possible that the oligodendrocyte lineage provides direct paracrine signals that contribute to the niche microenvironment through yet to be discovered mechanisms. Oligodendrocytes are generated from oligodendrocyte precursor cells (OPCs), which are generated in three waves in the developing brain: (1) OPCs emerge from the MGE around E12.5, (2) soon after the first wave, OPCs are generated from the LGE, and (3) soon after birth, OPCs are generated from the cortical VZ (Bergles & Richardson, 2015; Kessaris et al., 2006). After birth, MGE-derived OPCs are eliminated and the majority of oligodendrocytes in the adult brain are those derived from the cortical VZ (third wave), with a small contribution from LGE-derived oligodendrocytes (Tripathi et al., 2011). Oligodendrogenesis and myelination are protracted and arise from cumulative processes that occur throughout postnatal development and in adulthood (Monje, 2018). Ependymal cells are specialized glia that have epithelial characteristics and line the ventricles of the brain. Ependymal cells contribute to structural integrity of ventricles and play a functional role in the

V-SVZ niche as the source of signaling factors and by beating their cilia to propel CSF. For example, Noggin, an inhibitor of BMP signaling, exerts a pro-neurogenic effect on B cells, blocking astrocyte differentiation (Lim et al., 2000), and pigment epithelium-derived factor (PEDF) promotes B cell self-renewal (Ramirez-Castillejo et al., 2006). Ependymal cells are generated from RG between E14 and E16 (Spassky et al., 2005). Although ependymal cells are generated embryonically, multiciliated ependymal cells are not detectable until birth and are not widespread throughout the lateral ventricles until the first postnatal week (Spassky et al., 2005). Mature ependymal cell characteristics, such as Foxj1 expression, loss of Vcam1 expression, expansion of the apical domain, and pinwheel ultrastructural formation, emerge between late embryonic development and the first postnatal week, suggesting that ependymal cell differentiation largely occurs during postnatal development (Redmond et al., 2019).

Many niche cell types are generated postnatally, and even those that are generated embryonically must undergo an extended maturation period, which frequently lasts weeks into postnatal development (Figs. 2 and 3). Thus, SGZ and V-SVZ neurogenic niches reach their adult form long after NSCs have entered quiescence. This is especially true for V-SVZ NSCs, which become quiescent embryonically. The offset timing of NSC quiescence and niche establishment prompts multiple questions. How is NSC quiescence maintained before the adult niche has matured? How does niche maturation affect quiescent NSC function over early postnatal development? Do quiescent NSCs go through their own maturation process once they exit cell cycle and is this maturation process affected by the developing niche environment? To address these questions, future research must focus on the interaction between niche maturation and the progression of NSC properties during late embryonic and early postnatal development.

5. Mechanisms regulating establishment of the adult neural stem cell pool

Proper establishment of the adult NSC pool requires expansion of the precursor RG population, transition to adult NSC properties and maintenance of self-renewal capacity. Most studies have investigated how manipulations during embryonic development affect the number, proliferation and morphology of adult NSCs. However, few studies have focused on more specific manipulations during the period of adult NSC transformation in the SGZ (P3–P7) and the V-SVZ (E13.5–E15.5). We summarize broad cellular mechanisms that are likely critical for proper establishment of the adult neural stem cell pool, and give examples of molecular mechanisms that impact the development of adult NSCs. Both intrinsic and extrinsic molecular mechanisms regulate the establishment of adult NSC pools, but are often distinct for SGZ and V-SVZ NSCs.

The size of a stem cell population is regulated by its cell division dynamics. During cortical development, a period of symmetric self-renewing cell division first expands the RG population and is followed by a period of asymmetric cell division when neurogenesis begins. Altering the cell division mode or the overall rate of proliferation has long-term effects on the output of the NSC population. For example, if the cell division mode is switched to asymmetric division or differentiation prematurely, then the NSC pool will be smaller and fewer neurons will be generated. The number of adult NSCs in the SGZ and the V-SVZ decreases with age, suggesting that once the NSC pool transitions to its adult form, the NSC population loses the innate ability to expand (Encinas et al., 2011; Obernier et al., 2018). Therefore, the size of the adult NSC pool is largely dependent on the cell division dynamics of the precursor pool prior to entering quiescence. Molecular and environmental manipulations have begun to reveal mechanisms that regulate developmental NSC cell division and subsequently impact the establishment of the adult NSC pools. Shh is a morphogen critical for dorso-ventral patterning in the developing brain. Decreased Shh signaling in embryonic NSCs, through deletion of pathway components Sufu or Smoothened, impairs NSC proliferation in the developing DG, resulting in a smaller pool of quiescent adult RGLs (Han et al., 2008; Li et al., 2013; Noguchi et al., 2019). In contrast, reduced Shh signaling in embryonic cortical RG, as a result of Sufu deletion, induces expansion of the dorsal V-SVZ in the adult brain due to increased number and proliferation of B cells (Gomez et al., 2019). Surprisingly, despite increased B cell proliferation, Sufu deletion reduces the number of transit amplifying progenitor cells, resulting in lower levels of adult neurogenesis. Thus, Shh signaling is critical to the establishment of appropriate levels of adult neurogenesis in both the SGZ and V-SVZ, but affects the behavior of the two NSC pools differently. Erk1 and Erk2 are key players in the MAP kinase signaling cascade, which transduces signals from cell surface receptors to effect changes in transcription. Double knockout of Erk1 and Erk2 severely impairs DG morphogenesis due to reduced number and proliferation of RG in the DG ventricular zone at E16.5, resulting in reduced embryonic neurogenesis and near complete absence of postnatal NSCs or neurogenesis (Vithayathil, Pucilowska,

Goodnough, Atit, & Landreth, 2015). Alterations to both the maternal and early postnatal environment can impact adult NSC establishment. Modeling maternal diabetes by maternal leptin receptor haploinsufficiency (*Lepr*$^{db/+}$) or by maternal exposure to the circulating metabolite methylglyoxal, which is increased during maternal diabetes, results in offspring with depleted adult NSC pools in both the SGZ and V-SVZ (Yang et al., 2016). The reduced population of adult V-SVZ B cells is a result of a premature switch to neurogenesis in embryonic precursors, which stunts the expansion of the precursor pool. However, other interventions can increase the size of the adult NSC pool. Maternal infection during pregnancy results in an immediate inflammatory response, which can be modeled by maternal injection of IL-6. A single injection of IL-6 at E13.5 increases the size and proliferation of the adult V-SVZ NSC pool due to increased proliferation and self-renewal of embryonic precursors (Gallagher et al., 2013). It is unclear whether more embryonic RG transform into adult B cells or whether increased proliferation of adult B cells expands the adult NSC population. In contrast, IL-6 injection at E13.5 did not elicit a phenotype in the adult SGZ NSC pool, suggesting that some mechanisms controlling the establishment of the two adult NSC pools are time and/or niche specific.

The establishment of the V-SVZ B cell population is linked to the establishment of the ependymal cell population. The Geminin superfamily, composed of Geminin, GemC1 and Mcidas, control DNA replication and multiciliogenesis in cells throughout the body by antagonistically regulating centriole amplification—Geminin inhibits centriole amplification, while GemC1 and Mcidas promote centriole amplification (Balestrini, Cosentino, Errico, Garner, & Costanzo, 2010; McGarry & Kirschner, 1998; Pefani et al., 2011). The balance of Geminin superfamily members expressed in embryonic RG regulates the choice between ependymal vs B cell fate and can have long-term effects on the adult V-SVZ NSC pool. Overexpression of Mcidas or GemC1 in embryonic RG induces premature differentiation into ependymal cells at the expense of B cell generation (Kyrousi et al., 2015; Ortiz-Álvarez et al., 2019). In contrast, GemC1 deficiency in embryonic RG prevents the appearance of multiciliated ependymal cells during postnatal development and promotes B cell generation, increasing the number of B cells in the adult V-SVZ and the level of neurogenesis (Kyrousi et al., 2015; Lalioti et al., 2019). Similarly, overexpression of Geminin biases embryonic RG toward the B cell fate, increasing the number of RG that symmetrically divide to give rise to two B cells immediately preceding quiescence, a cell division mode which rarely occurs under endogenous conditions

(Lalioti et al., 2019; Ortiz-Álvarez et al., 2019). Multiple upstream signaling pathways impact the Geminin/GemC1/Mcidas regulation of fate choice between ependymal cells and B cells (Kyrousi, Lygerou, & Taraviras, 2017). For example, Gli3 knockout results in a mixed B cell/ependymal cell phenotype, suggesting that Shh signaling regulates proper establishment of the ependymal cell and B cell populations in the V-SVZ (Wang, Kane, Lee, & Ahn, 2014). In addition, Notch signaling prevents GemC1 and Mcidas from promoting the ependymal cell fate program in embryonic RGs (Kyrousi et al., 2015). Thus, the establishment of the adult V-SVZ NSC population is directly influenced by ependymal cell generation and is regulated by the Geminin superfamily members.

Developmental NSCs begin their transformation into adult NSCs by transitioning into a quiescent state. Although mechanisms regulating adult NSC re-entry into cell cycle from quiescence have been studied (Shin et al., 2015), mechanisms regulating developmental NSC exit from cell cycle into a quiescent state remain relatively unexplored. Adult NSCs maintain their quiescence largely due to extrinsic niche mechanisms, and disruption of these niche mechanisms often results in precocious entry into cell cycle and rapid depletion of the NSC pool (Urban et al., 2019). Because the adult niches are not fully formed or mature until after NSCs enter quiescence, it is possible that intrinsic mechanisms trigger quiescence entry during development. For example, cell cycle regulators are key regulators of adult NSC establishment. Cyclin-dependent kinase (CDK) inhibitors restrain cell cycle progression, often functioning as tumor suppressors. The CDK inhibitor p57 is required for the emergence of the quiescent B cell population in the V-SVZ. Knockout of p57 embryonically increases the proliferation of embryonic NSCs but reduces the size of the adult B cell population, resulting in lower levels of adult V-SVZ neurogenesis (Furutachi et al., 2015). In contrast, germline knockout of p21 or p27 results in increased RGL and progenitor cell proliferation in the adult SGZ, resulting in increased levels of neurogenesis (Andreu et al., 2015; Pechnick, Zonis, Wawrowsky, Pourmorady, & Chesnokova, 2008; Qiu et al., 2009). The cellular mechanism that results in increased RGL proliferation in the absence of p21 and p27 remains unclear because all three studies used germline knockout and only analyzed the adult brain. It is possible that p21 and p27 restrain the proliferation of both developmental and adult NSCs in the DG, but future studies will need to use temporally-restricted deletion of p21 and p27 to better define their role in NSC quiescence. The nuclear factor one (NFI) family of transcription factors, composed of NFIA, NFIB and NFIX, have been implicated in activating differentiation genes and repressing self-renewal genes in the developing hippocampus. Mice lacking any one of the NFI transcription factors exhibit hippocampal malformations (Barry et al., 2008; Heng et al., 2014; Piper et al., 2010). Adult NSCs in the DG of Nfix null mice are more proliferative and have impaired radial morphology and localization to the SGZ (Heng et al., 2014; Martynoga et al., 2013). The NFI transcription factors bind to enhancer regions specific to the quiescent adult NSC state, and NFIX is required and sufficient to induce aspects of quiescence in adult NSC cultures (Martynoga et al., 2013). Thus, NFIX regulates DG NSCs across development and may play a role in the establishment of the adult RGL pool, potentially through regulating proliferation/quiescence. However, temporally-restricted inactivation of NFIX has not been explored, and NFIX may have distinct effects during different stages of DG development. DNA methyltransferase 1 (DNMT1) is an enzyme responsible for maintenance of DNA methylation during DNA replication and is highly expressed in proliferating precursors in the embryonic DG. Tamoxifen-inducible Cre-mediated deletion of Dnmt1 in embryonic NSCs using *Nestin-CreER*^{T2} transgenic mice results in a dramatically smaller DG, which becomes apparent during early postnatal development, suggesting that the DNMT1 may play a role in DG development during the critical period of NSC transformation (Noguchi et al., 2016). Conditional knockout of Dnmt1 reduced the proliferation of postnatal NSCs, resulting in fewer RGLs and impaired neurogenesis in the adult (Noguchi et al., 2016). Dnmt1 knockdown in neonatal DG-derived NSC cultures recapitulated the decreased NSC proliferation observed in vivo and linked the proliferative defect to increased levels of p21 and p57 in the absence of DNMT1 (Noguchi et al., 2016). To date, most studies have used germline knockouts or conditional embryonic deletion of genes of interest, so it is unclear whether the mechanisms described above regulate proliferating embryonic precursors, transitioning NSCs, or both. Future studies will be needed to understand how molecular factors affect NSCs at different stages of development and how manipulating these factors during development ultimately affects the adult NSC pools.

6. Concluding remarks and future perspectives

Though most adult neurogenesis research in mammals has focused on mechanisms that regulate NSC behavior in the adult brain, recent research has begun to investigate the developmental establishment of adult NSCs and their niches. The field of adult NSC ontogeny is in its infancy, but has the potential to greatly impact our current concept of adult neurogenesis and NSC maintenance. The vast majority of embryonic NSCs do not survive beyond development, which suggests that the subpopulation of developmental NSCs that are maintained into adulthood possess a distinct capacity to be very long-lived. Understanding why some precursors transition into an adult NSC state while others do not will give us insight into mechanisms of NSC maintenance in the mammalian nervous system.

Multiple principles have already emerged from the burgeoning field of adult NSC ontogenesis. First, the adult NSC niche is established well after NSCs enter a quiescent state. Decades of work have shown that adult NSCs are extremely responsive to the surrounding niche environment, which provides both proliferation- and quiescence-inducing signals. Pro-quiescence signals dominate in the adult niches, resulting in a largely quiescent adult NSC population. Thus, one possibility was that the emergence of the adult NSC niche during development might trigger NSCs to enter quiescence, transitioning them to an adult NSC state. However, recent work in the SGZ and the V-SVZ indicates that NSCs in both compartments enter quiescence long before the adult niche is fully formed. The adult NSC niches are composed of many different cell types, including neurons, glia, and other non-neural cell types, which each have their own developmental timelines. Though most neurogenesis throughout the brain occurs embryonically, neuronal maturation and synapse development occurs largely postnatally. Gliogenesis occurs postnatally, followed by a maturation period that lasts well into the first month of life. Angiogenesis and microglial establishment continue to be dynamic throughout the first month of life. Collectively, this suggests that the adult NSC niches are not assembled in their adult form until around 1 month of age in mice, weeks after the restricted periods of development when NSCs enter long-term quiescence (E13.5–E15.5 in the V-SVZ and P3–P7 in the SGZ). Thus, intrinsic mechanisms and/or transient developmental niche mechanisms likely orchestrate the NSC transition to quiescence. However, the precise cellular and molecular mechanisms that trigger NSC entry into quiescence remain unknown and will be a major focus of future studies. It will also be interesting to explore how niche regulation of quiescent NSCs changes as niche maturation occurs across early postnatal development. NSCs in both the SGZ and V-SVZ go through an extended maturation period after they have entered quiescence, which is potentially impacted by the gradual formation of the adult neurogenic niche. Studying adult NSC establishment in parallel with adult niche formation will likely reveal synergistic interactions that link NSC function to its changing environment and may give us insight into how early life experience can have a long-term impact on adult neurogenesis. A second principle that has emerged from the study of adult NSC ontogenesis is that the adult SGZ and V-SVZ NSC populations are established through distinct developmental processes. Two separate models have been proposed—the "Continuous Model" for SGZ NSCs (Berg et al., 2019) and the "Set-Aside Model" for V-SVZ NSCs (Fuentealba et al., 2015; Furutachi et al., 2015)—to reflect major differences between SGZ and V-SVZ NSC establishment, including the timing of quiescence and whether NSC fate changes after entering quiescence (Fig. 1). In addition, a handful of studies have suggested that different molecular mechanisms regulate the establishment of SGZ and V-SVZ NSCs, showing that some mechanisms regulate the establishment of one adult NSC population but not the other, whereas other mechanisms regulate two populations in different ways. Differences in properties of adult NSCs in the SGZ and V-SVZ may be reflective of their distinct embryonic origins and trajectories across development. Gaining a deeper understanding of SGZ and V-SVZ adult NSC ontogeny as separate developmental phenomena may also provide insight into why adult neurogenesis in the DG appears to be more conserved in humans than adult olfactory bulb neurogenesis (Kempermann et al., 2018).

We are just beginning to unravel adult NSC ontogenesis and many unanswered questions remain. What triggers NSC entry into quiescence? Does an intrinsic clock initiate quiescence within a subpopulation of NSCs at a developmentally-specified time or do transient signals from the developing environment cause NSCs to enter quiescence? How is the size of the quiescent NSC pool determined? Can the establishment of the quiescent NSC pool be influenced by environmental factors and/or experience? Future studies should aim to better define and identify developmental precursors to adult NSCs. The cellular properties and molecular profiles of precursors to adult NSCs are not well-defined and it is unclear whether precursors to adult NSCs possess an innate identity that makes them very long-lived or whether long-term maintenance is an acquired state that occurs stochastically during transient stages of development. Though we have begun to identify precursors to adult RGLs in the SGZ, the identity of precursors to adult B cells remains unknown and may turn out to be heterogeneous. Defining precursor identity will be critical to developing targeted methods to specifically manipulate developmental precursors to adult NSCs. In addition, future studies should use what is known about the timeline of adult NSC ontogenesis to design temporally-specific experiments to interrogate mechanisms that regulate distinct developmental events, such as NSC pool expansion, generation of differentiated cell types, entry into quiescence, and quiescent NSC maturation. For example, recent work has identified the timing of NSC entry into long-term quiescence in both the SGZ and V-SVZ, but it remains unknown when the size of either adult NSC pool is established. A cyclical process of defining endogenous developmental processes, then interrogating their molecular mechanisms using cell type-specific and temporally-specific methodology, will drive forward adult NSC ontogenesis research.

Adult neurogenesis confers a unique layer of plasticity to specific regions of the mammalian brain by generating new neurons that integrate and modify the existing neural circuitry. Neurogenesis in the adult mammalian brain is wholly dependent on the long-term maintenance of a subset of embryonic NSCs, but we are just beginning to understand the developmental processes that promote NSC maintenance beyond development. Research investigating adult NSC ontogenesis will undoubtedly expand our understanding of NSC maintenance and will provide much needed insight into how developmental disorders and interventions can have long-lasting impacts on the adult NSC pool and subsequent neural plasticity in the adult mammalian brain.

Acknowledgments

We thank D.A. Berg for comments and members of the Ming and Song laboratories for discussion. This work was supported by grants from the National Institutes of Health (R35NS116843 to H.S. and R35NS097370 to G-l.M.).

References

Akdemir, E. S., Huang, A. Y., & Deneen, B. (2020). Astrocytogenesis: Where, when, and how. *F1000Res*, *9*.

Altman, J., & Bayer, S. A. (1990). Mosaic organization of the hippocampal neuroepithelium and the multiple germinal sources of dentate granule cells. *The Journal of Comparative Neurology*, *301*, 325–342.

Alvarez-Buylla, A., Kohwi, M., Nguyen, T. M., & Merkle, F. T. (2008). The heterogeneity of adult neural stem cells and the emerging complexity of their niche. *Cold Spring Harbor Symposia on Quantitative Biology*, *73*, 357–365.

Andreu, Z., Khan, M. A., González-Gómez, P., Negueruela, S., Hortigüela, R., San Emeterio, J., et al. (2015). The cyclin-dependent kinase inhibitor p27 kip1 regulates radial stem cell quiescence and neurogenesis in the adult hippocampus. *Stem Cells*, *33*, 219–229.

Ashton, R. S., Conway, A., Pangarkar, C., Bergen, J., Lim, K. I., Shah, P., et al. (2012). Astrocytes regulate adult hippocampal neurogenesis through ephrin-B signaling. *Nature Neuroscience*, *15*, 1399–1406.

Balestrini, A., Cosentino, C., Errico, A., Garner, E., & Costanzo, V. (2010). GEMC1 is a TopBP1-interacting protein required for chromosomal DNA replication. *Nature Cell Biology*, *12*, 484–491.

Bandeira, F., Lent, R., & Herculano-Houzel, S. (2009). Changing numbers of neuronal and non-neuronal cells underlie postnatal brain growth in the rat. *Proceedings of the National Academy of Sciences of the United States of America*, *106*, 14108–14113.

Barkho, B. Z., Song, H., Aimone, J. B., Smrt, R. D., Kuwabara, T., Nakashima, K., et al. (2006). Identification of astrocyte-expressed factors that modulate neural stem/progenitor cell differentiation. *Stem Cells and Development*, *15*, 407–421.

Barry, G., Piper, M., Lindwall, C., Moldrich, R., Mason, S., Little, E., et al. (2008). Specific glial populations regulate hippocampal morphogenesis. *The Journal of Neuroscience*, *28*, 12328–12340.

Batista-Brito, R., Close, J., Machold, R., & Fishell, G. (2008). The distinct temporal origins of olfactory bulb interneuron subtypes. *The Journal of Neuroscience*, *28*, 3966–3975.

Bayer, S. A. (1983). 3H-thymidine-radiographic studies of neurogenesis in the rat olfactory bulb. *Experimental Brain Research*, *50*, 329–340.

Bayer, S. A., & Altman, J. (1974). Hippocampal development in the rat: Cytogenesis and morphogenesis examined with autoradiography and low-level X-irradiation. *The Journal of Comparative Neurology*, *158*, 55–79.

Berg, D. A., Belnoue, L., Song, H., & Simon, A. (2013). Neurotransmitter-mediated control of neurogenesis in the adult vertebrate brain. *Development*, *140*, 2548–2561.

Berg, D. A., Bond, A. M., Ming, G. L., & Song, H. (2018). Radial glial cells in the adult dentate gyrus: What are they and where do they come from? *F1000Res*, *7*, 277.

Berg, D. A., Su, Y., Jimenez-Cyrus, D., Patel, A., Huang, N., Morizet, D., et al. (2019). A common embryonic origin of stem cells drives developmental and adult neurogenesis. *Cell*, *177*. 654–668 e615.

Bergles, D. E., & Richardson, W. D. (2015). Oligodendrocyte development and plasticity. *Cold Spring Harbor Perspectives in Biology*, *8*, a020453.

Bjornsson, C. S., Apostolopoulou, M., Tian, Y., & Temple, S. (2015). It takes a village: Constructing the neurogenic niche. *Developmental Cell*, *32*, 435–446.

Bonaguidi, M. A., Stadel, R. P., Berg, D. A., Sun, J., Ming, G. L., & Song, H. (2016). Diversity of neural precursors in the adult mammalian brain. *Cold Spring Harbor Perspectives in Biology*, *8*, a018838.

Bonaguidi, M. A., Wheeler, M. A., Shapiro, J. S., Stadel, R. P., Sun, G. J., Ming, G. L., et al. (2011). In vivo clonal analysis reveals self-renewing and multipotent adult neural stem cell characteristics. *Cell*, *145*, 1142–1155.

Bond, A. M., Ming, G. L., & Song, H. (2015). Adult mammalian neural stem cells and neurogenesis: Five decades later. *Cell Stem Cell*, *17*, 385–395.

Brill, M. S., Ninkovic, J., Winpenny, E., Hodge, R. D., Ozen, I., Yang, R., et al. (2009). Adult generation of glutamatergic olfactory bulb interneurons. *Nature Neuroscience*, *12*, 1524–1533.

Brill, M. S., Snapyan, M., Wohlfrom, H., Ninkovic, J., Jawerka, M., Mastick, G. S., et al. (2008). A dlx2- and pax6-dependent transcriptional code for periglomerular neuron specification in the adult olfactory bulb. *The Journal of Neuroscience*, *28*, 6439–6452.

Brunne, B., Zhao, S., Derouiche, A., Herz, J., May, P., Frotscher, M., et al. (2010). Origin, maturation, and astroglial transformation of secondary radial glial cells in the developing dentate gyrus. *Glia*, *58*, 1553–1569.

Bushong, E. A., Martone, M. E., & Ellisman, M. H. (2004). Maturation of astrocyte morphology and the establishment of astrocyte domains during postnatal hippocampal development. *International Journal of Developmental Neuroscience*, *22*, 73–86.

Caronia-Brown, G., Yoshida, M., Gulden, F., Assimacopoulos, S., & Grove, E. A. (2014). The cortical hem regulates the size and patterning of neocortex. *Development*, *141*, 2855–2865.

Ciric, T., Cahill, S. P., & Snyder, J. S. (2019). Dentate gyrus neurons that are born at the peak of development, but not before or after, die in adulthood. *Brain and Behavior: A Cognitive Neuroscience Perspective*, *9*, e01435.

Cole, J. D., Espinueva, D., Seib, D. R., Ash, A. M., Cooke, M. B., Cahill, S. P., et al. (2020). Adult-born hippocampal neurons undergo extended development and are morphologically distinct from neonatally-born neurons prolonged development of adult-born neurons. *The Journal of Neuroscience*, *40*, 5740–5756.
Del Bigio, M. R. (2010). Ependymal cells: Biology and pathology. *Acta Neuropathologica*, *119*, 55–73.
Delgado, R. N., & Lim, D. A. (2017). Maintenance of positional identity of neural progenitors in the embryonic and postnatal telencephalon. *Frontiers in Molecular Neuroscience*, *10*, 373.
Delgado, R. N., Lu, C., & Lim, D. A. (2016). Maintenance of neural stem cell regional identity in culture. *Neurogenesis (Austin)*, *3*, e1187321.
Delgado, R. N., Mansky, B., Ahanger, S. H., Lu, C., Andersen, R. E., Dou, Y., et al. (2020). Maintenance of neural stem cell positional identity by mixed-lineage leukemia 1. *Science*, *368*, 48–53.
Desmond, M. E., & Jacobson, A. G. (1977). Embryonic brain enlargement requires cerebrospinal fluid pressure. *Developmental Biology*, *57*, 188–198.
Doetsch, F., Garcia-Verdugo, J. M., & Alvarez-Buylla, A. (1997). Cellular composition and three-dimensional organization of the subventricular germinal zone in the adult mammalian brain. *The Journal of Neuroscience*, *17*, 5046–5061.
Encinas, J. M., Michurina, T. V., Peunova, N., Park, J. H., Tordo, J., Peterson, D. A., et al. (2011). Division-coupled astrocytic differentiation and age-related depletion of neural stem cells in the adult hippocampus. *Cell Stem Cell*, *8*, 566–579.
Falk, S., & Götz, M. (2017). Glial control of neurogenesis. *Current Opinion in Neurobiology*, *47*, 188–195.
Ferrón, S. R., Charalambous, M., Radford, E., McEwen, K., Wildner, H., Hind, E., et al. (2011). Postnatal loss of Dlk1 imprinting in stem cells and niche astrocytes regulates neurogenesis. *Nature*, *475*, 381–385.
Florio, M., & Huttner, W. B. (2014). Neural progenitors, neurogenesis and the evolution of the neocortex. *Development*, *141*, 2182–2194.
Fuentealba, L. C., Rompani, S. B., Parraguez, J. I., Obernier, K., Romero, R., Cepko, C. L., et al. (2015). Embryonic origin of postnatal neural stem cells. *Cell*, *161*, 1644–1655.
Furutachi, S., Miya, H., Watanabe, T., Kawai, H., Yamasaki, N., Harada, Y., et al. (2015). Slowly dividing neural progenitors are an embryonic origin of adult neural stem cells. *Nature Neuroscience*, *18*, 657–665.
Gallagher, D., Norman, A. A., Woodard, C. L., Yang, G., Gauthier-Fisher, A., Fujitani, M., et al. (2013). Transient maternal IL-6 mediates long-lasting changes in neural stem cell pools by deregulating an endogenous self-renewal pathway. *Cell Stem Cell*, *13*, 564–576.
Gebara, E., Bonaguidi, M. A., Beckervordersandforth, R., Sultan, S., Udry, F., Gijs, P. J., et al. (2016). Heterogeneity of radial glia-like cells in the adult hippocampus. *Stem Cells*, *34*, 997–1010.
Gerhardt, H., Ruhrberg, C., Abramsson, A., Fujisawa, H., Shima, D., & Betsholtz, C. (2004). Neuropilin-1 is required for endothelial tip cell guidance in the developing central nervous system. *Developmental Dynamics*, *231*, 503–509.
Giachino, C., Barz, M., Tchorz, J. S., Tome, M., Gassmann, M., Bischofberger, J., et al. (2014). GABA suppresses neurogenesis in the adult hippocampus through GABAB receptors. *Development*, *141*, 83–90.
Gilley, J. A., Yang, C. P., & Kernie, S. G. (2011). Developmental profiling of postnatal dentate gyrus progenitors provides evidence for dynamic cell-autonomous regulation. *Hippocampus*, *21*, 33–47.
Gomez, H. G., Noguchi, H., Castillo, J. G., Aguilar, D., Pleasure, S. J., & Yabut, O. R. (2019). Suppressor of fused regulates the proliferation of postnatal neural stem and precursor cells via a Gli3-dependent mechanism. *Biology Open*, *8*, bio039248.

Han, Y. G., Spassky, N., Romaguera-Ros, M., Garcia-Verdugo, J. M., Aguilar, A., Schneider-Maunoury, S., et al. (2008). Hedgehog signaling and primary cilia are required for the formation of adult neural stem cells. *Nature Neuroscience*, *11*, 277–284.
Heng, Y. H., McLeay, R. C., Harvey, T. J., Smith, A. G., Barry, G., Cato, K., et al. (2014). NFIX regulates neural progenitor cell differentiation during hippocampal morphogenesis. *Cerebral Cortex*, *24*, 261–279.
Hochgerner, H., Zeisel, A., Lönnerberg, P., & Linnarsson, S. (2018). Conserved properties of dentate gyrus neurogenesis across postnatal development revealed by single-cell RNA sequencing. *Nature Neuroscience*, *21*, 290–299.
Jang, M. H., Bonaguidi, M. A., Kitabatake, Y., Sun, J., Song, J., Kang, E., et al. (2013). Secreted frizzled-related protein 3 regulates activity-dependent adult hippocampal neurogenesis. *Cell Stem Cell*, *12*, 215–223.
Jang, M. H., Song, H., & Ming, G.-L. (2008). Regulation of adult neurogenesis by neurotransmitters. In F. H. Gage, G. Kempermann, & H. Song (Eds.), *Adult neurogenesis* CSHL Press.
Kempermann, G., Gage, F. H., Aigner, L., Song, H., Curtis, M. A., Thuret, S., et al. (2018). Human adult neurogenesis: Evidence and remaining questions. *Cell Stem Cell*, *23*, 25–30.
Kerloch, T., Clavreul, S., Goron, A., Abrous, D. N., & Pacary, E. (2019). Dentate granule neurons generated during perinatal life display distinct morphological features compared with later-born neurons in the mouse hippocampus. *Cerebral Cortex*, *29*, 3527–3539.
Kessaris, N., Fogarty, M., Iannarelli, P., Grist, M., Wegner, M., & Richardson, W. D. (2006). Competing waves of oligodendrocytes in the forebrain and postnatal elimination of an embryonic lineage. *Nature Neuroscience*, *9*, 173–179.
Kierdorf, K., Erny, D., Goldmann, T., Sander, V., Schulz, C., Perdiguero, E. G., et al. (2013). Microglia emerge from erythromyeloid precursors via Pu.1- and Irf8-dependent pathways. *Nature Neuroscience*, *16*, 273–280.
Kriegstein, A., & Alvarez-Buylla, A. (2009). The glial nature of embryonic and adult neural stem cells. *Annual Review of Neuroscience*, *32*, 149–184.
Kyrousi, C., Arbi, M., Pilz, G. A., Pefani, D. E., Lalioti, M. E., Ninkovic, J., et al. (2015). Mcidas and GemC1 are key regulators for the generation of multiciliated ependymal cells in the adult neurogenic niche. *Development*, *142*, 3661–3674.
Kyrousi, C., Lygerou, Z., & Taraviras, S. (2017). How a radial glial cell decides to become a multiciliated ependymal cell. *Glia*, *65*, 1032–1042.
Lalioti, M. E., Kaplani, K., Lokka, G., Georgomanolis, T., Kyrousi, C., Dong, W., et al. (2019). GemC1 is a critical switch for neural stem cell generation in the postnatal brain. *Glia*, *67*, 2360–2373.
Lehtinen, M. K., Bjornsson, C. S., Dymecki, S. M., Gilbertson, R. J., Holtzman, D. M., & Monuki, E. S. (2013). The choroid plexus and cerebrospinal fluid: Emerging roles in development, disease, and therapy. *The Journal of Neuroscience*, *33*, 17553–17559.
Lehtinen, M. K., Zappaterra, M. W., Chen, X., Yang, Y. J., Hill, A. D., Lun, M., et al. (2011). The cerebrospinal fluid provides a proliferative niche for neural progenitor cells. *Neuron*, *69*, 893–905.
Li, G., Fang, L., Fernandez, G., & Pleasure, S. J. (2013). The ventral hippocampus is the embryonic origin for adult neural stem cells in the dentate gyrus. *Neuron*, *78*, 658–672.
Lim, D. A., Tramontin, A. D., Trevejo, J. M., Herrera, D. G., Garcia-Verdugo, J. M., & Alvarez-Buylla, A. (2000). Noggin antagonizes BMP signaling to create a niche for adult neurogenesis. *Neuron*, *28*, 713–726.
Liu, X., Wang, Q., Haydar, T. F., & Bordey, A. (2005). Nonsynaptic GABA signaling in postnatal subventricular zone controls proliferation of GFAP-expressing progenitors. *Nature Neuroscience*, *8*, 1179–1187.

Lledo, P. M., Merkle, F. T., & Alvarez-Buylla, A. (2008). Origin and function of olfactory bulb interneuron diversity. *Trends in Neurosciences*, *31*, 392–400.
Ma, D. K., Bonaguidi, M. A., Ming, G. L., & Song, H. (2009). Adult neural stem cells in the mammalian central nervous system. *Cell Research*, *19*, 672–682.
Ma, S., & Huang, Z. (2015). Neural regulation of CNS angiogenesis during development. *Frontiers in Biology (Beijing)*, *10*, 61–73.
Ma, D. K., Jang, M. H., Guo, J. U., Kitabatake, Y., Chang, M. L., Pow-Anpongkul, N., et al. (2009). Neuronal activity-induced Gadd45b promotes epigenetic DNA demethylation and adult neurogenesis. *Science*, *323*, 1074–1077.
Ma, D. K., Ming, G. L., & Song, H. (2005). Glial influences on neural stem cell development: Cellular niches for adult neurogenesis. *Current Opinion in Neurobiology*, *15*, 514–520.
Marin-Padilla, M. (1985). Early vascularization of the embryonic cerebral cortex: Golgi and electron microscopic studies. *The Journal of Comparative Neurology*, *241*, 237–249.
Martynoga, B., Mateo, J. L., Zhou, B., Andersen, J., Achimastou, A., Urbán, N., et al. (2013). Epigenomic enhancer annotation reveals a key role for NFIX in neural stem cell quiescence. *Genes & Development*, *27*, 1769–1786.
Mathews, E. A., Morgenstern, N. A., Piatti, V. C., Zhao, C., Jessberger, S., Schinder, A. F., et al. (2010). A distinctive layering pattern of mouse dentate granule cells is generated by developmental and adult neurogenesis. *The Journal of Comparative Neurology*, *518*, 4479–4490.
Matsue, K., Minakawa, S., Kashiwagi, T., Toda, K., Sato, T., Shioda, S., et al. (2018). Dentate granule progenitor cell properties are rapidly altered soon after birth. *Brain Structure & Function*, *223*, 357–369.
McGarry, T. J., & Kirschner, M. W. (1998). Geminin, an inhibitor of DNA replication, is degraded during mitosis. *Cell*, *93*, 1043–1053.
Merkle, F. T., Fuentealba, L. C., Sanders, T. A., Magno, L., Kessaris, N., & Alvarez-Buylla, A. (2014). Adult neural stem cells in distinct microdomains generate previously unknown interneuron types. *Nature Neuroscience*, *17*, 207–214.
Merkle, F. T., Mirzadeh, Z., & Alvarez-Buylla, A. (2007). Mosaic organization of neural stem cells in the adult brain. *Science*, *317*, 381–384.
Merkle, F. T., Tramontin, A. D., García-Verdugo, J. M., & Alvarez-Buylla, A. (2004). Radial glia give rise to adult neural stem cells in the subventricular zone. *Proceedings of the National Academy of Sciences of the United States of America*, *101*, 17528–17532.
Ming, G. L., & Song, H. (2005). Adult neurogenesis in the mammalian central nervous system. *Annual Review of Neuroscience*, *28*, 223–250.
Ming, G. L., & Song, H. (2011). Adult neurogenesis in the mammalian brain: Significant answers and significant questions. *Neuron*, *70*, 687–702.
Mirzadeh, Z., Merkle, F. T., Soriano-Navarro, M., Garcia-Verdugo, J. M., & Alvarez-Buylla, A. (2008). Neural stem cells confer unique pinwheel architecture to the ventricular surface in neurogenic regions of the adult brain. *Cell Stem Cell*, *3*, 265–278.
Molnár, Z., Clowry, G. J., Šestan, N., Alzu'bi, A., Bakken, T., Hevner, R. F., et al. (2019). New insights into the development of the human cerebral cortex. *Journal of Anatomy*, *235*, 432–451.
Monje, M. (2018). Myelin plasticity and nervous system function. *Annual Review of Neuroscience*, *41*, 61–76.
Moreno-Estellés, M., González-Gómez, P., Hortigüela, R., Díaz-Moreno, M., San Emeterio, J., Carvalho, A. L., et al. (2012). Symmetric expansion of neural stem cells from the adult olfactory bulb is driven by astrocytes via WNT7A. *Stem Cells*, *30*, 2796–2809.
Namba, T., & Huttner, W. B. (2017). Neural progenitor cells and their role in the development and evolutionary expansion of the neocortex. *Wiley Interdisciplinary Reviews: Developmental Biology*, *6*.

Namba, T., Mochizuki, H., Onodera, M., Mizuno, Y., Namiki, H., & Seki, T. (2005). The fate of neural progenitor cells expressing astrocytic and radial glial markers in the postnatal rat dentate gyrus. *The European Journal of Neuroscience*, *22*, 1928–1941.

Namba, T., Mochizuki, H., Suzuki, R., Onodera, M., Yamaguchi, M., Namiki, H., et al. (2011). Time-lapse imaging reveals symmetric neurogenic cell division of GFAP-expressing progenitors for expansion of postnatal dentate granule neurons. *PLoS One*, *6*, e25303.

Nelson, B. R., Hodge, R. D., Daza, R. A., Tripathi, P. P., Arnold, S. J., Millen, K. J., et al. (2020). Intermediate progenitors support migration of neural stem cells into dentate gyrus outer neurogenic niches. *eLife*, *9*.

Nicola, Z., Fabel, K., & Kempermann, G. (2015). Development of the adult neurogenic niche in the hippocampus of mice. *Frontiers in Neuroanatomy*, *9*, 53.

Nikodemova, M., Kimyon, R. S., De, I., Small, A. L., Collier, L. S., & Watters, J. J. (2015). Microglial numbers attain adult levels after undergoing a rapid decrease in cell number in the third postnatal week. *Journal of Neuroimmunology*, *278*, 280–288.

Noguchi, H., Castillo, J. G., Nakashima, K., & Pleasure, S. J. (2019). Suppressor of fused controls perinatal expansion and quiescence of future dentate adult neural stem cells. *Elife*, *8*.

Noguchi, H., Murao, N., Kimura, A., Matsuda, T., Namihira, M., & Nakashima, K. (2016). DNA methyltransferase 1 is indispensable for development of the hippocampal dentate gyrus. *The Journal of Neuroscience*, *36*, 6050–6068.

Obernier, K., & Alvarez-Buylla, A. (2019). Neural stem cells: Origin, heterogeneity and regulation in the adult mammalian brain. *Development*, *146*.

Obernier, K., Cebrian-Silla, A., Thomson, M., Parraguez, J. I., Anderson, R., Guinto, C., et al. (2018). Adult neurogenesis is sustained by symmetric self-renewal and differentiation. *Cell Stem Cell*, *22*. 221-234.e228.

Ortega, F., Gascon, S., Masserdotti, G., Deshpande, A., Simon, C., Fischer, J., et al. (2013). Oligodendrogliogenic and neurogenic adult subependymal zone neural stem cells constitute distinct lineages and exhibit differential responsiveness to Wnt signalling. *Nature Cell Biology*, *15*, 602–613.

Ortega-Martínez, S., & Trejo, J. L. (2015). The postnatal origin of adult neural stem cells and the effects of glucocorticoids on their genesis. *Behavioural Brain Research*, *279*, 166–176.

Ortiz-Álvarez, G., Daclin, M., Shihavuddin, A., Lansade, P., Fortoul, A., Faucourt, M., et al. (2019). Adult neural stem cells and multiciliated ependymal cells share a common lineage regulated by the geminin family members. *Neuron*, *102*. 159–172.e157.

Palmer, T. D., Willhoite, A. R., & Gage, F. H. (2000). Vascular niche for adult hippocampal neurogenesis. *The Journal of Comparative Neurology*, *425*, 479–494.

Paul, A., Chaker, Z., & Doetsch, F. (2017). Hypothalamic regulation of regionally distinct adult neural stem cells and neurogenesis. *Science*, *356*, 1383–1386.

Pechnick, R. N., Zonis, S., Wawrowsky, K., Pourmorady, J., & Chesnokova, V. (2008). p21Cip1 restricts neuronal proliferation in the subgranular zone of the dentate gyrus of the hippocampus. *Proceedings of the National Academy of Sciences of the United States of America*, *105*, 1358–1363.

Pefani, D. E., Dimaki, M., Spella, M., Karantzelis, N., Mitsiki, E., Kyrousi, C., et al. (2011). Idas, a novel phylogenetically conserved geminin-related protein, binds to geminin and is required for cell cycle progression. *The Journal of Biological Chemistry*, *286*, 23234–23246.

Peretto, P., Dati, C., De Marchis, S., Kim, H. H., Ukhanova, M., Fasolo, A., et al. (2004). Expression of the secreted factors noggin and bone morphogenetic proteins in the subependymal layer and olfactory bulb of the adult mouse brain. *Neuroscience*, *128*, 685–696.

Pilz, G. A., Bottes, S., Betizeau, M., Jorg, D. J., Carta, S., Simons, B. D., et al. (2018). Live imaging of neurogenesis in the adult mouse hippocampus. *Science*, *359*, 658–662.

Piper, M., Barry, G., Hawkins, J., Mason, S., Lindwall, C., Little, E., et al. (2010). NFIA controls telencephalic progenitor cell differentiation through repression of the Notch effector Hes1. *The Journal of Neuroscience*, *30*, 9127–9139.

Qiu, J., Takagi, Y., Harada, J., Topalkara, K., Wang, Y., Sims, J. R., et al. (2009). p27Kip1 constrains proliferation of neural progenitor cells in adult brain under homeostatic and ischemic conditions. *Stem Cells*, *27*, 920–927.

Ramirez-Castillejo, C., Sanchez-Sanchez, F., Andreu-Agullo, C., Ferron, S. R., Aroca-Aguilar, J. D., Sanchez, P., et al. (2006). Pigment epithelium-derived factor is a niche signal for neural stem cell renewal. *Nature Neuroscience*, *9*, 331–339.

Redmond, S. A., Figueres-Oñate, M., Obernier, K., Nascimento, M. A., Parraguez, J. I., López-Mascaraque, L., et al. (2019). Development of ependymal and postnatal neural stem cells and their origin from a common embryonic progenitor. *Cell Reports*, *27*. 429-441.e423.

Reemst, K., Noctor, S. C., Lucassen, P. J., & Hol, E. M. (2016). The indispensable roles of microglia and astrocytes during brain development. *Frontiers in Human Neuroscience*, *10*, 566.

Robertson, P. L., Du Bois, M., Bowman, P. D., & Goldstein, G. W. (1985). Angiogenesis in developing rat brain: An in vivo and in vitro study. *Brain Research*, *355*, 219–223.

Save, L., Baude, A., & Cossart, R. (2019). Temporal embryonic origin critically determines cellular physiology in the dentate gyrus. *Cerebral Cortex*, *29*, 2639–2652.

Seki, T., Sato, T., Toda, K., Osumi, N., Imura, T., & Shioda, S. (2014). Distinctive population of Gfap-expressing neural progenitors arising around the dentate notch migrate and form the granule cell layer in the developing hippocampus. *The Journal of Comparative Neurology*, *522*, 261–283.

Shen, Q., Goderie, S. K., Jin, L., Karanth, N., Sun, Y., Abramova, N., et al. (2004). Endothelial cells stimulate self-renewal and expand neurogenesis of neural stem cells. *Science*, *304*, 1338–1340.

Shin, J., Berg, D. A., Zhu, Y., Shin, J. Y., Song, J., Bonaguidi, M. A., et al. (2015). Single-cell RNA-Seq with waterfall reveals molecular cascades underlying adult neurogenesis. *Cell Stem Cell*, *17*, 360–372.

Sierra, A., Encinas, J. M., Deudero, J. J., Chancey, J. H., Enikolopov, G., Overstreet-Wadiche, L. S., et al. (2010). Microglia shape adult hippocampal neurogenesis through apoptosis-coupled phagocytosis. *Cell Stem Cell*, *7*, 483–495.

Silva-Vargas, V., Maldonado-Soto, A. R., Mizrak, D., Codega, P., & Doetsch, F. (2016). Age-dependent niche signals from the choroid plexus regulate adult neural stem cells. *Cell Stem Cell*, *19*, 643–652.

Snyder, J. S. (2019). Recalibrating the relevance of adult neurogenesis. *Trends in Neurosciences*, *42*, 164–178.

Song, H., Stevens, C. F., & Gage, F. H. (2002). Astroglia induce neurogenesis from adult neural stem cells. *Nature*, *417*, 39–44.

Song, J., Sun, J., Olsen, R. H. J., Ming, G.-I., & Song, H. (2015). Neuronal circuitry mechanisms regulating adult mammalian neurogenesis. In *Cold spring harbor perspective in biology* (in press).

Song, J., Zhong, C., Bonaguidi, M. A., Sun, G. J., Hsu, D., Gu, Y., et al. (2012). Neuronal circuitry mechanism regulating adult quiescent neural stem-cell fate decision. *Nature*, *489*, 150–154.

Spassky, N., Merkle, F. T., Flames, N., Tramontin, A. D., Garcia-Verdugo, J. M., & Alvarez-Buylla, A. (2005). Adult ependymal cells are postmitotic and are derived from radial glial cells during embryogenesis. *The Journal of Neuroscience*, *25*, 10–18.

Sun, J., Bonaguidi, M. A., Jun, H., Guo, J. U., Sun, G. J., Will, B., et al. (2015). A septotemporal molecular gradient of sfrp3 in the dentate gyrus differentially regulates quiescent adult hippocampal neural stem cell activation. *Molecular Brain*, *8*, 52.

Tavazoie, M., Van der Veken, L., Silva-Vargas, V., Louissaint, M., Colonna, L., Zaidi, B., et al. (2008). A specialized vascular niche for adult neural stem cells. *Cell Stem Cell*, *3*, 279–288.

Taverna, E., Gotz, M., & Huttner, W. B. (2014). The cell biology of neurogenesis: Toward an understanding of the development and evolution of the neocortex. *Annual Review of Cell and Developmental Biology*, *30*, 465–502.

Tay, T. L., Savage, J. C., Hui, C. W., Bisht, K., & Tremblay, M. E. (2017). Microglia across the lifespan: From origin to function in brain development, plasticity and cognition. *The Journal of Physiology*, *595*, 1929–1945.

Tripathi, R. B., Clarke, L. E., Burzomato, V., Kessaris, N., Anderson, P. N., Attwell, D., et al. (2011). Dorsally and ventrally derived oligodendrocytes have similar electrical properties but myelinate preferred tracts. *The Journal of Neuroscience*, *31*, 6809–6819.

Urban, N., Blomfield, I. M., & Guillemot, F. (2019). Quiescence of adult mammalian neural stem cells: A highly regulated rest. *Neuron*, *104*, 834–848.

van Velthoven, C. T. J., & Rando, T. A. (2019). Stem cell quiescence: Dynamism, restraint, and cellular idling. *Cell Stem Cell*, *24*, 213–225.

Ventura, R. E., & Goldman, J. E. (2007). Dorsal radial glia generate olfactory bulb interneurons in the postnatal murine brain. *The Journal of Neuroscience*, *27*, 4297–4302.

Vicidomini, C., Guo, N., & Sahay, A. (2020). Communication, cross talk, and signal integration in the adult hippocampal neurogenic niche. *Neuron*, *105*, 220–235.

Vithayathil, J., Pucilowska, J., Goodnough, L. H., Atit, R. P., & Landreth, G. E. (2015). Dentate gyrus development requires ERK activity to maintain progenitor population and MAPK pathway feedback regulation. *The Journal of Neuroscience*, *35*, 6836–6848.

Wang, H., Kane, A. W., Lee, C., & Ahn, S. (2014). Gli3 repressor controls cell fates and cell adhesion for proper establishment of neurogenic niche. *Cell Reports*, *8*, 1093–1104.

Winkler, C. C., Yabut, O. R., Fregoso, S. P., Gomez, H. G., Dwyer, B. E., Pleasure, S. J., et al. (2018). The dorsal wave of neocortical oligodendrogenesis begins embryonically and requires multiple sources of sonic hedgehog. *The Journal of Neuroscience*, *38*, 5237–5250.

Yang, G., Cancino, G. I., Zahr, S. K., Guskjolen, A., Voronova, A., Gallagher, D., et al. (2016). A Glo1-methylglyoxal pathway that is perturbed in maternal diabetes regulates embryonic and adult neural stem cell pools in murine offspring. *Cell Reports*, *17*, 1022–1036.

Young, K. M., Fogarty, M., Kessaris, N., & Richardson, W. D. (2007). Subventricular zone stem cells are heterogeneous with respect to their embryonic origins and neurogenic fates in the adult olfactory bulb. *The Journal of Neuroscience*, *27*, 8286–8296.

Yuzwa, S. A., Borrett, M. J., Innes, B. T., Voronova, A., Ketela, T., Kaplan, D. R., et al. (2017). Developmental emergence of adult neural stem cells as revealed by single-cell transcriptional profiling. *Cell Reports*, *21*, 3970–3986.

Zhang, H., Kim, Y., Ro, E. J., Ho, C., Lee, D., Trapp, B. D., et al. (2020). Hippocampal neurogenesis and neural circuit formation in a cuprizone-induced multiple sclerosis mouse model. *The Journal of Neuroscience*, *40*, 447–458.

Zhang, Y., Liu, G., Guo, T., Liang, X. G., Du, H., Yang, L., et al. (2020). Cortical neural stem cell lineage progression is regulated by extrinsic signaling molecule sonic hedgehog. *Cell Reports*, *30*. 4490-4504.e4494.

CHAPTER THREE

The essential role of primary cilia in cerebral cortical development and disorders

Siling Liu[a], Mia X. Trupiano[a], Jeremy Simon[a], Jiami Guo[b], and E.S. Anton[a,*]

[a]UNC Neuroscience Center and the Department of Cell and Molecular Physiology, University of North Carolina School of Medicine, Chapel Hill, NC, United States
[b]Hotchkiss Brain Institute, Alberta Children's Hospital Research Institute, and the Department of Cell Biology and Anatomy, University of Calgary, Calgary, AB, Canada
*Corresponding author: e-mail address: anton@med.unc.edu

Contents

Abstract

Primary cilium, first described in the 19th century in different cell types and organisms by Alexander Ecker, Albert Kolliker, Aleksandr Kowalevsky, Paul Langerhans, and Karl Zimmermann (Ecker, 1844; Kolliker, 1854; Kowalevsky, 1867; Langerhans, 1876; Zimmermann, 1898), play an essential modulatory role in diverse aspects of nervous system development and function. The primary cilium, sometimes referred to as the cell's 'antennae', can receive wide ranging inputs from cellular milieu, including morphogens, growth factors, neuromodulators, and neurotransmitters. Its unique structural and functional organization bequeaths it the capacity to hyper-concentrate signaling machinery in a restricted cellular domain approximately one-thousandth the volume of cell soma. Thus enabling it to act as a signaling hub that integrates diverse developmental and homestatic information from cellular milieu to regulate the development and function of neural cells. Dysfunction of primary cilia contributes to the

Current Topics in Developmental Biology, Volume 142
ISSN 0070-2153
https://doi.org/10.1016/bs.ctdb.2020.11.003

pathophysiology of several brain malformations, intellectual disabilities, epilepsy, and psychiatric disorders. This review focuses on the most essential contributions of primary cilia to cerebral cortical development and function, in the context of neurodevelopmental disorders and malformations. It highlights the recent progress made in identifying the mechanisms underlying primary cilia's role in cortical progenitors, neurons and glia, in health and disease. A future challenge will be to translate these insights and advances into effective clinical treatments for ciliopathies.

1. Introduction

Primary cilia are microtubule based antennae-like sensory organelles extended by a vast majority of mammalian cells (Bloodgood, 2009; Fliegauf, Benzing, & Omran, 2007; Hilgendorf, Johnson, & Jackson, 2016; Reiter & Leroux, 2017; Sengupta, 2017). Primary cilia are anchored to the cell by mother centriole, the older of the pair of centrioles, which together with the pericentrosomal matrix, make up the centrosome complex. The primary cilia are structurally and functionally distinct from the motile cilia, that line epithelia and power fluid movement (Ringers, Olstad, & Jurisch-Yaksi, 2020). Disruption of the physiological function of primary cilia in humans lead to a broad range of multi-organ disease phenotypes, including obesity, renal, hepatic, and pancreatic cyst formation, situs abnormalities, congenital heart defects, anosmia, retinal degeneration, postaxial polydactyly, bronchiectasis, hypogonadism, infertility, hearing loss, facial anomalies, liver fibrosis, and brain malformations (Fliegauf et al., 2007; Hildebrandt, Benzing, & Katsanis, 2011; Lancaster & Gleeson, 2009; McIntyre et al., 2012; Mitchison & Valente, 2017; Park, Jang, & Lee, 2019; Reiter & Leroux, 2017; Sattar & Gleeson, 2011; Thomas, Boutaud, Reilly, & Benmerah, 2019; Valente, Rosti, Gibbs, & Gleeson, 2014). These disorders resulting from aberrant primary cilia function include Bardet-Biedl syndrome (BBS), nephronophthisis (NPHP), Senior-Loken syndrome (SNLS), Alstrom syndrome (ALMS), Meckel syndrome (MKS), Joubert syndrome (JBTS), Oral-facial-digital Type I (OFD 1), polycystic kidney diseases (PKD), Jeune asphyxiating thoracic dystrophy (JATD), Ellis van Creveld (EVC), and Leber congenital amaurosis (LCA) (see Reiter & Leroux, 2017 and Ringers et al., 2020 for comprehensive reviews of motile and non-motile ciliopathies).

Abnormalities in the formation, connectivity, and function of the central nervous system (CNS) underlie the neurological syndromes triggered by primary cilia malfunction (Caspary, Larkins, & Anderson, 2007; Doherty,

2009; Guo et al., 2015; Novarino, Akizu, & Gleeson, 2011; Parisi, 2019; Romani, Micalizzi, & Valente, 2013; Valente et al., 2014). These neurological syndromes include developmental delay, intellectual disabilities, autism spectrum disorder, mood disorders, epilepsy, abnormal respiratory rhythms, hypotonia, ataxia, oculomotor apraxia, and mirror movement synkinesis. The structural anomalies such as microcephaly, neuronal heterotopia, disrupted neuronal layer malformation, absent or reduced growth and decussation of axonal tracts (e.g., superior cerebellar peduncles (SCP), corticospinal tract (CST), corpus callosum (CC), and central pontine tracts), and cerebellar hypoplasia associated with these functional outcomes indicate the significance of physiological function of primary cilia in progenitors, neurons, and glia during human CNS development (Doherty, 2009; Guo et al., 2015; Juric-Sekhar, Adkins, Doherty, & Hevner, 2012; Parisi, 2019; Romani et al., 2013; Valente et al., 2014).

The human cerebral cortex forms as a result of coordinated unfolding of a series of inter-related developmental events (Kwan, Sestan, & Anton, 2012; Molnár et al., 2019; Fig. 1A). During early embryonic development, an undifferentiated sheet of pseudostratified neuroepithelial cells in telencephalic vesicles end their interkinetic nuclear migration, spanning the apical–basal axis, and transform into radial glial progenitors. Radial glial cells (RGCs), through their function as a source of neurogenic progenitors and neuronal migratory guides, provide a template for the formation of the cerebral cortex. RGCs in the pallium divide symmetrically and asymmetrically to expand the pool of progenitors and generate neurons, respectively. During the neurogenic period, asymmetric division of RGCs generates cortical neurons and neurogenic intermediate progenitors (IPs) or outer radial glial cells (oRGCs). Symmetric proliferation of IPs and oRGCs serve to rapidly expand the cortical neuronal population, particularly the upper layer neurons. The cell soma of radial progenitors remain apically positioned in the ventricular zone, with a long basal process extending toward the pial surface, thereby enabling the orderly generation and guidance of new neurons in the cerebral cortex. Migration of clonally related neurons along related radial glial basal processes leads to the formation of cortical radial columns (or radial units) and laminar organization of neurons. As neurogenesis nears its completion, progenitors either morphologically transform or undergo a final division to generate astrocytes. Oligodendrocytes are generated similarly from progenitors in the subpallium. In contrast, microglia, originating from early erythromyeloid progenitors (EMPs) in the extra-embryonic yolk sac, enter the CNS in association with waves of hematopoiesis, prior to the

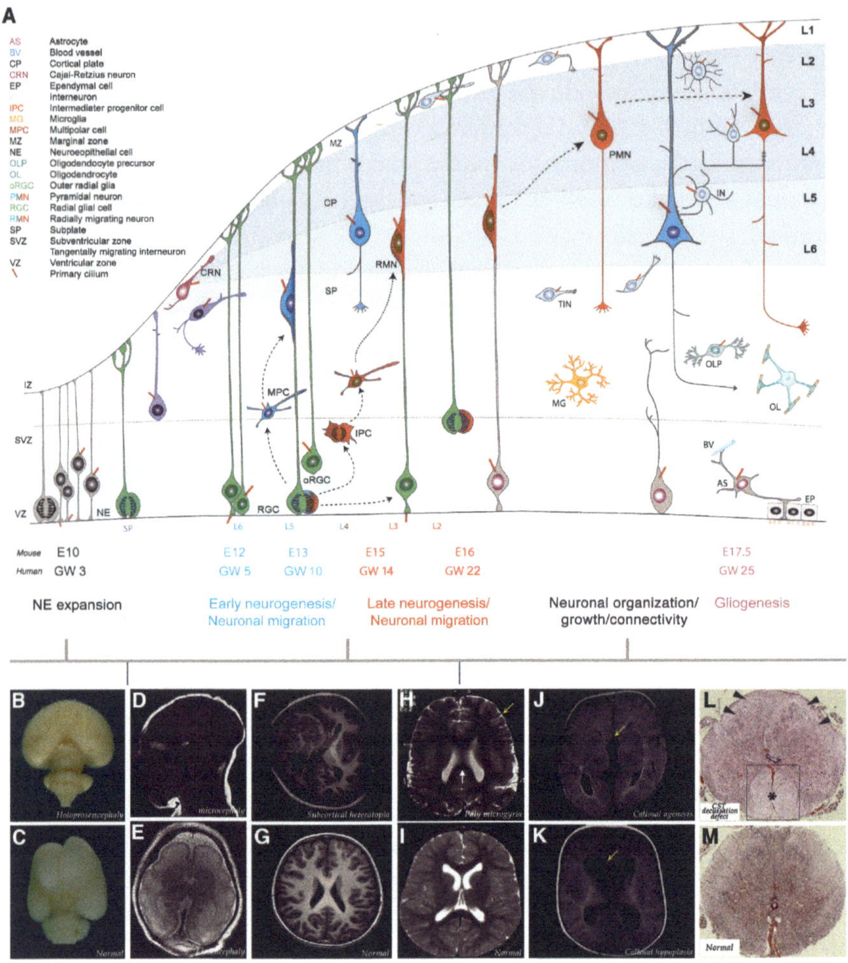

Fig. 1 Development of the cerebral cortex and cortical malformations associated with primary ciliary dysfunction. (A) During early development, symmetrically dividing neuroepithelial cells (NPs) in the ventricular zone (VZ; blue) assume radial glial identity. Radial glia cells (RGCs) divide symmetrically to expand the RGC pool or asymmetrically to generate neurons and other progenitors (intermediate progenitor cells [IPCs] or outer radial glial cells [oRGC]). Newborn neurons assume a multipolar morphology and then migrate away from the germinal zones, guided by the basal processes of the RGCs, to reach the mantle layers. The first-born neurons settle within the preplate (PP) to form the nascent cortical plate (CP). Continued addition of subsequently generated neurons then split the PP into the marginal zone (MZ) and the subplate (SP), leading to sequential formation of deep (L6 and L5; blue) and upper layers (L4, L3 and L2; red) neurons of the cerebral cortex. Intermediate progenitor cells (IPCs) or outer radial glial cells (oRGC) derived from radial glia undergo symmetric neurogenic divisions in the SVZ to generate upper layer neurons. Concurrent to radial migration of projection neurons, tangentially migrating interneurons originating in the ganglionic eminence invade

proliferation of neural progenitors (Thion, Ginhoux, & Garel, 2018). The developmental balance between different progenitor subtypes and the resultant differences in the generation and guidance of distinct classes of neurons and glia within the developing cerebral cortex enables the emergence of distinctly different cortical areas with characteristic patterns of neuronal and glial density and diversity. Once appropriately positioned, connectivity between the two main types of neurons, excitatory projection neurons with spiny dendrites and long axons and inhibitory interneurons with short axons and non-spiny dendrites, form the basic neuronal circuitry in the cerebral cortex. The distinct cellular architecture of these neuronal subtypes and their interactions with glia (i.e., astrocytes, oligodendroglia, and microglia) constrains the types of circuits formed and thus the excitatory/inhibitory (E/I) balance and the functional competence of these cortical circuits. Primary cilia play a diverse and essential modulatory role in these major developmental events underlying cerebral cortical formation and function (Fig. 1B–M).

The majority of the cellular components of the developing cerebral cortex, including neuroepithelial cells, progenitors, migrating neurons,

the cortical plate. At the end of neurogenesis, the radial glial scaffold is dismantled, progenitors become gliogenic, generate cortical and subependymal zone (SEZ) astrocytes (Ast), and give rise to a layer of ependymal cells (EL). Early embryonic microglial (MG) invasion into the developing cerebral wall and oligodendrocyte precursor (OLP) generation and differentiation into oligodendrocytes (OL) are not illustrated. AS, astrocyte; BV, blood vessel; CP, cortical plate; CRN, Cajal-Retzius neuron; EP, ependymal cell; IN, interneuron; IPC, intermediate progenitor cell; MG, microglia; MPC, multipolar cell; MZ, marginal zone; NPC, neuroepithelial progenitor cell; OLP, oligodendrocyte precursor; OL, oligodendrocyte; oRGC, outer radial glia; PMN, Pyramidal neuron; RGC, radial glial cell, RMN, radially migrating neuron; SP, subplate; SVZ, subventricular zone, TIN, tangentially migrating interneuron; VZ, ventricular zone. A timeline of mouse embryonic (E) day or human gestational week (GW) relevant to the initial stages of the respective developmental events are indicated. (B–M) Primary cilia dysfunction during distinct stages of cortical development contributes to specific cortical malformations. (B, C) Shh mutations, likely disrupting cilia mediated Shh signaling, in NPs lead to holoprosencephaly. (D–G) Primary cilia malfunction in RGCs or migrating neurons can promote microcephaly (D), lissencephaly (E), and subcortical heterotopia (F, G). Primary cilia dysfunction in postmigratory neurons contributes to poly microgyria (H[arrow], I), callosal agenesis (arrow, J) or hypoplasia (arrow, K), and CST decussation defects (L, M [asterisk in boxed area indicates enlarged, nondecussating anterior CST in HE stained cervical spinal cord]). Images are reproduced from Monuki and Walsh (2001) (B, C), Shamseldin et al. (2015) (D, E; *RTTN* mutation), Uzquiano et al. (2019) (F, G; *RPGRIP1L* mutation), Kheradmand Kia et al. (2012) (H, I; *RTTN* mutation), Putoux et al. (2011) (J, K; *KIF7* mutation), and Juric-Sekhar et al. (2012) (L, M; *OFD1* mutation). These mutations directly or indirectly disrupt primary cilia signaling and maintenance.

postmigratory neurons, and astrocytes, have active primary cilia (Fig. 2; Arellano, Guadiana, Breunig, Rakic, & Sarkisian, 2017; Bishop, Berbari, Lewis, & Mykytyn, 2007; Green & Mykytyn, 2014; Guemez-Gamboa, Coufal, & Gleeson, 2014; Sarkisian & Guadiana, 2015; Sengupta, 2017). Presence of primary cilia in mature oligodendrocytes and microglial cells has yet to be conclusively demonstrated. Primary cilia in the developing cortical cells are highly dynamic in terms of their ability to undergo changes in shape, length, and orientation during development (Fig. 2B; Higginbotham et al., 2012, 2013). Although the detection of primary cilia in different cortical cells with transmission or scanning electron microscopy analyses

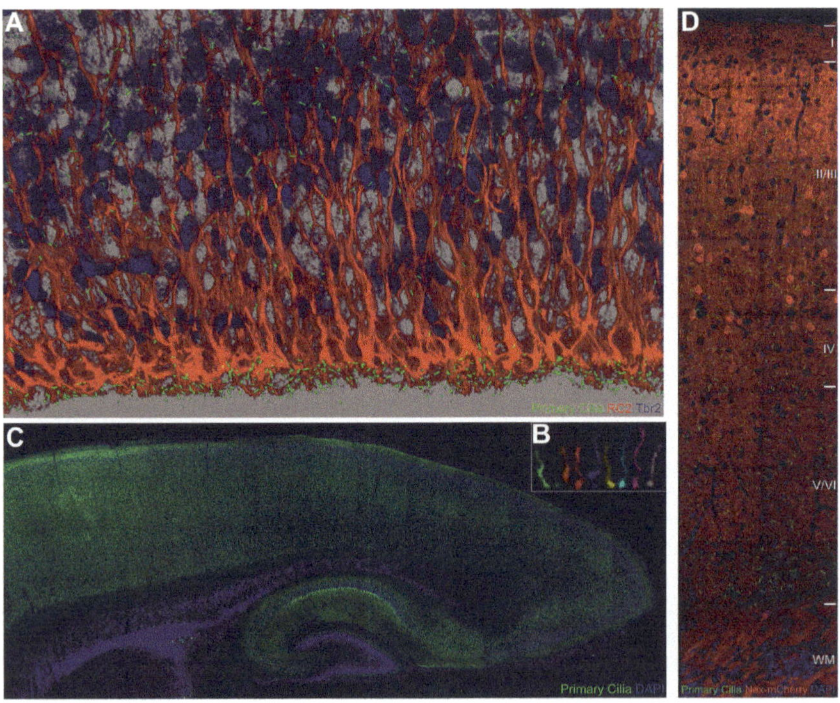

Fig. 2 Primary cilia in the embryonic and postnatal cerebral cortex. Primary cilia (green) in the ventricular zone of the developing mouse cerebral cortex (E16) were labeled with hGFAP-Cre induced Sstr3-GFP (A). Radial glia and intermediate progenitors were co-labeled with RC2 (red) and anti-Tbr2 (blue) antibodies, respectively. (B) Live imaging of a primary cilium in the embryonic cerebral wall (E16), over 5.5 h, indicates its dynamic nature. Images at different time points were pseudocolored to illustrate the changing shape, length, and orientation. (C) Projection neuronal primary cilia in postnatal brains (P7) were labeled with Nex-Cre induced Sstr3-GFP (B). Neurons across cortical layers display primary cilia of varying shapes and orientation (C).

or immunolabeling with pancilia-specific antibodies (e.g., anti-ACIII or Arl13b) do not reveal their functional diversity, primary cilia in each of these cell types are highly likely to be specialized to subserve cell-type specific functions during cortical development. Supporting this possibility, expression mapping of cilia-associated genes across different cell types during cerebral cortical development reveals distinct cell-type specific expression profiles (Loo et al., 2019; Fig. 3). How this molecular specification of primary cilia in different cortical cells translates into developmental stage- and cell type-specific functions during cerebral cortical formation remains open. However, the spectrum of primary cilia related developmental brain malformations indicate the significance of this sensory organelle for the appropriate development and differentiation of cortical progenitors, neurons, and glia and the resultant orderly progression of cerebral cortical formation and organization.

2. Progenitors, primary cilia, and associated brain developmental malformations

The organization of the ventricular niche where radial glial cells, intermediate progenitors, and outer radial glial progenitors actively form, divide, and interact with their daughter neuron or glial cells, depends on primary cilia signaling (Guo et al., 2015; Higginbotham et al., 2013; Lepanto, Badano, & Zolessi, 2016; Valente et al., 2014). In proliferating progenitors, primary cilia are disassembled during cell division and ciliary membrane remnants may preferentially associate with the mother centriole. During neurogenesis, asymmetric inheritance of the old mother centriole by daughter radial glial progenitors facilitates the maintenance of radial progenitor identity (Wang et al., 2009). Asymmetric inheritance of the primed mother centriole and thus, rapid extension of a primary cilium by one of the daughter cells is thought to provide differential competence for that daughter cell to respond to environmental cues in the ventricular niche (Basten & Giles, 2013; Delgehyr & Spassky, 2014). In dividing radial progenitors, WDR62 enables centrosome protein CEP170's localization to the basal body of the primary cilium, where CEP170 recruits a microtubule depolymerizing factor, KIF2A, to dismantle the cilium (Zhang et al., 2019). Disruption of this pathway and other WDR62 linked networks vital for cilia formation (e.g., WDR62-CPAP-IFT88) lead to microcephaly (Shohayeb et al., 2020; Zhang et al., 2019).

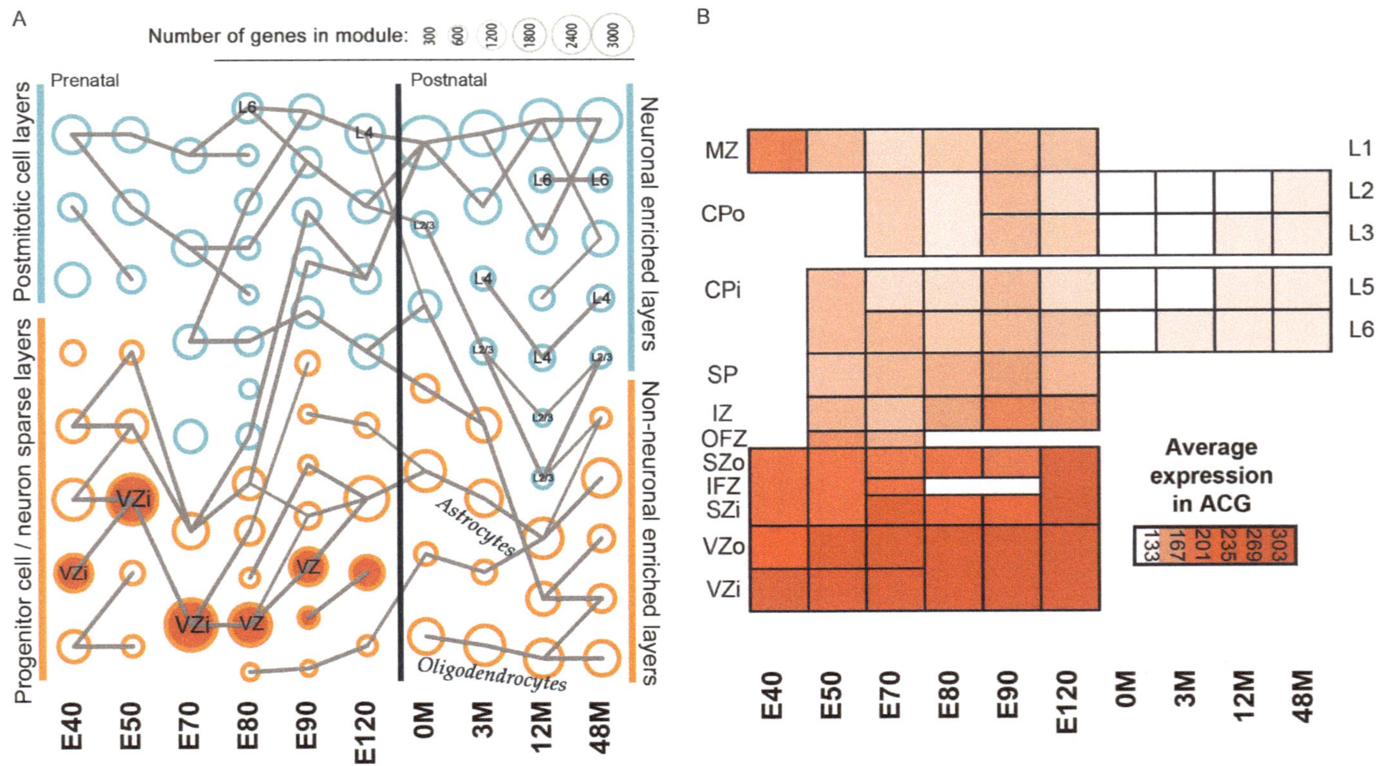

Fig. 3 Ciliopathy gene expression profile in the developing primate brain. (A) Gene expression levels of distinct populations of cells in individual cortical layers were profiled in rhesus monkey brain at 10 different pre and postnatal ages (E40, E50, E70, E80, E90, E120, 0M, 3M, 12M, and 48M; Bakken et al., 2016). Ciliopathy associated genes (confirmed and candidate) were significantly enriched among genes co-expressed in the developing cortex, primarily in the proliferative niche and new born neurons. Circle size indicates how many genes are in the module. Modules significantly enriched for risk genes associated with ciliopathies are in red ($P < 0.1$). Modules from adjacent ages with the most highly significant gene overlap are connected by gray lines. Though the number of ciliopathy genes overlapping with the other modules did not reach statistical significance, different ciliopathy genes are expressed in these modules (see Table 1). (B) Average expression pattern of cilia associated genes found in at least two enriched modules. Heat maps are organized by specific cortical layer and age as described in Bakken et al. (2016). MZ, marginal zone; CPo/CPi, outer/inner cortical plate; SP, subplate; IZ, intermediate zone; SZ, subventricular zone; VZo/VZi, outer/inner ventricular zone; L2-6, cortical neuronal layers 2–6.

Table 1 Ciliopathy and cilia-associated genes in gene expression modules of different cell types in the developing primate brain (related to Fig. 3).
a

Cortical gene expression modules	Total number of genes in module	% Ciliopathy genes in module	List of ciliopathy (confirmed and candidate) genes in module
BAKKEN2016_WGCNA_E40_M00	2484	2.254428341	ADCY6,AIPL1,AKAP9,ARF4,ARL2BP,ARL6,B9D1,BBS4,BBS7,CCNO, CEP104,CEP19,CP110,DCDC2,DNAH5,DNAI1,DNAL4,DNALI1,EFCAB7, HNF1B,HYDIN,INVS,IQCE,KIF24,LCA5,LRAT,LRRC49,LRRC6,MLF1, MST1,NEK1,NIN,NPHP3,NPHP4,OFD1,PACRG,PIH1D3,PKD1, POMGNT1,PTCH1,PTK7,PTPDC1,RAB3IP,RFX3,ROPN1L,RPE65, RSPH1,SPAG1,SPAG6,SSX2IP,TEKT1,TMEM107,TMEM17,TPRA1,TUBE1, TUBGCP6
BAKKEN2016_WGCNA_E40_M01	1621	1.480567551	CCDC28B,CEP170,CRB3,CTTN,DCTN1,KIF17,KIF3A,KIF3B,KIF3C, KIF5C,KIFAP3,MARK4,MCHR1,MXI1,NME7,OCRL,PRKACB,RAB11A, SDCCAG8,7-SEP,SNX10,TTBK2,USP9X,WDR60
BAKKEN2016_WGCNA_E40_M02	803	2.615193026	CCDC151,CETN2,GALNT11,GNAS,GSK3A,IFT140,IFT52,INPP5E, KATNB1,KIF2A,MAPRE1,ODF2,PDE6D,PEBP1,PPP2R1A,PRKAR2B, RAB11B,RAB11FIP3,TUBB,TUBB2B
BAKKEN2016_WGCNA_E40_M03	1271	1.41620771	ANO2,ARMC4,CNGB1,CYS1,DNAH6,GLI1,MAK,MORN3,PKHD1, PRKACA,RAB17,RDH12,RSG1,RSPH4A,SUFU,TTC21A,TTC25, ZMYND10
BAKKEN2016_WGCNA_E40_M04	596	1.006711409	KIF27,KIF7,MDM1,RSPH3,STK38L,TAPT1
BAKKEN2016_WGCNA_E40_M05	1107	1.445347787	C2CD3,CCDC22,CEP250,CEP89,CNTROB,GPR161,HTT,IFT122,MCRS1, MKKS,NUBP1,NUBP2,PAFAH1B1,RABL5,SSNA1,TCTN3,TULP3
BAKKEN2016_WGCNA_E40_M06	1863	1.986044015	AURKA,B9D2,BBS2,BBS9,CEP192,CEP72,CEP76,CEP97,CLUAP1,DNAL1, EVC2,FAM161A,FGFR1OP,GNAI2,GNAI3,HEATR2,IFT88,INPP5B, KIAA0586,MKS1,NEK9,ORC1,PLK1,POC1A,RAB23,RAB28,RILPL2, RTTN,SAV1,2- SEP,SMO,SYNE2, TCTN1,TMEM138,TMEM231,TTC17,WRAP73

Continued

Table 1 Ciliopathy and cilia-associated genes in gene expression modules of different cell types in the developing primate brain (related to Fig. 3).—Cont'd

a

Cortical gene expression modules	Total number of genes in module	% Ciliopathy genes in module	List of ciliopathy (confirmed and candidate) genes in module
BAKKEN2016_WGCNA_E40_M07	1603	4.304429195	AHI1,ALMS1,ANKS6,ARL13B,ASAP1,ATXN10,BBS10,BBS5,BUB1B, CC2D2A,CCDC14,CENPF,CENPJ,CEP120,CEP128,CEP135,CEP152, CEP290,CEP350,CEP63,CEP78,CETN3,CRB1,CROCC,CSPP1,DZIP1, EFHC1,FAM58A,FOXJ1,GLI3,HSPB11,IFT172,IFT57,IFT74,IFT80,IQCB1, IQUB,KCNJ13,KIF14,LZTFL1,MNS1,NEK2,PARD3,PCM1,PHF17,PIK3R4, PKD2,PLK4,POC1B,RFX2,RPGRIP1L,SASS6,SPA17,SPAG16,SPATA7, SPICE1,STIL,STK3,STK39,TEX9,TMEM216,TMEM67,TRIM32,TRIP11, TTC30A,TTC30B,TTC8,WDR19,WDR35
BAKKEN2016_WGCNA_E40_M08	1093	0.640439158	MAL,MERTK,RP2,TBX3,TUBA4A,WNK1,ZIC2
BAKKEN2016_WGCNA_E50_M00	3475	1.928057554	AHI1,AIPL1,ANKS6,ANO2,ARF4,ARL2BP,ARL6,ARMC4,BBS10,BBS7, CCNO,CEP290,CEP72,CEP97,CETN2,CNGB1,CP110,CSPP1,CYS1, DCDC2,DNAH5,DNAH6,GPR161,HNF1B,HYDIN,IFT140,INPP5E,INVS, IQCE,KCNJ13,KIF17,KIF27,LCA5,LRAT,LRRC6,MAK,MCHR1,MKKS, MLF1,MST1,NEK1,NPHP3,PKD1,PKHD1,PRKACA,PTPDC1,RAB17, RFX3,RPE65,RSG1,RSPH1,RSPH3,RSPH4A,SPAG6,SPATA7,SSX2IP, SUFU,TCTN3,TEKT1,TMEM67,TRIM32,TTC17,TTC21A,TTC30A, TUBE1,WDR35,ZMYND10
BAKKEN2016_WGCNA_E50_M01	1295	1.467181467	CEP19,EFCAB7,GALNT11,KIF2A,LRRC49,MORN3,NIN,NME7,PACRG, PRKACB,PRKAR2B,RAB3IP,RDH12,ROPN1L,SNX10,TAPT1,TMEM17, TTBK2,USP9X
BAKKEN2016_WGCNA_E50_M02	1304	1.303680982	AKAP9,ATXN10,CCDC28B,CEP104,CEP170,CRB3,CTTN,HSPB11,KIF3A, KIF3B,KIF3C,KIF5C,KIFAP3,MXI1,OCRL,RAB11A,SDCCAG8
BAKKEN2016_WGCNA_E50_M03	799	1.627033792	CCDC151,DCTN1,GSK3A,KATNB1,MAPRE1,MARK4,PDE6D,PEBP1, POMGNT1,PPP2R1A,RAB11B,TUBB,TUBB2B

BAKKEN2016_WGCNA_E50_M04	518	1.158301158	DNAL4,GNAS,IFT52,NUBP2,TUBGCP6,WRAP73
BAKKEN2016_WGCNA_E50_M05	1079	1.668211307	B9D2,C2CD3,CCDC22,CEP250,CEP89,CNTROB,FAM58A,GNAI2,HTT, IFT122,INPP5B,KIF7,MCRS1,NUBP1,PAFAH1B1,RAB11FIP3,RABL5, SSNA1,TUBA4A
BAKKEN2016_WGCNA_E50_M06	2916	3.429355281	ADCY6,ALMS1,ARL13B,ASAP1,AURKA,B9D1,BBS2,BBS4,BBS5,BBS9, BUB1B,CC2D2A,CCDC14,CENPF,CENPJ,CEP120,CEP128,CEP135, CEP152,CEP192,CEP350,CEP63,CEP76,CEP78,CETN3,CLUAP1,CRB1, CROCC,DNAI1,DNAL1,DNALI1,DZIP1,EFHC1,EVC2,FAM161A, FGFR1OP,FOXJ1,GLI1,GLI3,GNAI3,HEATR2,IFT172,IFT57,IFT74,IFT80, IFT88,IQCB1,IQUB,KIAA0586,KIF14,KIF24,LZTFL1,MDM1,MKS1,MNS1, NEK2,NEK9,NPHP4,ODF2,ORC1,PARD3,PCM1,PHF17,PIH1D3,PIK3R4, PKD2,PLK1,PLK4,POC1A,POC1B,PTCH1,PTK7,RAB23,RAB28,RFX2, RILPL2,RP2,RPGRIP1L,RTTN,SASS6,SAV1,2-SEP, SMO,SPA17,SPAG1,SPAG16,SPICE1,STIL,STK3,STK38L,STK39,SYNE2, TCTN1,TEX9,TMEM107,TMEM138,TMEM216,TMEM231,TPRA1, TRIP11,TTC30B,TTC8,TULP3,WDR19,ZIC2
BAKKEN2016_WGCNA_E50_M07	578	0.346020761	MAL,TBX3
BAKKEN2016_WGCNA_E50_M08	477	1.257861635	MERTK,OFD1,7-SEP,TTC25,WDR60,WNK1
BAKKEN2016_WGCNA_E70_M00	3085	1.458670989	ADCY6,ANKS6,ARL2BP,ARMC4,ASAP1,CEP72,CEP78,CEP89,CLUAP1, CP110,CRB3,DCDC2,DNAH5,DNAH6,EFHC1,HNF1B,HYDIN,INPP5E, INVS,KIF27,LRAT,MAL,MCHR1,MKKS,MLF1,MORN3,MST1,NEK1, NPHP3,PKD1,POMGNT1,PTK7,PTPDC1,ROPN1L,RPE65,SAV1,SPA17, SPAG1,TBX3,TRIM32,TTC17,TTC21A,TTC30B,TUBGCP6,WNK1
BAKKEN2016_WGCNA_E70_M01	1240	1.048387097	CEP19,GALNT11,IQCE,KIF3A,LRRC49,MAK,NIN,NME7,PRKACB, RAB3IP,RDH12,SNX10,TMEM17

Continued

Table 1 Ciliopathy and cilia-associated genes in gene expression modules of different cell types in the developing primate brain (related to Fig. 3).—Cont'd

a

Cortical gene expression modules	Total number of genes in module	% Ciliopathy genes in module	List of ciliopathy (confirmed and candidate) genes in module
BAKKEN2016_WGCNA_E70_M02	1405	0.996441281	CCDC28B,DCTN1,GSK3A,KIF2A,KIF3B,KIF3C,KIF5C,KIFAP3,OCRL, PEBP1,PPP2R1A,PRKAR2B,RAB11A,SDCCAG8
BAKKEN2016_WGCNA_E70_M03	1164	2.319587629	AKAP9,ARF4,ARL6,ATXN10,BBS10,BBS7,CEP104,CEP120,CEP170, CEP350,CSPP1,EFCAB7,FAM58A,HSPB11,IFT57,IFT80,MXI1,OFD1,RFX3, SPATA7,SSX2IP,TAPT1,TTBK2,TUBE1,USP9X,WDR35,WDR60
BAKKEN2016_WGCNA_E70_M04	1116	1.702508961	AIPL1,ANO2,CCNO,CEP290,CEP97,CNGB1,CYS1,KIF17,LRRC6,MARK4, PKHD1,PRKACA,RAB17,RSG1,RSPH4A,SPAG6,SUFU,TTC25,ZMYND10
BAKKEN2016_WGCNA_E70_M05	1315	2.053231939	B9D2,C2CD3,CCDC22,CEP250,CETN2,CNTROB,CTTN,DNAL4,GNAS, GPR161,HTT,IFT122,IFT140,IFT52,KATNB1,KIF7,MAPRE1,MCRS1, NPHP4,NUBP1,NUBP2,PDE6D,RAB11B,RSPH3,TUBB,TUBB2B,WRAP73
BAKKEN2016_WGCNA_E70_M06	3116	3.498074454	AHI1,ALMS1,ARL13B,AURKA,B9D1,BBS2,BBS4,BBS5,BBS9,BUB1B, CC2D2A,CCDC14,CCDC151,CENPF,CENPJ,CEP128,CEP135,CEP152, CEP192,CEP63,CEP76,CETN3,CRB1,CROCC,DNAI1,DNAL1,DNALI1, DZIP1,EVC2,FAM161A,FGFR1OP,FOXJ1,GLI1,GLI3,GNAI2,GNAI3, HEATR2,IFT172,IFT74,IFT88,INPP5B,IQCB1,IQUB,KCNJ13,KIAA0586, KIF14,KIF24,LCA5,LZTFL1,MDM1,MERTK,MKS1,MNS1,NEK2,NEK9, ODF2,ORC1,PACRG,PAFAH1B1,PARD3,PCM1,PHF17,PIH1D3,PIK3R4, PKD2,PLK1,PLK4,POC1A,POC1B,PTCH1,RAB11FIP3,RAB23,RAB28, RABL5,RFX2,RILPL2,RP2,RPGRIP1L,RSPH1,RTTN,SASS6,2-SEP,7-SEP, SMO,SPAG16,SPICE1,SSNA1,STIL,STK3, STK38L,STK39,SYNE2,TCTN1,TCTN3,TEKT1,TEX9,TMEM107, TMEM138,TMEM216,TMEM231,TMEM67,TPRA1,TRIP11,TTC30A,TTC8, TUBA4A,TULP3,WDR19,ZIC2

BAKKEN2016_WGCNA_E80_M00	3065	1.924959217	ADCY6,AIPL1,ANKS6,ANO2,ARL2BP,ARMC4,ASAP1,BBS9,CCDC151, CCNO,CEP135,CEP290,CEP350,CEP72,CEP78,CEP97,CNGB1,CP110, CRB3,CYS1,DNAH6,DNAL1,EFHC1,GLI1,HNF1B,HYDIN,INVS,KIF27, LRAT,LRRC6,MAL,MCHR1,MORN3,MST1,NIN,NPHP4,ODF2,PKD1, PKHD1,PRKACA,PTPDC1,RAB17,RFX3,RPE65,RSG1,RSPH3,RSPH4A, SAV1,SPAG1,SPAG6,SUFU,TCTN1,TCTN3,TMEM67,TPRA1,TTBK2, TTC21A,TTC25,WDR60
BAKKEN2016_WGCNA_E80_M01	936	0.854700855	KIFAP3,NME7,RAB3IP,RDH12,ROPN1L,SNX10,TAPT1,TMEM17
BAKKEN2016_WGCNA_E80_M02	434	1.152073733	AHI1,KIF17,MAK,OCRL,POMGNT1
BAKKEN2016_WGCNA_E80_M03	691	2.604920405	AKAP9,ARL6,ATXN10,CEP104,CEP170,CEP19,GALNT11,IFT57,KIF2A, KIF3A,LRRC49,MKKS,PRKACB,PRKAR2B,SPA17,SSX2IP,TRIM32, USP9X
BAKKEN2016_WGCNA_E80_M04	534	0.561797753	KIF3B,KIF3C,KIF5C
BAKKEN2016_WGCNA_E80_M05	364	0.549450549	CCDC28B,PEBP1
BAKKEN2016_WGCNA_E80_M06	834	2.757793765	ARF4,BBS10,BBS5,BBS7,CEP120,CEP76,CETN2,CSPP1,EFCAB7,FAM161A, HSPB11,IFT80,MDM1,MLF1,PHF17,PIK3R4,RAB11A,RAB23,RAB28, RPGRIP1L,SPATA7,TTC30A,TUBE1
BAKKEN2016_WGCNA_E80_M07	373	0.536193029	IQCE,TUBGCP6
BAKKEN2016_WGCNA_E80_M08	660	2.272727273	C2CD3,CEP250,CTTN,DCTN1,DNAL4,GSK3A,HTT,IFT122,IFT140, INPP5E,KATNB1,MARK4,NUBP2,PPP2R1A,RAB11B
BAKKEN2016_WGCNA_E80_M09	1096	1.551094891	B9D2,CCDC22,CEP89,CNTROB,FAM58A,GNAS,GPR161,IFT52,MAPRE1, MCRS1,NUBP1,PDE6D,SSNA1,TTC17,TUBB,TUBB2B,WRAP73
BAKKEN2016_WGCNA_E80_M10	835	2.395209581	ARL13B,CCDC14,CEP192,CETN3,GNAI3,HEATR2,IFT88,IQCB1,LCA5, NEK1,POC1B,RP2,RTTN,SASS6,7-SEP,STK38L, STK39,TRIP11,TTC8,WDR35

Continued

Table 1 Ciliopathy and cilia-associated genes in gene expression modules of different cell types in the developing primate brain (related to Fig. 3).—Cont'd

a

Cortical gene expression modules	Total number of genes in module	% Ciliopathy genes in module	List of ciliopathy (confirmed and candidate) genes in module
BAKKEN2016_WGCNA_E80_M11	371	0.808625337	MERTK,MXI1,RSPH1
BAKKEN2016_WGCNA_E80_M12	1952	3.790983607	ALMS1,AURKA,B9D1,BBS2,BBS4,BUB1B,CC2D2A,CENPF,CENPJ, CEP128,CEP152,CEP63,CLUAP1,CRB1,CROCC,DCDC2,DNAH5,DNAI1, DNALI1,DZIP1,EVC2,FGFR1OP,FOXJ1,GLI3,GNAI2,IFT172,IFT74, INPP5B,IQUB,KCNJ13,KIAA0586,KIF14,KIF24,KIF7,LZTFL1,MKS1,MNS1, NEK2,NEK9,NPHP3,ORC1,PACRG,PAFAH1B1,PARD3,PCM1,PIH1D3, PKD2,PLK1,PLK4,POC1A,PTCH1,PTK7,RAB11FIP3,RABL5,RFX2, RILPL2,2-SEP,SMO,SPAG16,SPICE1,STIL,STK3,SYNE2,TEKT1,TEX9, TMEM107,TMEM138,TMEM216,TMEM231,TTC30B,TUBA4A,TULP3, WDR19,ZIC2,ZMYND10
BAKKEN2016_WGCNA_E80_M13	296	1.351351351	OFD1,SDCCAG8,TBX3,WNK1
BAKKEN2016_WGCNA_E90_M00	3082	1.26541207	ADCY6,AIPL1,ANKS6,ANO2,ARL2BP,CEP192,CEP250,CEP89,CEP97, CNGB1,CP110,CRB3,CYS1,FAM161A,HNF1B,INPP5E,INVS,IQCE,KIF7, LRAT,MCHR1,MST1,NIN,NPHP4,OFD1,PKD1,PKHD1,PRKACA,RAB17, RFX3,RPE65,RSG1,SAV1,SDCCAG8,SUFU,TRIP11,TUBA4A,TUBGCP6
BAKKEN2016_WGCNA_E90_M01	1285	1.322957198	AKAP9,CEP170,CEP19,CEP72,GALNT11,KIF2A,KIF3A,KIFAP3,LRRC49, NME7,PRKACB,PRKAR2B,RAB3IP,SNX10,SSX2IP,TAPT1,TMEM17, USP9X
BAKKEN2016_WGCNA_E90_M02	941	0.743889479	KIF17,KIF3B,KIF3C,KIF5C,MKKS,OCRL,RDH12
BAKKEN2016_WGCNA_E90_M03	852	1.76056338	C2CD3,CCDC28B,DCTN1,GPR161,GSK3A,HTT,KATNB1,MARK4, PDE6D,PEBP1,PPP2R1A,RAB11A,TTC17,TUBB,TUBB2B

BAKKEN2016_WGCNA_E90_M04	923	1.19176598	CNTROB,CTTN,DNAL4,GNAS,HSPB11,IFT52,MCRS1,NUBP1,NUBP2,ODF2,RAB11B
BAKKEN2016_WGCNA_E90_M05	1043	2.396931927	ARF4,ARL6,ASAP1,ATXN10,BBS10,BBS5,BBS7,CEP104,CEP120,CEP76,CSPP1,EFCAB7,FAM58A,IFT57,IFT80,MDM1,MLF1,PIK3R4,RAB23,RAB28,SPATA7,STK39,TRIM32,TUBE1,WRAP73
BAKKEN2016_WGCNA_E90_M06	587	1.022146508	CEP290,CEP350,MXI1,PHF17,TTBK2,WDR60
BAKKEN2016_WGCNA_E90_M07	336	0.595238095	MAL,TBX3
BAKKEN2016_WGCNA_E90_M08	363	0.275482094	MERTK
BAKKEN2016_WGCNA_E90_M09	712	2.106741573	B9D2,CCDC22,CETN2,EVC2,GNAI2,IFT140,KIAA0586,MAPRE1,PAFAH1B1,POMGNT1,RAB11FIP3,RABL5,SSNA1,TCTN1,TMEM138
BAKKEN2016_WGCNA_E90_M10	1615	3.343653251	ARL13B,AURKA,BBS2,BUB1B,CC2D2A,CCDC14,CENPF,CENPJ,CEP152,CEP63,CETN3,CRB1,DZIP1,GLI1,GLI3,GNAI3,IFT74,IFT88,INPP5B,IQCB1,KIF14,LCA5,MKS1,MNS1,NEK1,NEK2,NEK9,NPHP3,ORC1,PARD3,PKD2,PLK1,PLK4,POC1A,POC1B,PTCH1,PTK7,PTPDC1,RILPL2,RP2,RTTN,SASS6,2-SEP,7-SEP,SMO,SPICE1,STIL,STK3,STK38L,SYNE2,TEX9,TMEM107,TMEM216,TULP3,ZIC2
BAKKEN2016_WGCNA_E90_M11	454	12.99559471	AHI1,ALMS1,ARMC4,B9D1,BBS4,BBS9,CCDC151,CCNO,CEP128,CEP135,CEP78,CLUAP1,CROCC,DCDC2,DNAH5,DNAH6,DNAI1,DNAL1,DNALI1,EFHC1,FGFR1OP,FOXJ1,HEATR2,HYDIN,IFT122,IFT172,IQUB,KCNJ13,KIF24,KIF27,LRRC6,LZTFL1,MAK,MORN3,PACRG,PCM1,PIH1D3,RFX2,ROPN1L,RPGRIP1L,RSPH1,RSPH3,RSPH4A,SPA17,SPAG1,SPAG16,SPAG6,TCTN3,TEKT1,TMEM231,TMEM67,TPRA1,TTC21A,TTC25,TTC30A,TTC30B,TTC8,WDR19,WDR35,ZMYND10
BAKKEN2016_WGCNA_E90_M12	248	0.403225806	WNK1

Continued

Table 1 Ciliopathy and cilia-associated genes in gene expression modules of different cell types in the developing primate brain (related to Fig. 3).—Cont'd

a

Cortical gene expression modules	Total number of genes in module	% Ciliopathy genes in module	List of ciliopathy (confirmed and candidate) genes in module
BAKKEN2016_WGCNA_E120_M00	3095	1.357027464	ADCY6,AHI1,AIPL1,ANO2,AURKA,C2CD3,CCDC14,CCDC22,CENPJ, CEP104,CEP135,CEP192,CEP350,CEP89,CEP97,CNGB1,CRB3,CTTN, CYS1,FAM161A,FAM58A,GPR161,IFT57,IFT80,KIF17,LRAT,MAPRE1, MCHR1,NIN,ODF2,PAFAH1B1,PKD1,POMGNT1,PRKACA,RAB17, RAB23,RPE65,SASS6,SSNA1,SUFU,SYNE2,TMEM17,TUBGCP6
BAKKEN2016_WGCNA_E120_M01	1673	1.434548715	ARF4,ASAP1,CEP19,CEP72,CEP76,GALNT11,INVS,KIF2A,KIF3A,KIFAP3, LRRC49,MKKS,NME7,PKHD1,PRKACB,PRKAR2B,RAB3IP,RDH12, SNX10,SSX2IP,STK39,TAPT1,TRIM32,USP9X
BAKKEN2016_WGCNA_E120_M02	1182	0.846023689	CCDC28B,IQCE,KIF3B,KIF3C,KIF5C,OCRL,PDE6D,PEBP1,TUBB, TUBB2B
BAKKEN2016_WGCNA_E120_M03	921	1.520086862	CEP250,CNTROB,DCTN1,GNAS,GSK3A,HTT,INPP5E,MARK4,MCRS1, NUBP1,NUBP2,PPP2R1A,RAB11B,WRAP73
BAKKEN2016_WGCNA_E120_M04	1167	1.199657241	AKAP9,ARL6,ATXN10,BBS7,CEP120,CEP170,EFCAB7,PHF17,PIK3R4, RAB11A,RAB28,TTBK2,TTC17,TUBE1
BAKKEN2016_WGCNA_E120_M05	594	1.346801347	DZIP1,KIF14,MAL,ORC1,PLK1,7-SEP,TBX3,TUBA4A
BAKKEN2016_WGCNA_E120_M06	2158	1.343836886	BUB1B,CENPF,CEP63,GLI1,GLI3,GNAI2,GNAI3,KIF7,MERTK,MXI1, NEK9,NPHP3,PACRG,PARD3,PKD2,POC1A,POC1B,PTCH1,PTK7, RAB11FIP3,RILPL2,RP2,2-SEP,SMO, STK3,STK38L,TMEM216,TPRA1,ZIC2

BAKKEN2016_WGCNA_E120_M07	1253	8.858739026	ALMS1,ANKS6,ARL13B,ARL2BP,ARMC4,B9D1,B9D2,BBS10,BBS2,BBS4,BBS5,BBS9,CC2D2A,CCDC151,CCNO,CEP128,CEP152,CEP290,CEP78,CETN2,CETN3,CLUAP1,CP110,CRB1,CROCC,CSPP1,DCDC2,DNAH5,DNAH6,DNAI1,DNAL1,DNAL4,DNALI1,EFHC1,EVC2,FGFR1OP,FOXJ1,HEATR2,HNF1B,HSPB11,HYDIN,IFT122,IFT140,IFT172,IFT52,IFT74,IFT88,INPP5B,IQCB1,IQUB,KATNB1,KCNJ13,KIAA0586,KIF24,KIF27,LCA5,LRRC6,LZTFL1,MAK,MDM1,MKS1,MLF1,MNS1,MORN3,MST1,NEK1,NEK2,NPHP4,OFD1,PCM1,PIH1D3,PLK4,PTPDC1,RABL5,RFX2,RFX3,ROPN1L,RPGRIP1L,RSG1,RSPH1,RSPH3,RSPH4A,RTTN,SAV1,SDCCAG8,SPA17,SPAG1,SPAG16,SPAG6,SPATA7,SPICE1,STIL,TCTN1,TCTN3,TEKT1,TEX9,TMEM107,TMEM138,TMEM231,TMEM67,TRIP11,TTC21A,TTC25,TTC30A,TTC30B,TTC8,TULP3,WDR19,WDR35,WDR60,ZMYND10
BAKKEN2016_WGCNA_E120_M08	398	0.251256281	WNK1
BAKKEN2016_WGCNA_0M_M00	2960	1.959459459	ADCY6,AIPL1,ALMS1,ANKS6,ANO2,ASAP1,B9D1,BBS4,CCDC151,CCNO,CENPF,CENPJ,CEP128,CEP135,CEP250,CEP350,CEP97,CNGB1,CP110,CTTN,DNAH6,FGFR1OP,HSPB11,IFT172,INPP5E,IQUB,KCNJ13,KIF24,LRAT,LRRC6,MAK,MCHR1,MNS1,MORN3,ORC1,PAFAH1B1,PIH1D3,PKD1,PLK4,POC1A,PRKACA,RAB11A,RAB11B,RAB3IP,RILPL2,RSPH4A,SAV1,SDCCAG8,SPAG1,SPAG6,SUFU,TMEM107,TMEM138,TMEM67,TTC17,USP9X,WDR60,ZMYND10
BAKKEN2016_WGCNA_0M_M01	3173	2.111566341	AHI1,ATXN10,AURKA,B9D2,C2CD3,CCDC22,CCDC28B,CEP19,CEP72,CEP76,DCTN1,DNAH5,DNAL4,FAM58A,GALNT11,GPR161,GSK3A,HTT,IFT122,IFT57,INVS,KATNB1,KIF17,KIF2A,KIF3A,KIF3B,KIF3C,KIF5C,KIFAP3,LRRC49,MARK4,MCRS1,MKKS,NEK2,NME7,NPHP4,NUBP1,NUBP2,OCRL,PDE6D,PEBP1,PHF17,PKHD1,PPP2R1A,PRKAR2B,RAB17,RABL5,RDH12,ROPN1L,RSPH1,RSPH3,SNX10,SPA17,SSX2IP,STIL,SYNE2,TCTN1,TCTN3,TEKT1,TMEM17,TRIM32,TTBK2,TUBA4A,TUBB,TUBB2B,TUBGCP6,WRAP73

Continued

Table 1 Ciliopathy and cilia-associated genes in gene expression modules of different cell types in the developing primate brain (related to Fig. 3).—Cont'd

a

Cortical gene expression modules	Total number of genes in module	% Ciliopathy genes in module	List of ciliopathy (confirmed and candidate) genes in module
BAKKEN2016_WGCNA_0M_M02	714	1.120448179	BBS9,CC2D2A,CYS1,DCDC2,DNAL1,GNAS,HNF1B,IQCE
BAKKEN2016_WGCNA_0M_M03	1079	1.853568119	AKAP9,ARF4,ARL6,BBS5,CEP192,CEP78,CETN2,CLUAP1,EFCAB7, EFHC1,FAM161A,IFT52,IFT80,LZTFL1,MLF1,PIK3R4,RAB28,RFX3, RPGRIP1L,SPATA7
BAKKEN2016_WGCNA_0M_M04	783	2.426564496	ARL2BP,BBS10,BBS7,BUB1B,CEP120,CEP170,CEP63,CSPP1,LCA5,MDM1, NIN,PRKACB,7-SEP,SPICE1,TAPT1,TTC30A,TTC8,TUBE1,WDR19
BAKKEN2016_WGCNA_0M_M05	1853	1.726929304	BBS2,CNTROB,CRB3,CROCC,DNAI1,EVC2,GLI3,GNAI2,GNAI3,IFT140, KIAA0586,MST1,NEK1,NEK9,NPHP3,PARD3,PKD2,POC1B,PTCH1,PTK7, RAB11FIP3,RFX2,RPE65,2-SEP,SMO,SPAG16,STK3,STK38L,TMEM216, TTC21A,TULP3,ZIC2
BAKKEN2016_WGCNA_0M_M06	526	1.330798479	DZIP1,GLI1,KIF7,MERTK,MXI1,ODF2,TBX3
BAKKEN2016_WGCNA_0M_M07	1353	3.17812269	ARL13B,ARMC4,CCDC14,CEP104,CEP152,CEP290,CEP89,CETN3,CRB1, DNALI1,FOXJ1,HEATR2,HYDIN,IFT74,IFT88,INPP5B,IQCB1,KIF14, KIF27,MAL,MAPRE1,MKS1,OFD1,PACRG,PCM1,PLK1,POMGNT1, PTPDC1,RAB23,RP2,RSG1,RTTN,SASS6,SSNA1,STK39,TEX9,TMEM231, TPRA1,TRIP11,TTC25,TTC30B,WDR35,WNK1
BAKKEN2016_WGCNA_3M_M00	2279	2.237823607	ADCY6,AIPL1,ALMS1,ANKS6,ANO2,ARMC4,B9D1,BBS4,CENPJ,CEP128, CEP135,CEP350,CEP78,CEP97,CLUAP1,CNGB1,CROCC,CSPP1,EFHC1, FOXJ1,GLI1,HNF1B,HYDIN,IFT74,INPP5B,KCNJ13,KIF14,KIF24,KIF27, KIF7,LRAT,LRRC6,MAK,MORN3,MST1,NIN,ORC1,PIH1D3,PKHD1, PLK1,PRKACA,PTK7,RAB11A,RAB11B,RAB3IP,SAV1,SDCCAG8,SUFU, TEX9,TTC21A,WDR19

BAKKEN2016_WGCNA_3M_M01	2318	1.98446937	AHI1,ARL6,ASAP1,ATXN10,C2CD3,CCNO,CYS1,DNAH6,DNAL1, DNAL4,GNAS,GSK3A,HTT,IFT57,INPP5E,INVS,IQCE,KATNB1,KIF3A, KIF3B,KIF3C,KIF5C,KIFAP3,LRRC49,MARK4,NME7,NPHP4,OCRL, PEBP1,POC1A,RAB17,RSPH1,RSPH3,SNX10,SSX2IP,STIL,TCTN3,TEKT1, TMEM17,TMEM67,TRIM32,TTBK2,TTC17,TUBA4A,TUBB,USP9X
BAKKEN2016_WGCNA_3M_M02	1448	1.726519337	B9D2,BBS9,CCDC151,CCDC22,CEP76,CETN2,DCTN1,FAM58A, FGFR1OP,GALNT11,IFT172,IFT80,LZTFL1,MKKS,NUBP1,NUBP2,PDE6D, ROPN1L,SPA17,SPAG1,SPAG6,SPATA7,TUBGCP6,WDR60,WRAP73
BAKKEN2016_WGCNA_3M_M03	584	2.739726027	AURKA,BBS7,BUB1B,CCDC28B,CENPF,CTTN,GPR161,IFT122,MCRS1, PHF17,PKD1,PPP2R1A,RDH12,SYNE2,TTC25
BAKKEN2016_WGCNA_3M_M04	678	2.212389381	CC2D2A,CEP72,DCDC2,DNAH5,HSPB11,KIF17,MCHR1,MLF1,NEK2, PRKAR2B,PTCH1,RFX3,RSPH4A,TUBB2B,ZMYND10
BAKKEN2016_WGCNA_3M_M05	1227	1.711491443	AKAP9,ARF4,ARL2BP,BBS10,BBS5,CEP120,CEP19,CEP192,EFCAB7, FAM161A,IFT52,KIF2A,MDM1,PAFAH1B1,PIK3R4,PLK4,RAB28,RABL5, RPGRIP1L,TTC8,WDR35
BAKKEN2016_WGCNA_3M_M06	1489	1.813297515	BBS2,CEP250,CRB1,DNAI1,DNALI1,EVC2,GLI3,GNAI2,GNAI3,KIAA0586, MXI1,NEK1,NEK9,NPHP3,PACRG,PARD3,PCM1,RAB11FIP3,RFX2, RSG1,2-SEP,SMO,STK38L,TMEM107,TMEM216,TULP3,ZIC2
BAKKEN2016_WGCNA_3M_M07	604	1.655629139	CNTROB,DZIP1,IFT140,IQUB,MERTK,MKS1,ODF2,RILPL2,TBX3, TPRA1
BAKKEN2016_WGCNA_3M_M08	1814	2.425578831	ARL13B,CCDC14,CEP104,CEP152,CEP170,CEP290,CEP63,CEP89,CETN3, CP110,CRB3,HEATR2,IFT88,IQCB1,LCA5,MAL,MAPRE1,MNS1,OFD1, PKD2,POC1B,POMGNT1,PRKACB,PTPDC1,RAB23,RP2,RPE65,RTTN, SASS6,7-SEP,SPAG16,SPICE1,SSNA1,STK3,STK39,TAPT1,TCTN1, TMEM138,TMEM231,TRIP11,TTC30A,TTC30B,TUBE1,WNK1

Continued

Table 1 Ciliopathy and cilia-associated genes in gene expression modules of different cell types in the developing primate brain (related to Fig. 3).—Cont'd

a

Cortical gene expression modules	Total number of genes in module	% Ciliopathy genes in module	List of ciliopathy (confirmed and candidate) genes in module
BAKKEN2016_WGCNA_12M_M00	2310	1.904761905	AIPL1,ANO2,ARMC4,AURKA,BBS4,BBS9,BUB1B,CENPJ,CEP128,CEP192, CEP350,CEP72,CEP78,CEP97,CLUAP1,CNGB1,DNAH6,GLI1,HNF1B, IFT140,IFT172,KIF14,KIF24,LRRC6,MAK,MKS1,MORN3,NEK1,NIN, ORC1,PIH1D3,PKD1,PKHD1,PLK1,POC1A,PRKACA,RAB11B,RAB3IP, SAV1,SDCCAG8,SSNA1,TCTN1,TMEM138,ZMYND10
BAKKEN2016_WGCNA_12M_M01	2025	1.827160494	AHI1,ANKS6,ASAP1,B9D2,C2CD3,CCDC151,CCDC22,CCNO,CYS1, DCTN1,DNAL4,FAM58A,GALNT11,GNAS,KATNB1,KIF3B,KIF3C,KIF5C, KIF7,KIFAP3,MCRS1,NPHP4,OCRL,PEBP1,PPP2R1A,RAB17,RSPH1, SNX10,SSX2IP,STIL,TCTN3,TRIM32,TTBK2,TTC17,TUBA4A,TUBGCP6, WRAP73
BAKKEN2016_WGCNA_12M_M02	588	2.040816327	CETN2,DNAH5,DNAL1,FGFR1OP,KIF17,MCHR1,PDE6D,PLK4,RABL5, ROPN1L,TEKT1,TUBB
BAKKEN2016_WGCNA_12M_M03	918	2.178649237	AKAP9,ARL6,CEP19,CEP76,EFCAB7,IFT57,INVS,KIF3A,LRRC49,LZTFL1, MKKS,PRKAR2B,RPGRIP1L,RSPH3,SPA17,SPAG6,SPATA7,TMEM17, USP9X,WDR60
BAKKEN2016_WGCNA_12M_M04	615	1.300813008	ADCY6,CCDC28B,DZIP1,GSK3A,HTT,INPP5E,IQCE,MARK4
BAKKEN2016_WGCNA_12M_M05	671	2.682563338	BBS7,CENPF,CTTN,GPR161,IFT122,INPP5B,MDM1,NME7,NUBP1, NUBP2,PHF17,RDH12,RSPH4A,SPAG1,SYNE2,TMEM67,TTC21A,TTC25
BAKKEN2016_WGCNA_12M_M06	289	0.692041522	MERTK,TBX3
BAKKEN2016_WGCNA_12M_M07	431	1.856148492	ATXN10,B9D1,CC2D2A,DCDC2,HSPB11,KCNJ13,NEK2,PTCH1

BAKKEN2016_WGCNA_12M_M08	1773	1.861252115	ALMS1,BBS2,CNTROB,CRB1,CROCC,DNAI1,DNALI1,EVC2,FOXJ1, GLI3,GNAI2,HYDIN,KIAA0586,NPHP3,ODF2,PACRG,PARD3, RAB11FIP3,RFX2,RILPL2,RSG1,2-SEP,SMO,SUFU,TMEM107,TMEM216, TMEM231,TPRA1,TUBB2B,TULP3,ZIC2
BAKKEN2016_WGCNA_12M_M09	1157	2.679343129	ARF4,ARL2BP,BBS10,BBS5,CEP120,CSPP1,EFHC1,FAM161A,IFT52,IFT74, IFT80,IQUB,KIF2A,LCA5,LRAT,MLF1,MNS1,PAFAH1B1,PIK3R4,RAB11A, RAB28,RFX3,RP2,SPICE1,TAPT1,TEX9,TTC30A,TTC8,TUBE1,WDR19, WDR35
BAKKEN2016_WGCNA_12M_M10	1664	2.584134615	ARL13B,CCDC14,CEP104,CEP135,CEP152,CEP170,CEP250,CEP290, CEP63,CEP89,CETN3,CP110,CRB3,GNAI3,HEATR2,IFT88,IQCB1,KIF27, MAL,MAPRE1,MST1,MXI1,NEK9,OFD1,PCM1,PKD2,POC1B,POMGNT1, PRKACB,PTK7,PTPDC1,RAB23,RPE65,RTTN,SASS6,7-SEP,SPAG16, STK3,STK38L,STK39,TRIP11,TTC30B,WNK1
BAKKEN2016_WGCNA_48M_M00	2109	2.038880986	ANO2,ARL2BP,BBS4,BBS7,BBS9,CENPJ,CEP135,CEP192,CEP97,CNGB1, CSPP1,EFCAB7,EFHC1,FGFR1OP,HNF1B,IQUB,KIAA0586,KIF14,KIF24, KIF27,LRAT,MNS1,NEK1,NME7,ORC1,PDE6D,PHF17,PIH1D3,PKHD1, PLK1,PLK4,RAB11A,RAB11B,RAB3IP,RPGRIP1L,SAV1,SDCCAG8,SPAG1, SUFU,TEX9,TMEM138,TMEM67,WDR19
BAKKEN2016_WGCNA_48M_M01	2217	2.390617952	AHI1,AIPL1,AKAP9,ARL6,ASAP1,C2CD3,CCNO,CEP19,CEP76,CYS1, DNAL1,DNAL4,FAM58A,GALNT11,GNAS,GSK3A,HTT,IFT57,INPP5E, INVS,IQCE,KATNB1,KIF3A,KIF3B,KIF3C,KIF5C,KIF7,KIFAP3,LRRC49, LZTFL1,MARK4,MORN3,OCRL,PEBP1,PRKACA,RAB17,ROPN1L, RSPH1,RSPH3,SNX10,SPA17,SPAG6,SPATA7,SSX2IP,STIL,TCTN3, TMEM17,TRIM32,TTBK2,TTC17,TTC25,TUBA4A,USP9X
BAKKEN2016_WGCNA_48M_M02	570	2.280701754	CEP78,CETN2,CLUAP1,DCDC2,DNAH5,DZIP1,IFT52,KIF17,LRRC6, MCHR1,PRKAR2B,RABL5,TEKT1
BAKKEN2016_WGCNA_48M_M03	1382	1.374819103	ADCY6,ANKS6,B9D2,CCDC151,CCDC22,CCDC28B,DCTN1,DNAH6, IFT140,IFT172,INPP5B,MCRS1,NPHP4,PKD1,POC1A,TTC21A,TUBB, TUBGCP6,WRAP73

Continued

Table 1 Ciliopathy and cilia-associated genes in gene expression modules of different cell types in the developing primate brain (related to Fig. 3).—Cont'd

a

Cortical gene expression modules	Total number of genes in module	% Ciliopathy genes in module	List of ciliopathy (confirmed and candidate) genes in module
BAKKEN2016_WGCNA_48M_M04	744	1.747311828	AURKA,CENPF,CTTN,GPR161,IFT122,MDM1,MKKS,NUBP1,NUBP2, PPP2R1A,RDH12,SYNE2,WDR60
BAKKEN2016_WGCNA_48M_M05	433	1.385681293	B9D1,BUB1B,CC2D2A,GLI1,NEK2,ZMYND10
BAKKEN2016_WGCNA_48M_M06	426	2.112676056	ARMC4,BBS2,CROCC,HSPB11,KCNJ13,MERTK,ODF2,PTCH1,TBX3
BAKKEN2016_WGCNA_48M_M07	1630	2.024539877	ALMS1,CEP250,CRB1,DNAI1,DNALI1,EVC2,FOXJ1,GLI3,GNAI2,GNAI3, MAPRE1,MXI1,NIN,NPHP3,PACRG,PARD3,PTK7,RAB11FIP3,RAB23, RFX2,RILPL2,RSG1,2-SEP,SMO,TCTN1,TMEM107,TMEM216,TMEM231, TRIP11,TTC30B,TUBB2B,TULP3,ZIC2
BAKKEN2016_WGCNA_48M_M08	719	1.529902643	ARF4,ATXN10,BBS5,IFT80,MLF1,PIK3R4,RAB28,RFX3,RSPH4A,TUBE1, WDR35
BAKKEN2016_WGCNA_48M_M09	812	3.448275862	ARL13B,BBS10,CCDC14,CEP120,CEP170,CEP290,CEP350,CEP63,CETN3, CP110,FAM161A,HYDIN,IFT74,IFT88,KIF2A,LCA5,MAK,PCM1,POC1B, RP2,RPE65,SASS6,7-SEP,SPICE1,STK38L,TAPT1,TTC30A,TTC8
BAKKEN2016_WGCNA_48M_M10	1399	1.858470336	CEP104,CEP128,CEP152,CEP72,CEP89,CNTROB,CRB3,HEATR2,IQCB1, MAL,MKS1,MST1,NEK9,OFD1,PAFAH1B1,PKD2,POMGNT1,PRKACB, PTPDC1,RTTN,SPAG16,SSNA1,STK3,STK39,TPRA1,WNK1

Modules with significant enrichment of ciliopathy genes are highlighted (adj P values are below)
BAKKEN2016_WGCNA_E40_M07 4.852e-7; BAKKEN2016_WGCNA_E50_M06 7.030e-9; BAKKEN2016_WGCNA_E70_M06 3.500e-7; BAKKEN2016_WGCNA_E80_M12 2.399e-6; BAKKEN2016_WGCNA_E90_M11 3.492e-29; BAKKEN2016_WGCNA_E120_M07 1.004e-41

b

Ciliopathy and cilia associated genes
ADCY3, ADCY6, AHI1, AIPL1, AK7, AKAP9, ALMS1, ANKRD26, ANKS3, ANKS6, ANO2, ARF4, ARL13B, ARL2, ARL2BP, ARL3, ARL6, ARMC4, ASAP1, ATXN10, AURKA, B9D1, B9D2, BBIP1, BBS1, BBS10, BBS12, BBS2, BBS4, BBS5, BBS7, BBS9, BUB1B, C10ORF107, C11ORF49, C21ORF2, C21ORF59, C2CD3, C2ORF71, C5orf42, C8orf37, CATIP, CBY1, CC2D2A, CCDC103, CCDC112, CCDC114, CCDC13, CCDC14, CCDC151, CCDC18, CCDC22, CCDC28B, CCDC37, CCDC39, CCDC40, CCDC63, CCDC65, CCDC66, CCDC88A, CCNO, CDK5RAP2, CDKL5, CENPF, CENPJ, CEP104, CEP120, CEP128, CEP131, CEP135, CEP152, CEP162, CEP164, CEP170, CEP19, CEP192, CEP250, CEP290, CEP350, CEP41, CEP63, CEP72, CEP76, CEP78, CEP83, CEP89, CEP97, CETN1, CETN2, CETN3, CFAP126, CFAP161, CFAP20, CFAP36, CFAP410, CFAP52, CFAP53, CFAP54, CFAP57, CFAP58, CFAP61, CLUAP1, CNGA1, CNGA2, CNGA4, CNGB1, CNTROB, CP110, CPLANE2, CRB1, CRB3, CROCC, CRX, CSNK1D, CSPP1, CTTN, CYS1, DCDC2, DCTN1, DDX59, DISC1, DNAAF1, DNAAF2, DNAAF3, DNAAF4, DNAAF5, DNAH1, DNAH10, DNAH11, DNAH2, DNAH5, DNAH6, DNAH9, DNAI1, DNAI2, DNAJB13, DNAL1, DNAL4, DNALI1, DPCD, DRC1, DRC3, DVL1, DYNC2H1, DYNC2LI1, DYNLT1, DYX1C1, DZIP1, EFCAB7, EFHC1, EFHC2, EVC, EVC2, EXOC4, FAM161A, FAM58A, FBF1, FGFR1OP, FIGNL1, FLCN, FOPNL, FOXJ1, FTO, FUZ, GALNT11, GAS8, GDI2, GLI1, GLI2, GLI3, GLIS2, GNAI2, GNAI3, GNAS, GPBAR1, GPR161, GSK3A, GUCY2D, HEATR2, HNF1B, HSPB11, HTR6, HTT, HYDIN, HYLS1, ICK, IFT122, IFT139, IFT140, IFT172, IFT20, IFT27, IFT43, IFT46, IFT52, IFT57, IFT74, IFT80, IFT81, IFT88, IMPDH1, INPP5B, INPP5E, INTU, INVS, IQCB1, IQCE, IQCG, IQUB, KATNB1, KCNJ13, KIAA0556, KIAA0586, KIAA0753, KIAA1217, KIF11, KIF14, KIF17, KIF19, KIF24, KIF27, KIF2A, KIF3A, KIF3B, KIF3C, KIF5C, KIF7, KIFAP3, KIZ, LCA5, LRAT, LRRC49, LRRC6, LUZP1, LZTFL1, MAK, MAL, MAPRE1, MARK4, MCHR1, MCIDAS, MCRS1, MDM1, MERTK, MKKS, MKS1, MLF1, MNS1, MOK, MORN3, MST1, MXI1, MYB, NEK1, NEK2, NEK4, NEK8, NEK9, NIN, NINL, NME5, NME7, NME8, NMNAT1, NOTO, NPHP1, NPHP2, NPHP3, NPHP4, NPHP5, NPHP7, NPHP9, NUBP1, NUBP2, OCRL, ODF2, OFD1, ORC1, OSR1, PACRG, PAFAH1B1, PARD3, PCARE, PCM1, PCNT, PDE6D, PEBP1, PHF17, PIBF1, PIFO, PIH1D3, PIK3R4, PKD1, PKD2, PKD1L1, PKD2L1, PKHD1, PLK1, PLK4, POC1A, POC1B, POC5, POMGNT1, PPP1R32, PPP2CA, PPP2R1A, PRKACA, PRKACB, PRKAR1A, PRKAR2A, PRKAR2B, PROSER3, PTCH1, PTK7, PTPDC1, RAB11A, RAB11B, RAB11FIP3, RAB17, RAB23, RAB28, RAB29, RAB3IP, RAB8A, RABL5, RD3, RDH12, RFX2, RFX3, RILPL1, RILPL2, ROPN1, ROPN1L, RP1, RP1L1, RP2, RPE65, RPGR, RPGRIP1, RPGRIP1L, RSG1, RSPH1, RSPH14, RSPH3, RSPH4A, RSPH9, RTTN, SASS6, SAV1, SAXO1, SCLT1, SDCCAG8, SEPT2, SEPT7, SLC47A2, SLC4A1, SMO, SNAP25, SNX10, SPA17, SPAG1, SPAG16, SPAG17, SPAG6, SPATA7, SPEF1, SPEF2, SPICE1, SSNA1, SSTR3, SSX2IP, STIL, STK3, STK36, STK38L, STK39, STOML3, STX3, SUFU, SYNE2, TAPT1, TBC1D31, TBC1D32, TBC1D7, TBX3, TCTEX1D2, TCTN1, TCTN2, TCTN3, TEKT1, TEKT2, TEKT3, TEKT4, TEKT5, TEX9, TMEM107, TMEM138, TMEM17, TMEM216 , TMEM231, TMEM237, TMEM67, TMEM80, TOPORS, TPRA1, TRAF3IP1, TRIM32, TRIP11, TTBK2, TTC17, TTC21A, TTC21B, TTC25, TTC26, TTC29, TTC30A, TTC30B, TTC8, TTLL3, TTLL5, TTLL6, TTLL9, TUB, TUBA1A, TUBA1C, TUBA4A, TUBB, TUBB2A, TUBB2B, TUBB3, TUBE1, TUBGCP6, TULP1, TULP3, ULK4, UNC119, USP33, USP9X, VHL, WDPCP, WDR19, WDR34, WDR35, WDR60, WDR62, WDR78, WNK1, WNK4, WRAP73, XPNPEP3, ZIC2, ZMYND10, ZNF423.

Names of the modules from different embryonic and postnatal ages, total number of genes in each module, % of cilia-related genes in the module, and the ciliary genes in each of the modules are listed in table (a). For a listing of all the genes in different modules, see (Bakken et al., 2016). All the cilia related genes used in this analysis are listed in table (b).

Table (b) compiled from Guo, J., Higginbotham, H., Li, J., Nichols, J., Hirt, J., Ghukasyan, V., & Anton, E. S. (2015). Developmental disruptions underlying brain abnormalities in ciliopathies. *Nature Communications, 6(7857)*. doi:10.1038/ncomms8857; Reiter, J. F., & Leroux, M. R. (2017). Genes and molecular pathways underpinning ciliopathies. *Nature Reviews. Molecular Cell Biology, 18(9)*, 533–547. doi:10.1038/nrm.2017.60.

Disruption of primary cilia activity in early radial progenitors via Arl13b deletion disrupt their apical-basal polarity, leading to drastic malformation of the cerebral cortex (Higginbotham et al., 2013). Shh mutations in humans, likely disrupting cilia mediated Shh signaling, lead to holoprosencephaly where defective early patterning of neuroepithelial cells results in the failure of the forebrain to develop into two cerebral cortical hemispheres (Fig. 1B and C; Monuki & Walsh, 2001; Nanni et al., 1999). Further, primary cilia in RGCs regulate the size of the surface of their ventricular apical domains via the mTORC1 pathway. Genetic deletion of primary cilia in RGCs (via conditional deletion of *Ift88* or *Kif3a* in RGCs) leads to misorientation of the mitotic spindle in radial glia, increased basal progenitors, dilation of brain ventricles (ventriculomegaly), and eventual hydrocephalus (Foerster et al., 2017). In contrast to RGCs, the dynamics and significance of primary cilia in intermediate progenitors and outer radial glia remains to be fully defined. In the postnatal germinal niche of the forebrain, the ventricular–subventricular zone, where there is ongoing progenitor proliferation and neurogenesis, the primary cilia is required for neurogenesis (Tong et al., 2014).

Several human ciliopathy mutations that disrupt ciliogenesis in progenitors, primarily via disrupting centrosome or centriolar functions, lead to primary microcephaly (Fig. 1D), where disrupted progenitor proliferation causes a reduction in the generation of appropriate complement of neurons and glia necessary to form a normal cerebral cortex. For example, mutations in centrosome related genes PCNT, PLK4, ASPM, WDR62, CPAP, DNYLT1, and C2CD3 disrupt appropriate patterns of ciliogenesis or cilia maintenance, post cell-division, thus disrupting the ability of progenitors to use primary cilia to sense essential environmental cues present in the CSF (e.g., SHH) or ventricular niche (e.g., IGF) necessary for progenitor differentiation (Bilguvar et al., 2010; Bober & Jackson, 2017; Boczek et al., 2018; Bond et al., 2002; Gabriel et al., 2016; Lehtinen et al., 2011; Li et al., 2011; Nicholas et al., 2010; Thauvin-Robinet et al., 2014). In addition, mutations in genes necessary for primary cilia structural maintenance, trafficking, or signaling, such as KIF2A, RPGRIP1L, Arl13b or Inpp5e, also disrupt the ventricular niche of the developing cerebral cortex and leads to microcephaly or subcortical heterotopia (Bielas et al., 2009; Broix et al., 2018; Cavallin et al., 2017; Higginbotham et al., 2013; Poirier et al., 2013; Uzquiano et al., 2019). Microcephaly in humans is often presented with hydrocephalus. Many of the genes disrupting primary cilia in progenitors can also affect motile cilia in ependymal cells lining the ventricles necessary for CSF flow. This may contribute to defective ependymal cell differentiation, ciliogenesis, and or motile ciliary beating frequency changes in these cells and thus hydrocephaly (Ringers et al., 2020).

Opposite of microcephaly, megalencephaly results from over proliferation of progenitors. Removal of CEP83 in RGCs disrupts distal appendage (DAP) assembly and thus impairs the anchoring of the centrosome to the apical membrane and primary ciliogenesis (Shao et al., 2020). CEP83 deficiency leads to megalencephaly and biallelic mutations in human *CEP83* cause intellectual disability (Failler et al., 2014). Furthermore, an activating mutation of AKT in humans cause hemimegalencephaly (Poduri et al., 2012). Primary cilia activation leads to phospho AKT localization at the base of primary cilium (Christensen, Morthorst, Mogensen, & Pedersen, 2017; Suizu et al., 2016). If and how the primary cilium is affected in human cortical progenitors with AKT mutations and how it contributes to brain overgrowth in these patients remains to be fully examined.

3. Migrating neurons and primary cilium in the developing cerebral cortex

Following their birth in the ventricular zone, newborn neurons migrate several thousand cell soma size distances through a complex cellular milieu to reach their final resting position in different cortical layers and areas. Both radially migrating projection neurons and tangentially migrating interneurons in the cerebral cortex display a dynamic primary cilium. Primary cilia change in length, rotate, branch, and appear to probe their surroundings in migrating inter neurons (Higginbotham et al., 2012). Cilia in neurons localize to the leading process during migration, but move closer to the cell soma during periods of pausing when they became much more dynamic in their movement and length. Electron tomography studies of interneuronal primary cilia further indicate that they cycle between elongated, surface-exposed and shorter, intracellular phases, in coordination with the cell migration stages (Baudoin et al., 2012). The localization and orientation of the primary cilium are also altered depending on the state of migration in migrating neuroblasts of the postnatal rostral migratory stream (Matsumoto et al., 2019). In contrast, signaling defective primary cilia in neurons do not display such dynamism (Higginbotham et al., 2012). Neurons with mutant cilia paused longer during migration and were unable to respond to micro gradients of guidance cues, suggesting that the reduction in cilia dynamics may reflect an inability to efficiently sense guidance cues in the migratory environment (Higginbotham et al., 2012).

Intriguingly, primary cilia activity modulates both radially and tangentially migrating neurons in the cerebral cortex (Baudoin et al., 2012; Guo et al., 2015; Higginbotham et al., 2012; Park et al., 2018). Radial migration involves close intercellular adhesive interactions with radial glial migratory guides, whereas tangential migration occurs for the most part in a cell-substrate independent manner. After they invade the developing cerebral wall, tangentially migrating neurons turn radially, along radial glia processes, to populate the emerging cortical layers (Yokota et al., 2007). Neuronal primary cilia are hypothesized to cell autonomously bind and transduce signals from gradients of guidance molecules, leading to changes in neuronal morphology and movement (Higginbotham et al., 2012). Shh signaling via primary cilia may enable tangentially migrating interneurons to turn radially and populate appropriate cortical plate regions (Baudoin et al., 2012). However, primary cilia may differentially affect the chemotaxic and haptotaxic mechanisms necessary for radial or tangential neuronal migration. Further, it remains untested if primary cilia and the growth cones in the leading processes of migrating neurons transduce different signals related to the directed movement of neurons.

Human genetic mutations of EML1 that perturb anterograde trafficking from the Golgi apparatus to primary cilia and thus the maintenance of primary cilia, lead to subcortical heterotopia, characterized by ectopic radial progenitor localization, disrupted neuronal migration, and aberrant placement of neurons away from their predestined cortical layers (Fig. 2F and G; Uzquiano et al., 2019). *RTTN* mutations associated with dysmorphic primary cilium can cause lissencephaly (Fig. 2E), a cortical neuronal migration defect resulting from an inability to properly maintain neuronal migration during cortical development (Kwan et al., 2012; Shamseldin et al., 2015). *MTOR* mutations affecting autophagy mediated ciliogenesis disrupts radial neuronal migration and leads to focal cortical dysplasia in FMCD patients with somatic mutations in *MTOR* (Di Nardo & Sahin, 2018; Park et al., 2019, 2018). MTOR mutant neurons without primary cilium or severely shortened primary cilium do not undergo normal radial migration and appropriate laminar placement when compared to wild type neurons in the same brain (Park et al., 2018), further demonstrating a role for primary cilia signaling in cortical neuronal migration.

Recent studies combining real time neuronal migration assays with live imaging of second messenger pathways in the ciliary compartment are beginning to provide new insights into how directional information from guidance cues in the developing cortex may be conveyed through the primary cilium to trigger changes in neuronal migration (Stoufflet et al., 2019). In embryonic, postnatal and adult migrating neurons, ciliary Adenylate Cyclase 3 (AC3) activity generates cAMP, which in turn activates centrosomal Protein Kinase A (PKA). Live-imaging with cAMP biosensors reveals a periodic cAMP hotspot at the centrosome, which appears to be necessary for centrosome/nucleus coupling and nucleokinesis of neurons. Delocalization of PKA from the centrosome leads to migratory defects. Primary cilia and its anchoring centrosome may form a cAMP-signaling unit to dynamically regulate neuronal migration (Stoufflet et al., 2019). Considering the cilium's proximity to the nucleus and its link to the centrosome-associated microtubule network, such primary cilia generated cAMP-signaling may promote rapid and efficient local and transcriptional signal transduction underlying nucleokinesis and directional migration of neurons.

4. Dendrites, axons, neuronal connectivity and primary cilium

As neurons arrive at their appropriate laminar locations, they extend dendrites and axons to form appropriate connections with other neurons in cortex and sub cerebral regions (Banker, 2018; Barnes & Polleux, 2009). Axonal pathway defects are a common feature in many human ciliopathies (Guo et al., 2015; Reiter & Leroux, 2017; Valente et al., 2014; Fig. 2J–M). Appropriate axonal growth and pathfinding are essential for the wiring of the developing brain (Chedotal & Richards, 2010; Engle, 2010). Joubert Syndrome Related disorders (JSRD), a subset of ciliopathies, are characterized by consistent and distinct axonal malformations (Brancati, Dallapiccola, & Valente, 2010; Engle, 2010; Parisi, 2009). An axonal tract malformation called the molar tooth sign (MTS) is a diagnostic feature of JSRD. MTS results from thickened, elongated, and horizontally misoriented superior cerebellar peduncles (SCPs) that fail to decussate in the midbrain (Brancati et al., 2010; Juric-Sekhar et al., 2012; Parisi, 2009; Sattar & Gleeson, 2011; Senocak, Oguz, Haliloglu, Topcu, & Cila, 2010). SCPs are formed by the axons of deep cerebellar nuclei (DCN). In addition to SCP malformation, absent or reduced decussation of the corticospinal tract (CST), corpus callosum (CC), and transverse pontine tracts are frequently observed in JSRD patients (Brancati et al., 2010; Engle, 2010; Hildebrandt et al., 2011; Jissendi-Tchofo et al., 2015; Juric-Sekhar et al., 2012; Poretti et al.,

2007; Romani et al., 2013). Consistent with these axonal anomalies and the resultant changes in brain wiring, JSRD patients are often diagnosed with delayed development, intellectual disabilities, autism spectrum disorder (ASD), muscular hypotonia, breathing difficulties, epilepsy, and ataxia (Akizu et al., 2013, 2014; Alvarez Retuerto et al., 2008; Novarino et al., 2011; Reiter & Leroux, 2017). Further, mutations in KIF7, a ciliary kinesin, leads to callosal agenesis or hypoplasia with associated intellectual disabilities (Putoux et al., 2011; Fig. 2J–K). Callosal axon malformations are also seen in ciliopathies such as Oral-facial-digital Syndrome resulting from mutations in OFD1 or TCTN3, necessary for centriole elongation and ciliogenesis or transition zone organization and SHH signaling, respectively (Guo et al., 2015; Thauvin-Robinet et al., 2013; Thomas et al., 2012). These diverse axonal phenotypes related to primary cilia malfunctions in humans strongly indicate the significance of ciliary signaling to axonal tract growth and connectivity.

During the formation of neuronal connectivity in the brain, axons are guided by tropic and trophic gradients of chemoattractant or chemorepellent cues toward their target. Axons extend in large tracts or fascicles and follow stereotyped pathways to reach their appropriate synaptic targets in distant brain regions. Such group extension of axons toward appropriate targets is necessary for neuronal wiring in the brain (Barnes & Polleux, 2009; Chedotal & Richards, 2010; Cionni et al., 2018; Engle, 2010; Paolino, Fenlon, Suárez, & Richards, 2018). The axon-dendrite polarization of neurites and initial extension of axons appear to be unaffected in the absence of primary cilia signaling (Guo et al., 2019). However, branching, fasciculation and crossing of axons and thus their accurate final projections and connectivity are disrupted in cilia mutants (Grochowsky & Gunay-Aygun, 2019; Guo et al., 2015; Parisi, 2019; Poretti, Huisman, Scheer, & Boltshauser, 2011; Qian, Song, & Ming, 2019; Thomas et al., 2019; Ware, Gunay-Aygun, & Hildebrandt, 2011). How does primary cilia signaling selectively modulate the fasciculation, decussation, and targeting of axons?

A motile growth cone that properly responds to guidance cues, and axon-axon interactions that enable axon sorting and fasciculation are both crucial to ensure the fidelity of axonal growth and navigation during development (Nishikimi, Oishi, & Nakajima, 2013; Raper & Mason, 2010; Wang & Marquardt, 2013; Yu & Bargmann, 2001). The axonal growth cone motility and direction relies on the interplay between filopodia that sense environmental cues for pathfinding, and lamellipodia that support persistent axonal outgrowth (Mattila & Lappalainen, 2008; Mejillano et al.,

2004). Growth cones of neurons with aberrant primary cilia show altered filopodia vs. lamellipodia balance (Guo et al., 2019). Further, adhesion molecules such as Pcdh17, essential to mediate axon-axon recognition (Hayashi et al., 2014; Hoshina et al., 2013), fail to localize to axon-axon contacts along cilia mutant axonal shafts (Guo et al., 2019). The combined disruption in axonal growth cone dynamics and axon-axon interactions may thus contribute to the axonal pathfinding and fasciculation defects evident in neurons with defective primary cilia (Guo et al., 2015, 2019). Some of the axonal tract defects in ciliopathies, particularly callosal malformations, may also involve abnormal guidepost cell positioning and associated inappropriate midline crossing (Laclef et al., 2015; Putoux et al., 2019).

Once they reach their appropriate targets, axons establish proper connections by extending terminal axon arbors (Courchet et al., 2013; Gibson & Ma, 2011; Kalil & Dent, 2014; Kalil, Li, & Hutchins, 2011). Upper layer cortical neurons with disrupted cilia grow a long primary axon with significantly reduced terminal arbors at targets in ipsi and contra lateral cortex, thus failing to sufficiently innervate their targets to form appropriate neuronal connections (Guo et al., 2019). Neuron intrinsic properties (e.g. neuronal activity, cAMP and Ca^{2+} gradients) are known to converge onto signaling nodes (e.g. PI3K/AKT/GSK3β, Rho GTPases) to shape axonal growth, branching and targeting (Kalil & Dent, 2014). But how do primary cilia convey such signals to axons?

The primary cilium, with a volume ~1/1000 of the soma (Phua, Lin, & Inoue, 2015), is home to diverse signaling machineries, including receptor tyrosine kinases (e.g., IGFR, EGFR, MET, PDGFR-α, Trk receptors $p75^{NTR}$ and TrkB; (Armato, Chakravarthy, Chiarini, PrÃ, & Whitfield, 2011; Christensen, Clement, Satir, & Pedersen, 2012) and GPCRs (e.g., Smo, Frizzled receptor FZD3, Sstr3, HTR6, NPY2R, NPY5R, DRD1, DRD2, MchR1, GPR161, GPR120 (Choi et al., 2011; Christensen et al., 2012; Green & Mykytyn, 2014; Hilgendorf et al., 2016; Mukhopadhyay & Rohatgi, 2014; Nichols, Floyd, Bruinsma, Narzinski, & Baranski, 2013; Omori et al., 2015; Schou, Pedersen, & Christensen, 2015), their downstream effectors (e.g., PI3 kinase (Franco et al., 2014), Inpp5b, Inpp5e (Conduit, Dyson, & Mitchell, 2012; Jacoby et al., 2009), Ocrl (Phua et al., 2015), AKT (Zhu, Shi, Wang, & Liao, 2009), AC3, 5 and 6 (Phua et al., 2015), and second messengers (e.g., PIP3, PIP2, Ca^{2+} and cAMP; (DeCaen, Delling, Vien, & Clapham, 2015; Delling, DeCaen, Doerner, Febvay, & Clapham, 2013). The primary cilium thus provides a unique environment where signaling components are highly concentrated in a small cellular domain to rapidly trigger and facilitate efficient signaling cascade activation and crosstalk (Gherman, Davis, & Katsanis, 2006; Ishikawa, Thompson, Yates, & Marshall, 2012; Kohli et al., 2017; Narita, Kozuka-Hata, Nonami, et al., 2012; Sigg et al., 2017; Van Dam et al., 2019, 2013; Mick et al., 2015).

Using optogenetic or chemogenetic methods to exert precise control of cilia-specific signaling events, the cellular and physiological impact of the signaling emanating selectively from the primary cilium on developing neurons and neural circuit formation are beginning to be understood (Guo et al., 2017, 2019). Optogenetic or chemogenetic manipulation of ciliary signaling events at multiple levels, including receptor activation (e.g., GPCRs), downstream effector activation (e.g., PI3K, AKT), and second messenger production (i.e., cAMP via ACIII), reveal that exogenously induced deregulation of ciliary signaling cascades rapidly alters growth cone behavior and filopodia/lamellipodia balance, to a large extent recapitulating the aberrant axonal growth cone behavior seen in primary cilia mutant neurons. Conversely, activation of the ciliary-phosphoinositide phosphatase (Inpp5e) pathway, that antagonizes PI3K/AKT signaling, induces an opposite effect, causing axons to retract filopodia and extend lamellipodia. These results suggest that disrupted primary cilia driven PI3K/AKT signaling homeostasis may be an underlying cause of aberrant axonal growth dynamics in cilia mutant neurons and that signals emanating from primary cilia have the capacity to remotely and rapidly regulate axonal growth behavior (Guo et al., 2015, 2019).

Once triggered, PI3K/AKT signaling co-opts downstream kinase cascades and second messengers (e.g. Ca^{2+}, cAMP, IP3, PIP3) to amplify and propagate signals for cell-wide, long-distance signaling (Civelli, 2012; Gavi, Shumay, Wang, & Malbon, 2006; Lemmon & Schlessinger, 2010). Ciliary GPCR signaling can induce axonal Ca^{2+} waves down to the growth cone (Guo et al., 2019). Localized activation of GPCR or PI3K/AKT signaling are known to trigger cell-wide Ca^{2+} waves through Ca^{2+} influx via Ca^{2+} channels on the plasma membrane, IP3-mediated Ca^{2+} release from ER internal stores, and Ca^{2+}/calmodulin signaling mediated by calcium binding proteins such as CaMKII (Averaimo & Nicol, 2014; Henle et al., 2011; Schneider et al., 2008; Zheng & Poo, 2007). Various Ca^{2+} channels localize to primary cilia (e.g., PKD1-L1 and 2-L1; DeCaen et al., 2015; Phua et al., 2015; Pablo, DeCaen, & Clapham, 2017). CaMKIIβ, known to amplify Ca^{2+} signals, is localized to the base of cilia (Puram et al., 2011). Furthermore, Shh-Smo signaling, likely mediated via primary cilia, can downregulate cAMP levels and protein kinase A activity in commissural neurons to allow Sema3-mediated repulsion of commissural axons at midline crossing (Parra & Zou, 2010). Even though the exact mechanisms that enable ciliary signaling to trigger axonal Ca^{2+} wave and its integration with axonal PI3K/AKT signaling waves are yet to be fully elucidated, signaling effectors downstream of PI3K/AKT, such as TSC/mTOR, GSK3β, and CREB, are known to regulate cytoskeleton organization, transcriptional and translational programs involved in neuronal morphogenesis and synaptic plasticity. Particularly, human patients and genetic models with mutations in PI3K/AKT signaling pathway components (e.g. PTEN, AKT1, TSC1/2) show aberrant axonal growth and disorganization of axon tracts (Choi et al., 2008; Geoffroy et al., 2015; Huang, Chen, & Page, 2016; Kwon et al., 2006; Shin et al., 2017; Tsai & Sahin, 2011). Elevated PI3K/AKT activity and altered downstream mTOR, GSK3β and CREB activity were evident in JSRD-related mouse neuronal cilia mutant models (Guo et al., 2017, 2019; Higginbotham et al., 2012). Further, transcriptional networks modulated by JSRD genes such as ARL13B and ZNF423 converge onto neuronal developmental programs including neuronal differentiation, cytoskeleton organization, and neural survival, potentially through AKT signaling network (Guo et al., 2019). These observations raise the intriguing hypothesis that PI3K/AKT signaling may serve as a converging node to integrate ciliary receptor signaling necessary for both rapid axonal growth modulation and long-term neuronal development programs.

Primary cilia signaling is known to modulate dendritic branching and spine formation in cortical neurons, possibly via cAMP and Shh-dependent pathways (Guadiana et al., 2013; Guo et al., 2017; Harwell et al., 2012; Lesiak, Brodsky, Cohenca, Croicu, & Neumaier, 2018). Disrupting primary cilia signaling via overexpression of 5HT6 and SSTR3 in cortical neurons, induces the formation of elongated and branched primary cilia, increases levels of intraflagellar transport proteins such as Kif3a, and reduces dendritic outgrowth and branching. These deficits are rescued when ciliary signaling capacity is restored with ACIII (Guadiana et al., 2013).

Intriguingly, both dendritic spines and cilia share a common function, i.e., to sense and transduce extracellular signals, with cilia being receptive to a broader range of extracellular cues than dendritic spines, which receive input signals primarily from presynaptic neurons and surrounding astrocytes or microglia (Nechipurenko, Doroquez, & Sengupta, 2013: Thion et al., 2018). Despite this commonality, if and how they influence each other remains an open question in neuronal functional biology. Intriguingly, synaptic integration that occurs in dendritic spines appears to require a functional primary cilium. Conditional deletion of cilia in hippocampal neurons induced defective dendritic refinement and synapse formation (Kumamoto et al., 2012). Importantly, unlike major axonal pathway defects detectable in MRI of ciliopathy patient brains, neuronal dendritic disruptions are yet to be systematically probed in genotyped ciliopathy patient brain tissue or organoid models.

5. Primary cilia's impact on neural circuit maintenance and function

In ciliopathies such JSRD, nearly two thirds of the affected individuals function in the intellectually disabled range, indicative of ongoing neural circuit malfunction (Grochowsky & Gunay-Aygun, 2019; Parisi, 2019). Psychomotor delay and epilepsy in subsets of ciliopathy patients further indicates neural circuit disruptions (Canning et al., 2018; Grochowsky & Gunay-Aygun, 2019). DISC1, a major susceptibility gene for psychiatric syndromes, regulates the formation and maintenance of primary cilia and targeting of dopamine D2 receptors to ciliary membrane (Marley & von Castro, 2012; Marley & von Zastrow, 2010). A diverse number of neuropsychiatric risk genes affect the formation or maintenance of primary cilia (Marley & von Castro, 2012). A decrease in the percentage of cells (olfactory neuronal precursors [ONP]) with primary cilia was noticed in schizophrenia (SCZ) and bipolar disorder (BD) patients (Muñoz-Estrada, Lora-Castellanos, Meza, Alarcón Elizaldes, & Benítez-King, 2017). Missense mutations in RTTN or AH1, resulting in short and dysmorphic primary cilia, cause polymicrogyria, characterized by abnormal cortical neuronal organization and circuits (Fig. 1H and I; Dixon-Salazar et al., 2004; Kheradmand Kia et al., 2012). Lithium, a commonly used antimanic mood stabilizer, elongates primary cilia (Miyoshi, Kasahara, Miyazaki, & Asanuma, 2009; Muñoz-Estrada et al., 2017) and is known to rescue brain malformations in mouse models of JSRD ciliopathy (Ferland et al., 2004). Interneurons (INs) with Arl13b deficient primary cilia ($Arl13b^{cKO\text{-}IN}$) display reduced synaptic bouton density and size (Guo et al., 2017). Patch-clamp electrophysiological recordings of interneuron-projection neuron circuits in brains with defective interneuronal primary cilia indicate a decrease in mIPSC frequency in projection neurons (Guo et al., 2017). In contrast, no changes in mIPSC amplitude, mEPSC frequency, or mEPSC amplitude were evident. The specific decrease in the inhibitory synaptic input onto projection neurons in cilia mutant mice ($Arl13b^{cKO\text{-}IN}$), leads to excitatory/inhibitory (E/I) synaptic imbalance in the IN-PN circuit. Intriguingly, accumulation and expression of somatostatin receptor 3 (Sstr3), a neurotransmitter receptor, is disrupted in the primary cilia of Arl13b deficient INs. Normal ciliary signaling involves ectocytosis of activated Sstr3 from ciliary tips (Nager et al., 2017). Disrupted ciliary targeting of neurotransmitter receptors such as Sstr3 may underlie cilia driven E/I imbalance in these mice. Consistent with this possibility, induced expression of Sstr3 in cilia mutant interneurons rescues their synaptic and E/I imbalance deficits (Guo et al., 2017). These observations indicate that primary cilia signaling is a non-synaptic signaling mechanism through which environmental signals can shape and refine interneuronal networks and this mechanism might be compromised in ciliopathy patients with neurobehavioral or epilepsy phenotypes.

6. Glial diversity (astrocytes, oligodendrocytes, microglia), function, and primary cilia in the developing cerebral cortex

Primary cilia functions and signaling in developing glial cells (astrocytes, oligodendendrocytes, and microglia) of the central nervous system remain for the most part uncharacterized. The vast majority of astrocytes in the postnatal cerebral cortex express an individual primary cilium (Kasahara, Miyoshi, Murakami, Miyazaki, & Asanuma, 2014). Primary cilia signaling is necessary for the neurogenic function of astrocyte like precursors in mature hippocampus (Breunig et al., 2008; Han et al., 2009). Altered functional states such as seizure activity shortens astroglial cilia length (Sterpka et al., 2020). In contrast, primary cilia are either absent or severely malformed in aberrantly proliferating, invasive astrocytes in astrocytomas or gliomas (Basten & Giles, 2013; Han et al., 2009; Han & Alvarez-Buylla, 2010; Moser, Fritzler, & Rattner, 2009; Sarkisian & Semple-Rowland, 2019; Seeger-Nukpezah & Golemis, 2012). Primary cilia are present in oligodendrocyte precursor cells and ciliary Shh signaling is critical for the generation of myelinating oligodendrocytes from these precursors (Bhattarai, 2015; Falcon-Urrutia, Carrasco, Lois, Palma, & Roth, 2015; Sanchez & Armstrong, 2018). Although loss of primary cilia in differentiated oligodendrocytes was noticed in vitro (Falcon-Urrutia et al., 2015), the presence of primary cilia in subsets of myelinating mature oligodendrocytes in vivo remains possible (Sanchez & Armstrong, 2018). Immunolabeling with common primary ciliary markers such as ACIII did not detect primary cilia in microglial cells. However, the lack of primary cilia in these cells remains to be fully validated with the use of other established ciliary markers and electron microscopy analysis. Further, comprehensive characterization of the presence and function of primary cilia in different glial cells of the cerebral cortex, spanning the entire spectrum of cortical development, is necessary to fully evaluate the functional impact of glial primary cilia in cerebral cortex.

7. Future directions: Primary cilia's role in brain development and neurodevelopmental disorders

The primary cilium is structured and positioned to be able to generate a fast and robust response to a wide range of extracellular signals present in the cellular milieu of the developing brain. The diffusion barrier at the base of the cilium allows a cell to maximize the local concentration of a plethora of signaling receptors and their downstream effectors within cilia. The design of the primary cilium enables high levels of second messengers to be sustained more easily within the narrow geometry of the cilium than in other regions of the cell where diffusion through a larger space could lead to a rapid diminishment of the signal. The primary cilium's proximity to the nucleus and its link to the centrosome-associated microtubule network may promote rapid and efficient intracellular signal transduction. Understanding in real time how signaling emanating from primary cilia is conveyed to far flung domains of a neuron, including, dendritic, spines, axonal growth cones, AIS, and nucleus (or similarly, between primary cilium and distinct cell domains in non-neural cell types of the CNS) will be essential to fully evaluate the cell biological significance of this sensory organelle in nervous system development and disease (Fig. 4). Further, understanding the cross-talk that occurs between intracellular signaling cascades that originate in distinct cellular compartments (e.g., primary cilium, growth cone, dendritic spine, glial end feet etc.) of neural cells during development will help define the primary cilia-specific pathways vital for normal CNS development.

Considering that the vast majority of ciliopathies have their origins in primary cilia malfunction during embryonic development (Fig. 3), better modeling of cilia dysfunction and rescue in relevant human models, such as iPSC derived brain organoids, or exploration of CRISPR/Cas9-based targeted editing of ciliary gene defects *in utero* during development will help open up new therapeutic strategies to address ciliopathies at their origin.

Centrosomes play an essential role in ciliogenesis and cilia maintenance. Often human mutations affecting primarily the centrosomes lead

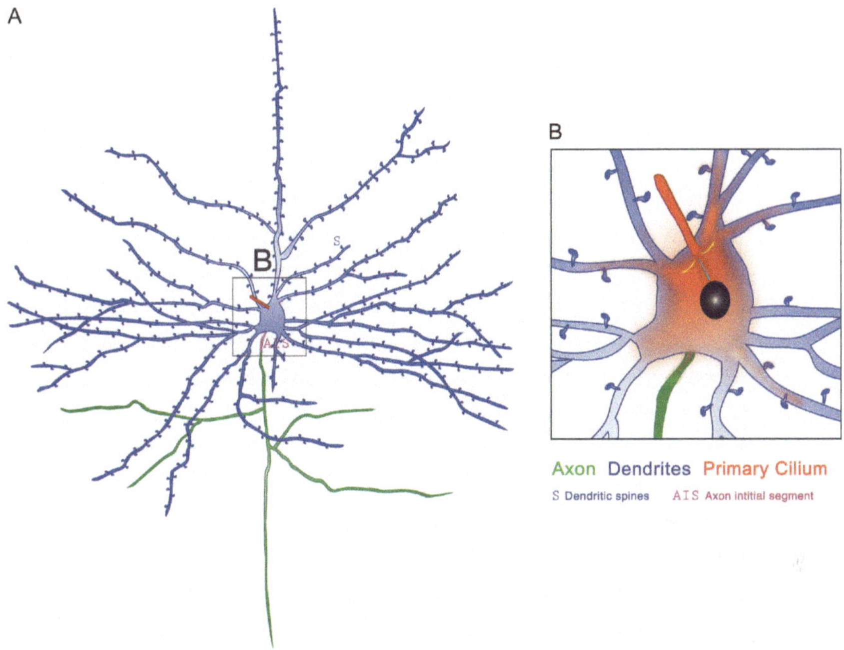

Fig. 4 Understanding the functional importance of primary cilia in neurons. Interrogating the selective effect of signaling emanating from primary cilia on axons and dendrites and defining how signaling spreads from the primary cilium to distal domains of neurons (red gradient) and to the nucleus (cyan arrow) to influence neuronal activity and behavior are challenges yet to be met. A similar dearth of knowledge exists for the importance of primary cilia signaling during glial cell development and behavior.

to secondary structural and functional changes in primary cilia, causing a spectrum of outcomes not entirely attributable to primary cilia signaling. Further development of unbiased screening assays for the discovery and interrogation of mutations that exclusively affect the signaling function of neuronal or glial primary cilium and its underpinning cilia-specific machinery will shed new light on the functional importance of primary cilia in the nervous system.

Neuronal or glial primary cilia may produce a "ciliary signaling signature" that elicits cilia-specific impact on neuronal or glial events such as calcium waves, second messenger cascades, cytoskeletal rearrangements, adhesion, and transcriptional regulation necessary for proper neuronal or glial generation, growth, differentiation, and interactions. In this regard, examining if primary cilia in distinct cell types are as molecularly specified as the cell types they subserve, will be highly informative. Future efforts to decipher the precise molecular mediators of communication between neuronal or glial cilia and their cellular environment in the context of brain development will enable us to decisively delineate how primary cilia convey environmental signals to modulate brain formation and organization. Since ciliary proteins, such as Arl13b, can also exert cilia-independent functions in neural cells (Ferent et al., 2019), a vigilant approach that excludes non-ciliary functions of cilia localized proteins should be part of this effort.

The neural circuit malfunctions associated with ciliopathies highlight the importance of investigating the possible participation of primary ciliary signaling in experience-driven refinement of cortical neuronal connectivity during critical periods or during learning and memory (Baraban et al., 2009; Chattopadhyaya et al., 2004; Southwell, Froemke, Alvarez-Buylla, Stryker, & Gandhi, 2010). Intriguingly, IGF1, a ligand for Igf1 R receptors in cilia (Higginbotham et al., 2013; Lehtinen et al., 2011), has been identified as a sensory experience regulated gene necessary to sculpt the synaptic connections of vasoactive intestinal peptide INs (Mardinly et al., 2016). In neurodegenerative disorders such as Huntington disease (HD), pathogenic polyQ expansion of huntingtin (HTT) protein causes centrosomal accumulation of PCM1 and abnormally long primary cilia in striatal cells, contributing to dysregulated circuit homeostasis in this disease (Keryer et al., 2011). It is thus conceivable that ciliary signaling is continuously required in mature neurons for the maintenance and modulation of neuronal network homeostasis. Moreover, the source and the molecular identity of the various environmental cues, including neurotransmitters or neuromodulators, that can initiate or modulate primary cilia signaling in mature neurons, and whether such signaling enables neurons to function cooperatively (Karnani et al., 2016) to modulate the dynamics of a circuit underlying specific functions or behavior are yet to be fully characterized. In this regard, the role of primary cilia signaling via ciliary melanocortin-4 receptors (MC4R) in hypothalamic neurons to regulate food intake and body weight provides an informative example (Siljee et al., 2018). Future efforts aimed at decisively resolving the above outlined issues will further our understanding of the fundamentally vital modulatory role primary cilia play in cerebral cortical formation, function, and disorders.

Acknowledgments

This research was supported by NIH grants MH060929 and NS to EA and the confocal imaging core of an NINDS institutional center core grant (5P30NS045892).

Declaration of competing financial interests

The authors declare no competing financial interests.

References

Akizu, N., Cantagrel, V., Schroth, J., Cai, N., Vaux, K., McCloskey, D., et al. (2013). AMPD2 regulates GTP synthesis and is mutated in a potentially treatable neurodegenerative brainstem disorder. *Cell, 154*(3), 505–517. https://doi.org/10.1016/j.cell.2013.07.005.

Akizu, N., Silhavy, J. L., Rosti, R. O., Scott, E., Fenstermaker, A. G., Schroth, J., et al. (2014). Mutations in CSPP1 lead to classical Joubert syndrome. *American Journal of Human Genetics, 94*(1), 80–86. https://doi.org/10.1016/j.ajhg.2013.11.015.

Alvarez Retuerto, A. I., Cantor, R. M., Gleeson, J. G., Ustaszewska, A., Schackwitz, W. S., Pennacchio, L. A., et al. (2008). Association of common variants in the Joubert syndrome gene (AHI1) with autism. *Human Molecular Genetics, 17*(24), 3887–3896. https://doi.org/10.1093/hmg/ddn291.

Arellano, J. I., Guadiana, S. M., Breunig, J. J., Rakic, P., & Sarkisian, M. R. (2017). Development and distribution of neuronal cilia in mouse neocortex. *The Journal of Comparative Neurology, 520*, 848–873. https://doi.org/10.1002/cne.22793.

Armato, U., Chakravarthy, B., Chiarini, A., PrÃ, I. D., & Whitfield, J. F. (2011). A paradigm-changing surprise from dentate gyrus granule cells—cilium-localized p75NTR may drive their progenitor cell proliferation. *Journal of Alzheimer's Disease & Parkinsonism, 1*(2), 1–3. https://doi.org/10.4172/2161-0460.1000e104. e104.

Averaimo, S., & Nicol, X. (2014). Intermingled cAMP, cGMP and calcium spatiotemporal dynamics in developing neuronal circuits. *Frontiers in Cellular Neuroscience, 8*, 376. https://doi.org/10.3389/fncel.2014.00376.

Bakken, T. E., Miller, J. A., Ding, S. L., Sunkin, S. M., Smith, K. A., Ng, L., et al. (2016). A comprehensive transcriptional map of primate brain development. *Nature, 535*(7612), 367–375. https://doi.org/10.1038/nature18637.

Banker, G. (2018). The development of neuronal polarity: A retrospective view. *International Journal of Neuroscience, 38*, 1867–1873. United States: 2018 the authors 0270-6474/18/381867-07$15.00/0.

Baraban, S. C., Southwell, D. G., Estrada, R. C., Jones, D. L., Sebe, J. Y., Alfaro-Cervello, C., et al. (2009). Reduction of seizures by transplantation of cortical GABAergic interneuron precursors into Kv1.1 mutant mice. *Proceedings of the National Academy of Sciences of the United States of America, 106*, 15472–15477.

Barnes, A. P., & Polleux, F. (2009). Establishment of axon-dendrite polarity in developing neurons. *Annual Review of Neuroscience, 32*, 347–381.

Basten, S. G., & Giles, R. H. (2013). Functional aspects of primary cilia in signaling, cell cycle and tumorigenesis. *Cilia, 2*(1), 6.

Baudoin, J.-P., Viou, L., Launay, P.-S., Luccardini, C., Espeso Gil, S., Kiyasova, V., et al. (2012). Tangentially migrating neurons assemble a primary cilium that promotes their reorientation to the cortical plate. *Neuron, 76*(6), 1108–1122. https://doi.org/10.1016/j.neuron.2012.10.027.

Bhattarai, S. R. (2015). *Characteristics of primary cilia and centrosomes in neuronal and glial lineages of the adult brain*. Denton, Texas: University of North Texas.

Bielas, S. L., Silhavy, J. L., Brancati, F., Kisseleva, M. V., Al-Gazali, L., Sztriha, L., et al. (2009). Mutations in INPP5E, encoding inositol polyphosphate-5-phosphatase E, link phosphatidyl inositol signaling to the ciliopathies. *Nature Genetics, 41*(9), 1032–1036.

Bilguvar, K., Ozturk, A. K., Louvi, A., Kwan, K. Y., Choi, M., Tatli, B., et al. (2010). Whole-exome sequencing identifies recessive WDR62 mutations in severe brain malformations. *Nature, 467*(7312), 207–210. https://doi.org/10.1038/nature09327.

Bishop, G. A., Berbari, N. F., Lewis, J., & Mykytyn, K. (2007). Type III adenylyl cyclase localizes to primary cilia throughout the adult mouse brain. *The Journal of Comparative Neurology*, *505*, 562–571. https://doi.org/10.1002/cne.21510.

Bloodgood, R. A. (2009). From central to rudimentary to primary: The history of an under-appreciated organelle whose time has come. The primary cilium. *Methods in Cell Biology*, *94*, 3–52. https://doi.org/10.1016/S0091-679X(08)94001-2.

Bober, M. B., & Jackson, A. P. (2017). Microcephalic osteodysplastic primordial dwarfism, type II: a clinical review. *Current Osteoporosis Reports*, *15*(2), 61–69. https://doi.org/10.1007/s11914-017-0348-1.

Boczek, N. J., Hopp, K., Benoit, L., Kraft, D., Cousin, M. A., Blackburn, P. R., et al. (2018). Characterization of three ciliopathy pedigrees expands the phenotype associated with biallelic C2CD3 variants. *European Journal of Human Genetics*, *26*, 1797–1809. https://doi.org/10.18926/AMO/53020.

Bond, J., Roberts, E., Mochida, G. H., Hampshire, D. J., Scott, S., Askham, J. M., et al. (2002). ASPM is a major determinant of cerebral cortical size. *Nature Genetics*, *32*, 316–320. United States.

Brancati, F., Dallapiccola, B., & Valente, E. (2010). Joubert syndrome and related disorders. *Orphanet Journal of Rare Diseases*, *5*(1), 20. https://doi.org/10.1186/1750-1172-5-20.

Breunig, J. J., Sarkisian, M. R., Arellano, J. I., Morozov, Y. M., Ayoub, A. E., Sojitra, S., et al. (2008). Primary cilia regulate hippocampal neurogenesis by mediating sonic hedgehog signaling. *Proceedings of the National Academy of Sciences of the United States of America*, *105*(35), 13127–13132. https://doi.org/10.1073/pnas.0804558105.

Broix, L., Asselin, L., Silva, C. G., Ivanova, E. L., Tilly, P., Gilet, J. G., et al. (2018). Ciliogenesis and cell cycle alterations contribute to KIF2A-related malformations of cortical development. *Human Molecular Genetics*, *27*, 224–238. England: The Author 2017. Published by Oxford University Press For Permissions, please email: journals.permissions@oup.com.

Canning, P., Park, K., Goncalves, J., Li, C., Howard, C. J., Sharpe, T. D., et al. (2018). CDKL family kinases have evolved distinct structural features and ciliary function. *Cell Reports*, *22*(4), 885–894. https://doi.org/10.1016/j.celrep.2017.12.083.

Caspary, T., Larkins, C. E., & Anderson, K. V. (2007). The graded response to sonic hedgehog depends on cilia architecture. *Developmental Cell*, *12*, 767–778. United States.

Cavallin, M., Bijlsma, E. K., El Morjani, A., Moutton, S., Peeters, E. A., Maillard, C., et al. (2017). Recurrent KIF2A mutations are responsible for classic lissencephaly. *Neurogenetics*, *18*, 73–79. United States.

Chattopadhyaya, B., Di Cristo, G., Higashiyama, H., Knott, G. W., Kuhlman, S. J., Welker, E., et al. (2004). Experience and activity-dependent maturation of perisomatic GABAergic innervation in primary visual cortex during a postnatal critical period. *Journal of Neuroscience*, *24*, 9598–9611.

Chedotal, A., & Richards, L. J. (2010). Wiring the brain: The biology of neuronal guidance. *Cold Spring Harbor Perspectives in Biology*, *2*(6), a001917. https://doi.org/10.1101/cshperspect.a001917.

Choi, Y. J., Di Nardo, A., Kramvis, I., Meikle, L., Kwiatkowski, D. J., Sahin, M., et al. (2008). Tuberous sclerosis complex proteins control axon formation. *Genes & Development*, *22*(18), 2485–2495. https://doi.org/10.1101/gad.1685008.

Choi, Y.-H., Suzuki, A., Hajarnis, S., Ma, Z., Chapin, H. C., Caplan, M. J., et al. (2011). Polycystin-2 and phosphodiesterase 4C are components of a ciliary A-kinase anchoring protein complex that is disrupted in cystic kidney diseases. *Proceedings of the National Academy of Sciences of the United States of America*, *108*(26), 10679–10684. https://doi.org/10.1073/pnas.1016214108.

Christensen, S. T., Clement, C. A., Satir, P., & Pedersen, L. B. (2012). Primary cilia and coordination of receptor tyrosine kinase (RTK) signalling. *The Journal of Pathology*, *226*(2), 172–184. https://doi.org/10.1002/path.3004.

Christensen, S. T., Morthorst, S. K., Mogensen, J. B., & Pedersen, L. B. (2017). Primary cilia and coordination of receptor tyrosine kinase (RTK) and transforming growth factor β (TGF-β) signaling. *Cold Spring Harbor Perspectives in Biology*. https://doi.org/10.1101/cshperspect.a028167.

Cionni, J. M., Wong, H. H.-W., Bressan, D., Kodama, L., Harris, W. A., & Holt, C. E. (2018). Axon-axon interactions regulate topographic optic tract sorting via CYFIP2-dependent WAVE complex function. *Neuron, 97*, 1078–1093. https://doi.org/10.1016/j.neuron.2018.01.027.

Civelli, O. (2012). Orphan GPCRs and neuromodulation. *Neuron, 76*(1), 12–21. https://doi.org/10.1016/j.neuron.2012.09.009.

Conduit, S. E., Dyson, J. M., & Mitchell, C. A. (2012). Inositol polyphosphate 5-phosphatases; new players in the regulation of cilia and ciliopathies. *FEBS Letters, 586*(18), 2846–2857. https://doi.org/10.1016/j.febslet.2012.07.037.

Courchet, J., Lewis, T. L., Jr., Lee, S., Courchet, V., Liou, D.-Y., Aizawa, S., et al. (2013). Terminal axon branching is regulated by the LKB1-NUAK1 kinase pathway via presynaptic mitochondrial capture. *Cell, 153*(7), 1510–1525. https://doi.org/10.1016/j.cell.2013.05.021.

DeCaen, P. G., Delling, M., Vien, T. N., & Clapham, D. E. (2015). Direct recording and molecular identification of the calcium channel of primary cilia. *Nature, 504*(7479), 315–318. https://doi.org/10.1038/nature12832.

Delgehyr, N., & Spassky, N. (2014). Interplay between primary cilia and cell cycle. *Medecine Sciences, 30*, 976–979. France: 2014 medecine/sciences - Inserm.

Delling, M., DeCaen, P. G., Doerner, J. F., Febvay, S., & Clapham, D. E. (2013). Primary cilia are specialized calcium signalling organelles. *Nature, 504*, 311–314.

Di Nardo, A., & Sahin, M. (2018). mTOR'ing across the cortex by chopping the cilia. *Neuron, 99*(1), 3–5. https://doi.org/10.1016/j.neuron.2018.06.042.

Dixon-Salazar, T., Silhavy, J. L., Marsh, S. E., Louie, C. M., Scott, L. C., Gururaj, A., et al. (2004). Mutations in the AHI1 gene, encoding jouberin, cause Joubert syndrome with cortical polymicrogyria. *American Journal of Human Genetics, 75*(6), 979–987. https://doi.org/10.1086/425985.

Doherty, D. (2009). Joubert Syndrome: Insight into brain development, cilium biology, and complex disease. *Seminars in Pediatric Neurology, 16*(3), 144–154. https://doi.org/10.1016/j.spen.2009.06.002.

Ecker, A. (1844). Flimmerbewegung im Gehörorgan von Petromyzon marinus. *Archiv für Anatomie, Physiologie und wissenschaftliche Medicin, 1*, 520–521.

Engle, E. C. (2010). Human genetic disorders of axon guidance. *Cold Spring Harbor Perspectives in Biology, 2*(3), a001784. https://doi.org/10.1101/cshperspect.a001784.

Failler, M., Gee, H., Krug, P., Joo, K., Halbritter, J., Belkacem, L., et al. (2014). Mutations of CEP83 cause infantile nephronophthisis and intellectual disability. *American Journal of Human Genetics, 94*(6), 905–915. https://doi.org/10.1016/j.ajhg.2014.05.002.

Falcon-Urrutia, P., Carrasco, C. M., Lois, P., Palma, V., & Roth, A. D. (2015). Shh signaling through the primary cilium modulates rat oligodendrocyte differentiation. *PLoS One, 10*(7). https://doi.org/10.1371/journal.pone.0133567.

Ferent, J., Constable, S., Gigante, E. D., Yam, P. T., Mariani, L. E., Legué, E., et al. (2019). The ciliary protein Arl13b functions outside of the primary cilium in Shh-mediated axon guidance. *Cell Reports, 29*(11), 3356–3366.e3. https://doi.org/10.1016/j.celrep.2019.11.015.

Ferland, R., Eyaid, W., Collura, R., Tully, L., Hill, R., Al-Nouri, D., et al. (2004). Abnormal cerebellar development and axonal decussation due to mutations in AHI1 in Joubert syndrome. *Nature Genetics, 36*(9), 1008–1013. https://doi.org/10.1038/ng1419.

Fliegauf, M., Benzing, T., & Omran, H. (2007). When cilia go bad: Cilia defects and ciliopathies. *Nature Reviews Molecular Cell Biology, 8*, 880–893. England.

Foerster, P., Daclin, M., Asm, S., Faucourt, M., Boletta, A., Genovesio, A., et al. (2017). mTORC1 signaling and primary cilia are required for brain ventricle morphogenesis. *Development, 144*(2), 201–210. https://doi.org/10.1242/dev.138271.

Franco, I., Gulluni, F., Campa, C. C., Costa, C., Margaria, J. P., Ciraolo, E., et al. (2014). PI3K class II alpha controls spatially restricted endosomal PtdIns3P and Rab11 activation to promote primary cilium function. *Developmental Cell, 28*(6), 647–658. https://doi.org/10.1016/j.devcel.2014.01.022.

Gabriel, E., Wason, A., Ramani, A., Gooi, L. M., Keller, P., Pozniakovsky, A., et al. (2016). CPAP promotes timely cilium disassembly to maintain neural progenitor pool. *The EMBO Journal, 35*(8), 803–819. https://doi.org/10.15252/embj.201593679.

Gavi, S., Shumay, E., Wang, H.-y., & Malbon, C. C. (2006). G-protein-coupled receptors and tyrosine kinases: Crossroads in cell signaling and regulation. *Trends in Endocrinology & Metabolism, 17*(2), 48–54. https://doi.org/10.1016/j.tem.2006.01.006.

Geoffroy, C. G., Lorenzana, A. O., Kwan, J. P., Lin, K., Ghassemi, O., Ma, A., et al. (2015). Effects of PTEN and nogo codeletion on corticospinal axon sprouting and regeneration in mice. *The Journal of Neuroscience : The Official Journal of the Society for Neuroscience, 35*(16), 6413–6428. https://doi.org/10.1523/jneurosci.4013-14.2015.

Gherman, A., Davis, E. E., & Katsanis, N. (2006). The ciliary proteome database: An integrated community resource for the genetic and functional dissection of cilia. *Nature Genetics, 38*(9), 961–962. https://doi.org/10.1038/ng0906-961.

Gibson, D. A., & Ma, L. (2011). Developmental regulation of axon branching in the vertebrate nervous system. *Development, 138*(2), 183–195. https://doi.org/10.1242/dev.046441.

Green, J. A., & Mykytyn, K. (2014). neuronal primary cilia: an under appreciated signaling and sensory organelle in the brain. *Neuropsychopharmacology, 39*, 244–245. https://doi.org/10.1038/npp.2013.203.

Grochowsky, A., & Gunay-Aygun, M. (2019). Clinical characteristics of individual organ system disease in non-motile ciliopathies. *Translational Science of Rare Diseases, 4*(1-2), 1–23. https://doi.org/10.3233/TRD-190033.

Guadiana, S. M., Semple-Rowland, S., Daroszewski, D., Madorsky, I., Breunig, J. J., Mykytyn, K., et al. (2013). Arborization of dendrites by developing neocortical neurons is dependent on primary cilia and type 3 adenylyl cyclase. *The Journal of Neuroscience, 33*(6), 2626–2638. https://doi.org/10.1523/jneurosci.2906-12.2013.

Guemez-Gamboa, A., Coufal, N. G., & Gleeson, J. G. (2014). Primary cilia in the developing and mature brain. *Neuron, 82*(3), 511–521.

Guo, J., Higginbotham, H., Li, J., Nichols, J., Hirt, J., Ghukasyan, V., et al. (2015). Developmental disruptions underlying brain abnormalities in ciliopathies. *Nature Communications, 6*(7857). https://doi.org/10.1038/ncomms8857.

Guo, J., Otis, J. M., Higginbotham, H., Monckton, C., Cheng, J., Asokan, A., et al. (2017). Primary cilia signaling shapes the development of interneuronal connectivity. *Developmental Cell, 42*(3), 286–300. e284.

Guo, J., Otis, J., Suciu, S., Catalano, C., Xing, L., Constable, S., et al. (2019). Primary cilia signaling promotes axonal tract development and is disrupted in Joubert syndrome-related disorders models. *Developmental Cell, 51*, 759–774. https://doi.org/10.1016/j.devel.2019.11.005.

Han, Y. G., & Alvarez-Buylla, A. (2010). Role of primary cilia in brain development and cancer. *Current Opinion in Neurobiology, 20*(1), 58–67. https://doi.org/10.1016/j.conb.2009.12.002.

Han, Y. G., Kim, H. J., Dlugosz, A. A., Ellison, D. W., Gilbertson, R. J., & Alvarez-Buylla, A. (2009). Dual and opposing roles of primary cilia in medulloblastoma development. *Nature Medicine, 15*(9), 1062–1065. https://doi.org/10.1038/nm.2020.

Harwell, C. C., Parker, P. R., Gee, S. M., Okada, A., McConnell, S. K., Kreitzer, A. C., et al. (2012). Sonic hedgehog expression in corticofugal projection neurons directs cortical microcircuit formation. *Neuron*, *73*(6), 1116–1126. https://doi.org/10.1016/j.neuron.2012.02.009.

Hayashi, S., Inoue, Y., Kiyonari, H., Abe, T., Misaki, K., Moriguchi, H., et al. (2014). Protocadherin-17 mediates collective axon extension by recruiting actin regulator complexes to interaxonal contacts. *Developmental Cell*, *30*(6), 673–687. https://doi.org/10.1016/j.devcel.2014.07.015.

Henle, S. J., Wang, G., Liang, E., Wu, M., Poo, M.m., & Henley, J. R. (2011). Asymmetric PI(3,4,5)P3 and Akt signaling mediates chemotaxis of axonal growth cones. *Journal of Neuroscience*, *31*(19), 7016–7027. https://doi.org/10.1523/jneurosci.0216-11.2011.

Higginbotham, H., Eom, T. Y., Mariani, L. E., Bachleda, A., Hirt, J., Gukassyan, V., et al. (2012). Arl13b in primary cilia regulates the migration and placement of interneurons in the developing cerebral cortex. *Developmental Cell*, *23*(5), 925–938.

Higginbotham, H., Guo, J., Yokota, Y., Umberger, N., Su, C., Li, J., et al. (2013). Arl13b-regulated cilia activities are essential for polarized radial glial scaffold formation. *Nature Neuroscience*, *16*(8), 1000–1007. https://doi.org/10.1038/nn.3451.

Hildebrandt, F., Benzing, T., & Katsanis, N. (2011). Ciliopathies. *The New England Journal of Medicine*, *364*(16), 1533–1543. https://doi.org/10.1056/nejmra1010172.

Hilgendorf, K. I., Johnson, C. T., & Jackson, P. K. (2016). The primary cilium as a cellular receiver: organizing ciliary GPCR signaling. *Current Opinion in Cell Biology*, *39*, 84–92. https://doi.org/10.1016/j.ceb.2016.02.008.

Hoshina, N., Tanimura, A., Yamasaki, M., Inoue, T., Fukabori, R., Kuroda, T., et al. (2013). Protocadherin 17 regulates presynaptic assembly in topographic corticobasal ganglia circuits. *Neuron*, *78*(5), 839–854. https://doi.org/10.1016/j.neuron.2013.03.031.

Huang, W. C., Chen, Y., & Page, D. T. (2016). Hyperconnectivity of prefrontal cortex to amygdala projections in a mouse model of macrocephaly/autism syndrome. *Nature Communications*, 7. https://doi.org/10.1038/ncomms13421, 13421.

Ishikawa, H., Thompson, J., Yates, J. R., 3rd, & Marshall, W. F. (2012). Proteomic analysis of mammalian primary cilia. *Current Biology*, *22*(5), 414–419. https://doi.org/10.1016/j.cub.2012.01.031.

Jacoby, M., Cox, J. J., Gayral, S., Hampshire, D. J., Ayub, M., Blockmans, M., et al. (2009). INPP5E mutations cause primary cilium signaling defects, ciliary instability and ciliopathies in human and mouse. *Nature Genetics*, *41*(9), 1027–1031. https://doi.org/10.1038/ng.427.

Jissendi-Tchofo, P., Severino, M., Nguema-Edzang, B., Toure, C., Soto Ares, G., & Barkovich, A. J. (2015). Update on neuroimaging phenotypes of mid-hindbrain malformations. *Neuroradiology*, *57*(2), 113–138. https://doi.org/10.1007/s00234-014-1431-2.

Juric-Sekhar, G., Adkins, J., Doherty, D., & Hevner, R. F. (2012). Joubert syndrome: Brain and spinal cord malformations in genotyped cases and implications for neurodevelopmental functions of primary cilia. *Acta Neuropathologica*, *123*(5), 695–709. https://doi.org/10.1007/s00401-012-0951-2.

Kalil, K., & Dent, E. W. (2014). Branch management: mechanisms of axon branching in the developing vertebrate CNS. *Nature Reviews. Neuroscience*, *15*(1), 7–18. https://doi.org/10.1038/nrn3650.

Kalil, K., Li, L., & Hutchins, B. I. (2011). Signaling mechanisms in cortical axon growth, guidance and branching. *Frontiers in Neuroanatomy*, *5*, 62. https://doi.org/10.3389/fnana.2011.00062.

Karnani, M. M., Jackson, J., Ayzenshtat, I., Tucciarone, J., Manoocheri, K., Snider, W. G., et al. (2016). Cooperative subnetworks of molecularly similar interneurons in mouse neocortex. *Neuron*, *90*(1), 86–100. https://doi.org/10.1016/j.neuron.2016.02.037.

Kasahara, K., Miyoshi, K., Murakami, S., Miyazaki, I., & Asanuma, M. (2014). Visualization of astrocytic primary cilia in the mouse brain by immunofluorescent analysis using the cilia marker Arl13b. *Acta Medica Okayama, 68*(6), 317–322. https://doi.org/10.18926/AMO/53020.

Keryer, G., Pineda, J. R., Liot, G., Kim, J., Dietrich, P., Benstaali, C., et al. (2011). Ciliogenesis is regulated by a huntingtin-HAP1-PCM1 pathway and is altered in Huntington disease. *The Journal of Clinical Investigation, 121*(11), 4372–4382. https://doi.org/10.1172/JCI57552.

Kheradmand Kia, S., Verbeek, E., Engelen, E., Schot, R., Poot, R. A., de Coo, I. F., et al. (2012). RTTN mutations link primary cilia function to organization of the human cerebral cortex. *American Journal of Human Genetics, 91*(3), 533–540. https://doi.org/10.1016/j.ajhg.2012.07.008.

Kohli, P., Höhne, M., Jüngst, C., Bertsch, S., Ebert, L. K., Schauss, A. C., et al. (2017). The ciliary membrane-associated proteome reveals actin-binding proteins as key components of cilia. *EMBO Reports, 18*(9), 1521–1535. https://doi.org/10.15252/embr.201643846.

Kolliker, A. (1854). *Manual of human microscopical anatomy (translation by George Busk of Mikroskopische Anatomie).* Philadelphia: Lippincott: Grambo and Company.

Kowalevsky, A. (1867). Entwickelungsgeschichte des *Amphioxus lanceolatus. Memoires de l'Academie Imperiale des Sciences de St-Petersbourg (ser VII), 11,* 1–17.

Kumamoto, N., Gu, Y., Wang, J., Janoschka, S., Takemaru, K.-I., Levine, J., et al. (2012). A role for primary cilia in glutamatergic synaptic integration of adult-born neurons. *Nature Neuroscience, 15*(3), 399–405. https://doi.org/10.1038/nn.3042.

Kwan, K., Sestan, N., & Anton, E. (2012). Transcriptional co-regulation of neuronal migration and laminar identity in the neocortex. *Development, 139*(9), 1535–1546. https://doi.org/10.1242/dev.069963.

Kwon, C.-H., Luikart, B. W., Powell, C. M., Zhou, J., Matheny, S. A., Zhang, W., et al. (2006). Pten regulates neuronal arborization and social interaction in mice. *Neuron, 50*(3), 377–388. https://doi.org/10.1016/j.neuron.2006.03.023.

Laclef, C., Anselme, I., Besse, L., Catala, M., Palmyre, A., Baas, D., et al. (2015). The role of primary cilia in corpus callosum formation is mediated by production of the Gli3 repressor. *Human Molecular Genetics, 24*(17), 4997–5014. https://doi.org/10.1093/hmg/ddv221.

Lancaster, M. A., & Gleeson, J. G. (2009). The primary cilium as a cellular signaling center: Lessons from disease. *Current Opinion in Genetics & Development, 19,* 220–229. https://doi.org/10.1016/J.GDE.2009.04.008.

Langerhans, P. (1876). Zur Anatomie des Amphioxus. *Archiv für Mikroskopische Anatomie, 12,* 290–348.

Lehtinen, M. K., Zappaterra, M. W., Chen, X., Yang, Y. J., Hill, A. D., Lun, M., et al. (2011). The cerebrospinal fluid provides a proliferative niche for neural progenitor cells. *Neuron, 69*(5), 893–905. https://doi.org/10.1016/j.neuron.2011.01.023.

Lemmon, M. A., & Schlessinger, J. (2010). Cell signaling by receptor tyrosine kinases. *Cell, 141*(7), 1117–1134. https://doi.org/10.1016/j.cell.2010.06.011.

Lepanto, P., Badano, J. L., & Zolessi, F. R. (2016). Neuron's little helper: The role of primary cilia in neurogenesis. *Neurogenesis, 3*(1). https://doi.org/10.1080/23262133.2016.1253363, e1253363.

Lesiak, A. J., Brodsky, M., Cohenca, N., Croicu, A. G., & Neumaier, J. F. (2018). Restoration of physiological expression of 5-HT6 receptor into the primary cilia of null mutant neurons lengthens both primary cilia and dendrites. *Molecular Pharmacology, 94*(1), 731–742. https://doi.org/10.1124/mol.117.111583.

Li, A., Saito, M., Chuang, J. Z., Tseng, Y. Y., Dedesma, C., Tomizawa, K., et al. (2011). Ciliary transition zone activation of phosphorylated Tctex-1 controls ciliary resorption, S-phase entry and fate of neural progenitors. *Nature Cell Biology, 13*(4), 402–411. https://doi.org/10.1038/ncb2218.

Loo, L., Simon, J. M., Xing, L., McCoy, E. S., Niehaus, J. K., Guo, J., et al. (2019). Single-cell transcriptomic analysis of mouse neocortical development. *Nature Communications*, *10*(1), 134.

Mardinly, A. R., Spiegel, I., Patrizi, A., Centofante, E., Bazinet, J. E., Tzeng, C. P., et al. (2016). Sensory experience regulates cortical inhibition by inducing IGF1 in VIP neurons. *Nature*, *531*(7594), 371–375. https://doi.org/10.1038/nature17187.

Marley, A., & von Castro, M. (2012). A simple cell-based assay reveals that diverse neuropsychiatric risk genes converge on primary cilia. *PLoS One*, 7. https://doi.org/10.1371/journal.pone.0046647.

Marley, A., & von Zastrow, M. (2010). DISC1 regulates primary cilia that display specific dopamine receptors. *PLoS One*, *5*(5). https://doi.org/10.1371/journal.pone.0010902, e10902.

Matsumoto, M., Sawada, M., García-González, D., Herranz-Pérez, V., Ogino, T., Bang Nguyen, H., et al. (2019). Dynamic changes in ultrastructure of the primary cilium in migrating neuroblasts in the postnatal brain. *The Journal of Neuroscience : The Official Journal of the Society for Neuroscience*, *39*(50), 9967–9988. https://doi.org/10.1523/JNEUROSCI.1503-19.2019.

Mattila, P. K., & Lappalainen, P. (2008). Filopodia: molecular architecture and cellular functions. *Nature Reviews. Molecular Cell Biology*, *9*(6), 446–454. https://doi.org/10.1038/nrm2406.

McIntyre, J. C., Davis, E. E., Joiner, A., Williams, C. L., Tsai, I. C., Jenkins, P. M., et al. (2012). Gene therapy rescues cilia defects and restores olfactory function in a mammalian ciliopathy model. *Nature Medicine*, *18*, 1423–1428. https://doi.org/10.1038/nm.2860.

Mejillano, M. R., Kojima, S., Applewhite, D. A., Gertler, F. B., Svitkina, T. M., & Borisy, G. G. (2004). Lamellipodial versus filopodial mode of the actin nanomachinery: Pivotal role of the filament barbed end. *Cell*, *118*(3), 363–373. https://doi.org/10.1016/j.cell.2004.07.019.

Mick, D. U., Rodrigues, R. B., Leib, R. D., Adams, C. M., Chien, A. S., Gygi, S. P., et al. (2015). Proteomics of primary cilia by proximity labeling. *Developmental Cell*, *35*(4), 497–512.

Mitchison, H., & Valente, E. (2017). Motile and non-motile cilia in human pathology: From function to phenotypes. *Journal of Pathology*, *241*, 294–309. https://doi.org/10.1002/path.4843.

Miyoshi, K., Kasahara, K., Miyazaki, I., & Asanuma, M. (2009). Lithium treatment elongates primary cilia in the mouse brain in in cultured cells. *Biochemical and Biophysical Research Communications*, *388*, 757–762. https://doi.org/10.1016/j.bbrc.2009.08.099.

Molnár, Z., Clowry, G. J., Šestan, N., Alzu'bi, A., Bakken, T., Hevner, R. F., et al. (2019). New insights into the development of the human cerebral cortex. *Journal of Anatomy*, *235*, 432–451. https://doi.org/10.1111/joa.13055.

Monuki, E., & Walsh, C. (2001). Mechanisms of cerebral cortical patterning in mice and humans. *Nature Neuroscience*, *4*, 1199–1206. https://doi.org/10.1038/nn752.

Moser, J. J., Fritzler, M. J., & Rattner, J. B. (2009). Primary ciliogenesis defects are associated with human astrocytoma/glioblastoma cells. *BMC Cancer*, *9*(448). https://doi.org/10.1186/1471-2407-9-448.

Mukhopadhyay, S., & Rohatgi, R. (2014). G-protein-coupled receptors, Hedgehog signaling and primary cilia. *Seminars in Cell and Developmental Biology*, *33*, 63–72. https://doi.org/10.1016/j.semcdb.2014.05.002.

Muñoz-Estrada, J., Lora-Castellanos, A., Meza, I., Alarcón Elizaldes, S., & Benítez-King, G. (2017). Primary cilia formation is diminished in schizophrenia and bipolar disorder: A possible marker for these psychiatric diseases. *Schizophrenia Research*, *195*, 412–420. https://doi.org/10.1016/j.schres.2017.08.055.

Nager, A. R., Goldstein, J. S., Herranz-Perez, V., Portran, D., Ye, F., Garcia-Verdugo, J. M., et al. (2017). An actin network dispatches ciliary GPCRs into extracellular vesicles to modulate signaling. *Cell*, *168*, 252–263.

Nanni, L., Ming, J. E., Bocian, M., Steinhaus, K., Bianchi, D. W., Die-Smulders, C., et al. (1999). The mutational spectrum of the sonic hedgehog gene in holoprosencephaly: SHH mutations cause a significant proportion of autosomal dominant holoprosencephaly. *Human Molecular Genetics*, *8*, 2479–2488. England.

Narita, K., Kozuka-Hata, H., Nonami, Y., et al. (2012). Proteomic analysis of multiple primary cilia reveals a novel mode of ciliary development in mammals. *Biology Open*, *1*(8), 815–825. https://doi.org/10.1242/bio.20121081.

Nechipurenko, I. V., Doroquez, D. B., & Sengupta, P. (2013). Primary cilia and dendritic spines: Different but similar signaling compartments. *Molecules and Cells*, *36*(4), 288–303. https://doi.org/10.1007/s10059-013-0246-z.

Nicholas, A. K., Khurshid, M., Desir, J., Carvalho, O. P., Cox, J. J., Thornton, G., et al. (2010). WDR62 is associated with the spindle pole and is mutated in human microcephaly. *Nature Genetics*, *42*(11), 1010–1014. https://doi.org/10.1038/ng.682.

Nichols, A. S., Floyd, D. H., Bruinsma, S. P., Narzinski, K., & Baranski, T. J. (2013). Frizzled receptors signal through G proteins. *Cellular Signalling*, *25*(6), 1468–1475. https://doi.org/10.1016/j.cellsig.2013.03.009.

Nishikimi, M., Oishi, K., & Nakajima, K. (2013). Axon guidance mechanisms for establishment of callosal connections. *Neural Plasticity*, *2013*(2), 1–7. https://doi.org/10.1155/2013/149060.

Novarino, G., Akizu, N., & Gleeson, J. G. (2011). Modeling human disease in humans: The ciliopathies. *Cell*, *147*(1), 70–79. https://doi.org/10.1016/j.cell.2011.09.014.

Omori, Y., Chaya, T., Yoshida, S., Irie, S., Tsujii, T., & Furukawa, T. (2015). Identification of G protein-coupled receptors (GPCRs) in primary cilia and their possible involvement in body weight control. *PLoS One*, *10*(6). https://doi.org/10.1371/journal.pone.0128422. e0128422-0128417.

Pablo, J. L., DeCaen, P. G., & Clapham, D. E. (2017). Progress in ciliary ion channel physiology. *The Journal of General Physiology*, *149*, 37–47. https://doi.org/10.1085/jgp.20161169.

Paolino, A., Fenlon, L. R., Suárez, R., & Richards, L. J. (2018). Transcriptional control of long-range cortical projections. *Current Opinion in Neurobiology*, *53*, 57–65. https://doi.org/10.1016/j.conb.2018.05.005.

Parisi, M. A. (2009). Clinical and molecular features of Joubert syndrome and related disorders. *American Journal of Medical Genetics. Part C, Seminars in Medical Genetics*, *151C*(4), 326–340. https://doi.org/10.1002/ajmg.c.30229.

Parisi, M. A. (2019). The molecular genetics of Joubert syndrome and related ciliopathies: The challenges of genetic and phenotypic heterogeneity. *Translational Science of Rare Diseases*, *4*, 25–49. https://doi.org/10.3233/TRD-190041.

Park, S. M., Jang, H. J., & Lee, J. H. (2019). Roles of primary cilia in the developing brain. *Frontiers of Cellular Neuroscience*, *13*(218). https://doi.org/10.3389/fncel.2019.00218.

Park, S., Lim, J., Ramakrishina, S., Kim, S., Kim, W., Lee, J., et al. (2018). Brain somatic mutations in *MTOR* disrupt neuronal ciliogenesis, Leading to focal cortical dyslamination. *Neuron*, *99*(1), 83–97. https://doi.org/10.1016/j.neuron.2018.05.039.

Parra, L. M., & Zou, Y. (2010). Sonic hedgehog induces response of commissural axons to Semaphorin repulsion during midline crossing. *Nature Neuroscience*, *13*, 29–35. United States.

Phua, S. C., Lin, Y.-C., & Inoue, T. (2015). An intelligent nano-antenna: Primary cilium harnesses TRP channels to decode polymodal stimuli. *Cell Calcium*, *58*(4), 415–422. https://doi.org/10.1016/j.ceca.2015.03.005.

Poduri, A., Evrony, G. D., Cai, X., Elhosary, P. C., Beroukhim, R., Lehtinen, M. K., et al. (2012). Somatic activation of AKT3 causes hemispheric developmental brain malformations. *Neuron, 74*(1), 41–48. https://doi.org/10.1016/j.neuron.2012.03.010.

Poirier, K., Lebrun, N., Broix, L., Tian, G., Saillour, Y., Boscheron, C., et al. (2013). Mutations in TUBG1, DYNC1H1, KIF5C and KIF2A cause malformations of cortical development and microcephaly. *Nature Genetics, 45*(6), 639–647. https://doi.org/10.1038/ng.2613.

Poretti, A., Boltshauser, E., Loenneker, T., Valente, E. M., Brancati, F., Il'yasov, K., et al. (2007). Diffusion tensor imaging in Joubert syndrome. *American Journal of Neuroradiology, 28*(10), 1929–1933. https://doi.org/10.3174/ajnr.A0703.

Poretti, A., Huisman, T. A., Scheer, I., & Boltshauser, E. (2011). Joubert syndrome and related disorders: Spectrum of neuroimagingfindings in 75 patients. *American Journal of Neuroradiology, 32*(8), 1459–1463. https://doi.org/10.3174/ajnr.A2517.

Puram, S. V., Kim, A. H., Ikeuchi, Y., Wilson-Grady, J. T., Merdes, A., Gygi, S. P., et al. (2011). A CaMKIIbeta signaling pathway at the centrosome regulates dendrite patterning in the brain. *Nature Neuroscience, 14*(8), 973–983. https://doi.org/10.1038/nn.2857.

Putoux, A., Baas, D., Paschaki, M., Morlé, L., Maire, C., Attié-Bitach, T., et al. (2019). Altered GLI3 and FGF8 signaling underlies acrocallosal syndrome phenotypes in Kif7 depleted mice. *Human Molecular Genetics, 28*(6), 877–887. https://doi.org/10.1093/hmg/ddy392.

Putoux, A., Thomas, S., Coene, K. L., Davis, E. E., Alanay, Y., Ogur, G., et al. (2011). KIF7 mutations cause fetal hydrolethalus and acrocallosal syndromes. *Nature Genetics, 43*(6), 601–606. https://doi.org/10.1038/ng.826.

Qian, X., Song, H., & Ming, G. L. (2019). Brain organoids: Advances, applications and challenges. *Development, 146*(8).

Raper, J., & Mason, C. (2010). Cellular strategies of axonal pathfinding. *Cold Spring Harbor Perspectives in Biology, 2*(9), a001933. https://doi.org/10.1101/cshperspect.a001933.

Reiter, J. F., & Leroux, M. R. (2017). Genes and molecular pathways underpinning ciliopathies. *Nature Reviews. Molecular Cell Biology, 18*(9), 533–547. https://doi.org/10.1038/nrm.2017.60.

Ringers, C., Olstad, E. W., & Jurisch-Yaksi, N. (2020). The role of motile cilia in the development and physiology of the nervous system. *Philosophical Transactions of the Royal Society of London. Series B, Biological Sciences, 375*(1792), 20190156. https://doi.org/10.1098/rstb.2019.0156.

Romani, M., Micalizzi, A., & Valente, E. M. (2013). Joubert syndrome: Congenital cerebellar ataxia with the molar tooth. *The Lancet Neurology, 12*(9), 894–905. https://doi.org/10.1016/S1474-4422(13)70136-4.

Sanchez, M. A., & Armstrong, R. C. (2018). Postnatal sonic hedgehog (Shh) responsive cells give rise to oligodendrocyte lineage cells during myelination and in adulthood contribute to remyelination. *Experimental Neurology, 299*, 122–136. United States: Published by Elsevier Inc.

Sarkisian, M. R., & Guadiana, S. M. (2015). Influences of primary cilia on cortical morphogenesis and neuronal subtype maturation. *Neuroscience, 21*, 136–151. https://doi.org/10.1177/1073858414531074.

Sarkisian, M. R., & Semple-Rowland, S. L. (2019). Emerging roles of primary cilia in glioma. *Frontiers in Cellular Neuroscience, 13*(55). https://doi.org/10.3389/fncel.2019.00055.

Sattar, S., & Gleeson, J. G. (2011). The ciliopathies in neuronal development: a clinical approach to investigation of Joubert syndrome and Joubert syndrome-related disorders. *Developmental Medicine and Child Neurology, 53*, 793–798. https://doi.org/10.1111/j.1469-8749.2011.04021.x.

Schneider, J. W., Gao, Z., Li, S., Farooqi, M., Tang, T. S., Bezprozvanny, I., et al. (2008). Small-molecule activation of neuronal cell fate. *Nature Chemical Biology*, *4*(7), 408–410. https://doi.org/10.1038/nchembio.95.

Schou, K. B., Pedersen, L. B., & Christensen, S. T. (2015). Ins and outs of GPCR signaling in primary cilia. *EMBO Reports*, *16*(9), 1099–1113. https://doi.org/10.15252/embr.201540530.

Seeger-Nukpezah, T., & Golemis, E. A. (2012). The extracellular matrix and ciliary signaling. *Current Opinion in Cell Biology*, *24*(5), 652–661. https://doi.org/10.1016/j.ceb.2012.06.002.

Sengupta, P. (2017). Cilia and sensory signaling: The journey from "animalcules" to human disease. *PLoS Biology*, *15*, e2002240.

Senocak, E. U., Oguz, K. K., Haliloglu, G., Topcu, M., & Cila, A. (2010). Structural abnormalities of the brain other than molar tooth sign in Joubert syndrome-related disorders. *Diagnostic and Interventional Radiology*, *16*(1), 3–6. https://doi.org/10.4261/1305-3825.DIR.2673-09.1.

Shamseldin, H., Alazami, A. M., Manning, M., Hashem, A., Caluseiu, O., Tabarki, B., et al. (2015). RTTN mutations cause primary microcephaly and primordial dwarfism in humans. *American Journal of Human Genetics*, *97*(6), 862–868. https://doi.org/10.1016/j.ajhg.2015.10.012.

Shao, W., Yang, J., He, M., Yu, X. Y., Lee, C. H., Yang, Z., et al. (2020). Centrosome anchoring regulates progenitor properties and cortical formation. *Nature*, *580*(7801), 106–112. https://doi.org/10.1038/s41586-020-2139-6.

Shin, J., Kim, M., Jung, H.-J., Cha, H. L., Suh-Kim, H., Ahn, S., et al. (2017). Characterization of developmental defects in the forebrain resulting from hyperactivated mTOR signaling by integrative analysis of transcriptomic and proteomic data. *Scientific Reports*, 7(1), 2826. https://doi.org/10.1038/s41598-017-02842-6.

Shohayeb, B., Ho, U., Yeap, Y. Y., Parton, R. G., Millard, S. S., Xu, Z., et al. (2020). The association of microcephaly protein WDR62 with CPAP/IFT88 is required for cilia formation and neocortical development. *Human Molecular Genetics*, *29*, 248–263. England: The Author(s) 2019. Published by Oxford University Press For Permissions, please email: journals.permissions@oup.com.

Sigg, M. A., Menchen, T., Lee, C., Johnson, J., Jungnickel, M. K., Choksi, S. P., et al. (2017). Evolutionary proteomics uncovers ancient associations of cilia with signaling pathways. *Developmental Cell*, *43*(6), 744–762.e11. doi.org/10.1016/j.devcel.2017.11.014.

Siljee, J. E., Wang, Y., Bernard, A. A., Ersoy, B. A., Zhang, S., Marley, A., et al. (2018). Subcellular localization of MC4R with ADCY3 at neuronal primary cilia underlies a common pathway for genetic predisposition to obesity. *Nature Genetics*, *50*, 180–185. https://doi.org/10.1038/s41588-017-0020-9.

Southwell, D. G., Froemke, R. C., Alvarez-Buylla, A., Stryker, M. P., & Gandhi, S. P. (2010). Cortical plasticity induced by inhibitory neuron transplantation. *Science*, *327*(5969), 1145–1148. https://doi.org/10.1126/science.1183962.

Sterpka, A., Yang, J., Strobel, M., Zhou, Y., Pauplis, C., & Chen, X. (2020). Diverged morphology changes of astrocytic and neuronal primary cilia under reactive insults. *Molecular Brain*, *13*, 28. doi.org/10.1186/s13041-020-00571-y.

Stoufflet, J., Chaulet, M., Doulazmi, M., Fouquet, C., Dubacq, C., Métin, C., et al. (2019). Primary cilium-dependent cAMP/PKA signaling at the centrosome regulates neuronal migration. *BioRxiv*. https://doi.org/10.1101/765925.

Suizu, F., Hirata, N., Kimura, K., Edamura, T., Tanaka, T., Ishigaki, S., et al. (2016). Phosphorylation-dependent Akt-inversin interaction at the basal body of primary cilia. *The EMBO Journal*, *35*(12), 1346–1363. https://doi.org/10.15252/embj.201593003.

Thauvin-Robinet, C., Lee, J. S., Lopez, E., Herranz-Perez, V., Shida, T., Franco, B., et al. (2014). The oral-facial-digital syndrome gene C2CD3 encodes a positive regulator of centriole elongation. *Nature Genetics*, *46*(8), 905–911. https://doi.org/10.1038/ng.3031.
Thauvin-Robinet, C., Thomas, S., Sinico, M., Aral, B., Burglen, L., Gigot, N., et al. (2013). OFD1 mutations in males: phenotypic spectrum and ciliary basal body docking impairment. *Clinical Genetics*, *84*(1), 86–90. https://doi.org/10.1111/cge.12013.
Thion, M., Ginhoux, F., & Garel, S. (2018). Microglia and early brain development: An intimate journey. *Science*, *362*(6411), 185–189. https://doi.org/10.1126/science.aat0474.
Thomas, S., Boutaud, L., Reilly, M. L., & Benmerah, A. (2019). Cilia in hereditary cerebral anomalies. *Biology of the Cell*, *111*, 217–231. https://doi.org/10.1111/boc.201900012.
Thomas, S., Legendre, M., Saunier, S., Bessieres, B., Alby, C., Bonniere, M., et al. (2012). TCTN3 mutations cause Mohr-Majewski syndrome. *American Journal of Human Genetics*, *91*(2), 372–378. https://doi.org/10.1016/j.ajhg.2012.06.017.
Tong, C. K., Han, Y.-G., Shah, J. K., Obernier, K., Guinto, C. D., & Alvarez-Buylla, A. (2014). Primary cilia are required in a unique subpopulation of neural progenitors. *Proceedings of the National Academy of Sciences of the United States of America*, *111*, 12438–12443. https://doi.org/10.1073/pnas.1321425111.
Tsai, P., & Sahin, M. (2011). Mechanisms of neurocognitive dysfunction and therapeutic considerations in tuberous sclerosis complex. *Current Opinion in Neurology*, *24*(2), 106–113. https://doi.org/10.1097/WCO.0b013e32834451c4.
Uzquiano, A., Cifuentes-Diaz, C., Jabali, A., Romero, D. M., Houllier, A., Dingli, F., et al. (2019). Mutations in the heterotopia gene Eml1/EML1 severely disrupt the formation of primary cilia. *Cell Reports*, *28*, 1596–1611. https://doi.org/10.1016/j.celrep.2019.06.096.
Valente, E. M., Rosti, R. O., Gibbs, E., & Gleeson, J. G. (2014). Primary cilia in neurodevelopmental disorders. *Nature Reviews. Neurology*, *10*(1), 27–36.
van Dam, T., Kennedy, J., van der Lee, R., de Vrieze, E., Wunderlich, K. A., Rix, S., et al. (2019). CiliaCarta: An integrated and validated compendium of ciliary genes. *PLoS One*, *14*(5), e0216705. doi.org/10.1371/journal.pone.0216705.
van Dam, T. J., Wheway, G., Slaats, G. G., SYSCILIA Study Group, Huynen, M. A., & Giles, R. H. (2013). The SYSCILIA gold standard (SCGSv1) of known ciliary components and its applications within a systems biology consortium. *Cilia*, *2*(1), 7. doi.org/10.1186/2046-2530-2-7.
Wang, L., & Marquardt, T. (2013). What axons tell each other: Axon-axon signaling in nerve and circuit assembly. *Current Opinion in Neurobiology*, *23*(6), 974–982. https://doi.org/10.1016/j.conb.2013.08.004.
Wang, X., Tsai, J. W., Imai, J. H., Lian, W. N., Vallee, R. B., & Shi, S. H. (2009). Asymmetric centrosome inheritance maintains neural progenitors in the neocortex. *Nature*, *461*(7266), 947–955. https://doi.org/10.1038/nature08435.
Ware, S., Gunay-Aygun, M., & Hildebrandt, F. (2011). Spectrum of clinical diseases caused by disorders of primary cilia. *Proceedings of the American Thoracic Society*, *8*, 444–450. https://doi.org/10.1513/pats.201103-025SD.
Yokota, Y., Ghashghaei, H., Han, C., Watson, H., Campbell, K., & Anton, E. (2007). Radial glial dependent and independent dynamics of interneuronal migration in the developing cerebral cortex. *PLoS One*, *2*(8), 794. https://doi.org/10.1371/journal.pone.0000794.
Yu, T. W., & Bargmann, C. I. (2001). Dynamic regulation of axon guidance. *Nature Neuroscience*, *4*(Supp), 1169–1176. https://doi.org/10.1038/nn748.
Zhang, W., Yang, S. L., Yang, M., Herrlinger, S., Shao, Q., Collar, J. L., et al. (2019). Modeling microcephaly with cerebral organoids reveals a WDR62-CEP170-KIF2A pathway promoting cilium disassembly in neural progenitors. *Nature Communications*, *10*(1), 2612. https://doi.org/10.1038/s41467-019-10497-2.

Zheng, J. Q., & Poo, M. M. (2007). Calcium signaling in neuronal motility. *Annual Review of Cell and Developmental Biology*, *23*, 375–404. https://doi.org/10.1146/annurev.cellbio.23.090506.123221.

Zhu, D., Shi, S., Wang, H., & Liao, K. (2009). Growth arrest induces primary-cilium formation and sensitizes IGF-1-receptor signaling during differentiation induction of 3T3-L1 preadipocytes. *Journal of Cell Science*, *122*(15), 2760–2768. https://doi.org/10.1242/jcs.046276.

Zimmermann, K. W. (1898). Beitrage zur Kenntniss einiger Drusen und Epithelien. *Arch. Mikrosk. Anat.*, *52*, 552–70610.

CHAPTER FOUR

New insights into the molecular mechanisms of axon guidance receptor regulation and signaling

Yixin Zang[†], Karina Chaudhari[†], and Greg J. Bashaw*

Department of Neuroscience, Perelman School of Medicine, University of Pennsylvania, Philadelphia, PA, United States

*Corresponding author: e-mail address: gbashaw@pennmedicine.upenn.edu

Contents

Abstract

As the nervous system develops, newly differentiated neurons need to extend their axons toward their synaptic targets to form functional neural circuits. During this highly dynamic process of axon pathfinding, guidance receptors expressed at the tips of motile axons interact with soluble guidance cues or membrane tethered molecules present in the environment to be either attracted toward or repelled away from the source of these cues. As competing cues are often present at the same location and during the same developmental period, guidance receptors need to be both spatially and

[†] These authors contributed equally to this work.

Current Topics in Developmental Biology, Volume 142
ISSN 0070-2153
https://doi.org/10.1016/bs.ctdb.2020.11.008

temporally regulated in order for the navigating axons to make appropriate guidance decisions. This regulation is exerted by a diverse array of molecular mechanisms that have come into focus over the past several decades and these mechanisms ensure that the correct complement of surface receptors is present on the growth cone, a fan-shaped expansion at the tip of the axon. This dynamic, highly motile structure is defined by a lamellipodial network lining the periphery of the growth cone interspersed with finger-like filopodial projections that serve to explore the surrounding environment. Once axon guidance receptors are deployed at the right place and time at the growth cone surface, they respond to their respective ligands by initiating a complex set of signaling events that serve to rearrange the growth cone membrane and the actin and microtubule cytoskeleton to affect axon growth and guidance. In this review, we highlight recent advances that shed light on the rich complexity of mechanisms that regulate axon guidance receptor distribution, activation and downstream signaling.

1. Introduction

More than 100 years ago, Ramon Y Cajal described the swellings at the tips of axons, which he named "growth cones" (Cajal, 1890). Based on observations of these structures in preparations of developing chick spinal cords, Cajal predicted that during development, growth cones would be dynamic and could respond to chemical cues released in the embryonic environment to be guided to the correct targets (Cajal, 1890). It was not until nearly a century later that conserved families of secreted and membrane-anchored cues and their neuronal receptors began to be identified (Tessier-Lavigne & Goodman, 1996). Initially, four major families of ligand receptor pairs were shown to influence axon growth and guidance in various contexts, including netrins and their Deleted in Colorectal Cancer (Dcc), neogenin and Uncoordinated-5 (Unc5) receptors (Kennedy, 2000), slits and their roundabout (Robo) receptors (Brose & Tessier-Lavigne, 2000), semaphorins and their plexin and neuropilin receptors (Pasterkamp & Kolodkin, 2003), and ephrins and their Eph receptors (Kullander & Klein, 2002) Fig. 1. The secreted and membrane-anchored cues were originally grouped into four rough categories, acting as attractants or repellents, at either short or long-range (Tessier-Lavigne & Goodman, 1996); however, these distinctions have blurred over the years, since it is now clear that many of these cues can trigger diverse and sometimes even opposite axon responses, depending on the receptor composition and intracellular properties of the responding growth cones (Bashaw & Klein, 2010). Nevertheless, in the majority of *in vivo* settings, slit-Robo, Sema-plexin and ephrin-Eph signaling result in axon repulsion, while netrin promotes attraction through its Dcc and neogenin

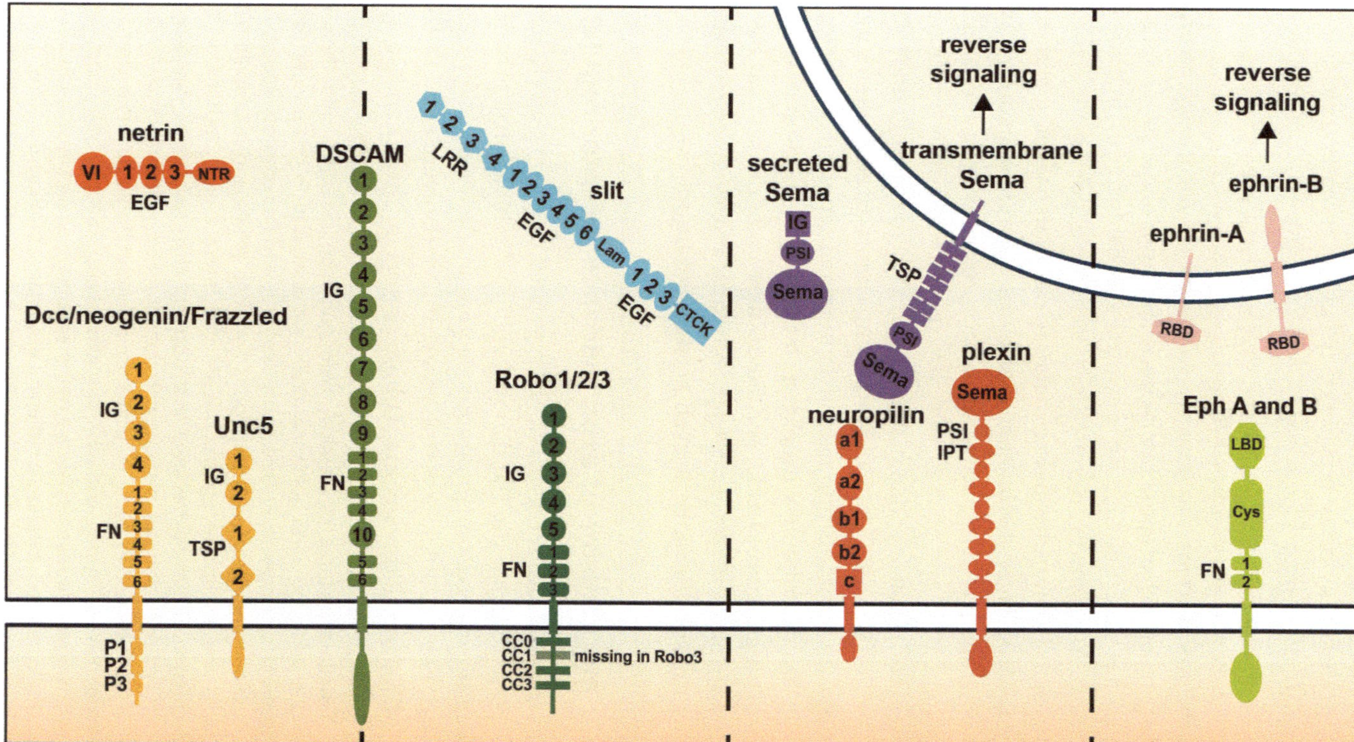

Fig. 1 See figure legend on next page.

receptors, and repulsion through its Unc5 family receptors. Additionally, it is now clear that many other families of ligand and receptors, including morphogens (BMP, Sonic Hedgehog, and Wnt), growth factors (VegF and Fgf) and their corresponding receptors also act to regulate axon guidance (Bashaw & Klein, 2010; Chedotal, 2019; Stoeckli, 2018).

In the first part of this review, we focus on regulatory mechanisms including receptor-receptor interactions and proteolytic processing where recent studies have shed light on how these mechanisms can diversify signaling outputs. Additionally, we discuss in detail the major advances that structural biological approaches have brought to our understanding of the diverse conformational landscape of surface receptor assemblies, and how in turn, this knowledge is driving insight into receptor activation mechanisms. In the second half of the chapter, we turn our attention toward intracellular signaling mechanisms, with a particular focus on advances in understanding the spatial and temporal control of signaling events. In addition to discussing how guidance receptors impinge on actin and microtubule regulatory proteins, we also describe new insights into the role of endosomes as signaling hubs in axon guidance. Lastly, we discuss new technologies that have been developed to allow investigators to probe spatial and temporal regulation and signaling in axon guidance at even greater resolution.

2. Mechanisms of axon guidance receptor regulation

Axon guidance receptors can be regulated at multiple levels, including at the transcriptional, the translational and post-translational/protein levels to ensure the right complement of receptors are deployed at the right time and place. For example, the netrin receptor Dcc (Leggere et al., 2016; Saito et al., 2016), and all three of the Robo receptors for slit (Chen, Gore, Long,

Fig. 1 Domain structures of classical axon guidance receptors and their ligands. A schematic depicting axon guidance molecules discussed in this review: netrin and its receptors Dcc, neogenin, Frazzled and Unc5, slit and its receptors Robo1, 2, and 3, transmembrane and secreted semaphorins and their plexin and neuropilin receptors, and ephrin and its Eph receptors. DSCAM can interact with both netrin and slit. Transmembrane Semas and ephrins can also signal in reverse as receptors. CTCK, C-terminal cysteine knot; Cys, cysteine-rich domains; EGF, epidermal growth factor-like domains; FN, fibronectin type III domains; IG, immunoglobulin-like domains; IPT, immunoglobulin-plexin-transcription domains; Lam, laminin-type domains; LRR, leucine-rich-repeat domains; NTR, the netrin module; PSI, plexin-semaphorin-integrin domain; RBD, receptor binding domains; LBD, ligand binding domains; TSP, thrombospondin type 1 domains.

Ma, & Tessier-Lavigne, 2008; Johnson, Junge, & Chen, 2019; Zhuang et al., 2019) are alternatively spliced to produce receptor isoforms that may have distinct activities. In addition, microRNA and RNA binding proteins modulate the translation of guidance receptors, including Robo1, ephrin-B1 and neuropilin 1 (Nrp1) to control their availability (Arvanitis, Jungas, Behar, & Davy, 2010; Hornberg, Cioni, Harris, & Holt, 2016; Yang et al., 2018). Post-translational modifications also play key roles in regulating the surface distribution of receptors. For example, differential regulation of Robo receptor trafficking at the growth cone at low levels before and high levels after midline crossing precisely times axonal responses to slit to ensure the correct formation of midline circuits (Alther, Domanitskaya, & Stoeckli, 2016; Gorla et al., 2019; Kinoshita-Kawada et al., 2019). In another example, glycosylation of the Wnt receptor Frizzled3 regulates its trafficking to the growth cone surface, which mediates the anterior turning of post-crossing commissural axons at the floor plate in the developing mouse spinal cord (Onishi & Zou, 2017). These mechanisms of regulated receptor splicing, translation and trafficking to control axon responsiveness during guidance have recently been reviewed in detail (Gorla & Bashaw, 2020). In this section of the review, we instead focus on the regulation of axon guidance receptors at the post-translational/protein level, paying particular attention to the ways these intricate regulatory events impinge on the activation and downstream signaling pathways of axon guidance receptors.

2.1 Receptor-coreceptor and receptor-receptor interactions

The disparity between the relatively small number of guidance receptor-ligand pairs available and the overwhelming complexity of connections in the mature nervous system creates a major developmental challenge. One mechanism that has emerged to solve this dilemma is to employ coreceptors, which are transmembrane or membrane-anchored proteins that directly bind to guidance receptors, but not their ligands, to modulate and diversify receptor outputs (Fig. 2). Additionally, receptors with distinct ligands can interact with each other to modify and modulate signaling activity. Binding to coreceptors or other receptors can induce additional downstream signaling events or activate the same effectors but at different levels, to adjust guidance receptors' responses to extracellular stimuli in a context-specific way. Such accurate spatial-temporal control of guidance events is paramount for the proper assembly of neural circuits.

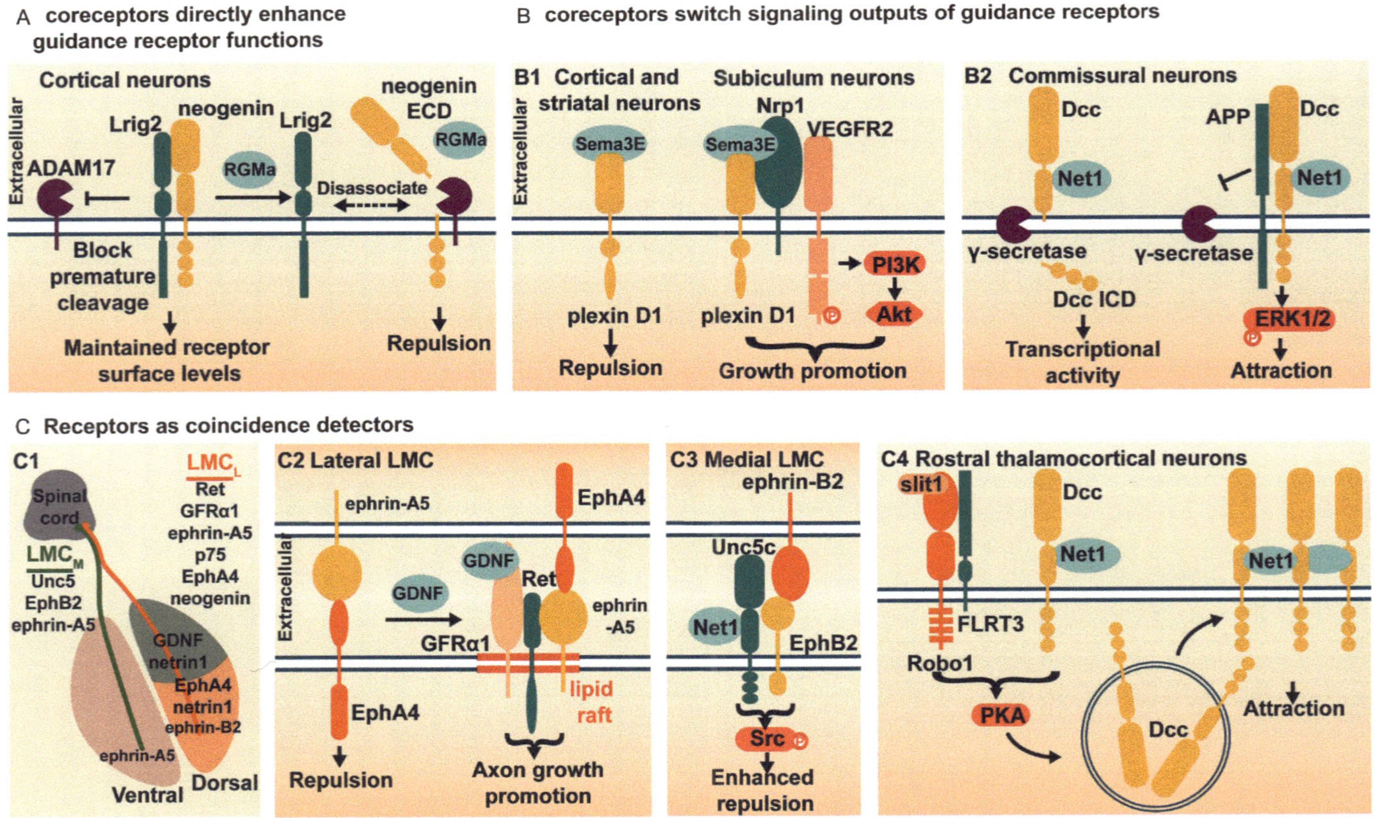

Fig. 2 See figure legend on opposite page.

Some coreceptors strengthen the signaling outputs of the guidance receptors with which they interact. For example, Leucine rich repeats and immunoglobulin like domains 2 (Lrig2) binds to neogenin in the developing mouse brain (Fig. 2A). Lrig2 is required for repulsive guidance molecule a (RGMa) induced growth cone collapse by blocking premature proteolytic processing of neogenin to maintain neogenin surface levels (van Erp et al., 2015). Binding of the RGMa induces dissociation of the Lrig2-neogenin complex, suggesting that ligand binding can regulate receptor-coreceptor interactions. Lrig2-dependent regulation of neogenin signaling controls cortical neuron migration and optic nerve regeneration, yet its significance in neogenin directed axon guidance in development remains to be seen.

Guidance receptors and their coreceptors can also undergo rapid and local phosphorylation or dephosphorylation, to allow for precise regulation and integration of guidance receptor signaling. In vertebrates, the divergent Robo3 receptor is essential for commissural axons to cross the midline. In *Robo3* knockout (KO) mouse embryos, dorsal commissural axons in the developing spinal cord display a complete failure of midline crossing, and initial explant culture and genetic analysis suggested that Robo3 promotes

Fig. 2 Regulation of axon guidance receptors via receptor-coreceptor interactions. (A) Coreceptors can enhance guidance receptor functions. In cortical neurons, in the absence of RGMa, Lrig2 binds to neogenin to block premature ADAM17-mediated proteolytic cleavage of the ECD of neogenin. Binding of RGMa initiates the dissociation of Lrig2 and neogenin, allowing neogenin ECD shedding and RGMa-induced neurite growth inhibition to occur. (B) Coreceptor binding can switch signaling outputs of guidance receptors. (B1) In cortical and striatal neurons, which do not express Nrp1, Sema3E acts as a repellent interacting with plexin D1. In subiculum neurons, Nrp1 recruits VEGFR2 to mediate a growth promoting response to Sema3E via activating the PI3K/Akt pathway. (B2) In spinal cord commissural neurons, APP binds to Dcc to inhibit γ-secretase dependent proteolytic cleavage of the Dcc ICD and its potential transcriptional activity, and to enhance netrin1-induced attraction via increasing the activation of ERK1/2. (C) Receptor-receptor interactions can also function as coincidence detectors to induce synergistic effects from two guidance cues. (C1) A schematic depicting the spinal cord, the lateral motor column spinal motor neurons that target the limb and the developing limb bud divided into ventral and dorsal halves. Expression of some genes either in the motor axons or the dorsal limb are shown. (C2) In medial LMCs that express EphA4, ephrin-A5 acts as a repellent. In lateral LMCs that express ephrinA5, co-detection of GDNF and EphA4 by the GFRα-Ret-ephrin-A5 receptor-coreceptor complex mediates axon growth promotion. (C3) Co-stimulation with netrin1 and ephrin-B2 leads to enhanced phosphorylation of Src kinase, which strengthens repulsion in medial LMCs mediated by the Unc5c-EphB2 complex. (C4) In rostral thalamocortical neurons, Robo1 binding to its coreceptor FLRT3 enables co-detection of slit1 and netrin1, which activates PKA to increase surface Dcc levels via enhanced vesicular transport, to facilitate attraction.

midline crossing by antagonizing Robo1-Robo2-slit mediated repulsion (Friocourt & Chedotal, 2017; Sabatier et al., 2004). Interestingly, Robo3's ability to inhibit repulsion is not, contrary to the initial models, due to competitive binding of slit, because mammalian Robo3 binds to slit2 with orders of magnitude lower affinity compared to Robo1 (Zelina et al., 2014). Moreover, *Robo3/slit1/slit2* triple KO and *Robo1/2/3* triple KO only partially rescue the Robo3-dependent loss of midline crossing, indicating that Robo3 likely functions in additional pathways independently of Robo1 and 2 to promote midline crossing (Jaworski, Long, & Tessier-Lavigne, 2010; Sabatier et al., 2004). Indeed, the Robo3 receptor forms a complex with Dcc to potentiate netrin-Dcc signaling, which requires tyrosine-phosphorylation of Robo3 in its conserved intracellular CC0 domain by Src-family kinases upon netrin1 stimulation (Zelina et al., 2014). *Robo3* mutants show migratory defects in precerebellar neurons, which can only be rescued by full-length Robo3 but not by Robo3 variants that cannot bind to Dcc or be tyrosine phosphorylated. This observation demonstrates the functional significance of Robo3 as a Dcc coreceptor; however, how netrin1 stimulation triggers Robo3 phosphorylation and what signaling machineries are assembled downstream remain unknown. Robo3 KO spinal cord explants also show significantly decreased axon outgrowth in response to bath application of netrin1 (Zelina et al., 2014). While this suggests that Robo3's ability to potentiate netrin1 signaling is important for axon outgrowth, whether Robo3 is also responsible for axon turning and axon guidance in this context has not been investigated.

Down syndrome cell adhesion molecule (DSCAM) is best known for its role in controlling dendritic self-avoidance through homophilic interactions mediated by a remarkable pool of diverse splice variants (Sachse et al., 2019). More recently another mode of action for Dscam has emerged from studies of *Drosophila* mechanosensory neurons (MSNs), where dephosphorylation of Dscam1 by the Receptor protein tyrosine phosphatase 69D (RPTP69D) is required cell-autonomously for the consolidation and extension of axon collaterals (Dascenco et al., 2015). Interestingly, RPTP69D's ability to interact with and regulate Dscam1's function is enhanced upon slit binding to Dscam1. Here slit might function independently of Robo, as Robo1 or Robo2 null mutants show no defect in MSNs axon collateral formation, although it remains possible that Robo1 and 2 might function redundantly. In addition, RPTP69D acts as a coreceptor for Robo3 to enhance its repulsive function by increasing Robo3 surface levels in *Drosophila* mushroom body small lateral neurons (Oliva et al., 2016). Interestingly, RPTP69D interacts with Robo3 in a dephosphorylation-independent pathway, which suggests that RPTPs have two distinct modes of function in brain development, acting as either phosphatases or coreceptors for different guidance receptors.

Instead of enhancing or inhibiting the function of guidance receptors, some coreceptors introduce novel functional outputs by switching responses to guidance cues. For example, Nrp1, the coreceptor for plexin D1, acts as a molecular gate to switch semaphorin 3E (Sema3E)-mediated repulsion to growth promotion in the subiculo-mammillary axons in descending forebrain axon tracts in mice (Chauvet et al., 2007). Nrp1 promotes growth by recruiting and inducing the phosphorylation of another transmembrane receptor, vascular endothelial growth factor receptor type 2 (VEGFR2), leading to the activation of the phosphatidylinositol-3 kinase (PI3K) pathway downstream of VEGFR2 (Bellon et al., 2010). Similarly, in the mouse visual system, Ng-CAM-related cell adhesion molecule (Nr-CAM) forms a complex with plexin A1 that shifts the Sema6D response from repulsion to attraction in the contralateral projecting RGCs to facilitate chiasm crossing (Kuwajima et al., 2012). Unexpectedly, the chiasm cells and the contralateral projecting RGCs both express plexin A1 and Nr-CAM, and their expression in both cell types is important for switching the repulsive response of Sema6D. Finally, in the developing mouse spinal cord, amyloid precursor protein (APP) can interact with Dcc to enhance the activation and phosphorylation of extracellular-regulated kinase 1/2 (ERK1/2), which are known Dcc effectors that mediate netrin-induced axon growth and turning (Rama et al., 2012). Biochemical experiments using rat neuroblastoma cell lysates overexpressing APP and Dcc indicates that APP might also inhibit γ-secretase cleavage of Dcc, a step essential for the transcriptional function of the *Drosophila* Dcc homologue Frazzled (Fra) (Neuhaus-Follini & Bashaw, 2015). Although yet to be tested *in vivo*, this observation raises the intriguing possibility that in some commissural neurons, APP could switch the Dcc response from the transcriptional regulation of target genes to the canonical Netrin-dependent regulation of the cytoskeleton.

Most studies to date only focus on the function of one ligand-receptor pair in a given guidance scenario. However, during normal development, most projecting axons encounter many different extracellular stimuli simultaneously at guidance choice points. To help navigating axons resolve and integrate the multitude of extracellular information, some receptors act as coincidence-detectors to elicit synergistic effects from two or more guidance signals, resulting in significantly elevated growth cone responses that are distinct from simple additive effects of two parallel events (for a review on the interaction between different types of axon guidance pathways, see Morales & Kania, 2017). Such coincidence-detection reduces targeting errors, increases fidelity, and also significantly reinforces correct guidance decisions. This is perhaps best illustrated in motor axon targeting in the vertebrate limb. During development, motor axons projecting from the medial and lateral regions of the spinal cord lateral motor column (LMC) bifurcate at the base of the limb bud to form ventral and dorsal limb nerves, respectively (Fig. 2C1). The RTK Ret acts as a coreceptor for both GFRα1, the receptor for glial-derived neurotrophin factor (GDNF), and ephrin-A5, the reverse signaling receptor for EphA4, and Ret is essential for the targeting of lateral LMC axons *in vivo* (Bonanomi et al., 2012). In growth cone turning assays, co-stimulation with low levels of both GDNF and EphA4 can promote turning, whereas low levels of either GDNF or EphA4 alone produce no response. GDNF stimulation can recruit Ret to lipid rafts, suggesting that GDNF-GFRα1 signaling potentiates ephrin-A5 reverse signaling by bringing the co-receptor Ret within close proximity to the GPI-anchored ephrin-A5. Furthermore, in stripe assays, ephrin-A5 application sensitizes netrin responses in chick LMC explants by enhancing the protein levels of both its repulsive receptor EphA4 and the netrin receptor neogenin, suggesting a mechanism where a repulsive guidance cue potentiates the responses of an attractive guidance cue (Croteau, Kao, & Kania, 2019). The intracellular domain of EphA4 is dispensable for this function, yet it remains to be seen if the extracellular domains or transmembrane domains of EphA4 can directly bind to neogenin or if EphA4 indirectly increases neogenin levels. In medial LMC axons, co-stimulation of both netrin1 and ephrin-B2 can induce complex formation between their respective receptors (Unc5c and EphB2) and amplify repulsion through greater and prolonged activation of their common downstream effector Src kinase (Poliak et al., 2015). However, Src inhibition only partially blocks this growth cone collapse evoked by simultaneous netrin1 and ephrin-B2 application, indicating that additional unknown mechanisms exist to synergize Unc5c and EphB2 signaling. Further illustrating the point, in rostral thalamocortical axons, simultaneous detection of slit1 and netrin1 activates attraction via synergistic action of Robo1, its coreceptor fibronectin leucine rich transmembrane protein 3 (FLRT3) and the netrin receptor Dcc (Leyva-Diaz et al., 2014) (Fig. 2C4). FLRT3 forms a complex with Robo1 and transduces slit1 and netrin1 co-stimulation into an increase in Dcc surface levels, likely by promoting PKA-dependent mobilization of intracellular pools of Dcc.

The control of guidance receptor signaling by direct binding to coreceptors provides an exciting model for how developing neurons expand the functional outputs of guidance receptors. These recent studies demonstrate that besides recruiting additional intracellular effectors (Fig. 2B, C3, C4), coreceptors can also modulate the proteolytic processing (Fig. 2A, B2) and the localization (Fig. 2C2) of guidance receptors. Being transmembrane or GPI-anchored affords coreceptors regulatory capacity on both sides of the plasma membrane, so that coreceptors are uniquely poised to regulate guidance receptor signaling in a context dependent manner.

2.2 Structural insights into axon guidance receptor signaling: dimerization, oligomerization and clustering

Receptor activation entails far more than a one-to-one binding between a receptor and a ligand, but instead often calls for the recruitment of other receptor molecules to form complexes or clusters. In the last decade, integrated structural studies that offer crystallographic analysis coupled with cell-based or *in vivo* functional assays, have provided valuable insights into the 3D organizations of receptor complexes and how these structural features modulate guidance receptor activation and function. Here, we focus on four prominent families of axon guidance receptors, the plexins, Robos, Ephs and netrin receptors Dcc/neogenin/Unc5, as examples to describe how clustering and the oligomerization state of guidance receptors at the surface of the cell plays a pivotal role in relaying information across the membrane.

Plexins form autoinhibitory *cis*-homodimers to prevent premature activation in the absence of semaphorin binding. Crystal structures of the Sema-binding regions of plexin B1, A2 and C1 in complex with their respective ligands Sema4D, 6A, and 7A show essentially identical domain architectures in which a Sema dimer binds to two plexin molecules (Janssen et al., 2010; Liu et al., 2010; Nogi et al., 2010). The formation of this semaphorin-plexin interface is essential for plexin-mediated membrane collapse in COS-7 cells and growth cone collapse in chick dorsal root ganglion neurons. PlexA1, A2 and A4 form autoinhibitory homodimers in "closed" ring-like structures, which are mediated by the intermolecular interface between the Sema domain located at the tip of one plexin A "ring" and the ECD 4 and 5 situated at the "stalk" of another plexin A receptor (Kong et al., 2016; Marita et al., 2015). This structure remained undetected until the crystal structures of the entire extracellular domains (ECDs) of plexin As were determined, underscoring the idea that structural data involving fragments of molecules should be interpreted with caution. Disrupting this plexin A-plexin A interaction in the absence of exogenous Sema stimulation, induces membrane collapse in COS-7 cells and dentate gyrus granule cell growth cones. Conversely, preserving this interaction eliminates collapse, supporting a model where plexin As are normally autoinhibited by *cis*-homodimerization (Kong et al., 2016). The amino acid residues at this interface are only highly conserved among vertebrate plexin As, but it remains possible that other classes of plexins employ similar autoinhibitory *cis*-interactions via alternative interfaces. In addition to preventing inappropriate activation of receptors, the pre-signaling association between two plexin As could potentially serve to position receptors within close proximity to each other to prepare for rapid and precise responses upon ligand binding.

The Robo receptor, on the other hand, autoinhibits its dimerization and downstream signaling by engaging in a compact "closed" conformation (Fig. 3B). The Ig4 domain in both Robo1 and 2 mediates receptor dimerization and is required for the Robo2 overexpression-induced membrane collapse phenotype in COS-7 cells (Aleksandrova et al., 2018; Yom-Tov et al., 2017). Crystal structures of the intact Robo2 ECD uncovered a "closed" hairpin-like conformation which blocks access to the Ig4 homodimerization interface (Barak et al., 2019). Additionally, it is also revealed that the Ig5 domain of Robo2 interacts with its Ig1 and 2 domains *in trans*, potentially mediating trans-inhibition via stabilizing the "closed" conformation of an opposing Robo2 receptor (Fig. 3B). These observations support a model in which slit2 binding to the Ig1 domain of Robo2 relieves trans-inhibition and shifts Robo2's conformation from "closed" to "open" to facilitate Robo2 homodimerization and signaling. This model is inferred by crystallographic analysis of superimposed slit-binding and trans-interacting Robo2 interfaces and has not been confirmed by co-crystals of Robo2 and slit2. An inhibitory trans-interaction between Robo receptors has also been observed in *Drosophila*, where Robo2 expressed on midline cells can bind to Robo1 expressed on crossing commissural axons to inhibit slit-mediated repulsion (Evans, Santiago, Arbeille, & Bashaw, 2015). However, whether the formation of this heterodimer also relies on the same Ig5-Ig1/2 binding interface remains unclear. While *Drosophila* Robo2 shares similarities with both vertebrate Robo1 and 2, equivalent trans-homo or -hetero dimers of Robo receptors have not been identified in vertebrates. The ECD of the divergent vertebrate Robo3 receptor, which has identical domain structure with that of Robo1 and 2, assumes a fully extended "open" conformation and exists exclusively in monomeric state in the absence of ligand binding (Pak et al., 2020).

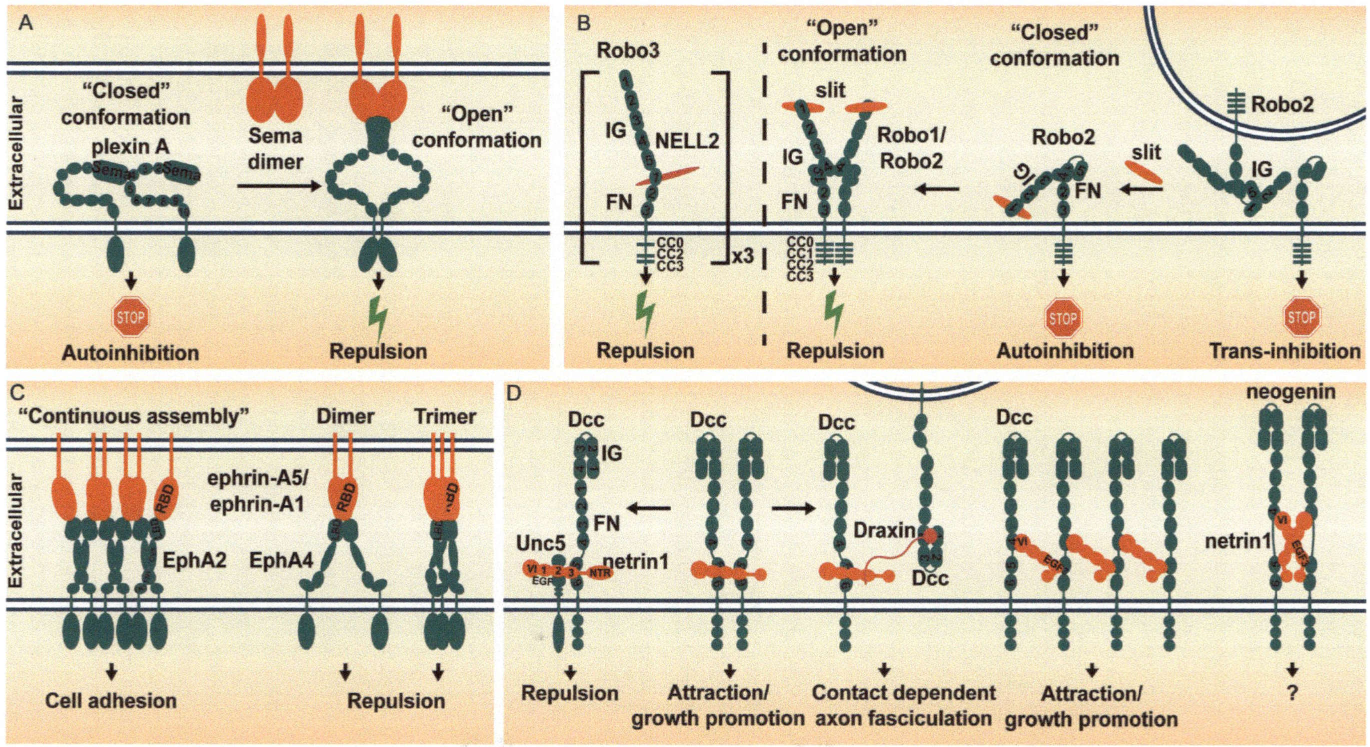

Fig. 3 See figure legend on next page.

Robo3 interacts with its ligand NELL2 via its FN1 domain. Exchanging the FN1 domain of Robo1 with that of Robo3 converts the predominantly dimeric Robo1, typically in a more compressed conformation, into a strictly monomeric molecule with an elongated "open" shape, indicating that the FN1 domains of Robos determine both their oligomerization and conformational characteristics. Future analysis that identifies the functionally relevant residues and interfaces within the FN1 domain of Robo3, that are not conserved in Robo 1 and 2, will provide important information on how different Robo receptors are regulated.

Different Eph receptors form disparate assemblies at the membrane to instigate distinct signaling pathways. Crystal structures suggest that full length EphA2 ECDs, either alone or in complex with the receptor binding domain of ephrin-A5 or -A1, could cluster to form a highly ordered oligomer that exists as a continuous array on the membrane (Himanen et al., 2010; Seiradake, Harlos, Sutton, Aricescu, & Jones, 2010). Restricting lateral movement of EphA2 using a physical barrier eliminates the formation of EphA2 arrays and blocks the cellular responses to ephrin-A1, as indicated by the decreased recruitment of signaling effectors and altered cytoskeletal morphology (Salaita et al., 2010). Thus, the ephrin-induced array-like arrangement of EphA2 is functionally important for signal propagation. However, large signaling arrays are not detected for all Ephs. EphA4 and EphB2, which mediate repulsion instead of cell-adhesion like EphA2, form smaller clusters

Fig. 3 The dynamic conformational landscape and multimerization state of axon guidance receptors at the membrane modulate their signaling outputs. (A) Plexin A receptors form an auto-inhibitory dimer, which can shift into an "open" shape upon binding to a Sema dimer, to mediate repulsion. (B) Slit binding can potentially break trans-inhibitory interactions between two Robo2 receptors on opposing cells, by inducing a conformational change of Robo2 from a "closed" auto-inhibitory conformation to an "open" conformation. The following *cis*-dimerization of Robo2 receptors generates a repulsive response. The Robo3 receptor instead assumes an elongated shape and interacts with its ligand NELL2 in a 3:3 trimer to mediate axon repulsion. For simplicity, the NELL2-Robo3 interaction is depicted as a monomeric interaction. (C) Upon binding to ephrin-A5 or -A1, EphA2 receptors cluster to form a continuous assembly at the membrane to mediate cell adhesion. EphA4 forms smaller multimers, such as dimers or trimers, to transduce repulsion. (D) Netrin binds to two Dcc receptors to generate attraction but can also mediate repulsion when the receptor binding interface on the EGF2/3 domains of netrin1 is occupied instead by UNC5. Draxin can interact with both netrin1 and Dcc to potentially stabilize trans-interaction between two Dcc receptors to mediate contact dependent axon fasciculation. netrin1 can also interact with the FN4 domains of Dcc and neogenin to form either an extended array or a 2:2 tetramer with the receptors, depending on the length of the linker region between the FN4 and 5 domains of the receptors.

such as dimers or trimers (Schaupp et al., 2014; Seiradake et al., 2013). It is currently unclear what mechanisms prevent the lateral expansion of these small multimers. One possibility is that the coalescence of small EphB2 oligomers into larger assemblies could terminate signaling, likely through accelerating endocytosis of large receptor aggregates (Ojosnegros et al., 2017).

Finally, netrin can employ distinct interfaces to form unique signaling complexes with different netrin receptors, and the composition of those complexes can determine the signaling outputs to be either attraction or repulsion. Several recent studies revealed three potential receptor binding sites on netrin1. First, crystal structures suggest that the EGF-1 and -2 domains of netrin1, when binding to the fibronectin type III (FN) 5 and 6 domains of Dcc, are required for chemoattractive multimerization of Dcc (Fig. 3D). Importantly, when the same binding site on netrin1 is occupied instead by Unc5A, the netrin response is switched from attraction to repulsion, as demonstrated in mouse spinal cord neurons (Finci et al., 2014). It is important to note that the netrin-Unc5 interaction still awaits structural characterization. Secondly, the interaction between the EGF-3 domain of netrin1 and the FN5 domain of the receptor is shared among Dcc and neogenin 1 (Neo1; (Finci et al., 2014; Xu et al., 2014). This binding interface is required for Dcc-mediated attraction, but its functional significance in Neo1 has not been demonstrated. Additionally, the novel Dcc ligand Draxin can also bind to netrin1 in the same region on EGF-3 as Dcc, with even higher affinity (Gao et al., 2015; Liu et al., 2018). This suggests that Draxin could function by connecting Dcc-bound netrin1 to an opposing Dcc receptor, to facilitate contact-dependent axon fasciculation (Liu et al., 2018). Lastly, the N-terminal laminin-like domain of netrin1 can bind to the FN4 domain of both Dcc and Neo1, yet may lead to the differential formation of either a continuous netrin1-Dcc assembly or a 2:2 netrin1-Neo1 tetramer, due to differences in the linker length connecting the FN4 and 5 domains of the receptors (Xu et al., 2014). The double mutant phenotype in the developing mouse spinal cord suggests that Neo1 and Dcc collaborate to mediate netrin1-dependent attraction in commissural axons. Whether and how differences in the structure of Dcc and Neo1 signaling complexes determine their relative contribution to midline crossing have not yet been explored.

Despite significant recent progress, our understanding of the structural characteristics of guidance receptor complexes is still largely incomplete. Indeed, continued progress is hindered by the challenge of crystalizing full-length transmembrane receptors, which are more likely to represent their native membrane-bound conformations than fragments of the receptors. Nevertheless, structures of the receptor ECD fragments in complex with their ligands have shed some light on the 3D organization of receptor signaling assemblies and suggest potential mechanisms to explain unique features of each axon guidance pathway. However, it has been difficult to test the significance of these assemblies *in vivo* due to a lack of genetic manipulation tools in vertebrates. Fortunately, the development of widely applicable new techniques for genome editing, such as the CRISPR/Cas-9 system, will enable investigators to test the functions of receptor assembly interfaces in relevant physiological contexts in the future.

2.3 Proteolytic processing of guidance receptors

Proteolytic processing controls the levels and functions of axon guidance receptors and is increasingly appreciated as one of the main mechanisms that regulates and diversifies their signaling outputs. Receptor cleavage can either downregulate surface levels of full-length receptors to terminate their signaling, or generate various receptor fragments that are active signaling molecules, or achieve both at the same time (Fig. 4). These receptor fragments can localize to different extracellular and intracellular spaces to initiate downstream pathways that are often distinct from the ones mediated by full-length receptors. As such, the contributing proteases act as molecular switches to change the neuron's responses to guidance cues. In this section of the review, we will focus on deciphering the involvement of proteases (often referred to as "sheddases"), which release receptor ECDs into extracellular space, and the intramembrane protease γ-secretase, which cleaves within the transmembrane domain to generate soluble receptor ICDs. It is important to note that in addition to functioning separately, sheddases and γ-secretase often target the same receptors in sequential cleavages, and in many cases are coupled with transcriptional activities of the receptor ICDs.

The β-site amyloid precursor protein cleaving enzyme 1 (BACE1), sometimes referred to as β-secretase, is best known for its central role in the proteolytic processing of APP (Lichtenthaler, Lemberg, & Fluhrer, 2018; Saftig & Bovolenta, 2015). BACE1, together with γ-secretase, cleaves APP to produce the pathogenic Aβ peptide, which forms the amyloid plaques that are pivotal in the pathogenesis of Alzheimer's disease. In addition to APP, BACE1 also cleaves the neural cell adhesion molecule close homolog of L1 (CHL1), which functions downstream of the Nrp1-plexin A receptor complex (Barao et al., 2015). In mouse embryonic thalamic

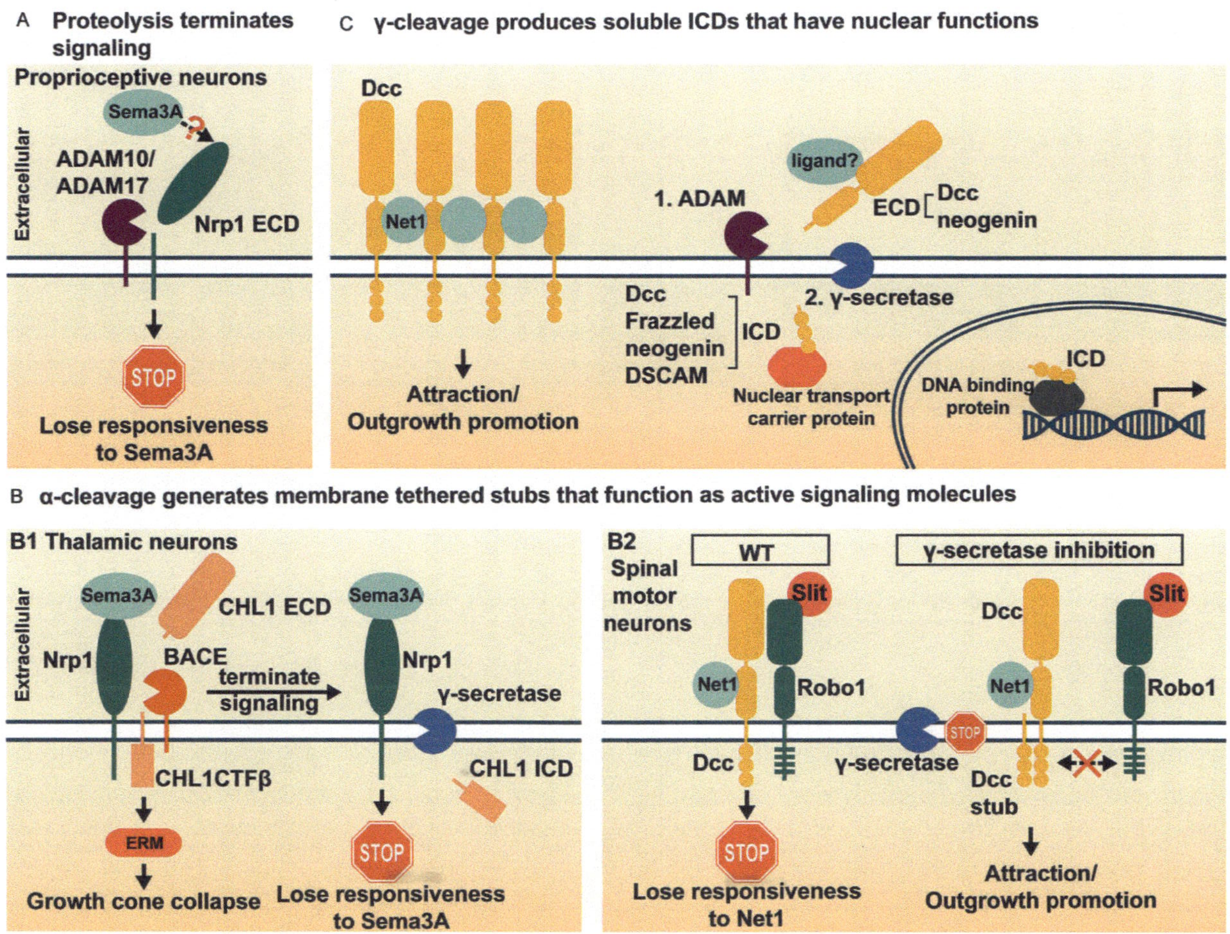

Fig. 4 See figure legend on next page.

neurons, BACE1 mediates Sema3A-induced growth cone collapse specifically via the membrane-tethered C-terminal fragment generated by BACE1 (CHL1CTFβ; Fig. 4B1). CHL1CTFβ recruits downstream effectors that can directly regulate actin polymerization in growth cone filopodia. Intriguingly, signaling is also terminated by proteolysis, when γ-secretase cleaves the CHL1 ICD containing the effector binding domain. How do neurons prevent premature γ-secretase cleavage of CHL1? The answer might lie in the differential subcellular localization of BACE1 (active in the trans-Golgi network and endosomes) and γ-secretase (active in endosomes and at the plasma membrane; (Lichtenthaler et al., 2018)). Aside from APP and CHL1, however, very few BACE1 substrates have been described, and even less is known about the functional significance of most BACE1-mediated cleavages. Candidate BACE1 substrates correspond to 19% of the neuronal proteins that are targeted by sheddases, as identified in a mass spectrometry-based screen, making BACE1 one of the major sheddases in mouse embryonic neurons (Kuhn et al., 2012).

Fig. 4 Receptor proteolytic processing regulates downstream signaling outputs. (A) Proteolytic cleavages of receptors can terminate their signaling by downregulating the surface levels of functional full-length receptors. To grow past a repulsive Sema3A barrier, Rat proprioceptive axons lose responsiveness to Sema3A as ADAM10 and ADAM17 cleave the Sema3A receptor Nrp1. It is unclear if this cleavage is induced by Sema-3A binding or if it happens constitutively without ligand stimulation. (B) ADAM-cleavage releases the ECDs of guidance receptors to form membrane-tethered C-terminal stubs that can function as active signaling molecules. (B1) In mouse thalamic neurons, CHL1 is recruited by Nrp1 upon Sema3A stimulation. BACE sheds the ECD of CHL1 to generate CHL1CTFβ which can interact with ezrin-radixin-moesin (ERM) proteins to mediate growth cone collapse. The signaling cascade is terminated when CHL1CTFβ is further cleaved by γ-secretase. (B2) In WT mouse spinal motor neurons, slit-activated Robo1 binds to full-length Dcc to silence netrin1 response. When γ-secretase function is inhibited, the accumulated Dcc stubs preferentially bind to full-length Dcc to protect them against slit-Robo1 mediated inhibition of netrin1 responsiveness. (C) γ-secretase cleavages of guidance receptors produce soluble ICD fragments that can function inside the nucleus. In canonical Dcc signaling pathways, netrin1 binding triggers the formation of a continuous netrin1-Dcc assembly which induces local cytoskeletal rearrangements to mediate attraction or axon growth promotion. In certain subpopulations of neurons, potential ligand binding initiates sequential proteolytic cleavages of Dcc/Frazzled, neogenin and DSCAM, first by ADAMs and then by γ-secretase. This process releases the ICD fragments that can translocate into the nucleus to regulate gene transcription. The function of ICDs might require assistance from nuclear transport carrier proteins (such as importin beta IPO5, which binds to the DSCAM ICD) and DNA binding proteins (such as the transcriptional regulator LMO4, which interacts with the Neo1 ICD).

ADAMs (a disintegrin and metalloproteases) are another family of transmembrane proteases. ADAMs are highly expressed in the developing nervous system with distinct spatial and temporal profiles, which suggests that they may have diverse roles in neurodevelopment (Hsia et al., 2019). ADAMs share a common domain structure consisting of the autoinhibitory pro-domain, the catalytic metalloprotease domain, the disintegrin domain, the cysteine-rich region, the EGF-like repeat domain (that is absent in ADAM10 and 17), the transmembrane domain and a shorter, less well-defined cytoplasmic tail (Lambrecht, Vanderkerken, & Hammad, 2018). In general, ADAMs function by cleaving the extracellular juxtamembrane region of type I transmembrane proteins, which leads to the shedding of the cleaved ECDs into extracellular space. For the purpose of this review, we will focus on discussing recent discoveries implicating ADAM10 and ADAM17, two of the most studied members of the ADAM family, in the modulation of axon guidance receptors. Phylogenetic and molecular evolution analyses demonstrate that a major diversification event separates ADAM10 and ADAM17 from all other members of the ADAM family, and as a result ADAM10 and ADAM17 are highly homologous, well-conserved and share many of the same substrates (Long et al., 2012). Unfortunately, global KOs of *Adam10* are embryonic lethal at E9.5 and global KOs of *Adam17* are perinatal lethal, and both mutants exhibit wide-spread developmental defects, complicating efforts to decipher their roles in neurodevelopment and to identify novel neuronal substrates (Hsia et al., 2019). In addition, compared to most other proteases, sheddases rely more on amino acid motifs or secondary structures to recognize their targets, rather than stringent substrate sequences (Lichtenthaler et al., 2018). Indeed, no consensus ADAM cleavage sequence has been identified to date (Lambrecht et al., 2018). This results in frequent failure in generating cleavage-resistant forms of ADAM targets, adding to the difficulties of predicting and verifying potential ADAM substrates. Conditional KO mutants for *Adam10* (Jorissen et al., 2010; Prox et al., 2013) and *Adam17* (Horiuchi et al., 2007), complemented by the development of novel proteomic strategies (Muller, Scilabra, & Lichtenthaler, 2016), will facilitate their study and help us gain mechanistic insights into their roles in specific axon guidance pathways.

Receptors of all four classical axon guidance cues-slits, netrins, semaphorins and ephrins-are physiological substrates for ADAMs during neurodevelopment, pointing to ADAM-mediated shedding as a universal mechanism in regulating guidance receptor signaling (Bai & Pfaff, 2011). In *Drosophila*, the ADAM10 homologue Kuzbanian cleaves Robo1 and is required cell-autonomously in neurons for slit-Robo1 mediated repulsion

(Coleman, Labrador, Chance, & Bashaw, 2010). Additionally, in cell lines, human ROBO1 undergoes sequential cleavages by ADAMs and γ-secretase to generate soluble ICDs that can localize to the nucleus, although the functions of the ICDs remain unexplored (Seki et al., 2010). Interestingly, the levels of ROBO1 ICDs are strongly enhanced by the ADAM17-specific activator phorbol 12-myristate 13-acetate (PMA), and also largely suppressed by the ADAM17-specific inhibitor TAPI-1, indicating that ADAM17 could be responsible for cleaving ROBO1 in this context. Although it remains unclear how ADAM cleavages facilitate Robo1 signaling, suppressing ADAM10 blocks ROBO1 cytosolic region-mediated recruitment of the Ras/Rac GEF Son of Sevenless (Sos), suggesting that the cytosolic domains of Robo1 are important for ADAM-dependent Robo1 signaling (Coleman et al., 2010). During rat development, proprioceptive axons (PAs) lose their sensitivity to the repellent Sema3A as ADAM10 and ADAM17 cleave its receptor Nrp1, and this process is required for the proper targeting of PAs in the spinal cord as they grow through a region of high Sema3A expression (Romi et al., 2014). Interestingly, in *Adam10* and *Adam17* double KO explants and spinal cords, PAs regain insensitivity to Sema3A after a one-day delay, indicating that Nrp1 cleavage by ADAMs might not be the only mechanism preventing Sema3A responsiveness in PAs. In primary rat cortical neuronal cultures, EphB2 is cleaved mainly by ADAM10 and then further processed by γ-secretase, although whether this proteolytic cascade regulates EphB2 in the context of axon guidance has not been investigated (Litterst et al., 2007). Instead of targeting the guidance receptor Eph, ADAMs regulate ephrin-Eph signaling in axon guidance by processing the ligand ephrin and the co-receptor neural cell adhesion molecule (NCAM). In HEK293 cells, the formation of the ephrin-A5-EphA3 complex activates ADAM10 (which is constitutively associated with EphA3) to cleave ephrin-A5 *in trans*, severing ephrin-A5-EphA3-mediated cell contact and enabling subsequent growth cone retraction (Janes et al., 2005). More recently, as shown in mouse cortical neurons, ephrin-A5 also induces ADAM10 cleavage of NCAM *in cis* (Brennaman, Moss, & Maness, 2014), which is required for ephrin-A5-EphA3-dependent growth cone collapse, possibly by terminating EphA3 clustering and signaling (Sullivan, Kumper, Temple, & Maness, 2016).

γ-Secretase is a multi-subunit intramembrane protease complex that mostly functions to release the ICDs of its target transmembrane proteins and can have major influences on the physiological signaling properties of guidance receptors (Bai & Pfaff, 2011). Multiple lines of evidence suggest that Dcc and neogenin are critically involved in orchestrating the netrin response in neurons (Chedotal, 2019), and both go through sequential proteolytic cleavages by ADAMs and γ-secretase (Fig. 4C). In such cleavage events, a membrane receptor is first processed by sheddases to release its ECD. The membrane-tethered stub, with only a small fragment of ECD remaining, can then successfully pass through a gate formed by the γ-secretase component nicastrin to enter the active site for cleavage (Wolfe, 2019). In *presenilin 1* (*Ps1*, the catalytic subunit of the γ-secretase complex) mutant mice, spinal motor neurons (MNs) ectopically cross the floor plate due to inappropriate netrin1 attraction (Bai et al., 2011). This netrin1 response is obtained through the accumulation of Dcc stubs in the absence of γ-secretase cleavages (Fig. 4B2). Dcc stubs might function by preferentially binding to full-length Dcc receptors in MNs through an unknown mechanism, thus shielding them against interacting with the Robo1 receptor, which can inhibit netrin1 responsiveness. The role of γ-secretase in spinal commissural neurons, which also employ netrin1-Dcc and slit-Robo1 as some of the major pathways to guide their axons across the floor plate, awaits further investigation. Spinal commissures suffer severe disorganization at the floor plate in *Ps1* KO embryos, but this phenotype has not been directly tied to the cleavage of Dcc or any other guidance molecules (Bai et al., 2011). However, a strikingly different signaling strategy operates in *Drosophila* commissural interneurons, where the Dcc homologue Fra is cleaved by γ-secretase to release its ICD as an active signaling molecule (Neuhaus-Follini & Bashaw, 2015). The Fra ICD can translocate to the nucleus to transcriptionally activate the expression of its target gene *commissureless*, which is required cell-autonomously in commissural neurons to facilitate commissural axon midline crossing (Neuhaus-Follini & Bashaw, 2015; Yang, Garbe, & Bashaw, 2009). Given that the P3 motif, which houses both the nuclear export signal and the transcriptional activation domain of Fra, is well conserved from invertebrates to vertebrates, it is tempting to speculate that similar transcriptional function might exist for Dcc as well. Indeed, *in vitro* evidence suggests that both Dcc and neogenin ICDs can localize to the nucleus, and in the case of neogenin, a role in regulating transcription has also been reported (Goldschneider, Rama, Guix, & Mehlen, 2008; Taniguchi, Kim, & Sisodia, 2003). Lastly, γ-secretase processing of neogenin is required for its function across different species, including the regulation of (1) neural tube elongation in zebrafish (Brown et al., 2019), (2) RGC axon targeting in the chick optic tectum (Banerjee et al., 2016), (3) cortical neuron migration and optic nerve regeneration in mouse (van Erp et al., 2015), and (4) transcriptional regulation of gene expression in human cancer cell lines (Goldschneider et al., 2008), attesting to the significance and the universality of regulated proteolysis for the proper signaling propagation of guidance receptors both during development and in disease pathogenesis.

In addition to Dcc and neogenin, the DSCAM ICD generated by γ-secretase cleavage can travel to the nucleus to influence the transcription of many genes involved in axon guidance and other aspects of neural circuit formation (Sachse et al., 2019). Forced expression of DSCAM ICD inhibits neurite outgrowth in primary mouse cortical neurons, yet it remains to be seen whether this inhibited outgrowth has physiological consequences. One perplexing aspect of such membrane-to-nucleus pathways is how proteolysis of a guidance receptor might interfere with downstream signaling outputs that only the full-length receptor can activate. One possibility involves activating the full-length receptor and the γ-secretase cleavage of the receptor with different ligands. Since netrins are not required in the transcriptional activation of *commissureless* (Yang et al., 2009), *Drosophila* commissural neurons might separate the function of full-length Fra and Fra ICD by activating γ-secretase cleavage with an alternative ligand, in a spatially and temporally regulated manner. Physical segregation of full-length receptors and receptor ICDs is another possible mechanism through which neurons could regulate guidance receptor functions. Future studies should address how receptor ICDs are retrogradely transported from the growth cone, where ligand stimulation is presumably detected, to the nucleus.

To further our understanding of the functional significance of proteolytic processing of axon guidance receptors, substantial future efforts should be directed toward uncovering the regulatory mechanisms that modulate the activity of the proteases themselves. ADAM10, ADAM17, BACE1 and γ-secretase are all broadly expressed in the developing nervous system and often target the same membrane receptors as proteolytic substrates to produce distinct amino and carboxy termini with variable functions. One strategy to ensure the specificity of proteases in regulating the precise cleavage of their substrates would be to spatially segregate the proteases into distinct subdomains on the plasma membrane or into different cellular compartments (Lichtenthaler et al., 2018). However, spatial segregation is unlikely to be the only means to regulate protease activity, since ADAM10 and BACE1, ADAM10 and γ-secretase, and BACE1 and γ-secretase can each interact and form proteolytically active multi-protease complexes in the developing mouse brain, suggesting physical separation might not hold true for all signaling pathways (Chen, Kim, et al., 2015; Wang, Wang, & Pei, 2018). It is interesting to note that γ-secretase could inhibit ADAM10 but enhance BACE1 processing of APP (Chen, Kim, et al., 2015). ADAM10, on the other hand, might facilitate BACE1 cleavage of CHL1 (Wang et al., 2018). Although the physiological significance of these regulatory interactions has yet come to light, these observations imply the existence of crosstalk within a functionally interconnected protease network.

Alternatively, neurons can also employ context-specific modulators to regulate the functions of proteases. It is well-established that interacting with different members of the tetraspanin C8 subgroup can differentially influence the trafficking, activity and substrate specificity of ADAM10, whereas those properties of ADAM17 are regulated by the two inactive rhomboid proteases iRhom1 and iRhom2 (Hsia et al., 2019; Lambrecht et al., 2018; Li et al., 2015; Vincent, 2016). Despite their high expression levels in the nervous system, little information is available on the involvement of tetraspanins and iRhoms in regulating axon guidance receptor function. Additionally, tissue inhibitor of matrix metalloproteinase-3 (TIMP3) inhibits ADAM17 activity and can induce neurite outgrowth in the hippocampus both in primary neurons and *in vivo*. However, whether the growth-promoting effect of TIMP3 is dependent upon modulating ADAM17 function has not been established (Gibb et al., 2015). Secreted Frizzled Related Proteins (Sfrps), which share similarities with TIMPs in their C-terminal netrin-related motif, can bind to and suppress ADAM10 cleavages of Notch and APP (Esteve et al., 2011). In *Sfrp*-null mouse embryos, RGC axon guidance and fasciculation defects are observed in both the developing retina and optic nerve (Marcos et al., 2015). Enhanced processing of N-cadherin and L1, two cell adhesion molecules that are also known ADAM10 substrates, correlates with the genetic ablation of *Sfrps* or the overexpression of ADAM10, yet causal relationships between specific ADAM10 substrates and RGC axon growth and guidance have not been established *in vivo*. The Robo receptors are attractive candidates, since inactivation of the slit-Robo pathway phenocopies the various defects that are present in *Sfrp* mutants (Marcos et al., 2015). Finally, ADAM10 might not be uniquely responsible for the axonal abnormalities observed in *Sfrp* mutants as it remains possible that Sfrps can also target and modulate ADAM17 function (Esteve et al., 2011; Marcos et al., 2015). Given the large number of existing physiological substrates for any of these proteases, global KOs of their essential modulators will likely result in altered proteolytic processing of several proteins and thus give rise to exceedingly complicated phenotypes. Future studies with conditional KO models for each of these proteases and their modulators, allowing close inspection of individual axon guidance events, will provide us with more insights into the many molecular mechanisms regulating proteolytic processing of guidance receptors.

3. Downstream signaling of axon guidance receptors

Thus far, we have discussed how axon guidance receptors are regulated, now we move on to how guidance receptors signal to the underlying growth cone cytoskeleton to control specific guidance decisions. Axon guidance receptors primarily induce growth cone turning by impinging on the axonal cytoskeleton to induce movement of the growth cone in a directional manner. A fundamental question is how these guidance receptors achieve specificity in signaling pathways in order to generate distinct functional outputs. Diversity in signaling can be achieved through different downstream targets. However, in many cases, a common set of effectors is recruited in response to a number of different guidance cues. In such instances, differences in signaling can be elicited via regulation of the spatiotemporal activation of an effector molecule. In previous years, much of the effort to elucidate signaling cascades downstream of guidance receptors has focused on RhoGTPases which play pivotal roles in axon growth and guidance. This work has been extensively covered in previous reviews and will not be discussed here (Niftullayev & Lamarche-Vane, 2019). In this section, we discuss recent insights into novel roles of certain actin and microtubule effectors in axon guidance, with a special emphasis on how their spatiotemporal activation can contribute to signaling specificity.

4. Actin binding proteins

The actin cytoskeleton is the primary driving force for growth cone guidance and exploration. When actin polymerization is inhibited using cytochalasin, neurons lose their ability to respond to guidance cues (Bentley & Toroian-Raymond, 1986). Actin remodeling drives the formation of lamellipodia and filopodia that typically underlie growth cone motility and protrusion. It seems reasonable then to infer that actin polymerization is associated with attractive guidance cues while actin depolymerizing agents are recruited by repulsive guidance cues. The reality, however, is more complicated. Many repulsive guidance receptors recruit downstream effectors that enhance actin polymerization, and recent studies (discussed below) highlight the nuanced and complex nature of these actin rearrangements and their importance in repulsive signaling. Aside from propelling the growth cone forward, actin effectors can also influence several other actin-dependent processes including membrane trafficking and endocytosis, that play important roles in regulating receptor signaling. Finally, actin effectors are also likely to act in a context-dependent manner, influenced by the repertoire of other signaling proteins present, and the cellular compartments within which they are activated. In this section, we highlight recent work describing the role of the Wave regulatory complex and Ena/VASP proteins, two major actin effector families, in axon guidance.

4.1 The WASP family of nucleation promoting factors

The actin-nucleating Arp2/3 complex is one of the most well-studied actin binding protein complexes. Upon activation, the Arp2/3 complex nucleates branched actin filaments and is important for the lamellipodial actin network although its role in filopodia formation remains controversial (Yang & Svitkina, 2011). The Arp2/3 complex is activated by nucleation promoting factors (NPFs) like the Wiskott-Aldrich Syndrome protein (WASP) family of NPFs, which consists of five subfamilies in mammals: (i) WASP and N-WASP (ii) the three isoforms of SCAR/WAVE (iii) WHAMM (iv) WASH and (v) JMY. For details on the cellular functions of the WASP family of NPFs, we refer the reader to reviews that explore this topic (Alekhina, Burstein, & Billadeau, 2017; Tyler, Allwood, & Ayscough, 2016). Here, we will highlight the recent work on the SCAR/WAVE subfamily in axon guidance.

The WAVE regulatory complex (WRC) consists of five different proteins: CYFIP/Sra1, Nap1/Kette, Abi, HSPC300/Brick1 and WAVE/SCAR. WAVE contains a VCA region that can bind to Arp2/3 and induce branched actin polymerization. In *Drosophila* and *C. elegans*, SCAR/WAVE has been implicated in axon guidance and targeting (Shakir et al., 2008; Stephan, Gohl, Fleige, Klambt, & Bogdan, 2011; Xu & Quinn, 2012) however the recruitment and activation of the WRC in response to guidance cues is still poorly understood. Recent work identified a unique binding site for the WRC known as WRC-interacting receptor sequence (WIRS; (Chen et al., 2014). The WIRS motif is present in various transmembrane proteins, including several axon guidance receptors. A number of studies have since followed up on the importance of this binding site in various cellular processes. In *Drosophila*, the WIRS motif present in neuroligins was found to be essential for the recruitment of the WRC to postsynaptic membranes at neuromuscular junctions to maintain normal synapse formation and synaptic transmission (Xing et al., 2018). In *C. elegans*, the synaptic cell adhesion protein SYG-1 requires its WIRS motif to interact with the WRC for proper axonal arborization and synapse assembly (Chia, Chen, Li, Rosen, & Shen, 2014). The WIRS-WRC interaction is also important for neogenin to regulate junctional stability in human epithelial cell lines (Lee et al., 2016). These WIRS-WRC interactions can restrict WRC-mediated actin assembly to desired subcellular locations and allow for high spatial and temporal specificity in directing actin polymerization. Since there are WIRS motifs in several axon guidance receptors including Robos and Dcc, it will be interesting to determine if they are functional and if their interaction with the WRC is regulated by the binding of their respective ligands. The WIRS motif has a conserved threonine or serine residue that is essential for its interaction with the WRC, and phosphorylation of various WIRS sites has been previously documented (Hornbeck et al., 2012). Further investigation is necessary to determine if phosphorylation can regulate the WIRS-WRC interaction. Another interesting question is whether in addition to recruiting the WRC, these WIRS-containing receptors can also influence its activity. The cytoplasmic tails of different protocadherin receptors were found to have differential effects on the activity of the WRC in a pyrene-actin polymerization assay (Chen et al., 2014). However, whether this phenomenon has physiological significance is still unclear. Altogether, identification of the WIRS motif has thus provided a direct link between axon guidance receptors and WRC-mediated actin dynamics, and is an important contribution to our understanding of WRC recruitment.

SCAR/WAVE proteins can also function in endocytosis of surface receptors. The WASP subfamily of Arp2/3 NPFs is traditionally accepted as the NPF responsible for regulating internalization events at the plasma membrane (Benesch et al., 2005; Merrifield, Qualmann, Kessels, & Almers, 2004; Tyler et al., 2016). However, in invertebrates like *Drosophila* and *Caenorhabditis*, SCAR/WAVE is a key regulator of endocytic events (Bai & Grant, 2015; Fricke et al., 2009; Giuliani et al., 2009; Patel & Soto, 2013; Shivas & Skop, 2012). New work suggests that this finding can be extended to vertebrate systems as well. In mouse hippocampal neurons, knocking down components of the WRC, but not N-WASP, impairs BDNF-mediated internalization of TrkB, suggesting that the WRC is the NPF responsible for BDNF-induced endocytosis of TrkB (Xu, Fu, Zhu, & Liu, 2016). However, TrkB does not contain a WIRS motif and there is no evidence for a direct interaction between TrkB and the WRC. Retrolinkin, an endosomal vesicle protein important for BDNF-TrkB trafficking (Fu et al., 2011), interacts with CYFIP, and knockdown of retrolinkin in hippocampal neurons shows a decrease in colocalization between WAVE1 and BDNF-activated TrkB, suggesting that retrolinkin functions in the recruitment of the WRC to TrkB. However, since no physical interaction between retrolinkin and TrkB has been reported, it remains unclear how the WRC is recruited to and activated by TrkB. The WRC-induced endocytosis of TrkB appears to be clathrin-independent, consistent with its clathrin-independent role in endocytosis of the interleukin-2 receptor (IL-2R; (Basquin et al., 2015). IL-2R directly interacts with the WRC via a WIRS motif present in its cytoplasmic tail and mutations in the WIRS motif disrupt this interaction and inhibit IL-2R endocytosis. Interestingly, IL-2R endocytosis also requires subsequent activation of N-WASP, providing a unique example of two Arp2/3 activators regulating different steps of endocytosis initiation in a temporally coordinated manner.

Taken together, these findings offer significant advances in understanding the mechanisms underlying recruitment and activation of the WRC, yet how these events are actively regulated in response to specific guidance cues remains largely unknown. Future studies should be directed toward uncovering if and how receptor-WRC interactions are modulated upon ligand binding. Additionally, while the different WASP subfamilies generally have distinct physiological functions, in some cases, they can act cooperatively to regulate a single actin-driven process. Thus, cooperativity between different WASP subfamilies may represent an additional mechanism to diversify individual receptor signaling pathways.

4.2 Ena/VASP proteins

The Ena/VASP family consists of actin-regulatory proteins that associate with the barbed ends of actin filaments and antagonize filament capping, thereby resulting in the generation of long, unbranched F-actin filaments. *Drosophila* and *C. elegans* each have a single Ena/VASP ortholog while mammals have three: Mena, VASP and EVL (Drees & Gertler, 2008). Ena/VASP family members are concentrated at the leading edge of lamellipodia and the tips of filopodia. They are thus perfectly poised to function as immediate modifiers of the actin cytoskeleton in response to guidance cues (Fig. 5). Ena acts in attractive signaling downstream of Dcc and netrin1, functioning to increase filopodia protrusion and elongation (Gitai, Yu, Lundquist, Tessier-Lavigne, & Bargmann, 2003; Lebrand et al., 2004; Yu, Hao, Lim, Tessier-Lavigne, & Bargmann, 2002). A number of studies have demonstrated the requirement for Ena in slit-Robo dependent repulsive signaling as well (Bashaw, Kidd, Murray, Pawson, & Goodman, 2000; Yu et al., 2002) and the *Drosophila* Robo receptor has been shown to directly interact with Ena (Bashaw et al., 2000).

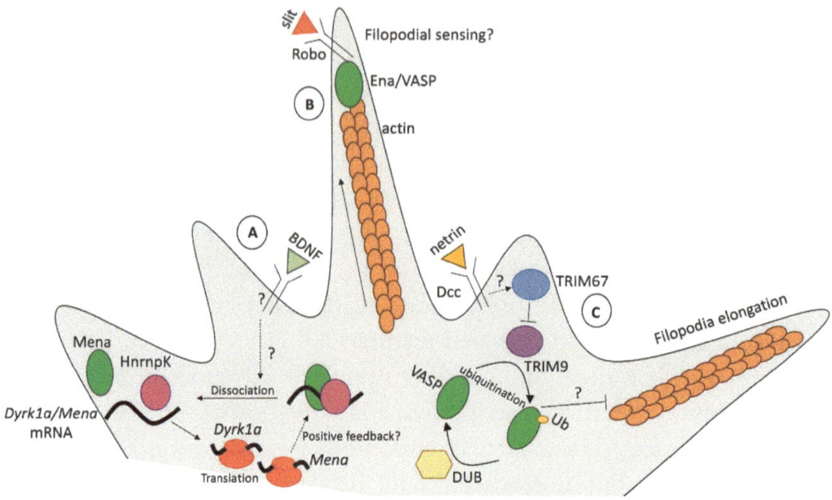

Fig. 5 Different roles of Ena/VASP proteins in axon guidance signaling. (A) BDNF stimulation induces dissociation of the complex of Mena, HnrnpK and *Dyrk1a* mRNA resulting in disinhibition of translation of *Dyrk1a*. Mena can also regulate translation of its own mRNA raising the potential for a positive feedback loop. (B) Robo recruits Ena/VASP proteins in response to slit stimulation to facilitate an initial extension of filopodia toward slit followed by retraction away from it possibly contributing to filopodial sensing of the environment. (C) TRIM9-mediated ubiquitination of VASP inhibits its function in filopodial formation although the mechanistic details are unclear. Netrin stimulation results in TRIM67-mediated inhibition of TRIM9 and activation of de-ubiquitinating (DUB) enzyme that removes the ubiquitin moiety from VASP resulting in its activation and promoting the growth of filopodia. Ub, ubiquitin; DUB, de-ubiquitinating enzyme.

Recent work to resolve the paradoxical role of Ena/VASP proteins in repulsive signaling has uncovered a novel response to slit in dorsal root ganglion axons (McConnell et al., 2016). When DRG axons encounter slit, they initially extend long filopodia toward the source of slit before retracting. This filopodial extension requires the activity of Ena/VASP proteins and is necessary for subsequent slit-induced retraction. Disrupting the Ena/VASP interaction site on Robo abolishes the extension of filopodia and the subsequent repulsion. In mouse embryos deficient for Ena/VASP proteins, DRG axons fail to bifurcate along the rostro-caudal axis and instead aberrantly extend into the dorsal midline similar to the phenotype seen in embryos lacking Robo or slit (McConnell et al., 2016). Together with the *in vitro* data, these observations strongly suggest that Ena/VASP proteins and Robo function together to guide DRG axons. slit-induced filopodia are likely to function in sensing the environment surrounding a growth cone, thereby facilitating a better resolution of guidance gradients (Fig. 5B).

Guidance cues can achieve rapid localized cytoskeletal changes by using molecular switches that allow for immediate regulation of cytoskeletal effector proteins. One such switch involves the rapid, reversible, non-degradative ubiquitination of VASP (Menon et al., 2015) Fig. 5C). The ubiquitination of VASP is mediated by TRIM9, a member of the tripartite motif (TRIM) family of E3 ubiquitin ligases. TRIM9 along with the invertebrate orthologs, MADD-2 and Asap. has previously been implicated in netrin-Dcc signaling (Hao et al., 2010; Morikawa, Kanamori, Yasunaga, & Emoto, 2011; Winkle et al., 2014). However, while *Trim9* loss of function phenocopies *netrin* and *Dcc* mutants in invertebrates (Hao et al., 2010; Morikawa et al., 2011), conditional *Trim9* knockout mice show thickening of the corpus callosum, a phenotype opposite to *Dcc* and *netrin* mutants. This suggests that TRIM9 function might have diverged in mammals (Menon et al., 2015). Mechanistically, TRIM9-mediated ubiquitination of VASP in cortical neurons, has no effect on VASP stability but results in relocalization of VASP away from filopodial tips (Fig. 5C). Although it is unclear how ubiquitination affects VASP localization, it appears that ubiquitination of VASP affects the rate of its dissociation from filopodial tips, which in turn reduces filopodial stability. Consistent with this idea, *Trim9* mutant growth cones show increased filopodial number and length. Upon netrin stimulation, VASP ubiquitination is reduced. This reduction is lost in the presence of a deubiquitinating enzyme (DUB) inhibitor, suggesting that VASP is de-ubiquitinated downstream of netrin signaling (Menon et al., 2015).

More recently, TRIM67 was found to competitively inhibit TRIM9-dependent ubiquitination of VASP (Boyer, McCormick, Menon, Urbina, & Gupton, 2020) Fig. 5C. Cortical neurons lacking TRIM67 show increased ubiquitination of VASP, presumably mediated by TRIM9. *In vivo* analysis of *Trim67* conditional mutant mice shows thinning of the corpus callosum similar to *netrin* and *Dcc* mutants, although the defects in *netrin and Dcc* are much more severe with complete agenesis of the corpus callosum. This might suggest compensatory mechanisms *in vivo*, or perhaps can be partly attributed to the use of whole animal knockouts for *netrin* and *Dcc*. Importantly, the ligase domain of TRIM67 appears to be important for the reduction in VASP ubiquitination. Is TRIM67 ubiquitinating TRIM9 to inhibit TRIM9 activity or is it acting indirectly via other substrates? Despite the contrasting phenotypes of *Trim9* and *Trim67* mutants *in vivo*, both *Trim67* and *Trim9* mutant cortical neurons have an increased growth cone area and number of basal filopodia, *in vitro*. Perhaps this reflects a continuing need for TRIM9 and TRIM67 downstream of netrin signaling to allow for repeated cycling of VASP between ubiquitinated and de-ubiquitinated states for successive rounds of filopodial attachment and detachment. This could represent a model similar to the cycling of RhoGTPases where dominant negative and constitutively active forms of RhoGTPases often show the same phenotype. This balance between TRIM9 and TRIM67 function may fine-tune VASP activity to an optimal level permitting filopodial sensing of the environment.

In addition to its canonical role as a cytoskeletal regulator, Mena, the mammalian ortholog of Ena, can function to regulate local translation both at the basal level and in a cue-dependent manner (Vidaki et al., 2017) Fig. 5A. Mena can bind to its target mRNAs, including the kinase *Dyrk1a*, through its association with the RNA-binding protein (RBP) hnrnpK. In neurons deficient in Ena/VASP proteins, both basal and BDNF-elicited local translation of *Dyrk1a* is impaired (Fig. 5A). Mena acts downstream of many axon guidance receptors that are known to regulate local translation, and thus can serve as a link to the translational machinery. This dual function of Mena places it in a pivotal position to coordinate the crosstalk between cytoskeletal reorganization and local protein synthesis downstream of guidance cues.

These recent studies are paving the way for understanding how a core set of actin-binding proteins can function downstream of both attractive and repulsive guidance receptors. The work on slit-induced filopodial extension in DRG axons, lends credence to the idea that actin polymerization events are essential for repulsive signaling and are more nuanced and complex than previously thought. Further, the discovery of novel functions of actin-binding proteins, namely the regulation of local translation by Mena, can substantially advance our understanding of how these proteins function in several different guidance pathways yet produce distinctive signaling outputs.

5. Microtubule associated proteins

It is widely accepted that the actin cytoskeleton is instrumental in driving growth cone motility in response to axon guidance cues. However, there is a rapidly growing body of evidence to suggest that microtubules (MTs) play more than just a passive role in axonal navigation and extension. The idea of MTs steering growth cones stems from one of the early observations that localized application of Taxol, a MT-stabilizing drug, induces attraction toward the site of application, while Nocodazole, a MT-destabilizer, induces repulsion away from it (Buck & Zheng, 2002). MTs contribute to axonal growth and guidance by advancing and stabilizing filopodia and by directing the asymmetric delivery of intracellular vesicles which contain new membrane components and receptor proteins necessary for membrane protrusion. Additionally, MTs are constantly switching between phases of growth and catastrophe, and this dynamic instability can help growth cones explore and navigate their surrounding environments. Here, we focus on recent work identifying the microtubule-associated proteins (MAPs) required for regulating MT dynamics to influence growth cone steering.

MTs are the stiffest of the cytoskeletal structures (Fletcher & Mullins, 2010). They are composed of α- and β-tubulin heterodimers that polymerize in a head-to-tail arrangement to give rise to protofilaments that in turn assemble to form polarized tubes. TUBB3, a neuronal isoform of β-tubulin, interacts with netrin receptors and plays a role in netrin-induced attraction and repulsion (Huang et al., 2018; Qu et al., 2013; Shao, Yang, Huang, Alarmanazi, & Liu, 2017). TUBB3 can interact with Dcc and this interaction is enhanced in the presence of netrin (Qu et al., 2013). *TUBB3* knockdown affects netrin-induced neurite outgrowth and attraction both *in vitro* and *in vivo* in chick spinal cords. Indeed, certain disease-associated missense mutations in TUBB3 that partially disrupt its interaction with Dcc, fail to rescue the netrin-dependent outgrowth and branching defects induced by *TUBB3* knockdown in cultured cortical neurons (Huang et al., 2018). Electroporation of these TUBB3 variants in spinal cords of chick embryos inhibits netrin-induced attraction of axons in an open-book turning assay. Because Dcc is absent in avians, it is likely that TUBB3 functions downstream of neogenin, a functional substitute for Dcc in chick; however, this requires further study. In mammals, *TUBB3* mutants show defects in formation of the corpus callosum and anterior commissure, phenotypes similar to that seen in *Dcc* and *netrin* mutants, further suggesting a role for TUBB3 in Dcc signaling (Fazeli et al., 1997; Serafini et al., 1996; Tischfield et al., 2010). Future studies should determine the interaction sites on Dcc that are responsible for binding to TUBB3, permitting manipulations that would specifically disrupt TUBB3 function in Dcc signaling rather than globally depleting TUBB3. Unc5 can also interact with TUBB3, although in contrast to Dcc, netrin stimulation reduces this interaction (Shao et al., 2017). The current model suggests that netrin stimulation causes TUBB3 to dissociate from Unc5 on the proximal side, allowing for collapse and repulsion away from netrin.

In addition to interacting with MTs directly, guidance receptors can also recruit MAPs that can bind to MTs and regulate MT dynamics. Several of these MAPs can bind to actin in addition to microtubules and can thus serve as points of crosstalk between the two cytoskeletal networks. Most MT bundles are sequestered in the central domain of the growth cone presumably restricted by actin retrograde flow, but a few extend further out into the periphery. These pioneer MTs can undergo crosslinking with peripherally located F-actin bundles that help guide them into filopodia. Subsequent microtubule capture and stabilization can facilitate directed growth. MT-actin interactions are crucial for achieving proper axon guidance (Coles & Bradke, 2015) and much effort has been directed toward identifying MAPs which have actin and MT crosslinking abilities. For example, XMAP215 and navigator1 (Nav1), two previously known MAPs, were recently shown to also directly interact with F-actin and regulate MT-actin interactions (Sanchez-Huertas et al., 2020; Slater et al., 2019). XMAP215 and Nav1 join the expanding repertoire of crosslinker proteins that includes cytoplasmic linker associated proteins (CLASPs), adenomatous poliposis coli (APC), spectraplakins and others. Although removal of several of these proteins has been shown to impair responsiveness to guidance cues *in vitro*, mechanistic insight into their recruitment and activation downstream of specific guidance receptors is relatively lacking.

MT studies are complicated by the fact that MAPs act in tip-stabilizing complexes, where problems of redundancy between complex components limit the extent of knowledge that can be gained from individual knockout studies. Further, MAPs can have roles independent of MT binding, as was recently reported for MAP6 in Sema3E signaling (Deloulme et al., 2015) and many MAPs can bind to actin as well as MTs, which complicates the interpretations from MAP knockout studies. Finally, some MAPs, like CLASPs, can both promote and inhibit MT elongation (Bearce, Erdogan, & Lowery, 2015) highlighting a complex regulation that may buffer against unwanted polymerization events. As such, reductionist approaches studying single MAPs in isolation are inherently limited. Additionally, with advancements in the fields of optogenetics and live-imaging microscopy, investigators can study the impact of localized MAP manipulations as individual growth cones are visualized in living tissue.

6. Spatiotemporal signaling

It is easy to imagine signal diversification when effectors can perform distinct functions downstream of different receptors, although many effectors have just a single known function. In such instances, differences in signaling can be achieved by regulating the spatiotemporal activation of these effectors. When and where in a cell, these effectors are active, can influence both the duration of their signaling and the protein assemblies with which they interact, resulting in differential signaling outputs (Fehrenbacher, Bar-Sagi, & Philips, 2009; Sugiyama, Fairn, & Antonescu, 2019). In this section we discuss spatial and temporal activation of downstream effectors of axon guidance, and novel tools for studying this local signaling.

6.1 Endosomal signaling

Endosomes play pivotal roles in axon pathfinding by regulating surface levels of axon guidance receptors (Fig. 6). Endocytosis is a well-documented mechanism for attenuating receptor signaling by decreasing the number of surface receptors and targeting them for degradation. In recent years however, several novel functions for endosomes have emerged, from serving as hotspots for local translation to functioning as platforms for diverse signaling cascades. Here, we focus on the role that endosomal signaling plays in axon guidance.

Endosomes can serve as major signaling hubs where continued ligand-receptor association leads to sustained signaling from the endosomal compartment, constituting a population known as "signaling endosomes." Signaling endosomes were traditionally viewed as long-distance, retrograde messengers that transmit information from the nerve terminals to the soma, an important component of neurotrophin signaling (reviewed in Barford, Deppmann, & Winckler, 2017). Now, the definition of signaling endosomes has expanded to include those that signal locally within the growth cone. Some receptors initiate different signaling cascades at the plasma membrane and in endosomes, thus requiring internalization to recruit specific effector molecules. There are several instances of guidance cue-induced signaling that is specifically initiated in endosomes. slit-induced endocytosis of the Robo receptor is required for the recruitment of its downstream effector SOS, a GEF for Rac1 activation (Chance & Bashaw, 2015) Fig. 6A. Disrupting endocytosis of Robo results in a failure to recruit SOS to Robo-induced cellular processes. Endosomal-specific Rac1 activation is also induced by the TrkA receptor, which is internalized upon NGF stimulation (Harrington et al., 2011). Activated Rac1 subsequently recruits cofilin which presumably severs actin filaments to enable the retrograde transport of these TrkA-NGF signaling endosomes to the soma to support neuronal survival (Fig. 6A). Finally, RhoA, another member of the RhoGTPase family is also activated on early endosomes upon NogoAΔ20

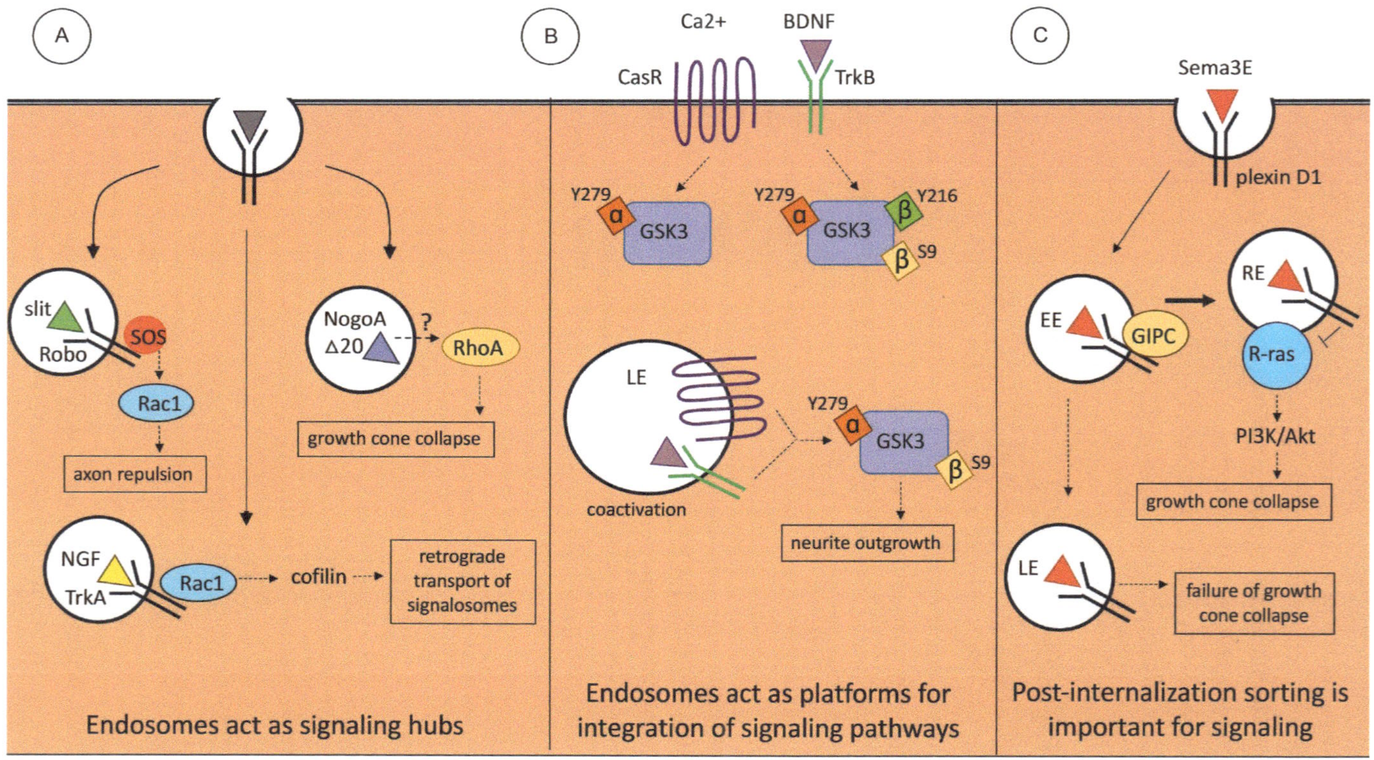

Fig. 6 See figure legend on opposite page.

stimulation (Joset, Dodd, Halegoua, & Schwab, 2010). The growth inhibitory fragment NogoAΔ20 is internalized via the pinocytic chaperone, pincher-dependent macroendocytosis, and generates signaling endosomes that are retrogradely transported to negatively regulate neuronal growth programs in the cell body. These signaling endosomes can also act locally to mediate growth cone collapse in response to NogoAΔ20 (Fig. 6A). Disrupting endocytosis of NogoAΔ20 abolishes RhoA activation and impairs NogoAΔ20-dependent growth cone collapse (Joset et al., 2010). Previously, endocytosis was thought to function primarily in repulsion and growth cone collapse, however a recent study has demonstrated a role for endocytosis in Shh-mediated attractive signaling. Shh induces endocytosis of its receptor, Boc, via an endocytic adapter, Numb, and this internalization is required for Shh-dependent axon turning (Ferent et al., 2019). One demonstrated role for Shh-mediated endocytosis is to facilitate Ptch1 internalization, permitting disinhibition of smoothened (Smo). However, it is tempting to speculate that Shh-induced endocytosis might function as more than just a means for Smo disinhibition, by actively regulating local signaling; perhaps even serving as a site for integration with the Shh-induced DOCK/ELMO/Rac1 pathway (Makihara et al., 2018).

Signaling endosomes are perfectly poised to serve as platforms for crosstalk between signaling pathways (Fig. 6B). Different guidance receptors internalized into a common vesicle can elicit signaling cascades that interact to generate an entirely novel output (Markworth et al., 2019). For example, CasR and TrkB can interact in a ligand-independent manner, and upon coactivation in chick sensory neurons, CasR and TrkB colocalize and co-traffic in Rab7-positive late endosomes. Both CasR and TrkB have a common downstream

Fig. 6 Endosomal signaling in axon guidance. (A) Some axon guidance receptors recruit effector molecules specifically at endosomal membranes and generate a local signaling output which is distinct from that occurring at the plasma membrane. Examples include Robo and SOS, TrkA and Rac1, and NogoA△20 and RhoA. (B) Endosomes can serve as platforms for the integration of signaling pathways. Independently, CasR and TrkB show different phosphorylation profiles of their common downstream target, GSK3. Upon coactivation, CasR and TrkB colocalize in late endosomes where CasR alters the GSK3 phosphorylation target sites of TrkB. (C) Post-internalization sorting into different endosomal populations can contribute to diversity in signaling as different endosomal pools have unique repertoires of effector molecules localized to them. Upon Sema3E binding, plexin D1 is internalized and further sorted into recycling endosomes by GIPC where it inactivates the R-ras population primarily expressed on the membranes of recycling endosomes. Missorting of plexin D1 into late endosomes fails to elicit inactivation of R-ras.

effector, GSK-3, which can be phosphorylated at multiple residues. When independently activated, CasR and TrkB show different phosphorylation profiles of GSK-3 and coactivation of CasR and TrkB results in a non-additive effect of GSK-3 phosphorylation, suggesting that CasR changes the GSK-3 phosphorylation target sites of TrkB (Fig. 6B). The authors hypothesized that GSK-3 cycles between active and inactive states. Supporting this idea, phosphorylation of tau, a downstream target of GSK-3, shows a cycling behavior when CasR and TrkB are coactivated but not when CasR is activated alone (Markworth et al., 2019). Since phosphorylation of tau can affect its ability to promote MT assembly (Lindwall & Cole, 1984), cycling between tau dephosphorylation and phosphorylation can potentially induce successive rounds of microtubule assembly and disassembly to regulate neurite outgrowth. It is unclear whether TrkB activation alone is also capable of eliciting the cycling behavior of phosphorylated tau and what that might mean for the biological significance of this integration of signaling. It remains to be seen whether CasR and TrkB interact directly to regulate crosstalk, and if late endosomes are actively contributing to the integration of their signaling or if they are simply acting as a platform for the localization of these receptors.

Once internalized, receptors can be sorted into recycling endosomes to be returned to multiple locations on the cell surface. This allows cells to fine tune the distribution of receptors and the extent of signaling. Post-internalization sorting is an important regulatory step in several axon guidance pathways and is vital for proper spatiotemporal signaling (Fig. 6C). Adaptor proteins are critical regulators of post-internalization sorting. Recently, an adaptor protein, HD-PTP, was found to be indispensable for ephrin-mediated growth cone collapse. HD-PTP, an accessory protein for the ESCRT complex, regulates post-endocytic sorting of EphB2 complexes by protecting them from lysosomal degradation and facilitating their recycling back to the plasma membrane (Lahaie et al., 2019). In some cases, recycling endosomes do not function solely to replenish surface receptors, but actively contribute to signaling cascades. For instance, active R-ras is detected primarily on the membranes of recycling endosomes. Plexin D1 is internalized in response to Sema3E and requires sorting by the adaptor protein GIPC, into recycling endosomes where it inactivates R-ras (Fig. 6C). Disrupting the interaction between plexin D1 and GIPC results in missorting of plexin D1 from recycling endosomes to late endosomes and loss of R-ras inactivation, along with a failure of Sema3E-induced growth cone collapse (Burk et al., 2017). A very interesting role for recycling endosomes in co-adaptation of ephrinA/EphA signaling was recently demonstrated in chick RGCs (Fiederling et al., 2017). Upon prolonged exposure to guidance signals, axons become less sensitive to those signals. For continued navigation, axons must adapt over time in order to overcome the decreased sensitivity to guidance signals. In this study, the authors use computational modeling and an *in vitro* "gap assay" to demonstrate that axons adapt to both forward and reverse ephrinA/EphA signals and this co-adaptation is based on vesicular trafficking. Upon internalization of ephrinAs and EphAs, enhanced cis signaling in recycling endosomes desensitizes growth cones to trans signals while resensitization is achieved by recycling of ephrinAs and EphAs back to the growth cone surface (Fiederling et al., 2017). These studies highlight the importance of post-endocytic sorting in axon guidance and lend credence to the idea that different endosomal populations have unique signaling profiles that can further contribute to signal diversification.

6.2 Tools for studying spatiotemporal signaling

Specific neuronal connections are established through the concerted action of several guidance receptor pathways that signal through common effector molecules whose dynamics are subject to rigorous spatial and temporal regulation. Thus, to dissect the molecular mechanisms involved in guidance receptor signaling, it is imperative to study the spatiotemporal signaling of these effectors together with their regulatory mechanisms. Unfortunately, the commonly used methods of studying signal transduction traditionally involve cell lysis or fixation, offering only limited spatial and temporal resolution. To overcome these challenges, many novel tools have been developed in the last decade to help visualize and manipulate these signaling molecules with greater precision. Here we provide a brief overview of these novel tools and refer readers to several excellent reviews for further information (Ross, Mehta, & Zhang, 2016; Rost, Schneider-Warme, Schmitz, & Hegemann, 2017).

Advancements in optical imaging have enabled the precise observation of signaling molecules at subcellular resolution and with high temporal specificity. Improvements in the design and sensitivity of genetically encoded fluorescent biosensors have made them powerful tools for monitoring effector activity in living cells. The most common biosensors are the fluorescence resonance energy transfer (FRET)-based biosensors, which are engineered by sandwiching the sensing unit between two fluorescent proteins that act as a FRET pair (Marx, 2017). FRET biosensors offer an advantage over single fluorescent protein biosensors because they are ratiometric, making them more reliable and robust. When targeted to specific subcellular compartments, these biosensors can provide valuable information on the functional compartmentalization of signaling in neurons. However, many fluorescent biosensors are limited by their reliance on visible light which penetrates poorly through tissues. Far-red and near-infrared sensors, which can penetrate deeper into tissues, may be more suitable for *in vivo* imaging (Chernov, Redchuk, Omelina, & Verkhusha, 2017).

With the advent of super-resolution (SR) microscopy, live imaging at nanoscale resolution is now possible. However, most SR techniques use very high intensity light which can cause photobleaching, phototoxicity and other adverse consequences, prohibiting long-term imaging of dynamic cellular events. To this end, super-resolution based on reversibly photoswitchable fluorescent proteins (RSFPs) is proving to be extremely useful. Reversible saturable optical fluorescence transition (RESOLFT) microscopy uses RSFPs that can undergo repeated cycles of photoactivation from a dark state to a fluorescent state, and enables live imaging for long periods with negligible photodamage (Kwon et al., 2015). Bioluminescence offers another strategy to avoid the problem of phototoxicity and many bioluminescence resonance energy transfer (BRET) sensors are being developed. While most luminescent proteins are dimmer, compromising resolution, Nano-lantern and other color variants enable high-speed multicolor luminescence in living cells (Takai et al., 2015). Other efforts are being directed at physically magnifying the sample itself in a new form of super-resolution microscopy known as expansion microscopy (Chen, Tillberg, & Boyden, 2015). This method uses a swellable polymer that expands such that labeled biomolecules within the sample can be separated to distances that enable their resolution with traditional microscopy.

In addition to being able to detect the spatiotemporal activation of effector molecules, it is imperative to develop tools that can precisely manipulate effector activity in time and space. Optical manipulation of specific effector activities allows for increasingly finer control in live cell studies. This can be achieved through the use of naturally occurring light-activated proteins like photoactivatable adenylyl cyclases (PACs) which can be optically activated to regulate cAMP levels in a spatiotemporal manner. Optical manipulation can also be achieved through the optical caging of proteins, wherein the target protein or its activity is blocked by a photolabile group that enables light-induced activation with high resolution. Alternatively, light-induced dimerization (LID) can be used to facilitate spatiotemporal activation of signaling by regulating protein interactions and effector recruitment. In LID, one component of the dimer is targeted to a specific subcellular compartment while the other component is expressed as a cytosolic protein. The dimerization induced by light facilitates rapid and precise recruitment of the cytosolic protein to the subcellular compartment. LID can also be used to achieve homo-oligomerization of proteins thereby allowing for finer control over receptor activation. This has been demonstrated for optoTrk receptors, which can recapitulate BDNF-TrkB signaling in response to light in cultured hippocampal neurons (Chang et al., 2014). Lastly, magnetic nanoparticles (MNPs) are another rapidly developing tool for manipulating single molecules with high spatiotemporal precision (Nimpf & Keays, 2017). MNPs can be used to control the spatial organization of cell surface receptors and induce signal transduction as reported for Notch receptors (Seo et al., 2016). However, technical considerations remain with regard to the delivery of these MNPs within cells and current research is geared toward enabling genetic encoding of MNPs.

There have been tremendous advancements in the fields of optogenetics, magnetogenetics and nanoscale microscopy. The availability of these novel tools for visualizing and manipulating effector molecules with high spatiotemporal precision, offers many exciting avenues of investigation. An ongoing effort is the design and optimization of these tools for various different effectors involved in guidance receptor signaling, and the improvement of existing technologies to make them more amenable to *in vivo* imaging.

7. Concluding remarks

In this chapter, we have sought to highlight recent advances and new concepts emerging from the investigation of axon guidance receptor regulation and signaling. When conserved families of guidance cues and receptors were first described, their functions were grouped into the two major categories of attraction and repulsion. They were further ascribed either short-range or long-range functions, or both. This proved to be a useful framework for thinking about axon guidance mechanisms, but not surprisingly this categorization represented an over-simplification of the diverse actions of these signaling proteins. In addition, while first described for their actions in the developing nervous system, it is now appreciated that these conserved families of signaling proteins regulate diverse aspects of neuronal morphogenesis, homeostasis and function.

Investigations into the mechanisms of receptor regulation over the past ten years have begun to highlight the myriad of ways in which interactions between receptors and coreceptors, as well as interactions between receptors for distinct guidance cues serve to diversify and modulate signaling outputs.

This has challenged the notion that guidance pathways only act in isolation to mediate discrete steps of the axon guidance process, and that it is only the net balance between attraction and repulsion that determines guidance outcomes. Instead, these pathways are often integrated and converge to yield distinct and sometimes synergistic effects on axonal responses. This new understanding of the complex cross talk between signaling pathways has been dramatically accelerated by the burgeoning body of structural work that sheds light on how receptor-ligand and receptor-receptor interactions control receptor activation. In addition to new roles for interactions between distinct guidance pathways, significant progress has been made in understanding how proteolytic processing of axon guidance receptors can regulate receptor availability and downstream signaling. One particularly interesting emerging theme is that receptors often undergo processive proteolysis to generate intracellular domains which can signal in new ways, including by acting as direct regulators of transcription. This suggests the intriguing possibility that axon guidance pathways used for building neuronal circuits could be repurposed to control other neuronal properties by directly regulating gene expression.

As we have seen for mechanisms of receptor regulation, the study of the complex mechanisms of signaling downstream of axon guidance receptor activation has also been a fertile area of investigation. One of the most challenging aspects of understanding axon guidance receptor signaling has been the finding that frequently the same signaling molecules can act in both axon attraction and axon repulsion. Higher resolution imaging and a finer focus on the spatial regulation of signaling events has begun to shed light on this paradox. For example, recent studies of Ena/VASP proteins point to the importance of local signaling within filopodia to allow growth cones to sample gradients of extracellular cues and reveal that an important aspect of the response to repellant cues is to first extend filopodia. As new tools become available to image the activity of signaling molecules in sub-domains of the growth cone, we expect that a clearer delineation of the roles of proteins that act in both attraction and repulsion will emerge. New studies of actin regulatory proteins such as Ena/VASP have also highlighted how these proteins can play distinct roles in the growth cone depending on how upstream pathways impinge upon them, and where within the growth cone they are activated. For instance, Ena/VASP can regulate filopodial extension in response to netrin or regulate local protein translation in response to BDNF. A clear challenge for the future will be to achieve an understanding of how these distinct activities are regulated spatially and temporally to confer distinct signaling outcomes.

Finally, it is important to recognize that without exception, conserved families of axon guidance proteins play broad roles in the development and function of multiple organ systems outside of the nervous system, and disruption of their signaling pathways is implicated in many human diseases. The broad and diverse roles of axon guidance signaling pathways creates an even stronger imperative to dissect their mechanisms of action. Thus, future studies of the regulation and function of axon guidance signaling pathways will not only increase our understanding of the development and function of the nervous system, they will also offer broad insights into development and disease.

Acknowledgments

We thank members of the Bashaw lab for thoughtful discussion and comments on this manuscript. We are also grateful to Engin Ozkan and Artur Kania for critical feedback on the review. Research in the Bashaw Lab is supported by grants from the National Science Foundation (IOS-1853719) and the National Institutes of Health (R35 NS097340).

References

Alekhina, O., Burstein, E., & Billadeau, D. D. (2017). Cellular functions of WASP family proteins at a glance. *Journal of Cell Science, 130*, 2235–2241.

Aleksandrova, N., Gutsche, I., Kandiah, E., Avilov, S. V., Petoukhov, M. V., Seiradake, E., et al. (2018). Robo1 forms a compact dimer-of-dimers assembly. *Structure, 26*(320-328), e324.

Alther, T. A., Domanitskaya, E., & Stoeckli, E. T. (2016). Calsyntenin 1-mediated trafficking of axon guidance receptors regulates the switch in axonal responsiveness at a choice point. *Development, 143*, 994–1004.

Arvanitis, D. N., Jungas, T., Behar, A., & Davy, A. (2010). Ephrin-B1 reverse signaling controls a posttranscriptional feedback mechanism via miR-124. *Molecular and Cellular Biology, 30*, 2508–2517.

Bai, G., Chivatakarn, O., Bonanomi, D., Lettieri, K., Franco, L., Xia, C., et al. (2011). Presenilin-dependent receptor processing is required for axon guidance. *Cell, 144*, 106–118.

Bai, Z., & Grant, B. D. (2015). A TOCA/CDC-42/PAR/WAVE functional module required for retrograde endocytic recycling. *Proceedings of the National Academy of Sciences of the United States of America, 112*, E1443–E1452.

Bai, G., & Pfaff, S. L. (2011). Protease regulation: The Yin and Yang of neural development and disease. *Neuron, 72*, 9–21.

Banerjee, P., Harada, H., Tassew, N. G., Charish, J., Goldschneider, D., Wallace, V. A., et al. (2016). Upsilon-secretase and LARG mediate distinct RGMa activities to control appropriate layer targeting within the optic tectum. *Cell Death and Differentiation, 23*, 442–453.

Barak, R., Yom-Tov, G., Guez-Haddad, J., Gasri-Plotnitsky, L., Maimon, R., Cohen-Berkman, M., et al. (2019). Structural principles in robo activation and auto-inhibition. *Cell, 177*(272-285), e216.

Barao, S., Gartner, A., Leyva-Diaz, E., Demyanenko, G., Munck, S., Vanhoutvin, T., et al. (2015). Antagonistic effects of BACE1 and APH1B-gamma-secretase control axonal guidance by regulating growth Cone collapse. *Cell Reports, 12*, 1367–1376.

Barford, K., Deppmann, C., & Winckler, B. (2017). The neurotrophin receptor signaling endosome: Where trafficking meets signaling. *Developmental Neurobiology*, *77*, 405–418.
Bashaw, G. J., Kidd, T., Murray, D., Pawson, T., & Goodman, C. S. (2000). Repulsive axon guidance: Abelson and enabled play opposing roles downstream of the roundabout receptor. *Cell*, *101*, 703–715.
Bashaw, G. J., & Klein, R. (2010). Signaling from axon guidance receptors. *Cold Spring Harbor Perspectives in Biology*, *2*, a001941.
Basquin, C., Trichet, M., Vihinen, H., Malarde, V., Lagache, T., Ripoll, L., et al. (2015). Membrane protrusion powers clathrin-independent endocytosis of interleukin-2 receptor. *The EMBO Journal*, *34*, 2147–2161.
Bearce, E. A., Erdogan, B., & Lowery, L. A. (2015). TIPsy tour guides: How microtubule plus-end tracking proteins (+TIPs) facilitate axon guidance. *Frontiers in Cellular Neuroscience*, *9*, 241.
Bellon, A., Luchino, J., Haigh, K., Rougon, G., Haigh, J., Chauvet, S., et al. (2010). VEGFR2 (KDR/Flk1) signaling mediates axon growth in response to semaphorin 3E in the developing brain. *Neuron*, *66*, 205–219.
Benesch, S., Polo, S., Lai, F. P., Anderson, K. I., Stradal, T. E., Wehland, J., et al. (2005). N-WASP deficiency impairs EGF internalization and actin assembly at clathrin-coated pits. *Journal of Cell Science*, *118*, 3103–3115.
Bentley, D., & Toroian-Raymond, A. (1986). Disoriented pathfinding by pioneer neurone growth cones deprived of filopodia by cytochalasin treatment. *Nature*, *323*, 712–715.
Bonanomi, D., Chivatakarn, O., Bai, G., Abdesselem, H., Lettieri, K., Marquardt, T., et al. (2012). Ret is a multifunctional coreceptor that integrates diffusible- and contact-axon guidance signals. *Cell*, *148*, 568–582.
Boyer, N. P., McCormick, L. E., Menon, S., Urbina, F. L., & Gupton, S. L. (2020). A pair of E3 ubiquitin ligases compete to regulate filopodial dynamics and axon guidance. *The Journal of Cell Biology*, *219*.
Brennaman, L. H., Moss, M. L., & Maness, P. F. (2014). EphrinA/EphA-induced ectodomain shedding of neural cell adhesion molecule regulates growth cone repulsion through ADAM10 metalloprotease. *Journal of Neurochemistry*, *128*, 267–279.
Brose, K., & Tessier-Lavigne, M. (2000). Slit proteins: Key regulators of axon guidance, axonal branching, and cell migration. *Current Opinion in Neurobiology*, *10*, 95–102.
Brown, S., Jayachandran, P., Negesse, M., Olmo, V., Vital, E., & Brewster, R. (2019). Rgma-induced Neo1 proteolysis promotes neural tube morphogenesis. *The Journal of Neuroscience*, *39*, 7465–7484.
Buck, K. B., & Zheng, J. Q. (2002). Growth cone turning induced by direct local modification of microtubule dynamics. *The Journal of Neuroscience*, *22*, 9358–9367.
Burk, K., Mire, E., Bellon, A., Hocine, M., Guillot, J., Moraes, F., et al. (2017). Post-endocytic sorting of Plexin-D1 controls signal transduction and development of axonal and vascular circuits. *Nature Communications*, *8*, 14508.
Cajal, R. (1890). A quelle époque apparaissent les expansions des cellules nerveuses de la moelle épinière du poulet. *Anatomischer Anzeiger*, *5*, 609–613.
Chance, R. K., & Bashaw, G. J. (2015). Slit-dependent endocytic trafficking of the robo receptor is required for son of sevenless recruitment and midline axon repulsion. *PLoS Genetics*, *11*, e1005402.
Chang, K. Y., Woo, D., Jung, H., Lee, S., Kim, S., Won, J., et al. (2014). Light-inducible receptor tyrosine kinases that regulate neurotrophin signalling. *Nature Communications*, *5*, 4057.
Chauvet, S., Cohen, S., Yoshida, Y., Fekrane, L., Livet, J., Gayet, O., et al. (2007). Gating of Sema3E/PlexinD1 signaling by neuropilin-1 switches axonal repulsion to attraction during brain development. *Neuron*, *56*, 807–822.

Chedotal, A. (2019). Roles of axon guidance molecules in neuronal wiring in the developing spinal cord. *Nature Reviews. Neuroscience*, *20*, 380–396.
Chen, B., Brinkmann, K., Chen, Z., Pak, C. W., Liao, Y., Shi, S., et al. (2014). The WAVE regulatory complex links diverse receptors to the actin cytoskeleton. *Cell*, *156*, 195–207.
Chen, Z., Gore, B. B., Long, H., Ma, L., & Tessier-Lavigne, M. (2008). Alternative splicing of the Robo3 axon guidance receptor governs the midline switch from attraction to repulsion. *Neuron*, *58*, 325–332.
Chen, A. C., Kim, S., Shepardson, N., Patel, S., Hong, S., & Selkoe, D. J. (2015). Physical and functional interaction between the alpha- and gamma-secretases: A new model of regulated intramembrane proteolysis. *The Journal of Cell Biology*, *211*, 1157–1176.
Chen, F., Tillberg, P. W., & Boyden, E. S. (2015). Optical imaging. Expansion microscopy. *Science*, *347*, 543–548.
Chernov, K. G., Redchuk, T. A., Omelina, E. S., & Verkhusha, V. V. (2017). Near-infrared fluorescent proteins, biosensors, and optogenetic tools engineered from phytochromes. *Chemical Reviews*, *117*, 6423–6446.
Chia, P. H., Chen, B., Li, P., Rosen, M. K., & Shen, K. (2014). Local F-actin network links synapse formation and axon branching. *Cell*, *156*, 208–220.
Coleman, H. A., Labrador, J. P., Chance, R. K., & Bashaw, G. J. (2010). The Adam family metalloprotease Kuzbanian regulates the cleavage of the roundabout receptor to control axon repulsion at the midline. *Development*, *137*, 2417–2426.
Coles, C. H., & Bradke, F. (2015). Coordinating neuronal actin-microtubule dynamics. *Current Biology*, *25*, R677–R691.
Croteau, L. P., Kao, T. J., & Kania, A. (2019). Ephrin-A5 potentiates netrin-1 axon guidance by enhancing neogenin availability. *Scientific Reports*, *9*, 12009.
Dascenco, D., Erfurth, M. L., Izadifar, A., Song, M., Sachse, S., Bortnick, R., et al. (2015). Slit and receptor tyrosine phosphatase 69D confer spatial specificity to axon branching via Dscam1. *Cell*, *162*, 1140–1154.
Deloulme, J. C., Gory-Faure, S., Mauconduit, F., Chauvet, S., Jonckheere, J., Boulan, B., et al. (2015). Microtubule-associated protein 6 mediates neuronal connectivity through Semaphorin 3E-dependent signalling for axonal growth. *Nature Communications*, *6*, 7246.
Drees, F., & Gertler, F. B. (2008). Ena/VASP: Proteins at the tip of the nervous system. *Current Opinion in Neurobiology*, *18*, 53–59.
Esteve, P., Sandonis, A., Cardozo, M., Malapeira, J., Ibanez, C., Crespo, I., et al. (2011). SFRPs act as negative modulators of ADAM10 to regulate retinal neurogenesis. *Nature Neuroscience*, *14*, 562–569.
Evans, T. A., Santiago, C., Arbeille, E., & Bashaw, G. J. (2015). Robo2 acts in trans to inhibit Slit-Robo1 repulsion in pre-crossing commissural axons. *eLife*, *4*, e08407.
Fazeli, A., Dickinson, S. L., Hermiston, M. L., Tighe, R. V., Steen, R. G., Small, C. G., et al. (1997). Phenotype of mice lacking functional deleted in colorectal cancer (Dcc) gene. *Nature*, *386*, 796–804.
Fehrenbacher, N., Bar-Sagi, D., & Philips, M. (2009). Ras/MAPK signaling from endomembranes. *Molecular Oncology*, *3*, 297–307.
Ferent, J., Giguere, F., Jolicoeur, C., Morin, S., Michaud, J. F., Makihara, S., et al. (2019). Boc acts via numb as a Shh-dependent endocytic platform for Ptch1 internalization and Shh-mediated axon guidance. *Neuron*, *102*(1157-1171), e1155.
Fiederling, F., Weschenfelder, M., Fritz, M., von Philipsborn, A., Bastmeyer, M., & Weth, F. (2017). Ephrin-A/EphA specific co-adaptation as a novel mechanism in topographic axon guidance. *eLife*, *6*.
Finci, L. I., Kruger, N., Sun, X., Zhang, J., Chegkazi, M., Wu, Y., et al. (2014). The crystal structure of netrin-1 in complex with DCC reveals the bifunctionality of netrin-1 as a guidance cue. *Neuron*, *83*, 839–849.

Fletcher, D. A., & Mullins, R. D. (2010). Cell mechanics and the cytoskeleton. *Nature*, *463*, 485–492.
Fricke, R., Gohl, C., Dharmalingam, E., Grevelhorster, A., Zahedi, B., Harden, N., et al. (2009). Drosophila Cip4/Toca-1 integrates membrane trafficking and actin dynamics through WASP and SCAR/WAVE. *Current Biology*, *19*, 1429–1437.
Friocourt, F., & Chedotal, A. (2017). The Robo3 receptor, a key player in the development, evolution, and function of commissural systems. *Developmental Neurobiology*, *77*, 876–890.
Fu, X., Yang, Y., Xu, C., Niu, Y., Chen, T., Zhou, Q., et al. (2011). Retrolinkin cooperates with endophilin A1 to mediate BDNF-TrkB early endocytic trafficking and signaling from early endosomes. *Molecular Biology of the Cell*, *22*, 3684–3698.
Gao, X., Metzger, U., Panza, P., Mahalwar, P., Alsheimer, S., Geiger, H., et al. (2015). A floor-plate extracellular protein-protein interaction screen identifies draxin as a secreted netrin-1 antagonist. *Cell Reports*, *12*, 694–708.
Gibb, S. L., Zhao, Y., Potter, D., Hylin, M. J., Bruhn, R., Baimukanova, G., et al. (2015). TIMP3 attenuates the loss of neural stem cells, mature neurons and neurocognitive dysfunction in traumatic brain injury. *Stem Cells*, *33*, 3530–3544.
Gitai, Z., Yu, T. W., Lundquist, E. A., Tessier-Lavigne, M., & Bargmann, C. I. (2003). The netrin receptor UNC-40/DCC stimulates axon attraction and outgrowth through enabled and, in parallel, Rac and UNC-115/AbLIM. *Neuron*, *37*, 53–65.
Giuliani, C., Troglio, F., Bai, Z., Patel, F. B., Zucconi, A., Malabarba, M. G., et al. (2009). Requirements for F-BAR proteins TOCA-1 and TOCA-2 in actin dynamics and membrane trafficking during Caenorhabditis elegans oocyte growth and embryonic epidermal morphogenesis. *PLoS Genetics*, *5*, e1000675.
Goldschneider, D., Rama, N., Guix, C., & Mehlen, P. (2008). The neogenin intracellular domain regulates gene transcription via nuclear translocation. *Molecular and Cellular Biology*, *28*, 4068–4079.
Gorla, M., & Bashaw, G. J. (2020). Molecular mechanisms regulating axon responsiveness at the midline. *Developmental Biology*, *466*, 12–21.
Gorla, M., Santiago, C., Chaudhari, K., Layman, A. A. K., Oliver, P. M., & Bashaw, G. J. (2019). Ndfip proteins target robo receptors for degradation and allow commissural axons to cross the midline in the developing spinal cord. *Cell Reports*, *26*(3298–3312), e3294.
Hao, J. C., Adler, C. E., Mebane, L., Gertler, F. B., Bargmann, C. I., & Tessier-Lavigne, M. (2010). The tripartite motif protein MADD-2 functions with the receptor UNC-40 (DCC) in netrin-mediated axon attraction and branching. *Developmental Cell*, *18*, 950–960.
Harrington, A. W., St Hillaire, C., Zweifel, L. S., Glebova, N. O., Philippidou, P., Halegoua, S., et al. (2011). Recruitment of actin modifiers to TrkA endosomes governs retrograde NGF signaling and survival. *Cell*, *146*, 421–434.
Himanen, J. P., Yermekbayeva, L., Janes, P. W., Walker, J. R., Xu, K., Atapattu, L., et al. (2010). Architecture of Eph receptor clusters. *Proceedings of the National Academy of Sciences of the United States of America*, *107*, 10860–10865.
Horiuchi, K., Kimura, T., Miyamoto, T., Takaishi, H., Okada, Y., Toyama, Y., et al. (2007). Cutting edge: TNF-alpha-converting enzyme (TACE/ADAM17) inactivation in mouse myeloid cells prevents lethality from endotoxin shock. *Journal of Immunology*, *179*, 2686–2689.
Hornbeck, P. V., Kornhauser, J. M., Tkachev, S., Zhang, B., Skrzypek, E., Murray, B., et al. (2012). PhosphoSitePlus: A comprehensive resource for investigating the structure and function of experimentally determined post-translational modifications in man and mouse. *Nucleic Acids Research*, *40*, D261–D270.

Hornberg, H., Cioni, J. M., Harris, W. A., & Holt, C. E. (2016). Hermes regulates axon sorting in the optic tract by post-trancriptional regulation of neuropilin 1. *The Journal of Neuroscience*, *36*, 12697–12706.
Hsia, H. E., Tushaus, J., Brummer, T., Zheng, Y., Scilabra, S. D., & Lichtenthaler, S. F. (2019). Functions of 'A disintegrin and metalloproteases (ADAMs)' in the mammalian nervous system. *Cellular and Molecular Life Sciences*, *76*, 3055–3081.
Huang, H., Yang, T., Shao, Q., Majumder, T., Mell, K., & Liu, G. (2018). Human TUBB3 mutations disrupt netrin attractive signaling. *Neuroscience*, *374*, 155–171.
Janes, P. W., Saha, N., Barton, W. A., Kolev, M. V., Wimmer-Kleikamp, S. H., Nievergall, E., et al. (2005). Adam meets Eph: An ADAM substrate recognition module acts as a molecular switch for ephrin cleavage in trans. *Cell*, *123*, 291–304.
Janssen, B. J., Robinson, R. A., Perez-Branguli, F., Bell, C. H., Mitchell, K. J., Siebold, C., et al. (2010). Structural basis of semaphorin-plexin signalling. *Nature*, *467*, 1118–1122.
Jaworski, A., Long, H., & Tessier-Lavigne, M. (2010). Collaborative and specialized functions of Robo1 and Robo2 in spinal commissural axon guidance. *The Journal of Neuroscience*, *30*, 9445–9453.
Johnson, V., Junge, H. J., & Chen, Z. (2019). Temporal regulation of axonal repulsion by alternative splicing of a conserved microexon in mammalian Robo1 and Robo2. *eLife*, *8*.
Jorissen, E., Prox, J., Bernreuther, C., Weber, S., Schwanbeck, R., Serneels, L., et al. (2010). The disintegrin/metalloproteinase ADAM10 is essential for the establishment of the brain cortex. *The Journal of Neuroscience*, *30*, 4833–4844.
Joset, A., Dodd, D. A., Halegoua, S., & Schwab, M. E. (2010). Pincher-generated Nogo-A endosomes mediate growth cone collapse and retrograde signaling. *The Journal of Cell Biology*, *188*, 271–285.
Kennedy, T. E. (2000). Cellular mechanisms of netrin function: Long-range and short-range actions. *Biochemistry and Cell Biology*, *78*, 569–575.
Kinoshita-Kawada, M., Hasegawa, H., Hongu, T., Yanagi, S., Kanaho, Y., Masai, I., et al. (2019). A crucial role for Arf6 in the response of commissural axons to Slit. *Development*, *146*.
Kong, Y., Janssen, B. J., Malinauskas, T., Vangoor, V. R., Coles, C. H., Kaufmann, R., et al. (2016). Structural basis for plexin activation and regulation. *Neuron*, *91*, 548–560.
Kuhn, P. H., Koroniak, K., Hogl, S., Colombo, A., Zeitschel, U., Willem, M., et al. (2012). Secretome protein enrichment identifies physiological BACE1 protease substrates in neurons. *The EMBO Journal*, *31*, 3157–3168.
Kullander, K., & Klein, R. (2002). Mechanisms and functions of Eph and ephrin signalling. *Nature Reviews. Molecular Cell Biology*, *3*, 475–486.
Kuwajima, T., Yoshida, Y., Takegahara, N., Petros, T. J., Kumanogoh, A., Jessell, T. M., et al. (2012). Optic chiasm presentation of Semaphorin6D in the context of Plexin-A1 and Nr-CAM promotes retinal axon midline crossing. *Neuron*, *74*, 676–690.
Kwon, J., Hwang, J., Park, J., Han, G. R., Han, K. Y., & Kim, S. K. (2015). RESOLFT nanoscopy with photoswitchable organic fluorophores. *Scientific Reports*, *5*, 17804.
Lahaie, S., Morales, D., Bagci, H., Hamoud, N., Castonguay, C. E., Kazan, J. M., et al. (2019). The endosomal sorting adaptor HD-PTP is required for ephrin-B:EphB signalling in cellular collapse and spinal motor axon guidance. *Scientific Reports*, *9*, 11945.
Lambrecht, B. N., Vanderkerken, M., & Hammad, H. (2018). The emerging role of ADAM metalloproteinases in immunity. *Nature Reviews. Immunology*, *18*, 745–758.
Lebrand, C., Dent, E. W., Strasser, G. A., Lanier, L. M., Krause, M., Svitkina, T. M., et al. (2004). Critical role of Ena/VASP proteins for filopodia formation in neurons and in function downstream of netrin-1. *Neuron*, *42*, 37–49.
Lee, N. K., Fok, K. W., White, A., Wilson, N. H., O'Leary, C. J., Cox, H. L., et al. (2016). Neogenin recruitment of the WAVE regulatory complex maintains adherens junction stability and tension. *Nature Communications*, 7, 11082.

Leggere, J. C., Saito, Y., Darnell, R. B., Tessier-Lavigne, M., Junge, H. J., & Chen, Z. (2016). NOVA regulates Dcc alternative splicing during neuronal migration and axon guidance in the spinal cord. *eLife*, *5*.
Leyva-Diaz, E., del Toro, D., Menal, M. J., Cambray, S., Susin, R., Tessier-Lavigne, M., et al. (2014). FLRT3 is a Robo1-interacting protein that determines Netrin-1 attraction in developing axons. *Current Biology*, *24*, 494–508.
Li, X., Maretzky, T., Weskamp, G., Monette, S., Qing, X., Issuree, P. D., et al. (2015). iRhoms 1 and 2 are essential upstream regulators of ADAM17-dependent EGFR signaling. *Proceedings of the National Academy of Sciences of the United States of America*, *112*, 6080–6085.
Lichtenthaler, S. F., Lemberg, M. K., & Fluhrer, R. (2018). Proteolytic ectodomain shedding of membrane proteins in mammals-hardware, concepts, and recent developments. *The EMBO Journal*, *37*.
Lindwall, G., & Cole, R. D. (1984). Phosphorylation affects the ability of tau protein to promote microtubule assembly. *The Journal of Biological Chemistry*, *259*, 5301–5305.
Litterst, C., Georgakopoulos, A., Shioi, J., Ghersi, E., Wisniewski, T., Wang, R., et al. (2007). Ligand binding and calcium influx induce distinct ectodomain/gamma-secretase-processing pathways of EphB2 receptor. *The Journal of Biological Chemistry*, *282*, 16155–16163.
Liu, Y., Bhowmick, T., Liu, Y., Gao, X., Mertens, H. D. T., Svergun, D. I., et al. (2018). Structural basis for draxin-modulated axon guidance and fasciculation by netrin-1 through DCC. *Neuron*, *97*(1261–1267), e1264.
Liu, H., Juo, Z. S., Shim, A. H., Focia, P. J., Chen, X., Garcia, K. C., et al. (2010). Structural basis of semaphorin-plexin recognition and viral mimicry from Sema7A and A39R complexes with PlexinC1. *Cell*, *142*, 749–761.
Long, J., Li, M., Ren, Q., Zhang, C., Fan, J., Duan, Y., et al. (2012). Phylogenetic and molecular evolution of the ADAM (A disintegrin and metalloprotease) gene family from Xenopus tropicalis, to Mus musculus, Rattus norvegicus, and Homo sapiens. *Gene*, *507*, 36–43.
Makihara, S., Morin, S., Ferent, J., Cote, J. F., Yam, P. T., & Charron, F. (2018). Polarized dock activity drives Shh-mediated axon guidance. *Developmental Cell*, *46*(410-425), e417.
Marcos, S., Nieto-Lopez, F., Sandonis, A., Cardozo, M. J., Di Marco, F., Esteve, P., et al. (2015). Secreted frizzled related proteins modulate pathfinding and fasciculation of mouse retina ganglion cell axons by direct and indirect mechanisms. *The Journal of Neuroscience*, *35*, 4729–4740.
Marita, M., Wang, Y., Kaliszewski, M. J., Skinner, K. C., Comar, W. D., Shi, X., et al. (2015). Class A plexins are organized as preformed inactive dimers on the cell surface. *Biophysical Journal*, *109*, 1937–1945.
Markworth, R., Adolfs, Y., Dambeck, V., Steinbeck, L. M., Lize, M., Pasterkamp, R. J., et al. (2019). Sensory axon growth requires spatiotemporal integration of CaSR and TrkB signaling. *The Journal of Neuroscience*, *39*, 5842–5860.
Marx, V. (2017). Probes: FRET sensor design and optimization. *Nature Methods*, *14*, 949–953.
McConnell, R. E., Edward van Veen, J., Vidaki, M., Kwiatkowski, A. V., Meyer, A. S., & Gertler, F. B. (2016). A requirement for filopodia extension toward slit during Robo-mediated axon repulsion. *The Journal of Cell Biology*, *213*, 261–274.
Menon, S., Boyer, N. P., Winkle, C. C., McClain, L. M., Hanlin, C. C., Pandey, D., et al. (2015). The E3 ubiquitin ligase TRIM9 is a filopodia off switch required for netrin-dependent axon guidance. *Developmental Cell*, *35*, 698–712.
Merrifield, C. J., Qualmann, B., Kessels, M. M., & Almers, W. (2004). Neural Wiskott Aldrich syndrome protein (N-WASP) and the Arp2/3 complex are recruited to sites of clathrin-mediated endocytosis in cultured fibroblasts. *European Journal of Cell Biology*, *83*, 13–18.

Morales, D., & Kania, A. (2017). Cooperation and crosstalk in axon guidance cue integration: Additivity, synergy, and fine-tuning in combinatorial signaling. *Developmental Neurobiology*, *77*, 891–904.

Morikawa, R. K., Kanamori, T., Yasunaga, K., & Emoto, K. (2011). Different levels of the Tripartite motif protein, anomalies in sensory axon patterning (Asap), regulate distinct axonal projections of Drosophila sensory neurons. *Proceedings of the National Academy of Sciences of the United States of America*, *108*, 19389–19394.

Muller, S. A., Scilabra, S. D., & Lichtenthaler, S. F. (2016). Proteomic substrate identification for membrane proteases in the brain. *Frontiers in Molecular Neuroscience*, *9*, 96.

Neuhaus-Follini, A., & Bashaw, G. J. (2015). The intracellular domain of the frazzled/dcc receptor is a transcription factor required for commissural axon guidance. *Neuron*, *87*, 751–763.

Niftullayev, S., & Lamarche-Vane, N. (2019). Regulators of Rho GTPases in the nervous system: Molecular implication in axon guidance and neurological disorders. *International Journal of Molecular Sciences*, *20*.

Nimpf, S., & Keays, D. A. (2017). Is magnetogenetics the new optogenetics? *The EMBO Journal*, *36*, 1643–1646.

Nogi, T., Yasui, N., Mihara, E., Matsunaga, Y., Noda, M., Yamashita, N., et al. (2010). Structural basis for semaphorin signalling through the plexin receptor. *Nature*, *467*, 1123–1127.

Ojosnegros, S., Cutrale, F., Rodriguez, D., Otterstrom, J. J., Chiu, C. L., Hortiguela, V., et al. (2017). Eph-ephrin signaling modulated by polymerization and condensation of receptors. *Proceedings of the National Academy of Sciences of the United States of America*, *114*, 13188–13193.

Oliva, C., Soldano, A., Mora, N., De Geest, N., Claeys, A., Erfurth, M. L., et al. (2016). Regulation of Drosophila brain wiring by neuropil interactions via a slit-Robo-RPTP signaling complex. *Developmental Cell*, *39*, 267–278.

Onishi, K., & Zou, Y. (2017). Sonic Hedgehog switches on Wnt/planar cell polarity signaling in commissural axon growth cones by reducing levels of Shisa2. *eLife*, *6*, e25269.

Pak, J. S., DeLoughery, Z. J., Wang, J., Acharya, N., Park, Y., Jaworski, A., et al. (2020). NELL2-Robo3 complex structure reveals mechanisms of receptor activation for axon guidance. *Nature Communications*, *11*, 1489.

Pasterkamp, R. J., & Kolodkin, A. L. (2003). Semaphorin junction: Making tracks toward neural connectivity. *Current Opinion in Neurobiology*, *13*, 79–89.

Patel, F. B., & Soto, M. C. (2013). WAVE/SCAR promotes endocytosis and early endosome morphology in polarized C. elegans epithelia. *Developmental Biology*, *377*, 319–332.

Poliak, S., Morales, D., Croteau, L.-P., Krawchuk, D., Palmesino, E., Morton, S., et al. (2015). Synergistic integration of netrin and ephrin axon guidance signals by spinal motor neurons. *eLife*, *4*, e10841.

Prox, J., Bernreuther, C., Altmeppen, H., Grendel, J., Glatzel, M., D'Hooge, R., et al. (2013). Postnatal disruption of the disintegrin/metalloproteinase ADAM10 in brain causes epileptic seizures, learning deficits, altered spine morphology, and defective synaptic functions. *Journal of Neuroscience*, *33*, 12915–12928.

Qu, C., Dwyer, T., Shao, Q., Yang, T., Huang, H., & Liu, G. (2013). Direct binding of TUBB3 with DCC couples netrin-1 signaling to intracellular microtubule dynamics in axon outgrowth and guidance. *Journal of Cell Science*, *126*, 3070–3081.

Rama, N., Goldschneider, D., Corset, V., Lambert, J., Pays, L., & Mehlen, P. (2012). Amyloid precursor protein regulates netrin-1-mediated commissural axon outgrowth. *The Journal of Biological Chemistry*, *287*, 30014–30023.

Romi, E., Gokhman, I., Wong, E., Antonovsky, N., Ludwig, A., Sagi, I., et al. (2014). ADAM metalloproteases promote a developmental switch in responsiveness to the axonal repellant Sema3A. *Nature Communications*, *5*, 4058.

Ross, B., Mehta, S., & Zhang, J. (2016). Molecular tools for acute spatiotemporal manipulation of signal transduction. *Current Opinion in Chemical Biology*, *34*, 135–142.
Rost, B. R., Schneider-Warme, F., Schmitz, D., & Hegemann, P. (2017). Optogenetic tools for subcellular applications in neuroscience. *Neuron*, *96*, 572–603.
Sabatier, C., Plump, A., Ma, L., Brose, K., Tamada, A., Murakami, F., et al. (2004). The divergent Robo family protein Rig-1/Robo3 is a negative regulator of slit responsiveness required for midline crossing by commissural axons. *Cell*, *117*, 157–169.
Sachse, S. M., Lievens, S., Ribeiro, L. F., Dascenco, D., Masschaele, D., Horre, K., et al. (2019). Nuclear import of the DSCAM-cytoplasmic domain drives signaling capable of inhibiting synapse formation. *The EMBO Journal*, *38*, e99669.
Saftig, P., & Bovolenta, P. (2015). Proteases at work: Cues for understanding neural development and degeneration. *Frontiers in Molecular Neuroscience*, *8*, 13.
Saito, Y., Miranda-Rottmann, S., Ruggiu, M., Park, C. Y., Fak, J. J., Zhong, R., et al. (2016). NOVA2-mediated RNA regulation is required for axonal pathfinding during development. *eLife*, *5*, e14371.
Salaita, K., Nair, P. M., Petit, R. S., Neve, R. M., Das, D., Gray, J. W., et al. (2010). Restriction of receptor movement alters cellular response: Physical force sensing by EphA2. *Science*, *327*, 1380–1385.
Sanchez-Huertas, C., Bonhomme, M., Falco, A., Fagotto-Kaufmann, C., van Haren, J., Jeanneteau, F., et al. (2020). The + TIP navigator-1 is an actin-microtubule crosslinker that regulates axonal growth cone motility. *The Journal of Cell Biology*, *219*, e201905199.
Schaupp, A., Sabet, O., Dudanova, I., Ponserre, M., Bastiaens, P., & Klein, R. (2014). The composition of EphB2 clusters determines the strength in the cellular repulsion response. *The Journal of Cell Biology*, *204*, 409–422.
Seiradake, E., Harlos, K., Sutton, G., Aricescu, A. R., & Jones, E. Y. (2010). An extracellular steric seeding mechanism for Eph-ephrin signaling platform assembly. *Nature Structural & Molecular Biology*, *17*, 398–402.
Seiradake, E., Schaupp, A., del Toro Ruiz, D., Kaufmann, R., Mitakidis, N., Harlos, K., et al. (2013). Structurally encoded intraclass differences in EphA clusters drive distinct cell responses. *Nature Structural & Molecular Biology*, *20*, 958–964.
Seki, M., Watanabe, A., Enomoto, S., Kawamura, T., Ito, H., Kodama, T., et al. (2010). Human ROBO1 is cleaved by metalloproteinases and gamma-secretase and migrates to the nucleus in cancer cells. *FEBS Letters*, *584*, 2909–2915.
Seo, D., Southard, K. M., Kim, J. W., Lee, H. J., Farlow, J., Lee, J. U., et al. (2016). A mechanogenetic toolkit for interrogating cell signaling in space and time. *Cell*, *165*, 1507–1518.
Serafini, T., Colamarino, S. A., Leonardo, E. D., Wang, H., Beddington, R., Skarnes, W. C., et al. (1996). Netrin-1 is required for commissural axon guidance in the developing vertebrate nervous system. *Cell*, *87*, 1001–1014.
Shakir, M. A., Jiang, K., Struckhoff, E. C., Demarco, R. S., Patel, F. B., Soto, M. C., et al. (2008). The Arp2/3 activators WAVE and WASP have distinct genetic interactions with Rac GTPases in Caenorhabditis elegans axon guidance. *Genetics*, *179*, 1957–1971.
Shao, Q., Yang, T., Huang, H., Alarmanazi, F., & Liu, G. (2017). Uncoupling of UNC5C with polymerized TUBB3 in microtubules mediates netrin-1 repulsion. *The Journal of Neuroscience*, *37*, 5620–5633.
Shivas, J. M., & Skop, A. R. (2012). Arp2/3 mediates early endosome dynamics necessary for the maintenance of PAR asymmetry in Caenorhabditis elegans. *Molecular Biology of the Cell*, *23*, 1917–1927.
Slater, P. G., Cammarata, G. M., Samuelson, A. G., Magee, A., Hu, Y., & Lowery, L. A. (2019). XMAP215 promotes microtubule-F-actin interactions to regulate growth cone microtubules during axon guidance in Xenopus laevis. *Journal of Cell Science*, *132*, jcs224311.

Stephan, R., Gohl, C., Fleige, A., Klambt, C., & Bogdan, S. (2011). Membrane-targeted WAVE mediates photoreceptor axon targeting in the absence of the WAVE complex in Drosophila. *Molecular Biology of the Cell*, *22*, 4079–4092.

Stoeckli, E. T. (2018). Understanding axon guidance: Are we nearly there yet? *Development*, *145*, dev151415.

Sugiyama, M. G., Fairn, G. D., & Antonescu, C. N. (2019). Akt-ing up just about everywhere: Compartment-specific Akt activation and function in receptor tyrosine kinase signaling. *Frontiers in Cell and Development Biology*, *7*, 70.

Sullivan, C. S., Kumper, M., Temple, B. S., & Maness, P. F. (2016). The neural cell adhesion molecule (NCAM) promotes clustering and activation of EphA3 receptors in GABAergic interneurons to induce Ras homolog gene family, member A (RhoA)/ Rho-associated protein kinase (ROCK)-mediated growth cone collapse. *The Journal of Biological Chemistry*, *291*, 26262–26272.

Takai, A., Nakano, M., Saito, K., Haruno, R., Watanabe, T. M., Ohyanagi, T., et al. (2015). Expanded palette of Nano-lanterns for real-time multicolor luminescence imaging. *Proceedings of the National Academy of Sciences of the United States of America*, *112*, 4352–4356.

Taniguchi, Y., Kim, S. H., & Sisodia, S. S. (2003). Presenilin-dependent "gamma-secretase" processing of deleted in colorectal cancer (DCC). *The Journal of Biological Chemistry*, *278*, 30425–30428.

Tessier-Lavigne, M., & Goodman, C. S. (1996). The molecular biology of axon guidance. *Science*, *274*, 1123–1133.

Tischfield, M. A., Baris, H. N., Wu, C., Rudolph, G., Van Maldergem, L., He, W., et al. (2010). Human TUBB3 mutations perturb microtubule dynamics, kinesin interactions, and axon guidance. *Cell*, *140*, 74–87.

Tyler, J. J., Allwood, E. G., & Ayscough, K. R. (2016). WASP family proteins, more than Arp2/3 activators. *Biochemical Society Transactions*, *44*, 1339–1345.

van Erp, S., van den Heuvel, D. M. A., Fujita, Y., Robinson, R. A., Hellemons, A., Adolfs, Y., et al. (2015). Lrig2 negatively regulates ectodomain shedding of axon guidance receptors by ADAM proteases. *Developmental Cell*, *35*, 537–552.

Vidaki, M., Drees, F., Saxena, T., Lanslots, E., Taliaferro, M. J., Tatarakis, A., et al. (2017). A requirement for Mena, an actin regulator, in local mRNA translation in developing neurons. *Neuron*, *95*(608-622), e605.

Vincent, B. (2016). Regulation of the alpha-secretase ADAM10 at transcriptional, translational and post-translational levels. *Brain Research Bulletin*, *126*, 154–169.

Wang, X., Wang, C., & Pei, G. (2018). alpha-secretase ADAM10 physically interacts with beta-secretase BACE1 in neurons and regulates CHL1 proteolysis. *Journal of Molecular Cell Biology*, *10*, 411–422.

Winkle, C. C., McClain, L. M., Valtschanoff, J. G., Park, C. S., Maglione, C., & Gupton, S. L. (2014). A novel Netrin-1-sensitive mechanism promotes local SNARE-mediated exocytosis during axon branching. *The Journal of Cell Biology*, *205*, 217–232.

Wolfe, M. S. (2019). Structure and function of the gamma-secretase complex. *Biochemistry*, *58*, 2953–2966.

Xing, G., Li, M., Sun, Y., Rui, M., Zhuang, Y., Lv, H., et al. (2018). Neurexin-Neuroligin 1 regulates synaptic morphology and functions via the WAVE regulatory complex in Drosophila neuromuscular junction. *eLife*, 7.

Xu, C., Fu, X., Zhu, S., & Liu, J. J. (2016). Retrolinkin recruits the WAVE1 protein complex to facilitate BDNF-induced TrkB endocytosis and dendrite outgrowth. *Molecular Biology of the Cell*, *27*, 3342–3356.

Xu, Y., & Quinn, C. C. (2012). MIG-10 functions with ABI-1 to mediate the UNC-6 and SLT-1 axon guidance signaling pathways. *PLoS Genetics*, *8*, e1003054.

Xu, K., Wu, Z., Renier, N., Antipenko, A., Tzvetkova-Robev, D., Xu, Y., et al. (2014). Structures of netrin-1 bound to two receptors provide insight into its axon guidance mechanism. *Science, 344*, 1275–1279.

Yang, L., Garbe, D. S., & Bashaw, G. J. (2009). A frazzled/DCC-dependent transcriptional switch regulates midline axon guidance. *Science, 324*, 944–947.

Yang, T., Huang, H., Shao, Q., Yee, S., Majumder, T., & Liu, G. (2018). miR-92 suppresses Robo1 translation to modulate slit sensitivity in commissural axon guidance. *Cell Reports, 24*, 2694–2708 e2696.

Yang, C., & Svitkina, T. (2011). Filopodia initiation: Focus on the Arp2/3 complex and formins. *Cell Adhesion & Migration, 5*, 402–408.

Yom-Tov, G., Barak, R., Matalon, O., Barda-Saad, M., Guez-Haddad, J., & Opatowsky, Y. (2017). Robo Ig4 is a dimerization domain. *Journal of Molecular Biology, 429*, 3606–3616.

Yu, T. W., Hao, J. C., Lim, W., Tessier-Lavigne, M., & Bargmann, C. I. (2002). Shared receptors in axon guidance: SAX-3/Robo signals via UNC-34/enabled and a netrin-independent UNC-40/DCC function. *Nature Neuroscience, 5*, 1147–1154.

Zelina, P., Blockus, H., Zagar, Y., Peres, A., Friocourt, F., Wu, Z., et al. (2014). Signaling switch of the axon guidance receptor Robo3 during vertebrate evolution. *Neuron, 84*, 1258–1272.

Zhuang, M., Li, X., Zhu, J., Zhang, J., Niu, F., Liang, F., et al. (2019). The m6A reader YTHDF1 regulates axon guidance through translational control of Robo3.1 expression. *Nucleic Acids Research, 47*, 4765–4777.

CHAPTER FIVE

Dorsal commissural axon guidance in the developing spinal cord

Sandy Alvarez[a,b], Supraja G. Varadarajan[c], and Samantha J. Butler[a,d,]*
[a]Department of Neurobiology, University of California, Los Angeles, CA, United States
[b]Molecular Biology Interdepartmental Doctoral Program, University of California, Los Angeles, CA, United States
[c]Department of Neurobiology, Stanford University, Palo Alto, CA, United States
[d]Eli and Edythe Broad Center of Regenerative Medicine and Stem Cell Research, University of California, Los Angeles, CA, United States
*Corresponding author: e-mail address: butlersj@ucla.edu

Contents

Abstract

Commissural axons have been a key model system for identifying axon guidance signals in vertebrates. This review summarizes the current thinking about the molecular and cellular mechanisms that establish a specific commissural neural circuit: the dI1 neurons in the developing spinal cord. We assess the contribution of long- and short-range

Current Topics in Developmental Biology, Volume 142
ISSN 0070-2153
https://doi.org/10.1016/bs.ctdb.2020.10.009

signaling while sequentially following the developmental timeline from the birth of dI1 neurons, to the extension of commissural axons first circumferentially and then contralaterally into the ventral funiculus.

1. Introduction

Our ability to think, feel and perform complex tasks depends on neurons making the correct connections with their target cells. The task of precisely wiring the nervous system begins during embryonic development when neurons extend axons into the embryonic environment, to navigate a specific path toward their synaptic targets. Errors in this process can lead to devastating deficits in functions as diverse as cognition, respiration, movement and speech. The process of axon guidance has been studied for over 100 years, beginning with Ramón y Cajal's seminal observation that the direction of axonal growth is mediated by the growth cone, a specialized, transient structure at the tip of the axon (Ramón y Cajal, 1995; Sotelo, 2002). Cajal described growth cones for axons projecting along early trajectories in the chicken embryonic spinal cord. He was thus also the first to note the importance of commissural axons, which project contralaterally, across midline structures. Commissural axons are a key organizing feature of the central nervous system (CNS) (Comer, Alvarez, Butler, & Kaltschmidt, 2019), with the capacity to carry and integrate information across the nervous system. While they arise from multiple populations of interneurons in the spinal cord and brain (Wentworth, 1984; Yaginuma, Homma, Kunzi, & Oppenheim, 1991), this review will focus specifically on the cellular and molecular mechanisms that specify and guide commissural axons extending from the dorsal interneuron (dI) 1 population of spinal neurons. dI1s mediate proprioception, our usually unconscious ability to sense the position and movement of our bodies in space (Yuengert et al., 2015).

dI1 commissural axons have been a key model system for identifying the molecular repertoire of signals that direct growth cones into their correct trajectory (Charron & Tessier-Lavigne, 2005; Chedotal, 2019; Ducuing, Gardette, Pignata, Tauszig-Delamasure, & Castellani, 2019). Guidance cues are generally classified as acting as either attractants or repellants, with short-range or long-range activities depending on the range over which they elicit a response (Dickson, 2002; Tessier-Lavigne & Goodman, 1996). Thus, long-range cues are usually diffusible, acting over hundreds of microns, while short-range cues are often contact-dependent, either membrane-bound or tethered to the extracellular matrix (ECM), with effects limited to a range of 1–2 cell diameters (Tessier-Lavigne & Goodman, 1996). While most studies have focused on the role of long-range cues guiding dI1 axons, interest has more recently shifted to considering the importance of local signals. This review will weigh the contribution of long- and short-range signaling while sequentially following the dI1 population over the developmental period where neural progenitor cells (NPCs) are specified as dI1s, and dI1 axons are directed into their trajectory, first circumferentially toward ventral midline, and then longitudinally in the spinocerebellar tract.

2. Specification of the dI1 population of dorsal spinal neurons

2.1 Cellular organization of the embryonic spinal cord

Dorsal spinal commissural neurons first arise in the neural tube, an early embryonic structure that is generated when the neural plate folds to form a cylindrical tube. Initially only a single cell thick, the neural tube is a pseudostratified epithelium comprised of rapidly dividing neural progenitor cells (NPCs). The early neural tube consists of two compartments: a medial ventricular zone (VZ), containing the nuclei of the NPCs and a more lateral marginal layer, containing the processes of the NPCs (Altman & Bayer, 1984; Butler & Bronner, 2015). These processes are radial, spanning the VZ with an apical attachment, on the lumenal side of the neuroepithelium, and a basal "endfeet" attachment on the basement membrane, which is also known as the pial surface (Fig. 1A). This architecture permits the cell bodies of the NPCs to migrate back and forth along a lateral-medial axis within the VZ, as a function of the cell cycle, with mitosis taking place on the apical surface. NPCs continue dividing until they are ready to differentiate into mature neuronal subtypes. Newly formed post-mitotic neurons lose their apical-basal attachments and migrate laterally out of the VZ to form a third layer, the mantle layer. This layer will ultimately become the gray matter of the adult spinal cord. As development proceeds, the marginal layer expands to contain the fiber tracts, eventually becoming the white matter of the spinal cord.

As NPCs divide, they are subdivided into discrete progenitor (p) domains, defined by combinatorial codes of transcription factors (Fig. 1B, C) (Briscoe, Pierani, Jessell, & Ericson, 2000; Jessell, 2000; Shirasaki & Pfaff, 2002). The identity of these progenitor domains is dependent on both intrinsic signaling

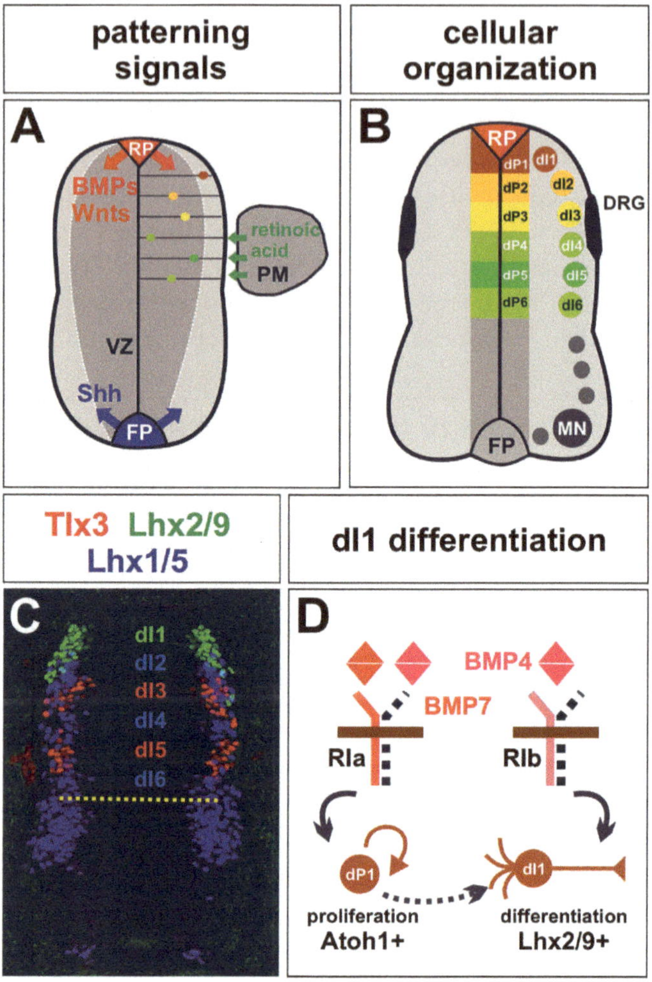

Fig. 1 Organization and specification of dI1s in the spinal cord. (A, B) Six domains of dorsal progenitor neurons (dP1–dP6) arise in the ventricular zone (VZ) in response to BMP/Wnt signals from the roof plate (RP) and retinoic acid (RA) from paraxial mesoderm (PM). This process results in six distinct classes of post-mitotic dorsal neurons (dI1-dI6). The ventral spinal cord is patterned by a gradient of Shh from the floor plate FP. (C) Dorsal interneurons can be distinguished by their expression of distinct complements of transcription factors (Andrews et al., 2017). (D) Model for the specification of dI1s. Both BMP4 and BMP7 can promote Atoh1^{+} dP1 patterning through BmprIa or BmprIb (chicken), but only BMP4 directs progenitors to differentiate as Lhx2^{+} dI1s through BmprIb (mouse and chicken) (Andrews et al., 2017).

from midline structures as well as extrinsic signaling from surrounding tissue, including the notochord and the paraxial mesoderm (Butler & Bronner, 2015). Briefly, ventral identity is specified by a putative gradient of sonic hedgehog (Shh) emanating from the floor plate (FP) at the ventral midline. Shh acts as a morphogen, with the concentration and duration of Shh signaling being decoded by signaling machinery in the primary cilia of NPCs on the apical side of the VZ (Briscoe & Ericson, 2001; Dessaud et al., 2007; Kong et al., 2015; Sasai & Briscoe, 2012). Sharp boundaries between the progenitor domains are cemented by cross-repressive interactions between the transcription factors that define domain identity (Alaynick, Jessell, & Pfaff, 2011; Briscoe et al., 2000). This process results in the specification of at least five progenitor domains in the ventral spinal cord: p0-3 and pMN, which give rise to the ventral interneurons and motor neurons (MNs) respectively. Similarly, the dorsal spinal cord contains at least six classes of dorsal interneurons, dI1-dI6, which are derived from six dP populations, dP1-dP6 (Fig. 1B, C). The identity and proliferative capacity of the dorsal-most NPCs (dP1-3P3) depends on signals from the roof plate (RP) at the dorsal midline (Lee, Dietrich, & Jessell, 2000), which include members of the bone morphogenetic protein (Bmp) and Wnt families (Andrews, Kong, Novitch, & Butler, 2019; Liem, Tremml, & Jessell, 1997). The more ventral-dorsal NPCs (dP4-dP6) arise independently from signals from the RP; their identity may be dependent on signals from the adjacent paraxial mesoderm (Le Dreau & Marti, 2013). This organization of the spinal cord results in the different laminae of the spinal cord containing neurons segregated according to their distinct physiological properties and functions (Rexed, 1954). In general, cells associated with control of motor functions are located in or adjacent to the ventral horns whereas cells mediating sensory activities are present within the dorsal horn.

2.2 Molecular specification of dI1 commissural neurons

The focus of this review is to assess the mechanisms that direct the formation of the dI1 commissural circuit. This process begins when the dorsal-most NPCs flanking the RP sequentially differentiate into dP1s and then dI1s. Multiple Wnts have a role in this process, including Wnt1 and Wnt3a, which are required for the formation of both the dI1s and dI2s (Muroyama, Fujihara, Ikeya, Kondoh, & Takada, 2002). Wnts have been proposed to act as mitogens that regulate the size of dI populations (Megason & McMahon, 2002), although their exact role remains unclear.

Bmp signaling is also required for the specification of the dI1 population (Lee et al., 2000; Lee, Mendelsohn, & Jessell, 1998; Wine-Lee et al., 2004). Multiple Bmps, including Bmp4, Bmp5, Bmp6, Bmp7 and Growth/Differentiation Factor (Gdf) 7 are secreted by the RP (Butler & Dodd, 2003; Liem et al., 1997) and are sufficient to promote some dI1 formation (Andrews et al., 2017; Gupta et al., 2018; Lee et al., 1998; Liem. et al., 1997). Bmps have been proposed to act collectively as spatial (Lee & Jessell, 1999) and/or temporal (Tozer, Le Dreau, Marti, & Briscoe, 2013) morphogen gradient(s), to specify dorsal cell fates by analogy with the models of Shh patterning in the ventral spinal cord (Briscoe & Ericson, 2001; Dessaud et al., 2007). However, it has alternatively hypothesized that different Bmps have specific effects on the induction of particular dorsal neural fates (Andrews et al., 2019; Le Dreau et al., 2012; Lee & Jessell, 1999).

Recent studies have distinguished between these models by methodically assessing the ability of RP-derived Bmps to direct dorsal spinal fates by either manipulating *Bmp* expression in chicken embryos *in vivo*, or culturing mouse (m) embryonic stem cells (ESCs) with different concentrations of recombinant Bmp proteins *in vitro* (Andrews et al., 2017). These studies unambiguously demonstrated that Bmps do not act as concentration-dependent morphogens. While altering the level of Bmps changed the effectiveness of NPC responses, i.e. the number of cells that were directed to a specific fate, the dose-dependent changes in dI identity predicted by the spatial or temporal morphogen models were not observed. Rather, each Bmp showed both distinct and reiterative activities directing RP and dI1-dI3 fates (Andrews et al., 2017; Lee & Jessell, 1999). dI1s appear to be most effectively specified by Bmp4 (Andrews et al., 2017; Duval et al., 2019) which acts reiteratively to first promote dP1 proliferation, and then direct these dP1s to differentiate into dI1 neurons (Fig. 1D). Bmps mediate their signal-specific activities by activating distinct type Bmp receptors (Bmprs) in dPs to permit differential progression through the cell cycle (Andrews et al., 2017; Panchision et al., 2001; Timmer, Wang, & Niswander, 2002; Yamauchi, Phan, & Butler, 2008). Activated type I Bmprs phosphorylate and thereby activate the R-Smads, the canonical intracellular mediators of Bmp signaling (Shi & Massague, 2003). Smad1 and Smad5 are the only R-Smads present in the developing spinal cord (Hazen et al., 2012) and it remains unresolved whether both Smad1 and Smad5 (Le Dreau et al., 2012) or Smad5 alone (Hazen et al., 2012) is critical for dI1 fate specification.

As in the ventral spinal cord, the consequence of dorsal signaling is to regulate codes of homeodomain and bHLH transcription factors that first define, discrete progenitor domains and then distinct classes of post-mitotic neurons (Andrews et al., 2019; Helms & Johnson, 2003; Zhuang & Sockanathan, 2006). The dorsal tube is demarcated by the expression of Pax3 (Goulding, Chalepakis, Deutsch, Erselius, & Gruss, 1991) and Pax7 (Jostes, Walther, & Gruss, 1990), which are the earliest known general markers of dorsal spinal identity. Work *in vitro* has suggested that the upregulation of Pax3/7 in dorsal tissue may depend on retinoic acid (RA) signaling from the paraxial mesoderm (Gupta, Kawaguchi, Mandric, Castellanos, & Butler, 2020). Subsequent Bmp signaling from the dorsal midline results in the cells immediately flanking the RP activating Atoh1 (Helms & Johnson, 1998), a bHLH transcription factor that is both necessary and sufficient for dP1 identity (Gowan et al., 2001). Atoh1 in turn upregulates a transcriptional cassette mediated by Lhx2 (Lh2a) and Lhx9 (Lh2b), which establishes the intraspinal trajectory of commissural neurons. Lhx2 and Lhx9 are initially coexpressed by all new-born dI1s as they start to extend axons (Lee et al., 1998). dI1s also concomitantly migrate ventrally within the mantle layer; on reaching the deep dorsal horn, they segregate into two subpopulations that differentially express either Lhx2 or Lhx9 (Lee et al., 1998; Wilson, Shafer, Lee, & Dodd, 2008). These subpopulations have distinct axial positions and axonal trajectories. The $Lhx2^+$ population (dI1c), has a medial position and projects axon contralaterally across the FP whereas the $Lhx9^+$ population (dI1i) has a more lateral position and projects axons ipsilaterally within the ventral funiculus (Avraham et al., 2009; Phan, Hazen, Frendo, Jia, & Butler, 2010; Wilson et al., 2008).

3. Guidance of dI1 axons away from the dorsal midline

The first decision made by all newly formed spinal neurons in the mantle zone, is to extend a growth cone into the marginal zone in a specific direction. Depending on their complement of guidance receptors, growth cones interpret signaling information in the neuroepithelial environment to guide them into their initial trajectory. Remarkably, all spinal commissural neurons initially extend axons ventrally thereby establishing one of the organizing architectural principles of the spinal cord: that contralaterally projecting axons only cross at the ventral midline (Altman & Bayer, 1984; Ramón y Cajal, 1995; Wentworth, 1984). The mechanisms that direct commissural axons ventrally have been most extensively assessed for the dI1 population, because it was one of the first vertebrate trajectories to be distinguishable by immunohistochemistry. dI1 axons were initially detected using antibodies against the glycoprotein, Tag1, which label the dorsal-most subset of commissural axons in their transverse trajectory in the spinal cord, i.e. as they extend up to and across the FP (Dodd, Morton, Karagogeos, Yamamoto, & Jessell, 1988). More recently, antibodies against robo3 have been shown to generally label all dorsal commissural axons (Sabatier et al., 2004; Tulloch, Teo, Carvajal, Tessier-Lavigne, & Jaworski, 2019), while dI1 axons can be unambiguously identified using a genetically encoded marker: the Atoh1::tau*gfp* transgenic mouse line (Hazen et al., 2012; Tran et al., 2013). dI1s are born relatively early, ~E10.5 in the mouse rostral spinal cord (Hazen et al., 2012), arising immediately flanking the RP. In addition to instructing cell fate specification, the RP is also thought to be a key source of axon guidance cues for dI1 axons. There are two complementary mechanistic possibilities: first, the RP is itself intrinsically repulsive and does not support axon growth, and second, the RP secretes a diffusible repellent that orients axons away it, thereby directing dI1 axons to grow ventrally. Supporting the latter model, *in vitro* tissue culture "reorientation" assays demonstrated that an ectopically placed RP explant is sufficient to reorient Tag1$^+$ axons away from the RP, as they extend within an explant of rat dorsal spinal cord (Augsburger, Schuchardt, Hoskins, Dodd, & Butler, 1999).

Multiple factors have been identified as RP repellents. Keratin sulfate, a sulfated proteoglycan specifically localized to the RP, is the key candidate for a short-range signal that locally renders the RP repulsive to axon growth (Snow et al., 1990b). *In vitro* stripe assays have shown that sulfated proteoglycans can repel elongating neurites in a concentration-dependent manner (Snow et al., 1990a). Both members of the Bmp family (Augsburger et al., 1999), and draxin, a novel secreted molecule (Islam et al., 2009), have been proposed to act as long-range guidance repellents for dI1 axons. In both cases, these factors were good candidates based on their localized presence in the RP, and their observed activities in *in vitro* assays. First, Bmps can phenocopy the ability of an ectopic RP to reorient rat Tag1$^+$ commissural axons *in vitro* (Augsburger et al., 1999); a heterodimer of Bmp7 and Gdf7 was subsequently observed to have the most potent reorienting activity (Butler & Dodd, 2003). These studies were amongst the first in the field to demonstrate that instructive growth factors can also act as axon guidance signals, with the striking implication that the Bmp family reiteratively specifies different cellular processes for dI1s, as they mature during development. Similarly, draxin was identified by its ability to prevent the outgrowth of Tag1$^+$ axons extending from a chicken dorsal spinal cord explant (Islam et al., 2009).

While both Bmps and draxin have dramatic *in vitro* activities consistent with a role mediating a long-range repellent, both factors have relatively modest loss of function phenotypes in mouse embryonic spinal cords *in vivo* (Butler & Dodd, 2003; Islam et al., 2009). It thus remains largely unresolved whether the polarizing activities observed *in vitro*, are required to guide dI1 axons away from the dorsal midline *in vivo*. Moreover, subsequent genetic studies assessing whether type I Bmp receptors (BmprI) have reiterative roles directing the formation of dI1 circuits (Yamauchi et al., 2008) did not robustly observe the randomized growth predicted for the loss of a chemorepellent. Similarly, the Smad family are not required for the putative reorienting activities of the Bmps *in vitro* (Hazen et al., 2012; Perron & Dodd, 2011). Rather, the most profound phenotypes observed *in vivo* after modulating the activity of the Bmp signaling pathway, were alterations in the rate at which dI1 axons grew. Specifically, decreasing signaling through the Bmp pathway accelerated dI1 axon growth (Phan et al., 2010; Yamauchi, Varadarajan, Li, & Butler, 2013), while slowed/stalled dI1 axon growth was observed when Bmp signaling was increased (Phan et al., 2010; Yamauchi et al., 2008). Together, these studies suggest that a key *in vivo* role of the Bmps is to set the rate at which dI1 axons grow through the dorsal spinal cord toward the FP. These studies thereby identified a novel "temporal" mode by which the process of axon guidance can be regulated (Phan et al., 2010). In contrast to the well described "spatial" cues which regulate the direction of axon growth, temporal guidance signals regulate rate at which axons grow toward their targets thereby ensuring that axons encounter subsequent spatial signals at a particular time in development. This process permits neural circuits to develop in concert with the rest of the developing embryo. If axons grow at the wrong speed, this process can become uncoupled; for example, accelerated dI1 axons make guidance errors on encountering the FP too early in development (Phan et al., 2010; Phan & Butler, 2013).

The temporal guidance activity of the Bmps appears to be mediated through Lim kinase (Limk) 1, a non-canonical effector of Bmp signaling that regulates actin dynamics (Wen et al., 2007). Limk1 is a direct inhibitor of cofilin, and they act together to control the rate of actin polymerization or "treadmilling". In the current model, the rate which dI1 axons first extend away from the RP is controlled by the Bmps binding to a heterodimeric complex of BmprIb (Yamauchi et al., 2013) and BmprII, which releases Limk1 into the cytosol (Lee-Hoeflich et al., 2004). Limk1 then phosphorylates and inactivates cofilin, which inhibits the depolymerization of actin

(Meberg & Bamburg, 2000) thereby slowing the rate of dI1 axons as they extend through the dorsal spinal cord.

4. Guidance of dI1 axons around the ventricular zone

The transverse path of dI1 axons in the developing spinal cord follows a circumferential dorsal to ventral route toward the FP at the ventral midline. Previous studies suggested this trajectory could be achieved mechanistically by a "push" from the RP and a "pull" from the FP (Fig. 2A) (Kaprielian, Imondi, & Runko, 2000). However, dI1 axons extend precisely around the border of VZ and avoid numerous potential exit points for pioneering axons, including the dorsal root entry zone (DREZ) and the lateral and motor exit points (Laumonnerie, Da Silva, Kania, & Wilson, 2014). Such complex pathfinding is likely to require instructive cues in addition to a simple combination of push/pull signaling. Supporting this hypothesis, netrin1 has been shown to play a crucial role both guiding commissural axons around the VZ and maintaining the CNS-PNS boundary (Laumonnerie, Tong, Alstermark, & Wilson, 2015; Moreno-Bravo, Roig Puiggros, Mehlen, & Chedotal, 2019; Varadarajan & Butler, 2017; Varadarajan et al., 2017). The netrins are a family of laminin-like proteins that were first characterized, in a seminal breakthrough in the early 1990s, as the prototypical vertebrate axon guidance cue (Kennedy, Serafini, de la Torre, & Tessier-Lavigne, 1994; Serafini et al., 1994). Netrin1 (from the Sanskrit *netr*, meaning "one who guides") was identified as a candidate for the FP attractant, whose existence was first hypothesized by Ramón y Cajal in 1890 (de Castro, López-Mascaraque, & De Carlos, 2007); see Section 5 for more details. Netrin1 mediates its activity through two classes of receptors: Deleted in Colon Cancer (Dcc) and the Unc5 family, a
ll members of the immunoglobulin (Ig) domain super family (Engelkamp, 2002; Keino-Masu et al., 1996; Leonardo et al., 1997; Moore, Tessier-Lavigne, & Kennedy, 2007). Dcc is thought to mediate the attractive properties of netrin1 (Fazeli et al., 1997), while the members of the Unc5 family, Unc5A-Unc5D, can mediate the repulsive properties of netrin1 (Dillon et al., 2007; Engelkamp, 2002; Leonardo et al., 1997; Williams et al., 2006).

Netrin1 has a complex distribution pattern in the mouse spinal cord: it is transcribed by both NPCs and the FP cells, with a sharp expression boundary in the dorsal spinal cord at the level of the DREZ (Fig. 2B, C). However, netrin1 protein does not remain in the VZ, rather it appears to be transported

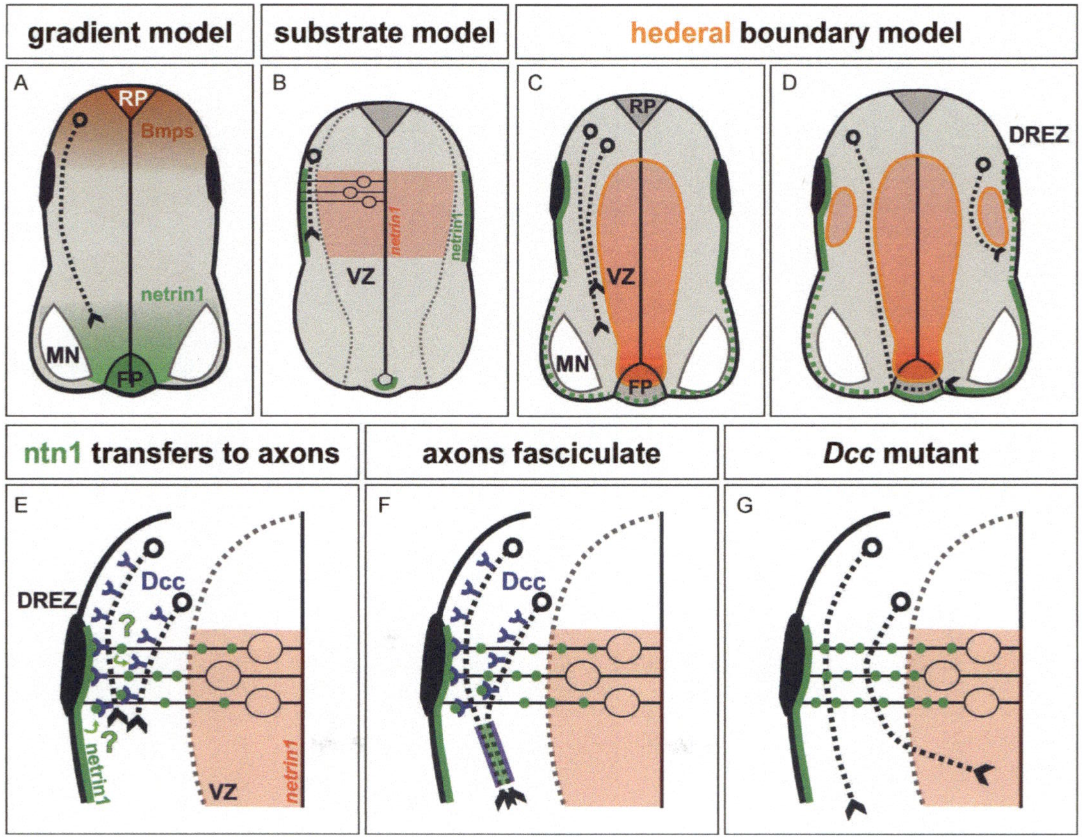

Fig. 2 See figure legend on opposite page.

along the radial processes of the NPCs (Fig. 2B, E). Netrin1 is then deposited on the pial surface, via the glial endfeet, and accumulates on commissural axons (Fig. 3A) (Dominici et al., 2017; Varadarajan et al., 2017). Pial-netrin1 deposition starts as early as E9.5 (Varadarajan & Butler, 2017), immediately before the pioneering dI1 axons commence their trajectory toward the FP (Fig. 2B).

4.1 Netrin1 prevents dI1 axons from innervating the VZ

As pioneering dI1 axons extend ventrally, they grow around the edge of the VZ, without invading it. This behavior is not unique to dI1 axons, rather all spinal axons avoid growing VZ, thereby establishing another architectural organizing principle of the spinal cord. Recent studies have shown that this remarkably homogenous behavior is mediated by NPC-derived netrin1. In the complete absence of netrin1, spinal axons, most notably the robo3$^+$ commissural axons, are profusely defasciculated and extend randomly, including into the VZ (Fig. 3B) (Varadarajan et al., 2017; Wu et al., 2019). This effect is

Fig. 2 Models for the long- and short-range activities of netrin1. (A) In the long-range gradient model, the RP-derived Bmps act as a chemorepellent "pushing" commissural axons away from the dorsal midline, while FP-derived netrin1 and Shh function as chemoattractants "pulling" commissural axons to the ventral midline. (B) In the short-range model, pial-netrin1 first orients early born commissural axons to extend ventrally. NPCs transcribe *netrin1* (red) and then deposit netrin1 protein (green) on the pial surface, where it may act as a growth substrate to promote axon extension (Varadarajan et al., 2017). (C) As development progresses, pre-crossing commissural axons extend into the ventral spinal cord, and no longer grow adjacent to pial-netrin1 (dotted green line). Rather, they project precisely around a "hederal" boundary of *netrin1* expressing NPCs (orange line). We have proposed that the netrin1 hederal boundary promotes directed axon fasciculation while preventing innervation of *netrin1* expressing cells. This activity permits commissural axons to grow around the VZ. (D) Commissural axons extend across the FP in a highly fasciculated bundle within a narrow corridor bounded by hederal-*netrin1* and pial-netrin1. Post-crossing commissural axons then turn rostrally to extend in the ventral funiculus, again growing adjacent to a pial-netrin1 substrate (solid green line). Concomitantly, a domain of *netrin1* expressing cells (red) emerges adjacent to the DREZ, which continue to sculpt axonal trajectories within the spinal cord. (E–G) We propose that the netrin1 produced by neural progenitors is transported to the pial surface in their progenitor endfeet, and then transfers from this pial-substrate to Dcc$^+$ commissural axons (E) (Varadarajan et al., 2017; Varadarajan & Butler, 2017). Dcc and netrin1 then interact *in cis* to promote the selective fasciculation and growth commissural axons around *netrin1* expressing NPCs (F). In absence of Dcc, netrin1 does accumulate on the pial surface, but does not transfer to commissural axons. These axons fail to fasciculate and grow randomly, including into the VZ (G).

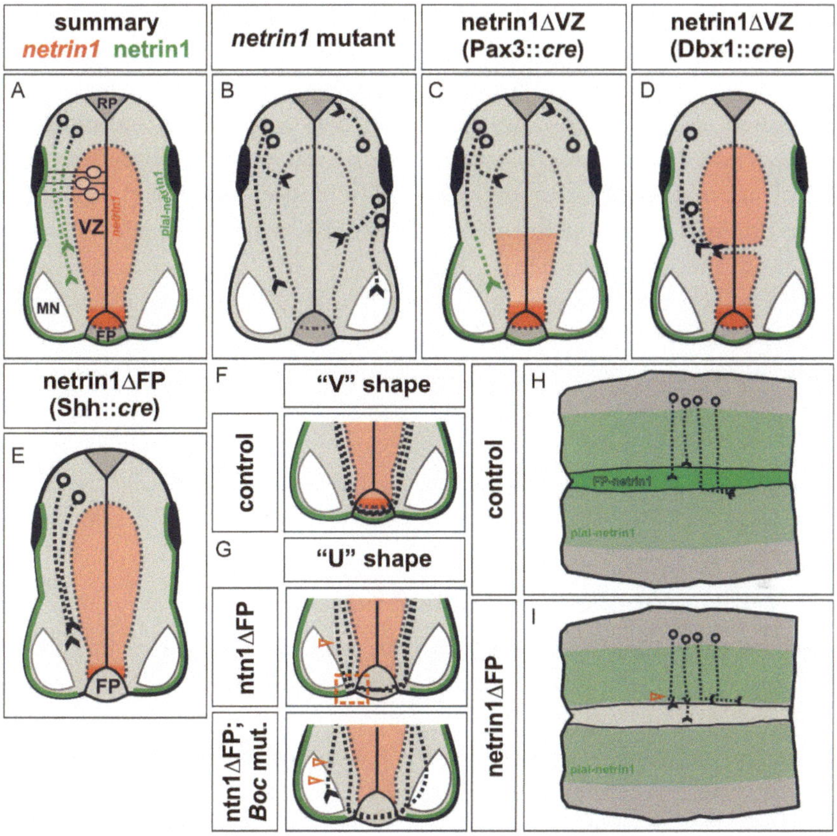

Fig. 3 Short- and long-range phenotypes observed in the absence of NPC- or FP-derived netrin1. (A) Summary of the distribution of netrin1 transcript (red) and protein (green). NPCs and FP cells express *netrin1*, while netrin1 protein is present on the pial surface and on commissural axons. (B) In *netrin1* loss of function mutants, spinal axons are highly defasciculated, extending into the VZ, and wandering across the motor column. (C, D) *Short range phenotypes for NPC-derived netrin1*: When *netrin1* expression is removed from either a large (C) or small (D) number of NPCs using conditional genetic approaches (C=Pax3::*cre* driver; D=Dbx1::*cre* driver) (Varadarajan et al., 2017), axons defasciculate specifically and locally in the region where netrin1 activity is absent. In particular, the introduction of ectopic *netrin1* on::off boundaries (D) locally reshapes axon trajectories in a manner consistent with the hederal boundary model. *Long-range phenotypes for NPC-derived netrin1*: none observed. (E–I) *Short-range phenotypes for FP-derived netrin1*: In netrin1ΔFP mice (Shh::*cre* driver), commissural axons project ventrally toward the FP (E) in a manner largely indistinguishable from controls. However, errors occur ~10 μm from the FP (yellow box, G), when axons reach the "off" edge of the ventral domain pial-netrin1, where they locally defasciculate (arrowheads, I). Axons start to cross the FP from this "off" edge, resulting in a laterally displaced "U" shaped trajectory (G), distinct from the "V" shape observed in controls (F). Control

(Continued)

phenocopied by the conditional removal of netrin1 from either the Pax3 domain (Fig. 3C) (Varadarajan et al., 2017; Yamauchi et al., 2017), i.e. all dorsal NPCs (netrin1ΔdVZ), or the Dbx1 domain, i.e. a specific subpopulation of NPCs (Fig. 3D). In netrin1ΔdVZ spinal cords, randomized axon growth is only observed in the dorsal spinal cord, where deposition of pial-netrin has been disrupted (Fig. 3C). This effect is very short-range: when netrin1 is removed from $Dbx1^+$ NPCs, axons only invade the narrow channel in the VZ where netrin1 has been depleted (Varadarajan et al., 2017) (Fig. 3D). Together, these studies identified that NPC-derived netrin1 establishes a boundary, that promotes the growth of fasciculated commissural axons around the VZ. This boundary activity of NPC-netrin1 is mediated by the Dcc receptor. In the absence of Dcc, netrin1 is still present at the pial surface but does not accumulate on commissural axons, which then grow in a profoundly defasciculated manner (Varadarajan et al., 2017) (Fig. 2F, G). Thus axonal-netrin1 appears to be key to establishing the VZ boundary.

4.2 Netrin1 supplies a "hederal" boundary

How mechanistically is the NPC-derived boundary of netrin1 established? The deposition of netrin1 on the pial surface suggests the formation of a local, attractive growth substrate in the dorsal spinal cord. And indeed, in the earliest stages of axogenesis, dorsal pial-netrin1 may act by haptotaxis, the directed growth of cells along an adhesive surface (Carter, 1965), instructing pioneering axons to grow ventrally, immediately adjacent to the $netrin1^+$ substrate (Fig. 2B). However, pial-netrin1 does not subsequently shape the transverse trajectory of commissural axons in the ventral

Fig. 3—Cont'd commissural axons then turn rostrally to project longitudinally beside the FP in the ventral funiculus (H). Most axons correctly make the rostral turn in netrin1ΔFP mice, however, we (S.A. and S.J.B., personal communication) and others (Moreno-Bravo et al., 2019) have observed that a subset of commissural axons fail to cross the FP, and turn ipsilaterally (I), resulting in a thinning of the FP commissure (Wu et al., 2019). It remains unresolved whether this ipsilateral turn is the result of axons following an ectopic hederal boundary created by the specific loss of netrin1 in the FP (I), or a direct requirement for FP-netrin1 in axon crossing. *Long-range phenotypes for FP-derived netrin1*: While we have not observed this phenotype, other reports (Wu et al., 2019) have shown that commissural axons are modestly defasciculated in the ventral spinal cord in the absence of FP-derived netrin1 (yellow arrowheads, G) i.e. ~100μm from the FP. This phenotype is significantly enhanced when netrin1ΔFP is combined with a *Boc* mutation (G), the non-canonical receptor that mediates the axon guidance activities of Shh (Wu et al., 2019).

spinal cord, rather commissural axons grow to precisely follow the boundary of the *netrin1*$^{+}$ VZ (Fig. 2C, D). Multiple conditional manipulations of *netrin1* expression (Varadarajan & Butler, 2017; Varadarajan et al., 2017) have shown that commissural axons will deviate from their trajectories to extend around ectopic borders of *netrin1*-expressing cells, as if these cells are supplying a boundary repellent (Fig. 3C, D, I).

We have proposed the hederal boundary model to reconcile these apparently distinct activities (Varadarajan & Butler, 2017) (Fig. 2C, D). This model is drawn from the analogy of a wall supporting a growing hedera (ivy plant); the wall provides a substrate for ivy to grow along, while preventing the ingrowth of the ivy into the wall. We propose that the hederal boundary is mediated by the ability of netrin1 to locally transfer, or be trafficked, from NPCs, to the pial surface and then onto Dcc^{+} axons (Fig. 2E, F). Axonal-netrin1 can both promote the directed fasciculated growth of commissural axons and prevent them from growing on *netrin1*-expressing cells in the VZ. Thus, early extending Dcc^{+} commissural axons are first oriented by the dorsal boundary of pial-netrin1 in the spinal cord. Commissural axons then accumulate netrin1; axonal-netrin1 acts to fasciculate axons together, and direct them around the *netrin1* expressing VZ. This model is supported by the observations that pial-netrin is not sufficient to guide *Dcc*$^{-}$ commissural axons (Fig. 2G) and that occasional axons growing aberrantly in the VZ in controls, have never accumulated netrin1 (Varadarajan et al., 2017).

The mechanism by which netrin1 becomes sequestered on axons is unresolved. Draxin may have a role in this process: the crystal structure of draxin either alone or in a netrin1-Dcc complex has shown that netrin1 and Dcc can bind to draxin on adjacent sites (Liu et al., 2018). Moreover, the sequestration of draxin results in the same pattern of randomized axon growth observed in *netrin1* and *Dcc* mutant spinal cords (Islam et al., 2009). Thus, draxin may bind to both molecules to facilitate the binding of netrin1 to Dcc^{+} axons, to promote fasciculation and directed growth.

Finally, the hederal activity of netrin1 appears to act over a very short-range. It remains unresolved whether netrin1 is always substrate bound, or whether there is some limited diffusion from the radial fibers (question marks, Fig. 2E). However, it is notable that there is no protein accumulation on the pial surface in the dorsal-most spinal cord, i.e. where NPCs are not expressing *netrin1*, suggesting very limited diffusivity, if any. Further studies examining the activities of membrane-bound forms of netrin1 *in vivo* are necessary to answer these questions.

4.3 Netrin1 maintains the CNS-PNS boundary

As dI1 axons begin their journey away from the RP, they must also avoid exiting the dorsal spinal cord from the DREZ, the entry point into the CNS for peripheral dorsal root ganglion (DRG) axons. The DRG begin innervating the spinal cord through the DREZ at E13.5 in mice (Yoshida, Han, Mendelsohn, & Jessell, 2006). Netrin1 plays a role preventing both commissural axons from leaving the spinal cord and sensory axons entering the spinal cord inappropriately. In *netrin1/Dcc* mutants, commissural axons aberrantly exit the spinal cord and grow into the periphery (Laumonnerie et al., 2014). The transient expression of netrin1 at E12.5 in a domain adjacent to the DREZ (Masuda et al., 2008; Watanabe et al., 2006), acts as a repellent boundary that prevents DRG axons from prematurely entering the spinal cord (Masuda et al., 2008; Varadarajan & Butler, 2017; Yoshida et al., 2006). Recent studies have suggested a role for cues emanating from the spinal meninges, which include the pial-substrate. Studies using dorsal spinal explants have shown that commissural axons are repelled by spinal meninges tissue while motor and sensory axons are attracted by meninges-derived cues (Suter, DeLoughery, & Jaworski, 2017). Is the meninges-bound cue that prevents commissural axons from leaving the spinal cord mediated by netrin1? The addition of netrin1 to the explant cultures complicates this interpretation, however there is support for this hypothesis from *in vivo* studies. In the absence of pial-netrin1, commissural axons, as well as precerebellar neurons in the hindbrain and pontine neurons, erroneously cross the CNS/PNS boundary and are no longer confined to the CNS (Dominici et al., 2017; Laumonnerie et al., 2014; Moreno-Bravo et al., 2018; Varadarajan et al., 2017; Yung et al., 2018).

5. Guidance of commissural axons toward and across the FP

On entering the ventral spinal cord, commissural axons detach from the circumferential edge of the spinal cord, to continue growing around the VZ, and toward the FP at the ventral midline. As commissural axons enter the ventral spinal cord, they become more progressively fasciculated until they project across the FP in very tightly bundled fascicle. The passage of commissural axons toward and across the FP has been the focus of intensive study, with multiple long- and short-range cues identified.

5.1 Long-range guidance signals from the FP

Netrin1 was initially identified as a compelling candidate for the "pull" attractant acting from the FP to draw commissural axons to the ventral midline. *Netrin1* is expressed at highest levels by FP cells (Kennedy et al., 1994), and COS cell aggregates expressing *netrin1* can mimic the ability of the FP to promote the fasciculated growth of commissural axons toward them (Kennedy et al., 1994; Placzek, Tessier-Lavigne, Yamada, Dodd, & Jessell, 1990; Serafini et al., 1994). Loss of function studies initially revealed that mice mutant for *netrin1* or *Dcc* exhibit a transient stall in $Tag1^+$ commissural axon growth, with axons failing to reach the FP (Fazeli et al., 1997; Serafini et al., 1996). Together, these studies suggested the model that FP-derived netrin1 is a long-range cue that acts by chemotaxis to guide commissural axons to the ventral midline (Fig. 2A).

However, more recent studies have cast doubt on this model. Studies first *in vitro* (Moore, Biais, & Sheetz, 2009) and then *in vivo* in Drosophila (Akin & Zipursky, 2016; Brankatschk & Dickson, 2006; Timofeev, Joly, Hadjieconomou, & Salecker, 2012) demonstrated that tethered netrin1 can effectively function as a guidance cue. Most recently, NPC-derived netrin1 has been shown to direct fasciculated commissural axon growth, while specifically depleting FP-derived netrin1 (netrin1ΔFP) results in surprisingly minimal effects on the passage of $robo3^+$ commissural axons toward the FP (Fig. 3E) (Dominici et al., 2017; Varadarajan et al., 2017; Wu et al., 2019). Studies further characterizing netrin1ΔFP embryos have found local defects (boxed region, Fig. 3F, G, see figure legend for details); in particular, some commissural axons now grow ipsilaterally immediately adjacent to the netrin1-depleted FP (Fig. 3H, I) (Moreno-Bravo et al., 2019). This phenotype is consistent with either a local role for FP-derived netrin1 promoting contralateral growth, or the secondary consequence of introducing a longitudinal hederal boundary, given that the netrin1ΔFP manipulation introduces an ectopic on:off *netrin* boundary along the edge of the FP (Fig. 3H, I).

Other candidates for the "pull" long-range attractant include the morphogen sonic hedgehog (Shh) (Charron, Stein, Jeong, McMahon, & Tessier-Lavigne, 2003), and vascular endothelial growth factor (Vegf) (Ruiz de Almodovar et al., 2011) which are both secreted by the FP. Localized sources of both Shh and Vegf can reorient commissural axons toward them (Charron et al., 2003; Ruiz de Almodovar et al., 2011), as has been observed for FP explants (Kennedy et al., 1994; Placzek et al.,

1990; Serafini et al., 1994) . Shh appears to act through a complex of smoothened (Charron et al., 2003) and brother of Cdo (Boc) (Okada et al., 2006). In the absence of these receptors, commissural axons defasciculate, and extend into the motor column. A similar phenotype is observed when Vegf, or its receptor fetal liver kinase 1 (Flk1), is specifically depleted from the FP or commissural neurons, respectively (Ruiz de Almodovar et al., 2011).

Together, these studies suggest that long-range activities supplied by FP-derived Shh and Vegf are necessary to direct fasciculated axon growth in the ventral spinal cord. Supporting this hypothesis, synergistic guidance interactions between Shh and netrin1 have been identified. These interactions were first observed *in vitro*: use of microfluidic guidance assays revealed that commissural growth cones can sense a shallow combined Shh/netrin1 gradient, but are nonresponsive to an equivalently shallow gradient of Shh or netrin1 alone (Sloan, Qasaimeh, Juncker, Yam, & Charron, 2015). This activity was recently observed *in vivo*: the combined loss of Boc and netrin1ΔFP results in significantly more defasciculation of robo3$^+$ axons into the motor column than is observed for the loss of Boc or netrin1ΔFP alone (Fig. 3F, G) (Wu et al., 2019). Thus, long-range signaling from the FP may be necessary for the progressive fasciculation of commissural axon bundles as they project to the FP.

5.2 Short-range axon guidance cues in the FP

As commissural growth cones reach the ventral midline, their extension slows to navigate the FP (Bak & Fraser, 2003), which is the source of many short-range guidance cues (Table 1). Commissural axons then extend contralaterally across the FP, and turn rostrally, to project longitudinally in the ventral funiculus. In mouse embryos that lack a FP, commissural axons project relatively normally to the ventral midline, but make errors navigating the FP, including inappropriately accumulating in the midline and turning randomly, both rostrally and caudally, after crossing the FP (Kadison & Kaprielian, 2004; Kadison, Murakami, Matise, & Kaprielian, 2006; Matise, Epstein, Park, Platt, & Joyner, 1998; Matise, Lustig, Sakurai, Grumet, & Joyner, 1999). These phenotypes are consistent with key role(s) for short-range signaling from the FP.

Multiple members of the Ig domain superfamily, including Tag1, L1 and Nrcam, appear to mediate contact-dependent interactions between commissural growth cones/axons and FP cells. Nrcam is expressed by FP cells

Table 1 Summary of molecules present in the FP that have been proposed to act as short-range commissural axon guidance cues.

Name	Model	Proposed function	References
Ngcam-related CAM (Nrcam)	Mouse, chicken	Promotes growth of commissural axons across the FP	Stoeckli and Landmesser (1995)
Slits	Mouse, chicken	Repels post-crossing commissural axons from the FP	Brose et al. (1999), Long et al. (2004)
Class 3 semaphorins	Mouse, chicken	Repels post-crossing commissural axons from the FP	Nawabi et al. (2010), Zou et al. (2000)
Fspondin	Rat, chicken	Promotes growth of commissural axons across the FP	Burstyn-Cohen et al. (1999)
Neuropilin2	Mouse	Promotes growth of commissural axons across the FP	Hernandez-Enriquez et al. (2015)
P84	Mouse	Role unresolved, may serve as a substrate for axon outgrowth.	Chuang and Lagenaur (1990)
Neurite outgrowth inhibitor (nogo) b (reticulon4b)	Mouse	Repels post-crossing commissural axons from the FP	Wang et al. (2017)
Dystroglycan	Mouse	Repels post-crossing commissural axons from the FP	Wright et al. (2012)
Nectin3	Mouse	Promotes for correct axon fasciculation, and rostral turn for post-crossing axons	Okabe et al. (2004)
MAM domain-containing glycosylphosphatidylinositol anchor (Mdga) 1	Chicken	Function has not been characterized	Fujimura, Iwashita, Matsuzaki, and Yamamoto (2006)
Mdga2	Chicken	Promotes rostral turn for post-crossing axons	Joset et al. (2011)

Continued

Table 1 Summary of molecules present in the FP that have been proposed to act as short-range commissural axon guidance cues.—cont'd

Name	Model	Proposed function	References
Synaptic cell adhesion molecules (Syncam)/nectin-like molecules (Necls)	Chicken	Promotes growth of commissural axons across the FP and rostral turn for post-crossing axons	Niederkofler, Baeriswyl, Ott, and Stoeckli (2010)
Plexina2	Chicken	Promotes growth of commissural axons across the FP and rostral turn for post-crossing axons	Andermatt et al. (2014)

(Stoeckli & Landmesser, 1995), while Tag1 (known as axonin1 as in chicken) (Dodd et al., 1988; Stoeckli et al., 1989), and L1 (known as Ngcam in chicken) (Krushel, Prieto, Cunningham, & Edelman, 1993; Moscoso & Sanes, 1995) have complementary distributions on commissural axons. Thus, Tag1 is expressed on commissural axons prior to crossing the FP, while L1 is present on post-crossing commissural axons growing longitudinally in the ventral funiculus (Dodd et al., 1988). This distribution suggested that Tag1/L1 controls the response of commissural axons to signals in the FP, such as Nrcam (Stoeckli & Landmesser, 1995). Together, these cues may interact to direct commissural axons across the FP and into their contralateral rostral turn, by preventing them from turning prematurely along an ipsilateral trajectory. *In vitro* studies have provided support for this model (Fitzli et al., 2000), while the loss of Tag1 or Nrcam *in vivo* results in some commissural axons failing to cross the midline and growing along the ipsilateral FP border (Stoeckli & Landmesser, 1995; Stoeckli, Sonderegger, Pollerberg, & Landmesser, 1997). These adhesion molecules may also regulate commissural axon fasciculation across the FP. In Tag1, L1 and Nrcam mutants, some commissural axons locally defasciculate as they approach the midline (Stoeckli & Landmesser, 1995).

Extracellular matrix (ECM) proteins have also been implicated in regulating the passage of commissural axons across the FP. These signals include netrin1, as discussed in Section 5.1 and F-spondin. F-spondin is produced by the FP and localizes to the basal lamina immediately underlying FP (Klar, Baldassare, & Jessell, 1992). F-spondin can promote outgrowth of commissural axons *in vitro,* while *in vivo* perturbations result in axon defasculation and turning defects (Burstyn-Cohen et al., 1999).

5.3 Robo/slit signaling changes commissural axon responsiveness

The slit/robo pathway is the most well studied, short-range molecular mechanism that regulates whether axons do, or do not, cross a midline structure. First identified in Drosophila, the slit glycoproteins mediate axon repulsion by binding to the roundabout (robo) receptors, also members of the Ig domain superfamily (Kidd et al., 1998; Nusslein-Volhard, Wieschaus, & Kluding, 1984; Seeger, Tear, Ferres-Marco, & Goodman, 1993; Tear, Seeger, & Goodman, 1993). Classic studies have led to a model in which axon responsiveness to a slit$^+$ midline is regulated by the presence or absence of robo receptors. Thus, robo$^+$ axons grow ipsilaterally to avoid the slit$^+$ midline, while robo$^-$ axons grow contralaterally across a slit$^+$ midline, because they are unable to detect the slit repellent. If contralaterally projecting axons subsequently become robo$^+$, this mechanism then prevents them from re-crossing the slit$^+$ midline.

Much of this evidence for this model has come from Drosophila studies (Blockus & Chedotal, 2014; Comer et al., 2019; Evans & Bashaw, 2010). Here, we will briefly review the role of the slit/robo pathway directing commissural axons in the spinal cord. In vertebrates, there are three homologues of the Drosophila robo1 gene - robo1, robo2, and robo3/Rig1 - which are present on commissural axons (Kidd et al., 1998; Sundaresan et al., 1998; Yuan et al., 1999) and three members of the slit family - slit1, slit2, and slit3 - that are secreted by the FP (Brose et al., 1999). Robo1 and robo2 are present at higher levels on the post-crossing segment of commissural axons (Long et al., 2004; Mambetisaeva, Andrews, Camurri, Annan, & Sundaresan, 2005; Sabatier et al., 2004). They are thus candidates for the repellent receptors that identify the slit1/2/3$^+$ FP as being repulsive, thereby preventing commissural axons from re-crossing the FP. *In vitro* studies have shown that the vertebrate slit proteins can bind robo1 and robo2 to mediate a repulsive interaction (Brose et al., 1999; Li et al., 1999). Genetic *in vivo* studies are more complex: slit triple knockouts exhibit the re-crossing and axon stalling phenotypes predicted for the loss of a repellent at the FP (Long et al., 2004). However, while robo1/2 double mutants display similar phenotypes, they are less severe than the slit triple mutants suggesting the existence of additional slit receptors (Jaworski, Long, & Tessier-Lavigne, 2010).

In contrast, robo3 is present on the pre-crossing segments of commissural axons and has been proposed to prevent pre-crossing commissural axons from responding prematurely to the slit^{+} midline (Sabatier et al., 2004). In the absence of robo3, commissural axons reach the FP, but then dramatically fail to cross it, rather turning ipsilaterally to grow alongside it, as if they are now responsive to slit signaling (Sabatier et al., 2004). The mechanism by which robo3 suppresses the response of commissural axons to slit remains unresolved. Robo3 has been proposed to function by interfering with the robo1/2 complex. However, neither robo1 nor robo2 are upregulated in pre-crossing axons in *robo3* mutants, suggesting that robo3 does not regulate *robo1/2* expression (Sabatier et al., 2004). Robo3 also does not have a high affinity for slit proteins, making it unlikely that robo3 competes with robo1/2 for slit binding (Mambetisaeva et al., 2005; Zelina et al., 2014). *In vitro* studies have suggested that robo3 can promote robo1/2 degradation (Li et al., 2014). Robo3 has also been proposed to prevent axons from invading the motor column, by acting through a novel secreted ligand, neural epidermal growth factor-like-like 2 (nell2) (Jaworski et al., 2015; Pak et al., 2020). Nell2 is present in the spinal cord motor column and central canal. While single mutants in *nell2* have no commissural axon guidance phenotype, adding a *robo3*$^{+/-}$ transheterozygous mutation, i.e. *nell2*$^{-/-}$ *robo3*$^{+/-}$, results in aberrant axon growth in the motor column, consistent with a role for this pathway mediating a repellent interaction in the motor column (Jaworski et al., 2015).

The distribution of robo3 in pre-crossing commissural axons is regulated by the transcriptional programs that define neuronal subtype that were discussed in Section 2. Lhx2, the key transcription factor directing dI1c neural identity, binds to the control region of the *robo3* gene. In *Lhx2/9* double mutants, *robo3* expression is lost from dI1c axons, which then fail to cross the midline. Together, these findings suggest that Lhx2 directly upregulates robo3 in the contralateraly-projecting dI1 axon population (Wilson et al., 2008). What downregulates robo3 in post-crossing axons? One candidate from cancer studies is miR-383, a microRNA that suppresses *robo3* expression with a distribution that is inversely correlated with robo3 (Han et al., 2015).

What regulates the distribution of robo1/2 on commissural axons? Multiple pathways have been shown to alter robo1/2 cell surface levels. The WAGR syndrome gene, proline-rich and gla domain 4 (Prrg4) can prevent robo1 from being trafficked to the cell surface *in vitro* (Justice, Barnum, & Kidd, 2017), while Rab guanine nucleotide dissociation inhibitor (GDI), ubiquitin-specific peptidase 33 (Usp33), and calsyntenin1 can stabilize or promote robo1 surface localization (Alther, Domanitskaya, & Stoeckli, 2016; Philipp et al., 2012; Yuasa-Kawada, Kinoshita-Kawada, Wu, Rao, & Wu, 2009). Nedd4-family interacting proteins, Ndfip1 and Ndfip2, have been shown to recruit the Nedd4 E3 ubiquitin ligases to target robo1 to endosomes for degradation by ubiquitylation. Loss of *Ndfip1/2 in vivo* results in increased *robo1* expression on commissural axons and reduced midline crossing (Gorla et al., 2019). The distribution of robo1/2 may also be controlled at the expression level. miR-92 can suppress the translation of robo1 protein in the spinal cord *in vivo* (Yang et al., 2018). Finally, intriguing recent studies have identified that the neuro-oncological ventral antigen (Nova) splicing factors can produce robo1/robo2 isoforms with different efficacies as repellents. Through the alternative splicing of a microexon, Nova1/2 may regulate the production of these isoforms at different times in development to control the timing by which commissural axons approach and cross the midline (Johnson, Junge, & Chen, 2019)

6. Commissural axons exit the FP and turn rostrally into the ventral funiculus

On exiting the FP, dI1 commissural axons sharply orthogonally to travel immediately beside the border of the FP (Bovolenta & Dodd, 1990). After extending for a short distance, dI1 axons fan out over an arcuate trajectory to extend longitudinally along the ventral funiculus toward the brain (Avraham et al., 2009; Kadison & Kaprielian, 2004). This trajectory is bounded along the dorsal-ventral axis, preventing commissural axons from extending back into the dorsal spinal cord on the contralateral side.

6.1 Semaphorin signaling mediates commissural axon exit from the FP

Commissural axons may be directed to exit from the spinal cord by the class3 semaphorins, present in the FP. Class3 semaphorins signal through a complex of neuropilin2 (Nrp2) and its co-receptor plexina1 (Pignata, Ducuing, & Castellani, 2016). While plexina1 is present specifically on commissural axons, Nrp2 is found on both axons and in the FP (Hernandez-Enriquez et al., 2015; Nawabi et al., 2010). *Nrp2* deficient mice exhibit a variety of pathfinding defects, including commissural axons stalling within the FP and growing aberrantly in their longitudinal trajectory along the contralateral FP border (Tran et al., 2013; Zou, Stoeckli, Chen, & Tessier-Lavigne, 2000). How is a premature response to FP-derived semaphorins prevented? *In vitro* studies using spinal cord explants, have shown that pre-crossing commissural axons are repelled by sema3b. However, this sensitivity to sema3b repulsive signaling is inhibited by FP-derived Npn2. One possibility is that FP-derived Nrp2 sequesters semaphorins, preventing repulsion of pre-crossing axons (Hernandez-Enriquez et al., 2015). It remains unresolved how this sequestration is lifted to permit post-crossing commissural axons to detect and then be repelled by FP-derived sema3b.

6.2 Long-range cues in the FP direct the rostral turn

Two long-range signaling pathways, both present in the FP, have been proposed to mediate the ability of post-crossing commissural axons to turn rostrally. A rostral-low, caudal-high Shh gradient has been shown to act as a repulsive signal (Bourikas et al., 2005; Yam et al., 2012), while a rostral-high, caudal-low Wnt gradient, can attract commissural axons rostrally (Aviles & Stoeckli, 2016; Lyuksyutova et al., 2003). The hedgehog-interacting protein (Hip) mediates the repulsive response to Shh (Bourikas et al., 2005). Disrupting this interaction *in vivo* in chicken embryos leads to some commissural axons projecting both rostrally and caudally. Thus, the initial attractive response of pre-crossing commissural axons to Shh appears to be modulated to repulsion in post-crossing axons. This modulation may be mediated by the 14-3-3 family of adaptor proteins. In both *in vivo* and *in vitro* studies in rodents, increasing the levels of 14-3-3 proteins switches axonal attraction to Shh to repulsion. Conversely, the loss of 14-3-3 proteins changes the response of commissural axons to Shh from repulsive to attractive (Yam et al., 2012).

The Wnt protein gradient is mediated in mouse embryos by Wnt4, acting through the frizzled3 receptor (Lyuksyutova et al., 2003). In the absence of frizzled3, commissural axons turn randomly after crossing the midline. Graded Wnt levels have not been identified in chicken embryos, rather graded Wnt5a and Wnt7a activity may be produced a rostral-high, caudal-low gradient of a Wnt antagonist: secreted frizzled-related protein (Sfrp1) (Domanitskaya et al., 2010). The gradient of Sfrp1 is established by the Shh gradient at the FP. Thus, in chicken, Shh controls the trajectory of post-crossing commissural axons both directly (Bourikas et al., 2005) and indirectly, by regulating Wnt activity (Domanitskaya et al., 2010).

6.3 Bounding the longitudinal commissural trajectory to the ventral funiculus

Finally, the Eph/ephrin signaling pathway has a key role creating the barrier that constrains post-crossing commissural axons to the ventral funiculus (Flanagan & Vanderhaeghen, 1998; Wilkinson, 2000). Many classic studies have identified the Eph/ephrin pathway as mediating short-range repulsive axon guidance decisions. In the mouse and chicken, ephrinB ligands are expressed in the dorsal spinal cord and FP, while the ephrin receptors, including EphB1 and EphA2, are expressed on distal segments of post-crossing commissural axons (Brittis, Lu, & Flanagan, 2002; Imondi, Wideman, & Kaprielian, 2000). Upregulation of Eph receptors, specifically EphA2, in post-crossing axons seems to be regulated by localized protein synthesis and expression (Brittis et al., 2002). Commissural axons turn at a boundary of ephrinB expression and project longitudinally (Imondi, 2000). Disturbing endogenous EphB-ephrinB interactions in cultured spinal cord explants results in growth of commissural axons into dorsal regions of the spinal cord where ephrinB is normally expressed (Imondi & Kaprielian, 2001).

7. Conclusions and speculations

This review has traced the current thinking about the molecular and cellular mechanisms that establish the early stages of dorsal commissural neural circuit, from the birth of dI1 neurons, to the mechanisms that direct their commissural axons first circumferentially and then contralaterally into the ventral funiculus. This is a very well-studied pathway, with clear, yet still puzzling, themes emerging. First, while the commissural trajectory itself is relatively simple: axons extend toward and across a midline, a multitude signals have been implicated in the guidance of this trajectory, with multiple signals often identified for each putative choice point. Why is it so complicated? Potential explanations include that many genes with overlapping redundancies are required to establish this trajectory with absolutely fidelity, or - less interestingly - that some of the identified factors are spurious, because this axonal pathway is easy to perturb through genetic manipulation. However, it is also notable that only a few of the genes required for formation of this neural circuit in vertebrates have highly penetrant, highly expressive loss of function phenotypes. For example, commissural axons are completely defasciculated in the absence of the *netrin1/Dcc* pathway, and they all project ipsilaterally in the absence of *robo3*. Thus, an alternative possibility is that only a few of the identified factors are "architects", critically required to establish the dI1 axonal circuit, while the other signals represent "sculptors", refining or modulating the response to the architect factors. Supporting this model, it is notable that commissural axons continually grow in a netrin1^{+} environment. We have suggested that pre-crossing axons extend around a "hederal" boundary of *netrin1*-expressing cells, including both NPCs and the FP; while post-crossing axons grow longitudinally within the pial-netrin1 substrate in the ventral spinal cord (Fig. 3H) (Varadarajan & Butler, 2017). It remains unresolved whether additional signals regulate these responses to netrin1, permitting pre-crossing commissural axons to avoid, or be unresponsive to, ventral pial-netrin1, while post-crossing commissural axons again use it as a growth substrate.

A second theme in this review, is a reassessment of the importance of long-range signaling. While *bona fide* long-range signals remain (Shh, Bmps), the key signaling pathways required to guide commissural axons appear to act principally by short-range mechanisms. What is their general cellular purpose? Clues may come from the involvement of the large number of Ig domain superfamily receptors, including Tag1, Dcc, the Unc5 and robo families, L1, Nrcam and Boc, and their predominant loss-of-function phenotype: modest to severe axon defasciculation. Thus, the critical "architect" activities of netrin1/robo3 may be to define the commissural axon trajectory as series of local fasciculation decisions to direct axons (1) ventrally and (2) ipsilaterally versus contralaterally. This model stands in contrast to the prevailing long-range model, whereby axons grow in discrete steps from one long-range directional guidepost signal to the next. In the short-range model, the "sculpting" activities could include mediating defasciculation and/or the strength of fasciculation as axons progressively bundle together to cross the FP, as well as directing the topographic position of axons within fascicles. This reassessment opens up many new avenues for mechanistic exploration and may explain why relatively little progress has been made using putatively long-range axon guidance cues as regenerative factors. Understanding how guidance signals act locally to direct fasciculation, may have more mechanistic significance for repairing damaged or diseased nerve tracts.

Acknowledgments

We would like to thank Orkun Akin, Greg Bashaw, James Briscoe, Lorenzo del Castillo, Bennett Novitch, Marc Tessier-Lavigne and Larry Zipursky for discussions. This work was supported by a Cota Robles fellowship and the Ruth L. Kirschstein National

Research Service Award GM007185 (S.A.) and grants from the Broad Stem Cell Research Center (UCLA), Rose Hills Foundation and the NIH/NINDS (NS085097, NS107509) (S.J.B.).

References

Akin, O., & Zipursky, S. L. (2016). Frazzled promotes growth cone attachment at the source of a Netrin gradient in the Drosophila visual system. *eLife*, *5*.

Alaynick, W. A., Jessell, T. M., & Pfaff, S. L. (2011). SnapShot: Spinal cord development. *Cell*, *146*. 178–178 e171.

Alther, T. A., Domanitskaya, E., & Stoeckli, E. T. (2016). Calsyntenin 1-mediated trafficking of axon guidance receptors regulates the switch in axonal responsiveness at a choice point. *Development (Cambridge, England)*, *143*, 994–1004.

Altman, J., & Bayer, S. A. (1984). *The development of the rat spinal cord*.

Andermatt, I., Wilson, N. H., Bergmann, T., Mauti, O., Gesemann, M., Sockanathan, S., et al. (2014). Semaphorin 6B acts as a receptor in post-crossing commissural axon guidance. *Development (Cambridge, England)*, *141*, 3709–3720.

Andrews, M. G., Del Castillo, L. M., Ochoa-Bolton, E., Yamauchi, K., Smogorzewski, J., & Butler, S. J. (2017). BMPs direct sensory interneuron identity in the developing spinal cord using signal-specific not morphogenic activities. *eLife*, *6*.

Andrews, M. G., Kong, J., Novitch, B. G., & Butler, S. J. (2019). New perspectives on the mechanisms establishing the dorsal-ventral axis of the spinal cord. *Current Topics in Developmental Biology*, *132*, 417–450.

Augsburger, A., Schuchardt, A., Hoskins, S., Dodd, J., & Butler, S. (1999). BMPs as mediators of roof plate repulsion of commissural neurons. *Neuron*, *24*, 127–141.

Aviles, E. C., & Stoeckli, E. T. (2016). Canonical wnt signaling is required for commissural axon guidance. *Developmental Neurobiology*, *76*, 190–208.

Avraham, O., Hadas, Y., Vald, L., Zisman, S., Schejter, A., Visel, A., et al. (2009). Transcriptional control of axonal guidance and sorting in dorsal interneurons by the Lim-HD proteins Lhx9 and Lhx1. *Neural Development*, *4*, 21.

Bak, M., & Fraser, S. E. (2003). Axon fasciculation and differences in midline kinetics between pioneer and follower axons within commissural fascicles. *Development*, *130*, 4999–5008.

Blockus, H., & Chedotal, A. (2014). The multifaceted roles of slits and robos in cortical circuits: From proliferation to axon guidance and neurological diseases. *Current Opinion in Neurobiology*, *27*, 82–88.

Bourikas, D., Pekarik, V., Baeriswyl, T., Grunditz, A., Sadhu, R., Nardo, M., et al. (2005). Sonic hedgehog guides commissural axons along the longitudinal axis of the spinal cord. *Nature Neuroscience*, *8*, 297–304.

Bovolenta, P., & Dodd, J. (1990). Guidance of commissural growth cones at the floor plate in embryonic rat spinal cord. *Development*, *109*, 435–447.

Brankatschk, M., & Dickson, B. J. (2006). Netrins guide Drosophila commissural axons at short range. *Nature Neuroscience*, *9*, 188–194.

Briscoe, J., & Ericson, J. (2001). Specification of neuronal fates in the ventral neural tube. *Current Opinion in Neurobiology*, *11*, 43–49.

Briscoe, J., Pierani, A., Jessell, T. M., & Ericson, J. (2000). A homeodomain protein code specifies progenitor cell identity and neuronal fate in the ventral neural tube. *Cell*, *101*, 435–445.

Brittis, P. A., Lu, Q., & Flanagan, J. G. (2002). Axonal protein synthesis provides a mechanism for localized regulation at an intermediate target. *Cell*, *110*, 223–235.

Brose, K., Bland, K. S., Wang, K. H., Arnott, D., Henzel, W., Goodman, C. S., et al. (1999). Slit proteins bind Robo receptors and have an evolutionarily conserved role in repulsive axon guidance. *Cell*, *96*, 795–806.

Burstyn-Cohen, T., Tzarfaty, V., Frumkin, A., Feinstein, Y., Stoeckli, E., & Klar, A. (1999). F-Spondin is required for accurate pathfinding of commissural axons at the floor plate. *Neuron*, *23*, 233–246.

Butler, S. J., & Bronner, M. E. (2015). From classical to current: Analyzing peripheral nervous system and spinal cord lineage and fate. *Developmental Biology*, *398*, 135–146.

Butler, S. J., & Dodd, J. (2003). A role for BMP heterodimers in roof plate-mediated repulsion of commissural axons. *Neuron*, *38*, 389–401.

Carter, S. B. (1965). Principles of cell motility: The direction of cell movement and cancer invasion. *Nature*, *208*, 1183–1187.

Charron, F., Stein, E., Jeong, J., McMahon, A. P., & Tessier-Lavigne, M. (2003). The morphogen sonic hedgehog is an axonal chemoattractant that collaborates with netrin-1 in midline axon guidance. *Cell*, *113*, 11–23.

Charron, F., & Tessier-Lavigne, M. (2005). Novel brain wiring functions for classical morphogens: A role as graded positional cues in axon guidance. *Development*, *132*, 2251–2262.

Chedotal, A. (2019). Roles of axon guidance molecules in neuronal wiring in the developing spinal cord. *Nature Reviews*, *20*, 380–396.

Chuang, W., & Lagenaur, C. F. (1990). Central nervous system antigen P84 can serve as a substrate for neurite outgrowth. *Developmental Biology*, *137*, 219–232.

Comer, J. D., Alvarez, S., Butler, S. J., & Kaltschmidt, J. A. (2019). Commissural axon guidance in the developing spinal cord: From Cajal to the present day. *Neural Development*, *14*, 9.

de Castro, F., López-Mascaraque, L., & De Carlos, J. A. (2007). Cajal: Lessons on brain development. *Brain Research Reviews*, *55*, 481–489.

Dessaud, E., Yang, L. L., Hill, K., Cox, B., Ulloa, F., Ribeiro, A., et al. (2007). Interpretation of the sonic hedgehog morphogen gradient by a temporal adaptation mechanism. *Nature*, *450*, 717–720.

Dickson, B. J. (2002). Molecular mechanisms of axon guidance. *Science*, *298*, 1959–1964.

Dillon, A. K., Jevince, A. R., Hinck, L., Ackerman, S. L., Lu, X., Tessier-Lavigne, M., et al. (2007). UNC5C is required for spinal accessory motor neuron development. *Molecular and Cellular Neurosciences*, *35*, 482–489.

Dodd, J., Morton, S. B., Karagogeos, D., Yamamoto, M., & Jessell, T. M. (1988). Spatial regulation of axonal glycoprotein expression on subsets of embryonic spinal neurons. *Neuron*, *1*, 105–116.

Domanitskaya, E., Wacker, A., Mauti, O., Baeriswyl, T., Esteve, P., Bovolenta, P., et al. (2010). Sonic hedgehog guides post-crossing commissural axons both directly and indirectly by regulating Wnt activity. *The Journal of Neuroscience: The Official Journal of the Society for Neuroscience*, *30*, 11167–11176.

Dominici, C., Moreno-Bravo, J. A., Puiggros, S. R., Rappeneau, Q., Rama, N., Vieugue, P., et al. (2017). Floor-plate-derived netrin-1 is dispensable for commissural axon guidance. *Nature*, *545*, 350–354.

Ducuing, H., Gardette, T., Pignata, A., Tauszig-Delamasure, S., & Castellani, V. (2019). Commissural axon navigation in the spinal cord: A repertoire of repulsive forces is in command. *Seminars in Cell & Developmental Biology*, *85*, 3–12.

Duval, N., Vaslin, C., Barata, T. C., Frarma, Y., Contremoulins, V., Baudin, X., et al. (2019). BMP4 patterns Smad activity and generates stereotyped cell fate organization in spinal organoids. *Development*, *146*.

Engelkamp, D. (2002). Cloning of three mouse Unc5 genes and their expression patterns at mid-gestation. *Mechanisms of Development*, *118*, 191–197.

Evans, T. A., & Bashaw, G. J. (2010). Axon guidance at the midline: Of mice and flies. *Current Opinion in Neurobiology*, *20*, 79–85.

Fazeli, A., Dickinson, S. L., Hermiston, M. L., Tighe, R. V., Steen, R. G., Small, C. G., et al. (1997). Phenotype of mice lacking functional deleted in colorectal cancer (Dcc) gene. *Nature*, *386*, 796–804.

Fitzli, D., Stoeckli, E. T., Kunz, S., Siribour, K., Rader, C., Kunz, B., et al. (2000). A direct interaction of axonin-1 with NgCAM-related cell adhesion molecule (NrCAM) results in guidance, but not growth of commissural axons. *The Journal of Cell Biology*, *149*, 951–968.

Flanagan, J. G., & Vanderhaeghen, P. (1998). The ephrins and Eph receptors in neural development. *Annual Review of Neuroscience*, *21*, 309–345.

Fujimura, Y., Iwashita, M., Matsuzaki, F., & Yamamoto, T. (2006). MDGA1, an IgSF molecule containing a MAM domain, heterophilically associates with axon- and muscle-associated binding partners through distinct structural domains. *Brain Research*, *1101*, 12–19.

Gorla, M., Santiago, C., Chaudhari, K., Layman, A. A. K., Oliver, P. M., & Bashaw, G. J. (2019). Ndfip proteins target robo receptors for degradation and allow commissural axons to cross the midline in the developing spinal cord. *Cell Reports*, *26*. 3298-3312 e3294.

Goulding, M. D., Chalepakis, G., Deutsch, U., Erselius, J. R., & Gruss, P. (1991). Pax-3, a novel murine DNA binding protein expressed during early neurogenesis. *The EMBO Journal*, *10*, 1135–1147.

Gowan, K., Helms, A. W., Hunsaker, T. L., Collisson, T., Ebert, P. J., Odom, R., et al. (2001). Crossinhibitory activities of Ngn1 and Math1 allow specification of distinct dorsal interneurons. *Neuron*, *31*, 219–232.

Gupta, S., Kawaguchi, R., Mandric, I., Castellanos, S., & Butler, S. J. (2020). *In vitro recapitulation of dorsal spinal cord development reveals Wnt signaling as a regulator of interneuron diversity*. in preparation.

Gupta, S., Sivalingam, D., Hain, S., Makkar, C., Sosa, E., Clark, A., et al. (2018). Deriving Dorsal Spinal Sensory Interneurons from Human Pluripotent Stem Cells. *Stem Cell Reports*, *10*, 390–405.

Han, S., Cao, C., Tang, T., Lu, C., Xu, J., Wang, S., et al. (2015). ROBO3 promotes growth and metastasis of pancreatic carcinoma. *Cancer Letters*, *366*, 61–70.

Hazen, V. M., Andrews, M. A., Umans, L., Crenshaw, E. B. 3rd, Zwijsen, A., & Butler, S. J. (2012). BMP receptor-activated Smads direct diverse functions during the development of the dorsal spinal cord. *Developmental Biology*, *367*, 216–227.

Helms, A. W., & Johnson, J. E. (1998). Progenitors of dorsal commissural interneurons are defined by MATH1 expression. *Development*, *125*, 919–928.

Helms, A. W., & Johnson, J. E. (2003). Specification of dorsal spinal cord interneurons. *Current Opinion in Neurobiology*, *13*, 42–49.

Hernandez-Enriquez, B., Wu, Z., Martinez, E., Olsen, O., Kaprielian, Z., Maness, P. F., et al. (2015). Floor plate-derived neuropilin-2 functions as a secreted semaphorin sink to facilitate commissural axon midline crossing. *Genes & Development*, *29*, 2617–2632.

Imondi, R., & Kaprielian, Z. (2001). Commissural axon pathfinding on the contralateral side of the floor plate: A role for B-class ephrins in specifying the dorsoventral position of longitudinally projecting commissural axons. *Development*, *128*, 4859–4871.

Imondi, R., Wideman, C., & Kaprielian, Z. (2000). Complementary expression of transmembrane ephrins and their receptors in the mouse spinal cord: A possible role in constraining the orientation of longitudinally projecting axons. *Development (CambridgeEngland)*, *127*, 1397–1410.

Islam, S. M., Shinmyo, Y., Okafuji, T., Su, Y., Naser, I. B., Ahmed, G., et al. (2009). Draxin, a repulsive guidance protein for spinal cord and forebrain commissures. *Science*, *323*, 388–393.

Jaworski, A., Long, H., & Tessier-Lavigne, M. (2010). Collaborative and specialized functions of Robo1 and Robo2 in spinal commissural axon guidance. *The Journal of Neuroscience: The Official Journal of the Society for Neuroscience*, *30*, 9445–9453.
Jaworski, A., Tom, I., Tong, R. K., Gildea, H. K., Koch, A. W., Gonzalez, L. C., et al. (2015). Operational redundancy in axon guidance through the multifunctional receptor Robo3 and its ligand NELL2. *Science (New YorkN.Y.)*, *350*, 961–965.
Jessell, T. M. (2000). Neuronal specification in the spinal cord: Inductive signals and transcriptional codes. *Nature Reviews. Genetics*, *1*, 20–29.
Johnson, V., Junge, H. J., & Chen, Z. (2019). Temporal regulation of axonal repulsion by alternative splicing of a conserved microexon in mammalian Robo1 and Robo2. *eLife*, *8*.
Joset, P., Wacker, A., Babey, R., Ingold, E. A., Andermatt, I., Stoeckli, E. T., et al. (2011). Rostral growth of commissural axons requires the cell adhesion molecule MDGA2. *Neural Development*, *6*, 22.
Jostes, B., Walther, C., & Gruss, P. (1990). The murine paired box gene, Pax7, is expressed specifically during the development of the nervous and muscular system. *Mechanisms of Development*, *33*, 27–37.
Justice, E. D., Barnum, S. J., & Kidd, T. (2017). The WAGR syndrome gene PRRG4 is a functional homologue of the commissureless axon guidance gene. *PLoS Genetics*, *13*, e1006865.
Kadison, S. R., & Kaprielian, Z. (2004). Diversity of contralateral commissural projections in the embryonic rodent spinal cord. *Journal of Comparative Neurology*, *472*, 411–422.
Kadison, S. R., Murakami, F., Matise, M. P., & Kaprielian, Z. (2006). The role of floor plate contact in the elaboration of contralateral commissural projections within the embryonic mouse spinal cord. *Developmental Biology*, *296*, 499–513.
Kaprielian, Z., Imondi, R., & Runko, E. (2000). Axon guidance at the midline of the developing CNS. *The Anatomical Record*, *261*, 176–197.
Keino-Masu, K., Masu, M., Hinck, L., Leonardo, E. D., Chan, S. S., Culotti, J. G., et al. (1996). Deleted in colorectal cancer (DCC) encodes a netrin receptor. *Cell*, *87*, 175–185.
Kennedy, T. E., Serafini, T., de la Torre, J. R., & Tessier-Lavigne, M. (1994). Netrins are diffusible chemotropic factors for commissural axons in the embryonic spinal cord. *Cell*, *78*, 425–435.
Kidd, T., Brose, K., Mitchell, K. J., Fetter, R. D., Tessier-Lavigne, M., Goodman, C. S., et al. (1998). Roundabout controls axon crossing of the CNS midline and defines a novel subfamily of evolutionarily conserved guidance receptors. *Cell*, *92*, 205–215.
Klar, A., Baldassare, M., & Jessell, T. M. (1992). F-spondin: A gene expressed at high levels in the floor plate encodes a secreted protein that promotes neural cell adhesion and neurite extension. *Cell*, *69*, 95–110.
Kong, J. H., Yang, L., Dessaud, E., Chuang, K., Moore, D. M., Rohatgi, R., et al. (2015). Notch activity modulates the responsiveness of neural progenitors to sonic hedgehog signaling. *Developmental Cell*, *33*, 373–387.
Krushel, L. A., Prieto, A. L., Cunningham, B. A., & Edelman, G. M. (1993). Expression patterns of the cell adhesion molecule Nr-CAM during histogenesis of the chick nervous system. *Neuroscience*, *53*, 797–812.
Laumonnerie, C., Da Silva, R. V., Kania, A., & Wilson, S. I. (2014). Netrin 1 and Dcc signalling are required for confinement of central axons within the central nervous system. *Development*, *141*, 594–603.
Laumonnerie, C., Tong, Y. G., Alstermark, H., & Wilson, S. I. (2015). Commissural axonal corridors instruct neuronal migration in the mouse spinal cord. *Nature Communications*, *6*, 7028.
Le Dreau, G., Garcia-Campmany, L., Rabadan, M. A., Ferronha, T., Tozer, S., Briscoe, J., et al. (2012). Canonical BMP7 activity is required for the generation of discrete neuronal populations in the dorsal spinal cord. *Development*, *139*, 259–268.

Le Dreau, G., & Marti, E. (2013). The multiple activities of BMPs during spinal cord development. *Cellular and Molecular Life Sciences: CMLS*, *70*, 4293–4305.

Lee, K. J., Dietrich, P., & Jessell, T. M. (2000). Genetic ablation reveals that the roof plate is essential for dorsal interneuron specification. *Nature*, *403*, 734–740.

Lee, K. J., & Jessell, T. M. (1999). The specification of dorsal cell fates in the vertebrate central nervous system. *Annual Review of Neuroscience*, *22*, 261–294.

Lee, K. J., Mendelsohn, M., & Jessell, T. M. (1998). Neuronal patterning by BMPs: A requirement for GDF7 in the generation of a discrete class of commissural interneurons in the mouse spinal cord. *Genes & Development*, *12*, 3394–3407.

Lee-Hoeflich, S. T., Causing, C. G., Podkowa, M., Zhao, X., Wrana, J. L., & Attisano, L. (2004). Activation of LIMK1 by binding to the BMP receptor, BMPRII, regulates BMP-dependent dendritogenesis. *The EMBO Journal*, *23*, 4792–4801.

Leonardo, E. D., Hinck, L., Masu, M., Keino-Masu, K., Ackerman, S. L., & Tessier-Lavigne, M. (1997). Vertebrate homologues of C. elegans UNC-5 are candidate netrin receptors. *Nature*, *386*, 833–838.

Li, H. S., Chen, J. H., Wu, W., Fagaly, T., Zhou, L., Yuan, W., et al. (1999). Vertebrate slit, a secreted ligand for the transmembrane protein roundabout, is a repellent for olfactory bulb axons. *Cell*, *96*, 807–818.

Li, L., Liu, S., Lei, Y., Cheng, Y., Yao, C., & Zhen, X. (2014). Robo3.1A suppresses slit-mediated repulsion by triggering degradation of Robo2. *Journal of Neuroscience Research*, *92*, 835–846.

Liem, K. F., Jr., Tremml, G., & Jessell, T. M. (1997). A role for the roof plate and its resident TGFbeta-related proteins in neuronal patterning in the dorsal spinal cord. *Cell*, *91*, 127–138.

Liu, Y., Bhowmick, T., Gao, X., Mertens, H. D. T., Svergun, D. I., Xiao, J., et al. (2018). Structural Basis for Draxin-Modulated Axon Guidance and Fasciculation by Netrin-1 through DCC. *Neuron*, *97*, 1261–1267. e1264.

Long, H., Sabatier, C., Ma, L., Plump, A., Yuan, W., Ornitz, D. M., et al. (2004). Conserved roles for Slit and Robo proteins in midline commissural axon guidance. *Neuron*, *42*, 213–223.

Lyuksyutova, A. I., Lu, C. C., Milanesio, N., King, L. A., Guo, N., Wang, Y., et al. (2003). Anterior-posterior guidance of commissural axons by Wnt-frizzled signaling. *Science*, *302*, 1984–1988.

Mambetisaeva, E. T., Andrews, W., Camurri, L., Annan, A., & Sundaresan, V. (2005). Robo family of proteins exhibit differential expression in mouse spinal cord and Robo-Slit interaction is required for midline crossing in vertebrate spinal cord. *Developmental Dynamics: An Official Publication of the American Association of the Anatomists*, *233*, 41–51.

Masuda, T., Watanabe, K., Sakuma, C., Ikenaka, K., Ono, K., & Yaginuma, H. (2008). Netrin-1 acts as a repulsive guidance cue for sensory axonal projections toward the spinal cord. *The Journal of Neuroscience*, *28*, 10380–10385.

Matise, M. P., Epstein, D. J., Park, H. L., Platt, K. A., & Joyner, A. L. (1998). Gli2 is required for induction of floor plate and adjacent cells, but not most ventral neurons in the mouse central nervous system. *Development*, *125*, 2759–2770.

Matise, M. P., Lustig, M., Sakurai, T., Grumet, M., & Joyner, A. L. (1999). Ventral midline cells are required for the local control of commissural axon guidance in the mouse spinal cord. *Development*, *126*, 3649–3659.

Meberg, P. J., & Bamburg, J. R. (2000). Increase in neurite outgrowth mediated by overexpression of actin depolymerizing factor. *The Journal of Neuroscience*, *20*, 2459–2469.

Megason, S. G., & McMahon, A. P. (2002). A mitogen gradient of dorsal midline Wnts organizes growth in the CNS. *Development*, *129*, 2087–2098.

Moore, S. W., Biais, N., & Sheetz, M. P. (2009). Traction on immobilized netrin-1 is sufficient to reorient axons. *Science (New York, N.Y.)*, *325*, 166.

Moore, S. W., Tessier-Lavigne, M., & Kennedy, T. E. (2007). Netrins and their receptors. *Advances in Experimental Medicine and Biology, 621*, 17–31.
Moreno-Bravo, J. A., Roig Puiggros, S., Blockus, H., Dominici, C., Zelina, P., Mehlen, P., et al. (2018). Commissural neurons transgress the CNS/PNS boundary in absence of ventricular zone-derived netrin 1. *Development, 145*.
Moreno-Bravo, J. A., Roig Puiggros, S., Mehlen, P., & Chedotal, A. (2019). Synergistic Activity of Floor-Plate- and Ventricular-Zone-Derived Netrin-1 in Spinal Cord Commissural Axon Guidance. *Neuron, 101*. 625–634 e623.
Moscoso, L. M., & Sanes, J. R. (1995). Expression of four immunoglobulin superfamily adhesion molecules (L1, Nr-CAM/Bravo, neurofascin/ABGP, and N-CAM) in the developing mouse spinal cord. *The Journal of Comparative Neurology, 352*, 321–334.
Muroyama, Y., Fujihara, M., Ikeya, M., Kondoh, H., & Takada, S. (2002). Wnt signaling plays an essential role in neuronal specification of the dorsal spinal cord. *Genes & Development, 16*, 548–553.
Nawabi, H., Briancon-Marjollet, A., Clark, C., Sanyas, I., Takamatsu, H., Okuno, T., et al. (2010). A midline switch of receptor processing regulates commissural axon guidance in vertebrates. *Genes & Development, 24*, 396–410.
Niederkofler, V., Baeriswyl, T., Ott, R., & Stoeckli, E. T. (2010). Nectin-like molecules/SynCAMs are required for post-crossing commissural axon guidance. *Development (Cambridge, England), 137*, 427–435.
Nusslein-Volhard, C., Wieschaus, E., & Kluding, H. (1984). Mutations affecting the pattern of the larval cuticle in Drosophila melanogaster: I. Zygotic loci on the second chromosome. *Wilhelm Roux's Archives of Developmental Biology, 193*, 267–282.
Okabe, N., Shimizu, K., Ozaki-Kuroda, K., Nakanishi, H., Morimoto, K., Takeuchi, M., et al. (2004). Contacts between the commissural axons and the floor plate cells are mediated by nectins. *Developmental Biology, 273*, 244–256.
Okada, A., Charron, F., Morin, S., Shin, D. S., Wong, K., Fabre, P. J., et al. (2006). Boc is a receptor for sonic hedgehog in the guidance of commissural axons. *Nature, 444*, 369–373.
Pak, J. S., DeLoughery, Z. J., Wang, J., Acharya, N., Park, Y., Jaworski, A., et al. (2020). NELL2-Robo3 complex structure reveals mechanisms of receptor activation for axon guidance. *Nature Communications, 11*, 1489.
Panchision, D. M., Pickel, J. M., Studer, L., Lee, S. H., Turner, P. A., Hazel, T. G., et al. (2001). Sequential actions of BMP receptors control neural precursor cell production and fate. *Genes & Development, 15*, 2094–2110.
Perron, J. C., & Dodd, J. (2011). Inductive specification and axonal orientation of spinal neurons mediated by divergent bone morphogenetic protein signaling pathways. *Neural Development, 6*, 36.
Phan, K. D., & Butler, S. J. (2013). Bilaterally Symmetric Populations of Chicken dI1 (Commissural) Axons Cross the Floor Plate Independently of Each Other. *PLoS One, 8*, e62977.
Phan, K. D., Hazen, V. M., Frendo, M., Jia, Z., & Butler, S. J. (2010). The bone morphogenetic protein roof plate chemorepellent regulates the rate of commissural axonal growth. *The Journal of Neuroscience, 30*, 15430–15440.
Philipp, M., Niederkofler, V., Debrunner, M., Alther, T., Kunz, B., & Stoeckli, E. T. (2012). RabGDI controls axonal midline crossing by regulating Robo1 surface expression. *Neural Development, 7*, 36.
Pignata, A., Ducuing, H., & Castellani, V. (2016). Commissural axon navigation: Control of midline crossing in the vertebrate spinal cord by the semaphorin 3B signaling. *Cell Adhesion & Migration, 10*, 604–617.
Placzek, M., Tessier-Lavigne, M., Yamada, T., Dodd, J., & Jessell, T. M. (1990). Guidance of developing axons by diffusible chemoattractants. *Cold Spring Harbor Symposia on Quantitative Biology, 55*, 279–289.

Ramón y Cajal, S. (1995). *Histology of the nervous system of man and vertebrates*. New York: Oxford University Press.
Rexed, B. (1954). A cytoarchitectonic atlas of the spinal cord in the cat. *Journal of Comparative Neurology*, *100*, 297–379.
Ruiz de Almodovar, C., Fabre, P. J., Knevels, E., Coulon, C., Segura, I., Haddick, P. C., et al. (2011). VEGF mediates commissural axon chemoattraction through its receptor Flk1. *Neuron*, *70*, 966–978.
Sabatier, C., Plump, A. S., Le, M., Brose, K., Tamada, A., Murakami, F., et al. (2004). The divergent Robo family protein rig-1/Robo3 is a negative regulator of slit responsiveness required for midline crossing by commissural axons. *Cell*, *117*, 157–169.
Sasai, N., & Briscoe, J. (2012). Primary cilia and graded Sonic Hedgehog signaling. *Wiley Interdisciplinary Reviews: Developmental Biology*, *1*, 753–772.
Seeger, M., Tear, G., Ferres-Marco, D., & Goodman, C. S. (1993). Mutations affecting growth cone guidance in Drosophila: Genes necessary for guidance toward or away from the midline. *Neuron*, *10*, 409–426.
Serafini, T., Colamarino, S. A., Leonardo, E. D., Wang, H., Beddington, R., Skarnes, W. C., et al. (1996). Netrin-1 is required for commissural axon guidance in the developing vertebrate nervous system. *Cell*, *87*, 1001–1014.
Serafini, T., Kennedy, T. E., Galko, M. J., Mirzayan, C., Jessell, T. M., & Tessier-Lavigne, M. (1994). The netrins define a family of axon outgrowth-promoting proteins homologous to C. elegans UNC-6. *Cell*, *78*, 409–424.
Shi, Y., & Massague, J. (2003). Mechanisms of TGF-beta signaling from cell membrane to the nucleus. *Cell*, *113*, 685–700.
Shirasaki, R., & Pfaff, S. L. (2002). Transcriptional codes and the control of neuronal identity. *Annual Review of Neuroscience*, *25*, 251–281.
Sloan, T. F., Qasaimeh, M. A., Juncker, D., Yam, P. T., & Charron, F. (2015). Integration of shallow gradients of Shh and Netrin-1 guides commissural axons. *PLoS Biology*, *13*, e1002119.
Snow, D. M., Lemmon, V., Carrino, D. A., Caplan, A. I., & Silver, J. (1990). Sulfated proteoglycans in astroglial barriers inhibit neurite outgrowth in vitro. *Experimental Neurology*, *109*, 111–130.
Snow, D. M., Steindler, D. A., & Silver, J. (1990). Molecular and cellular characterization of the glial roof plate of the spinal cord and optic tectum: A possible role for a proteoglycan in the development of an axon barrier. *Developmental Biology*, *138*, 359–376.
Sotelo, C. (2002). The chemotactic hypothesis of Cajal: A century behind. *Progress in Brain Research*, *136*, 11–20.
Stoeckli, E. T., & Landmesser, L. T. (1995). Axonin-1, Nr-CAM, and Ng-CAM play different roles in the in vivo guidance of chick commissural neurons. *Neuron*, *14*, 1165–1179.
Stoeckli, E. T., Lemkin, P. F., Kuhn, T. B., Ruegg, M. A., Heller, M., & Sonderegger, P. (1989). Identification of proteins secreted from axons of embryonic dorsal-root-ganglia neurons. *European Journal of Biochemistry*, *180*, 249–258.
Stoeckli, E. T., Sonderegger, P., Pollerberg, G. E., & Landmesser, L. T. (1997). Interference with axonin-1 and NrCAM interactions unmasks a floor-plate activity inhibitory for commissural axons. *Neuron*, *18*, 209–221.
Sundaresan, V., Roberts, I., Bateman, A., Bankier, A., Sheppard, M., Hobbs, C., et al. (1998). The DUTT1 gene, a novel NCAM family member is expressed in developing murine neural tissues and has an unusually broad pattern of expression. *Molecular and Cellular Neurosciences*, *11*, 29–35.
Suter, T. A. C. S., DeLoughery, Z. J., & Jaworski, A. (2017). Meninges-derived cues control axon guidance. *Developmental Biology*, *430*, 1–10.
Tear, G., Seeger, M., & Goodman, C. S. (1993). To cross or not to cross: A genetic analysis of guidance at the midline. *Perspectives on Developmental Neurobiology*, *1*, 183–194.

Tessier-Lavigne, M., & Goodman, C. S. (1996). The molecular biology of axon guidance. *Science*, *274*, 1123–1133.
Timmer, J. R., Wang, C., & Niswander, L. (2002). BMP signaling patterns the dorsal and intermediate neural tube via regulation of homeobox and helix-loop-helix transcription factors. *Development*, *129*, 2459–2472.
Timofeev, K., Joly, W., Hadjieconomou, D., & Salecker, I. (2012). Localized netrins act as positional cues to control layer-specific targeting of photoreceptor axons in Drosophila. *Neuron*, *75*, 80–93.
Tozer, S., Le Dreau, G., Marti, E., & Briscoe, J. (2013). Temporal control of BMP signalling determines neuronal subtype identity in the dorsal neural tube. *Development*, *140*, 1467–1474.
Tran, T. S., Carlin, E., Lin, R., Martinez, E., Johnson, J. E., & Kaprielian, Z. (2013). Neuropilin2 regulates the guidance of post-crossing spinal commissural axons in a subtype-specific manner. *Neural Development*, *8*, 15.
Tulloch, A. J., Teo, S., Carvajal, B. V., Tessier-Lavigne, M., & Jaworski, A. (2019). Diverse spinal commissural neuron populations revealed by fate mapping and molecular profiling using a novel Robo3(Cre) mouse. *Journal of Comparative Neurology*, *527*, 2948–2972.
Varadarajan, S. G., & Butler, S. J. (2017). Netrin1 establishes multiple boundaries for axon growth in the developing spinal cord. *Developmental Biology*, *430*, 177–187.
Varadarajan, S. G., Kong, J. H., Phan, K. D., Kao, T. J., Panaitof, S. C., Cardin, J., et al. (2017). Netrin1 produced by neural progenitors, not floor plate cells is required for axon guidance in the spinal cord. *Neuron*, *94*. 790–799 e793.
Wang, L., Yu, C., Wang, J., Leung, P., Ma, D., Zhao, H., et al. (2017). Nogo-B is the major form of Nogo at the floor plate and likely mediates crossing of commissural axons in the mouse spinal cord. *The Journal of Comparative Neurology*, *525*, 2915–2928.
Watanabe, K., Tamamaki, N., Furuta, T., Ackerman, S. L., Ikenaka, K., & Ono, K. (2006). Dorsally derived netrin 1 provides an inhibitory cue and elaborates the 'waiting period' for primary sensory axons in the developing spinal cord. *Development*, *133*, 1379–1387.
Wen, Z., Han, L., Bamburg, J. R., Shim, S., Ming, G. L., & Zheng, J. Q. (2007). BMP gradients steer nerve growth cones by a balancing act of LIM kinase and Slingshot phosphatase on ADF/cofilin. *The Journal of Cell Biology*, *178*, 107–119.
Wentworth, L. E. (1984). The development of the cervical spinal cord of the mouse embryo. II. A Golgi analysis of sensory, commissural, and association cell differentiation. *Journal of Comparative Neurology*, *222*, 96–115.
Wilkinson, D. G. (2000). Eph receptors and ephrins: Regulators of guidance and assembly. *International Review of Cytology*, *196*, 177–244.
Williams, M. E., Lu, X., McKenna, W. L., Washington, R., Boyette, A., Strickland, P., et al. (2006). UNC5A promotes neuronal apoptosis during spinal cord development independent of netrin-1. *Nature Neuroscience*, *9*, 996–998.
Wilson, S. I., Shafer, B., Lee, K. J., & Dodd, J. (2008). A molecular program for contralateral trajectory: Rig-1 control by LIM homeodomain transcription factors. *Neuron*, *59*, 413–424.
Wine-Lee, L., Ahn, K. J., Richardson, R. D., Mishina, Y., Lyons, K. M., & Crenshaw, E. B., 3rd. (2004). Signaling through BMP type 1 receptors is required for development of interneuron cell types in the dorsal spinal cord. *Development*, *131*, 5393–5403.
Wright, K. M., Lyon, K. A., Leung, H., Leahy, D. J., Ma, L., & Ginty, D. D. (2012). Dystroglycan organizes axon guidance cue localization and axonal pathfinding. *Neuron*, *76*, 931–944.
Wu, Z., Makihara, S., Yam, P. T., Teo, S., Renier, N., Balekoglu, N., et al. (2019). Long-range guidance of spinal commissural axons by netrin1 and Sonic hedgehog from midline floor plate cells. *Neuron*.

Yaginuma, H., Homma, S., Kunzi, R., & Oppenheim, R. W. (1991). Pathfinding by growth cones of commissural interneurons in the chick embryo spinal cord: A light and electron microscopic study. *Journal of Comparative Neurology*, *304*, 78–102.

Yam, P. T., Kent, C. B., Morin, S., Farmer, W. T., Alchini, R., Lepelletier, L., et al. (2012). 14-3-3 proteins regulate a cell-intrinsic switch from sonic hedgehog-mediated commissural axon attraction to repulsion after midline crossing. *Neuron*, *76*, 735–749.

Yamauchi, K., Phan, K. D., & Butler, S. J. (2008). BMP type I receptor complexes have distinct activities mediating cell fate and axon guidance decisions. *Development*, *135*, 1119–1128.

Yamauchi, K., Varadarajan, S. G., Li, J. E., & Butler, S. J. (2013). Type Ib BMP receptors mediate the rate of commissural axon extension through inhibition of cofilin activity. *Development*, *140*, 333–342.

Yamauchi, K., Yamazaki, M., Abe, M., Sakimura, K., Lickert, H., Kawasaki, T., et al. (2017). Netrin-1 Derived from the Ventricular Zone, but not the Floor Plate, Directs Hindbrain Commissural Axons to the Ventral Midline. *Scientific Reports*, *7*, 11992.

Yang, T., Huang, H., Shao, Q., Yee, S., Majumder, T., & Liu, G. (2018). miR-92 Suppresses Robo1 Translation to Modulate Slit Sensitivity in Commissural Axon Guidance. *Cell Reports*, *24*. 2694–2708.e2696.

Yoshida, Y., Han, B., Mendelsohn, M., & Jessell, T. M. (2006). PlexinA1 signaling directs the segregation of proprioceptive sensory axons in the developing spinal cord. *Neuron*, *52*, 775–788.

Yuan, W., Zhou, L., Chen, J. H., Wu, J. Y., Rao, Y., & Ornitz, D. M. (1999). The mouse SLIT family: Secreted ligands for ROBO expressed in patterns that suggest a role in morphogenesis and axon guidance. *Developmental Biology*, *212*, 290–306.

Yuasa-Kawada, J., Kinoshita-Kawada, M., Wu, G., Rao, Y., & Wu, J. Y. (2009). Midline crossing and Slit responsiveness of commissural axons require USP33. *Nature Neuroscience*, *12*, 1087–1089.

Yuengert, R., Hori, K., Kibodeaux, E. E., McClellan, J. X., Morales, J. E., Huang, T. P., et al. (2015). Origin of a Non-Clarke's Column Division of the Dorsal Spinocerebellar Tract and the Role of Caudal Proprioceptive Neurons in Motor Function. *Cell Reports*, *13*, 1258–1271.

Yung, A. R., Druckenbrod, N. R., Cloutier, J. F., Wu, Z., Tessier-Lavigne, M., & Goodrich, L. V. (2018). Netrin-1 Confines Rhombic Lip-Derived Neurons to the CNS. *Cell Reports*, *22*, 1666–1680.

Zelina, P., Blockus, H., Zagar, Y., Peres, A., Friocourt, F., Wu, Z., et al. (2014). Signaling switch of the axon guidance receptor Robo3 during vertebrate evolution. *Neuron*, *84*, 1258–1272.

Zhuang, B., & Sockanathan, S. (2006). Dorsal-ventral patterning: A view from the top. *Current Opinion in Neurobiology*, *16*, 20–24.

Zou, Y., Stoeckli, E., Chen, H., & Tessier-Lavigne, M. (2000). Squeezing axons out of the gray matter: A role for slit and semaphorin proteins from midline and ventral spinal cord. *Cell*, *102*, 363–375.

CHAPTER SIX

Molecular mechanisms that mediate dendrite morphogenesis

Julie L. Lefebvre[a,b,*]
[a]Program for Neurosciences and Mental Health, Hospital for Sick Children, Toronto, ON, Canada
[b]Department of Molecular Genetics, University of Toronto, Toronto, ON, Canada
*Corresponding author: e-mail address: julie.lefebvre@sickkids.ca

Contents

Current Topics in Developmental Biology, Volume 142
ISSN 0070-2153
https://doi.org/10.1016/bs.ctdb.2020.12.008

Abstract

Neurons develop dendritic morphologies that bear cell type-specific features in dendritic field size and geometry, branch placement and density, and the types and distributions of synaptic contacts. Dendritic patterns influence the types and numbers of inputs a neuron receives, and the ways in which neural information is processed and transmitted in the circuitry. Even subtle alterations in dendritic structures can have profound consequences on neuronal function and are implicated in neurodevelopmental disorders. In this chapter, I review how growing dendrites acquire their exquisite patterns by drawing examples from diverse neuronal cell types in vertebrate and invertebrate model systems. Dendrite morphogenesis is shaped by intrinsic and extrinsic factors such as transcriptional regulators, guidance and adhesion molecules, neighboring cells and synaptic partners. I discuss molecular mechanisms that regulate dendrite morphogenesis with a focus on five aspects of dendrite patterning: (1) Dendritic cytoskeleton and cellular machineries that build the arbor; (2) Gene regulatory mechanisms; (3) Afferent cues that regulate dendritic arbor growth; (4) Space-filling strategies that optimize dendritic coverage; and (5) Molecular cues that specify dendrite wiring. Cell type-specific implementation of these patterning mechanisms produces the diversity of dendrite morphologies that wire the nervous system.

1. Introduction

Dendrites develop through a dynamic, reiterative process of branch addition, extension, retraction and stabilization to establish their territory. While most neuronal cell types undergo similar progressions in dendrite development, they each form a characteristic morphology with particular features in dendritic arbor size and shape, branch complexity, targeting, and synaptic and ion channel patterning. This dendritic organization is a key determinant to a neuron's function (Lefebvre & Marocha, 2020). Dendrite location and branch density determine the types and number of inputs a neuron can receive. Dendrite size, synapse distribution and distances from the soma, as well as electrical properties, influence dendritic computations and transmission. Given the importance of dendritic form to function, studies of dendrite morphogenesis seek to understand how cell type-specific dendritic patterns are shaped by intrinsic programs and extrinsic cues, and how a diversity of dendritic morphologies is produced in nervous systems.

In this chapter, I review mechanisms of dendrite morphogenesis in invertebrate and vertebrate model systems with a focus on molecular players of cell type-specific patterning. Progress in elucidating cell type-specific factors is fuelled by the recent expansion of genetic tools that allow the profiling, monitoring and targeting of defined neuronal populations. I begin with an overview of the cell biological mechanisms that build the dendritic structure and relate their activities to branch patterning. I discuss cellular strategies and molecular pathways that pattern multiple aspects of dendritic field formation, including size, targeting, and dendrite coverage, drawing examples from a diversity of model systems (*Drosophila*, *C. elegans*, mice, chick, frog, and zebrafish) (Fig. 1). Where possible, I draw links between transcriptional programs, cell biological mechanisms, and recognition molecules to highlight the coordination of dendrite development.

2. The cell biology of dendrite morphogenesis

Dendrites form from a dynamic period of branch addition and extreme cellular growth. Dendrite morphogenesis requires a massive addition of cytoskeleton and plasma membrane, and delivery of proteins that respond to cues and that direct synapse formation. Dendrites differ in their microtubule organization and thus acquire dendrite-specific cytoskeletal effectors and secretory structures that play key roles in the development and compartmentalization of arbors. Here I review mechanisms of spatial regulation of microtubule and actin polymerization and the specialization of secretory trafficking within dendrites that together tune branch patterning.

2.1 The formation of the dendritic microtubule network

A major structural distinction between axons and dendrites is microtubule polarity, which specifies membrane and protein trafficking and thereby compartmentalizes their pre- and post-synaptic functions. Axons contain microtubules with plus-ends pointing out while dendrites have a mixed polarity or a predominant organization with minus-ends pointing out (Baas, Black, & Banker, 1989; Baas, Deitch, Black, & Banker, 1988). Axon polarity is established upon neuronal polarization by microtubule-organizing centers (MTOC) at the centrosome that anchor the minus-ends (Stiess et al., 2010). As other neurites differentiate into dendrites, plus- and minus-end distal microtubules are generated by a specialized mixed polarity of microtubule arrays. In vertebrate dendrites, the mixtures of microtubule polarity vary locally within the arbor and during maturation (Baas et al., 1988; Kollins, Bell, Butts, & Withers, 2009; Yau et al., 2016). In invertebrates, microtubules in immature dendrites have an initially mixed polarity but gradually orient to a predominantly minus-end-out arrangement (Harterink et al., 2018;

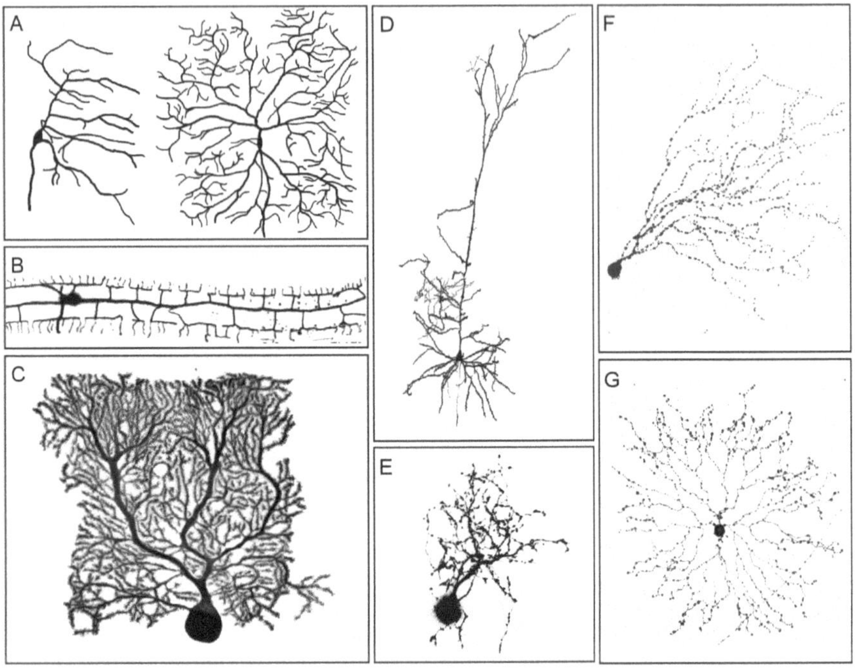

Fig. 1 Neuronal cell types that serve as key models for the study of dendrite morphogenesis. Inverted images of eight neuronal cell types labeled with fluorescent dyes or genetically-encoded fluorescent proteins. (A) Tracings of Class I (left) and IV (right) dendrite arborization (*da*) mechanosensory neurons of larval *Drosophila*. The *da* neurons are divided into four morphological classes that develop distinct dendritic arbors to innervate the epidermis and transduce different sensory modalities. Class I neurons have simple dendrites, while class IV neurons have complex space-filling arbors. (B) The *C. elegans* mechanosensory PVD neuron develops a candelabra-like branch pattern with a highly ordered arrangement of primary, secondary, and higher-order branches that innervate the length of the body wall. (C) A mouse Purkinje cell develops an elaborate dendritic arbor with a high branch complexity. The space-filling, non-overlapping arrangement is a characteristic of dendrite self-avoidance. (D) A mouse cortical pyramidal neuron with a large dendritic arbor that spans multiple cortical layers. Apical dendrites branch along the outer layer of the cortex and basal dendrites extend from the cell body. (E) A tectal neuron from a *Xenopus* tadpole. Live imaging studies have revealed factors that regulate dendrite dynamics and arborization in intact animals. (F) J-RGC is a retina ganglion cell (RGC) subtype with asymmetric dendrites that point ventrally along the retinal plane. (G) *En face* view of a mouse retinal starburst amacrine interneuron with a radial arrangement of dendrites, which are confined to a narrow plane. The uniform spacing of dendrites is dependent on dendrite self-avoidance. *Panel B: Adapted from Zou, W., Shen, A., Dong, X., Tugizova, M., Xiang, Y. K., & Shen, K. (2016). A multi-protein receptor-ligand complex underlies combinatorial dendrite guidance choices in C. elegans.* eLife, 5, e18345. *Panel D: Image courtesy of M. Scofield, Medical University of South Carolina. Panel E: Image courtesy of K. Haas, University of British Columbia. Some images are from Lefebvre, J. L., & Marocha, J. (2020). Chapter 12—Dendrite development: Vertebrates. In Rubenstein, J., Rakic, P., Chen, B., Kwan, K. Y., Kolodkin, A., & Anton, E. (Eds.),* Cellular migration and formation of axons and dendrites *(2nd ed.) (pp. 257–286). Academic Press.*

Hill et al., 2012; Stone, Roegiers, & Rolls, 2008). Dendritic microtubules also have a distinct signature of microtubule-associated proteins (MAPs) and post-translational modifications that influence dendritic branching and trafficking (Radler, Suber, & Spiliotis, 2020).

Microtubule polarity guides kinesin and dynein motors to transport dendrite-specific proteins and organelles, leading to the compartmentalization of post-synaptic neurotransmitter receptors and structures (Franker et al., 2016; Kapitein et al., 2010; Karasmanis et al., 2018; Petersen, Kaech, & Banker, 2014). The minus-end directed dynein motors biases cargo into dendrites, such as AMPA-type glutamate receptors in hippocampal neurons and ion channels in *Drosophila* sensory neurons (Kapitein et al., 2010; Zheng et al., 2008). Dendrites also require polarized transport to prevent entry of axonal cargo. For example in *C. elegans*, ectopic dendritic accumulation of plus-ended microtubules and axonal cargo is caused by the genetic deletion of the collapsin response mediator protein (CRMP) which regulates axonal sorting, and by deletion of kinesin-1, which normally slides plus-end-out microtubules out of the dendrite (Maniar et al., 2011; Yan et al., 2013).

Controlling when and where microtubule growth occurs determines branch formation and arbor patterning. Microtubules remodel extensively and are influenced by microtubule-associated proteins (MAPs) and plus-end tracking proteins (+TIPs) that stabilize or destabilize microtubules, modulate growth or shrinkage, and respond to receptor signaling (Kapitein & Hoogenraad, 2015). Microtubule nucleation occurs at non-centrosomal positions such as pre-existing microtubules and Golgi outposts, which are satellite secretory structures not present in axons (Ori-McKenney, Jan, & Jan, 2012; Yalgin et al., 2015). In *Drosophila* dendrite arborisation *(da)* neurons (Fig. 1A), Golgi outposts are located within distal dendrites and branchpoints where they regulate microtubule nucleation and minus-end growth (Ori-McKenney et al., 2012). Removal of Golgi outpost components does not disrupt all minus-end-out microtubules, suggesting that Golgi outposts are not essential for the formation or maintenance of a dendrite-specific cytoskeleton (Nguyen et al., 2014; Yang & Wildonger, 2020). In hippocampal neurons, microtubule nucleation is mediated by gamma tubulin and the associated ring complex (γ-TuRC), the HAUS/augmin complex, and modulated by the minus-end associated CAMSAP2 (Cunha-Ferreira et al., 2018; Yau et al., 2014). CAMSAP2 along with motor protein KIFC3 stabilize and immobilize microtubules at the minus end (Cao et al., 2020). Disruption of any of these components reduces dendrite branching and length. These components have conserved roles in invertebrate dendrite morphogenesis (Feng et al., 2019; Richardson et al., 2014; Yalgin et al., 2015). In *C. elegans*, dendrite morphogenesis of PVD neurons (Fig. 1B) is served by a non-centrosomal MTOC located in the growth cone of the extending primary dendrite (Liang et al., 2020). As the dendritic growth cone progresses and innervates its target region, the MTOC advances in a kinesin-1-dependent fashion. It generates a minus-end-out microtubule array toward the cell body and short plus-end-out microtubules toward the growing dendrite tip. Deciphering the mechanisms that spatially control microtubule initiation and polymerization remains a major direction for understanding how cell type-specific dendritic arbors are generated.

2.2 Regulation of dendrite initiation and patterning by the actin cytoskeleton

Dendritic outgrowth, retraction and stabilization is also determined by the spatiotemporal regulation of actin polymerization. A multitude of actin regulators control the dynamics and organization of actin filaments, and their roles in relation to dendrite formation have been captured in a diversity of cell types. Actin regulators include nucleators that promote linear F-actin elongation (e.g. formins, Enabled/VASP) or branched actin by nucleating new branches on existing filaments (e.g. the Arp2/3 complex) (Kwiatkowski et al., 2007). As illustrated in live imaging of *Drosophila da* dendrites, Arp2/3 complex is a major actin nucleator that initiates dendrite branchlet formation (Stürner et al., 2019). Dendrite branching is also enhanced by filament severing proteins (e.g. cofilin) that increase uncapped ends available for polymerization and F-actin turnover (Konietzny, Bar, & Mikhaylova, 2017; Rosario et al., 2012). Formins promote branch elongation and their activities are antagonized by formin inhibitors that shift from F-actin elongation to branched actin. Studies of the formin Daam1 and the formin inhibitor MTSS in Purkinje cell dendrites (Purkinje cell example is shown in Fig. 1C) provide a good example on how the balance between filament elongation and Arp2/3-dependent branching generates complex arbor morphologies (Kawabata Galbraith et al., 2018). In the absence of MTSS, Purkinje cells develop fewer branches and have a reduced dendrite complexity. Actomyosins also play critical roles in the dendrite cytoskeleton. As dense actin networks form, such as in lamellipodia, outward forces against the plasma membrane drive neurite protrusion and motility. Actomyosins capture and guide microtubule tips along actin filaments to facilitate microtubule invasion into a growing dendritic process (Withers & Wallace, 2020; Yoong et al., 2020). For example, myosin-6 promotes actin filament growth at the tips of a primary dendritic branch; the filaments then serve as tracks to guide microtubule polymerization and subdivide the tip into several branches (Yoong et al., 2020). As discussed below, many microtubule and actin regulators are transcriptionally regulated in a cell type-specific manner and serve as effectors of dendrite growth and patterning to contribute to the cell's characteristic morphology.

The spatial control of cytoskeletal remodeling by extrinsic cues –such as guidance factors, neuronal activity, and permissive substrates—is mediated through Rho-related GTPases and the WAVE regulatory complex. The Rho GTPases (e.g. RhoA, Rac1, and Cdc42) modulate the activities of actin regulators. Dendrite branching is positively regulated by Rac1 and Cdc42, while dendrite extension is inhibited by RhoA (Li, Van Aelst, & Cline, 2000; Nakayama, Harms, & Luo, 2000). Rho GTPases themselves are activated by guanine–nucleotide exchange factors (GEFs), and inactivated by GTPase activating proteins (GAPs). The diversity and cell type-specific expression patterns of GEFs and GAPs (Huang, Sun, Li, & Feng, 2017) suggest that they specify cytoskeletal responses to extracellular signals. The GEF Dock1 regulates Purkinje cell dendrite complexity by acting downstream of adhesion-G protein-coupled receptor BAI3 and regulating Rac1 activity (Lanoue et al., 2013). Activity-dependent dendrite arborization of pyramidal neurons is mediated by NMDA-receptor activation of Tiam1, a Rac1-specific GEF (Tolias et al., 2005). Rac1 also activates members of the Wilskott-Aldrich syndrome protein family (N-WAVE and WASP) and the WAVE Regulatory Complex (WRC) to drive Arp2/3-mediated actin polymerization. Loss of WAVE activity reduces dendritic branching (Stürner et al., 2019; Zou et al., 2018).

The WRC interacts with a diverse set of neuronal receptors and adhesion molecules, suggesting that receptor-WRC interactions instruct cytoskeletal changes to drive branch patterning (Chen et al., 2014). In *C. elegans*, the high-order branch patterning of the candelabra-like PVD neurons is achieved through local stimulation of F-actin assembly by the DMA-1 and HPO-30 receptors that directly recruit WRC and Tiam1 (Zou et al., 2018). Several receptors with WRC interaction motifs are implicated in dendrite patterning such as the protocadherins, ROBO and netrin receptors, neuroligins, and G-protein-coupled receptors (Chen et al., 2014). Functional interactions have yet to be tested in mammalian dendrite development. In the mouse cortex, Sema3A selectively promotes basal dendrite arborization through the Nrp1/PlxnA4 receptor complex and activation of the RhoGEF FARP2 and Rac1 (Danelon et al., 2020). These signaling components are dispensable for Sema3A-mediated repulsive events, illustrating how receptors produce divergent guidance responses through different intracellular signals. Local regulation of the actin cytoskeletal machinery by cell surface receptors could be a general mechanism for transducing guidance cues into dendritic patterns. Few examples have been elucidated to date. Understanding the spatiotemporal control of cytoskeletal remodeling responsible for morphogenesis remains a major gap in dendrite biology.

2.3 A satellite secretory pathway supports dendrite arborization

To facilitate the traffic of membranes and proteins to distantly located branches, dendrites have a satellite system of secretory structures that are not present in axons (Gardiol, Racca, & Triller, 1999; Horton & Ehlers, 2003; Pierce, Mayer, & McCarthy, 2001). Secretory pathway components include smooth and ribosome-bound endoplasmic reticulum (ER), ER exit sites, ER-to-Golgi intermediate compartment (ERGIC) and Golgi membranes. The importance for the secretory pathway in dendrite growth was demonstrated in a genetic screen in *Drosophila*, where mutations affecting several regulators of ER-to-Golgi trafficking reduced the growth of *da* dendrites but not of axons. For example, Sar1 initiates the formation of COPII-coated vesicles for forward trafficking from ER to Golgi. *Sar1* deletion leads to a decreased supply of plasma membrane to dendrites but not to axons (Ye et al., 2007). The reduction in dendritic growth is also observed in rodent pyramidal neurons following *Sar1* knockdown or other blockade of post-Golgi trafficking (Horton et al., 2005; Ye et al., 2007). In pyramidal neurons (example is shown in Fig. 1), the somatic Golgi apparatus is oriented towards and directs Golgi-derived vesicles to the long apical dendrites (Horton et al., 2005). Disruptions to Golgi polarity, but not to overall trafficking, impairs apical dendrite outgrowth, revealing that the addition of new membranes and subsequent dendrite elongation require polarized transport.

The satellite secretory system provides the local synthesis, processing and delivery of transmembrane proteins that might contribute to the highly arborized morphology of dendrites. In pyramidal neurons, the Golgi satellite secretory system extends throughout basal as well as apical dendrites, and glycosylates dendritic proteins en route to endosomes (Choy et al., 2014;

Mikhaylova, Bera, Kobler, Frischknecht, & Kreutz, 2016). Golgi outposts, which are particularly abundant at branch points and correlate with branch growth, also sort secreted and postsynaptic proteins such as BDNF and NMDA receptors (Horton & Ehlers, 2003; Horton et al., 2005; Jeyifous et al., 2009). Structures that are enriched for ER exit sites and Golgi outposts mediate the delivery of dendritic membrane proteins, such as AMPA receptors in response to activity (Cui-Wang et al., 2012; Evans, Gurung, Wilkinson, Stephens, & Henley, 2017). Manipulations that increase or decrease the formation of these satellite structures also enhance or reduce dendrite branching complexity, respectively, supporting an important role for satellite ER-to-Golgi transport for branch formation. As shown by live imaging studies, dendritic post-ER cargo transit through a variety of pathways, including long-range transport from somatic Golgi, Golgi-independent trafficking from the ERGIC and recycling endosomes to the plasma membrane, and through ER-Golgi satellites outposts distributed throughout the dendritic arbor (Bowen, Bourke, Hiester, Hanus, & Kennedy, 2017; Evans et al., 2017; Hanus et al., 2014; Mikhaylova et al., 2016). Thus, dendrite-specific secretory structures are conserved and involve a diversity of processing pathways. However the mechanisms by which they specify the rapid delivery of signaling receptors and thereby shape growing dendrites remain poorly understood.

3. Transcriptional control of dendrite morphogenesis

The morphological diversity of neuronal cell types is determined by transcriptional programs that specify branch growth patterns and responsiveness to extrinsic cues. Gene regulatory factors that influence dendrite patterning might function during neuronal specification, and thus relate to the neuron's overall identity. Or they might act during terminal differentiation and thus contribute to a specific dendritic or functional feature of the cell type (Enriquez et al., 2015; Hobert, 2008). Key concepts on how genetic programs dictate dendrite patterning and morphological diversity have come from studies on the four classes of dendrite arborization (*da*) neurons which develop distinct dendritic arbors that innervate the epidermis of larval *Drosophila* (Fig. 2A and B) (Grueber, Jan, & Jan, 2002). Gene deletion and misexpression studies in this model have been instrumental for identifying transcription factors and their gene targets that specify particular dendritic structures, as instructive factors are typically necessary and sufficient to regulate development of the features.

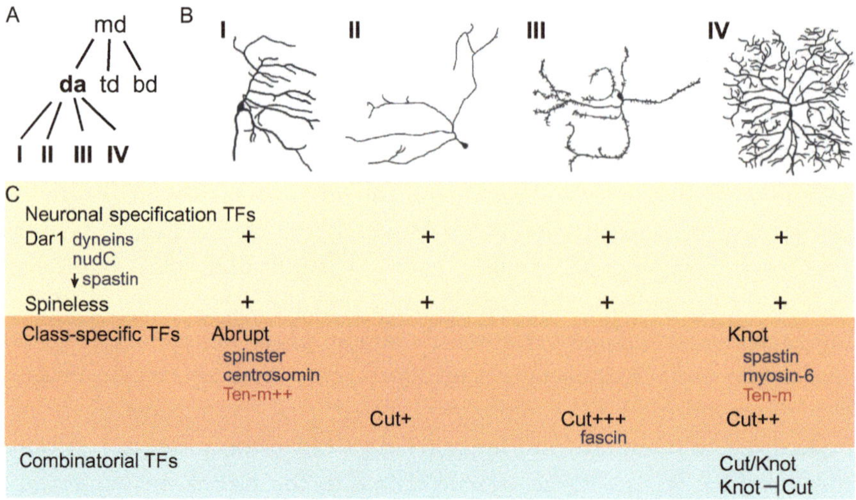

Fig. 2 Transcriptional control of *da* neuron dendrite morphology in *Drosophila*. (A) Lineage of sensory neurons in the larval peripheral nervous system. *Multidendritic* neurons comprise *da*, dendritic arborization neurons; *td*, tracheal dendrite neuron; and *bd*, bipolar dendritic neurons. (B) Characteristic dendritic morphologies of class I–IV *da*. neurons. (C) Summary of transcription factors that mediate pan-*da* or class-specific dendritic development. Known gene targets and effectors are listed, with cytoskeletal regulators in blue, and cell adhesion molecules in red. Differential expression levels are denoted by +.

3.1 Transcription factors influence dendrite shape through neuronal fate specification

The first concept is that transcription factors that endow a neuronal fate shared by multiple cell types regulate a common overall dendritic plan. All *da* classes are fated to a multipolar organization and extend two or three primary dendrites, which is dependent on the Krüppel-like transcription factor Dendritic arbor reduction 1 (Dar1) (Fig. 2C) (Wang et al., 2015). *Dar1* is necessary and sufficient for the multipolar dendritic morphology. Deletion of *Dar1* causes *da* neurons to become bipolar with a single primary dendrite and an axon, while misexpression of *Dar1* in bipolar and unipolar cell types induces multipolar dendritic outgrowth. Dar1 induces gene expression of dyneins and the nuclear distribution protein C (nudC) that regulate nuclear positioning, and suppresses the expression of microtubule-severing protein Spastin that normally limit microtubule-based branching (Wang et al., 2015; Ye et al., 2011). While *Dar1* is not required for axonal outgrowth, there are examples of factors that regulate both dendritic and axonal morphology. In the larval *Drosophila* CNS, Islet1 controls both axon targeting and dendrite positioning of motorneurons through regulation of expression of the Frazzled/DCC receptor that binds Netrin (Santiago & Bashaw, 2017).

A few examples have been described in mammalian dendrite morphogenesis. In the mouse cortex, the proneural transcription factor Neurogenin-2 specifies the unipolar dendritic morphology of pyramidal neurons and other traits of cortical glutamatergic fate such as radial migration properties (Hand et al., 2005). In the mouse retina, *Fezf1* is selectively expressed postmitotically in one of two starburst amacrine cell subtypes to regulate type-specific soma migration and dendrite targeting (Peng et al., 2020). Fezf1 mediates a binary fate choice by promoting the expression of genes related to one subtype, and repressing gene targets that are expressed in the other starburst subtype, including the atypical Rho GTPase *Rnd3* that mediates type-specific migration properties.

3.2 Class-specific traits are determined by class-specific transcription factors

Transcription factors Abrupt and Knot are exclusively expressed in class I and IV *da* neurons, respectively (Fig. 2C). When misexpressed in other *da* cell types, Abrupt and Knot induce ectopic branching features related to their endogenous class (Jinushi-Nakao et al., 2007; Li, Wang, Menut, & Gao, 2004; Sugimura, Satoh, Estes, Crews, & Uemura, 2004). Abrupt, a transcription factor of the Broad, Tramtrack, Bric-a-brac (BTB) family, produces a simple arbor morphology through gene targets that repress dendritic branching. One Abrupt target is *Centrosomin*, which encodes a centrosome-associated protein that acts at Golgi outposts to control microtubule nucleation and influence polymerization away from nascent branching to reduce branch number (Yalgin et al., 2015). By contrast, the helix-loop-helix (HLH) transcription factor Knot promotes dendrite branching required for the highly arborized class IV *da* dendrites by upregulating effectors of anterograde microtubule polymerization, such as myosin-6 (Yoong et al., 2020). Another Knot gene target is *Spastin*, which encodes a microtubule severing protein that generates microtubule breaks, allowing for new branch growth (Jinushi-Nakao et al., 2007). Abrupt and Knot also have common gene targets that shape dendritic arbors in either or both class I and IV (Hattori et al., 2013). Teneurin-m (Ten-m), a homophilic cell adhesion molecule, is a shared Abrupt and Knot effector but it is expressed more highly in class I *da* cells. As a result, growing class

I dendrites respond more strongly to, and orient their branch growth towards, Ten-m expression domains in the epidermis.

Transcription factors can also produce diverse morphologies through differences in expression levels and gene targets in multiple cell types (Fig. 2C). For example, the levels of Cut increase across II, IV and III classes; overexpression of *Cut* in a class II neuron produces a class IV morphology (Grueber, Jan, & Jan, 2003). Cut levels correlate with class-specific differences in the types and expression levels of effector proteins that control branch morphogenesis and stabilization. High Cut levels induce expression of Fascin, an actin-bundling protein that promotes formation of actin-rich dendritic spikes characteristic of class III neurons (Nagel et al., 2012; Sulkowski, Iyer, Kurosawa, Iyer, & Cox, 2011).

In the mouse retina, the Transcription factor T-box brain 1 (Tbr1) acts in four subtypes of retinal ganglion cells (RGC) to direct their dendrites to common target layers (Liu et al., 2018). In the absence of *Tbr1*, these RGCs misdirect their dendrites to lower retinal layers, while misexpression of Tbr1 in other RGC types re-directs their dendrites to upper layers. While a common Tbr1 program controls dendrite targeting, it also diverges through distinct gene targets to produce four distinct dendrite morphologies. As an example, one Tbr1-positive subtype exclusively expresses Cadherin 8 and Sorcs3, which are required for the final dendrite morphology (Liu et al., 2018). As highlighted in all of these examples, a central mechanism by which transcription factors specify dendrite morphogenesis is gene regulation of effectors that modify the dendritic cytoskeleton or mediate selective adhesion.

3.3 Combinatorial transcription factors regulate type-specific dendritic morphologies and wiring decisions

Diverse developmental outputs can be produced by transcription factor codes, where the activity of any one transcription factor varies according to the presence or absence of other transcription factors (Grueber & Ye, 2020). For instance, Knot and Cut regulate distinct and shared effectors in class IV *da* neurons that modulate the actin and microtubule cytoskeleton to produce the characteristic highly-branched complexity (Das et al., 2017). Similarly in *C. elegans*, class-specific dendrite patterning of mechanosensory neurons is regulated by combinatorial actions of transcription factors that tune the degree of branching and specify the sensory modality (O'Brien et al., 2017; Smith et al., 2013). In the *Drosophila* olfactory system, combinatorial codes involving seven transcription factors and co-factors were identified and shown to specify dendrite targeting of subclasses of olfactory projection neurons (PNs) to specific glomeruli (Komiyama, Johnson, Luo, & Jefferis, 2003; Komiyama & Luo, 2007).

Single-cell RNA sequencing (scRNASeq) offers a reverse approach to decipher how transcription factor codes confer distinct programs of dendritic wiring. For example in the *Drosophila* visual system, single-cell profiling of eight related neuronal morphological subtypes separated them into eight transcriptionally distinct clusters. Co-regulation of transcriptional factors among subtypes reflects shared dendritic wiring patterns, and different combinations uniquely define the eight subtypes (Kurmangaliyev, Yoo, LoCascio, & Zipursky, 2019). Although there is much to do to resolve and functionally test these transcriptional codes, single-cell profiling has the promise to uncover the transcriptional logic and repertoires of cell biological effectors and cell surface molecules that are responsible for wiring diverse neuron types during circuit assembly (Kurmangaliyev, Yoo, Valdes-Aleman, Sanfilippo, & Zipursky, 2020; Özel et al., 2020).

4. Regulation of dendritic arbor growth and size

Neuronal cell types develop a characteristic dendrite arbor size. The volume and branch density of the arbor delimits the numbers of presynaptic inputs that can be received, and influences the ways in which these signals are integrated and propagated. For instance, some cell types develop large arbors that receive inputs from many neurons and thereby converge multiple streams of information. By contrast, multiple cells bearing small dendritic arbors can divide information from a common afferent input into parallel pathways. In other cases, dendrite size adjusts to changes in the afferent territory size or number to preserve constant synaptic densities. Matching the dimensions of the dendritic arbor with the structural and functional demands of the circuit requires co-regulation of afferent connectivity and dendrite growth. The coordination between axonal and dendritic growth is also important as the animal grows. A developmental strategy to adjust dendrite size is to employ afferent-derived cues that promote or restrict dendrite elaboration.

4.1 Afferent-derived molecular cues influence dendrite size

Presynaptic afferents release neurotrophins and other growth factors to influence dendritic arbor size of target cells. Neurotrophins stimulate dendrite elaboration globally in several vertebrate neuronal cell types in vitro and in vivo, and can act in a short-range manner in vivo to control the growth of nearby dendrites (Horch & Katz, 2002; Lom, Cogen, Sanchez, Vu, & Cohen-Cory, 2002; McAllister, Lo, & Katz, 1995). Direct in vivo evidence that a pre-synaptic source of neurotrophin modulates dendrite size comes from a genetic study in the mouse cerebellum, where the extent of Purkinje cell dendrite arborization scales with the available Neurotrophin-3 (NT-3) secreted from parallel fiber afferents (Joo, Hippenmeyer, & Luo, 2014). In this example, Purkinje cell dendrites compete for afferent-derived NT-3 ligands to maximize their dendritic size. Sparse but not global knockout of the NT-3 receptor *TrkC* among Purkinje cells causes a marked reduction in total dendrite length and arbor size (Joo et al., 2014).

Afferent-derived growth factor signaling is another strategy to coordinate dendrite growth with axonal growth. As illustrated in the *Drosophila* visual system, this strategy is important for establishing a receptive field size that maintains a topographic transfer of sensory information. In one example, the dendritic field sizes of first-order neurons Dm8 and Tm20 are regulated by two factors, the TGF-beta superfamily ligand Activin and the Insulin-like peptide, which are secreted by their R7 and R8 photoreceptor inputs, respectively (Luo et al., 2020; Ting et al., 2014). Tm20 neurons receive one-to-one pre-synaptic connections from the R8 photoreceptors, while the wide-field Dm8 interneurons receive convergent inputs from ~16 R7 photoreceptors. During development, the R7 and R8 growth cones maintain a close contact with Dm8 and Tm20 dendrites, and secrete Activin to terminate dendrite growth and restrict dendritic field size. In the absence of the Activin receptor or the downstream effector Smad, the dendritic arbors of Dm8 and Tm20 expand into adjacent territories and synapse with excess photoreceptors. Interestingly, Activin-signaling is transmitted in an exquisite R7- or R8-specific manner that might result from direct interactions, because R7-specific Activin knockdown only affects Dm8 while R8-specific knockdown only affects Tm20. To grow a large arbor that receives multiple R7 inputs, the wide-field Dm8 requires the Insulin-like peptide-2 ligand to stimulate dendritic growth and synaptogenesis through the PI3K-mTORC1 pathway (Luo et al., 2020). Thus, the Dm8 dendritic field size is set by overlapping phases of 'grow' and 'stop-growing' signals derived from distinct pre-synaptic inputs. Antagonistic mechanisms likely operate in many contexts of dendrite morphogenesis: large arbors develop from longer phases of branch growth versus termination, and small arbors require an early onset of branch termination and stabilization to temper growth-inducing pathways.

4.2 Substrate-derived cues guide and anchor growing dendrites

Molecular cues positioned in the extracellular matrix or surrounding non-neural cells guide dendritic arbor growth to ensure that dendrites fill the available territory as the animal grows. Two examples of dendrite-substrate interactions are found in *C. elegans*. First, PVD mechanosensory neurons innervate the entire body wall with elaborate 'menorah-shaped' dendritic arbors (Fig. 1). PVD branch formation is elicited at designated points by cues derived from muscle and hypodermal cells that signal through a multi-protein ligand-receptor complex (Díaz-Balzac et al., 2016; Dong, Liu, Howell, & Shen, 2013; Salzberg et al., 2013; Zou et al., 2016). PVD neurons express the leucine-rich repeat-family (LRR) transmembrane receptor DMA-1 (Dendrite-morphogenesis-abnormal). This receptor binds to ligands SAX7 (an LI-CAM homolog) and MNR-1 (Menorin, a Fam151 homolog) derived from hypodermal cells, and to LECT2 (leukocyte cell-derived chemotaxin-2) secreted by muscle. The secondary and tertiary branch patterns are altered if any of these molecules are absent. Misexpression of these ligands in other cell types is sufficient to relocate PVD dendrites, revealing the instructive role for the SAX/MNR-1/LECT2/DMA-1 complex for dendritic branching and stabilization. In the second example, dendrites of amphid sensory neurons first anchor within a glial sheath at the nose tip and then elongate by retrograde migration of the cell body (Heiman & Shaham, 2009). Dendrite positioning and anchoring require dendrite-glia and dendrite-ECM interactions (Cebul, McLachlan, & Heiman, 2020).

4.3 Control of dendritic arbor size by neurotransmission

Neural transmission from presynaptic inputs can increase or decrease dendrite growth and complexity. The effect of activity depends on the neuron type or developmental stage. Manipulations of neurotransmission can result in global, cell-wide adaptations or local adjustments to dendritic growth. One manipulation that globally affects dendritic arbor size is inhibition of GABAergic transmission, which also increases excitatory drive. In the optic tectum of *Xenopus* tadpoles, tectal dendrites (see example in Fig. 1) receive glutamatergic inputs from retinal ganglion cells and inhibitory inputs from local interneurons. Blockade of inhibitory GABA transmission by overexpression of a dominant-negative GABA-A receptor subunit causes dendrite arbors to expand (Shen, Da Silva, He, & Cline, 2009). Similar increases are observed in rodent hippocampal neurons following GABA-A receptor inhibition (Wayman et al., 2006), and in gerbil auditory neurons upon decreased glycinergic transmission (Sanes & Hafidi, 1996; Sanes, Markowitz, Bernstein, & Wardlow, 1992). In the mouse cortex, long-range GABAergic neurons that project to the upper cortical layer stimulate dendrite branching and synaptogenesis of pyramidal neurons (Chen & Kriegstein, 2015). In this example, the effects of blocking GABA release are limited to nearby apical dendrites receiving inputs because outgrowth of distantly located basal dendrites is not affected.

Local modulation of dendritic outgrowth by neurotransmission shapes the arbors of nucleus laminaris neurons (NL) in the chick auditory brainstem. The bi-tufted NL neurons have two dendritic arbors that are similar in size and complexity, with one receiving excitatory inputs from the ipsilateral ear and the other from the contralateral ear. Denervation of one NL dendrite tuft, by transection of axonal inputs or application of tetrodotoxin (TTX), causes the denervated dendrite arbor to shrink (Deitch & Rubel, 1984; Wang & Rubel, 2012). The effect on dendrite size is not due to loss in afferent contact but rather due to loss of neurotransmission, because ablation of the cochlea two synapses away causes similar shrinkage. Differences in input strengths also differentially affect arbor growth: stimulated dendrites increased in arbor size and total branch length while unstimulated dendrites diminished in size (Sorensen & Rubel, 2006, 2011). This adaptive property allows dendrite arbor size to rapidly adjust to functional changes in presynaptic connectivity, thus ensuring optimal sound localization.

4.4 Activity-regulated dendrite dynamics and the synaptotropic hypothesis

The neurite dynamics and molecular signals that underlie activity-regulated dendrite outgrowth have been studied extensively in the developing optic tectum in *Xenopus* (Fig. 3A). Time-lapse imaging captured how glutamatergic synaptic activity first promotes dendrite arbor elaboration and then stabilization, as dendritic processes switch from responding to activity as 'grow' to 'stop-growing' signals. Dendrites of young tectal neurons first undergo a phase of rapid growth and branching, and filopodia form transient, activity-independent contacts with axonal processes through adhesion molecules such as Neurexin and Neuroligin-1 (Chen, Tari, She, & Haas, 2010; Hossain, Hewapathirane, & Haas, 2012). Dendritogenesis proceeds concurrently with synaptogenesis and the early glutamatergic transmission

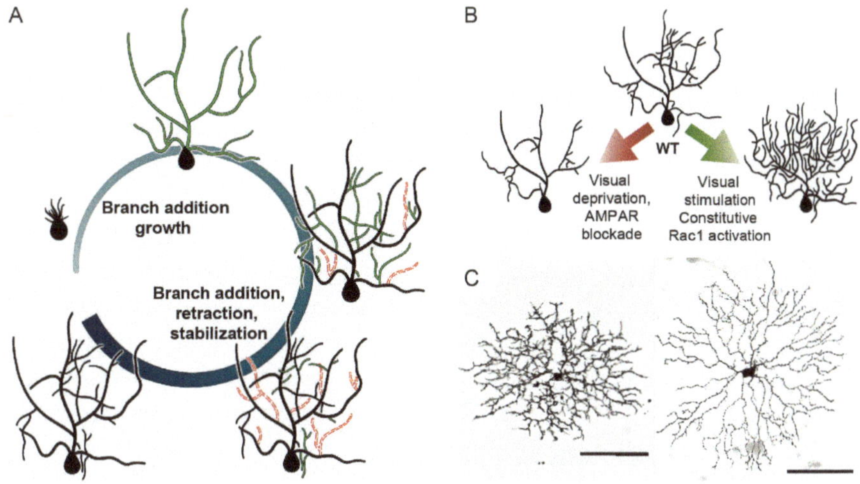

Fig. 3 Progression of dendrite growth, pruning and stabilization in the vertebrate visual system. (A) The trajectory of dendritic arbor development of optic tectal neurons in the *Xenopus* tadpole. Arbor development is governed by branch addition and lengthening (green), and then increasingly by branch elimination (red) or stabilization (black). (B) Manipulations to neural activity reveal experience-dependent structural plasticity of *Xenopus* tectal dendritic arbors. Dark-rearing or blockade of AMPA receptor-dependent synaptic maturation leads to reduced arbor size and complexity. Visual stimulation enhances dendritic size and complexity. (C) Starburst amacrine cell arborizations in mouse retina. At postnatal day 6 (left), starburst cells extend fine, exuberant branches that exhibit self-contacts. By P12 (right), the arbor is expanded but excess branches are eliminated. Starburst cell arborization is not influenced by neural activity. Scale bar 50μm. *Schematics in (A) are adapted from Lefebvre, J. L., & Marocha, J. (2020). Chapter 12—Dendrite development: Vertebrates. In Rubenstein, J., Rakic, P., Chen, B., Kwan, K.Y., Kolodkin, A., & Anton, E. (Eds.),* Cellular migration and formation of axons and dendrites *(2nd ed.) (pp. 257–286). Academic Press.*

through NMDA receptors promotes filopodia stabilization (Wu, Zou, Rajan, & Cline, 1999). Early periods of visual experience and NMDA-dependent transmission stimulates branch growth through activation of the cytoskeleton by Rho-GTPAses, Rac and Cdc42 (Fig. 3b) (Li, Aizenman, & Cline, 2002; Li et al., 2000; Sin, Haas, Ruthazer, & Cline, 2002). Once a particular arbor size is reached, tectal dendrites stabilize in response to visual activity and AMPA receptor-dependent transmission (Haas, Li, & Cline, 2006; Sin et al., 2002; Wu & Cline, 1998). Stop-growing responses are mediated by RhoA and CAMKII (Li et al., 2000; Wu & Cline, 1998). In the mouse brain as well, synaptogenesis influences Purkinje cell dendrite arborization through an initial stage of outgrowth to establish their territory. Arbor growth is then restricted as synapses from parallel fibers accumulate and stabilize through the Neurexin/Cerebellin1/GluD2 synaptic adhesion complex (Takeo et al., 2020). The interplay between dendrite branching and synapse formation forms the basis for the synaptotropic hypothesis: dendritic branch growth and stabilization is biased towards regions receiving more excitatory drive, as they are likely to encounter more synaptic partners (Cline & Haas, 2008; Vaughn, 1989).

Dendrite branch addition and stabilization are regulated by local calcium signaling. In live imaging of embryonic chick retinal ganglion cells, calcium fluxes were detected upon filopodia and axonal contacts, and transmitter-evoked release of local calcium correlated with filopodia stabilization (Lohmann & Bonhoeffer, 2008; Lohmann, Finski, & Bonhoeffer, 2005; Lohmann, Myhr, & Wong, 2002). Branch stabilization is influenced by local, but not global, fluxes of calcium. Reducing local calcium fluxes causes dendrite retraction, while inducing local calcium release promotes branch growth. Intracellular calcium activates calcium-dependent kinases, such as CaMKII that promotes filopodia extension and branching (Fink et al., 2003), and calcium-dependent transcription factors such as CREB, NeuroD, and MEF2 (Aizawa et al., 2004; Flavell et al., 2008; Hu, Chrivia, & Ghosh, 1999; Ince-Dunn et al., 2006; Redmond, Kashani, & Ghosh, 2002; Wayman et al., 2006; Zhou et al., 2006). These factors regulate the expression of numerous gene targets that influence dendrite development and synaptogenesis.

4.5 Sensory driven activity shapes dendrite growth in the somatosensory cortex

In some circuits, sensory experience reorganizes the dendritic arborization to maximize connections with stronger inputs and eliminate weaker ones. A key model for studying experience-dependent dendrite morphogenesis is the rodent somatosensory system, in which whisker stimulation is relayed to the cortex in a one whisker-one barrel topographic map (Petersen, 2007). Neurons within layer 4 of the somatosensory cortex, including spiny stellate neurons (L4 neurons), receive inputs from topographically arranged thalamocortical axon bundles that each transmits sensory information from one whisker. Developing L4 neurons cluster into discrete "barrel" structures surrounding each thalamocortical axon bundle and initially extend dendrites that are diffuse and receive multiple thalamic inputs. As L4 neurons receive sensory-driven thalamic activity, dendritic growth is re-oriented towards the center of one barrel receiving more active inputs. As a result, L4 neurons form barrel walls with asymmetrically-shaped dendritic arbors oriented towards the barrel center. Dendritic re-organization is dependent on activity: sensory deprivation, or global or cell-autonomous blockade of NMDA-dependent transmission during the neonatal period result in profuse branching and diminished barrels (Datwani, Iwasato, Itohara, & Erzurumlu, 2002; Espinosa, Wheeler, Tsien, & Luo, 2009; Iwasato et al., 2000, 1997). Live imaging of L4 dendrites in the neonatal mouse barrel cortex revealed that dendritic growth cones extend and branch towards the barrel hollow, where they stabilize in an NMDAR-dependent manner (Mizuno et al., 2014; Nakazawa, Mizuno, & Iwasato, 2018).

This example of experience-dependent remodeling requires transcription factors LIM homeobox 2 (Lhx2) and BTB domain containing 3 (BTBD3). Lhx2 directly regulates transcription of *Btbd3* in L4 neurons (Wang et al., 2017). BTBD3 translocation from the cytoplasm to the nucleus is induced by NMDAR-dependent stimulation (Matsui et al., 2013). BTBD3 also shapes dendritic remodeling and ocular dominance column formation in the ferret visual cortex, indicating a conserved role for experience-dependent structural plasticity in sensory cortex (Matsui et al., 2013). Deletion of *Lhx2* or *Btbd3* genes in L4 neurons in mice disrupts dendritic arbor asymmetry and barrel formation, while restoration of *Btbd3* in *Lhx2* mutant cortex rescues dendrite remodeling (Shetty et al., 2013; Wang et al., 2017). Loss of Lhx2 decreases the expression of developmental and activity-regulated genes involved in barrel cortex patterning, including *EphrinA5*, *Rorb* and *Lmo4*, suggesting a wider role for Lhx2 in terminal L4 neuron differentiation (Wang et al., 2017). Lhx2 and BTBD3 are proposed to provide L4 neurons competence to respond to sensory activity but the underlying gene targets are unknown. The next directions are to elucidate the gene regulatory programs and molecular mechanisms that drive sensory-dependent dendrite plasticity. New insights will allow us to better understand how afferent-target connectivity within circuits is spatially optimized to refine the point-point transmission of sensory information from the outside world.

5. Space-filling mechanisms to optimize dendritic field distribution

A common patterning feature in sensory systems is maximizing coverage of the receptive field with dendritic branches while minimizing branch overlap. The space-filling dendritic pattern arises from a "like repels like" mechanism: dendrites belonging to the same neuronal cell type (homotypic), and in some cases to the same neuron, employ contact-dependent repulsive interactions to re-direct growth towards the vacant territory. Early observations of homotypic repulsion come from ablation studies of mechanosensory neurons that innervate the leech epidermis and resemble the *Drosophila da* and the *C. elegans* PVD neurons. Neurons sensing a particular modality establish multiple neurite arbors across body segments that exhibit some or no overlap between neighboring arbors of the same type (e.g. nociceptive and touch neurons, respectively) (Blackshaw, Nicholls, & Parnas, 1982). They also exhibit a non-overlapping array of branches within the same arbor (Kramer & Kuwada, 1983). Ablation of a subset of touch neurons causes the remaining touch neuron to expand its arbor to cover the denervated territory (Blackshaw et al., 1982). Other cell types were unaffected, pointing to a homotypic basis for this adaptation. Moreover, ablation of an arbor subfield induced branch outgrowth to fill the vacant space while avoiding areas occupied by sibling branches arising from the same neuron (Kramer & Stent, 1985). Homotypic dendrite repulsion ensures that multiple arbors innervate the available epidermal territory with minimal overlap (tiling), and that sibling branches cover the available space (self-avoidance). I discuss subsequent genetic studies that led to identification of cellular and molecular mechanisms that produce three space-filling patterns: (a) Tiling; (b) Mosaics; and (c) Dendrite self-avoidance.

5.1 Dendritic tiling is achieved by mutual repulsion

In tiling, dendritic arbors cover the entire receptive surface with minimal overlap between homotypic neighbors (Fig. 4A). In mammals, the retinal circuitry is organized with a high degree of tiling among functionally distinct populations of retinal ganglion cells (RGCs) and bipolar cells (Lefebvre, Sanes, & Kay, 2015; Wassle, Peichl, & Boycott, 1981). This organization enables parallel retinal circuits to compute distinct aspects of visual information for any given point in the visual scene. We have learned key principles of dendritic tiling from cellular and genetic manipulations of mammalian retinal cells and the *Drosophila da* neurons, which also show tiling. First, tiling occurs through a mutual, dendro-dendritic repulsion between cell types belonging to the same functional class. In live imaging of RGCs in ferret retinal explants, contact-dependent repulsive behaviors were observed between dendrites of neighboring alpha ON-RGCs, but not with dendrites belonging to another type (Lohmann & Wong, 2001). Second, adjacent arbors constrain each other's territory growth through competitive and

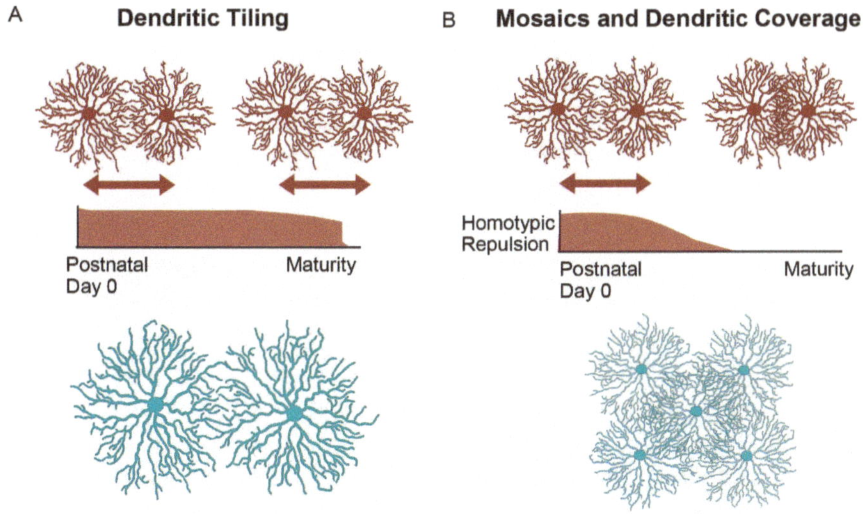

Fig. 4 Dendritic coverage, mosaics and tiling are controlled by homotypic repulsion. Homotypic repulsion refers to contact-dependent, repulsive interactions between developing neurites of the same cell type (homotypic). Developmental variations in the timing or extent of homotypic repulsion produce a range of dendritic coverage. (A) *Top,* Tiling. Contact-dependent repulsion between dendrite tips throughout development and maturation prevents overlap between arbors. *Bottom,* Tiling of neighboring dendritic territories ensures complete but nonredundant coverage of the innervation field. (B) *Top,* Mosaics and dendritic coverage. Homotypic dendrite repulsion between developing neurons generates a regular mosaic spacing between somata. Homotypic repulsion is attenuated, allowing growing arbors to extend over each other. *Bottom,* A transient period of homotypic repulsion arranges the neuronal population in a regular mosaic array, with overlapping dendritic arbors. Retinal cell types also exhibit a characteristic dendrite coverage factor, which is influenced by the timing of homotypic repulsion, as well as cell density and propensity for dendrite growth.

repulsive interactions. In *Drosophila,* laser ablation of class IV *da* neurons causes dendrites of neighboring class IV arbors to expand into denervated territories (Grueber, Ye, Moore, Jan, & Jan, 2003; Sugimura et al., 2003). The class IV neurons form boundaries as soon as dendritic tips meet, and these boundaries are maintained during larval growth through re-iterative contact and repulsion to ensure non-redundant innervation of the body wall. If additional homotypic cells are generated genetically and compete for the same territory, *da* neurons further partition their dendritic territories to prevent overlap. Similarly in the mouse retina, genetic modulations that increase or decrease the density of Type 7 cone bipolar cells lead to a corresponding decrease or increase in dendritic arbor size (Lee et al., 2011). In this case, bipolar cell dendrites initially overlap but undergo process refinement to achieve tiling. Together, these studies demonstrate the plasticity exhibited by dendrites to adjust their growth to local changes in cell density.

5.2 Regular spacing: Mosaics and dendritic coverage

Tiling is not the rule in the retina but rather an extreme example of regular cell spacing. Each distinct retinal cell type has a characteristic dendritic coverage and mosaic regularity, where the cell bodies are regularly distributed across the retina to ensure even spatial sampling (Reese & Keeley, 2015; Rockhill, Euler, & Masland, 2000). Dendritic coverage factor refers to the number of neurons and their dendritic fields that cover any point in the tissue. Neurons that tile perfectly have a coverage factor of ~1 and those with overlapping arbors have larger values (Reese & Keeley, 2015). Dendritic field size and mosaic regularity are patterned by acute phases of homotypic repulsion during development (Fig. 4B). For example, as horizontal retinal cells settle in their laminar position, they extend transient dendritic processes that exhibit homotypic repulsion and tiling to position their somas in a mosaic (Huckfeldt et al., 2009). Homotypic constraints subside, and arbors grow over each other to attain a coverage factor of ~6 (Reese & Keeley, 2015). Similarly, contact-dependent dendrite repulsion between young starburst amacrine cells establishes their mosaic but is attenuated to allow for extensive arbor overlap with a coverage factor of 25–35 (Galli-Resta, Resta, Tan, & Reese, 1997; Keeley, Whitney, Raven, & Reese, 2007; Ray et al., 2018).

The mosaic spacing of starburst and horizontal cells is mediated by the transmembrane Multiple EGF-like-domains -10 and -11 molecules (MEGF10 and MEGF11) (Kay, Chu, & Sanes, 2012). In *Megf10* or *Megf11* mutant retinas, horizontal and starburst cell densities are unaffected but the mosaics are abolished. Misexpressing *Megf10* in small patches of cells repels surrounding starburst cells. These experiments suggest that MEGF proteins serve as both ligand and receptor to produce contact-mediated repulsion between homotypic dendrites on all sides of the cell and lead to even spacing between somata (Kay et al., 2012; Ray et al., 2018).

As with MEGFs, the cell surface receptors that mediate homotypic repulsion are presumed to act in a cell type-specific manner. Few other receptors have been identified to date. The atypical Cadherin Flamingo was the first molecule identified to regulate dendritic tiling in a genetic screen for dendritic overgrowth in *Drosophila* larvae (Gao, Kohwi,

Brenman, Jan, & Jan, 2000; Grueber et al., 2002). *Da* neurons lacking *Flamingo* exhibit dendritic overlap at the dorsal midline and cross into the contralateral territory. Flamingo binds to the LIM domain protein Espina but the pathway by which Flamingo maintains homotypic repulsion is unclear (Kimura, Usui, Tsubouchi, & Uemura, 2006; Matsubara, Horiuchi, Shimono, Usui, & Uemura, 2011). Dendrite spacing might be achieved by other molecular strategies. For instance, dendrite repulsion is facilitated by attaching *da* class IV dendrites to the basement membrane in a two-dimensional plane through an adhesive complex involving Semaphorin-2B/PlexinB, Integrins, and the mTORC2–Tricornered complex (Emoto et al., 2004; Han et al., 2012; Kim, Shrestha, Blazeski, Mason, & Grueber, 2012; Koike-Kumagai, Yasunaga, Morikawa, Kanamori, & Emoto, 2009; Meltzer et al., 2016). When components of this pathway are abolished, dendrites are displaced and the probability of contact-dependent homotypic repulsion is reduced, resulting in significant overlap among dendritic fields and branches. Another strategy is to reduce adhesion between homotypic dendrites. This mechanism is proposed for mouse Dscam1 and Dscam1-like-dependent patterning of retinal mosaics, as several subpopulations form clumps of soma and dendrites when these genes are deleted (Fuerst et al., 2009; Fuerst, Koizumi, Masland, & Burgess, 2008; Garrett, Khalil, Walton, & Burgess, 2018).

5.3 Dendrite self-avoidance and self/non-discrimination

Dendrite self-avoidance between branches emanating from the *same* neuron produces a non-overlapping arrangement within the individual arbor territory and maximizes the area available for synaptic inputs. In contrast to tiling, repulsion is restricted to contacting sibling or "self" branches, allowing multiple neurons of the same type to overlap and innervate a shared territory. Dendrite self-avoidance has been described for a variety of neuronal cell types with highly-branched dendritic patterns, including the *da* neurons in *Drosophila*, the PVD neurons in *C. elegans*, and the RGCs, starburst and cerebellar Purkinje cells in mice (Grueber, Jan, et al., 2003; Matthews et al., 2007; Montague & Friedlander, 1989; Sdrulla & Linden, 2006; Smith, Watson, VanHoven, Colón-Ramos, & Miller, 2012).

To establish its characteristic candelabra-like branch patterns, PVD dendrites self-avoid through UNC-6/Netrin ligand and UNC-40/DCC; UNC-5 receptor signaling (Smith et al., 2012). Since the two PVD neurons do not overlap, a simple ligand-receptor pair acting cell autonomously is sufficient (Lefebvre et al., 2015). Interestingly, when a genetic duplication of PVD neurons forces them to compete for territory, they mutually repel into a tiled arrangement (Yip & Heiman, 2016). The tiling between ectopic PVD neurons is also dependent on UNC-6/UNC-40/UNC-5 signaling, illustrating how repulsive signaling pathways can produce homotypic repulsion in a variety of contexts. Similar mechanisms are proposed for Slit-Robo2-mediated spacing of cerebellar Purkinje dendrites, and for transmembrane Sema6A-PlexA2-mediated spacing of starburst cell dendrites (Gibson et al., 2014; Sun et al., 2013). PVD dendrite spacing is also regulated by MIG-14/Wntless, a multi-pass transmembrane protein that is essential for Wnt secretion but has a distinct role in dendrite self-avoidance (Liao, Li, Lee, Chien, & Pan, 2018). MIG-14 and UNC-6/Netrin act in parallel to induce dendrite retraction through modulation of Wasp/WAVE activity and F-actin assembly (Liao et al., 2018; Sundararajan et al., 2019). The intracellular signals that mediate dendrite self-avoidance in mice remain to be elucidated. They could regulate self-avoidance receptor delivery, such as LKB1-SIK-dependent trafficking of Robo2 to Purkinje dendrites (Kuwako & Okano, 2018), or regulate actin to modulate branch protrusion and retraction (Kawabata Galbraith et al., 2018).

Many neuronal cell types overlap extensively with homotypic neighbors, such as the radial starburst amacrine cells (example shown in Fig. 1) which form dendro-dendritic synaptic connections. In these scenarios, self-avoidance requires a molecular mechanism for dendrites to distinguish self from non-self discrimination. Two large families of recognition molecules have emerged as regulators of dendrite self/non-self discrimination: the *Drosophila* Dscam1s and the mammalian clustered Protocadherins (cPcdhs) (Fig. 5A). Although the IgSF-related Dscam1s and the Cadherin-related cPcdhs are divergent, they share several molecular features that set them apart from other cell adhesion families. First, the Dscam1s and cPcdhs generate extraordinary isoform diversity. *Dscam1* encodes more than 38,000 transmembrane molecules through alternative splicing (Schmucker et al., 2000). The mouse *cPcdh* genes, divided into the *Pcdh-alpha, -beta,* and *-gamma* clusters, encode 58 transmembrane isoforms expressed through alternative promoter choice (Wu & Maniatis, 1999). Second, neurons express multiple Dscam1 and cPcdh isoforms in stochastic and combinatorial patterns, and could each acquire a unique complement of isoforms (Kaneko et al., 2006; Miura, Martins, Zhang, Graveley, & Zipursky, 2013;

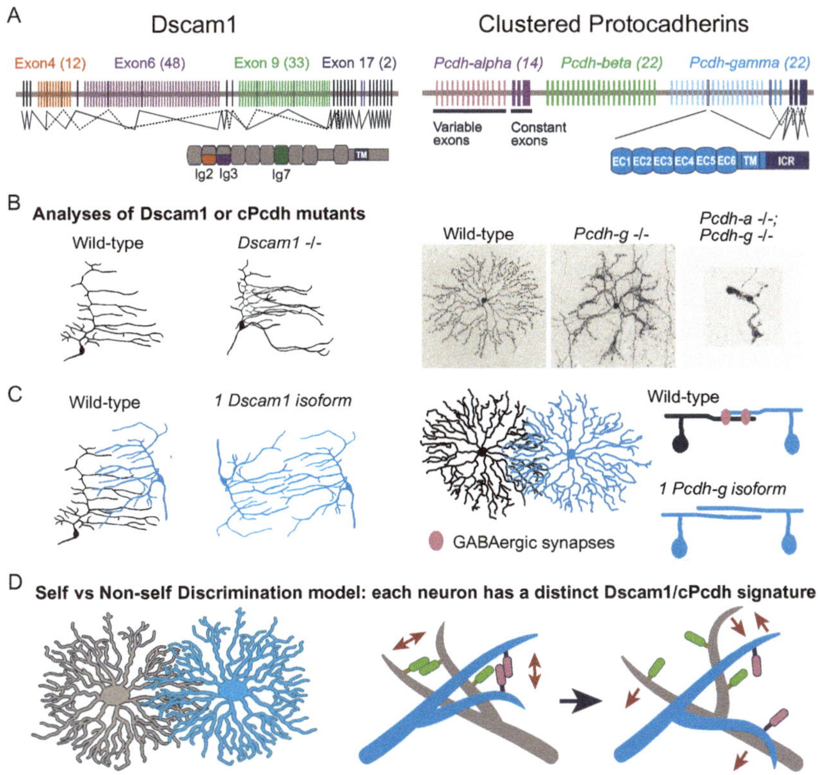

Fig. 5 Dendrite self/non-self discrimination is regulated by *Drosophila* Dscam1 and the clustered Protocadherins. (A) *Left*, *Dscam1* gene locus contains three large cassettes of alternate exons (number of alternate exons are in brackets). Exon 17 provides 2 alternate exons that encode transmembrane domains (TM), which determine axonal versus dendritic targeting. Dscam1 isoform expression is controlled at the splicing level, by probabilistic inclusion of alternate exons that each contribute to three variable Ig ectodomains. *Right*, The clustered Protocadherin locus (cPcdh) includes *alpha*, *beta*, and *gamma* clusters that encode 14, 22, and 22 isoforms respectively. Pcdh-α and -γ proteins are assembled by splicing one alternate variable exon encoding the extracellular and juxtamembrane sequences, to 3 constant exons that encode a common intracellular region (ICR). Each Pcdh-β protein is encoded by one *Pcdhb* variable exon. Choice of cPcdh expression is regulated at the level of individual promoters preceding each variable exon. Pcdh-γ proteins have six cadherin repeats (EC 1–6) encoded by variable extracellular sequences and a common intracellular region (ICR). (B) *Left*, The *da* neurons exhibit dendrite self-avoidance, as dendrites arising from the same neuron do not overlap. *Dscam1* mutant *da* neurons exhibit dendrite crossing and overlap, indicating a self-avoidance defect. *Right*, Starburst amacrine cells exhibit a radial, non-overlapping arrangement of dendrites. In the absence of the *Pcdhgs*, starburst cell dendrites form

(Continued)

Zhan et al., 2004). Third, the isoforms bind homophilically via their extracellular domains (Brasch et al., 2019; Meijers et al., 2007; Nicoludis et al., 2015; Rubinstein et al., 2015; Sawaya et al., 2008; Schreiner & Weiner, 2010; Thu et al., 2014; Wojtowicz et al., 2007). There are fewer cPcdhs but the oligomerization and homophilic selectivity of cPcdh multimers greatly expand the potential signatures.

Dscam1s and cPcdhs are widely expressed throughout the nervous system but roles in dendrite self-avoidance and self/non-self recognition have only been demonstrated for a few cell types. In *Drosophila*, Dscam1 is required for the non-overlapping branches arrangement of the four classes of *da* neurons as well as dendrite spacing of olfactory projection neurons (Hughes et al., 2007; Matthews et al., 2007; Soba et al., 2007; Zhu et al., 2006). In mice, the Pcdh-gamma cluster (*Pcdhg*) mediates dendrite self-avoidance of retinal starburst cells and cerebellar Purkinje cells (Lefebvre, Kostadinov, Chen, Maniatis, & Sanes, 2012). Manipulations that remove or alter *Dscam1* or *Pcdhg* isoform diversity produce dendritic

Fig. 5—Cont'd bundles and crossings. While removal of the *Pcdhas* does not affect dendrite self-avoidance, double *Pcdha; Pcdhg* mutant cells show severe defects suggesting cooperative, redundant roles for the cPcdhs in self-avoidance. (C) *Left*, Neurons of different *da* classes co-exist and overlap their dendritic fields. When both neurons are forced to express a single Dscam1 isoform, dendrite self-avoidance is rescued but dendrites of neighboring neurons show ectopic repulsion. *Right*, Dendritic arbors of starburst amacrine cells overlap extensively and form dendro-dendritic GABAergic synapses. Expression of a single Pcdhg isoform in starburst cells rescues dendrite self-avoidance but neighboring starburst dendrites are misaligned and have reduced GABAergic connections. (D) Model: Self/non-self discrimination ensures that homophilic recognition and contact-dependent repulsion is limited to sibling neurites expressing the same Dscam1 or cPcdh isoforms. Neighboring neurites are unlikely to express the same isoforms due to the stochastic and combinatorial nature of expression and therefore co-exist with other dendritic arbors. *Data summarized from Hughes, M. E., Bortnick, R., Tsubouchi, A., Bäumer, P., Kondo, M., Uemura, T., et al. (2007). Homophilic Dscam interactions control complex dendrite morphogenesis.* Neuron, 54, *417–427; Matthews, B. J., Kim, M. E., Flanagan, J. J., Hattori, D., Clemens, J. C., Zipursky, S. L., et al. (2007). Dendrite self-avoidance is controlled by Dscam.* Cell, 129, *593–604; Soba, P., Zhu, S., Emoto, K., Younger, S., Yang, S. J., Yu, H. H., et al. (2007). Drosophila sensory neurons require Dscam for dendritic self-avoidance and proper dendritic field organization.* Neuron, 54, *403–416. Panels in (B, right) are from Ing-Esteves, S., Kostadinov, D., Marocha, J., Sing, A. D., Joseph, K. S., Laboulaye, M. A., et al. (2018). Combinatorial effects of alpha- and gamma-protocadherins on neuronal survival and dendritic self-avoidance.* The Journal of Neuroscience : The Official Journal of the Society for Neuroscience, 38, *2713–2729.*

bundling and crossing defects, resulting in altered arbor morphology and incomplete dendritic field coverage (Fig. 5B). *Pcdhg* mutant starburst dendrites also form autapses (Kostadinov & Sanes, 2015), suggesting that self-avoidance prevents the formation of autapses but this idea remains to be tested in other neuronal types. In genetic manipulations that rendered all *da* Class IV or starburst cells to express an identical isoform, dendrite self-avoidance was rescued but neighboring cells recognized each other as self and avoided their territories or failed to form synapses (Fig. 5C). While Dscam1 isoforms provide a high degree of redundancy, isoform diversity is required for providing neurons within the population a robust mechanism to distinguish self from non-self (Hattori et al., 2009). Studies of starburst cell and Purkinje cells lacking one or both *Pcdha* and *Pcdhg* clusters also support cooperative roles for Pcdh-γ and Pcdh-α isoform diversity in dendrite self-avoidance (Ing-Esteves et al., 2018). Together, these studies support a self/non-self discrimination model where stochastic Dscam1 and cPcdh expression endow each neuron with a unique code for self-recognition. Self-avoidance is limited to sibling dendrites expressing matching isoforms, while dendrites of neighboring cells overlap because they express a different complement of Dscam1s/cPcdhs (Fig. 5D).

The extent to which the Dscam1s and cPcdhs mediate self/non-self discrimination in the CNS and the underlying signaling mechanisms remain unknown. Dscam1- and cPcdh-dependent self-avoidance control axon collateral formation and spacing (Chen et al., 2006, 2017; Hattori et al., 2007; Mountoufaris et al., 2017) but whether they mediate repulsive interactions in these cell types is not clear (Chen et al., 2006; Dascenco et al., 2015). cPcdh-dependent repulsive behaviors have not been demonstrated in vivo. Moreover, the molecular mechanisms that convert an initially homophilic cPcdh/Dscam1 adhesion event to neurite repulsion have not been identified. In pyramidal neurons, the Pcdh-γs promote dendrite branching and stabilization through homophilic adhesion and modulation of FAK/PKC/MARKCS and Rac1 signaling (Garrett, Schreiner, Lobas, & Weiner, 2012; Molumby, Keeler, & Weiner, 2016; Suo, Lu, Ying, Capecchi, & Wu, 2012). The diverse cPcdh and Dscam1 roles raise the interesting possibility that they provide recognition specificity but implement different intracellular signals to produce repulsion or adhesion, depending on the cell type or wiring context. Future efforts that define the cPcdh and Dscam1 interactomes, and that apply single cell techniques to spatially profile and manipulate isoform diversity will be instrumental for elucidating their mechanisms.

6. Regulation of dendrite targeting and synapse selectivity

The formation of precise connections between neurons unfolds in a stepwise fashion and requires multiple interactions between dendrites and their environment. Targeted dendrite growth is a major developmental strategy for promoting synapse specificity and segregating circuit components that process different streams of information (Lefebvre et al., 2015). Elucidation of dendrite targeting mechanisms has come mainly from studies of visual and olfactory systems, where axonal and dendritic growth and synaptic connectivity are concentrated to neuropil layers or columns that are devoid of cell somata. Here I discuss principles and mechanisms of dendrite targeting and synapse selectivity with a focus on the vertebrate retina (see reviews for more discussion of dendrite targeting in *Drosophila* (Corty, Matthews, & Grueber, 2009; Millard & Pecot, 2018; Sanes & Zipursky, 2010)).

Retinal neurons target their dendrites to synaptic layers called the outer plexiform layer and inner plexiform layer (IPL) (Fig. 6A). Bipolar cells receiving photoreceptor signals in the outer retina send axons to the IPL, where the ~15 bipolar subtypes terminate within narrow sublayers and synapse onto dendrites of amacrine interneurons and retinal ganglion cell (RGC) subtypes (~60 and ~40 subtypes, respectively) (Rheaume et al., 2018; Shekhar et al., 2016; Yan et al., 2020). The stratified connectivity between bipolar, amacrine and RGC subtypes within each sublayer creates microcircuits that compute different features of visual information. The IPL can be subdivided in the ON and OFF layers, where connections within the upper (or distal) layers respond to light "OFF" decrements and those in the lower "ON" half respond to light "ON." And the IPL can be further subdivided into >10 sublaminae with distinct anatomical and functional characteristics. Laminar target specificity within the IPL is hard-wired, as manipulations affecting neural activity in mice do not affect the laminar organization of dendrites (Kerschensteiner, Morgan, Parker, Lewis, & Wong, 2009).

6.1 Guidance cues and adhesion molecules target dendrites to layers

Laminar organization facilitates the recruitment of pre- and postsynaptic partners belonging to the same circuit or the exclusion of processes from inappropriate neurons. Targeted dendritic growth to specific IPL sublayers

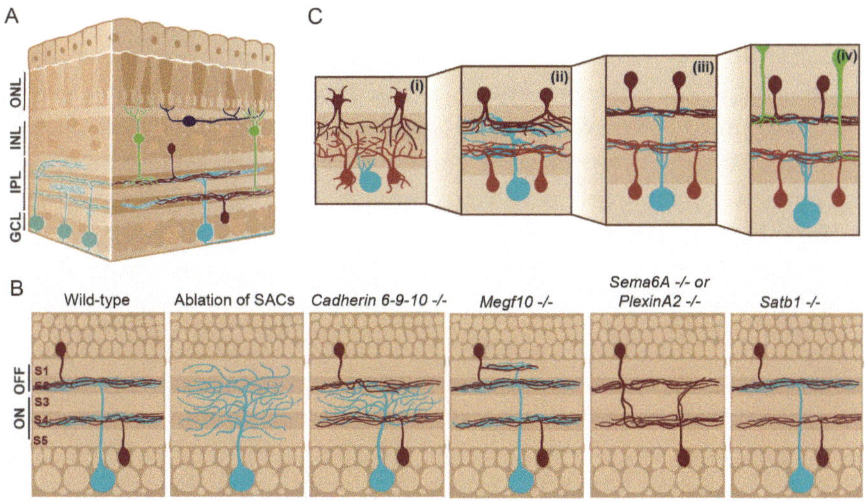

Fig. 6 Dendrite targeting in the mouse retina. (A) Cellular organization of the mouse retina. Neurons reside in three cellular layers: the outer nuclear layer (ONL), inner nuclear layer (INL), and ganglion cell layer (GCL). The ONL includes the rod and cone photoreceptors which form synapses onto horizontal cell (dark blue) and bipolar cell dendrites (green) in the outer plexiform layer. Bipolar cells and amacrine interneurons (dark red) reside in the INL and project to sublayers in the inner plexiform layer (IPL), where they synapse with RGC subtypes (cyan). Most amacrine cells have no axon and use their dendrites to make and receive synapses. Three RGC subtypes on the left (cyan) target their dendrites to different IPL sublaminae, including a bistratified ON-OFF DSGCs (ooDSGCs). The IPL on the right shows the direction-selective (DS) circuit, with two dendritic arbors of ooDSGCs receiving inputs from ON and OFF bipolar cells, and GABAergic inputs from ON and OFF starburst amacrine cells (dark red). (B) Summary of dendrite targeting defects in mutants affecting dendrite guidance and stabilization of DS retinal cell types. For simplicity, the IPL is subdivided into five sublayers, S1-S5. OFF layers reside in the upper half and ON layers in the lower half. ooDSGC dendrite stratification is disrupted upon genetic ablation of starburst amacrine cells, and in *Cadherin 6-9-10* triple mutants. MEGF10 stabilize starburst dendrites to form an IPL scaffold that guide ooDSGC dendrites. In *Megf10* mutants, starburst dendrites form ectopic layers and ooDSGC dendrites co-stratify with them because adhesive interactions are intact. In *Sema6A* or *PlexA2* mutants, OFF starburst dendrites misproject from S2 to the ON sublayer S4 due to loss of repulsion. In *Satb1* mutants, ooDSGCs are monostratified because they fail to stabilize an ON arbor in S4. (C) Development and laminar targeting of DS retinal cell types in the IPL. (i) Starburst amacrine cell dendrites (red) extend into the IPL and form a proto-IPL scaffold. (ii) ooDSGC target dendrites to the starburst scaffold. (iii) ooDSGC dendrites are stabilized and co-stratify with the ON and OFF starburst sublayers. (iv) ON and OFF bipolar cell axons terminate onto respective ON and OFF sublayers. *Adapted from Lefebvre, J. L., & Marocha, J. (2020). Chapter 12—Dendrite development: Vertebrates. In Rubenstein, J., Rakic, P., Chen, B., Kwan, K.Y., Kolodkin, A., & Anton, E. (Eds.),* Cellular migration and formation of axons and dendrites *(2nd ed.) (pp. 257–286). Academic Press.*

is critical for the assembly of retinal circuits. Some RGC types first grow dendrites broadly within the IPL and then prune their arbors to sublayers, while other RGC types restrict dendritic outgrowth to narrow sublayers (Kim, Zhang, Meister, & Sanes, 2010). Dendrite targeting requires various guidance strategies: (i) selective adhesion onto existing neurites in the nascent IPL; (ii) repulsion to prevent termination of dendrites from inappropriate sublayers; (iii) and mutual recognition or adhesion between cells that co-stratify (Lefebvre & Marocha, 2020). The cell surface recognition proteins guiding dendrite lamination vary by cell type, and include members of the immunoglobulin superfamily of cell adhesion molecules (IgSF), the Cadherin superfamily, and the Semaphorins. In the chick retina, different IgSF members provide instructive, homophilic cues for RGC and amacrine pairs to co-target an IPL sublayer. Misexpression of Sidekick-1, Sidekick-2, Dscam, DscamL, or Contactin-2 in the chick retina are sufficient to redirect dendrites to the corresponding IgSF-expressing sublayer (Yamagata & Sanes, 2008, 2012).

Repellent cues provided by the Semaphorins prevent the entry of dendrites into certain sublayers. Dendrites of amacrine subtypes destined for the OFF IPL express PlexinA2 and PlexinA4 receptors and are repelled by Semaphorin-6A localized within the ON half of the IPL (Fig. 6B) (Matsuoka, Nguyen-Ba-Charvet, et al., 2011; Sun et al., 2013). Instead of restricting their dendrites to a narrow OFF IPL sublayer, dopaminergic amacrine cells in retinas lacking Sema6A or PlexinA4 send errant dendrites to the ON IPL (Matsuoka, Nguyen-Ba-Charvet, et al., 2011). To prevent dendrites from extending out of the IPL, Sema5A and Sema5B expression within the cellular layers repel PlexinA1 or -A3-expressing neurites (Matsuoka, Chivatakarn, et al., 2011). Recognition molecules have the potential to cross-bind to each other and thus expand the combinatorial interactions and selectivity (Visser et al., 2015). The next challenges are to identify the repertoires and determine how they specify dendrite stratification and circuit assembly of multiple retinal neuron types.

6.2 Neurite scaffolds establish the first laminar cues to guide incoming dendrites

If retinal neurons require guidance cues in the developing IPL to target their dendrites, how are the first laminar cues established? As revealed by time-lapse imaging of fluorescently-labeled retinal cells in zebrafish, the IPL plexus emerges from amacrine cell neurites that first grow exuberantly and then stratify into two sublaminae (Godinho et al., 2005). These amacrine layers form scaffolds for the subsequent targeting of dendrites of RGCs and other amacrine subtypes. Some RGCs add dendritic layers in an inside-out manner while others confine dendritic growth towards particular sublayers (Mumm et al., 2006). RGC subtype-specific differences in dendrite growth are also observed in the mouse retina (Kim et al., 2010). Ablation studies in mice indicate that the early-born starburst amacrine cells provide a laminar scaffold that is essential for organizing the IPL (Duan et al., 2018). These studies suggest that amacrine cell dendrites form a proto-IPL by using homotypic, cell-surface cues to adhere to each other and self-assemble into a scaffold. RGC dendrites then use cues expressed by the amacrine scaffold for laminar positioning within the IPL (Fig. 6C) (Lefebvre & Marocha, 2020; Prigge & Kay, 2018).

In the mouse retina, the molecule that organizes the starburst amacrine cells into a proto-IPL is MEGF10, a transmembrane protein that mediates homotypic contacts between starburst amacrine cells (Kay et al., 2012; Ray et al., 2018). In the absence of *Megf10*, mutant starburst cells extend dendritic processes to ectopic layers. They still coalesce with neurites from their synaptic partners, the ON-OFF direction-selective RGCs (ooDSGC) and bipolar cells (Fig. 6B), indicating that starburst cells employ other cell adhesion molecules to attract RGC dendrites and bipolar terminals to their IPL sublayers. Starburst cells and their ooDSGC partners express type II Cadherin-6,-9, and -10, which mediate homophilic adhesive interactions to position ooDSGC dendrites along starburst sublaminae (Fig. 6B) (Duan et al., 2018). Direction-selective bipolar cell subtypes require Cadherin-8 or Cadherin-9 to fasciculate with the OFF or ON starburst layer, respectively (Duan, Krishnaswamy, De la Huerta, & Sanes, 2014). The IPL organizing activities of MEGF10 and the Cadherin molecules are essential for assembling the the bipolar inputs, starburst interneurons and the ooDSGCs into functional direction-selective circuits (Duan et al., 2014, 2018; Ray et al., 2018).

To recognize the laminar cues on target cells, incoming dendrites must be specified with their cell type-specific gene expression programs. The bi-stratified dendritic morphology of ooDSGCs is dependent on Satb1 (Special AT-rich sequence-binding protein-1), a global chromatin organizer and transcription factor (Peng et al., 2017). In contrast to wild-type ooDSGCs that bear two dendritic arbors, *Satb1* mutant ooDSGCs fail to stabilize branches in the ON layer and thus form the sole OFF arbor (Fig. 6b). Satb1 regulates the expression of Contactin 5, which stabilizes the ooDSGC dendrites in the ON layer through homophilic adhesion with their synaptic partners, the ON-starburst cells.

Another mechanism to regulate the spatial targeting of dendrites is through local and hierarchical interactions between growing dendrites, irrespective of target-derived cues. In *C. elegans*, the stereotyped organization of the amphid dendrites within the bundle that extends to the sensory nose tip is determined by differential adhesive interactions between neighboring dendrites during development (Yip & Heiman, 2018). Differential adhesion, through variations in expression levels or combinatorial actions of cell surface proteins, is a molecular strategy that also regulates neurite sorting and fine-scale positioning within layered or columnar neuropil structures (Barish et al., 2018; Schwabe, Borycz, Meinertzhagen, & Clandinin, 2014).

6.3 Synapse selectivity is determined by cell recognition molecules

Once dendrites reach their target location, they intermingle in dense neuropils that contain diverse potential partners. Within the mouse IPL for example, distinct RGC subtypes target the same sublayers but wire up into distinct circuits (Sanes & Masland, 2015). How do dendrites identify correct synaptic partners? Similar to dendrite targeting, synaptic selectivity is mediated by recognition molecules and selective adhesive interactions. An exquisite example was uncovered in the mouse retina, where the choice for an interneuron subtype to synapse with one of two potential RGC subtypes is mediated by the IgSF-related Sidekick proteins. Sidekick-1 (Sdk-1) and Sidekick-2 (Sdk-2) are differentially expressed by two functionally distinct RGC subtypes that have similar dendrite stratification patterns. Dendrites of the Sdk-1 + and Sdk-2 + expressing RGC populations intermingle with processes of an amacrine subtype that expresses Sdk-2 (Fig. 7A and B). Homophilic Sdk-2 interactions promote the synaptic connectivity between the Sdk-2 + RGCs and Sdk-2 + amacrine cells, while the mismatched Sdk-1 + AC and Sdk-2 + RGCs form few connections (Krishnaswamy, Yamagata, Duan, Hong, & Sanes, 2015). Similarly, homophilic interactions mediate the pairing of Sdk-1-expressing RGC and amacrine subtypes (Yamagata & Sanes, 2018). In *Sdk-2* mutants, RGC and amacrine dendrites remain confined to their IPL sublayer but show reduced synaptic transmission, suggesting that Sdk-2 mediates local synapse specificity (Fig. 7B and C). The Sdk-2+ amacrine-RGC pair underlies a circuit responsible for distinguishing global from local motion. Removal of *Sdk-2* leads to loss of this feature detection in mutant mice, demonstrating the importance for Sdk-mediated synapse selectivity in circuit wiring and function.

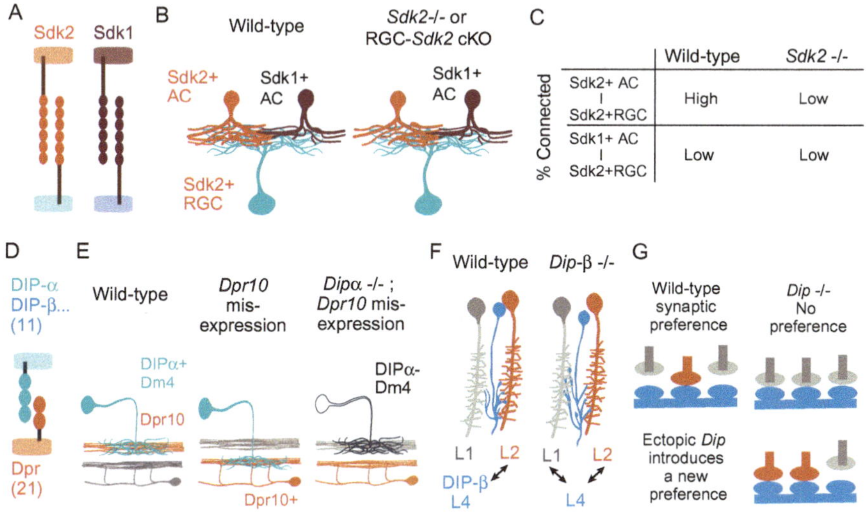

Fig. 7 Regulation of synapse selectivity by IgSF recognition molecules. (A) Sdk-1 and Sdk-2 mediate homophilic interactions to direct synapse partnering. (B, C) In the mouse retina, Sdk-2+ and Sdk-1+ amacrine subtypes send their dendrites to a common IPL sublayer but only the Sdk-2+ ACs (VG3-AC subtype, orange) selectively synapse with the Sdk2-expressing RGC subtype (W3-RGC, cyan). In the absence of *Sdk2*, dendrite targeting is not affected but synaptic connectivity is reduced. (D) DIP and Dpr pairs (from 11 and 21 members, respectively) mediate selective heterophilic interactions. (E) *Left,* In the *Drosphila* optic lobe, DIP-α-expressing Dm4 amacrine cell neurites (cyan) interact with Dpr10-expressing neuronal processes within medulla layer M3 (orange). *Middle,* Ecoptic expression of Dpr10 redirects Dm4 neurites. *Right,* Mistargeting is dependent on expression of DIP-α, as DIP-α-mutant Dm4 neurites fail to follow Dpr10 processes. (F) Within the lamina, the axon terminals of photoreceptors and the dendrites of lamina (L) neurons are organized into cartridges. L1 (gray) and L2 (orange) neurons receive inputs from photoreceptors. L4 (blue) form reciprocal connections with L2 in the proximal region, which is mediated by DIP-β and Dprs that have yet to be defined. In the absence of *Dip*-β, L4 dendrites overextend to distal regions of L4 and form ectopic synapses onto L1. (G) Summary of Dpr-DIP interactions for regulating synaptic selectivity. Dpr-DIP interactions establish a synaptic preference between appropriate partners (orange and blue). Preference for the correct synaptic partner is reduced when Dpr-DIP interactions are disrupted, revealing the intrinsic ability of neurons to synapse with other inputs. Ectopic Dpr-DIP interactions induce a preference for inappropriate synaptic partners. *Adapted from Xu, S., Xiao, Q., Cosmanescu, F., Sergeeva, A. P., Yoo, J., Lin, Y., et al. (2018). Interactions between the Ig-superfamily proteins DIP-α and Dpr6/10 regulate assembly of neural circuits.* Neuron, 100, *1369–1384.e6; Xu, C., Theisen, E., Maloney, R., Peng, J., Santiago, I., Yapp, C., et al. (2019). Control of synaptic specificity by establishing a relative preference for synaptic partners.* Neuron, 103, *865–877.e7.*

In *Drosophila*, two large IgSF subfamilies called Dpr (Defective proboscis retraction family; 21 members) and DIPs (Dpr-interacting proteins, 11 members) are proposed to determine synaptic selectivity (Fig. 7D). In vitro, each Dpr exhibits selective affinity for a particular subset of DIPs (Carrillo et al., 2015). Dprs and DIPs are differentially expressed by small subsets of neurons and in complementary patterns that suggest synaptic connectivity (Carrillo et al., 2015; Tan et al., 2015). In the fly visual system, Dpr-DIP pairs mediate heterophilic interactions that target neurites to optic layers or columns. For example, processes of DIPα-expressing Dm4 and Dm12 interneurons target the Dpr10-expressing layer in the medulla of the optic lobe. Misexpression of *Dpr10* redirects Dm4 and Dm12 processes to the ectopic Dpr10+ recipient layer, but the mistargeting is dependent on DIP-α expression (Fig. 7E) (Xu et al., 2018).

Dpr-DIPs also regulate synapse specificity within lamina cartridges. There, lamina neurons (L) L1-L2-L3 receive inputs from photoreceptor axon terminals, and L4 forms reciprocal connections with L2. Although L1-L3 dendrites are densely packed and contact each other extensively, they do not form reciprocal synapses, illustrating the high degree of synaptic specificity between L2 and L4 (Fig. 7F). DIP-β specifies the formation of reciprocal synapses between L2 and L4 neurons (Xu et al., 2019). *Dip-β* mutant L4 neurons form ectopic synapses onto inappropriate partners, suggesting that L4 neurons can synapse with multiple cell types but preferentially synapse onto DIP-β-expressing L2 neurons (Fig. 7G). Whether Dpr-DIP-mediated selectivity is the result of specific interactions between requisite pairs. or preferences based on relative affinities for multiple ligands remains unclear. Dendrite preference for particular synaptic inputs has also been observed for mouse RGCs (Tien, Soto, & Kerschensteiner, 2017). If the preferred bipolar cell partner is eliminated by genetic ablation, the RGC dendrites compensate by synapsing with the available inputs with no obvious changes to their structure or functional response. Thus, synaptic selectivity requires mechanisms to bias synaptic partnering but also allow plasticity for recruitment of alternate inputs to maintain proper outputs.

6.4 A model for multi-step wiring of dendrites into neural circuits

The wiring decisions that occur during circuit assembly are astoundingly complex. Studies of the mouse retina IPL are providing new insights into the coordination of multiple recognition events to govern dendritic wiring (Krishnaswamy et al., 2015; Lefebvre & Marocha, 2020). Dendrites of particular amacrine cells self-assemble into a proto-laminar scaffold and presents guidance and adhesive cues to stabilize growing dendrites to appropriate layers. Repulsive cues restrict dendrites within their target layers. Once dendrites reach their target zone, selective interactions are mediated by recognition molecules that bias connectivity between appropriate synaptic pairs. There is an extraordinary diversity of cell recognition molecules that can contribute to wiring specificity, including members of the Cadherin superfamily, Leucine-Rich Repeat (LRR) family, Neurexins, Teneurins, and classic axon guidance molecules such as the Ephrin/EPH and Semaphorin/Plexins (Honig & Shapiro, 2020; Sanes & Zipursky, 2020). Decoding the selective recognition events and molecular logic that underly synaptic specificity remains particularly challenging but is paramount to understanding neural circuit assembly. New methods to profile recognition molecules at the cell type or single cell level (Kurmangaliyev et al., 2020; Li et al., 2020), together with approaches that combine genetic, anatomical and functional assays of connectivity (Krishnaswamy et al., 2015) are providing the experimental access for dissecting mechanisms of dendritic wiring specificity.

7. Altered dendritic development in neurodevelopmental disorders (NDDs)

Altered dendritic development can lead to disruptions of neuronal connectivity and computations, which is thought to underlie cognitive dysfunctions associated with NDDs. Dendritic arborization defects are observed in many NDDs (Kulkarni & Firestein, 2012). Anatomical correlates include progressive dendrite atrophy in Down syndrome, an overabundance of immature spines and processes in Fragile X disorder, and reduced branching in autism spectrum disorders (ASDs) and schizophrenia (Becker, Armstrong, & Chan, 1986; Broadbelt, Byne, & Jones, 2002; Raymond, Bauman, & Kemper, 1996).

To define relationships between dendritic defects and NDD pathogenesis, several groups are modeling NDD-associated gene mutations using human induced pluripotent stem cells (hiPSCs) and animal models (Lefebvre & Marocha, 2020). For example, neurons derived from ASD-related Timothy syndrome patient hiPSCs, with mutations in the calcium channel subunit *CACNA1C*, show increased activity-dependent dendrite retraction (Krey et al., 2013). In studies of William's syndrome, hiPSC-derived cortical neurons develop dendritic phenotypes that are similar to those observed in post-mortem brain tissue (Chailangkarn et al., 2016). Another strategy is to model NDD-associated mutations to determine the developmental and cell biological basis for dendritic defects. For instance, *Dyrk1A* is located within the Down syndrome trisomy interval, and gene disruptions have also been identified in human ASD and intellectual disabilities (Park & Chung, 2013; van Bon et al., 2016). *Dyrk1A* encodes a kinase that regulates a myriad of processes, including microtubule and actin dynamics; disease-associated mutations lead to dendritic branching defects (Dang et al., 2018; Dowjat et al., 2007; Ori-McKenney et al., 2016). ASD and SCZ-linked Cntnap2 is a Contactin-associated protein that stabilizes dendrites of inhibitory neurons via interactions with CASK, a cytoskeletal scaffold protein (Gao et al., 2018). The levels and activity of CRMPs, which function in dendrite initiation and outgrowth, are also altered in patients and animals models from various NDDs (Quach, Honnorat, Kolattukudy, Khanna, & Duchemin, 2015). By studying how disease-associated mutations affect the structure and development of dendritic arbors in experimental models, the findings will advance a biological understanding of dendritic connectivity defects that underlie NDDs.

8. Conclusions and future directions

The nervous system requires a diversity of dendritic morphologies to organize neural circuits and fulfill a wide range of computations. We are just beginning to identify molecular pathways that pattern dendritic features for a handful of model neuronal cell types. As highlighted throughout the chapter, the next steps are to understand how transcriptional programs, extrinsic signals, and effectors are coordinated to regulate branch patterning. It also important to understand how these patterns influence neuronal computations and circuit functions. Such a detailed understanding will be achieved through further studies of the *Drosophila da* and *C. elegans* PVD mechanosensory neurons, as well as mouse retinal cells. Many examples from the mouse retina discussed in this chapter are the result of recent efforts to categorize retinal cell types, characterize their transcriptomic-morphological-functional properties, and apply genetic tools to probe their connectivity. This framework pioneered in invertebrates and in the mouse retina is being applied widely to study neural circuit assembly across model organisms. Single cell and spatial transcriptomics have the potential to reveal the gene expression programs that drive dendrite development and that vary according to cell lineage, spatial context or circuit activity. For instance genetic access to related neuronal subtypes will yield insights on how morphological diversification proceeds through the lineage, and lead to discovery of shared and distinct molecular pathways that specify dendrite patterns of each subtype. New technologies that induce complex gene or signaling manipulations, such as those mediated by CRISPR or optogenetics, and that report on cell biological processes such as cytoskeletal dynamics will be instrumental for interrogating the function of these pathways. New questions together with new tools will accelerate our understanding of the molecular mechanisms that regulate dendrite morphogenesis and that tune dendrite patterns for optimal neuronal function.

Acknowledgments

Wendy X. Wang assisted with the illustrations, and Samantha Ing-Esteves provided microscopy images. Julie L. Lefebvre is supported by a Canada Research Chair (Tier 2), a Sloan Fellowship in Neuroscience, an NSERC Discovery Grant (RGPIN-2016-06128) and a CIHR grant (PJT 148961).

References

Aizawa, H., Hu, S. C., Bobb, K., Balakrishnan, K., Ince, G., Gurevich, I., et al. (2004). Dendrite development regulated by CREST, a calcium-regulated transcriptional activator. *Science, 303*, 197–202.

Baas, P. W., Black, M. M., & Banker, G. A. (1989). Changes in microtubule polarity orientation during the development of hippocampal neurons in culture. *The Journal of Cell Biology, 109*, 3085–3094.

Baas, P. W., Deitch, J. S., Black, M. M., & Banker, G. A. (1988). Polarity orientation of microtubules in hippocampal neurons: Uniformity in the axon and nonuniformity in the dendrite. *Proceedings of the National Academy of Sciences of the United States of America, 85*, 8335–8339.

Barish, S., Nuss, S., Strunilin, I., Bao, S., Mukherjee, S., Jones, C. D., et al. (2018). Combinations of DIPs and Dprs control organization of olfactory receptor neuron terminals in Drosophila. *PLoS Genetics, 14*, e1007560.

Becker, L. E., Armstrong, D. L., & Chan, F. (1986). Dendritic atrophy in children with Down's syndrome. *Annals of Neurology, 20*, 520–526.

Blackshaw, S. E., Nicholls, J. G., & Parnas, I. (1982). Expanded receptive fields of cutaneous mechanoreceptor cells after single neurone deletion in leech central nervous system. *The Journal of Physiology, 326*, 261–268.

Bowen, A. B., Bourke, A. M., Hiester, B. G., Hanus, C., & Kennedy, M. J. (2017). Golgi-independent secretory trafficking through recycling endosomes in neuronal dendrites and spines. *eLife, 6*, e27362.

Brasch, J., Goodman, K. M., Noble, A. J., Rapp, M., Mannepalli, S., Bahna, F., et al. (2019). Visualization of clustered protocadherin neuronal self-recognition complexes. *Nature, 569*, 280–283.

Broadbelt, K., Byne, W., & Jones, L. B. (2002). Evidence for a decrease in basilar dendrites of pyramidal cells in schizophrenic medial prefrontal cortex. *Schizophrenia Research, 58*, 75–81.

Cao, Y., Lipka, J., Stucchi, R., Burute, M., Pan, X., Portegies, S., et al. (2020). Microtubule minus-end binding protein CAMSAP2 and Kinesin-14 motor KIFC3 control dendritic microtubule organization. *Current Biology, 30,* 899–908.e6.

Carrillo, R. A., Özkan, E., Menon, K. P., Nagarkar-Jaiswal, S., Lee, P. T., Jeon, M., et al. (2015). Control of synaptic connectivity by a network of Drosophila IgSF cell surface proteins. *Cell, 163,* 1770–1782.

Cebul, E. R., McLachlan, I. G., & Heiman, M. G. (2020). Dendrites with specialized glial attachments develop by retrograde extension using SAX-7 and GRDN-1. *Development, 147*. dev180448.

Chailangkarn, T., Trujillo, C. A., Freitas, B. C., Hrvoj-Mihic, B., Herai, R. H., Yu, D. X., et al. (2016). A human neurodevelopmental model for Williams syndrome. *Nature, 536,* 338–343.

Chen, B., Brinkmann, K., Chen, Z., Pak, C. W., Liao, Y., Shi, S., et al. (2014). The WAVE regulatory complex links diverse receptors to the actin cytoskeleton. *Cell, 156,* 195–207.

Chen, B. E., Kondo, M., Garnier, A., Watson, F. L., Püettmann-Holgado, R., Lamar, D. R., et al. (2006). The molecular diversity of Dscam is functionally required for neuronal wiring specificity in Drosophila. *Cell, 125,* 607–620.

Chen, J., & Kriegstein, A. R. (2015). A GABAergic projection from the zona incerta to cortex promotes cortical neuron development. *Science, 350,* 554–558.

Chen, W. V., Nwakeze, C. L., Denny, C. A., O'Keeffe, S., Rieger, M. A., Mountoufaris, G., et al. (2017). Pcdhalphac2 is required for axonal tiling and assembly of serotonergic circuitries in mice. *Science, 356,* 406–411.

Chen, S. X., Tari, P. K., She, K., & Haas, K. (2010). Neurexin-neuroligin cell adhesion complexes contribute to synaptotropic dendritogenesis via growth stabilization mechanisms in vivo. *Neuron, 67,* 967–983.

Choy, R. W., Park, M., Temkin, P., Herring, B. E., Marley, A., Nicoll, R. A., et al. (2014). Retromer mediates a discrete route of local membrane delivery to dendrites. *Neuron, 82,* 55–62.

Cline, H., & Haas, K. (2008). The regulation of dendritic arbor development and plasticity by glutamatergic synaptic input: A review of the synaptotrophic hypothesis. *The Journal of Physiology, 586,* 1509–1517.

Corty, M. M., Matthews, B. J., & Grueber, W. B. (2009). Molecules and mechanisms of dendrite development in Drosophila. *Development, 136,* 1049–1061.

Cui-Wang, T., Hanus, C., Cui, T., Helton, T., Bourne, J., Watson, D., et al. (2012). Local zones of endoplasmic reticulum complexity confine cargo in neuronal dendrites. *Cell, 148,* 309–321.

Cunha-Ferreira, I., Chazeau, A., Buijs, R. R., Stucchi, R., Will, L., Pan, X., et al. (2018). The HAUS complex is a key regulator of non-centrosomal microtubule organization during neuronal development. *Cell Reports, 24,* 791–800.

Danelon, V., Goldner, R., Martinez, E., Gokhman, I., Wang, K., Yaron, A., et al. (2020). Modular and distinct Plexin-A4/FARP2/Rac1 signalling controls dendrite morphogenesis. *The Journal of Neuroscience : The Official Journal of the Society for Neuroscience, 40,* 5413–5430.

Dang, T., Duan, W. Y., Yu, B., Tong, D. L., Cheng, C., Zhang, Y. F., et al. (2018). Autism-associated Dyrk1a truncation mutants impair neuronal dendritic and spine growth and interfere with postnatal cortical development. *Molecular Psychiatry, 23,* 747–758.

Das, R., Bhattacharjee, S., Patel, A. A., Harris, J. M., Bhattacharya, S., Letcher, J. M., et al. (2017). Dendritic cytoskeletal architecture is modulated by combinatorial transcriptional regulation in Drosophila melanogaster. *Genetics, 207,* 1401–1421.

Dascenco, D., Erfurth, M. L., Izadifar, A., Song, M., Sachse, S., Bortnick, R., et al. (2015). Slit and receptor tyrosine phosphatase 69D confer spatial specificity to axon branching via Dscam1. *Cell, 162,* 1140–1154.

Datwani, A., Iwasato, T., Itohara, S., & Erzurumlu, R. S. (2002). NMDA receptor-dependent pattern transfer from afferents to postsynaptic cells and dendritic differentiation in the barrel cortex. *Molecular and Cellular Neurosciences*, *21*, 477–492.

Deitch, J. S., & Rubel, E. W. (1984). Afferent influences on brain stem auditory nuclei of the chicken: Time course and specificity of dendritic atrophy following deafferentation. *The Journal of Comparative Neurology*, *229*, 66–79.

Díaz-Balzac, C. A., Rahman, M., Lázaro-Peña, M. I., Martin Hernandez, L. A., Salzberg, Y., Aguirre-Chen, C., et al. (2016). Muscle- and skin-derived cues jointly orchestrate patterning of somatosensory dendrites. *Current Biology*, *26*, 2379–2387.

Dong, X., Liu, O. W., Howell, A. S., & Shen, K. (2013). An extracellular adhesion molecule complex patterns dendritic branching and morphogenesis. *Cell*, *155*, 296–307.

Dowjat, W. K., Adayev, T., Kuchna, I., Nowicki, K., Palminiello, S., Hwang, Y. W., et al. (2007). Trisomy-driven overexpression of DYRK1A kinase in the brain of subjects with Down syndrome. *Neuroscience Letters*, *413*, 77–81.

Duan, X., Krishnaswamy, A., De la Huerta, I., & Sanes, J. R. (2014). Type II cadherins guide assembly of a direction-selective retinal circuit. *Cell*, *158*, 793–807.

Duan, X., Krishnaswamy, A., Laboulaye, M. A., Liu, J., Peng, Y. R., Yamagata, M., et al. (2018). Cadherin combinations recruit dendrites of distinct retinal neurons to a shared interneuronal scaffold. *Neuron*, *99*(1145–1154), e1146.

Emoto, K., He, Y., Ye, B., Grueber, W. B., Adler, P. N., Jan, L. Y., et al. (2004). Control of dendritic branching and tiling by the tricornered-kinase/furry signalling pathway in Drosophila sensory neurons. *Cell*, *119*, 245–256.

Enriquez, J., Venkatasubramanian, L., Baek, M., Peterson, M., Aghayeva, U., & Mann, R. S. (2015). Specification of individual adult motor neuron morphologies by combinatorial transcription factor codes. *Neuron*, *86*, 955–970.

Espinosa, J. S., Wheeler, D. G., Tsien, R. W., & Luo, L. (2009). Uncoupling dendrite growth and patterning: Single-cell knockout analysis of NMDA receptor 2B. *Neuron*, *62*, 205–217.

Evans, A. J., Gurung, S., Wilkinson, K. A., Stephens, D. J., & Henley, J. M. (2017). Assembly, secretory pathway trafficking, and surface delivery of kainate receptors is regulated by neuronal activity. *Cell Reports*, *19*, 2613–2626.

Feng, C., Thyagarajan, P., Shorey, M., Seebold, D. Y., Weiner, A. T., Albertson, R. M., et al. (2019). Patronin-mediated minus end growth is required for dendritic microtubule polarity. *The Journal of Cell Biology*, *218*, 2309–2328.

Fink, C. C., Bayer, K. U., Myers, J. W., Ferrell, J. E., Jr., Schulman, H., & Meyer, T. (2003). Selective regulation of neurite extension and synapse formation by the beta but not the alpha isoform of CaMKII. *Neuron*, *39*, 283–297.

Flavell, S. W., Kim, T. K., Gray, J. M., Harmin, D. A., Hemberg, M., Hong, E. J., et al. (2008). Genome-wide analysis of MEF2 transcriptional program reveals synaptic target genes and neuronal activity-dependent polyadenylation site selection. *Neuron*, *60*, 1022–1038.

Franker, M. A., Esteves da Silva, M., Tas, R. P., Tortosa, E., Cao, Y., Frias, C. P., et al. (2016). Three-step model for polarized sorting of KIF17 into dendrites. *Current Biology*, *26*, 1705–1712.

Fuerst, P. G., Bruce, F., Tian, M., Wei, W., Elstrott, J., Feller, M. B., et al. (2009). DSCAM and DSCAML1 function in self-avoidance in multiple cell types in the developing mouse retina. *Neuron*, *64*, 484–497.

Fuerst, P. G., Koizumi, A., Masland, R. H., & Burgess, R. W. (2008). Neurite arborization and mosaic spacing in the mouse retina require DSCAM. *Nature*, *451*, 470–474.

Galli-Resta, L., Resta, G., Tan, S. S., & Reese, B. E. (1997). Mosaics of islet-1-expressing amacrine cells assembled by short-range cellular interactions. *The Journal of Neuroscience : The Official Journal of the Society for Neuroscience*, *17*, 7831–7838.

Gao, F. B., Kohwi, M., Brenman, J. E., Jan, L. Y., & Jan, Y. N. (2000). Control of dendritic field formation in Drosophila: The roles of flamingo and competition between homologous neurons. *Neuron, 28*, 91–101.

Gao, R., Piguel, N. H., Melendez-Zaidi, A. E., Martin-de-Saavedra, M. D., Yoon, S., Forrest, M. P., et al. (2018). CNTNAP2 stabilizes interneuron dendritic arbors through CASK. *Molecular Psychiatry, 23*, 1832–1850.

Gardiol, A., Racca, C., & Triller, A. (1999). Dendritic and postsynaptic protein synthetic machinery. *The Journal of Neuroscience : The Official Journal of the Society for Neuroscience, 19*, 168–179.

Garrett, A. M., Khalil, A., Walton, D. O., & Burgess, R. W. (2018). DSCAM promotes self-avoidance in the developing mouse retina by masking the functions of cadherin superfamily members. *Proceedings of the National Academy of Sciences of the United States of America, 115*, E10216–E10224.

Garrett, A. M., Schreiner, D., Lobas, M. A., & Weiner, J. A. (2012). Gamma-protocadherins control cortical dendrite arborization by regulating the activity of a FAK/PKC/MARCKS signalling pathway. *Neuron, 74*, 269–276.

Gibson, D. A., Tymanskyj, S., Yuan, R. C., Leung, H. C., Lefebvre, J. L., Sanes, J. R., et al. (2014). Dendrite self-avoidance requires cell-autonomous slit/robo signalling in cerebellar purkinje cells. *Neuron, 81*, 1040–1056.

Godinho, L., Mumm, J. S., Williams, P. R., Schroeter, E. H., Koerber, A., Park, S. W., et al. (2005). Targeting of amacrine cell neurites to appropriate synaptic laminae in the developing zebrafish retina. *Development, 132*, 5069–5079.

Grueber, W. B., Jan, L. Y., & Jan, Y. N. (2002). Tiling of the Drosophila epidermis by multidendritic sensory neurons. *Development, 129*, 2867–2878.

Grueber, W. B., Jan, L. Y., & Jan, Y. N. (2003). Different levels of the homeodomain protein cut regulate distinct dendrite branching patterns of Drosophila multidendritic neurons. *Cell, 112*, 805–818.

Grueber, W. B., & Ye, B. (2020). Chapter 11—Dendrite development: Invertebrates. In J. Rubenstein, P. Rakic, B. Chen, K. Y. Kwan, A. Kolodkin, & E. Anton (Eds.), *Cellular migration and formation of axons and dendrites* (2nd ed., pp. 231–256). Academic Press.

Grueber, W. B., Ye, B., Moore, A. W., Jan, L. Y., & Jan, Y. N. (2003). Dendrites of distinct classes of Drosophila sensory neurons show different capacities for homotypic repulsion. *Current Biology, 13*, 618–626.

Haas, K., Li, J., & Cline, H. T. (2006). AMPA receptors regulate experience-dependent dendritic arbor growth in vivo. *Proceedings of the National Academy of Sciences of the United States of America, 103*, 12127–12131.

Han, C., Wang, D., Soba, P., Zhu, S., Lin, X., Jan, L. Y., et al. (2012). Integrins regulate repulsion-mediated dendritic patterning of drosophila sensory neurons by restricting dendrites in a 2D space. *Neuron, 73*, 64–78.

Hand, R., Bortone, D., Mattar, P., Nguyen, L., Heng, J. I., Guerrier, S., et al. (2005). Phosphorylation of Neurogenin2 specifies the migration properties and the dendritic morphology of pyramidal neurons in the neocortex. *Neuron, 48*, 45–62.

Hanus, C., Kochen, L., Tom Dieck, S., Racine, V., Sibarita, J. B., Schuman, E. M., et al. (2014). Synaptic control of secretory trafficking in dendrites. *Cell Reports, 7*, 1771–1778.

Harterink, M., Edwards, S. L., de Haan, B., Yau, K. W., van den Heuvel, S., Kapitein, L. C., et al. (2018). Local microtubule organization promotes cargo transport in C. elegans dendrites. *Journal of Cell Science, 131*. jcs223107.

Hattori, D., Chen, Y., Matthews, B. J., Salwinski, L., Sabatti, C., Grueber, W. B., et al. (2009). Robust discrimination between self and non-self neurites requires thousands of Dscam1 isoforms. *Nature, 461*, 644–648.

Hattori, D., Demir, E., Kim, H. W., Viragh, E., Zipursky, S. L., & Dickson, B. J. (2007). Dscam diversity is essential for neuronal wiring and self-recognition. *Nature, 449*, 223–227.

Hattori, Y., Usui, T., Satoh, D., Moriyama, S., Shimono, K., Itoh, T., et al. (2013). Sensory-neuron subtype-specific transcriptional programs controlling dendrite morphogenesis: Genome-wide analysis of abrupt and Knot/Collier. *Developmental Cell, 27*, 530–544.

Heiman, M. G., & Shaham, S. (2009). DEX-1 and DYF-7 establish sensory dendrite length by anchoring dendritic tips during cell migration. *Cell, 137*, 344–355.

Hill, S. E., Parmar, M., Gheres, K. W., Guignet, M. A., Huang, Y., Jackson, F. R., et al. (2012). Development of dendrite polarity in Drosophila neurons. *Neural Development*, 7, 34.

Hobert, O. (2008). Regulatory logic of neuronal diversity: Terminal selector genes and selector motifs. *Proceedings of the National Academy of Sciences of the United States of America, 105*, 20067–20071.

Honig, B., & Shapiro, L. (2020). Adhesion protein structure, molecular affinities, and principles of cell-cell recognition. *Cell, 181*, 520–535.

Horch, H. W., & Katz, L. C. (2002). BDNF release from single cells elicits local dendritic growth in nearby neurons. *Nature Neuroscience, 5*, 1177–1184.

Horton, A. C., & Ehlers, M. D. (2003). Dual modes of endoplasmic reticulum-to-Golgi transport in dendrites revealed by live-cell imaging. *The Journal of neuroscience : the official journal of the Society for Neuroscience, 23*, 6188–6199.

Horton, A. C., Racz, B., Monson, E. E., Lin, A. L., Weinberg, R. J., & Ehlers, M. D. (2005). Polarized secretory trafficking directs cargo for asymmetric dendrite growth and morphogenesis. *Neuron, 48*, 757–771.

Hossain, S., Hewapathirane, D. S., & Haas, K. (2012). Dynamic morphometrics reveals contributions of dendritic growth cones and filopodia to dendritogenesis in the intact and awake embryonic brain. *Developmental Neurobiology, 72*, 615–627.

Hu, S. C., Chrivia, J., & Ghosh, A. (1999). Regulation of CBP-mediated transcription by neuronal calcium signalling. *Neuron, 22*, 799–808.

Huang, G. H., Sun, Z. L., Li, H. J., & Feng, D. F. (2017). Rho GTPase-activating proteins: Regulators of Rho GTPase activity in neuronal development and CNS diseases. *Molecular and Cellular Neurosciences, 80*, 18–31.

Huckfeldt, R. M., Schubert, T., Morgan, J. L., Godinho, L., Di Cristo, G., Huang, Z. J., et al. (2009). Transient neurites of retinal horizontal cells exhibit columnar tiling via homotypic interactions. *Nature Neuroscience, 12*, 35–43.

Hughes, M. E., Bortnick, R., Tsubouchi, A., Bäumer, P., Kondo, M., Uemura, T., et al. (2007). Homophilic Dscam interactions control complex dendrite morphogenesis. *Neuron, 54*, 417–427.

Ince-Dunn, G., Hall, B. J., Hu, S. C., Ripley, B., Huganir, R. L., Olson, J. M., et al. (2006). Regulation of thalamocortical patterning and synaptic maturation by NeuroD2. *Neuron, 49*, 683–695.

Ing-Esteves, S., Kostadinov, D., Marocha, J., Sing, A. D., Joseph, K. S., Laboulaye, M. A., et al. (2018). Combinatorial effects of alpha- and gamma-Protocadherins on neuronal survival and dendritic self-avoidance. *The Journal of Neuroscience : the Official Journal of the Society for Neuroscience, 38*, 2713–2729.

Iwasato, T., Datwani, A., Wolf, A. M., Nishiyama, H., Taguchi, Y., Tonegawa, S., et al. (2000). Cortex-restricted disruption of NMDAR1 impairs neuronal patterns in the barrel cortex. *Nature, 406*, 726–731.

Iwasato, T., Erzurumlu, R. S., Huerta, P. T., Chen, D. F., Sasaoka, T., Ulupinar, E., et al. (1997). NMDA receptor-dependent refinement of somatotopic maps. *Neuron, 19*, 1201–1210.

Jeyifous, O., Waites, C. L., Specht, C. G., Fujisawa, S., Schubert, M., Lin, E. I., et al. (2009). SAP97 and CASK mediate sorting of NMDA receptors through a previously unknown secretory pathway. *Nature Neuroscience*, *12*, 1011–1019.

Jinushi-Nakao, S., Arvind, R., Amikura, R., Kinameri, E., Liu, A. W., & Moore, A. W. (2007). Knot/collier and cut control different aspects of dendrite cytoskeleton and synergize to define final arbor shape. *Neuron*, *56*, 963–978.

Joo, W., Hippenmeyer, S., & Luo, L. (2014). Neurodevelopment. Dendrite morphogenesis depends on relative levels of NT-3/TrkC signalling. *Science*, *346*, 626–629.

Kaneko, R., Kato, H., Kawamura, Y., Esumi, S., Hirayama, T., Hirabayashi, T., et al. (2006). Allelic gene regulation of Pcdh-alpha and Pcdh-gamma clusters involving both monoallelic and biallelic expression in single Purkinje cells. *The Journal of Biological Chemistry*, *281*, 30551–30560.

Kapitein, L. C., & Hoogenraad, C. C. (2015). Building the neuronal microtubule cytoskeleton. *Neuron*, *87*, 492–506.

Kapitein, L. C., Schlager, M. A., Kuijpers, M., Wulf, P. S., van Spronsen, M., MacKintosh, F. C., et al. (2010). Mixed microtubules steer dynein-driven cargo transport into dendrites. *Current Biology*, *20*, 290–299.

Karasmanis, E. P., Phan, C. T., Angelis, D., Kesisova, I. A., Hoogenraad, C. C., McKenney, R. J., et al. (2018). Polarity of neuronal membrane traffic requires sorting of kinesin motor cargo during entry into dendrites by a microtubule-associated Septin. *Developmental Cell*, *46*, 204–218.e7.

Kawabata Galbraith, K., Fujishima, K., Mizuno, H., Lee, S. J., Uemura, T., Sakimura, K., et al. (2018). MTSS1 regulation of actin-nucleating Formin DAAM1 in dendritic Filopodia determines final dendritic configuration of Purkinje cells. *Cell Reports*, *24*, 95–106.e9.

Kay, J. N., Chu, M. W., & Sanes, J. R. (2012). MEGF10 and MEGF11 mediate homotypic interactions required for mosaic spacing of retinal neurons. *Nature*, *483*, 465–469.

Keeley, P. W., Whitney, I. E., Raven, M. A., & Reese, B. E. (2007). Dendritic spread and functional coverage of starburst amacrine cells. *The Journal of Comparative Neurology*, *505*, 539–546.

Kerschensteiner, D., Morgan, J. L., Parker, E. D., Lewis, R. M., & Wong, R. O. (2009). Neurotransmission selectively regulates synapse formation in parallel circuits in vivo. *Nature*, *460*, 1016–1020.

Kim, M. E., Shrestha, B. R., Blazeski, R., Mason, C. A., & Grueber, W. B. (2012). Integrins establish dendrite-substrate relationships that promote dendritic self-avoidance and patterning in drosophila sensory neurons. *Neuron*, *73*, 79–91.

Kim, I. J., Zhang, Y., Meister, M., & Sanes, J. R. (2010). Laminar restriction of retinal ganglion cell dendrites and axons: Subtype-specific developmental patterns revealed with transgenic markers. *The Journal of Neuroscience : The Official Journal of the Society for Neuroscience*, *30*, 1452–1462.

Kimura, H., Usui, T., Tsubouchi, A., & Uemura, T. (2006). Potential dual molecular interaction of the Drosophila 7-pass transmembrane cadherin flamingo in dendritic morphogenesis. *Journal of Cell Science*, *119*, 1118–1129.

Koike-Kumagai, M., Yasunaga, K., Morikawa, R., Kanamori, T., & Emoto, K. (2009). The target of rapamycin complex 2 controls dendritic tiling of Drosophila sensory neurons through the tricornered kinase signalling pathway. *The EMBO Journal*, *28*, 3879–3892.

Kollins, K. M., Bell, R. L., Butts, M., & Withers, G. S. (2009). Dendrites differ from axons in patterns of microtubule stability and polymerization during development. *Neural Development*, *4*, 26.

Komiyama, T., Johnson, W. A., Luo, L., & Jefferis, G. S. (2003). From lineage to wiring specificity. POU domain transcription factors control precise connections of Drosophila olfactory projection neurons. *Cell*, *112*, 157–167.

Komiyama, T., & Luo, L. (2007). Intrinsic control of precise dendritic targeting by an ensemble of transcription factors. *Current Biology*, *17*, 278–285.
Konietzny, A., Bar, J., & Mikhaylova, M. (2017). Dendritic actin cytoskeleton: Structure, functions, and regulations. *Frontiers in Cellular Neuroscience*, *11*, 147.
Kostadinov, D., & Sanes, J. R. (2015). Protocadherin-dependent dendritic self-avoidance regulates neural connectivity and circuit function. *eLife*, *4*, e08964.
Kramer, A. P., & Kuwada, J. Y. (1983). Formation of the receptive fields of leech mechanosensory neurons during embryonic development. *The Journal of Neuroscience : The Official Journal of the Society for Neuroscience*, *3*, 2474–2486.
Kramer, A. P., & Stent, G. S. (1985). Developmental arborization of sensory neurons in the leech Haementeria ghilianii. II. Experimentally induced variations in the branching pattern. *The Journal of Neuroscience : The Official Journal of the Society for Neuroscience*, *5*, 768–775.
Krey, J. F., Pasca, S. P., Shcheglovitov, A., Yazawa, M., Schwemberger, R., Rasmusson, R., et al. (2013). Timothy syndrome is associated with activity-dependent dendritic retraction in rodent and human neurons. *Nature Neuroscience*, *16*, 201–209.
Krishnaswamy, A., Yamagata, M., Duan, X., Hong, Y. K., & Sanes, J. R. (2015). Sidekick 2 directs formation of a retinal circuit that detects differential motion. *Nature*, *524*, 466–470.
Kulkarni, V. A., & Firestein, B. L. (2012). The dendritic tree and brain disorders. *Molecular and Cellular Neurosciences*, *50*, 10–20.
Kurmangaliyev, Y. Z., Yoo, J., LoCascio, S. A., & Zipursky, S. L. (2019). Modular transcriptional programs separately define axon and dendrite connectivity. *eLife*, *8*, e50822.
Kurmangaliyev, Y. Z., Yoo, J., Valdes-Aleman, J., Sanfilippo, P., & Zipursky, S. L. (2020). Transcriptional programs of circuit assembly in the Drosophila visual system. *Neuron*, *108*, 1045–1057.
Kuwako, K. I., & Okano, H. (2018). The LKB1-SIK pathway controls dendrite self-avoidance in Purkinje cells. *Cell Reports*, *24*, 2808–2818.e4.
Kwiatkowski, A. V., Rubinson, D. A., Dent, E. W., Edward van Veen, J., Leslie, J. D., Zhang, J., et al. (2007). Ena/VASP is required for neuritogenesis in the developing cortex. *Neuron*, *56*, 441–455.
Lanoue, V., Usardi, A., Sigoillot, S. M., Talleur, M., Iyer, K., Mariani, J., et al. (2013). The adhesion-GPCR BAI3, a gene linked to psychiatric disorders, regulates dendrite morphogenesis in neurons. *Molecular Psychiatry*, *18*, 943–950.
Lee, S. C., Cowgill, E. J., Al-Nabulsi, A., Quinn, E. J., Evans, S. M., & Reese, B. E. (2011). Homotypic regulation of neuronal morphology and connectivity in the mouse retina. *The Journal of Neuroscience : The Official Journal of the Society for Neuroscience*, *31*, 14126–14133.
Lefebvre, J. L., Kostadinov, D., Chen, W. V., Maniatis, T., & Sanes, J. R. (2012). Protocadherins mediate dendritic self-avoidance in the mammalian nervous system. *Nature*, *488*, 517–521.
Lefebvre, J. L., & Marocha, J. (2020). Chapter 12—Dendrite development: Vertebrates. In J. Rubenstein, P. Rakic, B. Chen, K. Y. Kwan, A. Kolodkin, & E. Anton (Eds.), *Cellular migration and formation of axons and dendrites* (2nd ed., pp. 257–286). Academic Press.
Lefebvre, J. L., Sanes, J. R., & Kay, J. N. (2015). Development of dendritic form and function. *Annual Review of Cell and Developmental Biology*, *31*, 741–777.
Li, Z., Aizenman, C. D., & Cline, H. T. (2002). Regulation of rho GTPases by crosstalk and neuronal activity in vivo. *Neuron*, *33*, 741–750.
Li, J., Han, S., Li, H., Udeshi, N. D., Svinkina, T., Mani, D. R., et al. (2020). Cell-surface proteomic profiling in the Fly brain uncovers wiring regulators. *Cell*, *180*, 373–386.e15.
Li, Z., Van Aelst, L., & Cline, H. T. (2000). Rho GTPases regulate distinct aspects of dendritic arbor growth in Xenopus central neurons in vivo. *Nature Neuroscience*, *3*, 217–225.

Li, W., Wang, F., Menut, L., & Gao, F. B. (2004). BTB/POZ-zinc finger protein abrupt suppresses dendritic branching in a neuronal subtype-specific and dosage-dependent manner. *Neuron, 43*, 823–834.

Liang, X., Kokes, M., Fetter, R. D., Sallee, M. D., Moore, A. W., Feldman, J. L., et al. (2020). Growth cone-localized microtubule organizing center establishes microtubule orientation in dendrites. *eLife, 9*, e56547.

Liao, C. P., Li, H., Lee, H. H., Chien, C. T., & Pan, C. L. (2018). Cell-autonomous regulation of dendrite self-avoidance by the Wnt secretory factor MIG-14/Wntless. *Neuron, 98*, 320–334.e6.

Liu, J., Reggiani, J. D. S., Laboulaye, M. A., Pandey, S., Chen, B., Rubenstein, J. L. R., et al. (2018). Tbr1 instructs laminar patterning of retinal ganglion cell dendrites. *Nature Neuroscience, 21*, 659–670.

Lohmann, C., & Bonhoeffer, T. (2008). A role for local calcium signalling in rapid synaptic partner selection by dendritic filopodia. *Neuron, 59*, 253–260.

Lohmann, C., Finski, A., & Bonhoeffer, T. (2005). Local calcium transients regulate the spontaneous motility of dendritic filopodia. *Nature Neuroscience, 8*, 305–312.

Lohmann, C., Myhr, K. L., & Wong, R. O. (2002). Transmitter-evoked local calcium release stabilizes developing dendrites. *Nature, 418*, 177–181.

Lohmann, C., & Wong, R. O. (2001). Cell-type specific dendritic contacts between retinal ganglion cells during development. *Journal of Neurobiology, 48*, 150–162.

Lom, B., Cogen, J., Sanchez, A. L., Vu, T., & Cohen-Cory, S. (2002). Local and target-derived brain-derived neurotrophic factor exert opposing effects on the dendritic arborization of retinal ganglion cells in vivo. *The Journal of Neuroscience : The Official Journal of the Society for Neuroscience, 22*, 7639–7649.

Luo, J., Ting, C. Y., Li, Y., McQueen, P., Lin, T. Y., Hsu, C. P., et al. (2020). Antagonistic regulation by insulin-like peptide and activin ensures the elaboration of appropriate dendritic field sizes of amacrine neurons. *eLife, 9*, e50568.

Maniar, T. A., Kaplan, M., Wang, G. J., Shen, K., Wei, L., Shaw, J. E., et al. (2011). UNC-33 (CRMP) and ankyrin organize microtubules and localize kinesin to polarize axon-dendrite sorting. *Nature Neuroscience, 15*, 48–56.

Matsubara, D., Horiuchi, S. Y., Shimono, K., Usui, T., & Uemura, T. (2011). The seven-pass transmembrane cadherin flamingo controls dendritic self-avoidance via its binding to a LIM domain protein, Espinas, in Drosophila sensory neurons. *Genes & Development, 25*, 1982–1996.

Matsui, A., Tran, M., Yoshida, A. C., Kikuchi, S. S., Mami, U., Ogawa, M., et al. (2013). BTBD3 controls dendrite orientation toward active axons in mammalian neocortex. *Science, 342*, 1114–1118.

Matsuoka, R. L., Chivatakarn, O., Badea, T. C., Samuels, I. S., Cahill, H., Katayama, K., et al. (2011). Class 5 transmembrane semaphorins control selective mammalian retinal lamination and function. *Neuron, 71*, 460–473.

Matsuoka, R. L., Nguyen-Ba-Charvet, K. T., Parray, A., Badea, T. C., Chedotal, A., & Kolodkin, A. L. (2011). Transmembrane semaphorin signalling controls laminar stratification in the mammalian retina. *Nature, 470*, 259–263.

Matthews, B. J., Kim, M. E., Flanagan, J. J., Hattori, D., Clemens, J. C., Zipursky, S. L., et al. (2007). Dendrite self-avoidance is controlled by Dscam. *Cell, 129*, 593–604.

McAllister, A. K., Lo, D. C., & Katz, L. C. (1995). Neurotrophins regulate dendritic growth in developing visual cortex. *Neuron, 15*, 791–803.

Meijers, R., Puettmann-Holgado, R., Skiniotis, G., Liu, J. H., Walz, T., Wang, J. H., et al. (2007). Structural basis of Dscam isoform specificity. *Nature, 449*, 487–491.

Meltzer, S., Yadav, S., Lee, J., Soba, P., Younger, S. H., Jin, P., et al. (2016). Epidermis-derived Semaphorin promotes dendrite self-avoidance by regulating dendrite-substrate adhesion in Drosophila sensory neurons. *Neuron, 89*, 741–755.

Mikhaylova, M., Bera, S., Kobler, O., Frischknecht, R., & Kreutz, M. R. (2016). A dendritic Golgi satellite between ERGIC and Retromer. *Cell Reports*, *14*, 189–199.

Millard, S. S., & Pecot, M. Y. (2018). Strategies for assembling columns and layers in the Drosophila visual system. *Neural Development*, *13*, 11.

Miura, S. K., Martins, A., Zhang, K. X., Graveley, B. R., & Zipursky, S. L. (2013). Probabilistic splicing of Dscam1 establishes identity at the level of single neurons. *Cell*, *155*, 1166–1177.

Mizuno, H., Luo, W., Tarusawa, E., Saito, Y. M., Sato, T., Yoshimura, Y., et al. (2014). NMDAR-regulated dynamics of layer 4 neuronal dendrites during thalamocortical reorganization in neonates. *Neuron*, *82*, 365–379.

Molumby, M. J., Keeler, A. B., & Weiner, J. A. (2016). Homophilic protocadherin cell-cell interactions promote dendrite complexity. *Cell Reports*, *15*, 1037–1050.

Montague, P. R., & Friedlander, M. J. (1989). Expression of an intrinsic growth strategy by mammalian retinal neurons. *Proceedings of the National Academy of Sciences of the United States of America*, *86*, 7223–7227.

Mountoufaris, G., Chen, W. V., Hirabayashi, Y., O'Keeffe, S., Chevee, M., Nwakeze, C. L., et al. (2017). Multicluster Pcdh diversity is required for mouse olfactory neural circuit assembly. *Science*, *356*, 411–414.

Mumm, J. S., Williams, P. R., Godinho, L., Koerber, A., Pittman, A. J., Roeser, T., et al. (2006). In vivo imaging reveals dendritic targeting of laminated afferents by zebrafish retinal ganglion cells. *Neuron*, *52*, 609–621.

Nagel, J., Delandre, C., Zhang, Y., Förstner, F., Moore, A. W., & Tavosanis, G. (2012). Fascin controls neuronal class-specific dendrite arbor morphology. *Development*, *139*, 2999–3009.

Nakayama, A. Y., Harms, M. B., & Luo, L. (2000). Small GTPases Rac and rho in the maintenance of dendritic spines and branches in hippocampal pyramidal neurons. *The Journal of Neuroscience : The Official Journal of the Society for Neuroscience*, *20*, 5329–5338.

Nakazawa, S., Mizuno, H., & Iwasato, T. (2018). Differential dynamics of cortical neuron dendritic trees revealed by long-term in vivo imaging in neonates. *Nature Communications*, *9*, 3106.

Nguyen, M. M., McCracken, C. J., Milner, E. S., Goetschius, D. J., Weiner, A. T., Long, M. K., et al. (2014). Γ-tubulin controls neuronal microtubule polarity independently of Golgi outposts. *Molecular Biology of the Cell*, *25*, 2039–2050.

Nicoludis, J. M., Lau, S. Y., Scharfe, C. P., Marks, D. S., Weihofen, W. A., & Gaudet, R. (2015). Structure and sequence analyses of clustered protocadherins reveal antiparallel interactions that mediate homophilic specificity. *Structure*, *23*, 2087–2098.

O'Brien, B. M. J., Palumbos, S. D., Novakovic, M., Shang, X., Sundararajan, L., & Miller, D. M., 3rd. (2017). Separate transcriptionally regulated pathways specify distinct classes of sister dendrites in a nociceptive neuron. *Developmental Biology*, *432*, 248–257.

Ori-McKenney, K. M., Jan, L. Y., & Jan, Y. N. (2012). Golgi outposts shape dendrite morphology by functioning as sites of acentrosomal microtubule nucleation in neurons. *Neuron*, *76*, 921–930.

Ori-McKenney, K. M., McKenney, R. J., Huang, H. H., Li, T., Meltzer, S., Jan, L. Y., et al. (2016). Phosphorylation of β-tubulin by the down syndrome kinase, Minibrain/DYRK1a, regulates microtubule dynamics and dendrite morphogenesis. *Neuron*, *90*, 551–563.

Özel, M. N., Simon, F., Jafari, S., Holguera, I., Chen, Y. C., Benhra, N., et al. (2020). Neuronal diversity and convergence in a visual system developmental atlas. *Nature*.

Park, J., & Chung, K. C. (2013). New perspectives of Dyrk1A role in neurogenesis and neuropathologic features of down syndrome. *Experimental Neurobiology*, *22*, 244–248.

Peng, Y. R., James, R. E., Yan, W., Kay, J. N., Kolodkin, A. L., & Sanes, J. R. (2020). Binary fate choice between closely related interneuronal types is determined by a Fezf1-dependent postmitotic transcriptional switch. *Neuron*, *105*, 464–474.e6.

Peng, Y. R., Tran, N. M., Krishnaswamy, A., Kostadinov, D., Martersteck, E. M., & Sanes, J. R. (2017). Satb1 regulates Contactin 5 to pattern dendrites of a mammalian retinal ganglion cell. *Neuron, 95*, 869–883.e6.

Petersen, C. C. (2007). The functional organization of the barrel cortex. *Neuron, 56*, 339–355.

Petersen, J. D., Kaech, S., & Banker, G. (2014). Selective microtubule-based transport of dendritic membrane proteins arises in concert with axon specification. *The Journal of Neuroscience: the Official Journal of the Society for Neuroscience, 34*, 4135–4147.

Pierce, J. P., Mayer, T., & McCarthy, J. B. (2001). Evidence for a satellite secretory pathway in neuronal dendritic spines. *Current Biology, 11*, 351–355.

Prigge, C. L., & Kay, J. N. (2018). Dendrite morphogenesis from birth to adulthood. *Current Opinion in Neurobiology, 53*, 139–145.

Quach, T. T., Honnorat, J., Kolattukudy, P. E., Khanna, R., & Duchemin, A. M. (2015). CRMPs: Critical molecules for neurite morphogenesis and neuropsychiatric diseases. *Molecular Psychiatry, 20*, 1037–1045.

Radler, M. R., Suber, A., & Spiliotis, E. T. (2020). Spatial control of membrane traffic in neuronal dendrites. *Molecular and Cellular Neurosciences, 105*, 103492.

Ray, T. A., Roy, S., Kozlowski, C., Wang, J., Cafaro, J., Hulbert, S. W., et al. (2018). Formation of retinal direction-selective circuitry initiated by starburst amacrine cell homotypic contact. *eLife, 7*, e34241.

Raymond, G. V., Bauman, M. L., & Kemper, T. L. (1996). Hippocampus in autism: A Golgi analysis. *Acta Neuropathologica, 91*, 117–119.

Redmond, L., Kashani, A. H., & Ghosh, A. (2002). Calcium regulation of dendritic growth via CaM kinase IV and CREB-mediated transcription. *Neuron, 34*, 999–1010.

Reese, B. E., & Keeley, P. W. (2015). Design principles and developmental mechanisms underlying retinal mosaics. *Biological Reviews of the Cambridge Philosophical Society, 90*, 854–876.

Rheaume, B. A., Jereen, A., Bolisetty, M., Sajid, M. S., Yang, Y., Renna, K., et al. (2018). Single cell transcriptome profiling of retinal ganglion cells identifies cellular subtypes. *Nature Communications, 9*, 2759.

Richardson, C. E., Spilker, K. A., Cueva, J. G., Perrino, J., Goodman, M. B., & Shen, K. (2014). PTRN-1, a microtubule minus end-binding CAMSAP homolog, promotes microtubule function in Caenorhabditis elegans neurons. *eLife, 3*, e01498.

Rockhill, R. L., Euler, T., & Masland, R. H. (2000). Spatial order within but not between types of retinal neurons. *Proceedings of the National Academy of Sciences of the United States of America, 97*, 2303–2307.

Rosario, M., Schuster, S., Juttner, R., Parthasarathy, S., Tarabykin, V., & Birchmeier, W. (2012). Neocortical dendritic complexity is controlled during development by NOMA-GAP-dependent inhibition of Cdc42 and activation of cofilin. *Genes & Development, 26*, 1743–1757.

Rubinstein, R., Thu, C. A., Goodman, K. M., Wolcott, H. N., Bahna, F., Mannepalli, S., et al. (2015). Molecular logic of neuronal self-recognition through protocadherin domain interactions. *Cell, 163*, 629–642.

Salzberg, Y., Díaz-Balzac, C. A., Ramirez-Suarez, N. J., Attreed, M., Tecle, E., Desbois, M., et al. (2013). Skin-derived cues control arborization of sensory dendrites in Caenorhabditis elegans. *Cell, 155*, 308–320.

Sanes, D. H., & Hafidi, A. (1996). Glycinergic transmission regulates dendrite size in organotypic culture. *Journal of Neurobiology, 31*, 503–511.

Sanes, D. H., Markowitz, S., Bernstein, J., & Wardlow, J. (1992). The influence of inhibitory afferents on the development of postsynaptic dendritic arbors. *The Journal of Comparative Neurology, 321*, 637–644.

Sanes, J. R., & Masland, R. H. (2015). The types of retinal ganglion cells: Current status and implications for neuronal classification. *Annual Review of Neuroscience, 38*, 221–246.
Sanes, J. R., & Zipursky, S. L. (2010). Design principles of insect and vertebrate visual systems. *Neuron, 66*, 15–36.
Sanes, J. R., & Zipursky, S. L. (2020). Synaptic specificity, recognition molecules, and assembly of neural circuits. *Cell, 181*, 536–556.
Santiago, C., & Bashaw, G. J. (2017). Islet coordinately regulates motor axon guidance and dendrite targeting through the frazzled/DCC receptor. *Cell Reports, 18*, 1646–1659.
Sawaya, M. R., Wojtowicz, W. M., Andre, I., Qian, B., Wu, W., Baker, D., et al. (2008). A double S shape provides the structural basis for the extraordinary binding specificity of Dscam isoforms. *Cell, 134*, 1007–1018.
Schmucker, D., Clemens, J. C., Shu, H., Worby, C. A., Xiao, J., Muda, M., et al. (2000). Drosophila Dscam is an axon guidance receptor exhibiting extraordinary molecular diversity. *Cell, 101*, 671–684.
Schreiner, D., & Weiner, J. A. (2010). Combinatorial homophilic interaction between gamma-protocadherin multimers greatly expands the molecular diversity of cell adhesion. *Proceedings of the National Academy of Sciences of the United States of America, 107*, 14893–14898.
Schwabe, T., Borycz, J. A., Meinertzhagen, I. A., & Clandinin, T. R. (2014). Differential adhesion determines the organization of synaptic fascicles in the Drosophila visual system. *Current Biology, 24*, 1304–1313.
Sdrulla, A. D., & Linden, D. J. (2006). Dynamic imaging of cerebellar Purkinje cells reveals a population of filopodia which cross-link dendrites during early postnatal development. *Cerebellum, 5*, 105–115.
Shekhar, K., Lapan, S. W., Whitney, I. E., Tran, N. M., Macosko, E. Z., Kowalczyk, M., et al. (2016). Comprehensive classification of retinal bipolar neurons by single-cell transcriptomics. *Cell, 166*, 1308–1323.e30.
Shen, W., Da Silva, J. S., He, H., & Cline, H. T. (2009). Type a GABA-receptor-dependent synaptic transmission sculpts dendritic arbor structure in Xenopus tadpoles in vivo. *The Journal of Neuroscience : the Official Journal of the Society for Neuroscience, 29*, 5032–5043.
Shetty, A. S., Godbole, G., Maheshwari, U., Padmanabhan, H., Chaudhary, R., Muralidharan, B., et al. (2013). Lhx2 regulates a cortex-specific mechanism for barrel formation. *Proceedings of the National Academy of Sciences of the United States of America, 110*, E4913–E4921.
Sin, W. C., Haas, K., Ruthazer, E. S., & Cline, H. T. (2002). Dendrite growth increased by visual activity requires NMDA receptor and Rho GTPases. *Nature, 419*, 475–480.
Smith, C. J., O'Brien, T., Chatzigeorgiou, M., Spencer, W. C., Feingold-Link, E., Husson, S. J., et al. (2013). Sensory neuron fates are distinguished by a transcriptional switch that regulates dendrite branch stabilization. *Neuron, 79*, 266–280.
Smith, C. J., Watson, J. D., VanHoven, M. K., Colón-Ramos, D. A., & Miller, D. M. (2012). Netrin (UNC-6) mediates dendritic self-avoidance. *Nature Neuroscience, 15*, 731–737.
Soba, P., Zhu, S., Emoto, K., Younger, S., Yang, S. J., Yu, H. H., et al. (2007). Drosophila sensory neurons require Dscam for dendritic self-avoidance and proper dendritic field organization. *Neuron, 54*, 403–416.
Sorensen, S. A., & Rubel, E. W. (2006). The level and integrity of synaptic input regulates dendrite structure. *The Journal of Neuroscience : The Official Journal of the Society for Neuroscience, 26*, 1539–1550.
Sorensen, S. A., & Rubel, E. W. (2011). Relative input strength rapidly regulates dendritic structure of chick auditory brainstem neurons. *The Journal of Comparative Neurology, 519*, 2838–2851.

Stiess, M., Maghelli, N., Kapitein, L. C., Gomis-Ruth, S., Wilsch-Brauninger, M., Hoogenraad, C. C., et al. (2010). Axon extension occurs independently of centrosomal microtubule nucleation. *Science*, *327*, 704–707.

Stone, M. C., Roegiers, F., & Rolls, M. M. (2008). Microtubules have opposite orientation in axons and dendrites of Drosophila neurons. *Molecular Biology of the Cell*, *19*, 4122–4129.

Stürner, T., Tatarnikova, A., Mueller, J., Schaffran, B., Cuntz, H., Zhang, Y., et al. (2019). Transient localization of the Arp2/3 complex initiates neuronal dendrite branching in vivo. *Development*, *146*, dev171397.

Sugimura, K., Satoh, D., Estes, P., Crews, S., & Uemura, T. (2004). Development of morphological diversity of dendrites in Drosophila by the BTB-zinc finger protein abrupt. *Neuron*, *43*, 809–822.

Sugimura, K., Yamamoto, M., Niwa, R., Satoh, D., Goto, S., Taniguchi, M., et al. (2003). Distinct developmental modes and lesion-induced reactions of dendrites of two classes of Drosophila sensory neurons. *The Journal of Neuroscience : The Official Journal of the Society for Neuroscience*, *23*, 3752–3760.

Sulkowski, M. J., Iyer, S. C., Kurosawa, M. S., Iyer, E. P., & Cox, D. N. (2011). Turtle functions downstream of cut in differentially regulating class specific dendrite morphogenesis in Drosophila. *PLoS One*, *6*, e22611.

Sun, L. O., Jiang, Z., Rivlin-Etzion, M., Hand, R., Brady, C. M., Matsuoka, R. L., et al. (2013). On and off retinal circuit assembly by divergent molecular mechanisms. *Science*, *342*, 1241974.

Sundararajan, L., Smith, C. J., Watson, J. D., Millis, B. A., Tyska, M. J., & Miller, D. M., 3rd. (2019). Actin assembly and non-muscle myosin activity drive dendrite retraction in an UNC-6/netrin dependent self-avoidance response. *PLoS Genetics*, *15*, e1008228.

Suo, L., Lu, H., Ying, G., Capecchi, M. R., & Wu, Q. (2012). Protocadherin clusters and cell adhesion kinase regulate dendrite complexity through Rho GTPase. *Journal of Molecular Cell Biology*, *4*, 362–376.

Takeo, Y. H., Shuster, S. A., Jiang, L., Hu, M., Luginbuhl, D. J., Rülicke, T., et al. (2020). GluD2- and Cbln1-mediated competitive synaptogenesis shapes the dendritic Arbors of cerebellar Purkinje cells. *Neuron*, *109*. https://doi.org/10.1101/2020.06.14.151258.

Tan, L., Zhang, K. X., Pecot, M. Y., Nagarkar-Jaiswal, S., Lee, P. T., Takemura, S. Y., et al. (2015). Ig superfamily ligand and receptor pairs expressed in synaptic partners in Drosophila. *Cell*, *163*, 1756–1769.

Thu, C. A., Chen, W. V., Rubinstein, R., Chevee, M., Wolcott, H. N., Felsovalyi, K. O., et al. (2014). Single-cell identity generated by combinatorial homophilic interactions between alpha, beta, and gamma protocadherins. *Cell*, *158*, 1045–1059.

Tien, N. W., Soto, F., & Kerschensteiner, D. (2017). Homeostatic plasticity shapes cell-type-specific wiring in the retina. *Neuron*, *94*, 656–665.e4.

Ting, C. Y., McQueen, P. G., Pandya, N., Lin, T. Y., Yang, M., Reddy, O. V., et al. (2014). Photoreceptor-derived activin promotes dendritic termination and restricts the receptive fields of first-order interneurons in Drosophila. *Neuron*, *81*, 830–846.

Tolias, K. F., Bikoff, J. B., Burette, A., Paradis, S., Harrar, D., Tavazoie, S., et al. (2005). The Rac1-GEF Tiam1 couples the NMDA receptor to the activity-dependent development of dendritic arbors and spines. *Neuron*, *45*, 525–538.

van Bon, B. W., Coe, B. P., Bernier, R., Green, C., Gerdts, J., Witherspoon, K., et al. (2016). Disruptive de novo mutations of DYRK1A lead to a syndromic form of autism and ID. *Molecular Psychiatry*, *21*, 126–132.

Vaughn, J. E. (1989). Fine structure of synaptogenesis in the vertebrate central nervous system. *Synapse*, *3*, 255–285.

Visser, J. J., Cheng, Y., Perry, S. C., Chastain, A. B., Parsa, B., Masri, S. S., et al. (2015). An extracellular biochemical screen reveals that FLRTs and Unc5s mediate neuronal subtype recognition in the retina. *eLife*, *4*, e08149.

Wang, C. F., Hsing, H. W., Zhuang, Z. H., Wen, M. H., Chang, W. J., Briz, C. G., et al. (2017). Lhx2 expression in postmitotic cortical neurons initiates assembly of the thalamocortical somatosensory circuit. *Cell Reports*, *18*, 849–856.

Wang, Y., & Rubel, E. W. (2012). In vivo reversible regulation of dendritic patterning by afferent input in bipolar auditory neurons. *The Journal of neuroscience : the official journal of the Society for Neuroscience*, *32*, 11495–11504.

Wang, X., Zhang, M. W., Kim, J. H., Macara, A. M., Sterne, G., Yang, T., et al. (2015). The Krüppel-like factor Dar1 determines multipolar neuron morphology. *The Journal of Neuroscience : The Official Journal of the Society for Neuroscience*, *35*, 14251–14259.

Wassle, H., Peichl, L., & Boycott, B. B. (1981). Dendritic territories of cat retinal ganglion cells. *Nature*, *292*, 344–345.

Wayman, G. A., Impey, S., Marks, D., Saneyoshi, T., Grant, W. F., Derkach, V., et al. (2006). Activity-dependent dendritic arborization mediated by CaM-kinase I activation and enhanced CREB-dependent transcription of Wnt-2. *Neuron*, *50*, 897–909.

Withers, G. S., & Wallace, C. S. (2020). Transient lamellipodia predict sites of dendritic branch formation in hippocampal neurons. *Cell and Tissue Research*, *381*, 35–42.

Wojtowicz, W. M., Wu, W., Andre, I., Qian, B., Baker, D., & Zipursky, S. L. (2007). A vast repertoire of Dscam binding specificities arises from modular interactions of variable Ig domains. *Cell*, *130*, 1134–1145.

Wu, G. Y., & Cline, H. T. (1998). Stabilization of dendritic arbor structure in vivo by CaMKII. *Science*, *279*, 222–226.

Wu, Q., & Maniatis, T. (1999). A striking organization of a large family of human neural cadherin-like cell adhesion genes. *Cell*, *97*, 779–790.

Wu, G. Y., Zou, D. J., Rajan, I., & Cline, H. (1999). Dendritic dynamics in vivo change during neuronal maturation. *The Journal of Neuroscience : The Official Journal of the Society for Neuroscience*, *19*, 4472–4483.

Xu, C., Theisen, E., Maloney, R., Peng, J., Santiago, I., Yapp, C., et al. (2019). Control of synaptic specificity by establishing a relative preference for synaptic partners. *Neuron*, *103*, 865–877.e7.

Xu, S., Xiao, Q., Cosmanescu, F., Sergeeva, A. P., Yoo, J., Lin, Y., et al. (2018). Interactions between the Ig-superfamily proteins DIP-α and Dpr6/10 regulate assembly of neural circuits. *Neuron*, *100*, 1369–1384.e6.

Yalgin, C., Ebrahimi, S., Delandre, C., Yoong, L. F., Akimoto, S., Tran, H., et al. (2015). Centrosomin represses dendrite branching by orienting microtubule nucleation. *Nature Neuroscience*, *18*, 1437–1445.

Yamagata, M., & Sanes, J. R. (2008). Dscam and sidekick proteins direct lamina-specific synaptic connections in vertebrate retina. *Nature*, *451*, 465–469.

Yamagata, M., & Sanes, J. R. (2012). Expanding the Ig superfamily code for laminar specificity in retina: Expression and role of contactins. *The Journal of Neuroscience : The Official Journal of the Society for Neuroscience*, *32*, 14402–14414.

Yamagata, M., & Sanes, J. R. (2018). Expression and roles of the immunoglobulin superfamily recognition molecule Sidekick1 in mouse retina. *Frontiers in Molecular Neuroscience*, *11*, 485.

Yan, J., Chao, D. L., Toba, S., Koyasako, K., Yasunaga, T., Hirotsune, S., et al. (2013). Kinesin-1 regulates dendrite microtubule polarity in Caenorhabditis elegans. *eLife*, *2*, e00133.

Yan, W., Laboulaye, M. A., Tran, N. M., Whitney, I. E., Benhar, I., & Sanes, J. R. (2020). Mouse retinal cell atlas: Molecular identification of over sixty Amacrine cell types. *The Journal of Neuroscience : The Official Journal of the Society for Neuroscience*, *40*, 5177–5195.

Yang, S. Z., & Wildonger, J. (2020). Golgi outposts locally regulate microtubule orientation in neurons but are not required for the overall polarity of the dendritic cytoskeleton. *Genetics*, *215*, 435–447.

Yau, K. W., Schatzle, P., Tortosa, E., Pages, S., Holtmaat, A., Kapitein, L. C., et al. (2016). Dendrites in vitro and in vivo contain microtubules of opposite polarity and axon formation correlates with uniform plus-end-out microtubule orientation. *The Journal of Neuroscience : The Official Journal of the Society for Neuroscience*, *36*, 1071–1085.

Yau, K. W., van Beuningen, S. F., Cunha-Ferreira, I., Cloin, B. M., van Battum, E. Y., Will, L., et al. (2014). Microtubule minus-end binding protein CAMSAP2 controls axon specification and dendrite development. *Neuron*, *82*, 1058–1073.

Ye, B., Kim, J. H., Yang, L., McLachlan, I., Younger, S., Jan, L. Y., et al. (2011). Differential regulation of dendritic and axonal development by the novel Krüppel-like factor Dar1. *The Journal of Neuroscience : The Official Journal of the Society for Neuroscience*, *31*, 3309–3319.

Ye, B., Zhang, Y., Song, W., Younger, S. H., Jan, L. Y., & Jan, Y. N. (2007). Growing dendrites and axons differ in their reliance on the secretory pathway. *Cell*, *130*, 717–729.

Yip, Z. C., & Heiman, M. G. (2016). Duplication of a single neuron in C. elegans reveals a pathway for dendrite tiling by mutual repulsion. *Cell Reports*, *15*, 2109–2117.

Yip, Z. C., & Heiman, M. G. (2018). Ordered arrangement of dendrites within a C. elegans sensory nerve bundle. *eLife*, 7.

Yoong, L. F., Lim, H. K., Tran, H., Lackner, S., Zheng, Z., Hong, P., et al. (2020). Atypical myosin tunes dendrite arbor subdivision. *Neuron*, *106*, 452–467.e8.

Zhan, X. L., Clemens, J. C., Neves, G., Hattori, D., Flanagan, J. J., Hummel, T., et al. (2004). Analysis of Dscam diversity in regulating axon guidance in Drosophila mushroom bodies. *Neuron*, *43*, 673–686.

Zheng, Y., Wildonger, J., Ye, B., Zhang, Y., Kita, A., Younger, S. H., et al. (2008). Dynein is required for polarized dendritic transport and uniform microtubule orientation in axons. *Nature Cell Biology*, *10*, 1172–1180.

Zhou, Z., Hong, E. J., Cohen, S., Zhao, W. N., Ho, H. Y., Schmidt, L., et al. (2006). Brain-specific phosphorylation of MeCP2 regulates activity-dependent Bdnf transcription, dendritic growth, and spine maturation. *Neuron*, *52*, 255–269.

Zhu, H., Hummel, T., Clemens, J. C., Berdnik, D., Zipursky, S. L., & Luo, L. (2006). Dendritic patterning by Dscam and synaptic partner matching in the Drosophila antennal lobe. *Nature Neuroscience*, *9*, 349–355.

Zou, W., Dong, X., Broederdorf, T. R., Shen, A., Kramer, D. A., Shi, R., et al. (2018). A dendritic guidance receptor complex brings together distinct actin regulators to drive efficient F-actin assembly and branching. *Developmental Cell*, *45*, 362–375.e3.

Zou, W., Shen, A., Dong, X., Tugizova, M., Xiang, Y. K., & Shen, K. (2016). A multi-protein receptor-ligand complex underlies combinatorial dendrite guidance choices in C. elegans. *eLife*, *5*, e18345.

CHAPTER SEVEN

Wiring subcortical image-forming centers: Topography, laminar targeting, and map alignment

Kristy O. Johnson[a,b] **and Jason W. Triplett**[a,c,*]
[a]Center for Neuroscience Research, Children's National Research Institute, Washington, DC, United States
[b]Institute for Biomedical Sciences, The George Washington University School of Medicine, Washington, DC, United States
[c]Department of Pediatrics, The George Washington University School of Medicine, Washington, DC, United States
*Corresponding author: e-mail address: jtriplett@childrensnational.org

Contents

Abstract

Efficient sensory processing is a complex and important function for species survival. As such, sensory circuits are highly organized to facilitate rapid detection of salient stimuli and initiate motor responses. For decades, the retina's projections to image-forming centers have served as useful models to elucidate the mechanisms by which such exquisite circuitry is wired. In this chapter, we review the roles of molecular cues, neuronal activity, and axon-axon competition in the development of topographically ordered retinal ganglion cell (RGC) projections to the superior colliculus (SC) and dorsal lateral

Current Topics in Developmental Biology, Volume 142
ISSN 0070-2153
https://doi.org/10.1016/bs.ctdb.2020.10.004

geniculate nucleus (dLGN). Further, we discuss our current state of understanding regarding the laminar-specific targeting of subclasses of RGCs in the SC and its homolog, the optic tectum (OT). Finally, we cover recent studies examining the alignment of projections from primary visual cortex with RGCs that monitor the same region of space in the SC.

1. Introduction

Neurons in the visual system are connected in well-ordered arrays, allowing for efficient processing, modulation of information flow, and guiding behavioral responses. As such, this exquisite organization has served as an ideal model to understand the mechanisms underlying circuit development, the principles of which have been repeated in multiple brain regions and systems. Specifically, the pattern of connectivity between retinal ganglion cells (RGCs) and their image-forming targets, the superior colliculus (SC) and dorsal lateral geniculate nucleus (dLGN), have provided particularly rich insights. In these regions, RGC projections are organized topographically, such that the terminals of inputs reflect the order of their cell bodies, thereby retaining spatial information of the visual scene. In addition, inputs are organized laminarly, reflecting both eye-specific origins and functional subtypes. Finally, in both the SC and dLGN, feedback connections from the primary visual cortex (V1) are organized in alignment with RGC terminals monitoring the same region of visual space. The mechanisms underlying the establishment of eye-specific layers, especially in the dLGN and V1, are relatively well understood and have been covered elsewhere (Hensch & Quinlan, 2018; Murcia-Belmonte & Erskine, 2019). Thus, in this review, we will focus on recent advances in our understanding of the mechanisms of retinofugal topographic map formation, the establishment of functional laminar targeting, and the alignment of convergent V1 inputs with RGCs.

2. Mature organization of inputs in the SC and dLGN

As subcortical image-forming areas, both the SC and dLGN are organized topographically, such that the neighbor-neighbor relationships in the retina are maintained by their axon terminals in the target (Figs. 1A and 2A). RGCs project from a retina that can be divided along two Cartesian axes: temporal-nasal (T-N) and dorsal-ventral (D-V), which correspond to the azimuth (horizontal) and elevation (vertical) axes of the visual scene, respectively. The SC can similarly be divided such that the T-N and D-V retinal axes are mapped along the anterior-posterior (A-P) and lateral-medial (L-M)

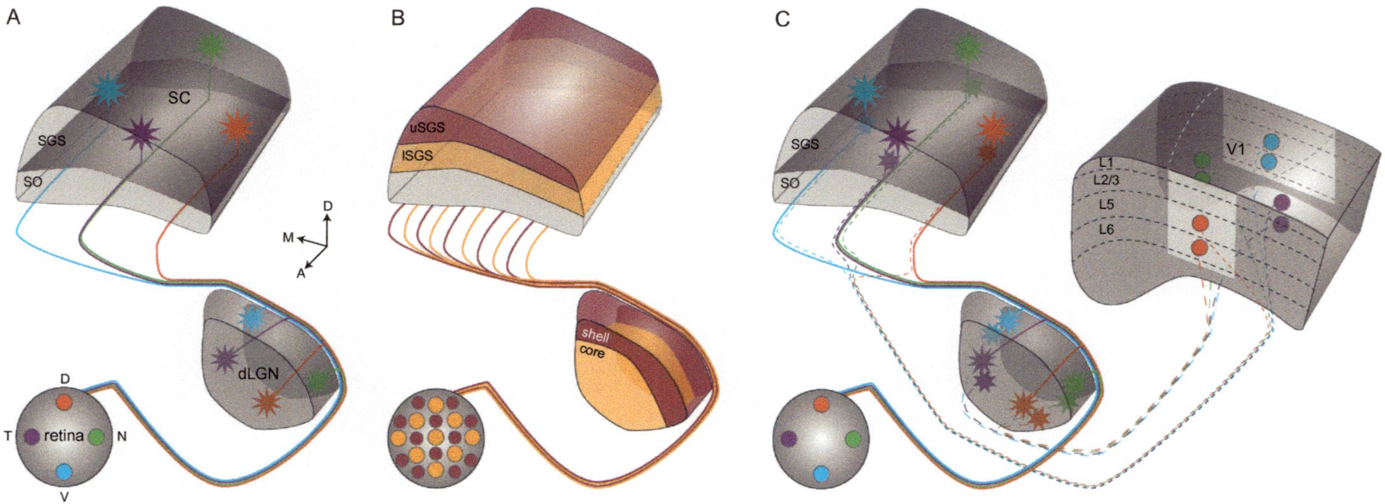

Fig. 1 Organization of subcortical image-forming circuits. Retinal ganglion cells (RGCs) send axons along the optic tract and terminate in multiple retinorecipient regions. In image-forming centers, the dorsal lateral geniculate nucleus (dLGN) and superior colliculus (SC), RGC terminations are organized topographically. (A) RGCs axons course along the optic tract on the dorsolateral side of the dLGN. Those terminating in the dLGN innervate into topographically appropriate regions, indicated by correlating colors. As RGC axons pass the dLGN, they begin to turn posteriorly and innervate the SC at in the stratum opticum (SO), ventral to their terminal located in the stratum griseum superficiale (SGS). (B) In addition to topography, RGCs are organized into parallel circuits, such that distinct subclasses of RGCs terminate in different sublaminae of the dLGN and SC. In each region, two broad termination strata have been identified: the core and shell in the dLGN, as well as the upper and lower SGS (uSGS, lSGS) in the SC. For simplicity, only two representative RGC subclasses are depicted here in yellow and maroon. (C) In addition to RGCs, both the dLGN and SC receive inputs from primary visual cortex (V1). The dLGN is primarily innervated by neurons from layer 6 (L6), while the SC receives L5 inputs. In each case, V1 projections terminate in topographic alignment with RGCs monitoring the same region of visual space, as depicted by correlating colors. *D, dorsal; N, nasal; V, ventral; T, temporal; A, anterior; P, posterior.*

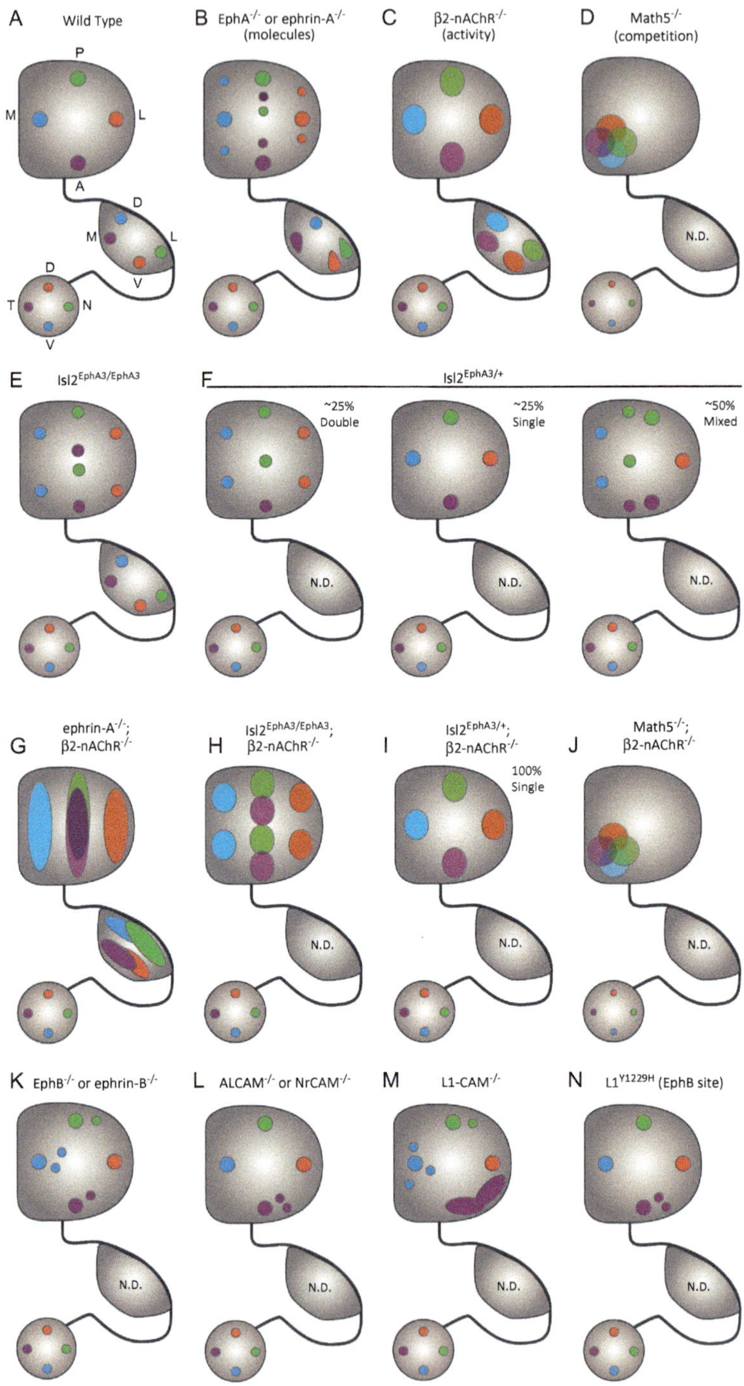

Fig. 2 See figure legend on opposite page.

axes of the SC, respectively. Thus, a two-dimensional visual scene is projected intact onto the SC, excepting the notable observation of a slight magnification of central visual space, corresponding to the anterior SC (Drager & Hubel, 1975). The dLGN is a more condensed structure than the SC, and thus the retinal axes are not as cleanly projected along two axes. Roughly, the T-N retinal axis project along the posterodorsomedial-anteroventrolateral dLGN, while the D-V retinal axis is found along the posteroventral-anterodorsal axis.

The SC and dLGN are comprised of sublaminae to which distinct subclasses of RGCs project their axons (Fig. 1B). The SC can be divided into multiple layers and sublayers along the D-V axis, where superficial layers process visual inputs and intermediate and lower layers mediate somatosenory, auditory, and motor functions (Ito & Feldheim, 2018).

Fig. 2—Cont'd Summary of genetic manipulations that result in topographic mapping errors. (A) In wild type mice, the temporal-nasal (T-N) axis of the retina projects along the lateral-medial (L-M) axis of the dLGN and the anterior-posterior (A-P) axis of the SC, while the dorsal-ventral (D-V) axis of the retina projects along the V-D axis of the dLGN and the L-M axis of the SC. (B) In mice lacking ephrin-A or EphA expression in both the retina and SC, ectopic termination zones (TZs) are observed along the A-P axis of the SC, while subtle broadening and shifts of TZ location are observed in the dLGN. (B) In mice lacking the β2 subunit of the nicotinic acetylcholine receptor (β2-nAChR), Stage II retinal waves are disrupted and TZs in both dLGN and SC are roughly in the appropriate topographic zone, but poorly refined. (D) In Math5 knockout mice, only ~5–20% of the normal number of RGCs are present, reducing axonal competition. As a result, RGC terminals are clustered in the anteromedial portion of the SC and topography is disrupted. Effects on organization in the dLGN have not been determined (N.D.) in these mutants. (E) In homozygous Islet2-EphA3 knock-in mice ($Isl2^{EphA3/EphA3}$), the retina's projection is duplicated along the A-P axis of the SC, but no change is observed in the dLGN. (F) In heterozygous $Isl2^{EphA3/+}$ mice, three different map organizations are observed: double (left), single (middle), and mixed (right). (G) In combination ephrin-A and β2-nAChR knockout mice, topographic order along the azimuth axes in both the dLGN and SC is completely lost. (H) When Stage II waves are disrupted in $Isl2^{EphA3/EphA3}$ mice, the retinocollicular projection is duplicated along the A-P axis of the SC, but TZs are broader. (I) When Stage II waves are disrupted in $Isl2^{EphA3/+}$ mice, heterogeneous map organizations are no longer observed. Instead, all maps appear single, though TZs are enlarged. (J) In combination Math5 and β2-nAChR knockout mice, RGCs remain clustered in the anteromedial SC. (K) Mutations in EphB or ephrin-B family members result in ectopic termination zones along the L-M axis of the SC. (L) Similarly, mutations in ALCAM and NrCAM drive disruptions along the L-M axis, though these tend to be concentrated in the anterior domain. (M) Deletion of L1-CAM results in dramatic disruptions in topography not restricted to either the A-P or L-M axis, with particularly large errors observed in anterolateral SC. (N) Deletion of a single EphB target site on L1-CAM results in errors only along the L-M axis, suggesting these molecules are in a linear pathway.

The superficial-most layer, the stratum griseum superficiale (SGS), receives retinal input and can be divided into upper and lower regions (uSGS, lSGS). Studies utilizing genetically-labeled RGCs revealed subtypes that terminate in each of these respective sublayers (Huberman et al., 2008, 2009; Kim, Zhang, Meister, & Sanes, 2010; Kim, Zhang, Yamagata, Meister, & Sanes, 2008; Siegert et al., 2009). Further, stochastic labeling of individual RGCs identified five "clusters" of RGC subclasses based on their terminal lamination pattern in the SC, suggesting that the uSGS and lSGS may be further subdivided (Hong, Kim, & Sanes, 2011). However, the precise localization of sublayers and their boundaries remains poorly defined in the mouse SC, a notable distinction from the zebrafish optic tectum (OT), where 10 synaptic laminae have been identified (Baier, 2013; Robles, Filosa, & Baier, 2013). Similarly, the sublayers targeted by RGC subclasses in the mouse dLGN are poorly defined compared to those in the ferret, cat, and monkey, where the high degree of stratification is apparent cytoarchitecturally. Thus far the mouse dLGN has been divided into "core" and "shell" regions, which receive inputs from distinct RGC subclasses and exhibit different functional properties (Seabrook, Burbridge, Crair, & Huberman, 2017).

In addition to RGCs, both the SC and dLGN receive feedback connections from visual cortex that play critical roles in both visual processing and execution of innate behaviors (Liang et al., 2015; Wang, Sarnaik, Rangarajan, Liu, & Cang, 2010). Neurons in Layer 5 (L5) of multiple visual cortical regions project to the SC (Wang & Burkhalter, 2013). Intriguingly, inputs from V1 and lateromedial area terminate in sublayers that overlap with retinal inputs (Fig. 1C), while those from other visual cortical areas terminate in more ventral (deep) sublayers. Importantly, the V1-SC projection is topographically organized and in register with the retinocollicular map. Similarly, the dLGN receives inputs from both L5 and L6 of V1, both of which are organized topographically and in alignment with retinal inputs and play critical roles in the development of proper connectivity in the thalamus (Hasse & Briggs, 2017).

3. Developmental timing of subcortical visual circuit wiring

The connectivity patterns of RGCs and V1 axons in the SC and dLGN are established in sequential, but overlapping, developmental windows. While we will discuss informative studies exploring the establishment of retinotectal connections in the fish, frogs, and chickens, we will focus on the mouse as a model organism to illustrate the timing of development. In both the dLGN and SC of mice, the bulk of retinal and cortical connections are established prior to eye opening (postnatal day 12–14 [P12–14]), though important experience-dependent processes are required for the maturation and maintenance of these connections (Hooks & Chen, 2020).

In the SC, RGC axons are the first to arrive, beginning as early as P0 and some precocious fibers appear embryonically (Triplett, 2014). Regardless of retinal origin location, the growth cones of all RGCs project to the posterior-most region of the SC. Following this, interstitial branches are dynamically formed and pruned at roughly topographically appropriate locations (Yates, Roskies, McLaughlin, & O'Leary, 2001). Over the first postnatal week, these branches eventually coalesce into a tightly packed terminal arbor, that is, positioned in the proper topographic location and within the appropriate sublamina of the SC. As the final refinement of retinocollicular axons is occurring, L5-V1 projections begin to innervate the SC, beginning at P6 (Triplett et al., 2009). Akin to RGCs, these axons go through a protracted period of branching and pruning, establishing adult-like topographic order and alignment with the retinal map by P12, just prior to eye opening. Importantly, development of L5-V1 connectivity is not entirely complete at this stage, as eye opening triggers a critical consolidation of synaptic inputs (Phillips et al., 2011).

Similar to the SC, the dLGN is first innervated by RGCs, which refine to a mature-like state by P8/9, and later by projections from V1, which refine during the second postnatal week. In contrast to the SC, as RGCs project into the dLGN, they do not appear to overshoot their future termination zone, but rather target it almost directly with minimal interstitial branching (Dhande et al., 2011). Intriguingly, while L6-V1 projections to the thalamus are observed at the ventromedial border of the dLGN as early as P2, they do not begin to innervate the dLGN until later (Seabrook, El-Danaf, Krahe, Fox, & Guido, 2013). This delay appears to be directed by RGCs, as L6-V1 axons precociously invade the dLGN in the absence of retinal input (Seabrook et al., 2013). In a peculiar twist to this mechanism, cortical input is required for proper RGC targeting to the dLGN (Shanks et al., 2016), suggesting highly orchestrated crosstalk between innervating axonal populations.

4. Mechanisms of retinofugal topographic map assembly

In the decades since Sperry's initial experiments examining the regrowth of RGC axons into the frog tectum (Sperry, 1943, 1947, 1963), the field has made substantial advances towards understanding the intricacies of the forces driving the formation of topographic maps of space. In addition to the molecular cues that Sperry postulated in his chemoaffinity hypothesis, clear evidence of a role for spontaneous waves of retinal activity has emerged in the past two decades. Further, inter-axon interactions, such as competition, play a role in mapping. Finally, the concept that these forces do not operate in a vacuum, but rather interact and influence one another, has recently been investigated.

4.1 Molecular cues

In his chemoaffinity hypothesis, Sperry postulated that specific "chemical tags" expressed by RGCs and neurons in the OT served to ensure proper connectivity (Sperry, 1963). His hypothesis was refined to postulate that gradients of molecules might be expressed along the cardinal axes of the retina and OT. As molecular and genetic tools were developed in subsequent decades, evidence in support of a mechanism akin to Sperry's truly insightful conclusion from his regeneration experiments has become "textbook" material.

4.1.1 Relative levels of repellant ephrin-A/EphA signals shape azimuth axis topography

The best demonstration of the influence of molecular gradients is in the mapping of the T-N axis of the retina onto the A-P axis of the SC/OT. There, counter-gradients of EphA receptor tyrosine kinases and their ligands, ephrin-As are found. The function of EphAs and ephrin-As has been reviewed in detail previously (Feldheim & O'Leary, 2010; Triplett & Feldheim, 2012), so we will briefly describe it here.

The clearest action of ephrin-A ligands expressed in the SC/OT is their role as repellant molecules. Elegant studies by the Boenhoffer lab leveraging the in vitro stripe assay demonstrated that temporal RGCs avoid posterior tectal membranes, an action mediated by a glycophosphoinosital (GPI)-linked protein (Walter, Henke-Fahle, & Bonhoeffer, 1987; Walter, Kern-Veits, Huf, Stolze, & Bonhoeffer, 1987). Subsequent work by many groups identified a family of GPI-linked molecules, later named ephrin-As, that are sufficient to replicate RGC avoidance results in the stripe assay and are expressed in high posterior to low anterior gradients in the SC/OT (Drescher et al., 1995). Intriguingly, these molecules are bound by a family of orphan receptor tyrosine kinases (Cheng & Flanagan, 1994;

Cheng, Nakamoto, Bergemann, & Flanagan, 1995), later named EphAs, that are expressed in high temporal to low nasal gradients in the retina.

The expression patterns of ephrin-As in the SC/tectum and EphAs in the retina are consistent with a mass-action repellent model that could drive topographic mapping. That is, RGCs from the temporal retina express high levels of EphA, are most repelled by the ephrin-A gradient, and terminate anteriorly. As RGCs progressively nasal in position express less EphA, they are less repelled by the ephrin-A gradient and terminate progressively posteriorly. Experimental evidence in support of this model came via both gain- and loss-of-function experiments. Indeed, global deletion of the ephrin-As expressed in the SC/OT (ephrin-A2, ephrin-A3, and ephrin-A5 in mouse) alone or in combination leads to topographic guidance errors of RGCs projecting to the SC and dLGN, as assessed by anatomical tracing (Fig. 2B) (Feldheim et al., 2000, 1998; Frisén et al., 1998; Pfeiffenberger, Yamada, & Feldheim, 2006). Further, deletion of EphA5, expressed in a gradient in the retina, phenocopies ephrin-A knockout experiments (Feldheim et al., 2004). Over-expression of ephrin-A3 sporadically in the tectum results in RGCs avoiding these areas, suggesting sufficiency for repellant action (Nakamoto et al., 1996).

Further studies suggest that the sum of ephrin-A/EphA signaling, rather than the action of specific molecules, results in topographic order. In Islet2-EphA3 knock-in mice ($Isl2^{EA3}$), the EphA3 receptor is expressed by $Isl2^{+}$ RGCs, which represent ~40–50% of all RGCs and are scattered across the retina in a salt-and-pepper fashion (Brown et al., 2000). Amazingly, in homozygous $Isl2^{EA3/EA3}$ mice, the retinocollicular map is duplicated both anatomically and physiologically along the azimuth axis (Fig. 2E) (Brown et al., 2000; Triplett et al., 2009). Since EphA3 is not normally expressed by RGCs, these findings suggest that the relative level of ephrin-A/EphA signaling shapes topography. Later computational modeling and combination knockout mouse models support this relative signaling mechanism (Bevins, Lemke, & Reber, 2011; Reber, Burrola, & Lemke, 2004). However, recent findings suggest a complicated interplay between ephrin-A/EphAs and other forces that was revealed in heterozygous $Isl2^{EA3/+}$ mice (Owens, Feldheim, Stryker, & Triplett, 2015), which we will discuss in detail below.

4.1.2 Mapping nasal axons: Reverse signaling and dual functionality

The ephrin-A/EphA repellant model of retinofugal topographic mapping raises a potential paradox: how are nasal neurons guided to terminate in the most repellant regions of the gradient? Three putative mechanisms to achieve innervation in ephrin-A rich areas have been put forth: (1) inter-axon competition for synaptic territory, (2) EphA-mediated repulsion, and (3) ephrin-A attraction. Here, we will focus on the latter two, as we will discuss competition in more detail later. In addition, we will discuss how recent findings affect our interpretation of these models and, potentially, uncover a novel ephrin-A/EphA-mediated mechanism.

The hypothesis that EphAs may function as repulsive cues in the guidance of RGC axons stems from their complementary expression patterns in both the retina and target structures and their ability to initiate signaling in the "reverse" direction (Triplett & Feldheim, 2012). Signaling through the ephrin-A/EphA complex is propagated in two directions: "forward" into the EphA-bearing cell and "reverse" into the ephrin-A-bearing cell. Support for activation of reverse signaling in retinofugal map formation stems from both loss- and gain-of-function experiments. Deletion of EphA7, which is expressed in a high anterior to low posterior gradient in the SC (but not in the retina), results in topographic guidance errors, suggesting an active role for collicular EphAs (Rashid et al., 2005). Further, guidance errors induced by over-expression of a constitutively-active EphA6 in RGCs are prevented by co-expression of ephrin-A5, suggesting reverse signaling may provide at minimum a counterbalancing force (Carreres et al., 2011). Curiously, ephrin-As are membrane attached and do not have an intracellular domain, so the mechanism by which reverse signaling occurs likely involves co-receptors. The neurotrophin receptors p75 and TrkB can interact with ephrin-As and have been suggested to act as co-receptors that mediate reverse signaling (Lim et al., 2008; Marler, Poopalasundaram, Broom, Wentzel, & Drescher, 2010; Poopalasundaram, Marler, & Drescher, 2011). Further, each is capable of modulating EphA/ephrin-A mediated guidance events such as terminal positioning in the posterior SC or interstitial branching. However, effects on overall topography are not as robust as global ephrin-A/EphA knockouts, raising the possibility that further co-receptors remain to be identified.

In other systems, ephrin-A/EphA interactions drive cell adhesion and axon tracking, raising the possibility that similar functions could be elicited in the formation of topographic maps by RGCs. Indeed, explant assays suggest that exogenous ephrin-A has differential effects on axon outgrowth depending on the location of explant origin (Hansen, Dallal, & Flanagan, 2004). As expected, high levels of ephrin-A inhibits outgrowth regardless of location, but moderate levels have a surprisingly growth-promoting effect on explants from nasal retina. Thus, under the right conditions in vivo nasal axons could be attracted, or at least exhibit increased growth, in the presence of high levels of ephrin-A. Alternatively, recent studies of growth cone responses of lateral motor column neurons suggest that ephrin-A5 enhances attraction to netrin-1, despite its known role as a repulsive cue (Croteau, Kao, & Kania, 2019; Poliak et al., 2015). These results raise the intriguing possibility that in the context of topographic mapping, ephrin-As may facilitate attraction via another as yet unidentified molecular cue.

The dissection of how ephrin-A/EphA interactions between RGCs and SC neurons was recently extended using a conditional knockout approach, which yielded surprising results that add a new understanding to how these molecules shape topography (Suetterlin & Drescher, 2014). In contrast to expectations derived from the EphA repellant model, deletion of ephrin-A5 specifically from RGCs does not result in dramatic disruptions in topography, particularly of nasal axons. Intriguingly, SC-specific deletion of ephrin-A5 does not alter the projection pattern of RGCs originating from temporal or central retina. However, when ephrin-A5 was conditionally-deleted from both retina and SC, topographic mapping errors were observed in all RGC populations, similar to global knockouts. These data suggest that axon-axon interactions mediated via ephrin-A5 play a critical role in mapping. Indeed, in vitro assays suggest that nasal axons (which express high levels of ephrin-As) can repel temporal RGCs. This target-independent function of ephrin-A5 represents a novel mechanism of action in the retinofugal system, though similar roles have been reported in other systems. Additionally, this study found that SC-specific deletion of ephrin-A5 results in topographic mapping errors by nasal RGCs, which may support an attractive role for ephrin-As on this population.

4.1.3 Diverse signaling pathways shape topography of the elevation axis

While ephrin-A/EphA signaling is a clear molecular driver of topographic order along the azimuth axis, a more diverse set of molecular pathways appears to influence mapping along the elevation axis. Interestingly, much of the topographic order along this axis, especially in the retina's projection to the SC/OT, is established prior to entering the target (Plas, Lopez, & Crair, 2005; Simon & O'Leary, 1991). Thus, many of the molecular regulators of mapping regulate the extent of interstitial branching along either side of the RGC axon, which ultimately dictates topographic order.

To begin, akin to the graded expression of EphAs/ephrin-As along the T-N and A-P axes, gradients of molecular cues have been described along the orthogonal axes. Indeed, the B-class of Ephs are expressed in a high ventral to low dorsal gradient in the retina, while their ligands, the ephrin-Bs are expressed in a high medial to low lateral gradient in the SC/OT (Braisted et al., 1997; Higenell, Han, Feldheim, Scalia, & Ruthazer, 2012; Holash et al., 1997). In contrast to the azimuth axis, these molecules primarily exert an attractive force, which shapes topography appropriately. In support of this, deletion of EphBs results in ectopic terminations lateral to the topographically appropriate region (Hindges, McLaughlin, Genoud, Henkemeyer, & O'Leary, 2002) (Fig. 2K). However, it should be noted that ectopic expression of ephrin-B1 appears to inhibit terminal stabilization (Mclaughlin, Hindges, Yates, & O'Leary, 2003; Thakar, Chenaux, & Henkemeyer, 2011), raising the possibility that these molecules may signal in a more complex manner, akin to their A-class family members.

Similar to EphBs/ephrin-Bs, molecules in the Wnt signaling pathway are expressed in gradients along the elevation axes and play a role in topographic mapping (Schmitt et al., 2005). Indeed, Wnt3 is expressed in a high medial to low lateral gradient in the SC, while one of its receptors, Ryk, is expressed in a high ventral to low dorsal gradient in the retina. Intriguingly, explant assays suggest that Wnt3 promotes the outgrowth of dorsal RGCs, while inhibiting and even repelling ventral RGCs. Thus, these signals may serve to counter-balance attractive/repulsive cues generated by ephrin-B/EphB signals. However, whether these signals converge on the same intracellular pathways remains unclear.

Finally, members of the immunoglobulin superfamily of cell adhesion molecules (CAMs) play critical roles in topographic mapping (Fig. 2L–N). First, anatomical tracing experiments in global knockout mice indicate that the L1 CAM is required for multiple aspects of RGC targeting, including sorting at the optic chiasm (Sretevan, Feng, Puré, & Reichardt, 1994), innervation of the SC, and topographic mapping (Demyanenko & Maness, 2003). Topographic targeting errors are observed along both the A-P and L-M axes of the SC, consistent with the uniform expression of L1 in both the retina and SC (Buhusi, Schlatter, Demyanenko, Thresher, & Maness, 2008; Demyanenko & Maness, 2003). Interestingly, similar experiments in knock-in mice harboring a targeted mutation of a tyrosine residue phosphorylated by EphB ($L1^{Y1229H}$) reveal topographic mapping errors specifically along the L-M axis of the SC, suggesting that L1 is a critical mediator of ephrin-B/EphB mediated guidance (Dai et al., 2012). Next, a related molecule, activated leukocyte CAM (ALCAM), is a binding partner of L1 expressed in overlapping patterns in the retinocollicular pathway.

Deletion of ALCAM results in topographic mapping errors that phenocopy those observed in $L1^{Y1229H}$ mice, suggesting that ALCAM may be critical for the EphB-mediated signaling via L1 (Buhusi et al., 2009). Lastly, neuron-glia related CAM (NrCAM) is a member of the L1 subfamily of CAMs, which is expressed in a similar pattern to L1 and ALCAM. Consistent with this, knockouts of NrCAM show similar L-M errors in topography (Dai et al., 2013). Intriguingly, EphB can phosphorylate NrCAM, which inhibits its ability to recruit ankyrin to the cell membrane, akin to its action on L1. Together, these data suggest that these related CAMs may signal in a complex downstream of EphBs to regulate topographic mapping.

4.2 Spontaneous activity

While molecular cues clearly play important roles in the establishment of retinofugal topography, experimental evidence suggests that they represent only part of the story. For instance, in the absence of any of the cues described above, the most common phenotype observed are ectopic TZs rather than a complete loss of organization. The coalescence of aberrant projections into clustered TZs is reminiscent of the organization of eye-specific inputs in the SC and dLGN, which is mediated in large part by spontaneous activity in the form of retinal waves. Indeed, many groups have investigated the role of retinal waves in the establishment of topography and have built a strong case in favor of them playing an instructive role in circuit formation (Ackman & Crair, 2014; Feller, 2009).

4.2.1 Stage II cholinergic waves instruct retinofugal topography

During development of the mammalian visual system, spontaneous neuronal activity propagates across the retina in a wave-like manner, such that neighboring clusters of RGCs have highly correlated patterns of activity. These waves progress through three developmental stages based on the mode of propagation: Stage I retinal waves occur during late gestation in rodents and are mediated through gap junctions; Stage II retinal waves are orchestrated by cholinergic starburst amacrine cells and occur from ~P0–10 in mice; and, Stage III waves are mediated via glutamatergic signaling driven by bipolar cells and modulated by amacrines and predominate from ~P10–15. A preponderance of evidence supports Stage II waves as the critical drivers of retinotopy, though Stage III waves are capable of compensating to some degree (Chandrasekaran, Plas, Gonzalez, & Crair, 2005).

Pharmacological manipulations confirmed that Stage II waves are mediated by the nicotinic acetylcholine receptor (nAChR), specifically those containing the β2 subunit (Feller, Wellis, Stellwagen, Werblin, & Shatz, 1996; Wong, Myhr, Miller, & Wong, 2000). Indeed, the pattern of spontaneous activity during the first postnatal week is dramatically altered, though not absent, in β2-nAChR$^{-/-}$ mice (Bansal et al., 2000; Burbridge et al., 2014; McLaughlin, Torborg, Feller, & O'Leary, 2003; Stafford, Sher, Litke, & Feldheim, 2009; Sun, Warland, Ballesteros, van der List, & Chalupa, 2008). As a result, RGC projections to the SC and dLGN are dramatically disrupted, failing to refine to condensed TZs (Fig. 2C) (Chandrasekaran et al., 2005; Grubb, Rossi, Changeux, & Thompson, 2003; McLaughlin, Torborg, et al., 2003). While more diffuse, TZs in both the SC and dLGN are centered in approximately the right topographic zone, reflecting the influence of molecular forces. As predicted, anatomical disruptions result in altered functional receptive fields of neurons in the SC, which are enlarged and elongated along the azimuth axis (Chandrasekaran et al., 2005; Mrsic-Flogel et al., 2005). Intriguingly, Stage II waves preferentially travel along the azimuth axis both in vitro (Stafford et al., 2009) and in vivo (Ackman, Burbridge, & Crair, 2012).

These functional findings support an instructive role for Stage II waves in circuit development, as do multiple other lines of evidence. First, the level of RGC activity in β2-nAChR$^{-/-}$ mice is similar to that of control mice, arguing against a simply permissive role for Stage II waves. Second, conditional deletion of β2-nAChR specifically from the retina, but not the midbrain, results in alterations to wave patterns and topographic order similar to those observed in global knockouts (Burbridge et al., 2014). Finally, using the tet-Off inducible expression system, Xu and colleagues were able to overexpress β2-nAChR specifically in RGCs in the background of a global knockout (Xu et al., 2011). This manipulation restored correlated activity in the developing retina, though waves were more spatially restricted. Intriguingly, topographic order was rescued in both the dLGN and SC for RGCs originating from the dorsonasal two thirds of the retina, but not the ventrotemporal third. As this region overlaps with ipsilaterally-projecting RGCs, the authors suspected an interaction and found that topography was rescued regardless of retinal origin location when binocular competition was eliminated via enucleation.

4.2.2 *Exogenously manipulated activity can shape retinofugal topography*

In order to further assess the instructive role activity plays in establishing topography, recent studies have leveraged techniques to manipulate the pattern of activity during development. In one study, the light-activatable channel rhodopsin (ChR2) was expressed in ~20% of RGCs under the control of the Thy1 promoter (Thy1-ChR2), allowing for stimulation via focused light exposure to the eyes (Zhang, Ackman, Xu, & Crair, 2011). A critical finding from these experiments was that synchronous activation of both eyes from ~P9–12 reduced, while asynchronous activation enhanced, eye-specific segregation in the SC and dLGN. Further, optogenetic activation of RGCs resulted in smaller TZs and partially rescued topographic deficits in β2-nAChR$^{-/-}$ mice.

Using a similar strategy, Munz and colleagues demonstrated that exogenously induced correlated firing stabilized RGC axonal branching in the OT, resulting in more compact terminal arbors and more efficient relay to tectal neurons (Munz et al., 2014). In this study, the authors took advantage of the observation that occasionally, a single RGC will project to the ipsilateral OT in *Xenopus laevis* tadpoles. By rearing tadpoles in varying visual environments or expressing ChR2 in these projections, they could be forced to fire synchronously or asynchronously with neighboring RGC fibers from the opposing eye. Intriguingly, asynchronous activation increased arbor dynamics whether the period of stimulation followed an epoch of darkness or synchronous stimulation, while synchronous activation only resulted in modest increases in dynamics when following a dark epoch.

Finally, manipulations of the pattern of visual experience can alter the topographic organization of RGC inputs to the OT in tadpoles. In normally-reared tadpoles, the predominant direction of optic flow is in the A-P direction as the tadpole swims forward. In a clever experiment, Hiramoto and Cline reared tadpoles in visual environments in which they were exposed to a natural stimulus (alternating white and black bars moving A to P) or the opposite (Hiramoto & Cline, 2014). Amazingly, 4 days of exposure to the natural direction of optic flow sharpened the topographic distribution of RGCs in the OT, while this sharpening did not occur in animals reared in the opposite environment nor in those dark-reared.

4.2.3 NMDARs play a critical role in activity-dependent map formation

The detection of coincident neuronal firing driven by spatially-correlated Stage II waves is critical for proper topographic map refinement, and several studies suggest that *N*-methyl-D-aspartate receptors (NMDARs) mediate this process. NMDARs are coincidence detectors that require both a membrane depolarization and glutamate to be activated and are thus well-suited for this task. Indeed, pharmacologic blockade of NMDAR function via topical application of MK-801 to the tectal surface results in disrupted retinotectal topography (Cline & Constantine-Paton, 1989). And, similar manipulations with MK-801 or AP5 in the rat SC results in topographic mapping errors (Simon, Prusky, O'Leary, & Constantine-Paton, 1992). Further, NMDARs play important roles in regulating the dynamics and stabilization of RGC terminal arbors (Ruthazer, Akerman, & Cline, 2003; Sin, Haas, Ruthazer, & Cline, 2002). Interestingly, NMDARs may mediate both synaptic strengthening and elimination in the SC during development. On one hand, NR2A-containing NMDARs are required for induction of long-term potentiation in the SC in juvenile mice (P15–17) (Zhao & Constantine-Paton, 2007). On the other hand, activation of NMDARs results in synaptic depression and synaptic elimination in the SC at earlier developmental stages (Colonnese & Constantine-Paton, 2006; Shi, Aamodt, Townsend, & Constantine-Paton, 2001).

As in other systems, NMDARs are critical for plasticity in the SC, which can be induced experimentally through lesion or manipulation of activity patterns. In a model of optic nerve crush, which induces regeneration and re-innervation of the goldfish OT, NMDAR function is required for the refinement of tectal receptive fields (Schmidt, 1990). Similarly, partial ablation of the SC results in map compression with compensating plasticity to maintain receptive field size (Finlay, Schneps, & Schneider, 1979; Pallas & Finlay, 1989), which is prevented when NMDAR activity is blocked (Huang & Pallas, 2001). As discussed above, the ability of exogenously synchronized activation to limit branch dynamics and optic flow-induced shifts in topographic position are dependent on signaling through NMDARs (Hiramoto & Cline, 2014; Munz et al., 2014). Together, these studies implicate NMDARs in playing critical roles in activity-dependent circuit development in the SC; however, it remains unclear whether NMDAR function is required solely on the post-synaptic side of the cleft or if molecules localized pre-synaptically may contribute as well.

4.3 Axon-axon competition

As mentioned previously, a putative mechanism by which nasal RGCs are guided to terminate in regions of highly repellant ephrin-A expression is biased competition. That is, RGC axons that express relatively lower levels of EphA receptor in any given region are less repelled and, thus, capable of establishing synapses in regions neighboring axons cannot. Rather than compete for synaptic territory with these neighboring axons, those originating from the nasal retina terminate in more posterior regions. Though some controversy exists, computational and genetic mouse models support a competitive mechanism in establishing retinocollicular topography.

4.3.1 Clues of competitive forces from molecular manipulations

Genetic mouse models in which the EphA/ephrin-A gradient has been altered support a competitive mechanism. As discussed above, single or combination knockout of EphA or ephrin-A molecules results in topographic mapping errors. Importantly, however, bulk labeling of RGCs reveals that the entire target region of either the dLGN or SC are completely filled (Feldheim et al., 2000), suggesting that RGCs establish similar numbers of synapses. Additionally, in $Isl2^{EphA3/EphA3}$ mice, $Isl2^{+}$ axons expressing exogenously high EphA levels terminate in regions that would be occupied by $Isl2^{-}$ RGCs from the anterior SC, effectively out-competing them for this territory (Brown et al., 2000). Indeed, genetically-labeled classes of $Isl2^{+}$ and $Isl2^{-}$ RGCs are sharply segregated into the anterior and posterior regions of the SC in $Isl2^{EphA3/EphA3}$ mice, respectively (Triplett et al., 2014). Intriguingly, this segregation is reflected in altered functionality in each sub-region, but not altered receptive field size (Kay & Triplett, 2017), again suggesting competitive forces help shape optimal RGC convergence.

4.3.2 Models of reduced RGC axon-axon competition reveal a role in topographic mapping

To more directly test the hypothesis that RGC axon-axon competition plays a role in establishing topography, models in which the total number of RGCs is reduced have been utilized. In one such approach, retinocollicular topography was assessed in mice lacking the transcription factor Math5 (Atoh7), which plays a critical role in RGC differentiation (Brown, Patel, Brzezinski, & Glaser, 2001). As a result, the RGC population in $Math5^{-/-}$ mice is only 5–20% of that found in controls. In strong support of competitive mechanisms, bulk labeling reveals that RGCs innervate only the anteromedial portion of the SC (Fig. 2D) (Triplett et al., 2011). Furthermore, both antero- and retrograde tracing experiments suggest that topographic order within this region is disrupted. These data are consistent with computational models that incorporate combinations of ephrin-A/EphA repulsive, ephrin-B/EphB attractive and competitive forces (Triplett et al., 2011). Of note, in an evaluation of multiple computational models of retinocollicular mapping based on experimental data, the Koulakov model developed in this study performed most efficiently (Hjorth, Sterratt, Cutts, Willshaw, & Eglen, 2014).

Consistent with these findings, a less dramatic reduction in RGC numbers yielded similar changes in topography, as well as altered eye-specific segregation. Using the Pax6α-Cre line, which is expressed in nasal and temporal "wedges" of the retina but not central retina, Maiorano and Hindges reduced axon-axon competition by crossing this line with mice harboring floxed alleles of Dicer (Maiorano & Hindges, 2013). As a result, retinal progenitors at the nasal and temporal poles undergo apoptosis, leading to an ~40% reduction in the number of RGCs by P8. However, these remaining RGC axons occupy a substantial proportion of the SC, suggesting a plasticity in response to reduced competition. In addition, the authors report disrupted topographic organization, as revealed by both anterograde and retrograde tracing. Finally, the reduction in competition also results in an alteration of eye-specific segregation, such that ipsilaterally-projecting RGCs occupy the superficial layers in the anterior- and posterior-most SC which are unoccupied by contralaterally-projecting RGCs.

Another approach at reducing RGC numbers yielded less convincing results in support of axon-axon competition. In that study, progenitor cells from a zebrafish donor embryo that expresses GFP driven by the Brn3c promotor were transplanted into either a wild type or *lakritz* mutant host (Gosse, Nevin, & Baier, 2008). Since *lakritz* mutants fail to develop due to a disruption in the Ath5 (Atoh7) transcription factor and Brn3c is expressed exclusively in RGCs, the authors could visualize single, labeled RGCs in an otherwise RGC-less fish. By determining the location of the posterior-most branch of the terminal arbor in the tectum as a function of RGC cell body location in the retina, the authors found that the organization in the *lakritz* background was identical to control. Intriguingly, when the same analysis was performed for the anterior-most branch, the fidelity of topographic order decreased, though order was still ostensibly present.

Consistent with these results, manipulations of neuronal activity suggest that comparative levels of synaptic vesicle release influence terminal arbor size. By expressing the tetanus toxin light chain (TeNT) in single RGCs, Ben Fredj and colleagues found that terminal arbor size was increased compared to control axons (Ben Fredj et al., 2010). However, when TeNT was expressed in a background of other vesicle release incapable RGCs, terminal size was similar to control axons in an unaltered background. Similar to other activity-dependent manipulations in the context of retinocollicular mapping, NMDARs functioned as mediators of this competition. Intriguingly, however, manipulations of Stage II waves in the context of reduced RGC numbers did not exert dramatic effects. In Math5$^{-/-}$;β2-nAChR$^{-/-}$ combination mutants, RGCs remained restricted to the anterior-medial SC (Fig. 2J) (Triplett et al., 2011). However, focal tracing was not performed in these animals, so the precise impact on topographic order remains unclear. Together, these data suggest that the influence of competitive forces may vary depending on organismal context.

4.4 Interactions between forces

Thus far, we have discussed the forces driving topographic map formation in isolation. However, molecular cues, spontaneous activity, and competition operate in overlapping temporal windows in vivo. Indeed, recent work has highlighted that these forces interact with one another during development, particularly ephrin-A/EphA signaling and Stage II waves.

4.4.1 Neuronal activity facilitates ephrin-A/EphA-mediated signaling

While activity has a clearly instructive role in its own right, evidence suggests that it is also required for efficient signaling of molecules that guide topographic mapping. To uncover this function, Nicol and colleagues developed an in vitro system in which they cultured retinal explants with SC slices (Nicol et al., 2007). When they blocked activity in this system, they found that ingrowing RGCs failed to prune branches in inappropriate topographic areas. Since they could manipulate the genotype of the retinal explant and SC explant independently, they were able to demonstrate that this was not due to a synaptic transmission mechanism. Further, in vitro growth cone collapse assays revealed that repulsion mediated by bath addition of ephrin-A5 was inhibited when action potentials were blocked with tetrodotoxin. Amazingly, they demonstrated that oscillations of cyclic AMP (cAMP) induced by uncaging could ameliorate this effect. Consistent with this, mice lacking functional adenylate cyclase 1 (AC1), which catalyzes the formation of cAMP, exhibit topographic mapping errors (Plas, Visel, Gonzalez, She, & Crair, 2004). And retina-specific, but not midbrain-specific, deletion of AC1 is sufficient to cause these errors (Dhande et al., 2012). Intriguingly, subsequent studies suggest that ephrin-As induce cAMP formation localized to lipid rafts which may be crucial for local guidance events (Averaimo et al., 2016). However, it remains unclear what role activity might play in the generation of these signals or if they act in parallel pathways.

In contrast to these studies, in vivo manipulations suggest that ephrin-A/EphA signaling and spontaneous activity are separable phenomena. By electroporating RGCs in utero, Benjumeda and colleagues were able to label axonal projections and express an inward-rectifying potassium channel (Kir2.1) to reduce activity (Benjumeda et al., 2013). In this system, the authors found that RGCs project to topographically appropriate regions in the SC, and co-expression of EphA6 results in an anteriorly-shifted termination zone even in Kir2.1$^+$ RGCs. Notably, terminal arbors of single RGCs were less dense when activity was reduced, suggesting the primary function of Stage II waves may be condensation of terminal branches rather than gross axonal guidance.

4.4.2 Interactions between molecular cues and neuronal activity

At the resolution of groups of neurons, both molecular cues and spontaneous activity should act in concert to shape an ordered topographic map of space. Indeed, in combination ephrin-A2$^{-/-}$;ephrin-A5$^{-/-}$;β2-nAChR$^{-/-}$ mice, topographic order along the azimuth axes of both the SC and dLGN is completely abolished (Fig. 2G) (Pfeiffenberger et al., 2006). Intriguingly, however, disruption of Stage II waves does not interfere with the duplication along the azimuth axis of the SC observed in Isl2$^{EphA3/EphA3}$ mice (Fig. 2H) (Triplett et al., 2009). Further, recent studies suggest that local clustering of similarly tuned visual neurons may exist in the SC (Ahmadlou & Heimel, 2015; Feinberg & Meister, 2015), which by necessity must disregard the precise local topography, suggesting that more complex interactions between these forces may exist. A recent study leveraging an animal model in which molecular and activity- driven forces were put in opposition revealed one potential mechanism by which such organizations may be achieved. In heterozygous Isl2$^{EphA3/+}$, the EphA3 receptor is over-expressed in Isl2$^+$ RGCs, but not to the same level as in homozygotes (Bevins et al., 2011; Brown et al., 2000; Reber et al., 2004). Thus, in temporal regions of the retina, the distinction between EphA levels in neighboring Isl2$^+$ and Isl2$^-$ RGCs is less sharp. In initial anatomical tracing experiments, it appeared that rather than a complete duplication of the retina's projection along the SC's A-P axis, as in Isl2$^{EphA3/EphA3}$ mice, the map was only partially duplicated in Isl2$^{EphA3/+}$ mice (Bevins et al., 2011; Brown et al., 2000). However, later experiments revealed a more complex situation.

In Isl2$^{EphA3/+}$ mice, the functional organization of the SC was found to be heterogeneous (Owens et al., 2015). That is, in ~25% of cases, the map is partially duplicated, as predicted by anatomical studies. Intriguingly, the functional map is similar to a wild type organization in ~25% of heterozygotes; and, in ~50% of cases the organization is "mixed," meaning some regions appeared to be singular, while other regions were duplicated (Fig. 2F). Such heterogeneity suggests a stochasticity in local map organizations, which may be dictated by opposing molecular and activity-dependent forces. In support of this, map organizations are not only heterogeneous between animals, but even between colliculi of the same animal, ruling out potential genetic background causes. Further, when Stage II waves are disrupted by crossing into the β2-nAChR$^{-/-}$ line, heterogeneity is eliminated and all maps appear singular (Fig. 2I). Finally, mathematical models in which the difference in EphA expression between Isl2^{+} and Isl2^{-} RGC populations could be altered replicates in vivo data. Together, these data suggest that in Isl2$^{EphA3/+}$ mice, molecules and activity are forced into precarious balance and each can dictate local map organization in a seemingly stochastic manner. Such a mechanism could allow for local rearrangements to prioritize synapse formation in functionally-relevant ways at the expense of topography.

5. Mechanisms of laminar-specific targeting

A critical feature of vision is the parallel nature of processing, in which different subclasses of the same neuronal type are tuned to distinct features of the visual scene. For example, recent functional and molecular fingerprinting analyses have identified upwards of 40 subclasses of RGCs (Baden et al., 2016; Rheaume et al., 2018). As this information is relayed to image-forming targets, terminations are localized to specific sublayers to facilitate efficient transfer of information (Inayat et al., 2015; Piscopo, El-Danaf, Huberman, & Niell, 2013; Shi et al., 2017). Thus, deciphering the developmental mechanisms by which these connections are established is a critical area of investigation. Current studies have focused on the contribution of topographic order, neuronal activity, and molecular cues to uncover these mechanisms.

5.1 Limited influence of topographic order

Despite the temporal overlap in the development of retinocollicular topography and laminar refinement of RGC subtypes, these two events are achieved via independent guidance mechanisms. To visualize subclass-specific arbors, multiple lines leveraging BAC transgenics to express GFP or Cre recombinase have been generated. For instance, in CB2-GFP mice a subset of "transient" OFF-aRGCs is labeled with GFP, which terminate in the core of the dLGN and the lSGS in the SC (Huberman et al., 2008). In contrast, in DRD4-GFP and TRHR-GFP mice, related, but distinct, direction-selective RGCs are labeled, and both terminate in the shell of the dLGN and uSGS of the SC (Huberman et al., 2009; Rivlin-Etzion et al., 2011). When these lines are crossed into an ephrin-A5$^{-/-}$ or ephrin-A2/A5$^{-/-}$ background, each still targets the appropriate sublamina (Sweeney, James, Sales, & Feldheim, 2015). Intriguingly, the authors of this study also found that ectopic TZs tend to be specific to either the uSGS or lSGS, implicating a misalignment of functional map organizations. In further support of their independence, DRD4-GFP and CB2-RGCs target the appropriate sublaminae in the SC of Isl2$^{EphA3/EphA3}$, despite a dramatic rearrangement into distinct subzones (Triplett et al., 2014). Together, these studies suggest that topography and laminar targeting are distinct events in the SC; however, whether the same is true in the dLGN remains unclear.

5.2 Activity-independent laminar targeting

Investigations in both the mouse SC and zebrafish OT suggest that laminar targeting does not rely on normal patterns of spontaneous or visually-evoked activity. By crossing CB2-GFP mice into the β2-nAChR background, Huberman and colleagues were able to assess the role of Stage II waves in lamina-specific targeting (Huberman et al., 2008). In this context, CB2-GFP^{+} RGCs terminate in the ventromedial core region, similar to the pattern observed in controls. Similarly, axon terminals are restricted to the lSGS in the SC; however, the "columnar" organization of CB2-GFP+ RGCs is disrupted in β2-nAChR$^{-/-}$ mice. Consistent with these findings, the exquisite lamination observed in the zebrafish OT is not altered when activity patterns are changed in a variety of ways. Nevin and colleagues leveraged a line in which expression of GFP was driven by Pou4f3, which labels ~40% of RGCs to assess retinotectal targeting (Nevin, Taylor, & Baier, 2008). Neither manipulation of visual experience by dark rearing nor pharmacologic blockade of activity nor blockade of vesicle release altered the lamination pattern of Pou4f3:GFP arbors in the OT. Together, these data suggest that a molecular, rather than activity-dependent mechanism is utilized to establish lamina-specific targeting.

5.3 Molecular regulators of lamina-specific targeting

While by no means as well understood as the forces controlling topographic mapping, molecular cues that regulate lamina-specific targeting in the OT have been identified. To begin, analysis of Pou4f3-labeled RGCs via a GAL4::UAS system revealed deficits in tectal targeting in zebrafish harboring a mutation of a subunit of type IV collagen (Col4a5) (Xiao & Baier, 2007). Importantly, immunohistochemical markers of tectal laminae are normal in these mutants, suggesting a direct guidance role. Interestingly, treatment with heparitinase phenocopies these results, suggesting that heparin sulfate proteoglycans play a critical role in laminar targeting. Indeed, subsequent studies demonstrated that a gradient of *slit1* originating from the basement membrane is necessary for targeting and that this action is mediated by the *robo2* receptor (Xiao et al., 2011). Finally, a forward genetic screen identified topoisomerase 2beta (Top2b) as a key regulator of laminar targeting (Nevin, Xiao, Staub, & Baier, 2011). Since Top2b regulates DNA organization and transcription, the molecular cues at the surface responsible for guidance errors observed in this mutant remain unclear. However, these cues are expressed by RGCs, since errors are observed in Top2b-deficient RGCs innervating a wild type OT, but lamination is unaltered in the reverse context.

5.4 Relationship between laminar targeting and development of tuning

The exquisite laminar organization of RGC subtype axons in the SC/OT and dLGN suggests its importance in relaying visual information. However, a recent study suggests that a properly laminated OT is not essential for establishing functional circuitry [Nikolaou & Meyer, 2015]. To test the role of lamination, these investigators utilized the *astray* (ast^{ti272z}) line, which disrupts *robo2* expression (Fricke, Lee, Geiger-Rudolph, Bonhoeffer, & Chien, 2001), resulting in altered laminar targeting in the OT (Xiao et al., 2011). Indeed, the targeting of multiple direction-selective RGCs is disrupted in ast^{ti272z} mutants, and the morphology of direction-selective tectal cells is also dramatically altered (Nikolaou & Meyer, 2015). Despite this, the expected repertoire of direction-selectivity in the tectum of older animals is intact; however, the rate of circuit development is delayed compared to controls. The authors further demonstrated that specific rescue of *robo2* in RGCs or tectal cells can serve as a positional cue for the dendrites or axons of the reciprocal population. Together, these data suggest that laminar targeting is required for efficient assembly of functional circuits but that other mechanisms may regulate the establishment of specific synapses.

6. Mechanisms of visual map alignment

In many visual, auditory, and somatosensory nuclei, neurons and their projections are organized topographically; in multisensory areas, these maps are aligned in order to facilitate integration and lower the threshold for detection of salient stimuli (Stein, Stanford, & Rowland, 2014). The SC is one such region (Basso & May, 2017), and investigations into the mechanisms by which L5-V1 inputs to the SC are aligned with RGCs has advanced our understanding of how these processes may occur in other modalities and regions.

6.1 Retinal instruction of corticocollicular projection alignment

Two models have been proposed to explain how RGC and V1 inputs could be aligned in the SC. First, since similar gradients of ephrin-As/EphAs are observed along the L-M axis in V1, alignment could be achieved by a gradient-matching mechanism. Second, since L5-V1 projections innervate the SC after the retinocollicular map is established and partially course through the retinorecipient region, they could be aligned by "reading" the existing map. Experiments in $Isl2^{EphA3/EphA3}$ mice were able to distinguish between these models. In contrast to the retinocollicular projection, the retinogeniculate projection in $Isl2^{EphA3/EphA3}$ mice is maintained as a single map (Fig. 2E), and the cortical representation of visual space is similar to controls (Triplett et al., 2009). Strikingly, anatomical tracing experiments demonstrate that the L5-V1 to SC projection in these mice is duplicated along the azimuth to align with the retinal map (Fig. 3C). Notably, L5-V1 inputs do appear to be capable of responding to the endogenous ephrin-A/EphA gradients in the SC, as rough topography is maintained when RGC input is reduced via enucleation or in $Math5^{-/-}$ mice (Fig. 3D) (Triplett et al., 2009).

The mechanism underlying retinal instruction of L5-V1 input alignment is not fully understood and could involve molecular cues, activity, or a combination thereof. To determine the role of Stage II waves, $Isl2^{EphA3/EphA3}$ mice were crossed into the $\beta2$-$nAChR^{-/-}$ background. In these combination mutants, the retinocollicular projection remains duplicated, but each termination zone is poorly refined (Fig. 2H). However, tracing experiments

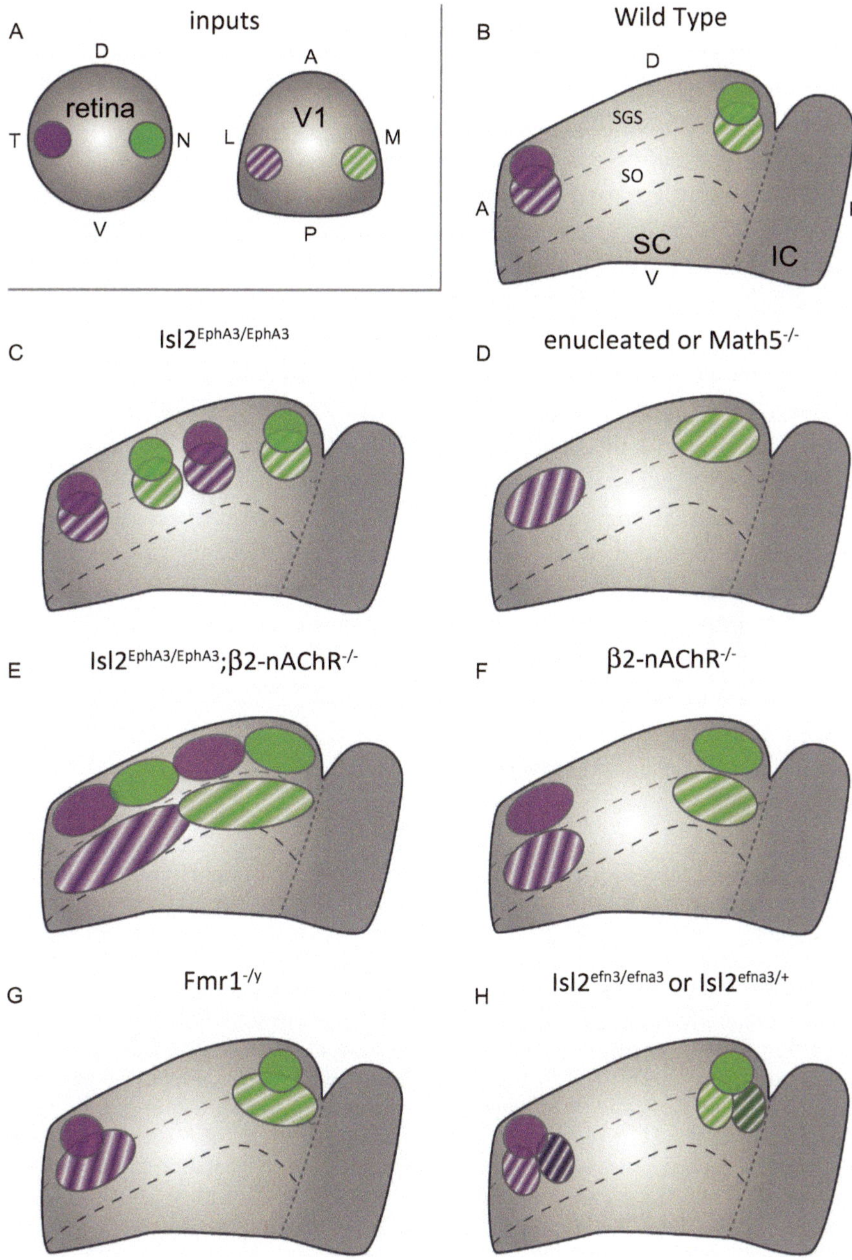

Fig. 3 See figure legend on next page.

suggest that L5-V1 neurons project to form a single, broad map that is effectively misaligned with the retinal map (Fig. 3E). Notably, however, overlap between L5-V1 TZs and retinal input is significantly decreased in β2-nAChR$^{-/-}$ mice (Fig. 3F) (Triplett et al., 2009), potentially preventing any retinal instruction via molecular cues.

The downstream pathways utilized in an activity-dependent mechanism of retinal instruction remain unclear. Interestingly, recently developed computational models of visual map alignment in the SC support an activity-dependent mechanism that could proceed via simple correlational firing between RGCs and L5-V1 or a Hebbian-like mechanism in which SC neurons integrate RGC and L5-V1 inputs (Tikidji-Hamburyan, El-Ghazawi, & Triplett, 2016). And, investigations in a mouse model of fragile X syndrome (Fmr1$^{-/y}$) implicate the fragile X mental retardation protein (FMRP) in the regulation of visual map alignment in the SC (Kay, Gabreski, & Triplett, 2018). Functional analyses in Fmr1$^{-/y}$ mice reveal increases in receptive field size, decreased direction-selectivity, and hyperactivity of orientation-selective neurons in the SC. Intriguingly, topography of RGC projections to the SC are unperturbed, but those from V1 are enlarged (Fig. 3G). These data suggest that FMRP may play a specific role in the development or maintenance of visual map alignment in the SC. Given its well described role in neuronal plasticity, FMRP may provide a link between neuronal activity and visual map alignment.

Fig. 3—Cont'd Summary of effects of manipulations that impact visual map alignment in the SC. (A & B) The superior colliculus (SC) receives inputs from the retina and primary visual cortex (V1) (A). RGCs from temporal retina and layer 5 (L5) neurons from lateral V1, which monitor the same region of space, terminate in the anterior portion of the SC (B). Likewise, RGCs from nasal retina and medial V1 terminate posteriorly. RGCs terminate in the stratum griseum superficiale (SGS), while V1 projections terminate more ventral, within the stratum opticum (SO) but overlapping RGC inputs. (C) In Isl2$^{EphA3/EphA3}$ mice, the retina's projection is duplicated along the anterior-posterior axis and V1 projections duplicate to maintain alignment. (D) When retinal input to the SC is reduced by enucleation or knockout of Math5, L5-V1 axons terminate in topographically appropriate regions, though termination zones (TZs) are broader. (E) When Stage II waves are disrupted in Isl2$^{EphA3/EphA3}$ mice, L5-V1 inputs fail to align with the duplicated retinal map. (F) In β2-nAChR knockout mice with disrupted waves, L5-V1 axons terminate in topographically appropriate region, though TZs are broader and less overlapped with retinal inputs. (G) In fragile X mice (Fmr1$^{-/y}$), retinocollicular topography is intact, but L5-V1 neurons have broader TZs. (H) In Islet2-ephrinA3 knock-in mice (Isl2^{efna3}), the retinal projection to the SC is topographically ordered, but the TZs of L5-V1 inputs are partially duplicated and slightly shifted.

In opposition to findings supporting an activity-dependent mechanism of retinal instruction of visual map alignment, recent data provides compelling support for the influence of ephrin-As expressed on RGC axons in the SC. To demonstrate this, Savier and colleagues generated an *Islet2-ephrin-A3* knock-in mouse model (Isl2^{efna3}) (Savier et al., 2017). Anatomical tracing experiments in these mice suggest that topography of the retinocollicular projection is organized similar to controls, as a single map of visual space along the azimuth axis. However, rather than a single TZ for each V1 labeling, as observed in controls, approximately 50% of TZs are partially- or fully-duplicated in Isl2$^{efna3/efna3}$ and Isl2$^{efna3/+}$ mice (Fig. 3F). Notably, the duplication observed in this context is distinct from that reported in Isl2$^{EphA3/EphA3}$ mice, appearing as a subtle displacement along the A-P axis rather than two well-separated TZs. Amazingly, L5-V1 topographic mapping errors are abrogated when the authors generated Isl2$^{EphA3/efna3}$ "double heterozygous" mice, which effectively blocks ephrin-A3 signaling by increasing *cis* interactions within RGC axons. Further, a novel 3-step computational model that incorporates alterations to the ephrin-A/EphA gradient based on cues transported along RGC axons replicates these in vivo findings (Savier et al., 2017).

7. Concluding remarks

Subcortical visual circuits have served as valuable models to elucidate fundamental mechanisms by which molecular cues, neuronal activity, and axon-axon competition drive brain wiring. In the 70+ years since Sperry's initial investigations into the regenerating retinotectal projection, we have achieved an almost complete understanding of the mechanisms by which topographic maps of visual space are established. However, several problems remain for the field to solve in order to provide a complete understanding of how these circuits are wired. First, the paradox of nasal axon mapping remains flummoxing. While support exists for multiple mechanisms, a unified model would solidify, and largely complete, our understanding of retinocollicular/retinotectal topographic map formation. Second, the mechanisms by which different subclasses of RGCs establish specific connections in the SC/OT and dLGN remains poorly understood. Surprisingly, in the decade or so since the generation of valuable tools to label different RGC subtypes in both fish and mice, only a few cues have been identified. One potential cause for this may be the very genetics that allow for labeling, since double or triple transgenic mice may be needed to assess the role of a candidate gene for each given line. Developing tools that allow for labeling different subtypes by a molecular means, such as viruses or electroporation, would allow for more rapid screening. Solving this mystery is critical, both for understanding how subcircuits are established and for aiding the development of regenerative therapies. Finally, elucidating the mechanisms by which topographic maps of space are aligned with one another presents a significant challenge. This is particularly true for cortical inputs to the dLGN, which play a critical role in processing. Further, while progress is being made in understanding the mechanisms of alignment in the SC, multiple cortical regions integrate sensory inputs, yet we have little understanding of how these circuits are assembled. Underscoring this is the prevalence of deficits in sensory processing and integration in neurodevelopmental disorders. As our recent work suggests, elucidating the organization of topographic maps in integrative centers may provide critical insights into the neural correlates of this dysfunction.

References

Ackman, J. B., Burbridge, T. J., & Crair, M. C. (2012). Retinal waves coordinate patterned activity throughout the developing visual system. *Nature*, *490*, 219–225.

Ackman, J. B., & Crair, M. C. (2014). Role of emergent neural activity in visual map development. *Current Opinion in Neurobiology*, *24*, 166–175.

Ahmadlou, M., & Heimel, J. A. (2015). Preference for concentric orientations in the mouse superior colliculus. *Nature Communications*, *6*, 6773.

Averaimo, S., Assali, A., Ros, O., Couvet, S., Zagar, Y., Genescu, I., et al. (2016). A plasma membrane microdomain compartmentalizes ephrin-generated cAMP signals to prune developing retinal axon arbors. *Nature Communications*, 7, 12896.

Baden, T., Berens, P., Franke, K., Rosón, M. R., Bethge, M., & Euler, T. (2016). The functional diversity of retinal ganglion cells in the mouse. *Nature*, *529*, 345–350.

Baier, H. (2013). Synaptic laminae in the visual system: Molecular mechanisms forming layers of perception. *Annual Review of Cell and Developmental Biology*, *29*, 385–416.

Bansal, A., Singer, J. H., Hwang, B. J., Xu, W., Beaudet, A., & Feller, M. B. (2000). Mice lacking specific nicotinic acetylcholine receptor subunits exhibit dramatically altered spontaneous activity patterns and reveal a limited role for retinal waves in forming ON and OFF circuits in the inner retina. *The Journal of Neuroscience*, *20*, 7672–7681.

Basso, M. A., & May, P. J. (2017). Circuits for action and cognition: A view from the superior colliculus. *Annual Review of Vision Science*, *3*, 197–226.

Ben Fredj, N., Hammond, S., Otsuna, H., Chien, C.-B., Burrone, J., & Meyer, M. P. (2010). Synaptic activity and activity-dependent competition regulates axon arbor maturation, growth arrest, and territory in the retinotectal projection. *The Journal of Neuroscience*, *30*, 10939–10951.

Benjumeda, I., Escalante, A., Law, C., Morales, D., Chauvin, G., Muca, G., et al. (2013). Uncoupling of EphA/ephrinA signaling and spontaneous activity in neural circuit wiring. *The Journal of Neuroscience*, *33*, 18208–18218.

Bevins, N., Lemke, G., & Reber, M. (2011). Genetic dissection of EphA receptor signaling dynamics during retinotopic mapping. *The Journal of Neuroscience*, *31*, 10302–10310.

Braisted, J. E., McLaughlin, T., Wang, H. U., Friedman, G. C., Anderson, D. J., & O'Leary, D. D. (1997). Graded and lamina-specific distributions of ligands of EphB receptor tyrosine kinases in the developing retinotectal system. *Developmental Biology, 191*, 14–28.

Brown, N. L., Patel, S., Brzezinski, J., & Glaser, T. (2001). Math5 is required for retinal ganglion cell and optic nerve formation. *Development, 128*, 2497–2508.

Brown, A., Yates, P. A., Burrola, P., Ortuño, D., Vaidya, A., Jessell, T. M., et al. (2000). Topographic mapping from the retina to the midbrain is controlled by relative but not absolute levels of EphA receptor signaling. *Cell, 102*, 77–88.

Buhusi, M., Demyanenko, G. P., Jannie, K. M., Dalal, J., Darnell, E. P. B., Weiner, J. A., et al. (2009). ALCAM regulates mediolateral retinotopic mapping in the superior colliculus. *The Journal of Neuroscience, 29*, 15630–15641.

Buhusi, M., Schlatter, M. C., Demyanenko, G. P., Thresher, R., & Maness, P. F. (2008). L1 interaction with ankyrin regulates mediolateral topography in the retinocollicular projection. *The Journal of Neuroscience, 28*, 177–188.

Burbridge, T. J., Xu, H.-p., Ackman, J. B., Ge, X., Zhang, Y., Ye, M.-J., et al. (2014). Visual circuit development requires patterned activity mediated by retinal acetylcholine receptors. *Neuron, 84*, 1049–1064.

Carreres, M. I., Escalante, A., Murillo, B., Chauvin, G., Gaspar, P., Vegar, C., et al. (2011). Transcription factor foxd1 is required for the specification of the temporal retina in mammals. *The Journal of Neuroscience, 31*, 5673–5681.

Chandrasekaran, A. R., Plas, D. T., Gonzalez, E., & Crair, M. C. (2005). Evidence for an instructive role of retinal activity in retinotopic map refinement in the superior colliculus of the mouse. *The Journal of Neuroscience, 25*, 6929–6938.

Cheng, H. J., & Flanagan, J. G. (1994). Identification and cloning of ELF-1, a developmentally expressed ligand for the Mek4 and Sek receptor tyrosine kinases. *Cell, 79*, 157–168.

Cheng, H. J., Nakamoto, M., Bergemann, A. D., & Flanagan, J. G. (1995). Complementary gradients in expression and binding of ELF-1 and Mek4 in development of the topographic retinotectal projection map. *Cell, 82*, 371–381.

Cline, H. T., & Constantine-Paton, M. (1989). NMDA receptor antagonists disrupt the retinotectal topographic map. *Neuron, 3*, 413–426.

Colonnese, M. T., & Constantine-Paton, M. (2006). Developmental period for N-methyl-D-aspartate (NMDA) receptor-dependent synapse elimination correlated with visuotopic map refinement. *The Journal of Comparative Neurology, 494*, 738–751.

Croteau, L. P., Kao, T. J., & Kania, A. (2019). Ephrin-A5 potentiates netrin-1 axon guidance by enhancing Neogenin availability. *Scientific Reports, 9*, 12009.

Dai, J., Buhusi, M., Demyanenko, G. P., Brennaman, L. H., Hruska, M., Dalva, M. B., et al. (2013). Neuron glia-related cell adhesion molecule (NrCAM) promotes topographic retinocollicular mapping. *PLoS One, 8*, e73000.

Dai, J., Dalal, J. S., Thakar, S., Henkemeyer, M., Lemmon, V. P., Harunaga, J. S., et al. (2012). EphB regulates L1 phosphorylation during retinocollicular mapping. *Molecular and Cellular Neurosciences, 50*, 201–210.

Demyanenko, G. P., & Maness, P. F. (2003). The L1 cell adhesion molecule is essential for topographic mapping of retinal axons. *The Journal of Neuroscience, 23*, 530–538.

Dhande, O. S., Bhatt, S., Anishchenko, A., Elstrott, J., Iwasato, T., Swindell, E. C., et al. (2012). Role of adenylate cyclase 1 in retinofugal map development. *The Journal of Comparative Neurology, 520*, 1562–1583.

Dhande, O. S., Hua, E. W., Guh, E., Yeh, J., Bhatt, S., Zhang, Y., et al. (2011). Development of single retinofugal axon arbors in normal and β2 knock-out mice. *The Journal of Neuroscience, 31*, 3384–3399.

Drager, U. C., & Hubel, D. H. (1975). Physiology of visual cells in mouse superior colliculus and correlation with somatosensory and auditory input. *Nature, 253*, 203–204.

Drescher, U., Kremoser, C., Handwerker, C., Löschinger, J., Noda, M., & Bonhoeffer, F. (1995). In vitro guidance of retinal ganglion cell axons by RAGS, a 25 kDa tectal protein related to ligands for Eph receptor tyrosine kinases. *Cell, 82*, 359–370.

Feinberg, E. H., & Meister, M. (2015). Orientation columns in the mouse superior colliculus. *Nature, 519*, 229–232.

Feldheim, D. A., Kim, Y. I., Bergemann, A. D., Frisén, J., Barbacid, M., & Flanagan, J. G. (2000). Genetic analysis of ephrin-A2 and ephrin-A5 shows their requirement in multiple aspects of retinocollicular mapping. *Neuron, 25*, 563–574.

Feldheim, D. A., Nakamoto, M., Osterfield, M., Gale, N. W., DeChiara, T. M., Rohatgi, R., et al. (2004). Loss-of-function analysis of EphA receptors in retinotectal mapping. *The Journal of Neuroscience, 24*, 2542–2550.

Feldheim, D. A., & O'Leary, D. D. (2010). Visual map development: Bidirectional signaling, bifunctional guidance molecules, and competition. *Cold Spring Harbor Perspectives in Biology, 2*, a001768.

Feldheim, D. A., Vanderhaeghen, P., Hansen, M. J., Frisén, J., Lu, Q., Barbacid, M., et al. (1998). Topographic guidance labels in a sensory projection to the forebrain. *Neuron, 21*, 1303–1313.

Feller, M. B. (2009). Retinal waves are likely to instruct the formation of eye-specific retinogeniculate projections. *Neural Development, 4*, 24.

Feller, M. B., Wellis, D. P., Stellwagen, D., Werblin, F. S., & Shatz, C. J. (1996). Requirement for cholinergic synaptic transmission in the propagation of spontaneous retinal waves. *Science, 272*, 1182–1187.

Finlay, B. L., Schneps, S. E., & Schneider, G. E. (1979). Orderly compression of the retinotectal projection following partial ablation in the newborn hamster. *Nature, 280*, 153–155.

Fricke, C., Lee, J.-S., Geiger-Rudolph, S., Bonhoeffer, F., & Chien, C.-B. (2001). Astray, a zebrafish roundabout homolog required for retinal axon guidance. *Science, 292*, 507–510.

Frisén, J., Yates, P. A., McLaughlin, T., Friedman, G. C., O'Leary, D. D., & Barbacid, M. (1998). Ephrin-A5 (AL-1/RAGS) is essential for proper retinal axon guidance and topographic mapping in the mammalian visual system. *Neuron, 20*, 235–243.

Gosse, N., Nevin, L., & Baier, H. (2008). Retinotopic order in the absence of axon competition. *Nature, 452*, 892–895.

Grubb, M. S., Rossi, F. M., Changeux, J.-P., & Thompson, I. D. (2003). Abnormal functional organization in the dorsal lateral geniculate nucleus of mice lacking the beta 2 subunit of the nicotinic acetylcholine receptor. *Neuron, 40*, 1161–1172.

Hansen, M. J., Dallal, G. E., & Flanagan, J. G. (2004). Retinal axon response to ephrin-as shows a graded, concentration-dependent transition from growth promotion to inhibition. *Neuron, 42*, 717–730.

Hasse, J. M., & Briggs, F. (2017). A cross-species comparison of corticogeniculate structure and function. *Visual Neuroscience, 34*, E016.

Hensch, T. K., & Quinlan, E. M. (2018). Critical periods in amblyopia. *Visual Neuroscience, 35*, E014.

Higenell, V., Han, S. M., Feldheim, D. A., Scalia, F., & Ruthazer, E. S. (2012). Expression patterns of Ephs and ephrins throughout retinotectal development in *Xenopus laevis*. *Developmental Neurobiology, 72*, 547–563.

Hindges, R., McLaughlin, T., Genoud, N., Henkemeyer, M., & O'Leary, D. (2002). EphB forward signaling controls directional branch extension and arborization required for dorsal-ventral retinotopic mapping. *Neuron, 35*, 475–487.

Hiramoto, M., & Cline, H. T. (2014). Optic flow instructs retinotopic map formation through a spatial to temporal to spatial transformation of visual information. *Proceedings of the National Academy of Sciences of the United States of America, 111*, E5105–E5113.

Hjorth, J. J. J., Sterratt, D. C., Cutts, C. S., Willshaw, D. J., & Eglen, S. J. (2014). Quantitative assessment of computational models for retinotopic map formation. *Developmental Neurobiology*, *75*, 641–666.
Holash, J. A., Soans, C., Chong, L. D., Shao, H., Dixit, V. M., & Pasquale, E. B. (1997). Reciprocal expression of the Eph receptor Cek5 and its ligand(s) in the early retina. *Developmental Biology*, *182*, 256–269.
Hong, Y. K., Kim, I.-J., & Sanes, J. R. (2011). Stereotyped axonal arbors of retinal ganglion cell subsets in the mouse superior colliculus. *The Journal of Comparative Neurology*, *519*, 1691–1711.
Hooks, B. M., & Chen, C. (2020). Circuitry underlying experience-dependent plasticity in the mouse visual system. *Neuron*, *106*, 21–36.
Huang, L., & Pallas, S. L. (2001). NMDA antagonists in the superior colliculus prevent developmental plasticity but not visual transmission or map compression. *Journal of Neurophysiology*, *86*, 1179–1194.
Huberman, A. D., Manu, M., Koch, S. M., Susman, M. W., Lutz, A. B., Ullian, E. M., et al. (2008). Architecture and activity-mediated refinement of axonal projections from a mosaic of genetically identified retinal ganglion cells. *Neuron*, *59*, 425–438.
Huberman, A. D., Wei, W., Elstrott, J., Stafford, B. K., Feller, M. B., & Barres, B. A. (2009). Genetic identification of an on-off direction-selective retinal ganglion cell subtype reveals a layer-specific subcortical map of posterior motion. *Neuron*, *62*, 327–334.
Inayat, S., Barchini, J., Chen, H., Feng, L., Liu, X., & Cang, J. (2015). Neurons in the most superficial lamina of the mouse superior colliculus are highly selective for stimulus direction. *The Journal of Neuroscience*, *35*, 7992–8003.
Ito, S., & Feldheim, D. A. (2018). The mouse superior colliculus: An emerging model for studying circuit formation and function. *Frontiers in Neural Circuits*, *12*, 10.
Kay, R. B., Gabreski, N. A., & Triplett, J. W. (2018). Visual subcircuit-specific dysfunction and input-specific mispatterning in the superior colliculus of fragile X mice. *Journal of Neurodevelopmental Disorders*, *10*, 23.
Kay, R. B., & Triplett, J. W. (2017). Visual neurons in the superior colliculus innervated by Islet2+ or Islet2− retinal ganglion cells display distinct tuning properties. *Frontiers in Neural Circuits*, *11*, 73.
Kim, I.-J., Zhang, Y., Meister, M., & Sanes, J. R. (2010). Laminar restriction of retinal ganglion cell dendrites and axons: Subtype-specific developmental patterns revealed with transgenic markers. *The Journal of Neuroscience*, *30*, 1452–1462.
Kim, I.-J., Zhang, Y., Yamagata, M., Meister, M., & Sanes, J. R. (2008). Molecular identification of a retinal cell type that responds to upward motion. *Nature*, *452*, 478–482.
Liang, F., Xiong, X. R., Zingg, B., Ji, X.-y., Zhang, L. I., & Tao, H. W. (2015). Sensory cortical control of a visually induced arrest behavior via corticotectal projections. *Neuron*, *86*, 755–767.
Lim, Y.-S., McLaughlin, T., Sung, T.-C., Santiago, A., Lee, K.-F., & O'Leary, D. D. (2008). p75(NTR) mediates ephrin-A reverse signaling required for axon repulsion and mapping. *Neuron*, *59*, 746–758.
Maiorano, N. A., & Hindges, R. (2013). Restricted perinatal degeneration induces retina reshaping and correlated structural rearrangement of the retinotopic map. *Nature Communications*, *4*, 1938.
Marler, K. J. M., Poopalasundaram, S., Broom, E. R., Wentzel, C., & Drescher, U. (2010). Pro-neurotrophins secreted from retinal ganglion cell axons are necessary for ephrinA-p75NTR-mediated axon guidance. *Neural Development*, *5*, 30.
Mclaughlin, T., Hindges, R., Yates, P. A., & O'Leary, D. D. (2003). Bifunctional action of ephrin-B1 as a repellent and attractant to control bidirectional branch extension in dorsal-ventral retinotopic mapping. *Development*, *130*, 2407–2418.

McLaughlin, T., Torborg, C. L., Feller, M. B., & O'Leary, D. D. M. (2003). Retinotopic map refinement requires spontaneous retinal waves during a brief critical period of development. *Neuron*, *40*, 1147–1160.

Mrsic-Flogel, T. D., Hofer, S. B., Creutzfeldt, C., Cloëz-Tayarani, I., Changeux, J.-P., Bonhoeffer, T., et al. (2005). Altered map of visual space in the superior colliculus of mice lacking early retinal waves. *The Journal of Neuroscience*, *25*, 6921–6928.

Munz, M., Gobert, D., Schohl, A., Poquerusse, J., Podgorski, K., Spratt, P., et al. (2014). Rapid Hebbian axonal remodeling mediated by visual stimulation. *Science*, *344*, 904–909.

Murcia-Belmonte, V., & Erskine, L. (2019). Wiring the binocular visual pathways. *International Journal of Molecular Sciences*, *20*, 3282.

Nakamoto, M., Cheng, H. J., Friedman, G. C., McLaughlin, T., Hansen, M. J., Yoon, C. H., et al. (1996). Topographically specific effects of ELF-1 on retinal axon guidance in vitro and retinal axon mapping in vivo. *Cell*, *86*, 755–766.

Nevin, L. M., Taylor, M. R., & Baier, H. (2008). Hardwiring of fine synaptic layers in the zebrafish visual pathway. *Neural Development*, *3*, 36.

Nevin, L. M., Xiao, T., Staub, W., & Baier, H. (2011). Topoisomerase IIbeta is required for lamina-specific targeting of retinal ganglion cell axons and dendrites. *Development*, *138*, 2457–2465.

Nicol, X., Voyatzis, S., Muzerelle, A., Narboux-Nême, N., Südhof, T. C., Miles, R., et al. (2007). cAMP oscillations and retinal activity are permissive for ephrin signaling during the establishment of the retinotopic map. *Nature Neuroscience*, *10*, 340–347.

Nikolaou, N., & Meyer, M. P. (2015). Lamination speeds the functional development of visual circuits. *Neuron*, *88*, 999–1013.

Owens, M. T., Feldheim, D. A., Stryker, M. P., & Triplett, J. W. (2015). Stochastic interaction between neural activity and molecular cues in the formation of topographic maps. *Neuron*, *87*, 1261–1273.

Pallas, S. L., & Finlay, B. L. (1989). Conservation of receptive-field properties of superior colliculus cells after developmental rearrangements of retinal input. *Visual Neuroscience*, *2*, 121–135.

Pfeiffenberger, C., Yamada, J., & Feldheim, D. A. (2006). Ephrin-As and patterned retinal activity act together in the development of topographic maps in the primary visual system. *The Journal of Neuroscience*, *26*, 12873–12884.

Phillips, M. A., Colonnese, M. T., Goldberg, J., Lewis, L. D., Brown, E. N., & Constantine-Paton, M. (2011). A synaptic strategy for consolidation of convergent visuotopic maps. *Neuron*, *71*, 710–724.

Piscopo, D. M., El-Danaf, R. N., Huberman, A. D., & Niell, C. M. (2013). Diverse visual features encoded in mouse lateral geniculate nucleus. *The Journal of Neuroscience*, *33*, 4642–4656.

Plas, D. T., Lopez, J. E., & Crair, M. C. (2005). Pretarget sorting of retinocollicular axons in the mouse. *The Journal of Comparative Neurology*, *491*, 305–319.

Plas, D. T., Visel, A., Gonzalez, E., She, W.-C., & Crair, M. C. (2004). Adenylate Cyclase 1 dependent refinement of retinotopic maps in the mouse. *Vision Research*, *44*, 3357–3364.

Poliak, S., Morales, D., Croteau, L. P., Krawchuk, D., Palmesino, E., Morton, S., et al. (2015). Synergistic integration of Netrin and ephrin axon guidance signals by spinal motor neurons. *eLife*, *4*, e10841.

Poopalasundaram, S., Marler, K. J. M., & Drescher, U. (2011). EphrinA6 on chick retinal axons is a key component for p75(NTR)-dependent axon repulsion and TrkB-dependent axon branching. *Molecular and Cellular Neurosciences*, *47*, 131–136.

Rashid, T., Upton, A. L., Blentic, A., Ciossek, T., Knöll, B., Thompson, I. D., et al. (2005). Opposing gradients of ephrin-As and EphA7 in the superior colliculus are essential for topographic mapping in the mammalian visual system. *Neuron*, *47*, 57–69.

Reber, M., Burrola, P., & Lemke, G. (2004). A relative signalling model for the formation of a topographic neural map. *Nature, 431*, 847–853.

Rheaume, B. A., Jereen, A., Bolisetty, M., Sajid, M. S., Yang, Y., Renna, K., et al. (2018). Single cell transcriptome profiling of retinal ganglion cells identifies cellular subtypes. *Nature Communications, 9*, 2759.

Rivlin-Etzion, M., Zhou, K., Wei, W., Elstrott, J., Nguyen, P. L., Barres, B. A., et al. (2011). Transgenic mice reveal unexpected diversity of on-off direction-selective retinal ganglion cell subtypes and brain structures involved in motion processing. *The Journal of Neuroscience, 31*, 8760–8769.

Robles, E., Filosa, A., & Baier, H. (2013). Precise lamination of retinal axons generates multiple parallel input pathways in the tectum. *The Journal of Neuroscience, 33*, 5027–5039.

Ruthazer, E. S., Akerman, C. J., & Cline, H. T. (2003). Control of axon branch dynamics by correlated activity in vivo. *Science, 301*, 66–70.

Savier, E., Eglen, S. J., Bathelemy, A., Perraut, M., Pfrieger, F. W., Lemke, G., et al. (2017). A molecular mechanism for the topographic alignment of convergent neural maps. *eLife, 6*, e20470.

Schmidt, J. T. (1990). Long-term potentiation and activity-dependent retinotopic sharpening in the regenerating retinotectal projection of goldfish: Common sensitive period and sensitivity to NMDA blockers. *The Journal of Neuroscience, 10*, 233–246.

Schmitt, A. M., Shi, J., Wolf, A. M., Lu, C.-C., King, L. A., & Zou, Y. (2005). Wnt–Ryk signalling mediates medial–lateral retinotectal topographic mapping. *Nature, 439*, 31–37.

Seabrook, T. A., Burbridge, T. J., Crair, M. C., & Huberman, A. D. (2017). Architecture, function, and assembly of the mouse visual system. *Annual Review of Neuroscience (Palo Alto, CA), 40*, 499–538.

Seabrook, T. A., El-Danaf, R. N., Krahe, T. E., Fox, M. A., & Guido, W. (2013). Retinal input regulates the timing of corticogeniculate innervation. *The Journal of Neuroscience, 33*, 10085–10097.

Shanks, J. A., Ito, S., Schaevitz, L., Yamada, J., Chen, B., Litke, A., et al. (2016). Corticothalamic axons are essential for retinal ganglion cell axon targeting to the mouse dorsal lateral geniculate nucleus. *The Journal of Neuroscience, 36*, 5252–5263.

Shi, J., Aamodt, S. M., Townsend, M., & Constantine-Paton, M. (2001). Developmental depression of glutamate neurotransmission by chronic low-level activation of NMDA receptors. *The Journal of Neuroscience, 21*, 6233–6244.

Shi, X., Barchini, J., Ledesma, H. A., Koren, D., Jin, Y., Liu, X., et al. (2017). Retinal origin of direction selectivity in the superior colliculus. *Nature Neuroscience, 20*, 550–558.

Siegert, S., Scherf, B. G., Del Punta, K., Didkovsky, N., Heintz, N., & Roska, B. (2009). Genetic address book for retinal cell types. *Nature Neuroscience, 12*, 1197–1204.

Simon, D. K., & O'Leary, D. D. (1991). Relationship of retinotopic ordering of axons in the optic pathway to the formation of visual maps in central targets. *The Journal of Comparative Neurology, 307*, 393–404.

Simon, D. K., Prusky, G. T., O'Leary, D. D., & Constantine-Paton, M. (1992). N-methyl-D-aspartate receptor antagonists disrupt the formation of a mammalian neural map. *Proceedings of the National Academy of Sciences of the United States of America, 89*, 10593–10597.

Sin, W. C., Haas, K., Ruthazer, E. S., & Cline, H. T. (2002). Dendrite growth increased by visual activity requires NMDA receptor and Rho GTPases. *Nature, 419*, 475–480.

Sperry, R. W. (1943). Effect of 180 degree rotation of the retinal field on visuomotor coordination. *The Journal of Experimental Zoology, 92*, 263–279.

Sperry, R. W. (1947). Nature of functional recovery following regeneration of the oculomotor nerve in amphibians. *The Anatomical Record, 97*, 293–316.

Sperry, R. W. (1963). Chemoaffinity in the orderly growth of nerve fiber patterns and connections. *Proceedings of the National Academy of Sciences of the United States of America*, *50*, 703–710.
Sretevan, D. W., Feng, L., Puré, E., & Reichardt, L. F. (1994). Embryonic neurons of the developing optic chiasm express L1 and CD44, cell surface molecules with opposing effects on retinal axon growth. *Neuron*, *12*, 957–975.
Stafford, B. K., Sher, A., Litke, A. M., & Feldheim, D. A. (2009). Spatial-temporal patterns of retinal waves underlying activity-dependent refinement of retinofugal projections. *Neuron*, *64*, 200–212.
Stein, B. E., Stanford, T. R., & Rowland, B. A. (2014). Development of multisensory integration from the perspective of the individual neuron. *Nature Reviews. Neuroscience*, *15*, 520–535.
Suetterlin, P., & Drescher, U. (2014). Target-independent ephrinA/EphA-mediated axon-axon repulsion as a novel element in retinocollicular mapping. *Neuron*, *84*, 740–752.
Sun, C., Warland, D. K., Ballesteros, J. M., van der List, D., & Chalupa, L. M. (2008). Retinal waves in mice lacking the beta2 subunit of the nicotinic acetylcholine receptor. *Proceedings of the National Academy of Sciences of the United States of America*, *105*, 13638–13643.
Sweeney, N. T., James, K. N., Sales, E. C., & Feldheim, D. A. (2015). Ephrin-As are required for the topographic mapping but not laminar choice of physiologically distinct RGC types. *Developmental Neurobiology*, *75*, 584–593.
Thakar, S., Chenaux, G., & Henkemeyer, M. (2011). Critical roles for EphB and ephrin-B bidirectional signalling in retinocollicular mapping. *Nature Communications*, *2*, 431.
Tikidji-Hamburyan, R. A., El-Ghazawi, T. A., & Triplett, J. W. (2016). Novel models of visual topographic map alignment in the superior colliculus. *PLoS Computational Biology*, *12*, e1005315.
Triplett, J. W. (2014). Molecular guidance of retinotopic map development in the midbrain. *Current Opinion in Neurobiology*, *24*, 7–12.
Triplett, J. W., & Feldheim, D. A. (2012). Eph and ephrin signaling in the formation of topographic maps. *Seminars in Cell & Developmental Biology*, *23*, 7–15.
Triplett, J. W., Owens, M. T., Yamada, J., Lemke, G., Cang, J., Stryker, M. P., et al. (2009). Retinal input instructs alignment of visual topographic maps. *Cell*, *139*, 175–185.
Triplett, J. W., Pfeiffenberger, C., Yamada, J., Stafford, B. K., Sweeney, N. T., Litke, A. M., et al. (2011). Competition is a driving force in topographic mapping. *Proceedings of the National Academy of Sciences of the United States of America*, *108*, 19060–19065.
Triplett, J. W., Wei, W., Gonzalez, C., Sweeney, N. T., Huberman, A. D., Feller, M. B., et al. (2014). Dendritic and axonal targeting patterns of a genetically-specified class of retinal ganglion cells that participate in image-forming circuits. *Neural Development*, *9*, 2.
Walter, J., Henke-Fahle, S., & Bonhoeffer, F. (1987). Avoidance of posterior tectal membranes by temporal retinal axons. *Development*, *101*, 909–913.
Walter, J., Kern-Veits, B., Huf, J., Stolze, B., & Bonhoeffer, F. (1987). Recognition of position-specific properties of tectal cell membranes by retinal axons in vitro. *Development*, *101*, 685–696.
Wang, Q., & Burkhalter, A. (2013). Stream-related preferences of inputs to the superior colliculus from areas of dorsal and ventral streams of mouse visual cortex. *The Journal of Neuroscience*, *33*, 1696–1705.
Wang, L., Sarnaik, R., Rangarajan, K., Liu, X., & Cang, J. (2010). Visual receptive field properties of neurons in the superficial superior colliculus of the mouse. *The Journal of Neuroscience*, *30*, 16573–16584.
Wong, W. T., Myhr, K. L., Miller, E. D., & Wong, R. O. (2000). Developmental changes in neurotransmitter regulation of correlated spontaneous retinal activity. *The Journal of Neuroscience*, *20*, 351–360.

Xiao, T., & Baier, H. (2007). Lamina-specific axonal projections in the zebrafish tectum require the type IV collagen Dragnet. *Nature Neuroscience, 10*, 1529–1537.
Xiao, T., Staub, W., Robles, E., Gosse, N. J., Cole, G. J., & Baier, H. (2011). Assembly of lamina-specific neuronal connections by slit bound to type IV collagen. *Cell, 146*, 164–176.
Xu, H.-p., Furman, M., Mineur, Y. S., Chen, H., King, S. L., Zenisek, D., et al. (2011). An instructive role for patterned spontaneous retinal activity in mouse visual map development. *Neuron, 70*, 1115–1127.
Yates, P. A., Roskies, A. L., McLaughlin, T., & O'Leary, D. D. (2001). Topographic-specific axon branching controlled by ephrin-As is the critical event in retinotectal map development. *The Journal of Neuroscience, 21*, 8548–8563.
Zhang, J., Ackman, J. B., Xu, H.-p., & Crair, M. C. (2011). Visual map development depends on the temporal pattern of binocular activity in mice. *Nature Neuroscience, 15*, 298–307.
Zhao, J.-P., & Constantine-Paton, M. (2007). NR2A −/− mice lack long-term potentiation but retain NMDA receptor and L-type Ca2+ channel-dependent long-term depression in the juvenile superior colliculus. *The Journal of Neuroscience, 27*, 13649–13654.

CHAPTER EIGHT

Synaptic recognition molecules in development and disease

Dhrubajyoti Chowdhury[a,†], Katherine Watters[a,b,†], and Thomas Biederer[a,*]
[a]Department of Neurology, Yale School of Medicine, New Haven, CT, United States
[b]Neuroscience Graduate Program, Tufts University School of Medicine, Boston, MA, United States
*Corresponding author: e-mail address: thomas.biederer@yale.edu

Contents

† These authors contributed equally.

Current Topics in Developmental Biology, Volume 142
ISSN 0070-2153
https://doi.org/10.1016/bs.ctdb.2020.12.009

Abstract

Synaptic connectivity patterns underlie brain functions. How recognition molecules control where and when neurons form synapses with each other, therefore, is a fundamental question of cellular neuroscience. This chapter delineates adhesion and signaling complexes as well as secreted factors that contribute to synaptic partner recognition in the vertebrate brain. The sections follow a developmental perspective and discuss how recognition molecules (1) guide initial synaptic wiring, (2) provide for the rejection of incorrect partner choices, (3) contribute to synapse specification, and (4) support the removal of inappropriate synapses once formed. These processes involve a rich repertoire of molecular players and key protein families are described, notably the Cadherin and immunoglobulin superfamilies, Semaphorins/Plexins, Leucine-rich repeat containing proteins, and Neurexins and their binding partners. Molecular themes that diversify these recognition systems are defined and highlighted throughout the text, including the neuron-type specific expression and combinatorial action of recognition factors, alternative splicing, and post-translational modifications. Methodological innovations advancing the field such as proteomic approaches and single cell expression studies are additionally described. Further, the chapter highlights the importance of choosing an appropriate brain region to analyze synaptic recognition factors and the advantages offered by laminated structures like the hippocampus or retina. In a concluding section, the profound disease relevance of aberrant synaptic recognition for neurodevelopmental and psychiatric disorders is discussed. Based on the current progress, an outlook is presented on research goals that can further advance insights into how recognition molecules provide for the astounding precision and diversity of synaptic connections.

1. Introduction

The synaptic connectivity patterns in the complex neuropil packed with neurons and glia are being revealed in stunning detail by connectomic reconstruction studies (Kasthuri et al., 2015; Motta et al., 2019). This chapter provides a molecular perspective on how these circuitry patterns are established in the vertebrate brain and reviews adhesion and signaling complexes that contribute to neuronal partner recognition during synapse development and refinement. A series of subcellular events assembles synapses in late prenatal and early postnatal stages, following axon guidance and dendrite differentiation (Biederer, Kaeser, & Blanpied, 2017; Bury & Sabo, 2016; Emperador-Melero & Kaeser, 2020; Südhof, 2018; Yoshihara, De Roo, & Muller, 2009). Filopodia enable axonal contacts with postsynaptic target cells, resulting in the reorganization of the actin cytoskeleton at these sites. In nascent presynaptic terminals, discrete exocytotic areas termed active zones are formed to establish the molecular machinery that couples calcium influx to synaptic vesicle fusion. The active zone precisely aligns with a specialized postsynaptic membrane domain of the target neuron, which is built through the recruitment of scaffold proteins into the postsynaptic density (PSD) and the sorting and stabilization of neurotransmitter receptors. The assembly of pre- and post-synaptic specializations is coordinated and can even be instructed in time and space by adhesion complexes, some of which include secreted molecules. The components of these *trans*-synaptic complexes often share similar extracellular domains yet engage in diverse cell–cell interactions and exhibit distinct dynamic properties and subcellular localizations (Apóstolo & de Wit, 2019; Benson & Huntley, 2012; Chamma & Thoumine, 2018; Missler, Südhof, & Biederer, 2012). Partner recognition between neurons during synapse development hence involves a rich repertoire of molecular players.

How molecular recognition specifies neuronal connectivity continues to be a question at the forefront of molecular and cellular neuroscience (Sanes & Zipursky, 2020). The cellular expression patterns of recognition molecules provide intriguing leads to ask how they contribute to this diversity (Favuzzi et al., 2019; Foldy et al., 2016; Paul et al., 2017). Moreover, proteomic analyses of synaptic surface proteins have begun to reveal the molecular complexity of these adhesion complexes (Cijsouw et al., 2018; Loh et al., 2016; Takano et al., 2020). Yet, even if each of these recognition factors would execute a different role to specify synapses, they could not individually account for the vast number of connections. Indeed, multiple molecular themes enhance the power of these factors to generate and diversify synaptic recognition patterns as summarized in Box 1 and Fig. 1. These themes are highlighted throughout this chapter.

This chapter follows a developmental perspective on synaptic recognition molecules in vertebrate systems. In Section 2, we discuss how these molecules guide the establishment of neuronal connectivity. Synaptic recognition not only helps to find the right partners, it also allows for rejection of incorrect partner choices as described in Section 3. Section 4 presents how synapse organizing adhesion molecules can act beyond the formation of synapses and contribute to the diversification of synapse types. Once synapses are formed, removing or pruning those connections recognized as inappropriate is a key developmental process and described in Section 5. The concluding Section 6 outlines the profound disease relevance of aberrant synaptic recognition, with a focus on neurodevelopmental and psychiatric disorders. The role of glial factors in synapse development is the focus of another chapter (reference chapter "Role of astrocytes in synapse formation

BOX 1 Molecular concepts at play to generate and diversify synaptic recognition

Regional expression variation: The families of synaptic recognition molecules have several members. Restricting their expression to select brain regions, or to subtypes of neurons and synapses, can increase their power to establish connectivity patterns. Case in point: The homophilic protein Cadherin-9 is expressed in the developing hippocampal CA3 and dentate gyrus and facilitates synapse formation between their neuronal populations (Williams et al., 2011).

Combinatorial expression: The concerted expression and function of recognition molecules can increase their individual ability to specify synaptic connections. Case in point: LRRTM and Neuroligin family members, which both individually engage Neurexins as presynaptic proteins, act together at postsynaptic sites of developing hippocampal neurons to control excitatory synapse number and glutamatergic transmission (Ko, Soler-Llavina, Fuccillo, Malenka, & Südhof, 2011; Siddiqui, Pancaroglu, Kang, Rooyakkers, & Craig, 2010).

Coincidence detection: Combinatorial assembly of multiple recognition factors into higher order adhesion complexes enables coincidence detection and validate neuronal partner choice. Case in point: Simultaneous binding of Latrophilin 3, FLRT3 and Teneurin-2 in a ternary complex is required for input-specific synapse formation in the hippocampal CA1 area (Sando, Jiang, & Südhof, 2019).

Temporally defined roles: Recognition molecules act during restricted temporal windows to enable proteins to act in different developmental contexts. Case in point: Pre- and post-synaptic Cadherins are required for synapse assembly in young neurons, but dispensable for synapse assembly in maturing neurons (Bozdagi, Valcin, Poskanzer, Tanaka, & Benson, 2004).

Post-transcriptional modifications: Alternative splicing profoundly increases the molecular diversity of synaptic recognition molecules. Case in point: Cell-type select utilization of six alternative splice sites in Neurexins gives rise to ~1300 isoforms that can differentially engage with a variety of ligands (Schreiner, Simicevic, Ahrne, Schmidt, & Scheiffele, 2015; Treutlein, Gokce, Quake, & Südhof, 2014).

Post-translational modifications: The range of recognition interactions can be increased through post-translational modifications that endow molecules with different binding properties. Case in point: Modification of the extracellular domain of presynaptic Neurexin-1 with heparan sulfate glycans promotes binding to postsynaptic LRRTM proteins (Roppongi et al., 2020; Zhang et al., 2018).

Subcellular targeting: Adhesive recognition molecules can interact with receptors/ligands on distinct target domains of neurites to guide inputs. Case in point: Netrin-G1 and Netrin-G2 are expressed in axons originating from different neuronal populations and restrict their cognate Netrin-G ligands NGL-1 and NGL-2 to subdendritic segments of hippocampal CA1 neurons (Matsukawa et al., 2014; Nishimura-Akiyoshi, Niimi, Nakashiba, & Itohara, 2007).

BOX 1 Molecular concepts at play to generate and diversify synaptic recognition—cont'd

Restrictive factors: Along with positive factors enabling synaptic partner recognition, negative factors can lower the rate of synapse formation, restrict inappropriately formed synapses, or tip the balance to remove synapses that only need to be formed transiently in development. Case in point: MDGA Ig superfamily members bind Neuroligins in *cis* to restrict interactions between Neuroligins and Neurexins in *trans* (Lee et al., 2013; Pettem, Yokomaku, Takahashi, Ge, & Craig, 2013).

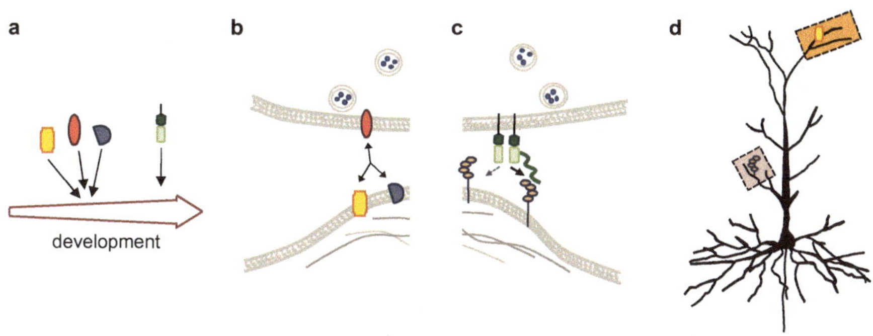

Fig. 1 Molecular themes that increase the coding power of synaptic recognition factors described in Box 1. (A) The diagram depicts recognition factors expressed during distinct developmental windows. These patterns include transient co-expression, providing for temporally defined functional cooperation. (B) The combinatorial expression of synaptic recognition factors that form multimeric complexes enables them to cooperate. (C) Post-translational modifications such as glycans (green line) can modulate and even enable recognition. (D) Recognition factors can exhibit subcellular localization to different domains of neurites. This cellular targeting defines their synaptic site of action.

and maturation" by Tan et al.). Throughout, we provide examples that highlight that it is crucial to choose an appropriate brain region to analyze synaptic recognition and how laminated structures like the retina and hippocampus provide an anatomical matrix that facilitates these studies. The scope of this chapter is unusually broad, and we would like to apologize to the colleagues whose work we could not reference here.

2. Wiring neuronal partners through adhesive recognition

The specificity of synaptic recognition is perhaps best illustrated by those regions of the central nervous system (CNS) that exhibit a laminar architecture of neuronal connectivity, providing a structural basis for the integration of inputs. The anatomical stratification of these brain regions facilitates studies of synaptic recognition and examples are depicted in Fig. 2.

The highly stereotyped laminar architecture of the retina has provided for pioneering studies to identify cell type-specific connectivity (Fig. 2A). It is organized into layers that contain either cell nuclei (the outer and inner nuclear layers and ganglion cell layer) or synaptic connections between projecting axons and dendrites (the outer and inner plexiform layers). In the outer plexiform layer, rod and cone photoreceptor cells form synapses with bipolar and horizontal cells. In the narrow inner plexiform layer of the mouse retina, the dendrites of >40 types of retinal ganglion cells (RGCs) receive inputs from >50 types of bipolar and amacrine cells (Macosko et al., 2015; Yan et al., 2020; Zeng & Sanes, 2017).

Another important brain region organized in a laminated manner is the hippocampus, where granule cells in the dentate gyrus, pyramidal neurons in the CA3 area, and pyramidal neurons in CA1 form three cell layers that are connected into a trisynaptic circuit (Fig. 2B). Additional projections add to these intrinsic connections as illustrated by pyramidal neurons in CA1. While their dendritic segments in the stratum radiatum receive inputs from Schaffer collateral axons of CA3 neurons, the more distal domains of apical dendrites of CA1 neurons in the stratum lacunosum-moleculare are contacted by axons of the temporoammonic branch of the perforant path from the entorhinal cortex. Specific partner recognition is required for all these connections, including subcellular input targeting.

The patterning of synaptic connectivity on a subcellular level is also exemplified by Purkinje cells, the principal output neurons of the cerebellum. The proximal and distal dendritic segments of a Purkinje cell receive glutamatergic innervation originating from the inferior olivary nucleus (one climbing fiber input) and cerebellar granule cells (thousands of parallel fiber inputs), respectively (Hirano, 2018; Kano, Watanabe, Uesaka, & Watanabe, 2018). In addition, Purkinje cells receive GABAergic inputs from local stellate and basket cells that form synapses on the shafts of their dendrites and soma, respectively.

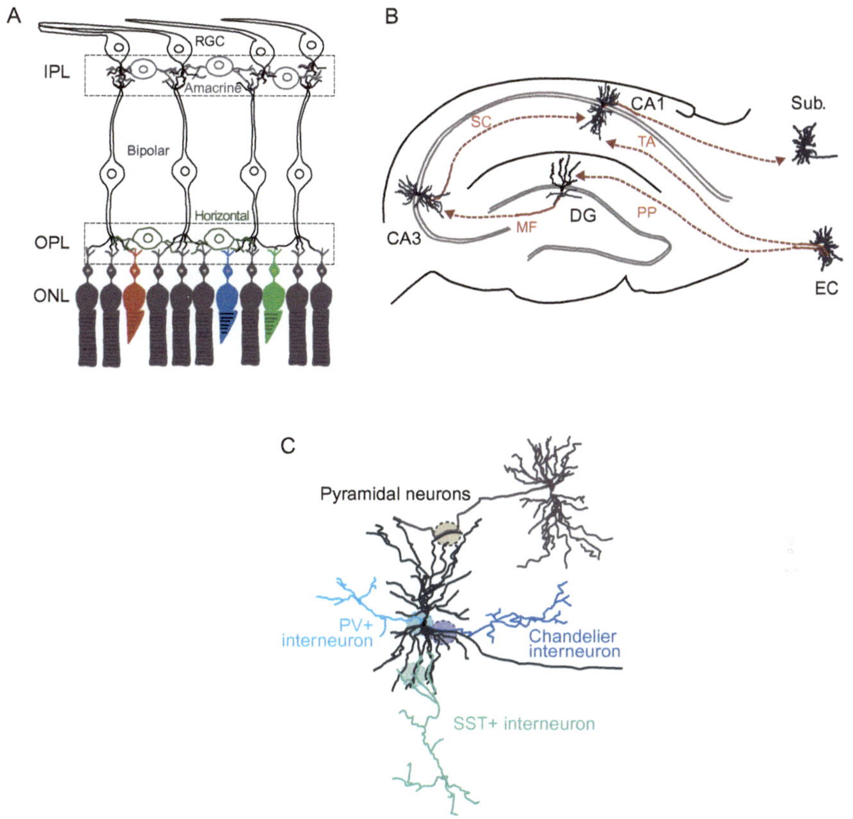

Fig. 2 Simplified illustrations of neuronal connectivity in select brain regions. (A) The retina exhibits exemplary laminar organization of cell types and synaptic connections. Rod (grey) and cone (red, green, blue) photoreceptors are located in the outer nuclear layer (ONL). Photoreceptors form synapses with bipolar cells (dark blue) and horizontal cells (dark green) in the outer plexiform layer (OPL). In the inner plexiform layer (IPL), neurites from retinal ganglion cells (RGCs, black), bipolar cells and amacrine cells (light grey) form synapses. (B) In the classic trisynaptic circuit of the hippocampus, dentate gyrus (DG) granule cells receive via the perforant path (PP) inputs from the entorhinal cortex (EC). DG neurons project via mossy fibers (MF) to CA3 pyramidal neurons, which project via Schaffer collaterals (SC) to CA1 pyramidal cells. In one of the additional projections, CA1 pyramidal neurons receive direct inputs from the EC via the temporoammonic (TA) pathway. CA1 pyramidal cells project to the subiculum (Sub) and EC (not shown). (C) Cortical pyramidal neurons (black) receive excitatory inputs from other pyramidal neurons (grey) and in addition subcellularly targeted inhibitory inputs from GABAergic neurons including Parvalbumin-positive (PV) interneurons, Somatostatin (SST)-positive interneurons, and Chandelier cells.

How connectivity patterns arise can not only be understood by analyzing target cells but also by comparing different populations of cells providing inputs to the same target neurons. For example, interneurons in the hippocampus and cortex make their inhibitory inputs to specific subcellular sites of pyramidal neurons, a targeting precision that plays central roles in controlling neuronal activity and network function (Cardin, 2018; Pelkey et al., 2017). Different types of GABAergic neurons synapse onto distinct subcellular compartments like soma and proximal dendrites, distal dendrites, or the axon initial segment, as shown for fast-spiking Parvalbumin (PV)-positive basket cells, Somatostatin-positive interneurons, and Chandelier cells, respectively (Fig. 2C).

In this section, we focus on major classes of recognition factors known to play key roles in establishing connectivity between distinct neuronal populations, including Cadherins, immunoglobulin superfamily molecules and leucine-rich repeat domain containing proteins. Roles of molecules in the recognition events that diversify synapse types, such as Neurexins and their partners, are described in Section 4. Fig. 3 shows proteins discussed below and their domain organization.

2.1 Classical Cadherins

The Cadherin (*Ca*lcium-*d*ependent ad*her*ent prote*in*) superfamily of single-pass transmembrane glycoproteins has more than 100 members in mammals and is characterized by two or more Extracellular Cadherin (EC) domains. Most Cadherins engage in homophilic interactions in *trans* mediated principally by the most membrane-distal EC domain (Brasch, Harrison, Honig, & Shapiro, 2012). While these interactions tend to be weak, Cadherins can be ordered into arrays for strong adhesion (Al-Amoudi, Diez, Betts, & Frangakis, 2007). Binding of Cadherin proteins to partners in *cis* further extends their interaction repertoire. The ~20 classical Cadherins, comprised of type I and II Cadherins differing in a motif in the most distal EC domain, share a highly similar intracellular sequence with an extended β-catenin-binding motif. Complex intracellular interactions provide means by which the strength of Cadherin-based adhesion can be adjusted (Brigidi & Bamji, 2011).

Classical Cadherins play diverse roles in the development and maintenance of synaptic circuits (Friedman, Benson, & Huntley, 2015). Most type I Cadherins, including the extensively studied N-Cadherin, are broadly expressed in the CNS, whereas type II Cadherins mostly exhibit distinct

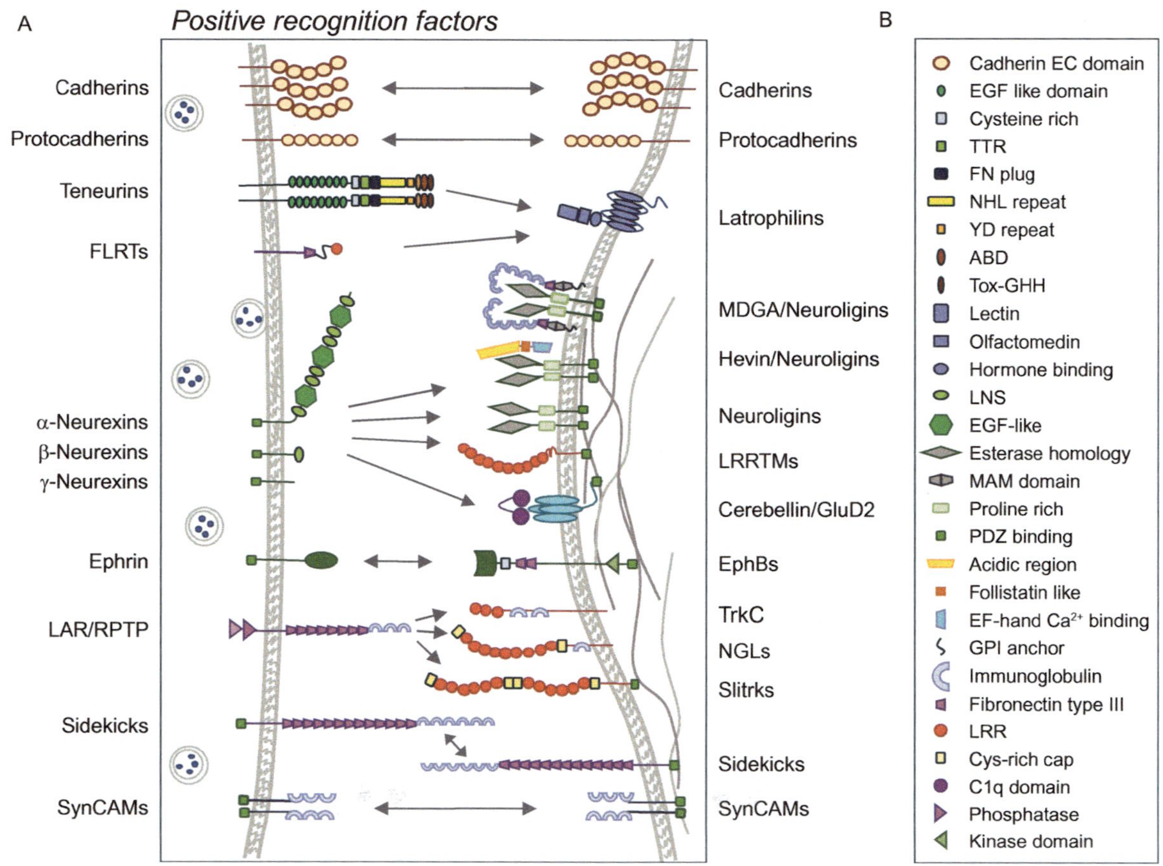

Fig. 3 See figure legend on next page.

expression patterns across areas of brain and spinal cord and are concentrated at nascent synaptic sites early in postnatal development. Cadherins provide for synaptic targeting across many brain regions. For example, blocking N-Cadherin function with antibodies in the developing chick optic tectum causes incoming retinal ganglion cell axons to overshoot their targets in a culture system (Inoue & Sanes, 1997). Similarly, an N-Cadherin zebrafish mutant shows defects in retinal lamination among several other neurite projection deficits of retinal cells (Masai et al., 2003). Moreover, N-Cadherin shapes connectivity by acting together with other family members like Cadherin-8, illustrating the importance of temporally correlated expression. This is exemplified by the topographic mapping of converging thalamic input streams to the barrels of the somatosensory cortex in the early postnatal period. While N-Cadherin becomes concentrated at thalamocortical synapses of the stream arising from ventral posterior medial nucleus (Huntley & Benson, 1999), Cadherin-8 is enriched at synapses of the other stream from the medial division of the posterior nucleus (Gil, Needleman, & Huntley, 2002) when these projections develop. Akin to the abovementioned findings from the chick tectum, the application of N-Cadherin function-blocking antibodies to organotypic co-cultures of thalamus and somatosensory cortex results in thalamic axons overshooting their targets in layer IV (Poskanzer, Needleman, Bozdagi, & Huntley, 2003). Another example for the role of classical Cadherins in target recognition is provided by Cadherin-6. Its deletion in mice impairs axon-target matching for a subset of RGCs whose axons fail to stop at their normal targets in the subcortical visual nuclei, and instead innervate inappropriate visual nuclei (Osterhout et al., 2011). In the cerebellum, Cadherin-7 is expressed in mossy fiber neurons of the pontine nucleus and their targets, the cerebellar granule neurons, but not in climbing fiber neurons, during the synaptogenic stage of development (Kuwako, Nishimoto,

Fig. 3 Recognition molecules that promote synaptic connectivity. (A) Factors promoting recognition and synaptic connections. Pre- and postsynaptic membranes are shown on the left and right, respectively. EphB, EphB receptors; FLRTs, fibronectin LRR transmembrane proteins; GluD2, glutamate receptor delta 2; LAR/RPTPs, LAR-type receptor protein tyrosine phosphatases; LRRTMs, leucine-rich repeat transmembrane proteins; MDGA, MAM domain-containing GPI-anchor proteins; NGLs, Netrin-G ligands; Slitrk, Slit- and Trk-like protein; SynCAMs, Synaptic Cell Adhesion Molecules; Trk, neurotrophin receptor-tyrosine kinase. (B) Domains utilized by the recognition molecules shown in (A). ABD, antibiotic-binding domain-like; C1q, complement component 1q domain; EC, extracellular cadherin domain; EGF, epidermal growth factor; FN, fibronectin; LNS, laminin-neurexin-sex hormone binding globulin domains; LRR, leucine-rich repeat; NHL, ncl-1, HT2A and lin-41 domain; TTR, transthyretin-related domain; YD, tyrosine aspartate repeat.

Kawase, Okano, & Okano, 2014). In agreement with a role in synaptic recognition, homophilic Cadherin-7 signaling induces presynaptic differentiation of pontine nucleus axons in co-culture assays and knockdown of Cadherin-7 in pontine nucleus neurons *in vivo* severely impairs their connectivity with granule cells in the developing cerebellum.

Apart from guiding long-range connectivity between brain regions, Cadherins also specify synaptic targeting of distinct neuronal populations within the same region. In the hippocampus, Cadherin-9 is specifically expressed in CA3 pyramidal neurons and dentate gyrus (DG) granule neurons (Bekirov, Needleman, Zhang, & Benson, 2002). Cadherin-9 knockdown from either CA3 or DG neurons exclusively reduces DG-CA3 mossy fiber synapse formation without affecting non-DG synapses (Williams et al., 2011). This is reminiscent of the role of Cadherin-8 in the development of synaptic laminae of the direction-selective retinal circuit within the inner plexiform layer, where different subtypes of bipolar cells specifically connect with their respective targets of starburst amacrine cells and retinal ganglion cells in appropriate sublaminae. Cadherin-8 and 9 are selectively expressed in bipolar cells and control the targeting of their axons to sublaminae in the inner plexiform layer by a heterophilic mechanism (Duan, Krishnaswamy, De la Huerta, & Sanes, 2014). Moreover, combinatorial interactions between six Cadherins (Cadherins 6, 7, 8, 9, 10 and 18) generate the appropriate connectivity between distinct bipolar cells, starburst amacrine cells and retinal ganglion cells to establish the complex direction-selective circuit of the mouse retina (Duan et al., 2018). Together, a wealth of data has established Cadherins as synaptic recognition factors throughout the brain and underlines the importance of their regional and temporal expression patterns, including their combinatorial expression, for circuit development.

2.2 Protocadherins

The largest subfamily within the Cadherin superfamily consists of Protocadherins, each containing six extracellular EC domains that are diverse among the isoforms and a short, constant cytoplasmic domain. Based on the genomic organization of the respective genes, Protocadherins are subdivided into gene clusters with 58 large variable exons in mice encoding α-, β-, and γ-Protocadherins, with multiple members within each cluster (Mountoufaris, Canzio, Nwakeze, Chen, & Maniatis, 2018; Wu & Maniatis, 1999). In addition, there are ~10 nonclustered δ-Protocadherins (Hulpiau & van Roy, 2009). Clustered Protocadherins are expressed in a combinatorial and stochastic fashion (Schreiner & Weiner, 2010;

Thu et al., 2014). Protocadherin proteins from the same and different clusters promiscuously form isoform-independent *cis* dimers through membrane proximal repeats with efficient *trans* binding between the dimers. Further, Protocadherins show *cis* interactions with classical Cadherins (Weiner & Jontes, 2013). Apart from multifaceted functions in nervous system development including neuronal survival and dendritic and axonal arborization (Lefebvre, 2017), clustered Protocadherins can generate cell-specific recognition codes. One example is in the mouse barrel cortex, where clustered Protocadherins are required for lineage-dependent postnatal reciprocal synaptic connections between excitatory layer IV neurons (Tarusawa et al., 2016). In hippocampal pyramidal neurons, conditional deletion of the atypical Protocadherin Celsr3 after the first postnatal week results in a ~50% decrease in the number of excitatory but not inhibitory synapses (Thakar et al., 2017). Similar to Dscams (see below), Protocadherins can also mediate self-/non-self-discrimination for self-avoidance and restriction of synapse formation as shown for γ-Protocadherins in starburst amacrine cells in the retina (Ing-Esteves et al., 2018). These studies show intriguing roles of this diverse protein family in synaptic recognition.

2.3 Immunoglobulin superfamily members

The first discovered calcium-independent adhesion molecules are the proteins of the immunoglobulin (Ig) superfamily (IgSF). They were identified in parallel to the Cadherin superfamily and are now known to have at least 500 members, with roughly half expressed in neurons. Many members of the IgSF have been implicated in cell-type-specific recognition throughout brain development. Their extracellular domains are comprised of multiple Ig domains that engage in homophilic as well as heterophilic binding (Verschueren et al., 2020; Wojtowicz et al., 2020), often in combination with fibronectin III domains. A prominent group within the Ig superfamily is the L1 family, consisting of L1 (also known as Neuron-glia Cell Adhesion Molecule NgCAM), Close Homolog of L1 (CHL1), Neuron-glia-related Cell Adhesion Molecule (NrCAM), and Neurofascins. The L1 family plays roles in subcellular-specific targeting in several circuits. In the mouse brain, Neurofascin-186 (NF186) is enriched at the axon initial segment (AIS) of several cell types, such as cerebellar Purkinje neurons, hippocampal granule cells and neocortical pyramidal cells. NF186 in Purkinje neurons binds Neuropilin-1 expressed on axons of GABAergic basket neurons to restrict formation of the complex pinceau synapses at the Purkinje AIS, named for their brush-like appearance (Ango et al., 2004; Telley et al., 2016). In the neocortex, instead of AIS-enriched NF186, pan-axonally expressed L1 is required for selective innervation of pyramidal neuron AIS by GABAergic Chandelier cells (Tai, Gallo, Wang, Yu, & Van Aelst, 2019). In both cases, anchoring the L1 family member by the cytoskeletal ankyrin G/spectrin complex is necessary for this localized innervation. The cooperation and coincidence detection by Ig recognition factors is illustrated by NrCAM and CHL1 in the sensory-motor circuit of the mouse spinal cord. They interact on GABApre interneurons with the Ig complex of Contactin 5 and Cntn-associated protein 4 (Caspr4) on proprioceptive sensory neurons to guide inhibitory synapse formation precisely at the axonal termini of sensory afferents (Ashrafi et al., 2014). Synaptic recognition includes interactions with astrocytic processes and a proximity-labeling approach performed in a partner-specific manner has recently mapped the proteome of astrocyte-neuron contacts in the cortex (Takano et al., 2020). NrCAM was one of the astrocyte-expressed proteins identified at these perisynaptic sites. Intriguingly, extracellular Ig interactions of astrocytic NrCAM are required for normal inhibitory synapse number and strength, while intracortical excitatory synapses are unaffected by NrCAM loss in astrocytes. This work advances our understanding of the roles that perisynaptic astrocyte contacts and NrCAM in particular play in synapse organization. It additionally highlights the power of targeted proteomic approaches to define synaptic recognition.

Studies in the retina have implicated other IgSF families in synaptic recognition, including the Down syndrome cell adhesion molecules (Dscams, comprised of Dscam and Dscam-like1), Sidekicks (Sdk1 and 2) and GPI-anchored Contactins (Cntns 1-6), which mostly bind homophilically. In the chick retina, isoforms of these proteins are expressed by largely non-overlapping subsets of bipolar cells, amacrine cells and retinal ganglion cells, with cells expressing the same molecule projecting to the same inner plexiform layer lamina (Yamagata & Sanes, 2008, 2012; Yamagata, Weiner, & Sanes, 2002). The isoform-specific adhesion between these proteins mediates lamina-specific connectivity in the retina, although interesting differences have been observed across molecules and species. One example are Dscams, which are required for synaptic lamination in the chick retina (Yamagata & Sanes, 2008). Their function in synapse specification does not appear to be conserved in the mammalian retina, and Dscam in the mouse retina facilitates neurite self-avoidance by counteracting cell type-specific adhesion by Cadherins and Protocadherins (Fuerst, Koizumi, Masland, & Burgess, 2008; Garrett, Khalil, Walton, & Burgess, 2018). Sidekicks are expressed in subsets of retinal neurons in chick and mice and are critical for laminar restriction of their neurites (Krishnaswamy, Yamagata, Duan, Hong, & Sanes, 2015; Yamagata & Sanes, 2018). Further, dendrites of mouse amacrine cells and retinal ganglion cells expressing Sdk1 but not Sdk2 arborize in the same stratum in the inner plexiform layer and this sublaminar restriction is disrupted in absence of Sdk1 (Yamagata & Sanes, 2018).

The Ig family of Synaptic Cell Adhesion Molecules (SynCAMs) is comprised of four members also known as Cadms (Cell adhesion molecules) or Nectin-like molecules (Biederer & Shrestha, 2015; Frei & Stoeckli, 2014). These vertebrate-specific proteins engage in homo- and heterophilic binding, with select pairwise interaction patterns between distinct members (Fogel et al., 2007; Kakunaga et al., 2005; Shingai et al., 2003; Thomas, Akins, & Biederer, 2008). SynCAM 1 is required and sufficient for excitatory synapse formation as shown in the hippocampal CA1 area (Robbins et al., 2010). SynCAM 1 also contributes to synaptic recognition in the CA3 area, where it organizes mossy fiber inputs to pyramidal neurons and is additionally required for mossy fibers to form synapses onto fast-spiking, PV-positive interneurons (Park et al., 2016). In the visual cortex, SynCAM 1 acts postsynaptically in PV-positive interneurons to promote their innervation by long-range thalamocortical inputs (Ribic, Crair, & Biederer, 2019) with implications for inhibitory network maturation and cortical plasticity (Ribic & Biederer, 2019). SynCAM 1 organizes synapses in the retina as well and contributes to cell-cell recognition in the outer plexiform layer. Here, it is expressed on mouse rod photoreceptor terminals and is required for their interactions with processes of horizontal cells and bipolar cell dendrites and the assembly of triadic rod ribbon synapses (Ribic, Liu, Crair, & Biederer, 2014).

The homophilic protein Kirrel3/Neph2 has provided insight into roles of Ig interactions in target recognition in the hippocampus. It is expressed by DG granule neurons and calbindin-positive GABAergic neurons in CA3 and regulates the development of mossy fiber synapses between these two cell types (Martin et al., 2015). With respect to hippocampal mossy fiber synapses, a proteomic approach that combined biochemical fractionation and FACS sorting has recently identified the Ig protein IgSF8 as a novel component of these large synaptic specializations (Apóstolo et al., 2020). IgSf8 is not only strongly enriched at mossy fiber synapses, it also acts as presynaptic organizer of their ultrastructure and connectivity. Methodologies that map the proteomes of different synapse types as in this study will advance our understanding of how recognition factors contribute to the staggering diversity of synapses across brain regions.

2.4 Leucine-rich repeat family proteins

The leucine-rich repeat (LRR) family of synaptic adhesion molecules is characterized by the presence of multiple consecutive LRR motifs in the extracellular domain which engage in diverse trans-synaptic interactions. LRR proteins have been implicated in all steps of circuit formation from neuronal migration and neurite outgrowth, to the formation and functional assembly of synaptic contacts (Schroeder & de Wit, 2018). The major subfamilies are Netrin-G ligands (NGLs, also called laminets), LRR Transmembrane proteins (LRRTMs), Slit and NTRK-like proteins (Slitrks), Fibronectin LRR Transmembrane proteins (FLRTs) and Synaptic Adhesion-Like Molecules (SALMs, also called LRFNs). LRR proteins provide examples for differential roles of family members in synaptic recognition and the specification of connectivity. Here, NGL-1 and -2 localize to the postsynaptic membrane and form trans-synaptic complexes selectively with GPI-anchored Netrin-G1 and -G2, respectively. In the neocortex and hippocampus of mice, Netrin-G1 and -G2 are distributed on different populations of developing axons (Nishimura-Akiyoshi et al., 2007). Their partners NGL-1 and -2 are concentrated in distinct segments within dendrites of these target areas that correspond to the termination zones of axons expressing Netrin-G1 or -G2 (Matsukawa et al., 2014). In Netrin-G1 and -G2 deficient mice, axonal pathfinding is normal, but the differential distribution of NGL-1 and -2 across dendritic segments is selectively disrupted. Consistent with a role of this subcellular targeting in synaptic recognition, NGL-2 loss selectively reduces the density of spines in the dendritic segment where CA1 pyramidal neurons receive Schaffer collateral inputs, while spine density on their distal dendrites is unaffected (DeNardo, de Wit, Otto-Hitt, & Ghosh, 2012). These studies provide an example for how differential subcellular targeting and adhesive interactions properties can be utilized to generate recognition codes.

The members of the FLRT family, FLRT1-3, were identified as high-affinity postsynaptic ligands of the Latrophilin family of adhesion-type G-protein coupled receptors (O'Sullivan et al., 2012). FLRT2 and FLRT3 show cell-type-specific expression patterns, with complementary and non-overlapping expression in the hippocampus. FLRT3 is highly expressed during the first 2 postnatal weeks in the principal cells of DG and CA3, and its conditional knockdown in the hippocampus of rat pups reduces spine density in DG granule cells and lowers the strength of perforant path inputs onto these cells (O'Sullivan et al., 2012). The FLRT ligands Latrophilin 2 and 3 localize to different dendritic domains of CA1 pyramidal neurons and are essential for synapse formation by entorhinal cortex afferents and Schaffer collateral axons in these strata, respectively (Anderson et al., 2017; Sando et al., 2019). Both pre- vs post-synaptic localizations of Latrophilins have been reported in different regions of the hippocampus, suggesting they may localize to both sides of synapses, perhaps in a synapse-type specific manner (Anderson et al., 2017; O'Sullivan et al., 2012; Sando et al., 2019).

2.5 Teneurins

Teneurins are large transmembrane proteins that play roles in dendrite morphogenesis, axon pathfinding, partner selection and synapse differentiation (Arac & Li, 2019). They form a family of four proteins comprised of a large C-terminal extracellular domain including eight epidermal growth factor (EGF) motifs, a single transmembrane region, and an N-terminal cytoplasmic domain. Teneurins form constitutive *cis* dimers through the membrane proximal EGF repeats and are involved in homophilic and heterophilic *trans* interactions that mediate target recognition. Here, trans-synaptic homophilic interactions of Teneurin-3 control targeting of axons from proximal CA1 neurons to their targets in the distal subiculum (Berns, DeNardo, Pederick, & Luo, 2018). Teneurins can also bind Latrophilins, and these heterophilic high-affinity *trans* interactions provide for synapse-specifying functions of Teneurins and are regulated by alternative splicing (Li et al., 2018; Silva et al., 2011). Intriguingly, simultaneous binding of Teneurin-2 and FLRT3 to Latrophilin 3 promotes Schaffer collateral-CA1 synapse formation (Sando et al., 2019). This coincidence detection via a ternary interaction, akin to a two-factor authentication protocol, highlights the coding power for target-dependent synapse specification provided by multimeric recognition complexes.

3. Restrictive recognition cues shape neuronal connectivity

In concert with the positive, synapse-promoting cell surface interactions described above, cell-surface interactions and molecules secreted from target and non-target cells can also prevent the inappropriate formation of synapses. Such restrictive recognition cues can act locally to refine connection specificity. Alternatively, soluble restrictive factors can act at a distance to the extent they are diffusible within the neuropil. Select restrictive factors discussed below are depicted in Fig. 4A.

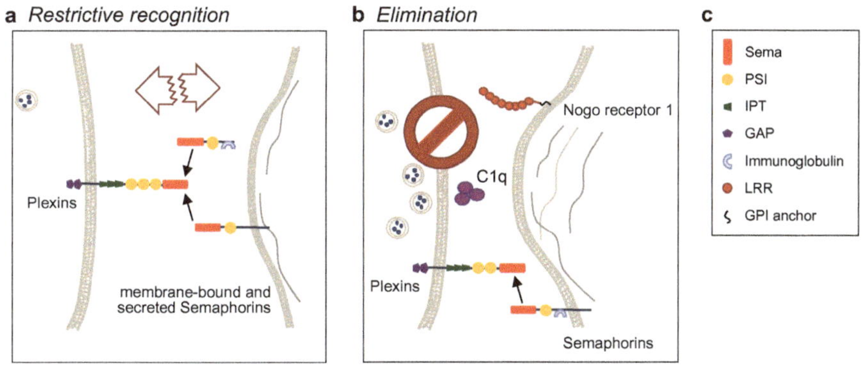

Fig. 4 Recognition molecules that restrict synaptic connectivity or participate in synapse elimination. (A) Factors restricting synaptic connections. (B) Factors that eliminate synapses once formed. Pre- and post-synaptic membranes are shown on the left and right, respectively. C1q, complement component 1q. (C) Domains utilized by the molecules in (A, B). GAP, GTPase activating domain; GPI, glycosylphosphatidylinositol; IPT, Immunoglobulin-like fold shared by Plexin and transcription factors, LRR, leucine-rich repeat; PSI, Plexin, Semaphorins and Integrin domain; Sema, Semaphorin domain.

3.1 Semaphorin–Plexin interactions

Some of the best understood restrictive recognition mechanisms are mediated by Semaphorin-Plexin signaling, which prevents mismatches in synaptic connections by inhibiting inappropriate target selection. This has been demonstrated in the hippocampus, retina, olfactory bulb, striatum and spinal cord (Pasterkamp, 2012; Yoshida, 2012). In the mouse hippocampal CA3 area, the transmembrane Semaphorins Sema6A and 6B, expressed on CA3 pyramidal neurons, interact with Plexin A4 on mossy fibers to control their lamina-restricted projection to the stratum lucidum of CA3 (Suto et al., 2007; Tawarayama, Yoshida, Suto, Mitchell, & Fujisawa, 2010). Sema6A and 6B are required for the repulsion of Plexin A4-expressing mossy fibers, preventing them from forming aberrant projections into stratum radiatum and stratum oriens. Semaphorin-Plexin signaling can also restrict connectivity by acting in the same cells as shown in hippocampal DG granule cells. Here, the transmembrane Semaphorin Sema5A is highly expressed in developing granule cells and signals through its receptor Plexin A2 co-expressed by these cells to suppress spinogenesis (Duan et al., 2014).

In the developing mouse retina, Sema6A acts as a repulsive cue to direct laminar termination away from inappropriate sublaminae. Sema6A and its receptor Plexin A4 are expressed in a complementary fashion in a subset of amacrine cells and of retinal ganglion cells in the inner plexiform layer that differ in their response to a luminance change (Matsuoka et al., 2011). Mice lacking Plexin A4 or Sema6A exhibit severe mistargeting of these cell projections to sublaminae of the inner plexiform layer. Similarly, Sema5A and Sema5B, acting through their receptors Plexin A1 and A3, constrain the neurites from multiple retinal neuron subtypes to the inner plexiform layer (Matsuoka et al., 2011). In the absence of these molecules, retinal neurons fail to correctly stratify in the inner plexiform layer and neurites become mistargeted to the outer portions of the developing retina. These results support that Semaphorins specify laminar stratification through restrictive recognition.

Semaphorins also restrict synaptic recognition in other neural circuits. In the mouse spinal cord, repulsive interactions between secreted Sema3E, expressed on the cutaneous maximus motor neurons, and PlexinD1, expressed on sensory afferents, prevents the formation of direct sensory-motor neuron synapses (Pecho-Vrieseling, Sigrist, Yoshida, Jessell, & Arber, 2009). Sema3E-Plexin D1 signaling also negatively regulates thalamo-striatal synapse formation selectively on direct pathway medium spiny neurons (Ding, Oh, Sabatini, & Gu, 2011). While these examples highlight the restrictive activities of Semaphorin–Plexin interactions during circuit assembly, it should be noted that depending on the context, these ligand-receptor systems can promote connectivity as shown, e.g., for Sema4D that enhances inhibitory synapse formation in the hippocampus (Acker, Wong, Kang, & Paradis, 2018). Semaphorins and their receptors additionally play important roles during later stages of circuit refinement that are discussed in Section 5.

3.2 Other restrictive recognition factors

The FLRT family member FLRT2 and Unc-5C, a receptor for Netrin axon guidance cues, are expressed in a strikingly complementary fashion within specific retinal sublaminae of the developing inner plexiform layer. Heterophilic repulsion between FLRT2 and Unc-5c has been proposed to mediate the laminar restriction of presynaptic starburst amacrine cells and a subset of ON-OFF direction-selective ganglion cells (Visser et al., 2015). Synapse-restricting recognition is provided in the hippocampus during early postnatal development by the GPI-anchored Nogo Receptor 1 (NgR1), which acts through its co-receptor TROY to limit excitatory synapse formation and maturation (Lee et al., 2008; Wills et al., 2012).

4. Beyond making connections: Creating synapse diversity

Synapses are structurally and functionally highly diverse. During their assembly, cell-surface adhesion molecules and secreted factors modulate and even instruct pre- and post-synaptic differentiation at nascent contact sites. Many synapse organizing membrane proteins have been identified based on their synaptogenic ability in a mixed co-culture assay of neurons and non-neuronal cells (Biederer & Scheiffele, 2007). The co-culture assay serves to identify the sufficiency of candidate molecules to induce synaptic specializations, however it is unable to differentiate between their activities across the steps of synapse formation, stabilization and maturation. Additional tests are required to dissociate these roles of synapse organizing proteins, including loss-of-function analyses and acute interference with their interactions. Since the structural and functional diversity of synapses is best studied in the complex environment of the brain, we prioritize in this Section *in vivo* over *in vitro* studies evaluating roles of these proteins.

As a prerequisite to understand the molecular basis of synapse diversity, much progress has been made to parse out the transcriptional diversity of neurons in the mouse brain (see for examples, Paul et al., 2017; Tasic et al., 2016). Recently, Rico and colleagues made significant progress by demonstrating that cell type-specific expression patterns of recognition factors not only create molecular fingerprints to assign neuronal identities but also underly their connectivity patterns (Favuzzi et al., 2019). Using cell sorting, RNA sequencing and mouse genetics, they identified recognition factors that are differentially expressed between interneuron types in the developing mouse cerebral cortex and that allow these neurons to target distinct subcellular compartments of pyramidal cells. This study revealed presynaptic molecular programs that specify the sites where synaptic inputs are formed (see also Section 4.5). Such innovative analyses of cellular expression patterns promise to advance new insights into the organizing principles of input-specific connectivity and microcircuit wiring. Below, we describe select recognition factors to highlight how they can diversify synapse types.

4.1 Synapse-type specific functions of Neurexin–Neuroligin complexes

Much of our current understanding of the creation of synapse diversity derives from the trans-synaptic interactions of presynaptic Neurexins

(Südhof, 2017). Mammalian Neurexins are encoded by three genes giving rise to Neurexin 1-3, each with three promoters that can drive transcription of a longer α-Neurexin and shorter β-Neurexin isoform, and a γ isoform that lacks most of the extracellular domain (Roppongi et al., 2020; Tabuchi & Südhof, 2002; Ushkaryov, Petrenko, Geppert, & Südhof, 1992; Yan et al., 2015). Mice lacking all isoforms of α-Neurexins have ultrastructurally normal synapses and unaltered numbers of excitatory synapses, but Ca^{2+}-triggered neurotransmitter release is severely impaired (Missler et al., 2003; Varoqueaux et al., 2006). Inhibitory synapse number is significantly reduced in Neurexin triple KO mice (Dudanova, Tabuchi, Rohlmann, Südhof, & Missler, 2007). Neuroligins were the first of the postsynaptic ligands identified for Neurexins and form a family of four members. Intriguingly, Neuroligin-1 and Neuroligin-2 localize to excitatory and inhibitory synapses, respectively, pointing to synapse-type specific roles of their complexes with Neurexins (Song, Ichtchenko, Südhof, & Brose, 1999; Varoqueaux, Jamain, & Brose, 2004). Mice lacking the three Neuroligins have impaired synaptic transmission but unaltered synapse number (Varoqueaux et al., 2006). Neurexins engage in multiple extracellular interactions in addition to binding Neuroligins, establishing Neurexins as presynaptic hub proteins (Südhof, 2017).

Neurexin/Neuroligin complexes are involved in the functional specification of synapses. In agreement with their selective localization to excitatory and inhibitory synapses, deletion of Neuroligin-1 or -2 impairs evoked excitatory or inhibitory synaptic transmission, respectively (Chanda, Hale, Zhang, Wernig, & Südhof, 2017; Chubykin et al., 2007). Studies of Neurexins in interneurons reveal an even more diverse picture. Conditional deletion of α and β isoforms of Neurexin 1, 2 and 3 from fast-spiking PV-positive interneurons in the prefrontal cortex results in a loss of the inhibitory synapses they form and a decrease in synaptic strength but no impairment in action potential-triggered Ca^{2+} influx. In contrast, pan-Neurexin deletion in Somatostatin-positive interneurons causes no synapse loss but a large decrease in action potential-triggered Ca^{2+} influx that also suppresses synaptic strength (Chen, Jiang, Zhang, Gokce, & Südhof, 2017). Neurexins hence perform distinct roles in specifying synaptic properties that depend on pre- and postsynaptic partner combinations.

4.2 SynCAM cell adhesion molecules

SynCAMs are synaptogenic Ig proteins and induce functional presynaptic specializations (Biederer et al., 2002; Czondor et al., 2013; Fogel et al., 2007). SynCAM 1 is primarily postsynaptic at excitatory hippocampal

CA1 synapses, with a smaller fraction present in the presynaptic membrane (Perez de Arce et al., 2015). Loss of SynCAM 1 in mice reduces excitatory synapse number and temporal control of its forebrain expression in mice has demonstrated that it is first sufficient to promote excitatory synapse number in the hippocampal CA1 area and then required to maintain them (Robbins et al., 2010). A postsynaptic role of SynCAM 1 in the specification of synaptic inputs was shown in the visual cortex of mice. Cell-type specific knockdown of SynCAM 1 in PV-positive interneurons revealed that it is required in these cells to receive long-range thalamocortical excitatory inputs while its loss in the same cells does not impact short-range intracortical excitatory synapses (Ribic et al., 2019). SynCAMs hence engage in postsynaptic recognition to promote and specify synaptic connectivity.

4.3 Protein tyrosine phosphatases

The Leukocyte common antigen-receptor protein tyrosine phosphatases (LAR-RPTPs) comprise of the PTPσ, PTPδ and LAR proteins that organize excitatory and inhibitory synapse assembly (Han, Jeon, Um, & Ko, 2016). LAR-RPTPs contribute to the specification of synapses in a target-dependent manner by binding to multiple distinct postsynaptic ligands, such as NGL-3, TrkC, Slitrks, Synaptic Adhesion-Like Molecules (SALMs) and Interleukin 1 receptor accessory protein-like 1 (IL1RAPL1) (Han et al., 2016; Lie, Li, Kim, & Kim, 2018). This target-dependent role in synapse specification is shown by results that PTPσ is required for excitatory synapse induction by Slitrk1 (Han et al., 2018), while PTPδ is required for IL1RAPL1-mediated excitatory presynaptic differentiation (Yoshida et al., 2011) as well as Slitrk3-mediated inhibitory presynaptic differentiation (Yim et al., 2013). Conditional deletion of individual or all LAR-RPTPs either globally or selectively in excitatory cortical and hippocampal neurons does not affect spontaneous or evoked AMPAR-mediated synaptic transmission but does impair NMDA-receptor-mediated responses (Kim et al., 2020; Sclip & Südhof, 2020). This indicates that LAR-RPTPs specify functional postsynaptic properties.

4.4 Ephrin-EphB receptors

An additional class of trans-synaptic organizers with a direct signaling capability are the EphB receptor tyrosine kinases and their ephrinB ligands, which mediate excitatory synapse development in the hippocampus and cortex (Henderson & Dalva, 2018). Here, EphB-ephrinB signaling regulates spinogenesis as well as the clustering of glutamate receptors (Dalva et al., 2000; Henkemeyer, Itkis, Ngo, Hickmott, & Ethell, 2003; Kayser, McClelland, Hughes, & Dalva, 2006; Segura, Essmann, Weinges, & Acker-Palmer, 2007). Postsynaptic EphB2 can simultaneously bind to its ephrinB ligands and NMDA receptors, controlling the mobility of these receptors (Dalva et al., 2000; Washburn, Xia, Zhou, Mao, & Dalva, 2020). Another role in specifying synaptic properties is provided by postsynaptic ephrinB3, which balances the extent to which glutamatergic synapses are formed on dendritic shafts vs spines (Aoto et al., 2007).

4.5 Synapse diversification involves secreted factors

Neurons locally secrete factors that contribute to synapse-type specific recognition. Wingless and Int-1 proteins (Wnts) induce presynaptic assembly (Sahores, Gibb, & Salinas, 2010; Umemori, Linhoff, Ornitz, & Sanes, 2004) and neuronal Pentraxins cluster ionotropic glutamate receptors at excitatory postsynaptic sites (Pelkey et al., 2015). Other secreted neuronal factors are Fibroblast growth factors (FGFs), and FGF-22 and FGF-7 differentially control synapse formation (Dabrowski, Terauchi, Strong, & Umemori, 2015; Terauchi et al., 2015; Umemori et al., 2004). In the hippocampus, they are highly expressed in CA3 pyramidal neurons during synaptogenesis and serve as retrograde presynaptic organizers of excitatory and inhibitory synapses, respectively, on CA3 pyramidal neurons. Another neuronally secreted factor that controls synaptic connectivity is the protein Complement Component 1q Subcomponent-like 3 (C1ql3). C1ql3 contains two globular domains originally identified in the protein C1q that assembles the initiating complex of the complement cascade in the immune system. C1ql family members bind the adhesion G protein-coupled receptor BAI3 (Bolliger, Martinelli, & Südhof, 2011; Sigoillot et al., 2015). Presynaptic deletion of C1ql3 in basolateral amygdala neurons causes a strong loss of their excitatory outputs to the prefrontal cortex (Martinelli et al., 2016).

Cerebellins (Cblns), another class of secreted C1q family members, are critical adaptors for multiple pre- and post-synaptic molecules. Cbln1 is secreted from cerebellar granule neurons and binds simultaneously to its postsynaptic receptor, the Glutamate Receptor Delta-2 (GluD2) on dendritic spines of Purkinje cells and a splice form of presynaptic Neurexin on cerebellar granule cell axons. Neurexin-Cbln1-GluD2 signaling leads to presynaptic vesicle accumulation as well as the postsynaptic accumulation of GluD2, illustrating coordinated pre- and postsynaptic differentiation by trans-synaptic interactions (Ito-Ishida et al., 2012; Matsuda et al., 2010).

Other examples for synapse-specifying secreted factors include Cbln4 that is secreted by Somatostatin-positive interneurons to bridge presynaptic Neurexin and GluD1 in layer 2/3 pyramidal neurons in the somatosensory cortex (Fossati et al., 2019). In addition, Cbln4 expressed by cortical Somatostatin-positive interneurons is required for their ability to innervate dendrites of pyramidal neurons, and exogenous expression of Cbln4 in PV-positive interneurons is sufficient to re-direct their inhibitory inputs to pyramidal neuron dendrites (Favuzzi et al., 2019). Cerebellins thereby serve as secreted 'match-makers' to spatially specify synapse formation.

Among astrocyte-secreted factors, Hevin/SPARCL1 was reported to bridge interaction-incompatible Neurexin 1α and a splice isoform of Neuroligin-1 at thalamocortical synapses (Singh et al., 2016). While validating the synaptogenic role of Hevin/SPARCL1, another study has used conditional knock-out of Neuroligins and Neurexins to show that Hevin acts independent of them in cultured neurons (Gan & Südhof, 2020). Synapse-type specific roles of Hevin for bridging Neurexin/Neuroligin complexes could resolve this discrepancy but remain to be tested.

4.6 Diversification of synaptic recognition proteins by alternative splicing

Alternative splicing of synaptic recognition molecules enhances their molecular diversity far beyond the limited number of genes, controlling their ectodomain interactions with synaptic ligands. Alternative splicing of cell adhesion molecules regulates synapse development as exemplified by Neurexins. Their alternative splicing at six canonical splice sites (SS1–SS6) in their ectodomains can generate over 1300 detectable isoforms (Schreiner et al., 2014; Treutlein et al., 2014). Single-cell profiling of mRNAs as well as ribosome-associated transcripts complemented by mass-spectrometric profiling of isoforms in the adult mouse brain have revealed hundreds of alternatively spliced Neurexin mRNAs with remarkable cell type-specificity and brain region-select regulation (Fuccillo et al., 2015; Furlanis, Traunmüller, Fucile, & Scheiffele, 2019; Schreiner et al., 2015). For example, in the mouse hippocampus, the $SS4^+$ Neurexin isoform is selectively expressed in GABAergic PV-positive interneurons while the $SS4^-$ isoform is the major isoform in glutamatergic pyramidal cells in the CA1 region (Nguyen et al., 2016). This posttranscriptional processing is extensively used to control synaptic recognition by Neurexins. One example is provided by LRRTMs and Latrophilins that bind Neurexins only when they lack an insert in splice site 4 ($SS4^-$) (Boucard, Ko, & Südhof, 2012; Etherton, Blaiss, Powell, & Südhof, 2009; Siddiqui et al., 2010), while

Cerebellin 1 only binds to $SS4^+$ Neurexins (Uemura et al., 2010). Alternative Neurexin splicing impacts synaptic composition and transmission as shown by studies in which the SS4 insert in Neurexin-3 was constitutively retained *in vivo*. This decreased postsynaptic levels of AMPARs in hippocampal neurons, a non-cell-autonomous phenotype shared by Neurexin-3 knock-out neurons (Aoto, Martinelli, Malenka, Tabuchi, & Südhof, 2013). These results highlight the need for profiling the cell-type-specific splicing patterns of other recognition molecules.

4.7 Post-translational modifications modulate and mediate synaptic recognition

Modifications such as phosphorylation, glycosylation, and palmitoylation are abundant in synaptic proteins and regulate their subcellular localization and protein–protein interactions. One example for the regulation of synaptic adhesion molecules with synaptogenic activity by posttranslational modification is provided by SynCAM 1. Site-specific N-glycosylation of its most membrane-distal Ig domain promotes trans-synaptic SynCAM interactions and is required for synapse induction (Fogel et al., 2010). Further, polysialic acid is attached to N-glycosylation sites of a subset of SynCAM 1 proteins in the developing mouse brain and this modification blocks homophilic SynCAM 1 binding (Galuska et al., 2010). Diverse cell-surface recognition codes can be created by glycan modifications as shown for the covalent attachment of long, heterogeneous chains of the glycosaminoglycan heparan sulfate to ectodomains of synapse organizers (Condomitti & de Wit, 2018). These heparan sulfate glycans are bound by heparan sulfate-binding proteins, thereby expanding interactions beyond adhesion of protein domains alone. Here, the glial-derived GPI-anchored Glypican 4 is modified by heparan sulfate and forms a glycan-dependent complex with postsynaptic LRRTM4 to promote excitatory synapse development (de Wit et al., 2013; Siddiqui et al., 2013). The Neurexin-1 extracellular domain is also modified by heparan sulfate and this modification is required for presynaptic differentiation induced by its postsynaptic Neuroligin and LRRTM ligands (Roppongi et al., 2020; Zhang et al., 2018).

4.8 Cooperation of co-expressed recognition molecules

A growing body of evidence demonstrates that multiple synapse organizing proteins are often present at the same synapse, with several of them known to share trans-synaptic partners. This sets the stage for coincidence detection of the presence of recognition factors. At hippocampal DG excitatory synapses, presynaptic Neurexin 1 forms a complex with presynaptic PTPσ that also involves the modification of Neurexin with heparan sulfate, and both proteins cooperate in synaptogenesis mediated by postsynaptic LRRTM4 (Roppongi et al., 2020). This shows that the themes of post-translational diversification and concerted function can be combined to control recognition systems. Combinatorial roles were additionally shown for LRRTMs and Neuroligins in early postnatal development. Whereas individual or combined knockdown of LRRTM1 and LRRTM2 in cultured hippocampal neurons does not affect number of excitatory synapses, the additional loss of Neuroligin-1 and Neuroligin-3 leads to an extensive reduction of synapse number in an activity-dependent manner (Ko et al., 2011). Moreover, N-Cadherin is required for the postsynaptic adhesion molecules Neuroligin-1, SynCAM 1 and LRRTM2 to promote presynaptic differentiation as well as to enable postsynaptic differentiation by Neurexin-1β (Stan et al., 2010; Yamagata, Duan, & Sanes, 2018). Further, development of inhibitory synapses is jointly controlled by the postsynaptic organizers Slitrk3 and Neuroligin-2 (Li et al., 2017). These examples provide intriguing evidence that cooperative action of recognition molecules specifies synapses in a manner that depends on the molecular makeup of their synaptic clefts.

5. Refining neuronal connectivity through eliminating synapses

Once neuronal connections are established, synapses that were inappropriately formed or only need to be present during a select developmental period are removed. This refinement is critical for experience-dependent neuronal network maturation and involves microglial, astrocytic, and neuronal recognition factors. Several molecules involved in synapse elimination are depicted in Fig. 4B.

Tailoring synaptic connectivity includes microglia that migrate to 'tagged' synapses and engulf presynaptic terminals (Ji, Akgul, Wollmuth, & Tsirka, 2013; Paolicelli et al., 2011; Schafer et al., 2012; Weinhard et al., 2018). Multiple immune system effector molecules have been implicated in synaptic elimination by microglia in the CNS. The roles of these pruning molecules are best understood in the mouse retinogeniculate system, which is well suited to investigate synaptic pruning due to the anatomical precision with which eye-specific RGC inputs are refined and the fact that pruning in this brain region occurs during a narrow postnatal window. Here, the complement cascade that serves in the innate immune system to tag debris for phagocytosis has been found to mark synapses for removal by microglia (Stevens et al., 2007). The abovementioned complement factor C1q is expressed in developing but not mature RGCs and localizes to the synapses between RGCs and their target neurons in the dorsolateral geniculate nucleus (dLGN). C1q knock-out mice have defects in synaptic refinement in the dLGN by P30, supporting that inappropriate retinal inputs were not properly pruned. Loss of synaptically localized C3, another complement protein, also results in decreased synaptic pruning in the retinogeniculate system (Schafer et al., 2012; Stevens et al., 2007). Other mechanisms that involve microglia are also in play in the dLGN. Following visual stimulation, microglia upregulate the cytokine TNF-associated weak inducer of apoptosis (TWEAK) and relay neurons increase expression of the TWEAK receptor, the Fibroblast growth factor-inducible protein, 14 kDa (Fn14). Neuronal Fn14 is required for the vision-dependent strengthening of bulbous spines contacted by RGCs when not bound by TWEAK. If Fn14 is bound by TWEAK at synapses proximal to microglia, their signaling decreases the number of bulbous spines via a mechanism distinct from phagocytic engulfment. Microglial TWEAK hence locally balances the refinement of dLGN inputs in a sensory experience-dependent manner (Cheadle, Rivera, Phelps, & Ennis, 2020).

Pruning in the hippocampal CA1 area involves the Triggering Receptor Expressed on Myeloid cells 2 (TREM2), an innate immune receptor that is required by microglia to refine excitatory inputs in CA1 (Filipello et al., 2018). In addition, the fractalkine receptor Cx3cr1, a chemokine receptor expressed by microglia, contributes to pruning in CA1. Mice lacking Cx3cr1 exhibit increased postsynaptic puncta density, CA1 dendritic spine density and mEPSC frequency, in agreement with an excess of excitatory synaptic sites due to a decrease in pruning (Paolicelli et al., 2011). The effect of Cx3cr1 on microglial-mediated pruning appears to be brain region-dependent, as its deletion in the visual cortex results in no change to synapse turnover (Lowery, Tremblay, Hopkins, & Majewska, 2017).

Microglia-dependent synapse elimination accounts for only part of retinogeniculate circuit refinement. Astrocytes and the factors they secrete add to the complexity of synapse removal (Chung et al., 2013; Vainchtein & Molofsky, 2020). Here, astrocyte-expressed phagocytic receptors (MEGF10 and MERTK) and recognition molecules (ephrinB1) contribute to synapse elimination in dLGN and hippocampal CA1, respectively (Chung et al., 2013; Koeppen et al., 2018).

Additionally, neuronally expressed molecules originally identified in the immune system, including the class I major histocompatibility complex (MHC I) and the secreted Pentraxins, homologs of a class of immune proteins recognizing antigens, participate in synapse removal (Bjartmar et al., 2006; Huh et al., 2000). Neuronal Semaphorins that restrict synaptic recognition in earlier development can also act to eliminate synapses once formed in order to refine connectivity (Riccomagno & Kolodkin, 2015). Semaphorin 3F signaling through the Neuropilin-2/Plexin A3 holoreceptor promotes the progressive elimination of synapses transiently formed by infrapyramidal mossy fiber axon collaterals on the basal dendrites of CA3 pyramidal cells in the maturing hippocampus (P25) (Bagri, Cheng, Yaron, Pleasure, & Tessier-Lavigne, 2003; Liu, Low, Jones, & Cheng, 2005). Similar to the elimination of excess synapses in other regions like the retinorecipient superior colliculus (Cheng et al., 2010) or the Plexin A3/A4-dependent stereotypic pruning of inputs by corticospinal tract axons (Low, Liu, Faulkner, Coble, & Cheng, 2008), this synaptic refinement process in the hippocampus precedes retraction of axons. Further, mice lacking the secreted Semaphorin 3F and its receptor Neuropilin-2 have normal spine density in DG granule cells and cortical layer V pyramidal neurons at P14 but higher density at P21 (Tran et al., 2009). This suggests a role for these recognition molecules in restricting synapse number in the maturing hippocampus, in agreement with the increase in the number of dendritic spines after acute deletion of Neuropilin-2 in adult cortical layer V neurons (Assous et al., 2019). Additional support for roles of Semaphorins in the negative control of synapse density comes from studies of the L1 Ig family member NrCAM, an obligate component of the Semaphorin 3F receptor complex Neuropilin-2/Plexin A3. Deletion of NrCAM results in increased spine number on apical dendrites of star pyramidal neurons in layer 4 of the mouse primary visual cortex at both P21 and P60 (Demyanenko et al., 2014). A classical process of synapse elimination occurs in the cerebellum and is also controlled by Semaphorin/Plexin recognition. Here, climbing fibers project from the contralateral inferior olive and synapse onto Purkinje cells (Hashimoto & Kano, 2005; Kano et al., 2018; Sassoe-Pognetto & Patrizi, 2017). During the first week of postnatal development in rodents, Purkinje cells in the cerebellum are innervated by multiple climbing fibers (Crepel, Mariani, & Delhaye-Bouchaud, 1976). However, by postnatal week three, only one of these original climbing fibers innervates a single Purkinje cell and all other climbing fiber inputs are removed (Chedotal & Sotelo, 1993; Crepel, Delhaye-Bouchaud, & Dupont, 1981;

Mariani & Changeux, 1981). In this process, Semaphorin 3A acts as a retrograde signal from Purkinje cells to Plexin A4 in climbing fibers to protect one synapse from elimination, whereas Semaphorin 7A facilitates elimination of climbing fiber synapses on Purkinje cells through Plexin C1 and the basement membrane-related protein Integrin β1 (Uesaka et al., 2014).

While it is important for circuit refinement to tag specific subsets of synapses for removal, it is conceivable that molecules present at the retained synapses serve to prevent pruning. Evidence exists that CD47, another immune system molecule, and its receptor SIRPα are among such factors. CD47 is detected at dLGN synapses during the peak period of their pruning and loss of CD47 or its receptor SIRPα results in a decrease in dLGN excitatory synapse number and increased microglial engulfment of presynaptic inputs. This in turn significantly impairs retinal innervation (Lehrman et al., 2018).

6. Aberrant synaptic recognition and brain disorders

Genome wide association studies (GWAS) and the characterization of *de novo* mutations in neuropsychiatric disorders strongly support that aberrations in synaptic adhesion molecules are associated with increased risk for neurodevelopmental and psychiatric disorders. While functional compensation can occur among synaptic adhesion molecules from the same or different gene families, *de novo* mutations in neuropsychiatric patients provide evidence that a change as small as a single amino acid substitution in a recognition molecule can impact social behaviors and cognitive functions. The notion that even minor disruptions in synaptic recognition perturb synapses and alter circuits is supported by studies in which disorder-linked mutations were introduced into synaptic adhesion molecules. Among the consequences are synapse-type specific alterations, changes in synaptic transmission, and improper connectivity, all of which can impair brain functions.

We focus here on autism spectrum disorders and schizophrenia that present during early and late brain development, respectively. They were selected because altered synapse number and connectivity patterns are part of the etiology of these disorders. Further, a wealth of human genetic data and results from animal models with disease-linked mutations or deletions in synaptic adhesion molecules are available. While the focus of this chapter lies on developmental aspects, it needs to be considered that phenotypes correlated with mutations in synaptic recognition factors could in part reflect their functions in the maturing and adult brain.

6.1 Autism spectrum disorders

Autism Spectrum Disorders (ASDs) are a group of developmental disorders characterized by deficits in social-emotional reciprocity, impairments in verbal and nonverbal communication, and repetitive patterns of behaviors and interests that start to manifest early in life. Specifically, ASD symptoms can often be diagnosed around the age of 2. These impairments are distinct from intellectual disability or a general developmental delay.

How do synapses come into play? ASD has a high comorbidity with epilepsy and has long been thought to involve altered synaptic connectivity. This agrees with the onset of its symptoms during the period of most intense synaptogenesis, as well as postmortem data showing increased dendritic spine density in prefrontal cortex (PFC) pyramidal neurons of ASD patients (Hutsler & Zhang, 2010). Patients with ASD also present improper excitatory and inhibitory (E/I) synaptic balance resulting in abnormal transmission and oscillatory anomalies on a brain-wide scale (Cornew, Roberts, Blaskey, & Edgar, 2012; Orekhova et al., 2007). A large body of evidence from GWAS and *de novo* mutation analyses points to malfunctions of synaptic molecules including recognition factors in the etiology of ASD (Geschwind & State, 2015; Sestan & State, 2018). Animal models where synaptic adhesion molecules associated with ASD either carry human disease-linked mutations or are deleted exhibit E/I synaptic imbalance phenotypes as reviewed below. Whether these E/I alterations are cause or consequence of ASD-linked aberrations is being discussed (Antoine, Langberg, Schnepel, & Feldman, 2019), but it can be considered that synaptic adhesion molecules have a role in the homeostatic stabilization of circuits and that disease-linked mutations impair this. Together, the altered expression of ASD-linked synaptic adhesion molecules can impact neuronal transmission and partner recognition as described in this Section. Susceptibility to these mutation effects appears to differ across brain regions and for candidate molecules listed here, the brain area or synapse type that has been characterized is stated.

*6.1.1 Neurexins (*NRXN *genes)*

Neurexins have been strongly implicated in ASD, and deletions and rare variants of Neurexin-1α are found in ASD patients (Gauthier et al., 2011; Schaaf et al., 2012; Südhof, 2017; Yan et al., 2008). Neurexin-1α knock-out mice have reduced spontaneous and evoked excitatory synaptic strength in the hippocampus (Etherton et al., 2009). Along with these synaptic changes, Neurexin-1α knock-outs display increased repetitive grooming behaviors, impaired nest-building, and impaired pre-pulse inhibition, behavioral phenotypes that are linked to neurodevelopmental disorders. Non-social cognitive defects were also observed in rats lacking Neurexin-1α (Esclassan, Francois, Phillips, Loomis, & Gilmour, 2015).

*6.1.2 Neuroligins (*NLGN *genes)*

Neuroligin family members are genetically associated with ASD, with mutations found in syndromic and non-syndromic ASD patients (Marro et al., 2019; Nakanishi et al., 2017; Südhof, 2017; Xu et al., 2014). One mutation in Neuroligin-3, R451C, is a highly penetrant missense mutation (Jamain et al., 2003). This mutation resides within the extracellular, cholinesterase-like domain and causes altered intracellular protein trafficking, resulting in lower surface expression (De Jaco et al., 2005). While no loss of excitatory or inhibitory synapse density has been found in Neuroligin-3 knock-out or R415C knock-in mouse lines, Neuroligin-3 R415C knock-in mice display increased inhibitory synaptic transmission in cell layer II/III of the somatosensory cortex that is not seen in the Neuroligin-3 knock-out mice, suggesting a pathological dominant negative effect (Tabuchi et al., 2007). In the somatosensory barrel cortex, *in vivo* spine imaging revealed an increased spine turnover rate in three-week-old Neuroligin-3 R451C knock-in mice (Isshiki et al., 2014). Neuroligin-3 R451C knock-in mice also display social novelty defects with decreased activity in the medial prefrontal cortex (mPFC), an area implicated in ASD. Local field potential recordings in the mPFC of Neuroligin-3 R451C knock-in mice revealed reduced gamma band activity as well as reduced gamma-to-theta amplitude coupling, indicative of inappropriate synaptic connectivity (Cao et al., 2018). *In vitro* patch-clamp recordings in the mPFC found reduced excitability of PV-positive interneurons, but not pyramidal neurons. Interestingly, oscillation-coupled excitation of mPFC PV-positive interneurons via optogenetics was able to rescue gamma-to-theta coupling in Neuroligin-3 R451C knock-in mice, as well as social novelty defects (Cao et al., 2018). Targeting the connectivity of interneurons could therefore be an entry point for the treatment of ASD. ASD-relevant effects of Neuroligin-3 mutations also manifest in the striatal subregion of the nucleus accumbens, where Neuroligin-3 deletion decreases inhibitory transmission and results in repetitive behaviors (Rothwell et al., 2014).

*6.1.3 Contactins (*CNTNAP *genes)*

Contactin Associated Protein-like 2 (CNTNAP2) shares extracellular domains with Neurexins and strong genetic data implicate it in ASD, with genetic variants and microdeletions in the *CNTNAP2* gene associated with ASD (Alarcon et al., 2008; Al-Murrani, Ashton, Aftimos, George, & Love, 2012; Arking et al., 2008; Poot et al., 2010). There is also evidence that a common genetic variant affects inter-region connectivity in the human PFC, with carriers of the risk variant having impaired functional connectivity and significant reductions in grey and white matter volume (Scott-Van Zeeland et al., 2010; Tan, Doke, Ashburner, Wood, & Frackowiak, 2010).

Loss of CNTNAP2 reduces dendritic spine density and *in vivo* imaging in the somatosensory cortex of mice shows that it contributes to an accelerated loss of spines, suggesting CNTNAP2 plays a role in the stabilization of excitatory synaptic connections (Gdalyahu et al., 2015; Varea et al., 2015). In the mPFC, a decrease in both excitatory and inhibitory synaptic inputs was found in cell layer II/III in CNTNAP2 knock-out mice (Lazaro et al., 2019). Further, analyses of local field potentials and unit spiking in awake CNTNAP2 knock-out mice found impairments in oscillations, in agreement with a reduction in coordinated neuronal population activity (Lazaro et al., 2019).

*6.1.4 SynCAMs (*CADM *genes)*

Two different missense mutations (H246N and Y251S) in the gene encoding SynCAM 1 have been identified in ASD patients (Zhiling et al., 2008). Both mutations occur in the immunoglobulin domain that is proximal to the cell membrane, which is required for its lateral *cis* interactions (Fogel, Stagi, Perez de Arce, & Biederer, 2011). These mutations render SynCAM 1 more susceptible to protease cleavage, alter its intracellular trafficking, and shorten dendrite length (Fujita et al., 2010; Zhiling et al., 2008). A biological concept-based analysis of ASD-linked SNPs that was cross-validated with patient gene expression data identified several disease-linked clusters, with a prominent cluster for adhesion that includes the gene encoding SynCAM 1 (Esteban, Tonellato, & Wall, 2020). SynCAM 2, a heterophilic binding partner of SynCAM 1, has also been implicated through GWAS in ASD, and is additionally linked to attention-deficit hyperactive disorder, cognitive processing speed and educational attainment (Albayrak et al., 2013; Casey et al., 2012; Davies et al., 2016; Ibrahim-Verbaas et al., 2016; Okbay et al., 2016). Mice in which SynCAM 1 is deleted exhibit lower dendritic spine density and a reduction in excitatory transmission in the hippocampal CA1 area (Robbins et al., 2010) as well as impaired connectivity and E/I balance in the CA3 region (Park et al., 2016). Loss of SynCAM 1 impacts the cortex, too, and reduces thalamocortical inputs to PV-positive interneurons and impedes inhibitory maturation (Ribic et al., 2019).

*6.1.5 Cadherins and protocadherins (*CDH *and* PCDH *genes)*

Single nucleotide polymorphisms (SNPs) close to the genes encoding Cadherin 8, 9 and 10 (*CDH8-10*) are strongly associated with ASD, as well as large deletions in Cadherin 13 (*CDH13*) (Lin, Frei, Kilander, Shen, & Blatt, 2016; Sanders et al., 2011; Wang et al., 2009). The Protocadherin family is also implicated in ASD. Multiple SNPs within the Protocadherin-α gene cluster show significant associations with autism (Anitha et al., 2013). CNVs in Protocadherin 9 (encoded by *PCDH9*) and homozygous deletions in Protocadherin 10 (*PCH10*) have been reported in ASD cases, as well as numerous Protocadherin 19 mutations in families with members diagnosed with epilepsy and mental retardation. Five of these mutations result in early stop codons and two are missense mutations that are predicted to affect calcium binding (Dibbens et al., 2008; Marshall et al., 2008; Morrow et al., 2008). Correspondingly, loss of Protocadherins in mice results in autism-relevant phenotypes. Protocadherin 19 knock-out male mice show abnormal sociability as well as increased grooming, while Protocadherin 10 knock-out male mice exhibit social novelty defects and abnormal gamma oscillations in the basolateral amygdala (Lim, Ryu, Kang, Noh, & Kim, 2019; Schoch et al., 2017).

6.2 Schizophrenia

Schizophrenia, a neurodevelopmental disorder also considered to be synaptic in pathology, presents later in adolescence in a time period that corresponds with the final maturation of the PFC. Postmortem studies of schizophrenia patients show decreased dendritic spine density in layer II/III in the PFC, pointing to a significant loss in synapse number (Garey et al., 1998; Glantz & Lewis, 2000) Schizophrenia patients also display altered PFC gamma band oscillations which are driven through PV-positive interneurons and are presumed to synchronize local cortical networks (McNally & McCarley, 2016). These lower gamma band oscillations in schizophrenia patients suggest chronically dysfunctional long-range synaptic transmission in the cortex (Chen et al., 2014; Grent-t-Jong et al., 2018; Grutzner et al., 2013).

While there has been significantly less success in identifying *de novo* mutations in proteins in schizophrenia patients as compared to ASD due to the challenging genetic heterogeneity of schizophrenia, GWAS studies have provided substantial progress. Genetic risk loci include synapse organizing proteins, suggesting that schizophrenia is a disorder that involves, at least in part, improper synaptic connectivity during development

(Schizophrenia Working Group of the Psychiatric Genomics, 2014). We highlight below representative schizophrenia-relevant synapse organizing proteins for which animal studies have been performed.

6.2.1 *Neurexins*

Neurexin-1α has been implicated in schizophrenia through copy number variant deletions and duplications (Gauthier et al., 2011; Kirov et al., 2009; Rujescu et al., 2009). As outlined for ASD, Neurexin-1α mice have reduced excitatory synapse strength in the hippocampus, as well as altered behaviors. Specifically, Neurexin-1α knock-out mice display impaired pre-pulse inhibition (PPI), a behavioral assay that measures sensory gating and attentive processing, which is also impaired in human schizophrenia patients (Etherton et al., 2009).

6.2.2 *Neuroligins*

Damaging missense mutations have been found in the *NLGN2* gene in a cohort of schizophrenia patients. One disease-linked mutant (R215H), fails to bind presynaptic Neurexin and disrupts GABA transmission in a reconstituted system (Sun et al., 2011). Introducing the R215H mutation into Neuroligin-2 in mice results in reduced miniature and evoked inhibitory post-synaptic currents and abnormal gamma oscillations in the PFC (Chen et al., 2020). Neuroligin-2 R215H knock-in mice also exhibit reduced inhibitory synaptic transmission in the hippocampus, with lower inhibitory synaptic marker density and impaired memory processes (Jiang et al., 2018). In the PPI test, Neuroligin-2 R215H mice have performed differently given the study; in one they display impaired PPI, in another enhanced, but this effect may be due to differences in the genetic makeup of the mice analyzed (Chen, Lee, Liao, & Chang, 2017; Jiang et al., 2018).

6.2.3 *Leucine-rich repeat transmembrane proteins* (LRRTM *genes)*

The postsynaptic adhesion molecule and Neurexin partner LRRTM1 is encoded by an imprinted gene (disease risk associated with paternal inheritance) and hypomethylation of the promoter significantly increases the risk of developing schizophrenia (Brucato, DeLisi, Fisher, & Francks, 2014; Francks et al., 2007; Ludwig et al., 2009). LRRTM1 is also linked to schizotypy in a non-clinical population (Leach, Prefontaine, Hurd, & Crespi, 2014). Altered excitatory presynaptic protein distribution in the CA1 but not the CA3 region of the hippocampus and a significant decrease in excitatory synapses in the CA1 stratum radiatum have been observed in LRRTM1 knock-out mice, indicating that LRRTM1 supports hippocampal connectivity (Linhoff et al., 2009;

Schroeder & de Wit, 2018; Takashima et al., 2011). To what extent these phenotypes involve interactions with Neurexins is currently unknown.

7. Outlook

As reviewed in this chapter, recognition molecules are now known to play critical roles in neuronal partner identification and the formation and specification of synapses. What are the next key questions? On a functional level, the roles of recognition during the sequential steps of partner contact and synapse assembly remain to be elucidated. As reviewed here, multiple examples exist for proteins that are required and sufficient for neuronal partner recognition. This is not the case for synapse formation, where several proteins are sufficient to drive this process, but none is required. Future studies can test whether select factors that control neuronal partner recognition switch roles after contact and work with synapse-organizing proteins to initiate synapse development. These recognition processes may include positive cooperation as well as competition between recognition factors, mechanisms that we are only beginning to grasp. From a molecular perspective, it will be important to delineate the stoichiometry, subsynaptic distribution and dynamic properties of recognition factors at different synapse types, aiming to reach single synapse resolution. These measurements will determine to what extent the relative abundance and dynamics of recognition factors guide the trajectories along which different synapse types emerge.

The design of future studies will benefit from comparing brain regions and neuron types as this provides opportunities to determine contextual functions of synaptic recognition factors. Region- and neuron-specific roles can be of high relevance for understanding disease processes, including why certain brain areas are more vulnerable to synaptic aberrations than others. An additional health-relevant goal will be to analyze the roles of synaptic recognition once development has been completed. Many of the molecules discussed here persist at mature synapses, indicating functions beyond development that may include synapse maintenance, synaptic plasticity, synapse-type specific control of network maturation, or circuit remodeling. A better understanding of how synaptic recognition shapes the mature CNS can generate leads for therapeutic intervention in disorders of the adult and aging brain and for healthy aging.

These new directions are bound to advance our knowledge of how recognition molecules provide for the precise connectivity of the CNS and the astounding structural and functional diversity of synapses.

Acknowledgments

The authors would like to thank Drs. Deanna Benson, George Huntley, Roman Giger, the Reviewer, and members of the Biederer group for their helpful input. Synaptic biology research in the Biederer group is supported by NIH grants R01 DA018928, R01 MH119826 and R21 NS109637 (to T.B.).

References

Acker, D. W. M., Wong, I., Kang, M., & Paradis, S. (2018). Semaphorin 4D promotes inhibitory synapse formation and suppresses seizures in vivo. *Epilepsia, 59*(6), 1257–1268. https://doi.org/10.1111/epi.14429.

Al-Amoudi, A., Diez, D. C., Betts, M. J., & Frangakis, A. S. (2007). The molecular architecture of cadherins in native epidermal desmosomes. *Nature, 450*(7171), 832–837. https://doi.org/10.1038/nature05994.

Alarcon, M., Abrahams, B. S., Stone, J. L., Duvall, J. A., Perederiy, J. V., Bomar, J. M., et al. (2008). Linkage, association, and gene-expression analyses identify CNTNAP2 as an autism-susceptibility gene. *American Journal of Human Genetics, 82*(1), 150–159. https://doi.org/10.1016/j.ajhg.2007.09.005.

Albayrak, O., Putter, C., Volckmar, A. L., Cichon, S., Hoffmann, P., Nothen, M. M., et al. (2013). Common obesity risk alleles in childhood attention-deficit/hyperactivity disorder. *American Journal of Medical Genetics. Part B, Neuropsychiatric Genetics, 162B*(4), 295–305. https://doi.org/10.1002/ajmg.b.32144.

Al-Murrani, A., Ashton, F., Aftimos, S., George, A. M., & Love, D. R. (2012). Amino-terminal microdeletion within the CNTNAP2 gene associated with variable expressivity of speech delay. *Case Reports in Genetics, 2012*, 172408. https://doi.org/10.1155/2012/172408.

Anderson, G. R., Maxeiner, S., Sando, R., Tsetsenis, T., Malenka, R. C., & Südhof, T. C. (2017). Postsynaptic adhesion GPCR latrophilin-2 mediates target recognition in entorhinal-hippocampal synapse assembly. *The Journal of Cell Biology, 216*(11), 3831–3846. https://doi.org/10.1083/jcb.201703042.

Ango, F., di Cristo, G., Higashiyama, H., Bennett, V., Wu, P., & Huang, Z. J. (2004). Ankyrin-based subcellular gradient of neurofascin, an immunoglobulin family protein, directs GABAergic innervation at purkinje axon initial segment. *Cell, 119*(2), 257–272. https://doi.org/10.1016/j.cell.2004.10.004.

Anitha, A., Thanseem, I., Nakamura, K., Yamada, K., Iwayama, Y., Toyota, T., et al. (2013). Protocadherin alpha (PCDHA) as a novel susceptibility gene for autism. *Journal of Psychiatry & Neuroscience, 38*(3), 192–198. https://doi.org/10.1503/jpn.120058.

Antoine, M. W., Langberg, T., Schnepel, P., & Feldman, D. E. (2019). Increased excitation-inhibition ratio stabilizes synapse and circuit excitability in four autism mouse models. *Neuron, 101*(4), 648–661.e644. https://doi.org/10.1016/j.neuron.2018.12.026.

Aoto, J., Martinelli, D. C., Malenka, R. C., Tabuchi, K., & Südhof, T. C. (2013). Presynaptic neurexin-3 alternative splicing trans-synaptically controls postsynaptic AMPA receptor trafficking. *Cell, 154*(1), 75–88. https://doi.org/10.1016/j.cell.2013.05.060.

Aoto, J., Ting, P., Maghsoodi, B., Xu, N., Henkemeyer, M., & Chen, L. (2007). Postsynaptic EphrinB3 promotes shaft glutamatergic synapse formation. *The Journal of Neuroscience, 27*(28), 7508–7519.

Apóstolo, N., & de Wit, J. (2019). Compartmentalized distributions of neuronal and glial cell-surface proteins pattern the synaptic network. *Current Opinion in Neurobiology, 57*, 126–133. https://doi.org/10.1016/j.conb.2019.01.025.

Apóstolo, N., Smukowski, S. N., Vanderlinden, J., Condomitti, G., Rybakin, V., Ten Bos, J., et al. (2020). Synapse type-specific proteomic dissection identifies IgSF8 as a hippocampal CA3 microcircuit organizer. *Nature Communications*, *11*(1). https://doi.org/10.1038/s41467-020-18956-x.

Arac, D., & Li, J. (2019). Teneurins and latrophilins: Two giants meet at the synapse. *Current Opinion in Structural Biology*, *54*, 141–151. https://doi.org/10.1016/j.sbi.2019.01.028.

Arking, D. E., Cutler, D. J., Brune, C. W., Teslovich, T. M., West, K., Ikeda, M., et al. (2008). A common genetic variant in the neurexin superfamily member CNTNAP2 increases familial risk of autism. *American Journal of Human Genetics*, *82*(1), 160–164. https://doi.org/10.1016/j.ajhg.2007.09.015.

Ashrafi, S., Betley, J. N., Comer, J. D., Brenner-Morton, S., Bar, V., Shimoda, Y., et al. (2014). Neuronal Ig/Caspr recognition promotes the formation of axoaxonic synapses in mouse spinal cord. *Neuron*, *81*(1), 120–129. https://doi.org/10.1016/j.neuron.2013.10.060.

Assous, M., Martinez, E., Eisenberg, C., Shah, F., Kosc, A., Varghese, K., et al. (2019). Neuropilin 2 signaling mediates corticostriatal transmission, spine maintenance, and goal-directed learning in mice. *The Journal of Neuroscience*, *39*(45), 8845–8859. https://doi.org/10.1523/JNEUROSCI.1006-19.2019.

Bagri, A., Cheng, H. J., Yaron, A., Pleasure, S. J., & Tessier-Lavigne, M. (2003). Stereotyped pruning of long hippocampal axon branches triggered by retraction inducers of the semaphorin family. *Cell*, *113*(3), 285–299. https://doi.org/10.1016/s0092-8674(03)00267-8.

Bekirov, I. H., Needleman, L. A., Zhang, W., & Benson, D. L. (2002). Identification and localization of multiple classic cadherins in developing rat limbic system. *Neuroscience*, *115*(1), 213–227. https://doi.org/10.1016/s0306-4522(02)00375-5.

Benson, D. L., & Huntley, G. W. (2012). Building and remodeling synapses. *Hippocampus*, *22*(5), 954–968. https://doi.org/10.1002/hipo.20872.

Berns, D. S., DeNardo, L. A., Pederick, D. T., & Luo, L. (2018). Teneurin-3 controls topographic circuit assembly in the hippocampus. *Nature*, *554*(7692), 328–333. https://doi.org/10.1038/nature25463.

Biederer, T., Kaeser, P. S., & Blanpied, T. A. (2017). Transcellular nanoalignment of synaptic function. *Neuron*, *96*(3), 680–696. https://doi.org/10.1016/j.neuron.2017.10.006.

Biederer, T., Sara, Y., Mozhayeva, M., Atasoy, D., Liu, X., Kavalali, E. T., et al. (2002). SynCAM, a synaptic adhesion molecule that drives synapse assembly. *Science*, *297*(5586), 1525–1531. https://doi.org/10.1126/science.1072356.

Biederer, T., & Scheiffele, P. (2007). Mixed-culture assays for analyzing neuronal synapse formation. *Nature Protocols*, *2*(3), 670–676. https://doi.org/10.1038/nprot.2007.92.

Biederer, T., & Shrestha, N. (2015). SynCAM proteins. In M. Caplan (Ed.), *Reference module in biomedical research* Elsevier.

Bjartmar, L., Huberman, A. D., Ullian, E. M., Renteria, R. C., Liu, X., Xu, W., et al. (2006). Neuronal pentraxins mediate synaptic refinement in the developing visual system. *The Journal of Neuroscience*, *26*(23), 6269–6281. https://doi.org/10.1523/JNEUROSCI.4212-05.2006.

Bolliger, M. F., Martinelli, D. C., & Südhof, T. C. (2011). The cell-adhesion G protein-coupled receptor BAI3 is a high-affinity receptor for C1q-like proteins. *Proceedings of the National Academy of Sciences of the United States of America*, *108*(6), 2534–2539. https://doi.org/10.1073/pnas.1019577108.

Boucard, A. A., Ko, J., & Südhof, T. C. (2012). High affinity neurexin binding to cell adhesion G-protein-coupled receptor CIRL1/latrophilin-1 produces an intercellular adhesion complex. *The Journal of Biological Chemistry*, *287*(12), 9399–9413. https://doi.org/10.1074/jbc.M111.318659.

Bozdagi, O., Valcin, M., Poskanzer, K., Tanaka, H., & Benson, D. L. (2004). Temporally distinct demands for classic cadherins in synapse formation and maturation. *Molecular and Cellular Neurosciences*, *27*(4), 509–521.

Brasch, J., Harrison, O. J., Honig, B., & Shapiro, L. (2012). Thinking outside the cell: How cadherins drive adhesion. *Trends in Cell Biology*, *22*(6), 299–310. https://doi.org/10.1016/j.tcb.2012.03.004.

Brigidi, G. S., & Bamji, S. X. (2011). Cadherin-catenin adhesion complexes at the synapse. *Current Opinion in Neurobiology*, *21*(2), 208–214. https://doi.org/10.1016/j.conb.2010.12.004.

Brucato, N., DeLisi, L. E., Fisher, S. E., & Francks, C. (2014). Hypomethylation of the paternally inherited LRRTM1 promoter linked to schizophrenia. *American Journal of Medical Genetics. Part B, Neuropsychiatric Genetics*, *165B*(7), 555–563. https://doi.org/10.1002/ajmg.b.32258.

Bury, L. A., & Sabo, S. L. (2016). Building a terminal: Mechanisms of presynaptic development in the CNS. *The Neuroscientist*, *22*(4), 372–391. https://doi.org/10.1177/1073858415596131.

Cao, W., Lin, S., Xia, Q. Q., Du, Y. L., Yang, Q., Zhang, M. Y., et al. (2018). Gamma oscillation dysfunction in mPFC leads to social deficits in Neuroligin 3 R451C knockin mice. *Neuron*, *97*(6), 1253–1260. https://doi.org/10.1016/j.neuron.2018.04.025.

Cardin, J. A. (2018). Inhibitory interneurons regulate temporal precision and correlations in cortical circuits. *Trends in Neurosciences*, *41*(10), 689–700. https://doi.org/10.1016/j.tins.2018.07.015.

Casey, J. P., Magalhaes, T., Conroy, J. M., Regan, R., Shah, N., Anney, R., et al. (2012). A novel approach of homozygous haplotype sharing identifies candidate genes in autism spectrum disorder. *Human Genetics*, *131*(4), 565–579. https://doi.org/10.1007/s00439-011-1094-6.

Chamma, I., & Thoumine, O. (2018). Dynamics, nanoscale organization, and function of synaptic adhesion molecules. *Molecular and Cellular Neurosciences*, *91*, 95–107. https://doi.org/10.1016/j.mcn.2018.04.007.

Chanda, S., Hale, W. D., Zhang, B., Wernig, M., & Südhof, T. C. (2017). Unique versus redundant functions of Neuroligin genes in shaping excitatory and inhibitory synapse properties. *The Journal of Neuroscience*, *37*(29), 6816–6836. https://doi.org/10.1523/JNEUROSCI.0125-17.2017.

Cheadle, L, Rivera, S. A., Phelps, J. S., Ennis, K. A., et al. (2020). Sensory experience engages microglia to shape neural connectivity through a non-phagocytic mechanism. *Neuron*, *108*(3), 451–468. https://doi.org/10.1016/j.neuron.2020.08.002.

Chedotal, A., & Sotelo, C. (1993). The 'creeper stage' in cerebellar climbing fiber synaptogenesis precedes the 'pericellular nest'- -ultrastructural evidence with parvalbumin immunocytochemistry. *Brain Research. Developmental Brain Research*, *76*(2), 207–220. https://doi.org/10.1016/0165-3806(93)90209-s.

Chen, J., Dong, B., Feng, X., Jiang, D., Chen, G., Long, C., et al. (2020). Aberrant mPFC GABAergic synaptic transmission and fear behavior in neuroligin-2 R215H knock-in mice. *Brain Research*, *1730*. https://doi.org/10.1016/j.brainres.2020.146671, 146671.

Chen, L. Y., Jiang, M., Zhang, B., Gokce, O., & Südhof, T. C. (2017). Conditional deletion of all neurexins defines diversity of essential synaptic organizer functions for neurexins. *Neuron*, *94*(3), 611–625.e614. https://doi.org/10.1016/j.neuron.2017.04.011.

Chen, C. H., Lee, P. W., Liao, H. M., & Chang, P. K. (2017). Neuroligin 2 R215H mutant mice manifest anxiety, increased prepulse inhibition, and impaired spatial learning and memory. *Frontiers in Psychiatry*, *8*, 257. https://doi.org/10.3389/fpsyt.2017.00257.

Chen, C. M., Stanford, A. D., Mao, X., Abi-Dargham, A., Shungu, D. C., Lisanby, S. H., et al. (2014). GABA level, gamma oscillation, and working memory performance in schizophrenia. *NeuroImage: Clinical*, *4*, 531–539. https://doi.org/10.1016/j.nicl.2014.03.007.

Cheng, T. W., Liu, X. B., Faulkner, R. L., Stephan, A. H., Barres, B. A., Huberman, A. D., et al. (2010). Emergence of lamina-specific retinal ganglion cell connectivity by axon arbor retraction and synapse elimination. *The Journal of Neuroscience*, *30*(48), 16376–16382. https://doi.org/10.1523/JNEUROSCI.3455-10.2010.

Chubykin, A. A., Atasoy, D., Etherton, M. R., Brose, N., Kavalali, E. T., Gibson, J. R., et al. (2007). Activity-dependent validation of excitatory versus inhibitory synapses by neuroligin-1 versus neuroligin-2. *Neuron*, *54*(6), 919–931. https://doi.org/10.1016/j.neuron.2007.05.029.

Chung, W. S., Clarke, L. E., Wang, G. X., Stafford, B. K., Sher, A., Chakraborty, C., et al. (2013). Astrocytes mediate synapse elimination through MEGF10 and MERTK pathways. *Nature*, *504*(7480), 394–400. https://doi.org/10.1038/nature12776.

Cijsouw, T., Ramsey, A. M., Lam, T. T., Carbone, B. E., Blanpied, T. A., & Biederer, T. (2018). Mapping the proteome of the synaptic cleft through proximity labeling reveals new cleft proteins. *Proteomes*, *6*(4). https://doi.org/10.3390/proteomes6040048.

Condomitti, G., & de Wit, J. (2018). Heparan sulfate proteoglycans as emerging players in synaptic specificity. *Frontiers in Molecular Neuroscience*, *11*. https://doi.org/10.3389/fnmol.2018.00014.

Cornew, L., Roberts, T. P., Blaskey, L., & Edgar, J. C. (2012). Resting-state oscillatory activity in autism spectrum disorders. *Journal of Autism and Developmental Disorders*, *42*(9), 1884–1894. https://doi.org/10.1007/s10803-011-1431-6.

Crepel, F., Delhaye-Bouchaud, N., & Dupont, J. L. (1981). Fate of the multiple innervation of cerebellar Purkinje cells by climbing fibers in immature control, x-irradiated and hypothyroid rats. *Brain Research*, *227*(1), 59–71. https://doi.org/10.1016/0165-3806(81)90094-8.

Crepel, F., Mariani, J., & Delhaye-Bouchaud, N. (1976). Evidence for a multiple innervation of Purkinje cells by climbing fibers in the immature rat cerebellum. *Journal of Neurobiology*, 7(6), 567–578. https://doi.org/10.1002/neu.480070609.

Czondor, K., Garcia, M., Argento, A., Constals, A., Breillat, C., Tessier, B., et al. (2013). Micropatterned substrates coated with neuronal adhesion molecules for high-content study of synapse formation. *Nature Communications*, *4*, 2252. https://doi.org/10.1038/ncomms3252.

Dabrowski, A., Terauchi, A., Strong, C., & Umemori, H. (2015). Distinct sets of FGF receptors sculpt excitatory and inhibitory synaptogenesis. *Development*, *142*(10), 1818–1830. https://doi.org/10.1242/dev.115568.

Dalva, M. B., Takasu, M. A., Lin, M. Z., Shamah, S. M., Hu, L., Gale, N. W., et al. (2000). EphB receptors interact with NMDA receptors and regulate excitatory synapse formation. *Cell*, *103*(6), 945–956. https://doi.org/10.1016/s0092-8674(00)00197-5.

Davies, G., Marioni, R. E., Liewald, D. C., Hill, W. D., Hagenaars, S. P., Harris, S. E., et al. (2016). Genome-wide association study of cognitive functions and educational attainment in UK Biobank. *Molecular Psychiatry*, *21*(6), 758–767. https://doi.org/10.1038/mp.2016.45.

De Jaco, A., Kovarik, Z., Comoletti, D., Jennings, L. L., Gaietta, G., Ellisman, M. H., et al. (2005). A single mutation near the C-terminus in alpha/beta hydrolase fold protein family causes a defect in protein processing. *Chemico-Biological Interactions*, *157-158*, 371–372. https://doi.org/10.1016/j.cbi.2005.10.057.

de Wit, J., O'Sullivan, M. L., Savas, J. N., Condomitti, G., Caccese, M. C., Vennekens, K. M., et al. (2013). Unbiased discovery of glypican as a receptor for LRRTM4 in regulating excitatory synapse development. *Neuron, 79*(4), 696–711. https://doi.org/10.1016/j.neuron.2013.06.049.

Demyanenko, G. P., Mohan, V., Zhang, X., Brennaman, L. H., Dharbal, K. E., Tran, T. S., et al. (2014). Neural cell adhesion molecule NrCAM regulates Semaphorin 3F-induced dendritic spine remodeling. *The Journal of Neuroscience, 34*(34), 11274–11287. https://doi.org/10.1523/JNEUROSCI.1774-14.2014.

DeNardo, L. A., de Wit, J., Otto-Hitt, S., & Ghosh, A. (2012). NGL-2 regulates input-specific synapse development in CA1 pyramidal neurons. *Neuron, 76*(4), 762–775. https://doi.org/10.1016/j.neuron.2012.10.013.

Dibbens, L. M., Tarpey, P. S., Hynes, K., Bayly, M. A., Scheffer, I. E., Smith, R., et al. (2008). X-linked protocadherin 19 mutations cause female-limited epilepsy and cognitive impairment. *Nature Genetics, 40*(6), 776–781. https://doi.org/10.1038/ng.149.

Ding, J. B., Oh, W. J., Sabatini, B. L., & Gu, C. (2011). Semaphorin 3E-Plexin-D1 signaling controls pathway-specific synapse formation in the striatum. *Nature Neuroscience, 15*(2), 215–223. https://doi.org/10.1038/nn.3003.

Duan, X., Krishnaswamy, A., De la Huerta, I., & Sanes, J. R. (2014). Type II cadherins guide assembly of a direction-selective retinal circuit. *Cell, 158*(4), 793–807. https://doi.org/10.1016/j.cell.2014.06.047.

Duan, X., Krishnaswamy, A., Laboulaye, M. A., Liu, J., Peng, Y. R., Yamagata, M., et al. (2018). Cadherin combinations recruit dendrites of distinct retinal neurons to a shared interneuronal scaffold. *Neuron, 99*(6), 1145–1154. https://doi.org/10.1016/j.neuron.2018.08.019.

Duan, Y., Wang, S. H., Song, J., Mironova, Y., Ming, G. L., Kolodkin, A. L., et al. (2014). Semaphorin 5A inhibits synaptogenesis in early postnatal- and adult-born hippocampal dentate granule cells. *eLife, 3*. https://doi.org/10.7554/eLife.04390.

Dudanova, I., Tabuchi, K., Rohlmann, A., Südhof, T. C., & Missler, M. (2007). Deletion of alpha-neurexins does not cause a major impairment of axonal pathfinding or synapse formation. *The Journal of Comparative Neurology, 502*(2), 261–274. https://doi.org/10.1002/cne.21305.

Emperador-Melero, J., & Kaeser, P. S. (2020). Assembly of the presynaptic active zone. *Current Opinion in Neurobiology, 63*, 95–103. https://doi.org/10.1016/j.conb.2020.03.008.

Esclassan, F., Francois, J., Phillips, K. G., Loomis, S., & Gilmour, G. (2015). Phenotypic characterization of nonsocial behavioral impairment in neurexin 1α knockout rats. *Behavioral Neuroscience, 129*(1), 74–85. https://doi.org/10.1037/bne0000024.

Esteban, F. J., Tonellato, P. J., & Wall, D. P. (2020). Enrichment of genomic variation in pathways linked to autism. *bioRxiv*. https://doi.org/10.1101/2020.10.19.346072.

Etherton, M. R., Blaiss, C. A., Powell, C. M., & Südhof, T. C. (2009). Mouse neurexin-1alpha deletion causes correlated electrophysiological and behavioral changes consistent with cognitive impairments. *Proceedings of the National Academy of Sciences of the United States of America, 106*(42), 17998–18003. https://doi.org/10.1073/pnas.0910297106.

Favuzzi, E., Deogracias, R., Marques-Smith, A., Maeso, P., Jezequel, J., Exposito-Alonso, D., et al. (2019). Distinct molecular programs regulate synapse specificity in cortical inhibitory circuits. *Science, 363*(6425), 413–417. https://doi.org/10.1126/science.aau8977.

Filipello, F., Morini, R., Corradini, I., Zerbi, V., Canzi, A., Michalski, B., et al. (2018). The microglial innate immune receptor TREM2 Is required for synapse elimination and normal brain connectivity. *Immunity, 48*(5), 979–991.e978. https://doi.org/10.1016/j.immuni.2018.04.016.

Fogel, A. I., Akins, M. R., Krupp, A. J., Stagi, M., Stein, V., & Biederer, T. (2007). SynCAMs organize synapses through heterophilic adhesion. *The Journal of Neuroscience*, *27*(46), 12516–12530. https://doi.org/10.1523/JNEUROSCI.2739-07.2007.
Fogel, A. I., Li, Y., Giza, J., Wang, Q., Lam, T. T., Modis, Y., et al. (2010). N-glycosylation at the SynCAM (Synaptic Cell Adhesion Molecule) immunoglobulin interface modulates synaptic adhesion. *The Journal of Biological Chemistry*, *285*(45), 34864–34874. https://doi.org/10.1074/jbc.M110.120865.
Fogel, A. I., Stagi, M., Perez de Arce, K., & Biederer, T. (2011). Lateral assembly of the immunoglobulin protein SynCAM 1 controls its adhesive function and instructs synapse formation. *The EMBO Journal*, *30*(23), 4728–4738. https://doi.org/10.1038/emboj.2011.336.
Foldy, C., Darmanis, S., Aoto, J., Malenka, R. C., Quake, S. R., & Südhof, T. C. (2016). Single-cell RNAseq reveals cell adhesion molecule profiles in electrophysiologically defined neurons. *Proceedings of the National Academy of Sciences of the United States of America*, *113*(35), E5222–E5231. https://doi.org/10.1073/pnas.1610155113.
Fossati, M., Assendorp, N., Gemin, O., Colasse, S., Dingli, F., Arras, G., et al. (2019). Trans-synaptic signaling through the Glutamate Receptor Delta-1 mediates inhibitory synapse formation in cortical pyramidal neurons. *Neuron*, *104*(6), 1081–1094. https://doi.org/10.1016/j.neuron.2019.09.027.
Francks, C., Maegawa, S., Lauren, J., Abrahams, B. S., Velayos-Baeza, A., Medland, S. E., et al. (2007). LRRTM1 on chromosome 2p12 is a maternally suppressed gene that is associated paternally with handedness and schizophrenia. *Molecular Psychiatry*, *12*(12), 1129–1139. 1057. https://doi.org/10.1038/sj.mp.4002053.
Frei, J. A., & Stoeckli, E. T. (2014). SynCAMs extend their functions beyond the synapse. *The European Journal of Neuroscience*, *39*(11), 1752–1760. https://doi.org/10.1111/ejn.12544.
Friedman, L. G., Benson, D. L., & Huntley, G. W. (2015). Cadherin-based transsynaptic networks in establishing and modifying neural connectivity. *Current Topics in Developmental Biology*, *112*, 415–465. https://doi.org/10.1016/bs.ctdb.2014.11.025.
Fuccillo, M. V., Foldy, C., Gokce, O., Rothwell, P. E., Sun, G. L., Malenka, R. C., et al. (2015). Single-cell mRNA profiling reveals cell-type-specific expression of Neurexin isoforms. *Neuron*, *87*(2), 326–340. https://doi.org/10.1016/j.neuron.2015.06.028.
Fuerst, P. G., Koizumi, A., Masland, R. H., & Burgess, R. W. (2008). Neurite arborization and mosaic spacing in the mouse retina require DSCAM. *Nature*, *451*(7177), 470–474. https://doi.org/10.1038/nature06514.
Fujita, E., Dai, H., Tanabe, Y., Zhiling, Y., Yamagata, T., Miyakawa, T., et al. (2010). Autism spectrum disorder is related to endoplasmic reticulum stress induced by mutations in the synaptic cell adhesion molecule, CADM1. *Cell Death & Disease*, *1*. https://doi.org/10.1038/cddis.2010.23.
Furlanis, E., Traunmüller, L., Fucile, G., & Scheiffele, P. (2019). Landscape of ribosome-engaged transcript isoforms reveals extensive neuronal-cell-class-specific alternative splicing programs. *Nature Neuroscience*, *22*(10), 1709–1717. https://doi.org/10.1038/s41593-019-0465-5.
Galuska, S. P., Rollenhagen, M., Kaup, M., Eggers, K., Oltmann-Norden, I., Schiff, M., et al. (2010). Synaptic cell adhesion molecule SynCAM 1 is a target for polysialylation in postnatal mouse brain. *Proceedings of the National Academy of Sciences of the United States of America*, *107*(22), 10250–10255. https://doi.org/10.1073/pnas.0912103107.
Gan, K. J., & Südhof, T. C. (2020). SPARCL1 promotes excitatory but not inhibitory synapse formation and function independent of Neurexins and Neuroligins. *The Journal of Neuroscience*, *40*(42), 8088–8102. https://doi.org/10.1523/JNEUROSCI.0454-20.2020.
Garey, L. J., Ong, W. Y., Patel, T. S., Kanani, M., Davis, A., Mortimer, A. M., et al. (1998). Reduced dendritic spine density on cerebral cortical pyramidal neurons in schizophrenia. *Journal of Neurology, Neurosurgery, and Psychiatry*, *65*(4), 446–453.

Garrett, A. M., Khalil, A., Walton, D. O., & Burgess, R. W. (2018). DSCAM promotes self-avoidance in the developing mouse retina by masking the functions of cadherin superfamily members. *Proceedings of the National Academy of Sciences of the United States of America*, *115*(43), E10216–E10224. https://doi.org/10.1073/pnas.1809430115.

Gauthier, J., Siddiqui, T. J., Huashan, P., Yokomaku, D., Hamdan, F. F., Champagne, N., et al. (2011). Truncating mutations in NRXN2 and NRXN1 in autism spectrum disorders and schizophrenia. *Human Genetics*, *130*(4), 563–573. https://doi.org/10.1007/s00439-011-0975-z.

Gdalyahu, A., Lazaro, M., Penagarikano, O., Golshani, P., Trachtenberg, J. T., & Geschwind, D. H. (2015). The autism related protein contactin-associated protein-Like 2 (CNTNAP2) stabilizes new spines: An in vivo mouse study. *PLoS One*, *10*(5). https://doi.org/10.1371/journal.pone.0125633, e0125633.

Geschwind, D. H., & State, M. W. (2015). Gene hunting in autism spectrum disorder: On the path to precision medicine. *Lancet Neurology*, *15*, S1474–S4422. https://doi.org/10.1016/S1474-4422(15)00044-7.

Gil, O. D., Needleman, L., & Huntley, G. W. (2002). Developmental patterns of cadherin expression and localization in relation to compartmentalized thalamocortical terminations in rat barrel cortex. *The Journal of Comparative Neurology*, *453*(4), 372–388. https://doi.org/10.1002/cne.10424.

Glantz, L. A., & Lewis, D. A. (2000). Decreased dendritic spine density on prefrontal cortical pyramidal neurons in schizophrenia. *Archives of General Psychiatry*, *57*(1), 65–73. https://doi.org/10.1001/archpsyc.57.1.65.

Grent-t-Jong, T., Gross, J., Goense, J., Wibral, M., Gajwani, R., Gumley, A. I., et al. (2018). Resting-state gamma-band power alterations in schizophrenia reveal E/I-balance abnormalities across illness-stages. *eLife*, 7. https://doi.org/10.7554/eLife.37799, e37799.

Grutzner, C., Wibral, M., Sun, L., Rivolta, D., Singer, W., Maurer, K., et al. (2013). Deficits in high-(>60 Hz) gamma-band oscillations during visual processing in schizophrenia. *Frontiers in Human Neuroscience*, 7. https://doi.org/10.3389/fnhum.2013.00088.

Han, K. A., Jeon, S., Um, J. W., & Ko, J. (2016). Emergent synapse organizers: LAR-RPTPs and their companions. *International Review of Cell and Molecular Biology*, *324*, 39–65. https://doi.org/10.1016/bs.ircmb.2016.01.002.

Han, K. A., Ko, J. S., Pramanik, G., Kim, J. Y., Tabuchi, K., Um, J. W., et al. (2018). PTPsigma drives excitatory presynaptic assembly via various extracellular and intracellular mechanisms. *The Journal of Neuroscience*, *38*(30), 6700–6721. https://doi.org/10.1523/JNEUROSCI.0672-18.2018.

Han, K. A., Woo, D., Kim, S., Choii, G., Jeon, S., Won, S. Y., et al. (2016). Neurotrophin-3 regulates synapse development by modulating TrkC-PTPsigma synaptic adhesion and intracellular signaling pathways. *The Journal of Neuroscience*, *36*(17), 4816–4831. https://doi.org/10.1523/JNEUROSCI.4024-15.2016.

Hashimoto, K., & Kano, M. (2005). Postnatal development and synapse elimination of climbing fiber to Purkinje cell projection in the cerebellum. *Neuroscience Research*, *53*(3), 221–228. https://doi.org/10.1016/j.neures.2005.07.007.

Henderson, N. T., & Dalva, M. B. (2018). EphBs and ephrin-Bs: Trans-synaptic organizers of synapse development and function. *Molecular and Cellular Neurosciences*, *91*, 108–121. https://doi.org/10.1016/j.mcn.2018.07.002.

Henkemeyer, M., Itkis, O. S., Ngo, M., Hickmott, P. W., & Ethell, I. M. (2003). Multiple EphB receptor tyrosine kinases shape dendritic spines in the hippocampus. *The Journal of Cell Biology*, *163*(6), 1313–1326. https://doi.org/10.1083/jcb.200306033.

Hirano, T. (2018). Purkinje neurons: Development, morphology, and function. *Cerebellum*, *17*(6), 699–700. https://doi.org/10.1007/s12311-018-0985-7.

Huh, G. S., Boulanger, L. M., Du, H., Riquelme, P. A., Brotz, T. M., & Shatz, C. J. (2000). Functional requirement for class I MHC in CNS development and plasticity. *Science*, *290*(5499), 2155–2159. https://doi.org/10.1126/science.290.5499.2155.

Hulpiau, P., & van Roy, F. (2009). Molecular evolution of the cadherin superfamily. *The International Journal of Biochemistry & Cell Biology*, *41*(2), 349–369. https://doi.org/10.1016/j.biocel.2008.09.027.

Huntley, G. W., & Benson, D. L. (1999). Neural (N)-cadherin at developing thalamocortical synapses provides an adhesion mechanism for the formation of somatopically organized connections. *The Journal of Comparative Neurology*, *407*(4), 453–471.

Hutsler, J. J., & Zhang, H. (2010). Increased dendritic spine densities on cortical projection neurons in autism spectrum disorders. *Brain Research*, *1309*, 83–94. https://doi.org/10.1016/j.brainres.2009.09.120.

Ibrahim-Verbaas, C. A., Bressler, J., Debette, S., Schuur, M., Smith, A. V., Bis, J. C., et al. (2016). GWAS for executive function and processing speed suggests involvement of the CADM2 gene. *Molecular Psychiatry*, *21*(2), 189–197. https://doi.org/10.1038/mp.2015.37.

Ing-Esteves, S., Kostadinov, D., Marocha, J., Sing, A. D., Joseph, K. S., Laboulaye, M. A., et al. (2018). Combinatorial effects of alpha- and gamma-protocadherins on neuronal survival and dendritic self-avoidance. *The Journal of Neuroscience*, *38*(11), 2713–2729. https://doi.org/10.1523/JNEUROSCI.3035-17.2018.

Inoue, A., & Sanes, J. R. (1997). Lamina-specific connectivity in the brain: Regulation by N-cadherin, neurotrophins, and glycoconjugates. *Science*, *276*(5317), 1428–1431. https://doi.org/10.1126/science.276.5317.1428.

Isshiki, M., Tanaka, S., Kuriu, T., Tabuchi, K., Takumi, T., & Okabe, S. (2014). Enhanced synapse remodelling as a common phenotype in mouse models of autism. *Nature Communications*, *5*, 4742. https://doi.org/10.1038/ncomms5742.

Ito-Ishida, A., Miyazaki, T., Miura, E., Matsuda, K., Watanabe, M., Yuzaki, M., et al. (2012). Presynaptically released Cbln1 induces dynamic axonal structural changes by interacting with GluD2 during cerebellar synapse formation. *Neuron*, *76*(3), 549–564. https://doi.org/10.1016/j.neuron.2012.07.027.

Jamain, S., Quach, H., Betancur, C., Rastam, M., Colineaux, C., Gillberg, I. C., et al. (2003). Mutations of the X-linked genes encoding neuroligins NLGN3 and NLGN4 are associated with autism. *Nature Genetics*, *34*(1), 27–29. https://doi.org/10.1038/ng1136.

Ji, K., Akgul, G., Wollmuth, L. P., & Tsirka, S. E. (2013). Microglia actively regulate the number of functional synapses. *PLoS One*, *8*(2), e56293. https://doi.org/10.1371/journal.pone.0056293.

Jiang, D. Y., Wu, Z., Forsyth, C. T., Hu, Y., Yee, S. P., & Chen, G. (2018). GABAergic deficits and schizophrenia-like behaviors in a mouse model carrying patient-derived neuroligin-2 R215H mutation. *Molecular Brain*, *11*(1), 31. https://doi.org/10.1186/s13041-018-0375-6.

Kakunaga, S., Ikeda, W., Itoh, S., Deguchi-Tawarada, M., Ohtsuka, T., Mizoguchi, A., et al. (2005). Nectin-like molecule-1/TSLL1/SynCAM3: A neural tissue-specific immunoglobulin-like cell-cell adhesion molecule localizing at non-junctional contact sites of presynaptic nerve terminals, axons and glia cell processes. *Journal of Cell Science*, *118*(Pt. 6), 1267–1277. https://doi.org/10.1242/jcs.01656.

Kano, M., Watanabe, T., Uesaka, N., & Watanabe, M. (2018). Multiple phases of climbing fiber synapse elimination in the developing cerebellum. *Cerebellum*, *17*(6), 722–734. https://doi.org/10.1007/s12311-018-0964-z.

Kasthuri, N., Hayworth, K. J., Berger, D. R., Schalek, R. L., Conchello, J. A., Knowles-Barley, S., et al. (2015). Saturated reconstruction of a volume of neocortex. *Cell*, *162*(3), 648–661. https://doi.org/10.1016/j.cell.2015.06.054.

Kayser, M. S., McClelland, A. C., Hughes, E. G., & Dalva, M. B. (2006). Intracellular and trans-synaptic regulation of glutamatergic synaptogenesis by EphB receptors. *The Journal of Neuroscience*, *26*(47), 12152–12164. https://doi.org/10.1523/JNEUROSCI.3072-06.2006.

Kim, K., Shin, W., Kang, M., Lee, S., Kim, D., Kang, R., et al. (2020). Presynaptic PTPsigma regulates postsynaptic NMDA receptor function through direct adhesion-independent mechanisms. *eLife, 9*, e54224. https://doi.org/10.7554/eLife.54224.
Kirov, G., Rujescu, D., Ingason, A., Collier, D. A., O'Donovan, M. C., & Owen, M. J. (2009). Neurexin 1 (NRXN1) deletions in schizophrenia. *Schizophrenia Bulletin, 35*(5), 851–854. https://doi.org/10.1093/schbul/sbp079.
Ko, J., Soler-Llavina, G. J., Fuccillo, M. V., Malenka, R. C., & Südhof, T. C. (2011). Neuroligins/LRRTMs prevent activity- and Ca2+/calmodulin-dependent synapse elimination in cultured neurons. *The Journal of Cell Biology, 194*(2), 323–334. https://doi.org/10.1083/jcb.201101072.
Koeppen, J., Nguyen, A. Q., Nikolakopoulou, A. M., Garcia, M., Hanna, S., Woodruff, S., et al. (2018). Functional consequences of synapse remodeling following astrocyte-specific regulation of Ephrin-B1 in the adult hippocampus. *The Journal of Neuroscience, 38*(25), 5710–5726. https://doi.org/10.1523/JNEUROSCI.3618-17.2018.
Krishnaswamy, A., Yamagata, M., Duan, X., Hong, Y. K., & Sanes, J. R. (2015). Sidekick 2 directs formation of a retinal circuit that detects differential motion. *Nature, 524*(7566), 466–470. https://doi.org/10.1038/nature14682.
Kuwako, K. I., Nishimoto, Y., Kawase, S., Okano, H. J., & Okano, H. (2014). Cadherin-7 regulates mossy fiber connectivity in the cerebellum. *Cell Reports, 9*(1), 311–323. https://doi.org/10.1016/j.celrep.2014.08.063.
Lazaro, M. T., Taxidis, J., Shuman, T., Bachmutsky, I., Ikrar, T., Santos, R., et al. (2019). Reduced prefrontal synaptic connectivity and disturbed oscillatory population dynamics in the CNTNAP2 model of autism. *Cell Reports, 27*(9), 2567–2578.e2566. https://doi.org/10.1016/j.celrep.2019.05.006.
Leach, E. L., Prefontaine, G., Hurd, P. L., & Crespi, B. J. (2014). The imprinted gene LRRTM1 mediates schizotypy and handedness in a nonclinical population. *Journal of Human Genetics, 59*(6), 332–336. https://doi.org/10.1038/jhg.2014.30.
Lee, K., Kim, Y., Lee, S. J., Qiang, Y., Lee, D., Lee, H. W., et al. (2013). MDGAs interact selectively with neuroligin-2 but not other neuroligins to regulate inhibitory synapse development. *Proceedings of the National Academy of Sciences of the United States of America, 110*(1), 336–341. https://doi.org/10.1073/pnas.1219987110.
Lee, H., Raiker, S. J., Venkatesh, K., Geary, R., Robak, L. A., Zhang, Y., et al. (2008). Synaptic function for the Nogo-66 receptor NgR1: Regulation of dendritic spine morphology and activity-dependent synaptic strength. *The Journal of Neuroscience, 28*(11), 2753–2765. https://doi.org/10.1523/JNEUROSCI.5586-07.2008.
Lefebvre, J. L. (2017). Neuronal territory formation by the atypical cadherins and clustered protocadherins. *Seminars in Cell & Developmental Biology, 69*, 111–121. https://doi.org/10.1016/j.semcdb.2017.07.040.
Lehrman, E. K., Wilton, D. K., Litvina, E. Y., Welsh, C. A., Chang, S. T., Frouin, A., et al. (2018). CD47 protects synapses from excess microglia-mediated pruning during development. *Neuron, 100*(1), 120–134. https://doi.org/10.1016/j.neuron.2018.09.017.
Li, J., Han, W., Pelkey, K. A., Duan, J., Mao, X., Wang, Y. X., et al. (2017). Molecular dissection of Neuroligin 2 and Slitrk3 reveals an essential framework for GABAergic synapse development. *Neuron, 96*(4), 808–826. https://doi.org/10.1016/j.neuron.2017.10.003.
Li, J., Shalev-Benami, M., Sando, R., Jiang, X., Kibrom, A., Wang, J., et al. (2018). Structural basis for Teneurin function in circuit-wiring: A toxin motif at the synapse. *Cell, 173*(3), 735–748.e715. https://doi.org/10.1016/j.cell.2018.03.036.
Lie, E., Li, Y., Kim, R., & Kim, E. (2018). SALM/Lrfn family synaptic adhesion molecules. *Frontiers in Molecular Neuroscience, 11*. https://doi.org/10.3389/fnmol.2018.00105.
Lim, J., Ryu, J., Kang, S., Noh, H. J., & Kim, C. H. (2019). Autism-like behaviors in male mice with a Pcdh19 deletion. *Molecular Brain, 12*(1). https://doi.org/10.1186/s13041-019-0519-3.

Lin, Y. C., Frei, J. A., Kilander, M. B., Shen, W., & Blatt, G. J. (2016). A subset of autism-associated genes regulate the structural stability of neurons. *Frontiers in Cellular Neuroscience*, *10*. https://doi.org/10.3389/fncel.2016.00263.

Linhoff, M. W., Lauren, J., Cassidy, R. M., Dobie, F. A., Takahashi, H., Nygaard, H. B., et al. (2009). An unbiased expression screen for synaptogenic proteins identifies the LRRTM protein family as synaptic organizers. *Neuron*, *61*(5), 734–749. https://doi.org/10.1016/j.neuron.2009.01.017.

Liu, X. B., Low, L. K., Jones, E. G., & Cheng, H. J. (2005). Stereotyped axon pruning via plexin signaling is associated with synaptic complex elimination in the hippocampus. *The Journal of Neuroscience*, *25*(40), 9124–9134. https://doi.org/10.1523/JNEUROSCI.2648-05.2005.

Loh, K. H., Stawski, P. S., Draycott, A. S., Udeshi, N. D., Lehrman, E. K., Wilton, D. K., et al. (2016). Proteomic analysis of unbounded cellular compartments: Synaptic clefts. *Cell*, *166*(5), 1295–1307. https://doi.org/10.1016/j.cell.2016.07.041.

Low, L. K., Liu, X. B., Faulkner, R. L., Coble, J., & Cheng, H. J. (2008). Plexin signaling selectively regulates the stereotyped pruning of corticospinal axons from visual cortex. *Proceedings of the National Academy of Sciences of the United States of America*, *105*(23), 8136–8141. https://doi.org/10.1073/pnas.0803849105.

Lowery, R. L., Tremblay, M. E., Hopkins, B. E., & Majewska, A. K. (2017). The microglial fractalkine receptor is not required for activity-dependent plasticity in the mouse visual system. *Glia*, *65*(11), 1744–1761. https://doi.org/10.1002/glia.23192.

Ludwig, K. U., Mattheisen, M., Muhleisen, T. W., Roeske, D., Schmal, C., Breuer, R., et al. (2009). Supporting evidence for LRRTM1 imprinting effects in schizophrenia. *Molecular Psychiatry*, *14*(8), 743–745. https://doi.org/10.1038/mp.2009.28.

Macosko, E. Z., Basu, A., Satija, R., Nemesh, J., Shekhar, K., Goldman, M., et al. (2015). Highly parallel genome-wide expression profiling of individual cells using nanoliter droplets. *Cell*, *161*(5), 1202–1214. https://doi.org/10.1016/j.cell.2015.05.002.

Mariani, J., & Changeux, J. P. (1981). Ontogenesis of olivocerebellar relationships. I. Studies by intracellular recordings of the multiple innervation of Purkinje cells by climbing fibers in the developing rat cerebellum. *The Journal of Neuroscience*, *1*(7), 696–702.

Marro, S. G., Chanda, S., Yang, N., Janas, J. A., Valperga, G., Trotter, J., et al. (2019). Neuroligin-4 regulates excitatory synaptic transmission in human neurons. *Neuron*, *103*(4), 617–626. https://doi.org/10.1016/j.neuron.2019.05.043.

Marshall, C. R., Noor, A., Vincent, J. B., Lionel, A. C., Feuk, L., Skaug, J., et al. (2008). Structural variation of chromosomes in autism spectrum disorder. *American Journal of Human Genetics*, *82*(2), 477–488. https://doi.org/10.1016/j.ajhg.2007.12.009.

Martin, E. A., Muralidhar, S., Wang, Z., Cervantes, D. C., Basu, R., Taylor, M. R., et al. (2015). The intellectual disability gene Kirrel3 regulates target-specific mossy fiber synapse development in the hippocampus. *eLife*, *4*, e09395. https://doi.org/10.7554/eLife.09395.

Martinelli, D. C., Chew, K. S., Rohlmann, A., Lum, M. Y., Ressl, S., Hattar, S., et al. (2016). Expression of C1ql3 in discrete neuronal populations controls efferent synapse numbers and diverse behaviors. *Neuron*, *91*(5), 1034–1051. https://doi.org/10.1016/j.neuron.2016.07.002.

Masai, I., Lele, Z., Yamaguchi, M., Komori, A., Nakata, A., Nishiwaki, Y., et al. (2003). N-cadherin mediates retinal lamination, maintenance of forebrain compartments and patterning of retinal neurites. *Development*, *130*(11), 2479–2494. https://doi.org/10.1242/dev.00465.

Matsuda, K., Miura, E., Miyazaki, T., Kakegawa, W., Emi, K., Narumi, S., et al. (2010). Cbln1 is a ligand for an orphan glutamate receptor delta2, a bidirectional synapse organizer. *Science*, *328*(5976), 363–368. https://doi.org/10.1126/science.1185152.

Matsukawa, H., Akiyoshi-Nishimura, S., Zhang, Q., Lujan, R., Yamaguchi, K., Goto, H., et al. (2014). Netrin-G/NGL complexes encode functional synaptic diversification. *The Journal of Neuroscience, 34*(47), 15779–15792. https://doi.org/10.1523/JNEUROSCI.1141-14.2014.

Matsuoka, R. L., Chivatakarn, O., Badea, T. C., Samuels, I. S., Cahill, H., Katayama, K., et al. (2011). Class 5 transmembrane semaphorins control selective mammalian retinal lamination and function. *Neuron, 71*(3), 460–473. https://doi.org/10.1016/j.neuron.2011.06.009.

Matsuoka, R. L., Nguyen-Ba-Charvet, K. T., Parray, A., Badea, T. C., Chedotal, A., & Kolodkin, A. L. (2011). Transmembrane semaphorin signalling controls laminar stratification in the mammalian retina. *Nature, 470*(7333), 259–263. https://doi.org/10.1038/nature09675.

McNally, J. M., & McCarley, R. W. (2016). Gamma band oscillations: A key to understanding schizophrenia symptoms and neural circuit abnormalities. *Current Opinion in Psychiatry, 29*(3), 202–210. https://doi.org/10.1097/YCO.0000000000000244.

Missler, M., Südhof, T. C., & Biederer, T. (2012). Synaptic cell adhesion. *Cold Spring Harbor Perspectives in Biology, 4*(4). https://doi.org/10.1101/cshperspect.a005694, a005694.

Missler, M., Zhang, W., Rohlmann, A., Kattenstroth, G., Hammer, R. E., Gottmann, K., et al. (2003). Alpha-neurexins couple Ca2+ channels to synaptic vesicle exocytosis. *Nature, 423*(6943), 939–948. https://doi.org/10.1038/nature01755.

Morrow, E. M., Yoo, S. Y., Flavell, S. W., Kim, T. K., Lin, Y., Hill, R. S., et al. (2008). Identifying autism loci and genes by tracing recent shared ancestry. *Science, 321*(5886), 218–223. https://doi.org/10.1126/science.1157657.

Motta, A., Berning, M., Boergens, K. M., Staffler, B., Beining, M., Loomba, S., et al. (2019). Dense connectomic reconstruction in layer 4 of the somatosensory cortex. *Science, 366*-(6469). https://doi.org/10.1126/science.aay3134.

Mountoufaris, G., Canzio, D., Nwakeze, C. L., Chen, W. V., & Maniatis, T. (2018). Writing, reading, and translating the clustered Protocadherin cell surface recognition code for neural circuit assembly. *Annual Review of Cell and Developmental Biology, 34*, 471–493. https://doi.org/10.1146/annurev-cellbio-100616-060701.

Nakanishi, M., Nomura, J., Ji, X., Tamada, K., Arai, T., Takahashi, E., et al. (2017). Functional significance of rare neuroligin 1 variants found in autism. *PLoS Genetics, 13*(8). https://doi.org/10.1371/journal.pgen.1006940, e1006940.

Nguyen, T. M., Schreiner, D., Xiao, L., Traunmuller, L., Bornmann, C., & Scheiffele, P. (2016). An alternative splicing switch shapes neurexin repertoires in principal neurons versus interneurons in the mouse hippocampus. *eLife, 5*, e22757. https://doi.org/10.7554/eLife.22757.

Nishimura-Akiyoshi, S., Niimi, K., Nakashiba, T., & Itohara, S. (2007). Axonal netrin-Gs transneuronally determine lamina-specific subdendritic segments. *Proceedings of the National Academy of Sciences of the United States of America, 104*(37), 14801–14806. https://doi.org/10.1073/pnas.0706919104.

Okbay, A., Beauchamp, J. P., Fontana, M. A., Lee, J. J., Pers, T. H., Rietveld, C. A., et al. (2016). Genome-wide association study identifies 74 loci associated with educational attainment. *Nature, 533*(7604), 539–542. https://doi.org/10.1038/nature17671.

Orekhova, E. V., Stroganova, T. A., Nygren, G., Tsetlin, M. M., Posikera, I. N., Gillberg, C., et al. (2007). Excess of high frequency electroencephalogram oscillations in boys with autism. *Biological Psychiatry, 62*(9), 1022–1029. https://doi.org/10.1016/j.biopsych.2006.12.029.

Osterhout, J. A., Josten, N., Yamada, J., Pan, F., Wu, S. W., Nguyen, P. L., et al. (2011). Cadherin-6 mediates axon-target matching in a non-image-forming visual circuit. *Neuron, 71*(4), 632–639. https://doi.org/10.1016/j.neuron.2011.07.006.

O'Sullivan, M. L., de Wit, J., Savas, J. N., Comoletti, D., Otto-Hitt, S., Yates, J. R., 3rd, et al. (2012). FLRT proteins are endogenous latrophilin ligands and regulate excitatory synapse development. *Neuron*, *73*(5), 903–910. https://doi.org/10.1016/j.neuron.2012.01.018.

Paolicelli, R. C., Bolasco, G., Pagani, F., Maggi, L., Scianni, M., Panzanelli, P., et al. (2011). Synaptic pruning by microglia is necessary for normal brain development. *Science*, *333*(6048), 1456–1458. https://doi.org/10.1126/science.1202529.

Park, K. A., Ribic, A., Laage Gaupp, F. M., Coman, D., Huang, Y., Dulla, C. G., et al. (2016). Excitatory synaptic drive and feedforward inhibition in the hippocampal CA3 circuit are regulated by SynCAM 1. *The Journal of Neuroscience*, *36*(28), 7464–7475. https://doi.org/10.1523/JNEUROSCI.0189-16.2016.

Pasterkamp, R. J. (2012). Getting neural circuits into shape with semaphorins. *Nature Reviews. Neuroscience*, *13*(9), 605–618. https://doi.org/10.1038/nrn3302.

Paul, A., Crow, M., Raudales, R., He, M., Gillis, J., & Huang, Z. J. (2017). Transcriptional architecture of synaptic communication delineates GABAergic neuron identity. *Cell*, *171*(3), 522–539.e520. https://doi.org/10.1016/j.cell.2017.08.032.

Pecho-Vrieseling, E., Sigrist, M., Yoshida, Y., Jessell, T. M., & Arber, S. (2009). Specificity of sensory-motor connections encoded by Sema3e-Plxnd1 recognition. *Nature*, *459*(7248), 842–846. https://doi.org/10.1038/nature08000.

Pelkey, K. A., Barksdale, E., Craig, M. T., Yuan, X., Sukumaran, M., Vargish, G. A., et al. (2015). Pentraxins coordinate excitatory synapse maturation and circuit integration of parvalbumin interneurons. *Neuron*, *85*(6), 1257–1272. https://doi.org/10.1016/j.neuron.2015.02.020.

Pelkey, K. A., Chittajallu, R., Craig, M. T., Tricoire, L., Wester, J. C., & McBain, C. J. (2017). Hippocampal GABAergic inhibitory interneurons. *Physiological Reviews*, *97*(4), 1619–1747. https://doi.org/10.1152/physrev.00007.2017.

Perez de Arce, K., Schrod, N., Metzbower, S. W. R., Allgeyer, E., Kong, G. K., Tang, A. H., et al. (2015). Topographic mapping of the synaptic cleft into adhesive nanodomains. *Neuron*, *88*(6), 1165–1172. https://doi.org/10.1016/j.neuron.2015.11.011.

Pettem, K. L., Yokomaku, D., Takahashi, H., Ge, Y., & Craig, A. M. (2013). Interaction between autism-linked MDGAs and neuroligins suppresses inhibitory synapse development. *The Journal of Cell Biology*, *200*(3), 321–336. https://doi.org/10.1083/jcb.201206028.

Poot, M., Beyer, V., Schwaab, I., Damatova, N., Van't Slot, R., Prothero, J., et al. (2010). Disruption of CNTNAP2 and additional structural genome changes in a boy with speech delay and autism spectrum disorder. *Neurogenetics*, *11*(1), 81–89. https://doi.org/10.1007/s10048-009-0205-1.

Poskanzer, K., Needleman, L. A., Bozdagi, O., & Huntley, G. W. (2003). N-cadherin regulates ingrowth and laminar targeting of thalamocortical axons. *The Journal of Neuroscience*, *23*(6), 2294–2305.

Ribic, A., & Biederer, T. (2019). Emerging roles of synapse organizers in the regulation of critical periods. *Neural Plasticity*, *2019*. https://doi.org/10.1155/2019/1538137, 1538137.

Ribic, A., Crair, M. C., & Biederer, T. (2019). Synapse-selective control of cortical maturation and plasticity by parvalbumin-autonomous action of SynCAM 1. *Cell Reports*, *26*(2), 381–393. https://doi.org/10.1016/j.celrep.2018.12.069.

Ribic, A., Liu, X., Crair, M. C., & Biederer, T. (2014). Structural organization and function of mouse photoreceptor ribbon synapses involve the immunoglobulin protein Synaptic Cell Adhesion Molecule 1. *The Journal of Comparative Neurology*, *522*(4), 900–920. https://doi.org/10.1002/cne.23452.

Riccomagno, M. M., & Kolodkin, A. L. (2015). Sculpting neural circuits by axon and dendrite pruning. *Annual Review of Cell and Developmental Biology*, *31*, 779–805. https://doi.org/10.1146/annurev-cellbio-100913-013038.

Robbins, E. M., Krupp, A. J., Perez de Arce, K., Ghosh, A. K., Fogel, A. I., Boucard, A., et al. (2010). SynCAM 1 adhesion dynamically regulates synapse number and impacts plasticity and learning. *Neuron*, *68*(5), 894–906. https://doi.org/10.1016/j.neuron.2010.11.003.

Roppongi, R. T., Dhume, S. H., Padmanabhan, N., Silwal, P., Zahra, N., Karimi, B., et al. (2020). LRRTMs organize synapses through differential engagement of Neurexin and PTPsigma. *Neuron*, *106*(1), 108–125.e112. https://doi.org/10.1016/j.neuron.2020.01.003.

Rothwell, P. E., Fuccillo, M. V., Maxeiner, S., Hayton, S. J., Gokce, O., Lim, B. K., et al. (2014). Autism-associated neuroligin-3 mutations commonly impair striatal circuits to boost repetitive behaviors. *Cell*, *158*(1), 198–212. https://doi.org/10.1016/j.cell.2014.04.045.

Rujescu, D., Ingason, A., Cichon, S., Pietilainen, O. P., Barnes, M. R., Toulopoulou, T., et al. (2009). Disruption of the neurexin 1 gene is associated with schizophrenia. *Human Molecular Genetics*, *18*(5), 988–996. https://doi.org/10.1093/hmg/ddn351.

Sahores, M., Gibb, A., & Salinas, P. C. (2010). Frizzled-5, a receptor for the synaptic organizer Wnt7a, regulates activity-mediated synaptogenesis. *Development*, *137*(13), 2215–2225. https://doi.org/10.1242/dev.046722.

Sanders, S. J., Ercan-Sencicek, A. G., Hus, V., Luo, R., Murtha, M. T., Moreno-De-Luca, D., et al. (2011). Multiple recurrent de novo CNVs, including duplications of the 7q11.23 Williams syndrome region, are strongly associated with autism. *Neuron*, *70*(5), 863–885. https://doi.org/10.1016/j.neuron.2011.05.002.

Sando, R., Jiang, X., & Südhof, T. C. (2019). Latrophilin GPCRs direct synapse specificity by coincident binding of FLRTs and teneurins. *Science*, *363*(6429). https://doi.org/10.1126/science.aav7969.

Sanes, J. R., & Zipursky, S. L. (2020). Synaptic specificity, recognition molecules, and assembly of neural circuits. *Cell*, *181*(6), 1434–1435. https://doi.org/10.1016/j.cell.2020.05.046.

Sassoe-Pognetto, M., & Patrizi, A. (2017). The Purkinje cell as a model of synaptogenesis and synaptic specificity. *Brain Research Bulletin*, *129*, 12–17. https://doi.org/10.1016/j.brainresbull.2016.10.004.

Schaaf, C. P., Boone, P. M., Sampath, S., Williams, C., Bader, P. I., Mueller, J. M., et al. (2012). Phenotypic spectrum and genotype-phenotype correlations of NRXN1 exon deletions. *European Journal of Human Genetics*, *20*(12), 1240–1247. https://doi.org/10.1038/ejhg.2012.95.

Schafer, D. P., Lehrman, E. K., Kautzman, A. G., Koyama, R., Mardinly, A. R., Yamasaki, R., et al. (2012). Microglia sculpt postnatal neural circuits in an activity and complement-dependent manner. *Neuron*, *74*(4), 691–705. https://doi.org/10.1016/j.neuron.2012.03.026.

Schizophrenia Working Group of the Psychiatric Genomics Consortium. (2014). Biological insights from 108 schizophrenia-associated genetic loci. *Nature*, *511*(7510), 421–427. https://doi.org/10.1038/nature13595.

Schoch, H., Kreibich, A. S., Ferri, S. L., White, R. S., Bohorquez, D., Banerjee, A., et al. (2017). Sociability deficits and altered amygdala circuits in mice lacking Pcdh10, an autism associated gene. *Biological Psychiatry*, *81*(3), 193–202. https://doi.org/10.1016/j.biopsych.2016.06.008.

Schreiner, D., Nguyen, T. M., Russo, G., Heber, S., Patrignani, A., Ahrne, E., et al. (2014). Targeted combinatorial alternative splicing generates brain region-specific repertoires of neurexins. *Neuron*, *84*(2), 386–398. https://doi.org/10.1016/j.neuron.2014.09.011.

Schreiner, D., Simicevic, J., Ahrne, E., Schmidt, A., & Scheiffele, P. (2015). Quantitative isoform-profiling of highly diversified recognition molecules. *eLife*, *4*, e07794. https://doi.org/10.7554/eLife.07794.

Schreiner, D., & Weiner, J. A. (2010). Combinatorial homophilic interaction between gamma-protocadherin multimers greatly expands the molecular diversity of cell adhesion. *Proceedings of the National Academy of Sciences of the United States of America*, *107*(33), 14893–14898. https://doi.org/10.1073/pnas.1004526107.

Schroeder, A., & de Wit, J. (2018). Leucine-rich repeat-containing synaptic adhesion molecules as organizers of synaptic specificity and diversity. *Experimental & Molecular Medicine*, *50*(4). https://doi.org/10.1038/s12276-017-0023-8.

Sclip, A., & Südhof, T. C. (2020). LAR receptor phospho-tyrosine phosphatases regulate NMDA-receptor responses. *eLife*, *9*, e53406. https://doi.org/10.7554/eLife.53406.

Scott-Van Zeeland, A. A., Abrahams, B. S., Alvarez-Retuerto, A. I., Sonnenblick, L. I., Rudie, J. D., Ghahremani, D., et al. (2010). Altered functional connectivity in frontal lobe circuits is associated with variation in the autism risk gene CNTNAP2. *Science Translational Medicine*, *2*(56), 56ra80. https://doi.org/10.1126/scitranslmed.3001344.

Segura, I., Essmann, C. L., Weinges, S., & Acker-Palmer, A. (2007). Grb4 and GIT1 transduce ephrinB reverse signals modulating spine morphogenesis and synapse formation. *Nature Neuroscience*, *10*(3), 301–310. https://doi.org/10.1038/nn1858.

Sestan, N., & State, M. W. (2018). Lost in translation: Traversing the complex path from genomics to therapeutics in autism spectrum disorder. *Neuron*, *100*(2), 406–423. https://doi.org/10.1016/j.neuron.2018.10.015.

Shingai, T., Ikeda, W., Kakunaga, S., Morimoto, K., Takekuni, K., Itoh, S., et al. (2003). Implications of nectin-like molecule-2/IGSF4/RA175/SgIGSF/TSLC1/SynCAM1 in cell-cell adhesion and transmembrane protein localization in epithelial cells. *The Journal of Biological Chemistry*, *278*(37), 35421–35427. https://doi.org/10.1074/jbc.M305387200.

Siddiqui, T. J., Pancaroglu, R., Kang, Y., Rooyakkers, A., & Craig, A. M. (2010). LRRTMs and neuroligins bind neurexins with a differential code to cooperate in glutamate synapse development. *The Journal of Neuroscience*, *30*(22), 7495–7506. https://doi.org/10.1523/JNEUROSCI.0470-10.2010.

Siddiqui, T. J., Tari, P. K., Connor, S. A., Zhang, P., Dobie, F. A., She, K., et al. (2013). An LRRTM4-HSPG complex mediates excitatory synapse development on dentate gyrus granule cells. *Neuron*, *79*(4), 680–695. https://doi.org/10.1016/j.neuron.2013.06.029.

Sigoillot, S. M., Iyer, K., Binda, F., Gonzalez-Calvo, I., Talleur, M., Vodjdani, G., et al. (2015). The secreted protein C1QL1 and its receptor BAI3 control the synaptic connectivity of excitatory inputs converging on cerebellar Purkinje cells. *Cell Reports*, *10*(5), 820–832. https://doi.org/10.1016/j.celrep.2015.01.034.

Silva, J. P., Lelianova, V. G., Ermolyuk, Y. S., Vysokov, N., Hitchen, P. G., Berninghausen, O., et al. (2011). Latrophilin 1 and its endogenous ligand Lasso/teneurin-2 form a high-affinity transsynaptic receptor pair with signaling capabilities. *Proceedings of the National Academy of Sciences of the United States of America*, *108*(29), 12113–12118. https://doi.org/10.1073/pnas.1019434108.

Singh, S. K., Stogsdill, J. A., Pulimood, N. S., Dingsdale, H., Kim, Y. H., Pilaz, L. J., et al. (2016). Astrocytes assemble thalamocortical synapses by bridging NRX1alpha and NL1 via Hevin. *Cell*, *164*(1-2), 183–196. https://doi.org/10.1016/j.cell.2015.11.034.

Song, J. Y., Ichtchenko, K., Südhof, T. C., & Brose, N. (1999). Neuroligin 1 is a postsynaptic cell-adhesion molecule of excitatory synapses. *Proceedings of the National Academy of Sciences of the United States of America*, *96*(3), 1100–1105. https://doi.org/10.1073/pnas.96.3.1100.

Stan, A., Pielarski, K. N., Brigadski, T., Wittenmayer, N., Fedorchenko, O., Gohla, A., et al. (2010). Essential cooperation of N-cadherin and neuroligin-1 in the transsynaptic control of vesicle accumulation. *Proceedings of the National Academy of Sciences of the United States of America*, *107*(24), 11116–11121.

Stevens, B., Allen, N. J., Vazquez, L. E., Howell, G. R., Christopherson, K. S., Nouri, N., et al. (2007). The classical complement cascade mediates CNS synapse elimination. *Cell*, *131*(6), 1164–1178.

Südhof, T. C. (2017). Synaptic neurexin complexes: A molecular code for the logic of neural circuits. *Cell*, *171*(4), 745–769. https://doi.org/10.1016/j.cell.2017.10.024.

Südhof, T. C. (2018). Towards an understanding of synapse formation. *Neuron*, *100*(2), 276–293. https://doi.org/10.1016/j.neuron.2018.09.040.

Sun, C., Cheng, M. C., Qin, R., Liao, D. L., Chen, T. T., Koong, F. J., et al. (2011). Identification and functional characterization of rare mutations of the neuroligin-2 gene (NLGN2) associated with schizophrenia. *Human Molecular Genetics*, *20*(15), 3042–3051. https://doi.org/10.1093/hmg/ddr208.

Suto, F., Tsuboi, M., Kamiya, H., Mizuno, H., Kiyama, Y., Komai, S., et al. (2007). Interactions between plexin-A2, plexin-A4, and semaphorin 6A control lamina-restricted projection of hippocampal mossy fibers. *Neuron*, *53*(4), 535–547. https://doi.org/10.1016/j.neuron.2007.01.028.

Tabuchi, K., Blundell, J., Etherton, M. R., Hammer, R. E., Liu, X., Powell, C. M., et al. (2007). A neuroligin-3 mutation implicated in autism increases inhibitory synaptic transmission in mice. *Science*, *318*(5847), 71–76. https://doi.org/10.1126/science.1146221.

Tabuchi, K., & Südhof, T. C. (2002). Structure and evolution of neurexin genes: Insight into the mechanism of alternative splicing. *Genomics*, *79*(6), 849–859. https://doi.org/10.1006/geno.2002.6780.

Tai, Y., Gallo, N. B., Wang, M., Yu, J. R., & Van Aelst, L. (2019). Axo-axonic innervation of neocortical pyramidal neurons by GABAergic Chandelier cells requires AnkyrinG-associated L1CAM. *Neuron*, *102*(2), 358–372.e359. https://doi.org/10.1016/j.neuron.2019.02.009.

Takano, T., Wallace, J. T., Baldwin, K. T., Purkey, A. M., Uezu, A., Courtland, J. L., et al. (2020). Chemico-genetic discovery of astrocytic control of inhibition in vivo. *Nature*. https://doi.org/10.1038/s41586-020-2926-0.

Takashima, N., Odaka, Y. S., Sakoori, K., Akagi, T., Hashikawa, T., Morimura, N., et al. (2011). Impaired cognitive function and altered hippocampal synapse morphology in mice lacking Lrrtm1, a gene associated with schizophrenia. *PLoS One*, *6*(7), e22716. https://doi.org/10.1371/journal.pone.0022716.

Tan, G. C., Doke, T. F., Ashburner, J., Wood, N. W., & Frackowiak, R. S. (2010). Normal variation in fronto-occipital circuitry and cerebellar structure with an autism-associated polymorphism of CNTNAP2. *NeuroImage*, *53*(3), 1030–1042. https://doi.org/10.1016/j.neuroimage.2010.02.018.

Tarusawa, E., Sanbo, M., Okayama, A., Miyashita, T., Kitsukawa, T., Hirayama, T., et al. (2016). Establishment of high reciprocal connectivity between clonal cortical neurons is regulated by the Dnmt3b DNA methyltransferase and clustered protocadherins. *BMC Biology*, *14*(1), 103. https://doi.org/10.1186/s12915-016-0326-6.

Tasic, B., Menon, V., Nguyen, T. N., Kim, T. K., Jarsky, T., Yao, Z., et al. (2016). Adult mouse cortical cell taxonomy revealed by single cell transcriptomics. *Nature Neuroscience*, *19*(2), 335–346. https://doi.org/10.1038/nn.4216.

Tawarayama, H., Yoshida, Y., Suto, F., Mitchell, K. J., & Fujisawa, H. (2010). Roles of semaphorin-6B and plexin-A2 in lamina-restricted projection of hippocampal mossy fibers. *The Journal of Neuroscience*, *30*(20), 7049–7060. https://doi.org/10.1523/JNEUROSCI.0073-10.2010.

Telley, L., Cadilhac, C., Cioni, J. M., Saywell, V., Jahannault-Talignani, C., Huettl, R. E., et al. (2016). Dual function of NRP1 in axon guidance and subcellular target recognition in cerebellum. *Neuron*, *91*(6), 1276–1291. https://doi.org/10.1016/j.neuron.2016.08.015.

Terauchi, A., Timmons, K. M., Kikuma, K., Pechmann, Y., Kneussel, M., & Umemori, H. (2015). Selective synaptic targeting of the excitatory and inhibitory presynaptic organizers FGF22 and FGF7. *Journal of Cell Science, 128*(2), 281–292. https://doi.org/10.1242/jcs.158337.

Thakar, S., Wang, L., Yu, T., Ye, M., Onishi, K., Scott, J., et al. (2017). Evidence for opposing roles of Celsr3 and Vangl2 in glutamatergic synapse formation. *Proceedings of the National Academy of Sciences of the United States of America, 114*(4), E610–E618. https://doi.org/10.1073/pnas.1612062114.

Thomas, L. A., Akins, M. R., & Biederer, T. (2008). Expression and adhesion profiles of SynCAM molecules indicate distinct neuronal functions. *The Journal of Comparative Neurology, 510*(1), 47–67. https://doi.org/10.1002/cne.21773.

Thu, C. A., Chen, W. V., Rubinstein, R., Chevee, M., Wolcott, H. N., Felsovalyi, K. O., et al. (2014). Single-cell identity generated by combinatorial homophilic interactions between alpha, beta, and gamma protocadherins. *Cell, 158*(5), 1045–1059. https://doi.org/10.1016/j.cell.2014.07.012.

Tran, T. S., Rubio, M. E., Clem, R. L., Johnson, D., Case, L., Tessier-Lavigne, M., et al. (2009). Secreted semaphorins control spine distribution and morphogenesis in the postnatal CNS. *Nature, 462*(7276), 1065–1069. https://doi.org/10.1038/nature08628.

Treutlein, B., Gokce, O., Quake, S. R., & Südhof, T. C. (2014). Cartography of neurexin alternative splicing mapped by single-molecule long-read mRNA sequencing. *Proceedings of the National Academy of Sciences of the United States of America, 111*(13), E1291–E1299. https://doi.org/10.1073/pnas.1403244111.

Uemura, T., Lee, S. J., Yasumura, M., Takeuchi, T., Yoshida, T., Ra, M., et al. (2010). Trans-synaptic interaction of GluRdelta2 and Neurexin through Cbln1 mediates synapse formation in the cerebellum. *Cell, 141*(6), 1068–1079. https://doi.org/10.1016/j.cell.2010.04.035.

Uesaka, N., Uchigashima, M., Mikuni, T., Nakazawa, T., Nakao, H., Hirai, H., et al. (2014). Retrograde semaphorin signaling regulates synapse elimination in the developing mouse brain. *Science, 344*(6187), 1020–1023. https://doi.org/10.1126/science.1252514.

Umemori, H., Linhoff, M. W., Ornitz, D. M., & Sanes, J. R. (2004). FGF22 and Its close relatives are presynaptic organizing molecules in the mammalian brain. *Cell, 118*(2), 257–270.

Ushkaryov, Y. A., Petrenko, A. G., Geppert, M., & Südhof, T. C. (1992). Neurexins: Synaptic cell surface proteins related to the alpha-latrotoxin receptor and laminin. *Science, 257*(5066), 50–56.

Vainchtein, I. D., & Molofsky, A. V. (2020). Astrocytes and microglia: In sickness and in health. *Trends in Neurosciences, 43*(3), 144–154. https://doi.org/10.1016/j.tins.2020.01.003.

Varea, O., Martin-de-Saavedra, M. D., Kopeikina, K. J., Schurmann, B., Fleming, H. J., Fawcett-Patel, J. M., et al. (2015). Synaptic abnormalities and cytoplasmic glutamate receptor aggregates in contactin associated protein-like 2/Caspr2 knockout neurons. *Proceedings of the National Academy of Sciences of the United States of America, 112*(19), 6176–6181. https://doi.org/10.1073/pnas.1423205112.

Varoqueaux, F., Aramuni, G., Rawson, R. L., Mohrmann, R., Missler, M., Gottmann, K., et al. (2006). Neuroligins determine synapse maturation and function. *Neuron, 51*(6), 741–754. https://doi.org/10.1016/j.neuron.2006.09.003.

Varoqueaux, F., Jamain, S., & Brose, N. (2004). Neuroligin 2 is exclusively localized to inhibitory synapses. *European Journal of Cell Biology, 83*(9), 449–456. https://doi.org/10.1078/0171-9335-00410.

Verschueren, E., Husain, B., Yuen, K., Sun, Y., Paduchuri, S., Senbabaoglu, Y., et al. (2020). The immunoglobulin superfamily receptome defines cancer-relevant networks associated with clinical outcome. *Cell, 182*(2), 329–344.e319. https://doi.org/10.1016/j.cell.2020.06.007.

Visser, J. J., Cheng, Y., Perry, S. C., Chastain, A. B., Parsa, B., Masri, S. S., et al. (2015). An extracellular biochemical screen reveals that FLRTs and Unc5s mediate neuronal subtype recognition in the retina. *eLife*, *4*. https://doi.org/10.7554/eLife.08149, e08149.

Wang, K., Zhang, H., Ma, D., Bucan, M., Glessner, J. T., Abrahams, B. S., et al. (2009). Common genetic variants on 5p14.1 associate with autism spectrum disorders. *Nature*, *459*(7246), 528–533. https://doi.org/10.1038/nature07999.

Washburn, H. R., Xia, N. L., Zhou, W., Mao, Y. T., & Dalva, M. B. (2020). Positive surface charge of GluN1 N-terminus mediates the direct interaction with EphB2 and NMDAR mobility. *Nature Communications*, *11*(1), 570. https://doi.org/10.1038/s41467-020-14345-6.

Weiner, J. A., & Jontes, J. D. (2013). Protocadherins, not prototypical: A complex tale of their interactions, expression, and functions. *Frontiers in Molecular Neuroscience*, *6*. https://doi.org/10.3389/fnmol.2013.00004.

Weinhard, L., di Bartolomei, G., Bolasco, G., Machado, P., Schieber, N. L., Neniskyte, U., et al. (2018). Microglia remodel synapses by presynaptic trogocytosis and spine head filopodia induction. *Nature Communications*, *9*(1), 1228. https://doi.org/10.1038/s41467-018-03566-5.

Williams, M. E., Wilke, S. A., Daggett, A., Davis, E., Otto, S., Ravi, D., et al. (2011). Cadherin-9 regulates synapse-specific differentiation in the developing hippocampus. *Neuron*, *71*(4), 640–655. https://doi.org/10.1016/j.neuron.2011.06.019.

Wills, Z. P., Mandel-Brehm, C., Mardinly, A. R., McCord, A. E., Giger, R. J., & Greenberg, M. E. (2012). The nogo receptor family restricts synapse number in the developing hippocampus. *Neuron*, *73*(3), 466–481. https://doi.org/10.1016/j.neuron.2011.11.029.

Wojtowicz, W. M., Vielmetter, J., Fernandes, R. A., Siepe, D. H., Eastman, C. L., Chisholm, G. B., et al. (2020). A human IgSF cell-surface interactome reveals a complex network of protein-protein interactions. *Cell*, *182*(4), 1027–1043. https://doi.org/10.1016/j.cell.2020.07.025.

Wu, Q., & Maniatis, T. (1999). A striking organization of a large family of human neural cadherin-like cell adhesion genes. *Cell*, *97*(6), 779–790. https://doi.org/10.1016/s0092-8674(00)80789-8.

Xu, X., Xiong, Z., Zhang, L., Liu, Y., Lu, L., Peng, Y., et al. (2014). Variations analysis of NLGN3 and NLGN4X gene in Chinese autism patients. *Molecular Biology Reports*, *41*(6), 4133–4140. https://doi.org/10.1007/s11033-014-3284-5.

Yamagata, M., Duan, X., & Sanes, J. R. (2018). Cadherins interact with synaptic organizers to promote synaptic differentiation. *Frontiers in Molecular Neuroscience*, *11*. https://doi.org/10.3389/fnmol.2018.00142.

Yamagata, M., & Sanes, J. R. (2008). Dscam and Sidekick proteins direct lamina-specific synaptic connections in vertebrate retina. *Nature*, *451*(7177), 465–469. https://doi.org/10.1038/nature06469.

Yamagata, M., & Sanes, J. R. (2012). Expanding the Ig superfamily code for laminar specificity in retina: Expression and role of contactins. *The Journal of Neuroscience*, *32*(41), 14402–14414. https://doi.org/10.1523/JNEUROSCI.3193-12.2012.

Yamagata, M., & Sanes, J. R. (2018). Expression and roles of the immunoglobulin superfamily recognition molecule Sidekick1 in mouse retina. *Frontiers in Molecular Neuroscience*, *11*. https://doi.org/10.3389/fnmol.2018.00485.

Yamagata, M., Weiner, J. A., & Sanes, J. R. (2002). Sidekicks: Synaptic adhesion molecules that promote lamina-specific connectivity in the retina. *Cell*, *110*(5), 649–660. https://doi.org/10.1016/s0092-8674(02)00910-8.

Yan, W., Laboulaye, M. A., Tran, N. M., Whitney, I. E., Benhar, I., & Sanes, J. R. (2020). Mouse retinal cell atlas: Molecular identification of over sixty amacrine cell types. *The Journal of Neuroscience*, *40*(27), 5177–5195. https://doi.org/10.1523/JNEUROSCI.0471-20.2020.

Yan, J., Noltner, K., Feng, J., Li, W., Schroer, R., Skinner, C., et al. (2008). Neurexin 1α structural variants associated with autism. *Neuroscience Letters*, *438*(3), 368–370. https://doi.org/10.1016/j.neulet.2008.04.074.

Yan, Q., Weyn-Vanhentenryck, S. M., Wu, J., Sloan, S. A., Zhang, Y., Chen, K., et al. (2015). Systematic discovery of regulated and conserved alternative exons in the mammalian brain reveals NMD modulating chromatin regulators. *Proceedings of the National Academy of Sciences of the United States of America*, *112*(11), 3445–3450. https://doi.org/10.1073/pnas.1502849112.

Yim, Y. S., Kwon, Y., Nam, J., Yoon, H. I., Lee, K., Kim, D. G., et al. (2013). Slitrks control excitatory and inhibitory synapse formation with LAR receptor protein tyrosine phosphatases. *Proceedings of the National Academy of Sciences of the United States of America*, *110*(10), 4057–4062. https://doi.org/10.1073/pnas.1209881110.

Yoshida, Y. (2012). Semaphorin signaling in vertebrate neural circuit assembly. *Frontiers in Molecular Neuroscience*, *5*. https://doi.org/10.3389/fnmol.2012.00071.

Yoshida, T., Yasumura, M., Uemura, T., Lee, S. J., Ra, M., Taguchi, R., et al. (2011). IL-1 receptor accessory protein-like 1 associated with mental retardation and autism mediates synapse formation by trans-synaptic interaction with protein tyrosine phosphatase delta. *The Journal of Neuroscience*, *31*(38), 13485–13499. https://doi.org/10.1523/JNEUROSCI.2136-11.2011.

Yoshihara, Y., De Roo, M., & Muller, D. (2009). Dendritic spine formation and stabilization. *Current Opinion in Neurobiology*, *19*(2), 146–153. https://doi.org/10.1016/j.conb.2009.05.013.

Zeng, H., & Sanes, J. R. (2017). Neuronal cell-type classification: Challenges, opportunities and the path forward. *Nature Reviews. Neuroscience*, *18*(9), 530–546. https://doi.org/10.1038/nrn.2017.85.

Zhang, P., Lu, H., Peixoto, R. T., Pines, M. K., Ge, Y., Oku, S., et al. (2018). Heparan sulfate organizes neuronal synapses through neurexin partnerships. *Cell*, *174*(6), 1450–1464. https://doi.org/10.1016/j.cell.2018.07.002.

Zhiling, Y., Fujita, E., Tanabe, Y., Yamagata, T., Momoi, T., & Momoi, M. Y. (2008). Mutations in the gene encoding CADM1 are associated with autism spectrum disorder. *Biochemical and Biophysical Research Communications*, *377*(3), 926–929. https://doi.org/10.1016/j.bbrc.2008.10.107.

CHAPTER NINE

Role of astrocytes in synapse formation and maturation

Christabel X. Tan[a,†], Caley J. Burrus Lane[a,b,†], and Cagla Eroglu[a,b,c,d,*]
[a]Department of Cell Biology, Duke University Medical Center, Durham, NC, United States
[b]Department of Neurobiology, Duke University Medical Center, Durham, NC, United States
[c]Duke Institute for Brain Sciences, Durham, NC, United States
[d]Regeneration Next Initiative, Duke University, Durham, NC, United States
[*]Corresponding author e-mail address: cagla.eroglu@duke.edu

Contents

[†] These authors contributed equally.

Current Topics in Developmental Biology, Volume 142
ISSN 0070-2153
https://doi.org/10.1016/bs.ctdb.2020.12.010

Abstract

Astrocytes are the most abundant glial cells in the mammalian brain and directly participate in the proper functioning of the nervous system by regulating ion homeostasis, controlling glutamate reuptake, and maintaining the blood-brain barrier. In the last two decades, a growing body of work also identified critical roles for astrocytes in regulating synaptic connectivity. Stemming from the observation that functional and morphological development of astrocytes occur concurrently with synapse formation and maturation, these studies revealed that both developmental processes are directly linked. In fact, astrocytes both physically contact numerous synaptic structures and actively instruct many aspects of synaptic development and function via a plethora of secreted and adhesion-based molecular signals. The complex astrocyte-to-neuron signaling modalities control different stages of synaptic development such as regulating the initial formation of structural synapses as well as their functional maturation. Furthermore, the synapse-modulating functions of astrocytes are evolutionarily conserved and contribute to the development and plasticity of diverse classes of synapses and circuits throughout the central nervous system. Importantly, because impaired synapse formation and function is a hallmark of many neurodevelopmental disorders, deficits in astrocytes are likely to be major contributors to disease pathogenesis. In this chapter, we review our current understanding of the cellular and molecular mechanisms by which astrocytes contribute to synapse development and discuss the bidirectional secretion-based and contact-mediated mechanisms responsible for these essential developmental processes.

1. Introduction

Astrocytes are the most abundant glial cells in the mammalian brain. When these glial cells were first identified in the late 1800s, their complex morphologies captivated the attention of those who observed them and called into question their underlying functions in the central nervous system (CNS) (Kettenmann & Verkhratsky, 2008). The name astrocyte, or "star cell," was coined by Michael von Lenhossek in the late 1800s and aptly describes the intricate and elaborate morphology of these cells (Lenhossek, 1893). Radiating out from the astrocyte's soma are several thick branches that arborize into thousands of thin, filamentous perisynaptic astrocytic processes, giving the astrocyte its characteristic bush-like appearance (Fig. 1A). More than a century after their initial characterization, advanced imaging techniques revealed that astrocytes tile the entire brain parenchyma by forming non-overlapping, but tightly packed territories (Bushong et al., 2002; Halassa et al., 2007). For example, in the rodent cortex, an astrocyte can envelope four to eight neuronal cell bodies and up to 600 dendrites (Halassa et al., 2007). While they morphologically resemble rodent astrocytes, human astrocytes are dramatically larger and have approximately ten-fold more processes than

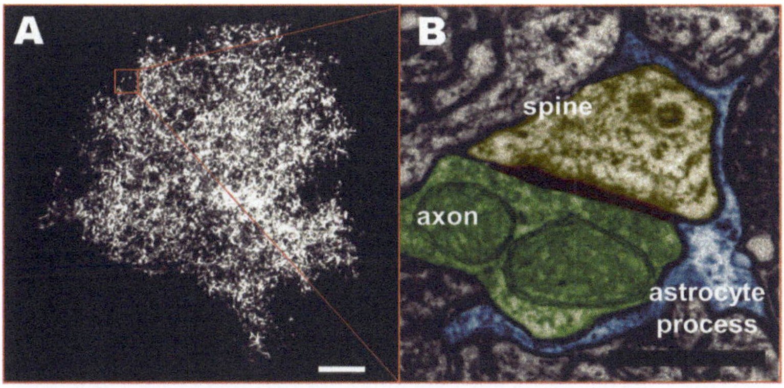

Fig. 1 Astrocytes are integral components of the tripartite synapse. (A) A protoplasmic mouse cortical astrocyte (P21, *in vivo*) gains its morphological complexity due to arborization of primary and secondary branches that end in thousands of perisynaptic astrocytic processes. Scale bar: 20 μm. Image by Christabel X. Tan (Eroglu Lab, Duke University). (B) An electron micrograph of the tripartite synapse in the mouse visual cortex. Perisynaptic astrocyte process (blue) ensheathes the pre-synaptic axon (green) and post-synaptic spine (yellow). Scale bar: 0.5μm. Image courtesy of W. Christopher Risher (Marshall University).

rodent astrocytes (Oberheim et al., 2006). Human astrocytes also tile within gray matter structures, although their larger size and more extensive processes leads to greater process overlap between neighboring cells (Oberheim et al., 2009).

Astrocytes were initially thought to be inert cells that held the brain together and provided a matrix for brain architecture and neuron distribution. Later, astrocytes were viewed as support cells necessary for maintaining neuronal function (Volterra & Meldolesi, 2005). These functions include neurotransmitter uptake, recycling and release (Anderson & Swanson, 2000), as well as ion (particularly K^+ buffering) and water homeostasis (Mederos, González-Arias, & Perea, 2018; Olsen et al., 2015). More recently, astrocytes have been recognized as active players at the synapse, directly regulating synapse formation, elimination and function (Chung, Allen, & Eroglu, 2015; Clarke & Barres, 2013; Perea, Navarrete, & Araque, 2009). Through their extensive branching networks, murine astrocytes can contact nearly 100,000 synapses (Halassa et al., 2007) and human astrocytes can contact close to 2 million synapses (Bushong et al., 2002; Oberheim et al., 2006). Therefore, many synapses throughout the brain are comprised not only by their neuronal elements i.e. a presynaptic axon and postsynaptic dendrite, but also are closely juxtaposed by a perisynaptic astrocyte process that plays critical structural and functional roles at the synapse. To capture this intimate cellular association, the term "tripartite synapse" (Fig. 1B) was coined by Haydon and colleagues (Araque et al., 1999).

2. Astrocyte maturation occurs concurrently with the peak period of synaptogenesis

The closely intertwined relationship between astrocytes and neurons begins very early in development, as both astrocytes and neurons are born from the same neuroepithelial stem cells known as radial glial cells. In the murine cortex, radial glial cells first generate precursors for excitatory neurons between embryonic day (E)12 to birth. Right before birth, Notch (Sun et al., 2001; Tanigaki et al., 2001) and BMP (Nakashima et al., 2001) trigger an intrinsic, temporally-regulated shift from neurogenesis to astrogenesis (Qian et al., 2000). Consequently, astrogenesis peaks immediately after birth and is largely completed by the end of the first postnatal week (Kriegstein & Alvarez-Buylla, 2009; Sauvageot & Stiles, 2002). These newly born astrocytes then undergo a period of postnatal developmental maturation, which can be characterized by both functional and morphological changes in these cells (Fig. 2). For instance, astrocytes upregulate their expression of potassium channel Kir4.1, which is essential for their regulation of ion and osmolarity homeostasis (Olsen et al., 2015). Apart from regulating ion homeostasis, astrocytes also participate in a number of key functions including maintaining the blood-brain barrier, as well as regulating glutamate reuptake and metabolism

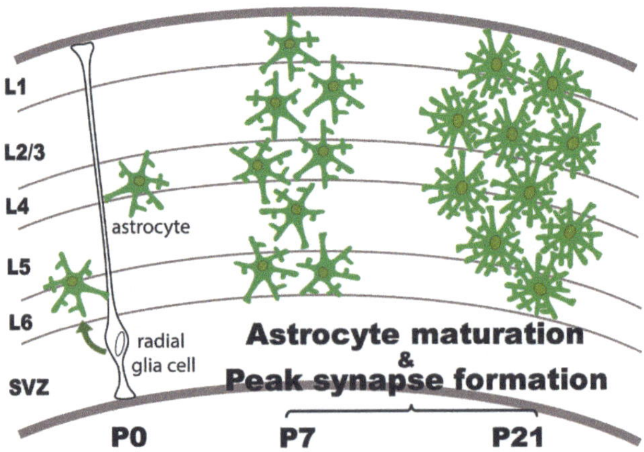

Fig. 2 Astrocyte morphological and functional maturation coincides with the peak period of synaptogenesis during mouse cortical development.

at the synapse. Key genes involved in these processes are similarly turned on during the same peak window of astrocyte maturation (Kimelberg & Nedergaard, 2010).

Functional maturation of astrocytes is closely intertwined with synaptic development and influenced by activity of underlying circuits (Hasel et al., 2017). For example, *in vitro* studies showed that the astrocytic expression of glutamate transporter subtype GLT1, a protein that is critical for removal of synaptically released extracellular glutamate, is dependent on co-culture with neurons (Swanson et al., 1997). In turn, neurons are dependent on GLT1-mediated uptake of neurotransmitter glutamate by astrocytes, so that glutamate-induced excitotoxicity can be prevented (Rothstein et al., 1996).

Concurrent with this functional development, astrocytes also undergo significant morphological changes that are essential for allowing astrocyte-synapse contacts to form. Interestingly, in the developing rodent cortex the period of astrocyte morphological arborization/maturation (postnatal day (P)7-21) closely coincides with the period of accelerated synaptogenesis (Farhy-Tselnicker & Allen, 2018; Stogsdill et al., 2017). The intricate spatial and temporal overlap of astrocyte maturation and synaptogenesis led to the hypothesis that astrocytes directly participate in synaptogenesis by delivering signals that induce different stages of synapse assembly and maturation. In support of this hypothesis, over the last two decades astrocytes have been found to directly contribute to formation of nascent synapses as well as their structural and functional maturation. In this chapter, we will discuss our current understanding of how astrocytes contribute to synapse formation and maturation. Specifically, we will review the astrocyte-secreted molecules and associated neuronal mechanisms which drive synaptogenesis, as well as the astrocyte-neuron contact-mediated mechanisms that contribute to synapse development.

3. Astrocyte-secreted proteins regulate synapse formation

Synaptogenesis is thought to be initiated when a presynaptic axon and a target postsynaptic dendrite contact each other. Cell adhesion molecules that transverse the synapse then anchor the pre and postsynaptic compartments together, allowing their stabilization and growth. The presynaptic terminal accumulates neurotransmitter-filled vesicles and vesicle-release machinery, while the postsynaptic terminal contains various proteinaceous scaffolds that tether neurotransmitter receptors (Schikorski & Stevens, 1997; Walmsley, Alvarez, & Fyffe, 1998). In the case of excitatory synapses, the postsynaptic terminal is often localized to a dendritic spine, which morphologically matures from a thin, filopodia-like structure to a mushroom-like shape. This structure of a dendritic spine is postulated to be important for maximizing efficient neurotransmission and signal processing (Adrian et al., 2014; Nimchinsky, Sabatini, & Svoboda, 2002). Functionally, immature synapses, which are often post-synaptically silent because they contain only NMDA and lack AMPA receptors, mature as they recruit AMPA receptor subunits allowing the excitatory synapse to be electrically active (Hall & Ghosh, 2008; Hestrin, 1992).

The formation and maturation of synapses occur at a time period closely overlapping with the birth and morphological and functional maturation of astrocytes. Furthermore, synaptogenesis does not proceed properly without the direct involvement of astrocytes. In pivotal studies by the Barres Lab, it was shown that the addition of astrocyte feeder inserts, or adding astrocyte-conditioned media (ACM) to purified rat retinal ganglion cell (RGC) neuron cultures is sufficient to induce a robust increase both in the number of synapses formed by the RGCs, as well as the functional maturation of these synapses. These results demonstrated that astrocytes secrete factors that drive synaptogenic processes (Pfrieger & Barres, 1997; Ullian et al., 2001). Following these seminal discoveries, several astrocyte-secreted factors have been identified in the past two decades to contribute to myriad aspects of synaptogenesis. It is important to note that the synaptogenic effects of astrocytes are not only specific to RGCs. In fact, astrocytes have been shown to regulate the synapse formation between neurons from many brain regions, including the cortex, hippocampus, superior colliculus, and cerebellum (Araujo, Carpi-Santos and Gomes, 2019; Elmariah et al., 2005; Eroglu et al., 2009; Stogsdill et al., 2017; Ullian et al., 2001). Furthermore, astrocytes regulate the formation and function of multiple types of synapses, such as intracortical, thalamocortical, cholinergic, and GABAergic synapses (Albrecht et al., 2012; Christopherson et al., 2005; Eroglu et al., 2009; Hughes, Elmariah, & Balice-Gordon, 2010; Kucukdereli et al., 2011; López-Murcia, Terni, & Llobet, 2015). Importantly, the synaptogenic effects of astrocytes are evolutionarily conserved. Astrocytes and astrocyte-like glia from species ranging from *C. elegans* to humans are critical for regulating synapse formation and function (Allen et al., 2012; Christopherson et al., 2005; Colón-Ramos, Margeta, & Shen, 2007; Freeman, 2015; Johnson et al., 2007).

In the following section, we will summarize some of the pivotal studies that identified these synapse inducing factors that are secreted by astrocytes and determined how they control synapse formation (Fig. 3A).

3.1 Thrombospondins

The thrombospondin family proteins (TSPs) are the first astrocyte-secreted synaptogenic factors identified, and are critical for increasing the number of synapses made between RGCs *in vitro* (Christopherson et al., 2005). TSPs (five isoforms exist in mammals made by five different genes) are large, oligomeric extracellular matrix (ECM) proteins which modify cell-cell and cell-matrix interactions in many body systems by interacting with cell-surface receptors, cytokines, and other ECM proteins (Adams, 2001; Bornstein, 2001). TSPs 1 through 4 are expressed in the CNS and these TSPs are greatly enriched in astrocytes compared to neurons (Zhang et al., 2014). Interestingly, the peak of TSP1 and TSP2 expression coincides with the initiation of postnatal excitatory synaptogenesis, P5to P7 in rodents; whereas, TSP3 and TSP4 are expressed by astrocytes in adulthood (Zhang et al., 2014).

In a candidate screen to identify astrocyte-secreted synaptogenic factors, Christopherson, Ullian and colleagues applied TSPs to RGC neuronal

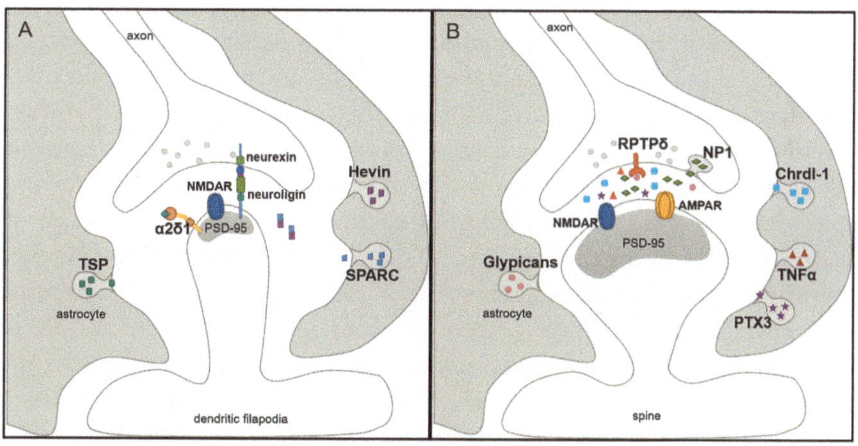

Fig. 3 Astrocyte-secreted factors involved in synapse formation and maturation. (A) Astrocyte-secreted Thrombospondin binds to post-synaptic receptor α2δ-1 to induce synaptogenesis. In a separate pathway, astrocytic Hevin mediates neurexin-neuroligin interaction critical for synapse formation. This pathway is antagonized by SPARC through an unknown mechanism. (B) Astrocyte secreted Glypicans, Chordin-like1, TNFα and Pentraxin-3 regulate synapse maturation.

cultures, and found that TSP1 or TSP2 greatly increases the number of excitatory synaptic structures. Electron microscopy analyses revealed that the synapses induced by TSP treatment are structurally identical to those formed when RGCs were exposed to a feeder layer of astrocytes; however, TSP-induced synapses are post-synaptically silent because they lack postsynaptic AMPA receptors. However, these synapses are capable of release and reuptake of glutamate. Using knockout mice, they also showed that both TSP1 and TSP2 are required for early synapse formation *in vivo,* as TSP1 and TSP2 double knockout mice have significantly reduced number of intracortical synapses at P8 and P21 (Christopherson et al., 2005).

Astrocyte-secreted TSPs mediate their synaptogenic effects through their interactions with a neuronal receptor, calcium channel subunit α2δ-1 (or Gabapentin receptor). The conserved EGF-like domain of TSPs are synaptogenic by themselves as they bind to the von Willebrand factor A (VWFA) domain of α2δ-1 to induce synapse formation (Eroglu et al., 2009). Genetic knockdown of α2δ-1 from RGC cultures prohibits TSP-induced synapse formation. Likewise, the addition of gabapentin, an anti-epileptic/anti-neuropathic pain drug that binds to α2δ-1, occludes the binding between TSP and α2δ-1. Thus co-application of gabapentin inhibits the ability of TSPs to induce synapse formation (Eroglu et al., 2009). α2δ-1 is also required for TSP-induced synaptogenesis in cultured cortical neurons, as TSP cannot induce synapse formation in cortical cultures derived from α2δ-1 knockout (KO) mice (Risher et al., 2018).

Even though α2δ-1 is present both at the pre- and post-synaptic sites, only post-synaptic α2δ-1 is required for TSP-induced synaptogenesis. *In vivo,* deleting α2δ-1 from mice results in a significant decrease in VGlut1 positive (VGlut1+) intracortical synapses within the visual cortex at P21, and this synaptic deficit is still maintained by P40 (Risher et al., 2018). No effect on thalamocortical synapses expressing VGlut2+ was observed. Not only were there fewer intracortical synapses present in α2δ-1 KO mice, but the maturity of dendritic spines was significantly impaired, indicating a critical role for α2δ-1 in both the formation and maturation of excitatory synaptic structures. Electrophysiological recordings revealed a sharp decrease in the frequency of miniature excitatory postsynaptic currents (mEPSCs) from α2δ-1 KO visual cortex pyramidal neurons, with no change in the amplitude, when compared to their littermate wildtype siblings, consistent with the finding of reduced synapse numbers (Risher et al., 2018).

α2δ-1 was first identified and characterized as a calcium channel subunit (Arikkath & Campbell, 2003; Klugbauer et al., 1999); however, its role in

TSP-mediated synaptogenesis is likely to be independent of calcium-channel function. When applied to RGCs, calcium channel blockers had no effect on the ability of TSPs to induce synapse formation (Eroglu et al., 2009). Instead, α2δ-1 signals through GTPase Rac1 to develop and mature nascent synapses and dendritic spines (Risher et al., 2018). Knockdown of Rac1 in cortical neurons prevents the synaptogenic effect of TSP2, which can be rescued by introducing fast-cycling Rac1 that mimics Rac1's active GTPase function. In addition to its interaction with α2δ-1, TSP1 has also been shown to interact with the cell adhesion protein Neuroligin-1 (NL1), and knockdown of NL1 from cultured hippocampal astrocyte-neuron mixed cultures prevented the ability of TSP1 to increase synaptic density (Xu, Xiao, & Xia, 2010).

In a recent study, it was shown that both TSP1 and α2δ-1 expression are enriched in the sublayers of the retinal synaptic zone where RGCs receive synapses, the inner plexiform layer, and critically regulate retinal synapse formation (Koh et al., 2018). Interestingly, TSP1 specifically regulated formation and function of synapses onto the on-off direction-selective RGCs, a critical component of the direction selective circuit of the retina. This specificity was conferred by a second TSP receptor, β1-integrin, which is enriched in on-off direction-selective RGCs. TSP1/β1-integrin binding in conjunction with α2δ-1 promotes excitatory synaptogenesis uniquely onto these direction-selective RGCs, thereby conferring both cell-type and circuit specificity to TSP1-induced synapse formation in the retina (Koh et al., 2019).

In addition to its developmental roles, thrombospondins may also regulate synaptic structure and function in adult CNS. In agreement with this, TSP1/β-integrin signaling regulates excitation/inhibitory balance in adult spinal cord cultures by upregulating glycinergic receptors and reducing AMPA receptor surface expression (Hennekinne et al., 2013). Even more recently, astrocyte-mediated TSP1/α2δ-1 signaling in the adult striatum was demonstrated to regulate excitatory synapse activity. Activated medium spiny neurons release GABA which bind to astrocytic $GABA_B$ receptors, resulting in an increase in intracellular calcium signaling and TSP1 mRNA abundance in astrocytes. This increase in TSP1 gene expression correlates with increased excitatory synapse formation and function in the adult striatum, augmenting medium spiny neuron firing and these effects can be blocked by gabapentin treatment. The outcome of this stimulation is behavioral outcomes resembling hyperactivity and decreased attention in mice (Nagai et al., 2019).

3.2 Hevin/SPARCL1

The interaction between TSP/α2δ-1 largely mediates the formation of excitatory intracortical synapses, which express presynaptic VGlut1 (Risher et al., 2018). However, another astrocyte-secreted factor Hevin/SPARCL1, which has also been identified using the RGC neuron culture system, regulates the formation of thalamocortical synapses. Hevin is a glycoprotein that is known to be localized to synaptic clefts (Lively, Ringuette, & Brown, 2007), and is expressed predominantly by astrocytes (Eroglu et al., 2009). Hevin is involved in the formation of VGlut2+ synapses, such as those made between RGCs *in vitro* and between RGCs and their superior colliculus target neurons *in vivo* (Kucukdereli et al., 2011; Risher et al., 2014). When applied to glia-free RGC cultures, like TSPs, Hevin also induces the formation of pre-synaptically active but post-synaptically silent synapses (Kucukdereli et al., 2011) and promotes a sharp increase in NMDA receptor-mediated currents (Singh et al., 2016). In purified cortical and thalamic neuron cultures, Hevin treatment robustly increases thalamocortical synapse formation without inducing any changes in the number of intracortical synapses (Risher et al., 2014). *In vivo*, lack of Hevin in mice leads to close to a 35% reduction specifically in retinocollicular and thalamocortical synapses and a reduction in cortical dendritic spine maturity during early postnatal development (Kucukdereli et al., 2011; Risher et al., 2014).

The loss of VGlut2+ thalamocortical synapses occurs alongside an increase in the number of VGlut1+ intracortical synapses in *Hevin* knockout mice. Investigation into the three-dimensional ultrastructure of dendrites in early postnatal mouse visual cortex revealed that during normal development, many dendritic spines are initially contacted with more than one presynaptic axon terminal. These "Spines with Multiple Excitatory Contacts," or SMECs, are prevalent at P14 but largely disappear by P25. SMECs often contain a VGlut2+ terminal alongside a VGlut1+ terminal. When Hevin is deleted, many SMECs remain by P25. Together, these results strongly suggest a role for Hevin not only in the formation of thalamocortical synapses, but also in the refinement of thalamic and cortical inputs on SMECs, an unexpected step in the maturation of synaptic structures during development (Risher et al., 2014).

Hevin regulates the formation and maturation of neuronal synapses by interacting simultaneously with postsynaptic neuroligins and presynaptic neurexins (Singh et al., 2016), both of which are encoded by families of cell-adhesion molecules that are critically involved in synapse formation

(Chih, Engelman, & Scheiffele, 2005). Canonically, postsynaptic neuroligins directly interact with presynaptic neurexins to bridge pre- and postsynaptic structures and aid in the development and maturation of synapses (Craig & Kang, 2007). However, certain neuroligins and neurexins, such as postsynaptic NL1β and presynaptic Nrxn1α, have only a weak affinity for each other and therefore do not bind together effectively. Astrocyte-secreted Hevin directly interacts with both Nrxn1α and NL1β and serves as a linking bridge between them to facilitate their interaction (Singh et al., 2016). The trans-synaptic bridging of Nrxn1α and NL1β by Hevin results in the recruitment of postsynaptic scaffolding proteins such as Homer-1 and thus the stabilization and maturation of thalamocortical synapses (Fig. 3). In a recent study, addition of Hevin into mixed cortical glia-neuron co-cultures was shown to increase excitatory synapse number and NMDA receptor current, independent of its interaction with neurexins and neuroligins (Gan & Südhof, 2020) In cortical neuron only cultures, Hevin does not induce intracortical synaptogenesis (Singh et al., 2016). Therefore, it is highly likely that this apparent effect of exogenous Hevin on synaptogenesis is mainly due to Hevin's effect on glia. Future studies to determine how Hevin signals to astrocytes or other glial cells are needed.

3.3 SPARC

Importantly, not all astrocyte-secreted factors are pro-synaptogenic. Astrocytes secrete a close homolog of Hevin/Sparcl1, called SPARC (Secreted Protein Acidic Rich in Cysteine). SPARC antagonizes Hevin's effects on synapse formation in RGC cultures and treating purified RGC neuron cultures with ACM from which SPARC had been depleted resulted in a significant increase in the number of excitatory RGC synapses formed (Kucukdereli et al., 2011). It is unclear how this antagonistic effect is achieved. A direct interaction between Hevin and SPARC was proposed; however, SPARC does not co-immunoprecipitate with Hevin and did not co-immunodeplete in Hevin-depleted mouse ACM, suggesting that SPARC does not directly interact with Hevin. Therefore, it was hypothesized that SPARC blocks Hevin-induced synapse formation via direct competition with an yet unknown neuronal receptor for Hevin (Kucukdereli et al., 2011).

Knockout (KO) of SPARC *in vivo* enhances the formation of thalamocortical synapses within the superior colliculus at P14. Additionally, recordings from mEPSCs from hippocampal neurons from SPARC KO mice displayed both greater mEPSC frequency and amplitude, indicating enhanced numbers and postsynaptic function of SPARC KO synapses (Jones et al., 2011). SPARC was proposed to be involved in the recruitment of AMPAR subunits GluA1 and GluA2 to developing synapses, as hippocampal neurons grown on a feeder layer of SPARC KO astrocytes had greater surface expression of both GluA1 and GluA2 (Jones et al., 2011).

Interestingly, SPARC also negatively regulates the maturation and refinement of cholinergic synapses in addition to thalamocortical synapses. For instance, treating cultured neonatal rat superior cervical ganglion neurons with SPARC decreased the density of cholinergic synapses formed, once again indicating a role for SPARC in antagonizing synapse formation. SPARC treatment can also cause the retraction and elimination of already-formed synapses, suggesting that SPARC may serve to regulate synaptic connectivity after synaptogenesis has occurred (López-Murcia et al., 2015). In cholinergic micro-cultures generated from neonatal rat superior cervical ganglion, SPARC treatment also increased spontaneous neurotransmission but decreased the number of vesicles available for presynaptic release, indicative of an immature synaptic phenotype (Albrecht et al., 2012). Taken together, these results suggest that astrocyte-secreted SPARC antagonizes the formation and prevents the maturation of excitatory and cholinergic presynaptic terminals.

4. Astrocyte-secreted factors regulate postsynaptic function and maturation

The astrocyte-secreted factors we have discussed so far are responsible for regulating the formation of excitatory synaptic structures and enhancing surface expression of only NMDARs (Fig. 3A). However, it was shown that ACM is capable of promoting formation of synapses which were also post-synaptically active (containing AMPARs), strongly indicating that other factors are required for the maturation of ("unsilencing") structural synapses induced by factors such as TSP and Hevin. The classical view of synapse development suggests that some synapses are first formed without postsynaptic AMPARs, and rather only contain NMDARs. The insertion of AMPAR subunits is needed to make them functional. Moreover, during development the predominant AMPAR subunits at post-synapses switch from GluA1 to GluA2 (Brill & Huguenard, 2008). This developmental switch decreases the entry of calcium into the postsynaptic neuron and limits calcium-dependent plasticity mechanisms, allowing for the closure of critical periods. Initially believed to be a neuron-intrinsic process, recent studies have identified several astrocyte-secreted molecules that are involved in this step of synapse maturation via the recruitment of AMPAR subunits to developing excitatory synapses (Fig. 3B).

4.1 Glypicans

The first astrocyte-secreted factors discovered to unsilence synapses were glypicans, a family of heparan sulfate proteoglycans. There are six known members of the glypican family in mammals. Glypican 4 (Gpc4) and Glypican 6 (Gpc6) are both predominantly expressed and secreted by cortical astrocytes during postnatal development, although glypican expression is also high in hippocampal and cerebellar astrocytes. Interestingly, *in vivo* temporal expression of these two glypicans suggest that they might have different but complementary roles in synapse formation and function. Astrocytic Gpc4 expression peaks between P6-14 when synaptogenesis first begins while Gpc6 expression peaks between P14-21 when synapses mature (Allen et al., 2012). The addition of Gpc4 or Gpc6, which share significant structural homology, to glia-free RGC cultures induced a 2.5-fold increase in surface GluA1 expression (Allen et al., 2012). A substantial increase in mEPSC amplitude accompanied this increase in GluA1 expression, confirming that Gpc4 and Gpc6 can enhance the electrical activity of RGC synapses. In addition to unsilencing RGC synapses via GluA1 insertion, Gpc4 and 6 treatment also led to a significant increase in the total number of synapses formed, suggesting that astrocyte-secreted glypicans can induce synapse formation in the absence of other factors such as TSP and Hevin. Consistent with these *in vitro* results, *Gpc4* KO mice display reductions in mEPSC amplitude within the CA1 region of the hippocampus, concurrent with a decline in the total number of excitatory synapses containing GluA1 subunits (Allen et al., 2012). *Dallylike*, the *Drosophila* homologue of mammalian Gpc4, has been shown to play similar roles in synaptic development, demonstrating a conservation of these mechanisms across species (Johnson et al., 2006).

Interestingly, Gpc4 regulates postsynaptic GluA1 expression via a presynaptic mechanism. Astrocyte-secreted Gpc4 induces the expression and secretion of neuronal pentraxin-1 (NP-1) from the pre-synapse. NP-1 binds to and stabilizes AMPARs on dendritic surfaces (Farhy-Tselnicker et al., 2017). Treating glia-free RGC cultures with Gpc4 significantly increases the amount of NP1 within the culture media. On the other hand, siRNA-mediated knockdown of *NP1* in glia-free RGC cultures precludes the increase in surface GluA1 expression resulting from Gpc4 treatment and prevents excitatory synapse formation. Similarly, conditional knockout of Gpc4 from astrocytes in mice increases the number of NP1 puncta on neurons within the superior colliculus, due to diminished secretion of NP1 and enhanced retention of the protein within the pre-synapse. Additional experiments demonstrated that Gpc4 mediates these effects by binding to its neuronal receptor, RPTPδ (Farhy-Tselnicker et al., 2017). Blocking RPTPδ in RGCs prevents Gpc4-mediated increase both in GluA1 surface accumulation and in NP1 expression. Thus, the interaction between astrocyte secreted Gpc4 and its neuronal RPTPδ receptor allows for the upregulation of NP1 secretion, resulting in the enhanced recruitment of GluA1 subunits to the synapse. Currently, the specific mechanism by which astrocyte secreted Gpc6 unsilences synapses remains unknown.

4.2 TNFα

Astrocytes also secrete TNFα, a proinflammatory cytokine that controls the synaptic surface expression of AMPAR subunits. Treating cultured hippocampal neurons with TNFα results in a two-fold increase in surface AMPAR expression, and is also accompanied by an increase in mEPSC frequency (Beattie et al., 2002). Further work clarified that the TNFα-induced increase in AMPAR subunit expression is specific to GluA1, as TNFα treatment has no effect on GluA2 expression (Stellwagen et al., 2005). These effects are mediated by astrocyte-secreted TNFα's interaction with its neuronal receptor TNFR1. Although TNFα treatment increases mEPSC frequency and GluA1 synaptic expression, TNFα treatment leads to a decrease in mIPSC frequency and GABAaR expression, demonstrating that TNFα could play a role both in the maturation of excitatory synapses and the negative regulation of inhibitory synapses (Stellwagen et al., 2005). Furthermore, astrocyte-secreted TNFα is important for homeostatic synaptic scaling, which is the compensatory upregulation of synaptic activity that occurs in response to prolonged periods of decreased activity (Stellwagen & Malenka, 2006). When synaptic activity has been diminished, TNFα enhances synaptic activity via the increase in surface AMPAR expression to compensate for the overall decrease in activity. Therefore, astrocyte secreted TNFα is important not only for the developmental recruitment and expression of GluA1 at excitatory synapses, but also serves to regulate synaptic strength in response to altered patterns in synaptic activity after the development of synaptic circuitry is largely complete.

4.3 Chordin-like 1

In addition to regulating the insertion of GluA1 AMPAR subunits into developing synapses, astrocytes also play a critical role in the maturational switch that happens as GluA1 subunits are replaced by calcium-impermeable GluA2 subunits. Interestingly, this switch, which typically happens between P7-P16 in the mouse brain, is not directly regulated by glypicans. Instead, Chordin-Like 1 (Chrdl-1), a member of the chordin family of proteins, which is enriched in and secreted by developing astrocytes, was found to induce a significant increase in the surface clustering of GluA2 subunits on RGC dendrites when administered to glia-free RGC cultures (Blanco-Suarez et al., 2018). The mean frequency, but not amplitude, of mEPSCs recorded from RGCs treated with Chrdl-1 was also increased. Knockout of Chrdl-1 from mice substantially reduces the amount of GluA2 subunits colocalized with excitatory synapses at P14, further demonstrating a role for astrocyte-secreted Chrdl-1 in recruiting GluA2 subunits to the synapse *in vivo*. Additionally, *Chrdl-1* KO mice displayed a reduction in their total number of VGlut2+ thalamocortical synapses within the primary visual cortex, although no difference in the number of VGlut1+ intracortical synapses was observed, suggesting an important role for Chrdl-1 in the formation of thalamocortical synapses.

Synaptic maturation, in part via the recruitment of GluA2 subunits, results both in the refinement of synaptic circuitry along with the closure of critical periods of plasticity. In the visual cortex, there is a critical period during early development during which monocular enucleation or other forms of visual deprivation result in the substantial functional plasticity-dependent rewiring of patterns of ocular dominance (Levelt & Hubener, 2012). After the critical period window closes, however, the synaptic connections are no longer able to efficiently reorganize. Intriguingly, *Chrdl-1* KO mice display a significantly protracted critical period following monocular enucleation, as evidenced by enhanced synaptic remodeling after vision deprivation (Blanco-Suarez et al., 2018). This enhanced plasticity may be due to the relative immaturity of synapses formed in the absence of astrocytic Chrdl-1, in part owing to their deficit in GluA2 subunit surface expression. Thus, astrocytes participate not only in the formation and maturation of synapses themselves, but also in the regulation of critical periods.

4.4 Pentraxin-3

Pentraxin-3 (PTX3) has also recently been identified as an astrocyte-secreted protein that specifically enhances formation and function of developing excitatory synapses via the recruitment of AMPAR subunits (Fossati et al., 2019). PTX3 is a soluble innate immunity pattern recognition molecule secreted by astrocytes mainly during late embryonic and early postnatal development. Treating cultured hippocampal neurons with PTX3 increased both the frequency and amplitude of mEPSCs without changing the number of glutamatergic synapses or dendritic spines themselves. PTX3 treatment also enhanced the number of AMPAR subunits at the synapse. No changes in mRNA levels of NP1 or Gpc4 were detected when neurons were treated with PTX3, suggesting that PTX3 invokes the increase in postsynaptic AMPARs via a mechanism distinct from glypicans and NP-1. Instead, PTX3, which is known to interact with the extracellular matrix (ECM) and perineuronal nets, relies on these interactions to mediate its synapse-maturing effects. Specifically, destruction of the perineuronal net architecture by hyaluronidase disrupts PTX3-induced insertion of AMPAR subunits at the post-synapse surface leading to the conclusion that PTX3 enhances AMPAR recruitment to the synapse via the remodeling of perineuronal nets (Fossati et al., 2019). Interestingly, TSP1, which is known both to increase excitatory synaptogenesis and interact with the ECM, is a negative regulator of PTX3, blocking PTX3's ability to upregulate AMPAR subunit expression (Christopherson et al., 2005; Fossati et al., 2019). This interaction suggests that astrocyte-secreted molecules not only act upon neuronal mechanisms, but can also bidirectionally regulate one another.

4.5 Astrocyte-secreted factors regulate inhibitory synapse formation

Considerable progress has been made over the past two decades deciphering the mechanisms by which astrocytes control excitatory synapse formation and maturation. However, astrocytes have also been shown to contribute to inhibitory synapse formation (Elmariah et al., 2005; Hughes et al., 2010). Proper regulation of inhibitory synaptogenesis is also crucial for the development and maintenance of excitatory-inhibitory balance throughout the brain, deficits in which have been implicated in many neurological conditions including autism spectrum disorders (Nelson & Valakh, 2015; Rubenstein & Merzenich, 2003). Both astrocytes and astrocyte-conditioned media are able to increase the length and arborization of GABAergic axons, the density of GABAergic synaptic structures and the surface expression of postsynaptic GABA receptors (Diniz et al., 2014; Elmariah et al., 2005;

Hughes et al., 2010). However, the identity of the astrocytic factors that directly control inhibitory synapse formation are not yet known.

TGF-β1 is a factor present in ACM that is required for inhibitory synaptogenesis. When TGF-β1 was pharmacologically inhibited with SB-431542, ACM no longer induces inhibitory synapse formation (Diniz et al., 2014). Although TGF-β1 was also shown to act through D-serine to promote excitatory synapse formation, different pathways are involved for promoting inhibitory synaptogenesis (Diniz et al., 2012, 2014). However, these studies were conducted in neuron-glia mixed cultures, thus it is not known if TGF-β1 controls inhibitory synapse formation through a direct or indirect mechanism. The interplay between TGF-β1 and CaMKII was shown to be critical for inhibitory synaptogenesis, as blocking CaMKII with its inhibitor KN-93 prevented TGF-β1's effect on inhibitory but not excitatory synaptogenesis. TGF-β1 also increased the expression of neuroligin-2, which is presumed to be located at the post-synapse at inhibitory synapses, and knockdown of neuroligin-2 abolished the synaptogenic effect of TGF-β1 (Diniz et al., 2014). Other signaling pathways, such as brain-derived neurotrophic factor (BDNF) signaling, have also been implicated in astrocytic promotion of GABA receptor clustering and inhibitory synapse formation, although the details of this mechanism have yet to be fully elucidated (Elmariah et al., 2005).

5. Astrocyte-secreted factors regulate presynaptic release properties

Initial studies using a RGC neuron-only culture system utilized fractionation of glia-conditioned media (GCM) and assays to determine which fractions produced an increase in synaptic activity. Normally very few synapses are formed between RGCs in the absence of glia. Using this screening method, cholesterol and the ApoE particles carrying cholesterol were found to be one of the first astrocyte-secreted synaptogenic factors that increase synaptic transmission (Mauch et al., 2001). Neurons require cholesterol pre-synaptically for synaptic vesicle release and post-synaptically for the clustering and stability of neurotransmitter receptors (Allen, Halverson-Tamboli and Rasenick, 2007; Thiele et al., 2000). However, neurons do not synthesize a significant amount of cholesterol themselves. Rather, lipid synthesis and release from astrocytes provides neurons with the substantial majority of cholesterol used by neurons. Applying cholesterol to RGC neuron-only cultures substantially increased the frequency and size of mEPSCs

(Goritz, Mauch, & Pfrieger, 2005; Mauch et al., 2001). Cholesterol also plays an important role in the maturation of RGC neurites. Under normal conditions *in vitro*, RGCs extend multiple immature neurite processes but are not able to form mature, MAP2-positive dendrites. Treating RGCs with GCM and cholesterol induced the relocation of MAP2 from the soma to the dendrites and the clustering of glutamate receptor subunits to dendritic surfaces, indicating dendrite maturation (Goritz et al., 2005).

Recently, cholesterol synthesis in astrocytes *in vivo* has been shown to be critical for synapse maturation. Cholesterol synthesis in astrocytes relies on sterol regulatory element binding proteins (SREBPs), which are transcription factors responsible for the activation of genes involved in lipid synthesis. SREBP transcription factors are post-transcriptionally activated by SREBP Cleavage-Activating Protein (SCAP) (Camargo et al., 2012; Matsuda et al., 2001). Thus, astrocytic cholesterol synthesis can be disrupted by reducing SREBP levels via the conditional deletion of SCAP in mice (*SCAP* cKO). Surprisingly, *SCAP* cKO led to a greater density of hippocampal synapses *in vivo*. However, the maturity of the dendritic spines hosting these synapses was impaired in *SCAP* cKO mice (van Deijk et al., 2017). The resulting deficits in paired-pulse ratio, long-term potentiation, and presynaptic vesicle cluster size in *SCAP* cKO mice suggested that the impaired astrocyte cholesterol synthesis predominantly disturbs presynaptic structure and function. Taken together these findings reveal that astrocyte-secreted cholesterol is critical for neurite maturation and presynaptic function.

These studies which we summarized above revealed the presence of dynamic signaling modalities between astrocytes and neurons that are mediated by a milieu of astrocyte-secreted factors. Starting with RGCs and now extended to human induced pluripotent stem cell derived neurons, the development of neuron only culture systems were critical to identify and characterize astrocyte-secreted factors that control synapse formation. Studying the effects of secreted factors in neuron-glia mixed cultures can be misleading as secreted molecules can affect synapse formation by directly signaling to neurons or indirectly by modifying the synaptogenic functions of astrocytes or a combination of both.

6. Astrocyte-neuron contact-dependent mechanisms regulating synapse formation and maturation

6.1 Astrocyte processes physically contact onto synapses

Astrocytic processes closely ensheathe synaptic structures (Fig. 1B) and this intimate juxtapositioning between astrocytes and neurons was first noted by

Remak (1836) and Virchow (1858). However, the extent to which this close association between astrocytic processes and synaptic structures occur was not fully appreciated until Ventura and Harris used serial electron microscopy to visualize astrocyte morphology and its relationship with neurons in three-dimension (Ventura & Harris, 1999). Their quantification of adult rat astrocyte processes at hippocampal CA1 synapses revealed that the average distance between astrocytes to their closest contiguous synapse is 0.42 μm. More importantly, almost all mature synapses imaged had perisynaptic astrocyte processes enveloping the axon-dendrite interface while immature/transient spines (small macular, thin spines) were less likely to be contacted by astrocyte processes (Ventura & Harris, 1999; Witcher, Kirov, & Harris, 2007), highlighting a possible functional importance for astrocyte-neuron contacts in synapse formation and maturation.

Astrocyte-neuron contact is conserved throughout evolution. In *C. elegans*, ventral cephalic sheath cells (glial cells morphologically similar to vertebrate astrocytes) express netrin at the cell surface which interacts with receptors on pre-synaptic interneurons at the nerve ring to direct synaptogenesis and spatially coordinate precise circuit assembly (Colón-Ramos et al., 2007). Neuron-glia interactions mediated by a surface glycoprotein called neuroglian restricts abnormal axonal branching and dendritic arborization in *D. melanogaster* embryos (Yamamoto et al., 2006). More recently, the Freeman lab discovered a distinct glial subtype in adult *D. melanogaster* known as "astrocytic glia" with a ramified structure remarkably similar to mammalian glia. These fly astrocytes make close contacts with the neuropil, where processes exclusively wrap around the synapses and not the neuronal cell body (Doherty et al., 2009; Stork et al., 2014). Finally, serial section transmission electron microscopy of human CA1 stratum radiatum revealed that perisynaptic astrocyte processes also ensheathe synapses in the human brains, especially larger, more stable synapses (Witcher et al., 2010)—a striking similarity to that observed previously in murine models (Witcher et al., 2007). It is important to note that despite the evolutionarily conserved features of astrocytes and their synapse association, as the evolution progressed these cells became larger in size and more complex permitting them to interact with an incredible number of synapses at the same time. For example, a human cortical protoplasmic astrocyte is ten times larger than its mouse counterpart and is estimated to interact with up to two million synapses at the same time (Oberheim et al., 2009).

The tripartite synapse is not unique to the hippocampus, but observed in all brain regions (Araque et al., 1999; Grosche et al., 1999; Halassa et al., 2007; Perea et al., 2009; Santello, Calì, & Bezzi, 2012). In addition, astrocytes have been shown to contact neuronal soma, axons and dendrites, but appear to preferentially contact onto dendritic spines, likely through cell adhesion-based mechanisms. Serial EM reconstruction of murine hippocampus revealed the presence of cell adhesion molecules concentrated at cell junctions ("puncta adherentia") between dendritic spines and neighboring astrocyte processes (Spacek & Harris, 1998), while 2-photon microscopy tracking of astrocyte process dynamics revealed that ephrin-mediated transient contacts between perisynaptic astrocyte processes and dendritic protrusions correlate with enhanced spine stability and maturity (Nishida & Okabe, 2007). This physical contact between astrocytes and synapses drives downstream signaling pathways and regulates both excitatory and inhibitory synaptogenesis. For example, hippocampal astrocytes contact synapses via integrin receptors to activate the protein kinase C signaling pathway to mediate excitatory synapse formation (Hama et al., 2004) while Bergmann glia directly bind to axons of stellate interneurons through cell adhesion mediated by cell adhesion molecule L1 (CHL1) to guide axonal branching onto dendrites of Purkinje cells in the cerebellum, actively directing GABAergic circuit formation (Ango et al., 2008). Strikingly, synaptic activation modulates intracellular calcium changes in astrocytes and in turn influences neuronal activity, highlighting that close functional communication between astrocytes and neurons is necessary for bi-directional signaling at the tripartite synapse (Grosche et al., 1999). Having emphasized the importance of physical contact between perisynaptic astrocyte processes with the pre- and postsynaptic components, in the next section we will discuss some of the cell-adhesion based mechanisms underlying contact dependent synapse formation and maturation (Fig. 4).

6.2 Neurexin/neuroligin

Neurexins and neuroligins are well-studied cell adhesion molecules with critical roles in synapse formation and functional maturation. It has been widely accepted that postsynaptic neuroligins bind to presynaptic neurexins, forming a trans-synaptic complex, which in turn interacts with a diverse array of intracellular proteins to initiate pre- and postsynaptic differentiation and maturation (Craig & Kang, 2007; Graf et al., 2004; Südhof, 2008). Due to alternative splicing, there are thousands of neurexin isoforms, which are differentially expressed in specific subsets of neurons (Nguyen et al., 2016). Therefore, research into neurexin-neuroligin signaling has largely been limited to understanding how the different permutations of neuronal neurexins

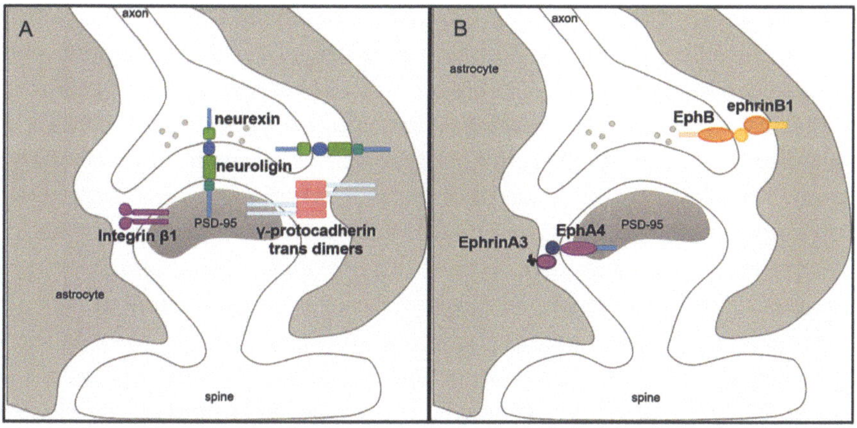

Fig. 4 Astrocytic cell adhesion molecules involved in synapse formation and maturation. (A) Astrocyte-neuron interactions mediated by neurexin/neuroligin, integrin β1 and γ-protocadherin transcellular interactions positively regulate synapse formation and maturation. (B) Conversely, ephrin:Eph cell-cell interaction between astrocytes and neurons negatively regulates synaptogenesis.

and neuroligins regulate synapse numbers and/or function in different circuits and brain regions. Importantly, non-neuronal sources of neuroligin are sufficient to induce formation of presynaptic structures by binding to axonal neurexins (Scheiffele et al., 2000). Overexpression of neuroligin in human embryonic kidney (HEK) 293 cells (1) promotes synapsin accumulation and induces synaptic vesicles in axons (Scheiffele et al., 2000); (2) aligns the pre and postsynaptic sites (Fu et al., 2003) and (3) regulates post-synapse differentiation by recruiting post-synaptic scaffolding proteins like PSD95 and assembling NMDA receptors at nucleation sites directly opposite the presynaptic terminal (Chih et al., 2005).

Do neuroligins play important roles in non-neuronal cells? In fact, it has been repeatedly demonstrated through transcriptomic analysis of FACS sorted or ribosome tagged astrocytes that astrocytes not only express endogenous neuroligins-1, 2 and 3 but that these astrocytic neuroligins are expressed at comparable or higher levels compared to neurons all throughout life (Boisvert et al., 2018; Srinivasan et al., 2016; Zhang et al., 2014)

Stogsdill et al. (2017) demonstrated that astrocytic neuroligins are critical for both astrocyte morphogenesis and synaptogenesis during cortical development. In particular, when neuroligin-2 was specifically eliminated in a sparse population of astrocytes in the primary visual cortex, the neuroligin-2 depleted astrocytes were far less elaborate and the density of excitatory synapses within the domains of knockout astrocytes was strongly diminished (Stogsdill et al., 2017). This was surprising, because it was previously suggested that neuroligin-2 expression is only neuronal and specific to inhibitory synapses, colocalizing with inhibitory postsynaptic markers such as gephryin and $GABA_AR\beta2/3$ (Varoqueaux, Jamain, & Brose, 2004). In fact, neuroligin-2 interacts with both gephyrin and collybistin, two post-synaptic proteins specific to inhibitory synapses (Poulopoulos et al., 2009). Loss of astrocytic neuroligin-2 did not affect the inhibitory synapse numbers (Stogsdill et al., 2017). Stogsdill et al. also tested the functional consequences of eliminating astrocytic neuroligin-2 on synapses through electrophysiological recordings from pyramidal neurons in layer 5 of the visual cortex. They found that loss of astrocytic neuroligin-2 results in a decrease in both the amplitude and frequency of miniature excitatory postsynaptic currents, demonstrating a novel role for astrocytic neuroligin-2 in the regulation of excitatory synapse formation and function. On the other hand, the same manipulation also significantly increased the frequency of miniature inhibitory postsynaptic currents. Taken together these findings revealed that astrocytic neuroligin-2 is critical for maintaining excitation/inhibition balance in the developing cortex, through local regulation of excitatory synapse numbers and inhibitory synaptic activity.

Another important outcome of these studies was the fact that neuroligin-neurexin adhesions between perisynaptic astrocyte processes and presynaptic neurexins molecularly couple astrocyte morphological maturation and synaptogenesis. These two vital neuroanatomical processes occur during the first 3 postnatal weeks of murine cortical development. Knocking down the expression of neuroligins 1-3 in astrocytes using short hairpin RNA (shRNA) resulted in a drastic reduction of astrocyte complexity when co-cultured onto wildtype neurons. Knocking down neurexin expression in neurons phenocopied the loss of astrocyte complexity, confirming that transcellular adhesion between astrocytes and neurons regulate astrocyte morphological maturation. *In vivo* experiments revealed that silencing astrocytic neuroligins resulted in a similar decrease in astrocyte stellate morphology (quantified by astrocyte territory volume and astrocyte neuropil infiltration volume). However, the individual astrocytic neuroligins were found to have non-overlapping functions and their effects on astrocyte neuropil infiltration volume varied over development. For example, astrocytic neuroligin-1 was critical in early astrocyte morphological development (P7); whereas astrocytic neuroligin-3 only had a robust effect on astrocyte complexity in later stages of postnatal development (P21). Astrocytic neuroligin-2 was required at both timepoints. These findings suggest that not only does astrocytic neuroligin-neuronal neurexin interaction bidirectionally regulate both astrocyte morphology and synapse formation, but that the different permutations of neuroligin-neurexin interactions at the tripartite synapse could potentially temporally instruct various steps of synapse formation, maturation and function.

6.3 Integrins

Integrins are a family of 24 heterodimeric proteins that serve as cell adhesion receptors that not only mediate cell-cell interactions but also mediate cell-extracellular matrix (cell-ECM) attachments. The function of integrins as bidirectional signaling receptors between astrocytes and neurons is best demonstrated through the Leyton lab's careful dissection of how interaction between neuronal Thy1 and astrocytic αvβ3 integrin heterodimers regulates formation of focal contact sites and stress fiber formation in astrocytes as well as inhibiting neurite outgrowth and retraction of neuronal processes (Avalos et al., 2009; Hermosilla et al., 2008; Herrera-Molina et al., 2012; Leyton et al., 2001; Maldonado et al., 2017). However, the roles of integrin-based cell adhesion mechanisms mediating astrocyte-neuron contact during synapse formation and function is less explored.

Hama et al. (2004) cultured astrocytes on top of individual neurons already exposed to media containing astrocyte secreted factors to determine the effects of astrocyte-neuron contact (Hama et al., 2004). The authors found that astrocyte contact improved dye uptake in active presynaptic terminals threefold, and that upon depolarization stimulation, astrocyte contact resulted in enhanced presynaptic function (Hama et al., 2004). These results suggested that astrocyte contact increases the number of neurotransmitter release sites and alters neurotransmitter release properties. This was also supported by electrophysiological data where astrocyte contact increased the amplitude of autaptic EPSC in response to paired pulse stimuli, as well as increased frequency of asynchronous EPSCs (Hama et al., 2004). Interestingly, the authors also found that limited, local astrocyte contact was sufficient to induce an increase in synaptogenesis throughout the neuron due to the activation of protein kinase C (PKC activation). PKC activation was only abolished by echistatin, an Arg-Gly-Asp (RGD) sequence containing peptide inhibitor for integrin β1 and β3 (Hama et al., 2004), suggesting that neuronal integrin β receptors are activated upon astrocyte contact and signal amplification through PKC pathway leads to global and long-lasting increases in synapse formation.

6.4 Protocadherins

Protocadherins form a subgroup of cadherin superfamily of cell adhesion molecules that mediate Ca^{2+}-dependent transcellular adhesion. Protocadherins are predominantly expressed in the CNS and have 6-7 extracellular cadherin (EC) domains which have low sequence homology to classical cadherin EC domains. Instead, they have RGD motif similar to integrin ligands (Morishita et al., 2006).

Protocadherin-mediated astrocyte-neuron contact on the other hand, has been demonstrated by the Weiner lab to be a critical controller of dendrite arborization (Molumby et al., 2017) and synaptogenesis (Garrett & Weiner, 2009). The seminal paper by Garrett and Weiner (2009) carefully characterized γ-protocadherin in the developing spinal cord and found that not only did astrocytes express all but 2 γ-protocadherins, but that these proteins are co-localized with ezrin at the perisynaptic astrocytic processes and came into close contact with the pre- and post-synaptic compartments (Garrett & Weiner, 2009). More critically, they demonstrated that astrocyte specific knockout of γ-protocadherin resulted significant decrease in excitatory and inhibitory synaptogenesis in vitro and in vivo, particularly during the early stages of synapse formation (Garrett & Weiner, 2009). The number of excitatory and inhibitory synapses in astrocytic γ-protocadherin knockout mice was significantly reduced when compared to control embryonic spinal cords (E15 and E17) but recovered by P0, unlike neuron specific γ-protocadherin knockout where reduction in synapse number was even more significant and sustained (Garrett & Weiner, 2009). In follow up studies in the developing cortex, it was revealed that disruption of astrocyte-neuron γ-protocadherin homophilic trans-interactions by knocking out γ-protocadherin or overexpressing mismatching γ-protocadherin specifically in astrocytes led to significant reductions in dendrite complexity (fewer branching and reduced dendrite length) in wildtype neurons (Molumby, Keeler, & Weiner, 2016). Interestingly, neuronal γ-protocadherin makes cis-interactions with neuroligin-1, sequestering neuroligin-1 and preventing the formation of neuroligin/neurexin trans-synaptic interactions required for synaptic organization at the pre and post synapse (Molumby et al., 2017). As described above, astrocytes also express neuroligins, and astrocytic neuroligin-2 binds with neuronal neurexins to mediate both astrocyte morphological maturation and synaptogenesis (Stogsdill et al., 2017). Future studies investigating the crosstalk between astrocytic protocadherins and neuroligins could determine how different cell adhesion pathways cooperatively work to link the tripartite synapse compartments to fine tune synapse formation and maturation.

6.5 Ephrin/Eph

Eph receptors are a class of transmembrane tyrosine kinase receptors which, by binding to their ephrin ligands, are critical regulators of short-range cell-cell communication underlying key processes during neurodevelopment such as neuron migration (Villar-Cerviño et al., 2013), axon guidance and the establishment of neural circuits (Paixão et al., 2013), as well as synaptogenesis (Kayser, Nolt, & Dalva, 2008). There are two subclasses of ephrins, ephrin-A and ephrin-B, both of which have an Eph receptor-binding domain and can be distinguished by their intracellular domain: Ephrin-As are anchored to the membrane by glycosylphosphatidylinositol (GPI) linkage while ephrin-Bs have intracellular PDZ domain (Lisabeth, Falivelli, & Pasquale, 2013). Canonically, ephrins bind to their Eph receptors which promote phosphorylation of tyrosine residues at both the juxtamembrane region and kinase domains, leading to signal amplification and the recruitment of downstream effector proteins that result in actin cytoskeletal organization (Kania & Klein, 2016; Murai & Pasquale, 2003). This is known as ephrin:Eph forward signaling. However, ephrin/Eph signaling is much more complex (and versatile) as the ephrin ligand is also capable of signal transduction in Eph:ephrin reverse signaling. For example, Eph receptor binding induces Tyr phosphorylation at the intracellular domain of ephrin-Bs which can then recruit Src kinases and other effectors (Kania and Klein, 2016).

The bidirectionality of Eph/ephrin signaling is best exemplified by crosstalk between perisynaptic astrocytic processes and dendritic spines. While it was recognized that glial cells expressed ephrin A ligand mRNA and protein at significant levels (Davenport et al., 1998), the function of astrocytic ephrins was unknown until it was demonstrated that astrocytic ephrinA3 bound to EphA4 receptors on hippocampal dendrites to induce spine retractions (Murai et al., 2003). In addition, both EphA4 knockout and eliminating EphA4 kinase activity resulted in significantly longer and misshapened spines (Murai et al., 2003). Even more significantly, EphA4 tyrosine phosphorylation occurs at basal levels in the adult hippocampus, strongly suggesting that the ephrinA3:EphA4 forward signaling only occurred when dendritic spines came into close contact with astrocytic processes, perhaps to counteract other spine enlargement signals during remodeling and serves as a mechanism to maintain proper brain circuitry (Murai et al., 2003). The binding of astrocytic ephrinA3 with neuronal EphA4 receptors also has functional consequences in astrocytes. EphA4:ephrinA3 reverse signaling significantly reduces cell surface expression of these glutamate transporter GLAST and GLT-1 and limits glutamate uptake at the synapse (Carmona et al., 2009).

Additionally, astrocytes also express the whole array of EphA and EphB receptors (Nestor et al., 2007; Zhuang et al., 2010). Activation of astrocytic EphA receptors induces persistent outgrowth of astrocytic processes in organotypic hippocampal slices, significantly decreasing the frequency of spontaneous intracellular calcium oscillations, resulting in a decrease in glutamate release that promotes extrasynaptic NMDA receptor activity (Nestor et al., 2007). It has also been demonstrated that activity dependent release of neuronal ephrinB3 activates astrocytic EphB3 and EphA4 to increase D-serine release, thereby modulating synaptic potentiation (Zhuang et al., 2010). All of these examples show that Eph/ephrin signaling plays a very critical role in regulating synaptic plasticity at the tripartite synapse. However, does astrocyte/neuron Eph/ephrin signaling play a role in synaptogenesis?

Careful characterization of ephrin and Eph mRNA expression in P0.5 murine brains by *in situ* hybridization revealed that these molecules are present during early cortical development (Liebl et al., 2003). The enrichment of ephrins and Eph receptors in the neuroepithelium and subventricular zone where radial glial cells reside also led to the speculation that Eph/ephrin signaling might be critical for astrocyte proliferation and migration (Liebl et al., 2003). The first direct evidence that Eph/ephrin signaling was critical for astrocyte-neuron interaction during synapse formation came from a series of two-photon time-lapse imaging experiments performed on hippocampal astrocytes and neurons isolated from P4-5 mice (Nishida & Okabe, 2007). A subset of cultured astrocytes was virally transduced with GFP while rhodamine-dextran (dye) was electroporated into pyramidal cells and the cells were imaged from DIV 3-7. Nishida and Okabe found that dendritic filopodia that frequently contacted astrocytes persisted longer and had an increased tendency to form spines independent of the size of filopodia, maturity of neuron or concentration of astrocyte-secreted factors (Nishida & Okabe, 2007). Modulating endogenous ephrinA3:EphA4 signaling by (1) introducing kinase-inactive EphA4 that sequesters endogenous ephrinA3 or by (2) introducing of ephrinA3/Fc that constitutively activates EphA4 both resulted in a significant decrease in the persistence of dendritic filopodia, especially filopodia which were newly formed (astrocyte-contact naive). These results demonstrated that the ephrin/Eph signaling pathway bridges transient astrocyte-dendrite contact critical to stabilization and maturation of newly formed dendrites into spines during synaptogenesis (Fig. 4B).

More recently, the Ethell lab has identified astrocytic ephrin B1 as an important negative regulator of synaptogenesis in the adult hippocampus. *In vivo* analyses of astrocyte-specific ephrinB1 conditional knockout mice revealed an increase in spine density and number of excitatory synapses specifically in the stratum radiatum (Koeppen et al., 2018). However, this did not result in increased synaptic function since Koeppen and colleagues unexpectedly found significant impairments in post-synaptic neuronal firing. Characterization of the spines showed an increase in immature spines which had significantly reduced AMPA receptor expression when normalized to synaptic PSD95 (Koeppen et al., 2018), strongly suggesting that loss of astrocytic ephrin B1 resulted in an increase in silent/immature synapses. Importantly, the authors demonstrated that astrocytic ephrin B1 bound to EphB receptor containing synaptosomes, suggesting that ephrinB1:EphB reverse signaling is critical for engulfment of presynaptic boutons and is an important mechanism through which astrocytes prune away immature synapses (Koeppen et al., 2018). This has important implications in synapse formation during learning and memory. Using a fear conditioning paradigm, astrocyte-specific ephrin B1 conditional knockout mice had improved contextual memory recall compared to wildtype which could be attributed to increased spine density and increased spine clustering, particularly on activated (cFos positive) neurons (Nguyen et al., 2020). The authors also found upregulation in the number of excitatory synapses through immunostaining as well as increases in both evoked excitatory responses and excitatory post-synaptic strength in trained knockout mice and not in naive knockout mice (Nguyen et al., 2020). Mice overexpressing astrocytic ephrin B1 not only had no changes in spine density and spine clustering on activated neurons after learning, but had a lower baseline of spines compared to wildtype, which was concordant with the significant impairments in contextual memory recall (Nguyen et al., 2020). Taken together, astrocytic ephrin B1 is poised to modulate excitatory synaptogenesis and is pivotal in the formation of new circuits during learning. Future studies are needed to also explore if astrocytic ephrin B1 plays a critical role in regulating synapse formation and maturation during development.

Early observations of the brain already highlighted the close juxtaposition of astrocytes and neurons at the synapse. The significance of astrocyte-neuron contact however, was not fully appreciated until the studies summarized above. The variety of molecular mechanisms, as well as the numerous receptor/ligand combinations within each signaling pathway described could underly intricate regional and temporal differences in astrocyte-neuron interactions which the field has only started to uncover. It is also possible that the additive effect of different contact-dependent mechanisms is required at the tripartite synapse to ensure lasting physical contact between astrocytes and neurons beyond development.

7. Conclusions and future directions

Here we have summarized the mounting evidence that emerged from a number studies detailing how astrocytes participate in the construction and remodeling of synaptic circuits. These findings clearly demonstrate that a dynamic bi-directional communication between astrocytes and neurons underlie all essential elements of synaptic development. However, there are still important outstanding questions:

1) How is the expression and secretion of synaptogenic factors from astrocytes regulated? We currently do not know whether astrocytes provide synapse modulating cues with spatial precision. Do astrocytes have polarity that ensures distribution of synapse modulating factors to distinct synapses where they are needed? Does local activity of synapses control secretion of astrocytic factors or modify the cell-adhesions between astrocytes and neurons. Future studies investigating the subcellular organization of astrocytes utilizing super-resolution microscopy techniques will be valuable to answer this long-standing question.
2) How do astrocyte-synapse adhesions control formation of excitatory and inhibitory synapses? Our current knowledge on astrocyte function at tripartite synapses point towards a critical role for astrocyte-synapse adhesions in controlling the local excitation/inhibition balance. This suggests the presence of intricate adhesion-based signaling mechanisms that control synapse formation. Future studies to define these mechanisms would benefit from identification of the molecular composition of astrocyte-synapse junctions. In fact, a recent study used a novel chemogenetic *in vivo* approach to proteomically map the astrocyte-neuron junctions as they exist *in vivo* (Takano et al., 2020). Extracellular biotin protein ligase variants were targeted to astrocyte-neuron interfaces through cell-type specific viral transduction and contact mediated complementation to achieve *in vivo* biotinylation of astrocyte-synapse contacts. Following biochemical isolation of biotinylated proteins, the authors used high resolution mass spectroscopy to identify 118 proteins representing a high-confidence perisynaptic proteome (Takano et al., 2020). Neuronal Cell Adhesion Molecule (NrCAM) was identified to be specifically localized to astrocyte-inhibitory synapse contacts through homophilic NrCAM interactions between astrocytes and dendrites. Loss of either astrocytic or neuronal NrCAM dramatically reduces inhibitory synapse number and function and induces overgrowth of astrocyte processes (Takano et al., 2020), a phenotype that is opposite of what was observed with astrocytic neuroligin 2 knockouts (Stogsdill et al., 2017).

3) Do astrocytes gain specialized functions based on their neuronal neighbors to properly support underlying synaptic connectivity and function? Mounting evidence points towards regional specialization of astrocytes (Bayraktar et al., 2015; Ben Haim & Rowitch, 2016; Boisvert et al., 2018). This specialization is likely to be critical for proper activity of brain regions and drive activity-dependent signaling between astrocytes and neurons both during development and in the adult CNS. Thus, future studies focusing on how circuit activity and architecture alter astrocytes will be fruitful.
4) How does astrocyte-neuron cross-talk shape behaviors? Recent studies in Drosophila, zebrafish and mice demonstrated that astrocytes respond to neuromodulators with increases in intracellular calcium concentration (Ma et al., 2016; Martín et al., 2015; Martin-Fernandez et al., 2017; Mu et al., 2019; Navarrete & Araque, 2008). In Drosophila, this activation leads to inhibition of dopaminergic neuron firing and alterations in behavior (Ma et al., 2016). In mice, elevating astrocytic calcium transients control attention as well as learning and memory attention (Adamsky et al., 2018; Kol et al., 2020; Mederos et al., 2018; Nagai et al., 2019) and diminishing them triggers repetitive grooming (Yu et al., 2018). Taken together, this growing body of work from multiple labs reveal the importance of these astrocytic calcium transients in different species and under distinct behavioral paradigms. Optimization and specialization of novel genetically encoded indicators and chemicogenetic tools for astrocytes will surely enable further discoveries and in-depth analyses of astrocyte-neuron communication in the future.

Astrocyte dysfunction is highly linked to human brain disorders. Therefore, answering these questions are not only important to unravel how our brains form and function, but also will be of importance for developing therapeutic strategies to combat neurological disease.

References

Adams, J. C. (2001). Thrombospondins: Multifunctional regulators of cell interactions. *Annual Review of Cell and Developmental Biology*, *17*(1), 25–51. https://doi.org/10.1146/annurev.cellbio.17.1.25.

Adamsky, A., et al. (2018). Astrocytic activation generates de novo neuronal potentiation and memory enhancement. *Cell*, *174*(1), 59–71.e14. https://doi.org/10.1016/j.cell.2018.05.002.

Adrian, M., et al. (2014). Barriers in the brain: Resolving dendritic spine morphology and compartmentalization. *Frontiers in Neuroanatomy*, *8*, 1–12. https://doi.org/10.3389/fnana.2014.00142. 142. Frontiers Media S.A.

Albrecht, D., et al. (2012). SPARC prevents maturation of cholinergic presynaptic terminals. *Molecular and Cellular Neuroscience*, *49*(3), 364–374. https://doi.org/10.1016/j.mcn.2012.01.005.

Allen, J. A., Halverson-Tamboli, R. A., & Rasenick, M. M. (2007). Lipid raft microdomains and neurotransmitter signalling. *Nature Reviews. Neuroscience*, *8*(2), 128–140. https://doi.org/10.1038/nrn2059.

Allen, N. J., et al. (2012). Astrocyte glypicans 4 and 6 promote formation of excitatory synapses via GluA1 AMPA receptors. *Nature*, *486*(7403), 410–414. https://doi.org/10.1038/nature11059.

Anderson, C. M., & Swanson, R. A. (2000). Astrocyte glutamate transport: Review of properties, regulation, and physiological functions. *Glia*, 1–14. https://doi.org/10.1002/1098-1136(200010)32:1<1::AID-GLIA10>3.0.CO;2-W. John Wiley & Sons, Ltd.

Ango, F., et al. (2008). Bergmann glia and the recognition molecule CHL1 organize GABAergic axons and direct innervation of Purkinje cell dendrites. *PLoS Biology*, *6*(4), e103. https://doi.org/10.1371/journal.pbio.0060103.

Araque, A., et al. (1999). Tripartite synapses: Glia, the unacknowledged partner. *Trends in Neurosciences*, 208–215. https://doi.org/10.1016/S0166-2236(98)01349-6.

Araujo, A. P. B., Carpi-Santos, R., & Gomes, F. C. A. (2019). The role of astrocytes in the development of the cerebellum. *Cerebellum*, *18*, 1017–1035. https://doi.org/10.1007/s12311-019-01046-0.

Arikkath, J., & Campbell, K. P. (2003). Auxiliary subunits: Essential components of the voltage-gated calcium channel complex. *Current Opinion in Neurobiology*, *13*(3), 298–307. https://doi.org/10.1016/S0959-4388(03)00066-7.

Avalos, A. M., et al. (2009). Neuronal Thy-1 induces astrocyte adhesion by engaging syndecan-4 in a cooperative interaction with αvβ3 integrin that activates PKCα and RhoA. *Journal of Cell Science*, *122*(19), 3462–3471. https://doi.org/10.1242/jcs.034827.

Bayraktar, O. A., et al. (2015). Astrocyte development and heterogeneity. *Cold Spring Harbor Perspectives in Biology*, 7(1). https://doi.org/10.1101/cshperspect.a020362.

Beattie, E. C., et al. (2002). Control of synaptic strength by glial TNFα. *Science*, *295*(5563), 2282–2285. https://doi.org/10.1126/science.1067859.

Ben Haim, L., & Rowitch, D. H. (2016). Functional diversity of astrocytes in neural circuit regulation. *Nature Reviews Neuroscience*, 31–41. https://doi.org/10.1038/nrn.2016.159.

Blanco-Suarez, E., et al. (2018). Astrocyte-secreted chordin-like 1 drives synapse maturation and limits plasticity by increasing synaptic GluA2 AMPA receptors. *Neuron*, *100*(5), 1116–1132.e13. https://doi.org/10.1016/j.neuron.2018.09.043.

Boisvert, M. M., et al. (2018). The aging astrocyte transcriptome from multiple regions of the mouse brain. *Cell Reports*, *22*(1), 269–285. https://doi.org/10.1016/j.celrep.2017.12.039.

Bornstein, P. (2001). Thrombospondins as matricellular modulators of cell function. *Journal of Clinical Investigation*, *107*, 929–934. https://doi.org/10.1172/JCI12749.

Brill, J., & Huguenard, J. R. (2008). Sequential changes in AMPA receptor targeting in the developing neocortical excitatory circuit. *Journal of Neuroscience*, *28*(51), 13918–13928. https://doi.org/10.1523/JNEUROSCI.3229-08.2008.

Bushong, E. A., et al. (2002). Protoplasmic astrocytes in CA1 stratum radiatum occupy separate anatomical domains. *Journal of Neuroscience*, *22*(1), 183–192. https://doi.org/10.1523/jneurosci.22-01-00183.2002.

Camargo, N., et al. (2012). High-fat diet ameliorates neurological deficits caused by defective astrocyte lipid metabolism. *FASEB Journal*, *26*(10), 4302–4315. https://doi.org/10.1096/fj.12-205807.

Carmona, M. A., et al. (2009). Glial ephrin-A3 regulates hippocampal dendritic spine morphology and glutamate transport. *Proceedings of the National Academy of Sciences of the United States of America*, *106*(30), 12524–12529. https://doi.org/10.1073/pnas.0903328106.

Chih, B., Engelman, H., & Scheiffele, P. (2005). Control of excitatory and inhibitory synapse formation by neuroligins. *Science*, *307*(5713), 1324–1328. https://doi.org/10.1126/science.1103773.

Christopherson, K. S., et al. (2005). Thrombospondins are astrocyte-secreted proteins that promote CNS synaptogenesis. *Cell*, *120*(3), 421–433. https://doi.org/10.1016/j.cell.2004.12.020.

Chung, W. S., Allen, N. J., & Eroglu, C. (2015). Astrocytes control synapse formation, function, and elimination. *Cold Spring Harbor Perspectives in Biology*, 7(9). https://doi.org/10.1101/cshperspect.a020370.

Clarke, L. E., & Barres, B. A. (2013). Emerging roles of astrocytes in neural circuit development. *Nature Reviews Neuroscience*, *14*(5), 311–321. https://doi.org/10.1038/nrn3484.

Colón-Ramos, D. A., Margeta, M. A., & Shen, K. (2007). Glia promote local synaptogenesis through UNC-6 (netrin) signaling in C. elegans. *Science*, *318*(5847), 103–106. https://doi.org/10.1126/science.1143762.

Craig, A. M., & Kang, Y. (2007). Neurexin-neuroligin signaling in synapse development. *Current Opinion in Neurobiology*, *17*(1), 43–52. https://doi.org/10.1016/j.conb.2007.01.011.

Davenport, R. W., et al. (1998). Cellular localization of ephrin-A2, ephrin-A5, and other functional guidance cues underlies retinotopic development across species. *Journal of Neuroscience*, *18*(3), 975–986. https://doi.org/10.1523/jneurosci.18-03-00975.1998.

Diniz, L. P., et al. (2012). Astrocyte-induced synaptogenesis is mediated by transforming growth factor β signaling through modulation of d-serine levels in cerebral cortex neurons. *Journal of Biological Chemistry*, *287*(49), 41432–41445. https://doi.org/10.1074/jbc.M112.380824.

Diniz, L. P., et al. (2014). Astrocytic control of neural circuit formation: Highlights on TGF-beta signaling. *Neurochemistry International*, *78*, 18–27. https://doi.org/10.1016/j.neuint.2014.07.008.

Doherty, J., et al. (2009). Ensheathing glia function as phagocytes in the adult Drosophila brain. *Journal of Neuroscience*, *29*(15), 4768–4781. https://doi.org/10.1523/JNEUROSCI.5951-08.2009.

Elmariah, S. B., et al. (2005). Astrocytes regulate inhibitory synapse formation via Trk-mediated modulation of postsynaptic GABAA receptors. *Journal of Neuroscience*, *25*(14), 3638–3650. https://doi.org/10.1523/JNEUROSCI.3980-04.2005.

Eroglu, Ç., et al. (2009). Gabapentin receptor α2δ-1 is a neuronal thrombospondin receptor responsible for excitatory CNS synaptogenesis. *Cell*, *139*(2), 380–392. https://doi.org/10.1016/j.cell.2009.09.025.

Farhy-Tselnicker, I., & Allen, N. J. (2018). Astrocytes, neurons, synapses: A tripartite view on cortical circuit development. *Neural Development*, *13*, 1–12. https://doi.org/10.1186/s13064-018-0104-y. 7.

Farhy-Tselnicker, I., et al. (2017). Astrocyte-secreted glypican 4 regulates release of neuronal pentraxin 1 from axons to induce functional synapse formation. *Neuron*, *96*(2), 428–445.e13. https://doi.org/10.1016/j.neuron.2017.09.053.

Fossati, G., et al. (2019). Pentraxin 3 regulates synaptic function by inducing AMPA receptor clustering via ECM remodeling and β1-integrin. *The EMBO Journal*, *38*(1), 1–20. https://doi.org/10.15252/embj.201899529.

Freeman, M. R. (2015). Drosophila central nervous system glia. *Cold Spring Harbor Perspectives in Biology*, 7, 1–14. a020552. Cold Spring Harbor Laboratory Press. https://doi.org/10.1101/cshperspect.a020552.

Fu, Z., et al. (2003). Functional excitatory synapses in HEK293 cells expressing neuroligin and glutamate receptors. *Journal of Neurophysiology*, *90*(6), 3950–3957. https://doi.org/10.1152/jn.00647.2003.

Gan, K. J., & Südhof, T. C. (2020). SPARCL1 promotes excitatory but not inhibitory synapse formation and function independent of neurexins and neuroligins. *Journal of Neuroscience*, *40*(42), 8088–8102. https://doi.org/10.1523/JNEUROSCI.0454-20.2020.

Garrett, A. M., & Weiner, J. A. (2009). Control of CNS synapse development by γ-protocadherin-mediated astrocyte-neuron contact. *Journal of Neuroscience*, *29*(38), 11723–11731. https://doi.org/10.1523/JNEUROSCI.2818-09.2009.

Goritz, C., Mauch, D. H., & Pfrieger, F. W. (2005). Multiple mechanisms mediate cholesterol-induced synaptogenesis in a CNS neuron. *Molecular and Cellular Neuroscience*, *29*(2)), 190–201. https://doi.org/10.1016/j.mcn.2005.02.006.

Graf, E. R., et al. (2004). Neurexins induce differentiation of GABA and glutamate postsynaptic specializations via neuroligins. *Cell*, *119*(7), 1013–1026. https://doi.org/10.1016/j.cell.2004.11.035.

Grosche, J., et al. (1999). Microdomains for neuron-glia interaction: Parallel fiber signaling to Bergmann glial cells. *Nature Neuroscience*, *2*(2), 139–143. https://doi.org/10.1038/5692.

Halassa, M. M., et al. (2007). Synaptic islands defined by the territory of a single astrocyte. *Journal of Neuroscience*, *27*(24), 6473–6477. https://doi.org/10.1523/JNEUROSCI.1419-07.2007.

Hall, B. J., & Ghosh, A. (2008). Regulation of AMPA receptor recruitment at developing synapses. *Trends in Neurosciences*, 82–89. https://doi.org/10.1016/j.tins.2007.11.010.

Hama, H., et al. (2004). PKC signaling mediates global enhancement of excitatory synaptogenesis in neurons triggered by local contact with astrocytes. *Neuron*, *41*(3), 405–415. https://doi.org/10.1016/S0896-6273(04)00007-8.

Hasel, P., et al. (2017). Neurons and neuronal activity control gene expression in astrocytes to regulate their development and metabolism. *Nature Communications*, *8*(1), 1–18. https://doi.org/10.1038/ncomms15132.

Hennekinne, L., et al. (2013). Differential control of thrombospondin over synaptic glycine and AMPA receptors in spinal cord neurons. *Journal of Neuroscience*, *33*(28), 11432–11439. https://doi.org/10.1523/JNEUROSCI.5247-12.2013.

Hermosilla, T., et al. (2008). Direct Thy-1/αVβ3 integrin interaction mediates neuron to astrocyte communication. *Biochimica et Biophysica Acta. Molecular Cell Research*, *1783*(6), 1111–1120. https://doi.org/10.1016/j.bbamcr.2008.01.034.

Herrera-Molina, R., et al. (2012). Astrocytic αvβ3 integrin inhibits neurite outgrowth and promotes retraction of neuronal processes by clustering thy-1. *PLoS One*, 7(3), e34295. https://doi.org/10.1371/journal.pone.0034295.

Hestrin, S. (1992). Developmental regulation of NMDA receptor-mediated synaptic currents at a central synapse. *Nature*, *357*(6380), 686–689. https://doi.org/10.1038/357686a0.

Hughes, E. G., Elmariah, S. B., & Balice-Gordon, R. J. (2010). Astrocyte secreted proteins selectively increase hippocampal GABAergic axon length, branching, and synaptogenesis. *Molecular and Cellular Neuroscience*, *43*(1), 136–145. https://doi.org/10.1016/j.mcn.2009.10.004.

Johnson, K. G., et al. (2006). The HSPGs syndecan and dallylike bind the receptor phosphatase LAR and exert distinct effects on synaptic development. *Neuron*, *49*(4), 517–531. https://doi.org/10.1016/j.neuron.2006.01.026.

Johnson, M. A., et al. (2007). Functional neural development from human embryonic stem cells: Accelerated synaptic activity via astrocyte coculture. *Journal of Neuroscience*, *27*(12), 3069–3077. https://doi.org/10.1523/JNEUROSCI.4562-06.2007.

Jones, E. V., et al. (2011). Astrocytes control glutamate receptor levels at developing synapses through SPARC—β-integrin interactions. *Journal of Neuroscience*, *31*(11), 4154–4165. https://doi.org/10.1523/JNEUROSCI.4757-10.2011.

Kania, A., & Klein, R. (2016). Mechanisms of ephrin-Eph signalling in development, physiology and disease. *Nature Reviews Molecular Cell Biology*, *17*, 240–256. https://doi.org/10.1038/nrm.2015.16.

Kayser, M. S., Nolt, M. J., & Dalva, M. B. (2008). EphB receptors couple dendritic filopodia motility to synapse formation. *Neuron*, *59*(1), 56–69. https://doi.org/10.1016/j.neuron.2008.05.007.

Kettenmann, H., & Verkhratsky, A. (2008). Neuroglia: The 150 years after. *Trends in Neurosciences*, *31*(12), 653–659. https://doi.org/10.1016/j.tins.2008.09.003.

Kimelberg, H. K., & Nedergaard, M. (2010). Functions of astrocytes and their potential as therapeutic targets. *Neurotherapeutics*, 7(4), 338–353. https://doi.org/10.1016/j.nurt.2010.07.006.

Klugbauer, N., et al. (1999). Molecular diversity of the calcium channel α2δ subunit. *Journal of Neuroscience*, *19*(2), 684–691. https://doi.org/10.1523/jneurosci.19-02-00684.1999.

Koeppen, J., et al. (2018). Functional consequences of synapse remodeling following astrocyte-specific regulation of ephrin-B1 in the adult hippocampus. *Journal of Neuroscience*, *38*(25), 5710–5726. https://doi.org/10.1523/JNEUROSCI.3618-17.2018.

Koh, S., et al. (2018). Subretinal human umbilical tissue-derived cell transplantation preserves retinal synaptic connectivity and attenuates müller glial reactivity. *Journal of Neuroscience*, *38*(12), 2923–2943. https://doi.org/10.1523/JNEUROSCI.1532-17.2018.

Koh, S., et al. (2019). Thrombospondin-1 promotes circuit-specific synapse formation via β1-integrin. *bioRXiv*. https://doi.org/10.1101/866590.

Kol, A., et al. (2020). Astrocytes contribute to remote memory formation by modulating hippocampal–cortical communication during learning. *Nature Neuroscience*, *23*(10), 1229–1239. https://doi.org/10.1038/s41593-020-0679-6.

Kriegstein, A., & Alvarez-Buylla, A. (2009). The glial nature of embryonic and adult neural stem cells. *Annual Review of Neuroscience*, *32*, 149–184. https://doi.org/10.1146/annurev.neuro.051508.135600.

Kucukdereli, H., et al. (2011). Control of excitatory CNS synaptogenesis by astrocyte-secreted proteins hevin and SPARC. *Proceedings of the National Academy of Sciences of the United States of America*, *108*(32), E440–E449. https://doi.org/10.1073/pnas.1104977108.

Lenhossek, M. (1893). *Der feinere Bau des Nervensystems im Lichte neuester Forschung*. Berlin: Fischer's Medicinische Buchhandlung.

Levelt, C. N., & Ḧubener, M. (2012). Critical-period plasticity in the visual cortex. *Annual Review of Neuroscience*, *35*, 309–330. https://doi.org/10.1146/annurev-neuro-061010-113813.

Leyton, L., et al. (2001). Thy-1 binds to integrin β3 on astrocytes and triggers formation of focal contact sites. *Current Biology*, *11*(13), 1028–1038. https://doi.org/10.1016/S0960-9822(01)00262-7.

Liebl, D. J., et al. (2003). mRNA expression of ephrins and Eph receptor tyrosine kinases in the neonatal and adult mouse central nervous system. *Journal of Neuroscience Research*, *71*(1), 7–22. https://doi.org/10.1002/jnr.10457.

Lisabeth, E. M., Falivelli, G., & Pasquale, E. B. (2013). Eph receptor signaling and ephrins. *Cold Spring Harbor Perspectives in Biology*, *5*(9), a009159. https://doi.org/10.1101/cshperspect.a009159.

Lively, S., Ringuette, M. J., & Brown, I. R. (2007). Localization of the extracellular matrix protein SC1 to synapses in the adult rat brain. *Neurochemical Research*, *32*(1), 65–71. https://doi.org/10.1007/s11064-006-9226-4.

López-Murcia, F. J., Terni, B., & Llobet, A. (2015). SPARC triggers a cell-autonomous program of synapse elimination. *Proceedings of the National Academy of Sciences of the United States of America*, *112*(43), 13366–13371. https://doi.org/10.1073/pnas.1512202112.

Ma, Z., et al. (2016). Neuromodulators signal through astrocytes to alter neural circuit activity and behaviour. *Nature*, *539*(7629), 428–432. https://doi.org/10.1038/nature20145.

Maldonado, H., et al. (2017). Astrocyte-to-neuron communication through integrin-engaged Thy-1/CBP/Csk/Src complex triggers neurite retraction via the RhoA/ROCK pathway. *Biochimica et Biophysica Acta. Molecular Cell Research*, *1864*(2), 243–254. https://doi.org/10.1016/j.bbamcr.2016.11.006.

Martín, R., et al. (2015). Circuit-specific signaling in astrocyte-neuron networks in basal ganglia pathways. *Science*, *349*(6249), 730–734. https://doi.org/10.1126/science.aaa7945.

Martin-Fernandez, M., et al. (2017). Synapse-specific astrocyte gating of amygdala-related behavior. *Nature Neuroscience*, *20*(11), 1540–1548. https://doi.org/10.1038/nn.4649.

Matsuda, M., et al. (2001). SREBP cleavage-activating protein (SCAP) is required for increased lipid synthesis in liver induced by cholesterol deprivation and insulin elevation. *Genes and Development*, *15*(10), 1206–1216. https://doi.org/10.1101/gad.891301.

Mauch, D. H., et al. (2001). CNS synaptogenesis promoted by glia-derived cholesterol. *Science*, *294*(5545), 1354–1357. https://doi.org/10.1126/science.294.5545.1354.

Mederos, S., González-Arias, C., & Perea, G. (2018). Astrocyte–neuron networks: A multilane highway of signaling for homeostatic brain function. *Frontiers in Synaptic Neuroscience*, *10*, 45. https://doi.org/10.3389/fnsyn.2018.00045.

Molumby, M. J., Keeler, A. B., & Weiner, J. A. (2016). Homophilic protocadherin cell-cell interactions promote dendrite complexity. *Cell Reports*, *15*(5), 1037–1050. https://doi.org/10.1016/j.celrep.2016.03.093.

Molumby, M. J., et al. (2017). γ-Protocadherins interact with neuroligin-1 and negatively regulate dendritic spine morphogenesis. *Cell Reports*, *18*(11), 2702–2714. https://doi.org/10.1016/j.celrep.2017.02.060.

Morishita, H., et al. (2006). Structure of the cadherin-related neuronal receptor/protocadherin-α first extracellular cadherin domain reveals diversity across cadherin families. *Journal of Biological Chemistry*, *281*(44), 33650–33663. https://doi.org/10.1074/jbc.M603298200.

Mu, Y., et al. (2019). Glia accumulate evidence that actions are futile and suppress unsuccessful behavior. *Cell*, *178*(1), 27–43.e19. https://doi.org/10.1016/j.cell.2019.05.050.

Murai, K. K., & Pasquale, E. B. (2003). Eph'ective signaling: Forward, reverse and crosstalk. *Journal of Cell Science*, *116*, 2823–2832. https://doi.org/10.1242/jcs.00625.

Murai, K. K., et al. (2003). Control of hippocampal dendritic spine morphology through ephrin-A3/EphA4 signaling. *Nature Neuroscience*, *6*(2), 153–160. https://doi.org/10.1038/nn994.

Nagai, J., et al. (2019). Hyperactivity with disrupted attention by activation of an astrocyte synaptogenic cue. *Cell*, *177*, 1280–1292.e20. https://doi.org/10.1016/j.cell.2019.03.019.

Nakashima, K., et al. (2001). BMP2-mediated alteration in the developmental pathway of fetal mouse brain cells from neurogenesis to astrocytogenesis. *Proceedings of the National Academy of Sciences of the United States of America*, *98*(10), 5868–5873. https://doi.org/10.1073/pnas.101109698.

Navarrete, M., & Araque, A. (2008). Endocannabinoids mediate neuron-astrocyte communication. *Neuron*, *57*(6), 883–893. https://doi.org/10.1016/j.neuron.2008.01.029.

Nelson, S. B., & Valakh, V. (2015). Excitatory/inhibitory balance and circuit homeostasis in autism spectrum disorders. *Neuron*, *87*(4), 684–698. https://doi.org/10.1016/j.neuron.2015.07.033.

Nestor, M. W., et al. (2007). Plasticity of neuron-glial interactions mediated by astrocytic EphARs. *Journal of Neuroscience*, *27*(47), 12817–12828. https://doi.org/10.1523/JNEUROSCI.2442-07.2007.

Nguyen, T. M., et al. (2016). An alternative splicing switch shapes neurexin repertoires in principal neurons versus interneurons in the mouse hippocampus. *eLife*, *5*. https://doi.org/10.7554/eLife.22757.001.

Nguyen, A. Q., et al. (2020). Astrocytic ephrin-B1 controls synapse formation in the hippocampus during learning and memory. *Frontiers in Synaptic Neuroscience*, *12*, 1–18. https://doi.org/10.3389/fnsyn.2020.00010. 10.

Nimchinsky, E. A., Sabatini, B. L., & Svoboda, K. (2002). Structure and function of dendritic spines. *Annual Review of Physiology*, *64*(1), 313–353. https://doi.org/10.1146/annurev.physiol.64.081501.160008.

Nishida, H., & Okabe, S. (2007). Direct astrocytic contacts regulate local maturation of dendritic spines. *Journal of Neuroscience*, *27*(2), 331–340. https://doi.org/10.1523/JNEUROSCI.4466-06.2007.

Oberheim, N. A., et al. (2006). Astrocytic complexity distinguishes the human brain. *Trends in Neurosciences*, *29*(10), 547–553. https://doi.org/10.1016/j.tins.2006.08.004.

Oberheim, N. A., et al. (2009). Uniquely hominid features of adult human astrocytes. *Journal of Neuroscience*, *29*(10), 3276–3287. https://doi.org/10.1523/JNEUROSCI.4707-08.2009.

Olsen, M. L., et al. (2015). New insights on astrocyte ion channels: Critical for homeostasis and neuron-glia signaling. *Journal of Neuroscience*, *35*(41), 13827–13835. https://doi.org/10.1523/JNEUROSCI.2603-15.2015.

Paixão, S., et al. (2013). EphrinB3/EphA4-mediated guidance of ascending and descending spinal tracts. *Neuron*, *80*, 1407–1420. https://doi.org/10.1016/j.neuron.2013.10.006.

Perea, G., Navarrete, M., & Araque, A. (2009). Tripartite synapses: Astrocytes process and control synaptic information. *Trends in Neurosciences*, 421–431. https://doi.org/10.1016/j.tins.2009.05.001. Elsevier Current Trends.

Pfrieger, F. W., & Barres, B. A. (1997). Synaptic efficacy enhanced by glial cells in vitro. *Science*, *277*(5332), 1684–1687. https://doi.org/10.1126/science.277.5332.1684.

Poulopoulos, A., et al. (2009). Neuroligin 2 drives postsynaptic assembly at perisomatic inhibitory synapses through gephyrin and collybistin. *Neuron*, *63*(5), 628–642. https://doi.org/10.1016/j.neuron.2009.08.023.

Qian, X., et al. (2000). Timing of CNS cell generation: A programmed sequence of neuron and glial cell production from isolated murine cortical stem cells. *Neuron*, *28*(1), 69–80. https://doi.org/10.1016/S0896-6273(00)00086-6.

Remak, R. (1836). *Observationes anatomicae et microscopicae systematis nervosi structura*. Available at: https://www.networkglia.eu/sites/networkglia.eu/files/downloads/Remak-Observationes_ anatomicae_et_microscopicae_systematis_nervosi_structura.pdf.

Risher, W. C., et al. (2014). Astrocytes refine cortical connectivity at dendritic spines. *eLife*, *3*, 1–24. https://doi.org/10.7554/eLife.04047.

Risher, W. C., et al. (2018). Thrombospondin receptor α2δ-1 promotes synaptogenesis and spinogenesis via postsynaptic Rac1. *The Journal of Cell Biology*, *217*(10), 3747–3765. https://doi.org/10.1083/jcb.201802057.

Rothstein, J. D., et al. (1996). Knockout of glutamate transporters reveals a major role for astroglial transport in excitotoxicity and clearance of glutamate. *Neuron*, *16*(3), 675–686. https://doi.org/10.1016/S0896-6273(00)80086-0.

Rubenstein, J. L. R., & Merzenich, M. M. (2003). Model of autism: increased ratio of excitation/inhibition in key neural systems. *Genes Brain & Behavior*, *2*(5), 255–267. https://doi.org/10.1046/j.1601-183X.2003.00037.x.

Santello, M., Calì, C., & Bezzi, P. (2012). Gliotransmission and the tripartite synapse. *Advances in Experimental Medicine and Biology*, *970*, 307–331. https://doi.org/10.1007/978-3-7091-0932-8_14.

Sauvageot, C. M., & Stiles, C. D. (2002). Molecular mechanisms controlling cortical gliogenesis. *Current Opinion in Neurobiology*, 244–249. https://doi.org/10.1016/S0959-4388(02)00322-7. Elsevier Ltd.

Scheiffele, P., et al. (2000). Neuroligin expressed in nonneuronal cells triggers presynaptic development in contacting axons. *Cell*, *101*(6), 657–669. https://doi.org/10.1016/S0092-8674(00)80877-6.

Schikorski, T., & Stevens, C. F. (1997). Quantitative ultrastructural analysis of hippocampal excitatory synapses. *Journal of Neuroscience*, *17*(15), 5858–5867. https://doi.org/10.1523/jneurosci.17-15-05858.1997.

Singh, S. K., et al. (2016). Astrocytes assemble thalamocortical synapses by bridging NRX1α and NL1 via Hevin. *Cell*, *164*(1–2), 183–196. https://doi.org/10.1016/j.cell.2015.11.034.

Spacek, J., & Harris, K. M. (1998). Three-dimensional organization of cell adhesion junctions at synapses and dendritic spines in area CA1 of the rat hippocampus. *Journal of Comparative Neurology*, *393*(1), 58–68. https://doi.org/10.1002/(SICI)1096-9861(19980330)393:1<58::AID-CNE6>3.0.CO;2-P.

Srinivasan, K., et al. (2016). Untangling the brain s neuroinflammatory and neurodegenerative transcriptional responses. *Nature Communications*, 7(1), 1–16. https://doi.org/10.1038/ncomms11295.

Stellwagen, D., & Malenka, R. C. (2006). Synaptic scaling mediated by glial TNF-α. *Nature*, *440*(7087), 1054–1059. https://doi.org/10.1038/nature04671.

Stellwagen, D., et al. (2005). Differential regulation of AMPA receptor and GABA receptor trafficking by tumor necrosis factor-α. *Journal of Neuroscience*, *25*(12), 3219–3228. https://doi.org/10.1523/JNEUROSCI.4486-04.2005.

Stogsdill, J. A., et al. (2017). Astrocytic neuroligins control astrocyte morphogenesis and synaptogenesis. *Nature*, *551*(7679), 192–197. https://doi.org/10.1038/nature24638.

Stork, T., et al. (2014). Neuron-Glia interactions through the heartless fgf receptor signaling pathway mediate morphogenesis of drosophila astrocytes. *Neuron*, *83*(2), 388–403. https://doi.org/10.1016/j.neuron.2014.06.026.

Südhof, T. C. (2008). Neuroligins and neurexins link synaptic function to cognitive disease. *Nature*, 903–911. https://doi.org/10.1038/nature07456. Nature Publishing Group.

Sun, Y., et al. (2001). Neurogenin promotes neurogenesis and inhibits glial differentiation by independent mechanisms. *Cell*, *104*(3), 365–376. https://doi.org/10.1016/S0092-8674(01)00224-0.

Swanson, R. A., et al. (1997). Neuronal regulation of glutamate transporter subtype expression in astrocytes. *Journal of Neuroscience*, *17*(3), 932–940. https://doi.org/10.1523/jneurosci.17-03-00932.1997.

Takano, T., et al. (2020). Chemico-genetic discovery of astrocytic control of inhibition in vivo. *Nature*, *588*, 296–302. https://doi.org/10.1038/s41586-020-2926-0.

Tanigaki, K., et al. (2001). Notch1 and Notch3 instructively restrict bFGF-responsive multipotent neural progenitor cells to an astroglial fate. *Neuron*, *29*(1), 45–55. https://doi.org/10.1016/S0896-6273(01)00179-9.

Thiele, C., et al. (2000). Cholesterol binds to synaptophysin and is required for biogenesis of synaptic vesicles. *Nature Cell Biology*, *2*(1), 42–49. https://doi.org/10.1038/71366.

Ullian, E. M., et al. (2001). Control of synapse number by glia. *Science*, *291*(5504), 657–661. https://doi.org/10.1126/science.291.5504.657.

van Deijk, A. L. F., et al. (2017). Astrocyte lipid metabolism is critical for synapse development and function in vivo. *Glia*, *65*(4), 670–682. https://doi.org/10.1002/glia.23120.

Varoqueaux, F., Jamain, S., & Brose, N. (2004). Neuroligin 2 is exclusively localized to inhibitory synapses. *European Journal of Cell Biology*, *83*(9), 449–456. https://doi.org/10.1078/0171-9335-00410.

Ventura, R., & Harris, K. M. (1999). Three-dimensional relationships between hippocampal synapses and astrocytes. *Journal of Neuroscience, 19*(16), 6897–6906. https://doi.org/10.1523/jneurosci.19-16-06897.1999.

Villar-Cerviño, V., et al. (2013). Contact repulsion controls the dispersion and final distribution of Cajal-Retzius cells. *Neuron*, 77(3), 457–471. https://doi.org/10.1016/j.neuron.2012.11.023.

Virchow, R. (1858). *Die cellularpathologie in ihrer bertundung auf physiologische und pathologische gewebelehre*. Berlin: Hirshwald.

Volterra, A., & Meldolesi, J. (2005). Astrocytes, from brain glue to communication elements: The revolution continues. *Nature Reviews Neuroscience*, 626–640. https://doi.org/10.1038/nrn1722. Nature Publishing Group.

Walmsley, B., Alvarez, F. J., & Fyffe, R. E. W. (1998). Diversity of structure and function at mammalian central synapses. *Trends in Neurosciences*, 81–88. https://doi.org/10.1016/S0166-2236(97)01170-3. Elsevier Ltd.

Witcher, M. R., Kirov, S. A., & Harris, K. M. (2007). Plasticity of perisynaptic astroglia during synaptogenesis in the mature rat hippocampus. *Glia, 55*(1), 13–23. https://doi.org/10.1002/glia.20415.

Witcher, M. R., et al. (2010). Three-dimensional relationships between perisynaptic astroglia and human hippocampal synapses. *58*(5), 572–587. https://doi.org/10.1002/glia.20946.

Xu, J., Xiao, N., & Xia, J. (2010). Thrombospondin 1 accelerates synaptogenesis in hippocampal neurons through neuroligin 1. *Nature Neuroscience, 13*(1), 22–24. https://doi.org/10.1038/nn.2459.

Yamamoto, M., et al. (2006). Control of axonal sprouting and dendrite branching by the Nrg-Ank Complex at the neuron-glia Interface. *Current Biology, 16*(16), 1678–1683. https://doi.org/10.1016/j.cub.2006.06.061.

Yu, X., et al. (2018). Reducing astrocyte calcium signaling in vivo alters striatal microcircuits and causes repetitive behavior. *Neuron, 99*, 1170–1187.e9. https://doi.org/10.1016/j.neuron.2018.08.015.

Zhang, Y., et al. (2014). An RNA-sequencing transcriptome and splicing database of glia, neurons, and vascular cells of the cerebral cortex. *Journal of Neuroscience, 34*(36), 11929–11947. https://doi.org/10.1523/JNEUROSCI.1860-14.2014.

Zhuang, Z., et al. (2010). EphrinBs regulate D-serine synthesis and release in astrocytes. *Journal of Neuroscience, 30*(47), 16015–16024. https://doi.org/10.1523/JNEUROSCI.0481-10.2010.

CHAPTER TEN

Development of motor circuits: From neuronal stem cells and neuronal diversity to motor circuit assembly

Julia L. Meng[a,b] and Ellie S. Heckscher[a,b,c,d,*]
[a]Department of Molecular Genetics and Cell Biology, University of Chicago, Chicago, IL, United States
[b]Program in Cell and Molecular Biology, University of Chicago, Chicago, IL, United States
[c]Committee on Development, Regeneration, and Stem Cell Biology, University of Chicago, Chicago, IL, United States
[d]Grossman Institute for Neuroscience, University of Chicago, Chicago, IL, United States
[*]Corresponding author: e-mail address: heckscher@uchicago.edu

Contents

Current Topics in Developmental Biology, Volume 142
ISSN 0070-2153
https://doi.org/10.1016/bs.ctdb.2020.11.010

Abstract

In this review, we discuss motor circuit assembly starting from neuronal stem cells. Until recently, studies of neuronal stem cells focused on how a relatively small pool of stem cells could give rise to a large diversity of different neuronal identities. Historically, neuronal identity has been assayed in embryos by gene expression, gross anatomical features, neurotransmitter expression, and physiological properties. However, these definitions of identity are largely unlinked to mature functional neuronal features relevant to motor circuits. Such mature neuronal features include presynaptic and postsynaptic partnerships, dendrite morphologies, as well as neuronal firing patterns and roles in behavior. This review focuses on recent work that links the specification of neuronal molecular identity in neuronal stem cells to mature, circuit-relevant identity specification. Specifically, these studies begin to address the question: to what extent are the decisions that occur during motor circuit assembly controlled by the same genetic information that generates diverse embryonic neuronal diversity? Much of the research addressing this question has been conducted using the Drosophila larval motor system. Here, we focus largely on Drosophila motor circuits and we point out parallels to other systems. And we highlight outstanding questions in the field. The main concepts addressed in this review are: (1) the description of temporal cohorts—novel units of developmental organization that link neuronal stem cell lineages to motor circuit configuration and (2) the discovery that temporal transcription factors expressed in neuronal stem cells control aspects of circuit assembly by controlling the size of temporal cohorts and influencing synaptic partner choice.

1. Introduction

Every time you move, feel a touch, or sense your body move, you rely on the activity of neurons in your spinal cord. These neurons are organized into motor circuits. When motor circuits are dysfunctional or develop abnormally this can lead to debilitating perceptual abnormalities such as allodynia, or movement disorders like paraplegia (Zeilig, Enosh, Rubin-Asher, Lehr, & Defrin, 2012). Understanding how motor circuits develop is a fundamental question in developmental neurobiology, which has implications for various fields including robotics, evolution, and medicine.

Motor circuit development starts with neurogenesis and culminates with the assembly of functional motor circuits. Motor circuit assembly is a multi-step process. First, neuronal stem cells divide to generate neurons. These neurons inherit gene expression from their stem cell parent. Then dendrites and axons must navigate through a dynamic, three-dimensional space; dendritic and axonal arbors must be elaborated; synaptic partnerships must be formed; neurotransmitter phenotypes must be established; and electrical properties must be acquired. Each of these events represents a key developmental decision that every neuron must make. The field of developmental neurobiology seeks to understand the cellular and molecular processes that control these motor circuit assembly decisions.

In this review, we discuss recent research on the topic of motor circuit assembly starting from the production of neurons from neuronal stem cells. Until recently, neuronal stem cells were studied in the context of generating the large diversity of neurons found in the motor system. Specifically, in this review, we address the question: to what extent are the decisions that occur during motor circuit assembly controlled by information that generates a diversity of neurons from a given neuronal stem cell? Much of the research addressing this question has been conducted using the Drosophila larval motor system. Consequently, we focus largely on Drosophila motor circuits in this review. Additionally, we point out parallels to other systems and highlight outstanding questions in the field.

2. Motor circuits

Motor circuits are unique from circuits found elsewhere in the CNS. They are the only circuits able to generate movement. Motor circuits are also unique in that they are required to detect and process multiple different types of somatosensory stimuli in parallel. For example, they can simultaneously detect touch, heat, and even light (exteroception) as well as self-movement (interoception) (Im & Galko, 2012; Koch, Acton, & Goulding, 2018; Kohsaka, Guertin, & Nose, 2017).

Anatomically, motor circuits are composed of two fundamental neuron types. (1) Motor neurons send axons from the CNS out to the periphery, and they control the activity of muscles (Stifani, 2014). Patterned muscle activity moves animals. (2) Interneurons have both dendrites and axons that reside within the CNS. Interneurons carry out a wide variety of functions—processing sensory information, relaying that information to the brain or to motor neurons, and generating different patterns of motor neuron firing, which drives various movements. In the motor system of many organisms, interneurons are by far the most abundant neuron type, and interneurons are the most morphologically and functionally diverse (Bikoff, 2019; Osseward & Pfaff, 2019). However, due to their relative experimental inaccessibility interneurons are less well-understood compared to motor neurons.

Conceptually, motor circuits are fundamental functional modules. That is, each circuit performs a specific computation. One motor circuit can be disrupted without perturbing the function of other motor circuits (Clark, McCumsey, Lopez-Darwin, Heckscher, & Doe, 2016), and individual motor circuits can be combined in different ways to generate complex outputs (Blitz & Nusbaum, 2012; Masson et al., 2020; Ohyama et al., 2015). Historically, motor circuits were studied using sharp electrodes in aquatic animals with large neuronal bodies, such as leech, lamprey, and lobster (Friesen, Poon, & Stent, 1976; Marder & Bucher, 2001). These animals are long-lived, have relatively few progeny, and have genomes not readily amenable to genetic manipulations. Together, these traits make it difficult to study circuit development in these systems. Consequently, our understanding of motor circuit assembly was extremely limited. In recent years the developmental genetic model Drosophila has become a prominent model to identify and study the function of motor circuits (Clark, Zarin, Carreira-Rosario, & Doe, 2018; Tastekin & Louis, 2017). This has been enabled by technological innovation—highly specific driver lines (Jenett et al., 2012; Li et al., 2014; Manning et al., 2012; Pfeiffer et al., 2008), genetically-encoded calcium indicators (Dana et al., 2019; Vaadia et al., 2019), optogenetics (Simpson & Looger, 2018), and connectomics (Schneider-Mizell et al., 2016). Now, many motor circuits have been identified at cellular and synaptic resolution. Some motor circuits generate patterned forward and reverse crawling (Fushiki et al., 2016; Kohsaka et al., 2017, 2019; Takagi et al., 2017; Zarin, Mark, Cardona, Litwin-Kumar, & Doe, 2019). These circuits can generate crawling or crawling-like motor neuron firing patterns without input from the central brain (Berni, Pulver, Griffith, & Bate, 2012) or input from sensory neurons (Lemon et al., 2015; Pulver et al., 2015) (Fig. 1). Motor circuits in the nerve cord can be controlled by neurons descending from the central brain such as Goro neurons for escape movement or MDN neurons for backward crawling (Fig. 1) (Carreira-Rosario et al., 2018; Namiki, Dickinson, Wong, Korff, & Card, 2018). Other circuits in the Drosophila nerve cord include those that

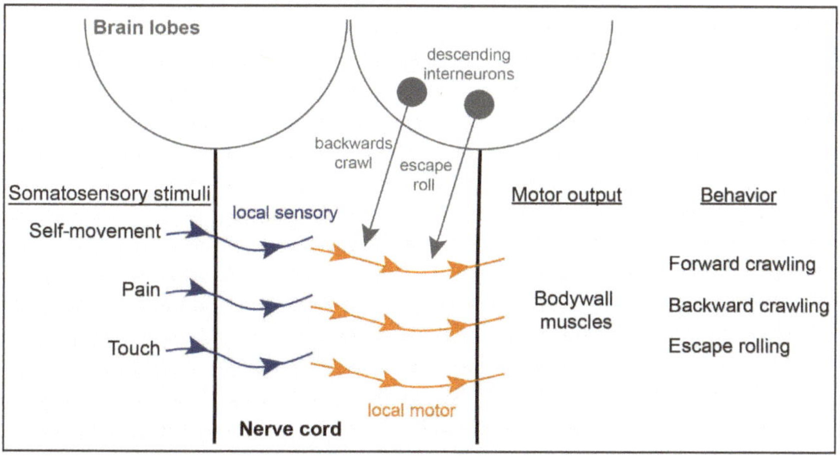

Fig. 1 Anatomical and functional definition of motor circuits. Illustration of the Drosophila larval nerve cord (black) and motor circuits (depicted as lines with nodes representing direction of circuit information flow). Sensory processing circuits, in blue to the left, relay somatosensory stimuli such as self-movement, pain, and touch. Motor circuits, in yellow to the right, can receive input from sensory neurons or descending neurons from the brain. Motor circuits control the patterned activity of bodywall muscles, which in turn generate patterned behaviors such as forward crawling, backward crawling, and escape rolling.

process different types of somatosensory stimuli (e.g., noxious stimuli, proprioceptive stimuli, light touch, etc.) (Fig. 1) (Babski et al., 2019; Burgos et al., 2018; Heckscher et al., 2015; Jovanic et al., 2019; Miroschnikow et al., 2018; Ohyama et al., 2015; Tastekin et al., 2018). These sensory processing circuits relay somatosensory information to other nerve cord circuits or to the central brain. One of the first circuits to be identified in the Drosophila nerve cord contains a set of Even-skipped Lateral neurons (ELs) (Heckscher et al., 2015). This EL circuit maintains left-right symmetrical muscle contraction during Drosophila larval crawling. Notably, when ELs are ablated many aspects of movement are normal, such as the timing of muscle contraction. Thus, even without ELs other circuits still function normally, demonstrating the modular nature of motor circuits. Ultimately, the identification of motor circuits in Drosophila larvae has opened the door for study of their assembly using a powerful developmental genetic toolkit.

3. Neuronal stem cells in the motor system

Neuronal stem cells, like other stem cells, divide asymmetrically to self-renew and to generate a daughter that will go on to make a post-mitotic neuronal cell (Knoblich, 2010). Neuronal stem cells change their potency over time, such that at different points in development the same stem cell produces different neurons. The set of neurons generated from a single neuronal stem cell is referred to as a lineage (Fig. 2C). Lineages are stereotyped from animal to animal and are composed of sets of non-identical neurons

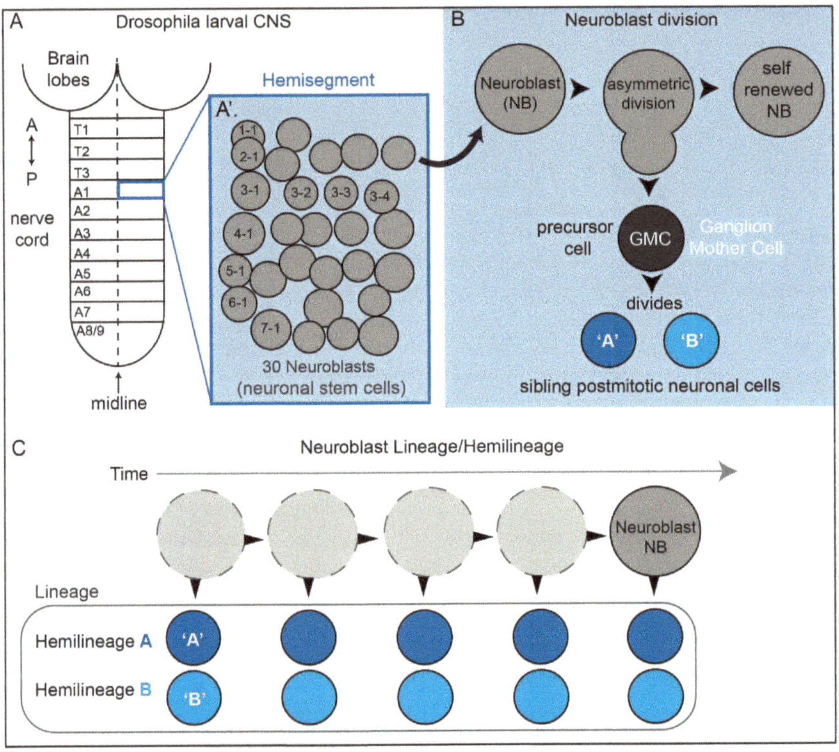

Fig. 2 Developmental organization of neuronal stem cell lineages. (A) Illustration of the Drosophila larvae Central Nervous System. The nerve cord is composed of repeating segments along the anterior–posterior (A–P) axis. There are three thoracic segments (T1–T3), seven abdominal segments (A1–A7), and two terminal segments (A8/9). Each segment can be further broken down into left-right symmetrical hemisegments that are separated by the midline (dotted line). (A′) Illustration of a hemisegment, which contains the same 30 neuronal stem cells called neuroblasts (NB), in gray. Each neuroblast is numerically named based on stereotyped row and column position (e.g., NB in row "1" and column "1" is named "1-1"). (B) Illustration depicting that neuroblasts divide asymmetrically to self-renew. Asymmetric divisions can generate precursor cells, dark gray, called a Ganglion Mother Cell (GMC) that generates two sibling postmitotic neuronal cells, "A" and "B." (C) Illustration of a single dividing Neuroblast over time, lighter gray circles with a dashed perimeter represent the Neuroblast at previous time points. The "A" and "B" neurons born from GMCs populate separate hemilineages A and B, respectively.

(Hartenstein & Stollewerk, 2015). Neuronal stem cell lineages can be considered fundamental developmental units of the nervous system.

In Drosophila, neuronal stem cells are called "neuroblasts." Nerve cord neuroblasts are among the most-intensely studied and best-characterized neuronal stem cells in the field (Birkholz, Rickert, Berger, Urbach, & Technau, 2013; Birkholz, Rickert, Nowak, Coban, & Technau, 2015; Broadus et al., 1995; Goodman & Doe, 1993). The Drosophila nerve cord is composed of repeating segments along the anterior–posterior (A–P) axis (Fig. 2A) (Technau, Rogulja-Ortmann, Berger, Birkholz, & Rickert, 2014). Segmentally repeating units are morphologically and functionally similar. Each segment can be further broken down into left-right, symmetrical hemisegments that are separated by the midline (Fig. 2A). Each hemisegment contains the same 30 neuroblasts (Fig. 2A') (Broadus et al., 1995; Doe, 1992). During embryogenesis, one set of 30 neuroblasts generates approximately 300 neurons (Bossing, Udolph, Doe, & Technau, 1996; Heckscher et al., 2014; Schmidt et al., 1997), each of which is thought to have a unique morphology (Bossing & Technau, 1994; Bossing et al., 1996; Rickert, Kunz, Harris, Whitington, & Technau, 2011; Schmid, Chiba, & Doe, 1999; Schmidt et al., 1997). Neurons produced during embryogenesis from nerve cord neuroblasts populate larval motor circuits.

In the Drosophila nerve cord, neuroblast lineages are often subdivided into two hemilineages. Hemilineages have been suggested to be an organizational unit of motor circuits. In the nerve cord, neuroblasts often divide asymmetrically to self-renew and bud off a ganglion mother cell (GMC) (Fig. 2B) (Goodman & Doe, 1993). The GMC divides once to produce two post-mitotic daughter cells (Fig. 2B). One of the two GMC daughters will become an "A" cell and will become part of the A hemilineage. The other GMC daughter will become a "B" cell and will populate the B hemilineage (Fig. 2C) (Truman, Moats, Altman, Marin, & Williams, 2010). In the adult Drosophila nerve cord, a comprehensive study showed that activation of distinct hemilineages of neurons generate distinct motor behaviors, with a majority of behaviors being unique (Harris, Pfeiffer, Rubin, & Truman, 2015). Behavior was used as a proxy for underlying circuit function and organization. Notably, caution must be taken to avoid over-interpreting circuit structure from behavior. Nonetheless, this study supports the idea that hemilineages produce groups of neurons that belong to specific motor circuits. Thus, a currently held view in the field is that neuroblast lineages can be considered fundamental developmental units of the motor system *and* also fundamental functional units of motor circuits.

4. From neuronal diversity to motor circuit assembly

Historically, the stem cell field has focused on understanding how early events in neurogenesis generate a diversity of different types of neurons (Jessell, 2000; Pearson & Doe, 2004). The terms neuronal diversity, neuronal fate, and neuronal identity have been used nearly interchangeably. Here, we will use the term identity to refer to all three. Past studies have provided a wealth of information regarding molecular programs that contribute to diverse neuronal identities (Doe, 2017). In this review, we divide these molecular programs into four classes: (1) Anterior–posterior diversity is regulated by *Hox transcription factors* (Fig. 3A) (Estacio-Gómez & Díaz-Benjumea, 2014). (2) Within a single hemisegment, neuroblasts become different from each other based on *Spatial patterning transcription factors* (Fig. 3B) (Skeath, 1999; Skeath & Thor, 2003). (3) Over time, a single neuroblast can generate a diversity of neuronal identities (Fig. 3C) (a.k.a., temporal identities), which is regulated by the activity of *Temporal transcription factors* (Fig. 3C') (Grosskortenhaus, Robinson, & Doe, 2006; Isshiki, Pearson, Holbrook, & Doe, 2001; Pearson & Doe, 2003; Tran & Doe, 2008). (4) An additional layer of diversity is achieved at the level of GMC daughter cells by *Notch signaling* (Fig. 3C") (Skeath & Doe, 1998; Truman et al., 2010). Thus, we have a basic framework for understanding how neuronal diversity is generated.

Neuronal identity has been measured by many criteria—embryonic neuronal gene expression, gross anatomical features, neurotransmitter expression, and physiological properties (Allan & Thor, 2015; Lacin & Truman, 2016; Rickert, Lüer, Vef, & Technau, 2018). A major gap in our knowledge, however, is that these definitions of identity are largely unlinked to mature neuronal features like presynaptic and postsynaptic partnerships, dendrite morphologies, neuronal firing patterns, or roles in behavior. The field needs to include mature neuronal features in the definition of neuronal identity if it hopes to link knowledge about the generation of neuronal diversity to motor circuit assembly. Throughout this review chapter, we will refer to historical concepts of identity as "molecular identity." The concept of identity that includes circuit-relevant mature neuronal features we denote as "functional identity." A major current focus of the field is to precisely define how regulators of molecular identity impact functional identity and therefore motor circuit assembly.

In this review, we describe two types of experiments—those in which researchers document circuit membership of neurons from a given stem

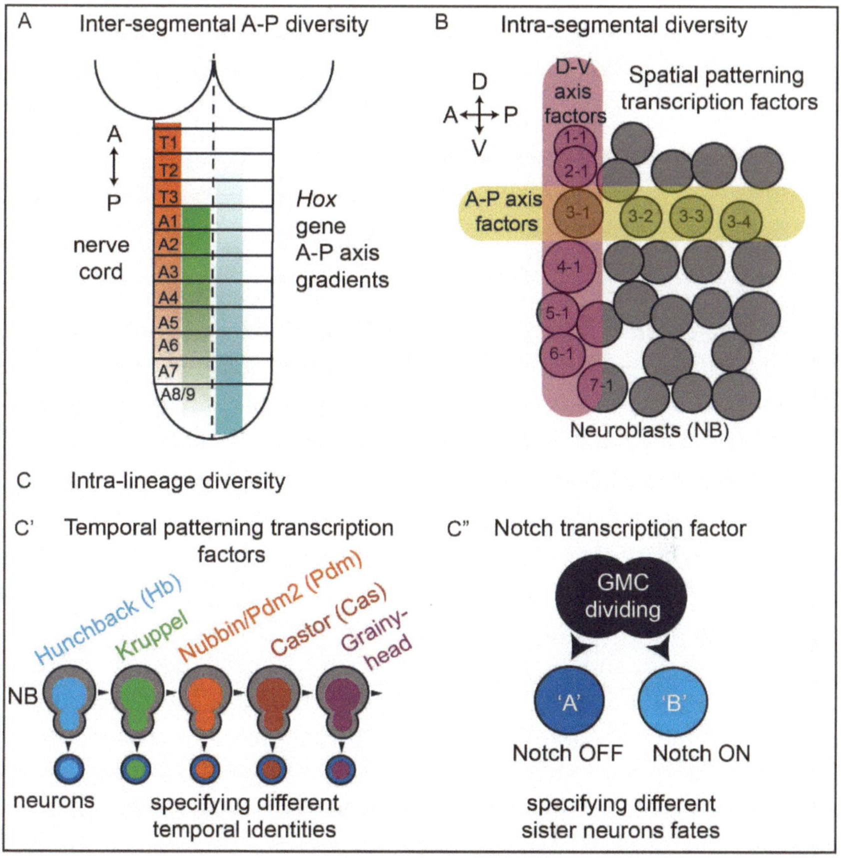

Fig. 3 Molecular programs contributing to diverse neuronal identities (A) Illustration of representative *Hox* gene gradients, different colored bars, along the A-P axis of the Drosophila larval nerve cord. (B) Illustration of representative neuroectoderm spatial transcription factors expressed along the A-P axis, green bar, and D-V axis, pink bar. The intersection, pink and green overlayed, of these spatial transcription factors specify neuroblast identity. (C) Illustration of intra-lineage diversity factors. (C′) Illustration of a single neuroblast (NB) dividing and expressing a series of temporal patterning transcription factors called temporal transcription factors, labeled and color coded, found in most, if not all, neuroblasts. Neuronal progeny inherit temporal transcription factor expressed during division. (C") Illustration of a GMC dividing to produce two sibling neuronal cells, where one is Notch ON whereas the other is Notch OFF. Notch ON/OFF promotes different molecular fates in the sibling cells.

cell lineage, and those in which researchers manipulate regulators of molecular identity and interrogate their impact on motor circuit assembly. These experiments provide insight into the lineage-based organization of motor circuits and the molecular processes giving rise to this organization.

5. What are the relationships between neuronal stem cell lineage and motor circuits?

A required first step in understanding how motor circuits arise from neuronal stem cells is to describe the circuit membership of neurons from a given neuronal stem cell lineage. In this section, we discuss examples of a recently described, neuronal stem cell lineage-based, developmental unit called a *temporal cohort* (Fig. 4A). Temporal cohorts are defined by the following three criteria: *Temporal cohorts are sets of neurons that come from one neuronal stem cell, are contiguously born, and have similar function at the circuit level.* This unit links neuronal stem cell development and circuit membership. Below, we highlight examples of temporal cohorts, put these examples into a larger context, and point out important areas for future study.

5.1 NB3-3 generates early-born EL and late-born EL temporal cohorts

In the Drosophila larval nerve cord, a pioneering study used the neuroblast NB3-3 lineage as a model to investigate the relationship between neuronal stem cell lineages and motor circuits (Wreden et al., 2017). NB3-3 generates a single motor neuron followed by a series of interneurons characterized by the expression of the transcription factor Even-skipped (Eve) and a lateral cell body position in the nerve cord (Fig. 4B) (Schmid et al., 1999). These interneurons are called "ELs" (Fujioka, Emi-Sarker, Yusibova, Goto, & Jaynes, 1999). Wreden et al. showed that ELs could be subdivided into early-born and late-born sets of neurons based on the expression of an enhancer called 11F02 (Wreden et al., 2017). For each set of ELs, circuit membership was assessed anatomically and functionally. Optogenetic activation of early-born ELs triggered escape rolling behavior, and optogenetic activation of late-born ELs resulted in abnormal left-right crawling (Wreden et al., 2017). Connectomic data showed at least a subset of early-born ELs receive synapses directly and indirectly from chordotonal (sound/vibration sensitive) sensory neurons, and late-born ELs receive synapses directly and indirectly from proprioceptive sensory neurons (He et al., 2019; Heckscher et al., 2015; Wreden et al., 2017). Calcium imaging confirmed that only early-born ELs were activated by sound stimuli, which activate chordotonal sensory neurons (Wreden et al., 2017). Taken together, this study demonstrated that NB3-3 generates ELs that contribute to two circuits. Early-born ELs populate a fast escape circuit that processes chordotonal input, and

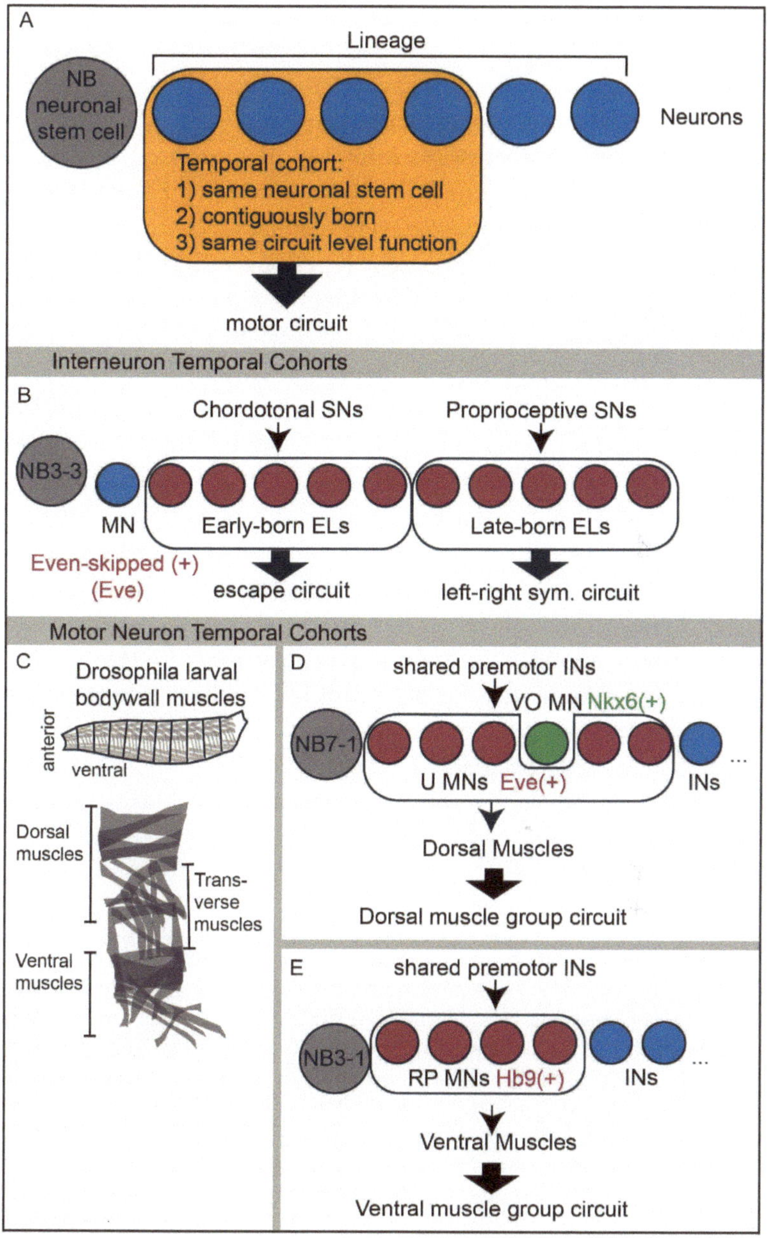

Fig. 4 Temporal cohorts are an organizational unit linking neuronal stem cell lineages to motor circuit membership. (A) Illustration depicting the criteria that must be met for a group of neurons to be considered a temporal cohort. (B) Illustration of the NB3-3 lineage, which after producing a single motor neuron, generates an Eve(+) early-born

(Continued)

late-born ELs populate a circuit that refines movement and processes proprioceptive stimuli (Fig. 4B). In both circuits, ELs are somatosensory processing interneurons. Thus, it appears that the NB3-3 lineage endows ELs with the capacity to process sensory information regardless of circuit functional identity. Furthermore, information associated with neuronal birth-timing segregates ELs into different circuits. Assembling motor circuits according to these rules provides a developmental mechanism for generating motor circuits able to process different types of sensory stimuli in parallel.

Wreden et al. was the first study to suggest that lineage-related neurons could be subdivided into temporal cohorts. Specifically, this suggested that early-born ELs belonged to one temporal cohort, and late-born ELs belonged to a second temporal cohort. To put this finding into context, it has been known for many years that there can be abrupt changes during lineage progression with respect to some neuronal features, such as morphology, gene expression, and neurotransmitter phenotype. For example, during lineage progression a neuroblast can abruptly change from producing projection interneurons to producing local interneurons (Schmid et al., 1999), or from producing Eve(+) to Eve(−) neurons (Pearson & Doe, 2004). The observation that lineages can be organized into temporal cohorts demonstrated that there can also be abrupt changes in the types of synaptic partnerships formed by neurons in a lineage. Of note, temporal cohort borders cannot necessarily be predicted based on differences in other features such as expression of well-established marker genes. This is illustrated by the example of ELs, which all share Eve expression yet contribute to two distinct circuits.

Fig. 4—Cont'd EL and late-born EL temporal cohorts. Early-born ELs receive innervation from chordotonal sensory neurons. Late-born ELs receive innervation from proprioceptive sensory neurons. (C) Illustration of the Drosophila larvae body wall musculature. Each body wall hemisegment contains 30 muscles. Each muscle is organized into three anatomical and functional groups: dorsal, transverse, and ventral. (D) Illustration of the NB7–1 lineage, which generates the Eve(+) U motor neurons, Nkx6(+) VO motor neuron, and a number, denoted by "...," of interneurons. The U motor neuron temporal cohort is not contiguously born because it is disrupted by the recently discovered VO motor neuron, which projects to the ventral oblique muscles, yet also shares premotor INs with several U MNs. The U motor neuron temporal cohort receives shared innervation from premotor interneurons and all U motor neurons synapse onto dorsal muscles. (E) Illustration of the NB3-1 lineage, which generates the Hb9(+) RP motor neurons followed by a number, denoted by "...," of interneurons. The RP motor neuron temporal cohort receives shared innervation from premotor interneurons and all RP motor neurons synapse onto ventral muscles.

5.2 Motor neuron temporal cohorts in Drosophila larvae

There are examples of motor neuron temporal cohorts. In Drosophila larvae, in each hemisegment, 35 motor neurons synapse onto 30 body wall muscles (Landgraf, Bossing, Technau, & Bate, 1997) (Fig. 4C). Body wall muscles are organized into three anatomical and functional groups—dorsal, transverse, and ventral (Fig. 4C) (Bate, 1993; Snodgrass, 1935). Each muscle group contracts at a distinct phase of larval crawling (Heckscher, Lockery, & Doe, 2012). Therefore, it is likely that sets of motor neurons share circuit membership if they synapse onto muscles found within a particular muscle group.

Dorsal and ventral muscle groups are innervated by motor neuron temporal cohorts. For the ventral muscle group, NB3-1 generates four consecutively born, RP motor neurons (RP1, RP4, RP3, RP5) (Fig. 4E) (Tran & Doe, 2008). Single neuron labeling shows each RP motor neuron forms unique and stereotyped synapses onto a ventral muscle (Choi, Park, & Griffith, 2004; Landgraf et al., 1997; Meng, Wang, Carrillo, & Heckscher, 2020; Zarin et al., 2019). Furthermore, connectomic analysis shows that RP motor neurons share common premotor interneuron synaptic inputs (Zarin et al., 2019). For the dorsal muscle group, NB7-1 generates five U motor neurons (U1–U5) (Fig. 4D) (Isshiki et al., 2001). Single cell labeling approaches showed that each U motor neuron forms a unique and stereotyped synapse onto a muscle in the dorsal muscle group (Landgraf et al., 1997; Meng, Marshall, Lobb-Rabe, & Heckscher, 2019; Seroka, Yazejian, Lai, & Doe, 2020). U motor neurons share many premotor interneuron inputs (Zarin et al., 2019). Both NB7-1 and NB3-1 generate interneurons after the window of motor neuron production. While little is known about the molecular identity or morphology of later born interneurons generated in the NB7-1 and NB3-1 lineages, from limited morphology data it is unlikely that all lineage related neurons populate a single circuit (Mark et al., 2019; Schmid et al., 1999). In summary, the Drosophila motor system contains at least two temporal cohorts of motor neurons that are defined both by muscle synaptic outputs and by presynaptic inputs.

An attractive model is that each circuit controlling a muscle group (dorsal, ventral, and transverse) is populated by a different motor neuron temporal cohort. In support of this idea, DiI clonal data of Drosophila neuroblasts lineages shows at least one lineage generates multiple motor neurons that project to the transverse muscle group (Schmid et al., 1999). Notably, also based on these DiI data, it is likely that not all motor neurons are organized into temporal cohorts because some lineages generate only one motor neuron (Schmid et al., 1999). Currently, it is unclear why some lineages make multiple motor neurons and others just a single motor neuron and yet others make no motor neurons at all.

5.3 Temporal cohorts are born contiguously or in a tight developmental time window

It is possible that temporal cohorts are not always comprised of contiguously-born neurons, but instead comprised of neurons born in a limited developmental time window. This idea is supported by a recent study that refuted the long-held perception that NB7-1 generates the five U motor neurons consecutively (Seroka et al., 2020). Seroka et al. showed that after the third-born U motor neuron, but before the fourth-born U motor neuron, an additional motor neuron is produced by NB7-1 (Fig. 4D). The newly identified NB7-1 motor neuron, VO, projects to ventral oblique muscles, meaning it is not a member of the dorsal muscle circuit (Seroka et al., 2020). This finding reveals a break in the previously identified U motor neuron temporal cohort. Notably, the VO motor neuron shares many presynaptic, premotor neuron inputs with U motor neurons (Zarin et al., 2019). This observation raises questions about how motor circuits evolve over time. Overall, these findings suggest that temporal cohorts are not always made up of neurons that are contiguously born, but can instead be comprised of neurons that are born in a tight developmental time window.

5.4 Motor neuron temporal cohorts in adult Drosophila

Beyond larval motor circuits, there is evidence of temporal cohorts of leg motor neurons in the Drosophila adult motor system. Clonal analysis revealed that out of 11 neuroblast lineages producing leg motor neurons, the majority are produced by 2 neuroblasts (Baek & Mann, 2009). In one lineage, LinA, birth dating of motor neurons showed a striking correlation between birth time and muscle targeting. All early-born LinA motor neurons synapse onto muscles (mostly long tendons) in the proximal tibia or proximal femur portion of the leg, whereas nearly all later-born LinA motor neurons synapse onto muscles (mostly reductors) in the distal tibia or distal femur portion of the leg. A second lineage, LinB, produces a group of late-born LinB motor neurons that synapse onto the coxa levator muscle. Each group of motor neurons in LinA (early-born and late-born) and LinB (late-born) had its own field of overlapping dendrites. Overlapping dendritic fields suggest that these neurons may receive common premotor input, and therefore they are predicted to be functionally recruited together. These findings provide evidence of leg motor neuron temporal cohorts. Therefore, temporal cohorts may be a widespread developmental strategy in motor systems throughout Drosophila life stages.

5.5 In many systems, motor circuits are specified sequentially

In the mouse spinal cord, there are examples of temporal cohort-like sets of neurons. These neurons are generated from stem cells within a small progenitor domain of the developing spinal cord, and these neurons perform analogous functions in different circuits. For example, Renshaw cells and Ia interneurons are sequentially produced by stem cells in the p1 progenitor domain (Benito-Gonzalez & Alvarez, 2012; Stam et al., 2012). Renshaw cells contribute to a motor neuron feedback circuit, whereas Ia interneurons contribute to a stretch reflex circuit. Despite participation in distinct circuits, Renshaw cells and Ia interneurons perform analogous functions—directly synapsing onto motor neurons and terminating firing (Hultborn, 2006). In addition, for extensor and flexor premotor interneurons, many of which originate from the same progenitor domain, time of neurogenesis is correlated with spatial and inferred functional segregation (Tripodi, Stepien, & Arber, 2011). Thus, Renshaw cells and Ia interneurons are examples of temporal-cohort like units in the spinal cord.

In the vertebrate spinal cord, lineage tracing studies are limited. In general, neuronal origins are traced to only progenitor domains that contain a large number of individual self-renewing stem cells (Sanes & Lichtman, 1999; Woodworth, Girskis, & Walsh, 2017). This means in spinal cords it is not possible to distinguish the origin of temporal cohort-like sets of neurons. They may originate from neuronal stem cells by one of two ways, based on either a stem cell heterogeneity model or a common stem cell model (Ravanelli & Appel, 2015). The heterogeneity model predicts that different stem cells are activated at different developmental times, and each stem cell gives rise to a distinctive temporal cohort-like set of neurons. The common model, analogous to the Drosophila neuroblast system, predicts that one stem cell divides to generate different neurons over time to make one or more temporal cohorts. Therefore, it is unknown if there are convergent or homologous cell biological mechanisms generating temporal cohort-like sets of neurons in the vertebrate spinal cord with respect to the Drosophila nerve cord. Nonetheless, the observation that there are temporal cohort-like groups of neurons in evolutionarily distant species points to the fundamental importance of temporally defined cohorts as a developmental strategy in the assembly of motor circuits.

The need for temporal specification of subsets of neurons in motor system development is likely driven by the fact that motor circuits develop sequentially. Indirect evidence for sequential motor circuit assembly comes from behavioral observations in species as diverse as humans and flies. For example, using ultrasound, reflex movements are first detected in human fetuses at 7 weeks post fertilization and then more complex behaviors such as hand-mouth coordination appear around 27 weeks (de Vries & Fong, 2006; Geertsen, Willerslev-Olsen, Lorentzen, & Nielsen, 2017). Similarly, in Drosophila embryos, the earliest motor behaviors involve forward peristalsis waves followed by more refined behaviors such as self-righting (Crisp, Evers, Fiala, & Bate, 2008). Work in zebrafish has pushed beyond behavioral observations to show that early-born neurons contribute to circuits for fast escape, whereas later-born neurons contribute to circuits for refined movements (Fetcho & McLean, 2010; Jay & McLean, 2019; McLean & Fetcho, 2009). These observations directly show that there is sequential assembly of motor circuits rather than just sequential expression of behavior from a set of pre-assembled circuits. Analogously, Wreden et al., found that in Drosophila larvae, early-born EL interneurons contribute to circuits for fast escape, whereas later-born EL interneurons contribute to circuits for proprioceptive refinement of movements (Wreden et al., 2017). This observation may point to a conserved developmental principle. Temporal cohorts or cohort-like groups of neurons may underlie motor circuit assembly across species, despite their separation by hundreds of millions of years of evolution.

5.6 Remaining questions about the lineage-based organization of motor circuits

There remain many questions about the relationships between stem cells and motor circuit configuration. Despite NB3-3 being currently the most completely characterized lineage, several key questions still remain. Can one EL be part of multiple temporal cohorts? NB3-3 makes additional neurons during larval stages. To what adult circuits do they contribute? How do the circuits to which NB3-3 contribute vary along the anterior–posterior axis of the larvae? In other insects, what circuits do NB3-3 neurons contribute to? Beyond NB3-3, to what extent are there additional examples of other interneuron temporal cohorts? For example, Saaghi neurons and Basin neurons, which synapse onto ELs, have been proposed to be temporal cohorts (Wreden et al., 2017). What is the relationship between neurons from different lineages that synapse with each other? For example, do temporal cohorts of early-born neurons preferentially wire together? Do early-born neurons from multiple lineages promote fast escape movements? Do all lineages synapse with each other, or is there some selectivity?

5.7 Conclusion

Work reviewed in this section provides evidence that temporal cohorts are fundamental, lineage-based organizational units of the Drosophila motor system. Temporal cohorts link neuron birth time within a neuronal stem cell lineage to motor circuit membership. Sequential development of motor circuits appears to be an ancient feature of motor system development. Thus, the observed link between neuronal birth time and functional circuit assembly may be representative of an ancient, widely-occurring strategy driving motor circuit assembly.

6. Why is neuron birth timing an important factor in motor circuit assembly?

Once stem cell to circuit relationships are identified, the next step in understanding circuit assembly is to identify the underlying cellular and molecular mechanisms that give rise to these relationships. The identification of temporal cohorts as an organizational unit of Drosophila motor circuits revealed a correlation between neuronal birth timing, within a single neuroblast lineage, and neuronal circuit membership. This finding raised the question of how those relationships are established.

During stem cell division, there are several simultaneously changing factors which could influence the circuit membership of the resulting neuron. One factor is the dynamic, instructive environment—earlier-born neurons and later-born neurons might encounter different physical substrates or receive different signaling cues. A second factor is that earlier-born neurons have more time for cell biological processes to occur, such as axon outgrowth. Third, the chromatin landscape within the nucleus of a neuronal stem cell can change over time. Finally, earlier-born and later-born neurons have different molecular profiles, which are controlled by temporal transcription factors. Determining the precise role of each factor associated with neuronal birth time is crucial for understanding how motor circuits assemble. In this section, we review studies that elucidate the roles of temporal transcription factors in Drosophila motor circuit assembly.

6.1 To what extent do temporal transcription factors control mature neuronal functional identity?

As neuronal stem cells divide, they sequentially express temporal transcription factors (Pearson & Doe, 2004). In Drosophila nerve cord neuroblasts, the following temporal transcription factors are expressed in neuroblasts in this sequence: Hunchback (Hb), Kruppel, Nubbin/Pdm2 (Pdm), Castor (Cas), and Grainy head (Fig. 3C') (Brody & Odenwald, 2000; Isshiki et al., 2001; Kambadur et al., 1998). Loss-of-function and misexpression experiments showed that temporal transcriptions collectively specify embryonic molecular temporal identity of neuronal progeny (Grosskortenhaus et al., 2006; Isshiki et al., 2001; Pearson & Doe, 2004; Tran & Doe, 2008). For example, when Hunchback is lost, expression of molecular markers for first-born neuronal progeny are lost (Isshiki et al., 2001). When Hunchback expression is prolonged in neuroblasts, more neurons express first-born molecular identity markers, and fewer neurons express later-born molecular identity markers (Isshiki et al., 2001). Limited anatomical evidence collected in embryos suggested that neurons with early-born molecular marker expression have initial axonal trajectories similar to those of early-born neurons (Isshiki et al., 2001; Pearson & Doe, 2003). Therefore collectively, there was strong evidence in the field to show that embryonic *molecular identity* could be regulated by temporal transcription factors.

The relationship between temporal transcription factor-induced changes in embryonic molecular identity and mature functional identity, and therefore circuit assembly, was far less clear. Notably, no data had been collected on mature features such as synaptic partnerships or dendrite morphology. Despite limited evidence, however, a widely-held model was that temporal transcription factors acting in combination with neuroblast specific spatial factors act upstream to control complex regulatory pathways controlling all aspects of neuronal identity, including mature neuronal features (Allan & Thor, 2015). This is an important model to test because, if true, it means that the information that controls all of a neuron's developmental decisions is present prior to the neuron's time of birth. It would mean that circuit assembly could be predicted and controlled by programs of gene expression in the neuroblast and would suggest that the environment in which motor circuits assemble is relatively stable and permissive. Finally, because temporal transcription factors are at the top of a molecular cascade, it would provide a molecular entry point for identifying important downstream factors that control motor circuit assembly.

Three recent studies, discussed below, directly probed the role of temporal transcription factors in controlling mature features of motor neurons

(Meng et al., 2019; Meng et al., 2020; Seroka & Doe, 2019). These studies used NB7-1 and NB3-1 lineage-specific gene expression of temporal transcription factors. This allowed the authors to rule out any non-lineage specific effects and to grow animals to mature larval stages. The manipulations in the recent studies generated the same embryonic phenotypes as seen in previous studies that focused only on embryonic stages (Grosskortenhaus et al., 2006; Isshiki et al., 2001; Tran & Doe, 2008). What sets the recent studies apart from previous studies is that they track motor neurons into mature larval stages. Collectively, these recent studies use neuromuscular junction fillet preparations, single cell tracing techniques, labeling with marker genes, confocal imaging, electrophysiology, and calcium imaging. They assess how temporal transcription factors regulate neuronal mature features such as anatomical and functional synaptic connectivity. Together they provide insight into how temporal transcription factors impact motor circuit assembly. They show: (1) temporal transcription factor expression in neuroblasts regulates motor neuron temporal cohort size and composition and (2) temporal transcription factor-induced changes in embryonic marker gene expression do not always accurately predict motor neuron-to-muscle synaptic partnerships.

6.2 Temporal transcription factor expression in neuroblasts regulates motor neuron temporal cohort size and composition

Recent studies support a model in which temporal transcription factors act as molecular switches in the neuroblast to regulate circuit membership of neuronal progeny. Bi-directional manipulation (i.e., precocious or prolonged) of temporal transcription factor expression in NB7-1 bi-directionally modulates the number of motor neurons produced by NB7-1. It does so without altering the total number of neurons in the lineage (Meng et al., 2019; Seroka & Doe, 2019). Prolonged expression of Hb in NB7-1 generates extra motor neurons that form functional synapses on muscles in the dorsal muscle group (Meng et al., 2019) (Fig. 5A and B). Hb expression in post-mitotic neurons does not have these effects (Meng et al., 2019). Forcing NB7-1 to precociously transition out of the Hb expression window by expressing the Hb switching factor Seven-up decreases the number of motor neurons synapsing on dorsal muscles (Fig. 5C) (Meng et al., 2019). Additionally, when Hb expression is prolonged in NB3-1, there is an increase in the number of motor neurons synapsing on the muscles in the ventral muscle group (Meng et al., 2020). Together these data demonstrate that Hb expression in neuroblasts is a potent regulator of motor neuron temporal cohort size.

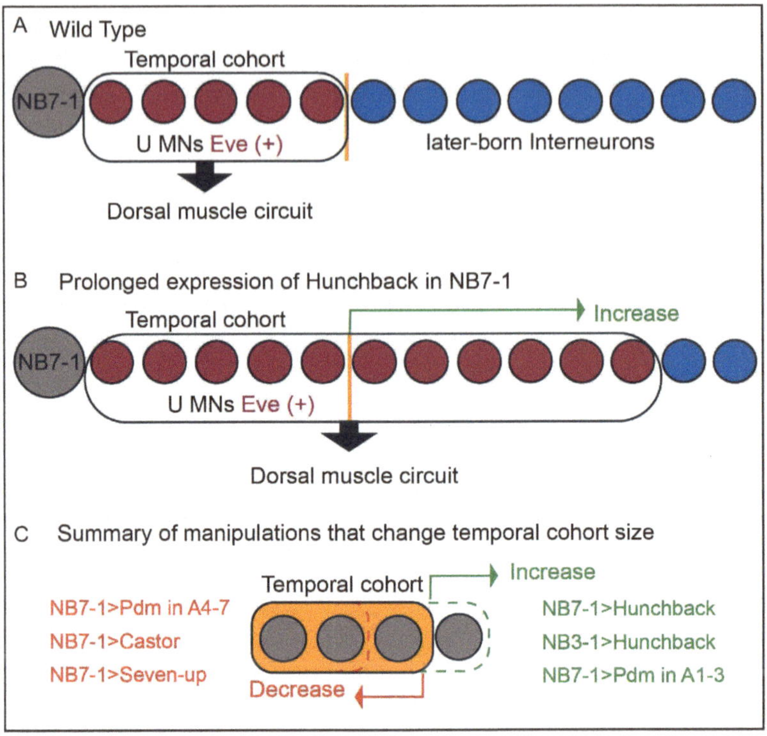

Fig. 5 Temporal transcription factors bi-directionally modulate motor neuron temporal cohorts from the NB7-1 and NB3-1 lineage. (A) Illustration of the NB7-1 lineage with the Eve(+) U motor neuron temporal cohort with the border highlighted by a yellow bar. In NB7-1. (B) Illustration showing that prolonged expression of Hunchback in NB7-1 leads to an increase, green arrow extended beyond the yellow bar, in the number of Eve(+) U motor neurons that project to dorsal muscles. (C) Illustration depicts manipulations, in green, that increase, green dotted line, the size of temporal cohorts, in yellow, and manipulations, in red, that decrease, red dotted line, the size of temporal cohorts, in yellow. For precocious expression of Pdm in NB7-1 (NB7-1 > Pdm), there are differences based on location, abdominal segments A1–A3 versus A4-7.

Other temporal transcription factors also regulate the size of motor neuron temporal cohorts. After the Hb expression window, Pdm and Castor are expressed (Pearson & Doe, 2004). Precocious expression of Castor in NB7-1 decreases the number of motor neurons that populate the dorsal muscle circuit (Fig. 5C). Precocious expression of Pdm in NB7-1 can either increase or decrease the number of U motor neurons that populate the dorsal muscle circuit, depending on which anterior–posterior segment is examined (Fig. 5C) (Meng et al., 2020). The observation that the effect of Pdm varies with anterior–posterior position suggests there may be a synergistic interaction with A-P patterning Hox genes. In summary, there is now strong evidence that multiple temporal transcription factors can modulate the proportion of neurons from a lineage that populate a given motor circuit.

When temporal transcription factor expression alters the size of a motor neuron temporal cohort, it does not linearly scale the proportion of neuronal functional identities within the temporal cohort. That is, a temporal cohort, which is twice as large as wild-type would not contain twice as many early-born and late-born motor neurons as wild-type. Instead, neurons with certain functional identities are overrepresented and other functional identities are underrepresented, depending on the specific temporal transcription factor manipulated (Meng et al., 2019, 2020; Seroka & Doe, 2019). For example, when Hb expression is prolonged in NB7-1, there is a dramatic increase in the number of synaptic branches onto the early U muscle targets and a significant decrease in the number of synaptic branches onto late U muscle targets (Meng et al., 2019; Seroka & Doe, 2019). These studies show that when temporal transcription factor manipulations change motor neuron temporal cohort size, they also change the composition of neurons within the temporal cohort.

6.3 Temporal transcription factor-induced changes in embryonic marker gene expression do not accurately predict motor neuron-to-muscle synaptic partnerships

The finding that temporal transcription factors impact motor circuit wiring was exciting, in part because it raised the possibility that synaptic partnerships might be predicted and controlled by programs of gene expression in the neuroblast. However, two of the recent studies show that this is an overly simplistic explanation (Meng et al., 2019; Seroka & Doe, 2019). When a single temporal transcription factor, Hb, is manipulated, there are mismatches between early marker gene expression and mature features of single neurons. For example, Meng et al. showed that when Hb expression is prolonged in NB7-1, many neurons express markers for the first-born, U1 embryonic molecular identity, but many of these neurons target muscles that were normally innervated by the U2 and U3 motor neurons (Meng et al., 2019). Likewise, Seroka and Doe showed that most neurons with U1-like dendritic morphologies expressed a molecular marker normally found only in U3–U5 motor neurons (Seroka & Doe, 2019). Together these studies demonstrate that when Hb expression is manipulated, embryonic molecular markers do not accurately predict mature features.

There are two lessons from these observations. (1) They demonstrate that manipulation of Hb in neuroblasts can control a neuron's decision to exit the nerve cord and a neuron's decision of where to branch off from an axon bundle. However, Hb cannot predictably control individual motor neuron synaptic partners. Consequently, in some cases temporal transcription factors *influence* rather than completely transform neuronal functional identity. Additional, not yet identified information must help neurons decide which specific muscle to innervate. (2) These observations provide a cautionary tale for the field. They demonstrate that in some cases embryonic molecular markers do not predict (or control) all mature neuronal features. However, notably, manipulation of Pdm and Cas in NB7-1 *does* change embryonic molecular identity and functional identity in a highly predictable way (Meng et al., 2020). Together these studies demonstrate that the extent to which mature features can be predicted by embryonic marker gene expression must be determined on a case-by-case basis.

6.4 In some cases, the CNS can compensate for temporal transcription factor-induced changes

As described above, temporal transcription factor manipulation results in two phenotypes—over innervated muscles and under innervated muscles. How does the motor system respond to these changes?

For over-innervated muscles, there is evidence for functional compensation. When Hb expression is prolonged in NB3-1, several ventral muscles get extra synaptic branches. Electrophysiological recordings show a significant increase in the number of vesicles released during stimulated synaptic activity. This is consistent with the idea that the total number of functional synapses is increased (Meng et al., 2020). However, the overall *muscle response* to nerve stimulation (EPSP) remains at wild-type levels. These data show functional compensation occurs. Although the specific cell biological mechanism for compensation is unclear, it is likely to be a form of synaptic homeostasis (Frank, James, & Müller, 2020).

There is evidence that under-innervated muscles receive anatomical compensation. When Hb expression is prolonged in NB7-1 or NB3-1, late-born U motor neurons and late-born RP motor neurons appear to be missing. Late-born U motor neuron and late-born RP motor neuron muscle targets are under-innervated. However, these muscles do not entirely lack synapses. Instead, what little synaptic branches are seen appear to be abnormally placed branches from motor neurons innervating other muscles. This phenotype is also seen on the RP3 motor neuron target when

RP3 is laser ablated during development (Chang & Keshishian, 1996). This phenotype is seen on U motor neuron muscle targets when U motor neurons are genetically ablated during development by expression of pro-apoptotic genes (Meng et al., 2020). Taken together these data support the idea that when muscles lack innervation due to loss of their normal motor neuron partner, other motor neurons excessively branch to compensate for the loss of innervation.

Anatomical compensation of under-innervated muscles is not sufficient to recover normal muscle physiology. When Hb expression is prolonged in NB3-1, several ventral muscles get fewer synaptic branches. Electrophysiological recordings of these under-innervated muscles show their overall response to nerve stimulation is significantly reduced in comparison to controls (Meng et al., 2020). These data show that temporal transcription factor manipulations have long-lasting functional effects.

In summary, manipulation of temporal transcription factors in neuroblasts leads to permanent changes in synaptic partnerships, thus triggering different types of plasticity and having long-lasting effects on physiology of the motor system.

6.5 Remaining questions about how temporal transcription factors impact motor circuit assembly

The observation that temporal transcription factors are potent regulators of motor neuron mature features raises the question of how manipulation of temporal transcription factors impacts interneuron mature features. There are two major gaps in our knowledge. (1) When motor neuron dendrite morphology is transformed by temporal transcription factor manipulation, how are premotor interneuron-to-motor neuron input synapses impacted? (2) How do temporal transcription factor manipulations impact axons and synaptic partner choice of interneurons whose axons do not leave the CNS? These questions must be addressed because motor neurons develop in a different context in comparison to interneurons. Interneurons form synaptic connections within the CNS, which is a dynamic environment constantly changing due to ongoing neurogenesis. Larval motor neurons acquire many of their mature features while projecting into a periphery that contains large, pre-existing muscles. Answering these questions will further elucidate the logic of motor circuit assembly.

The observation that temporal transcription factors can profoundly influence motor circuit assembly in Drosophila larvae raises questions about the impact of temporal transcription factors in other contexts.

(1) In the spinal cord, temporal programs drive neuronal molecular identity (Philippidou & Dasen, 2013; Sagner & Briscoe, 2019). However, the specific underlying molecular mechanisms are currently poorly-understood. Consequently, the degree to which stem cell intrinsic temporal transcriptional programs influence motor circuit assembly remains un-tested in the spinal cord. (2) Temporal transcription factors have been identified in other Drosophila circuits (Allan & Thor, 2015; Doe, 2017; Rossi, Fernandes, & Desplan, 2017; Stratmann & Thor, 2017; Sullivan, Warren, & Doe, 2019), but their impact on mature neuronal features and circuit assembly in these contexts is un-characterized. (3) It would be valuable to see the degree to which there are alterations in temporal transcription factor expression in neuroblasts in other insects. This is because all insect motor systems are generated from the same basic set of 30 neuroblasts (Doe, 1992; Doe, Kuwada, & Goodman, 1985; Shepherd & Bate, 1990; Tamarelle, Haget, & Ressouches, 1985; Truman & Ball, 1998). Yet, insect body plans have dramatically changed during the course of 300 million years of evolution (Niven, Graham, & Burrows, 2008). To accommodate changes in body plan, the duration of neuroblast division is altered (Biffar & Stollewerk, 2014; Truman & Ball, 1998). How temporal transcription factor expression changes over evolutionary time scales and how expression changes might impact diversification of motor system wiring are open questions.

The finding that temporal transcription factors strongly influence motor circuit assembly demonstrates that this class of transcription factors are at the top of a transcriptional hierarchy. A major unsolved question is what are the relevant programs of gene expression downstream of temporal transcription factors that regulate motor circuit assembly.

6.6 Conclusions

In this section, we reviewed studies that reveal temporal transcription factors expressed in neuronal stem cells can control circuit membership of motor neuron progeny. So far these studies reveal that transcription factor expression profile is more important than the time of neuronal birth for determining neuronal circuit membership. This indicates that synaptic partnerships between motor axons and muscles are controlled by lineage intrinsic programs.

7. Role of Hox in generating anterior–posterior (A–P) motor circuit diversity

Motor circuits are found repetitively along the anterior–posterior (A–P) axis in vertebrate spinal cord and Drosophila nerve cord.

These segmentally repeated circuits are not identical; instead they are specialized to control A–P region specific movements and to process A–P region specific somatosensory stimuli (Jung & Dasen, 2015; Venkatasubramanian & Mann, 2019). Despite A–P variation in spinal cord and nerve cord, the same set of neuronal stem cells is found at all A–P levels (Doe, 1992; Francius et al., 2013). It is still poorly understood how neuronal stem cells are modulated to generate A–P specific motor circuits.

The first major gap in our understanding of how stem cells generate A–P specific motor circuits is that we lack examples of how stem cell lineage to circuit relationships vary along the A–P axis. For example, in Drosophila larvae, in the first abdominal segment, NB3-3 generates early-born ELs and late-born ELs (Wreden et al., 2017). In thoracic segments, NB3-3 generates fewer ELs, and these ELs express early-born markers (Schmid et al., 1999, 1997; Tsuji, Hasegawa, & Isshiki, 2008). Unlike in the first abdominal segment where EL circuit membership information is known (Wreden et al., 2017), nothing is known about the circuit membership of thoracic ELs. Characterizing segment specific variation in stem-cell circuit relationships is a critical first step in understanding how stem cells are modulated to make A–P specific circuits.

The second major gap is that we lack an understanding of how *Hox* genes modulate neuronal stem cells to generate A–P specific motor circuits. *Hox* genes contribute to the specialization of many body regions along the A–P axis and have been extensively studied in the context of vertebrate motor neuron development and Drosophila neuronal diversity (Fig. 3A) (reviewed in Philippidou & Dasen, 2013). A major theme that emerges is that *Hox* genes can act alone or in combination over different stages in development. In Drosophila larvae, there is evidence that *Hox* gene expression is key to generating region-specific motor circuits. Evidence to support this idea comes from studies of larval behavior after manipulation of *Hox* gene expression (Dixit, VijayRaghavan, & Bate, 2008). Segment specific behavioral changes were observed indicating that underlying circuitry must also have been changed. However, it is still poorly understood how manipulation of Hox transcription factors impact mature neuronal features relevant for motor circuit assembly.

In Drosophila, one way in which Hox genes are likely to impact circuit assembly is by modulating the number of temporal cohorts produced by a given neuroblast. For many neuroblast lineages, Hox genes are known to impart A–P segmental differences as measured by neuron number and neuron gene expression profile (Technau et al., 2014). In the case of NB3-3, Hox gene activity can regulate the number of late-born neurons produced in thoracic segments. When *Hox* genes *Antennapedia* and *Abdominal-A* are manipulated, thoracic NB3-3 neuroblasts abnormally produce extra ELs with late-born marker gene expression (Tsuji et al., 2008). This observation supports the idea that one way in which Hox genes impact circuit assembly is by modulating the number of temporal cohorts produced by a given neuroblast. However, functional studies and additional examples are still lacking.

There remain many additional questions about the role of Hox genes in modulating NB3-3 to generate diverse motor circuits. These questions exemplify some of the major open questions in the field. (1) How does *Hox* gene manipulation alter mature EL neuronal features, such as synaptic partnerships and synaptic physiology? (2) If late-born ELs are generated in the thorax, what neurons do they synapse with? (3) Most *Hox* gene manipulations, including work done in NB3-3, lack temporal and cell type specificity. What is the spatiotemporal nature of *Hox* gene activity?

8. Open questions

8.1 To what extent does Notch contribute to motor circuit assembly?

Notch signaling is a likely source of control over motor circuit assembly. Notch signaling early in development contributes to neuronal molecular identity by acting in GMCs to generate neurons with A or B identities (Truman et al., 2010). There is mounting evidence that Notch signaling also regulates motor circuit membership. Evidence to support this idea comes from Mark et al. (2019). They identified seven neuroblast lineages in the TEM connectome spanning the first abdominal (A1) segment of the first instar Drosophila larval nerve cord. They found that hemilineages tend to project their neurites to either the sensory or motor region of the neuropil and neurons within one hemilineage project to a specific sub-region. Notch ON/OFF is correlated with neurite projection such that Notch ON neurons project to motor regions whereas Notch OFF hemilineages project to sensory regions. They investigated whether Notch ON/OFF is responsible for the differences in hemilineage projection sites. When Notch was misexpressed throughout an entire neuroblast lineage, no detectable change in lineage size was observed, however all neurons projected to the motor region of the nerve cord. Therefore, Notch controls the projection fate of neurons. Since morphology was characterized at the level of whole nerve cords, it remains unclear what role Notch has on synaptic partnership choice. Nonetheless, these data support the idea that Notch signaling could be a major regulator of motor circuit configuration.

8.2 To what extent do spatial factors contribute to motor circuit assembly?

Spatial patterning transcription factors are another likely source of control over motor circuit assembly. Within each nerve cord hemisegment, the 30 neuroblasts are arranged in a stereotyped grid-like fashion and have a unique molecular identity (Fig. 2A). The structured grid and neuroblast identity are controlled by the intersection of spatial factors expressed along the A–P and D–V axis in the neuroectoderm (Fig. 3B) (Skeath & Thor, 2003). An example of this control comes from work manipulating the spatial factor gooseberry-distal (gsb), which is normally expressed in the neuroectoderm row 5 (Skeath, Zhang, Holmgren, Carroll, & Doe, 1995). Loss of gsb causes loss of row 5 neuroblast identities, whereas ectopic expression of gsb leads to extra row 5 neuroblast identities (Skeath et al., 1995). While these factors are sufficient to specify neuroblast identity, it is unclear the extent to which they influence functional identity and motor circuit assembly.

A recent study suggested that spatial patterning transcription factors in neuroblasts limit or shape temporal transcription factor genomic binding by controlling the chromatin landscape. Recent work by Sen et al. used targeted DamID to identify genomic binding sites of Hunchback (Sen, Chanchani, Southall, & Doe, 2019). They found that Hunchback targets significantly different genes between the two different neuroblast lineages (Sen et al., 2019). This finding suggests that regulation by temporal transcription factors is dependent on other neuroblast-specific factors. Future work is needed to test this hypothesis and interrogate the role that spatial patterning factors play in circuit assembly.

9. Conclusion

The main concepts addressed in this review are: (1) the definition of temporal cohorts as novel organizational units in development that link neuronal stem cell lineages to motor circuit configuration and (2) the discovery that temporal patterning factors expressed in neuronal stem cells control aspects of circuit assembly by controlling the size of temporal cohorts and influencing synaptic partner choice. The recent work in Drosophila larval neuromuscular synapses presented in this review supports the model that intrinsic transcriptional programs control circuit assembly.

As the field's understanding of motor circuit architecture underlying the wide range of motor behavior increases, it will become feasible to interrogate the extent to which intrinsic transcriptional programs control synaptic partnerships throughout the motor system. Future work will need to begin to link key factors acting downstream of temporal patterning factors to specific decisions a neuron must make to establish functional synaptic partnerships. Understanding the extent to which developmental events are controlled by transcriptional cascades will undoubtedly reveal core, ancient developmental strategies underlying motor circuit assembly and reveal flexible developmental strategies, which may be shaped by evolutionary time.

Acknowledgments

We apologize to colleagues whose work we did not discuss or have the space to cite. We thank Jake Henderson and Kyle Lathem for valuable discussions and comments. Work in the Heckscher lab is funded by T32 GM007183 (J.L.M.), NSF GRFP (DGE-1746045) (J.L.M.), NIH R01-NS105748 (E.S.H.), University of Chicago MGCB start-up funds (E.S.H.).

References

Allan, D. W., & Thor, S. (2015). Transcriptional selectors, masters, and combinatorial codes: Regulatory principles of neural subtype specification. *Wiley Interdisciplinary Reviews: Developmental Biology*, *4*(5), 505–528.

Babski, H., Jovanic, T., Surel, C., Yoshikawa, S., Zwart, M. F., Valmier, J., et al. (2019). A GABAergic Maf-expressing interneuron subset regulates the speed of locomotion in Drosophila. *Nature Communications*, *10*(1), 1–17.

Baek, M., & Mann, R. S. (2009). Lineage and birth date specify motor neuron targeting and dendritic architecture in adult Drosophila. *Journal of Neuroscience*, *29*(21), 6904–6916.

Bate, M. (1993). In M. Bate, & A. Martinez Arias (Eds.), *The mesoderm and its derivatives* (pp. 1013–1090).

Benito-Gonzalez, A., & Alvarez, F. J. (2012). Renshaw cells and Ia inhibitory interneurons are generated at different times from p1 progenitors and differentiate shortly after exiting the cell cycle. *Journal of Neuroscience*, *32*(4), 1156–1170.

Berni, J., Pulver, S. R., Griffith, L. C., & Bate, M. (2012). Autonomous circuitry for substrate exploration in freely moving drosophila larvae. *Current Biology*, *22*(20), 1861–1870.

Biffar, L., & Stollewerk, A. (2014). Conservation and evolutionary modifications of neuroblast expression patterns in insects. *Developmental Biology*, *388*(1), 103–116.

Bikoff, J. B. (2019). Interneuron diversity and function in the spinal motor system. *Current Opinion in Physiology*, *8*, 36–43. Elsevier Ltd.

Birkholz, O., Rickert, C., Berger, C., Urbach, R., & Technau, G. M. (2013). Neuroblast pattern and identity in the Drosophila tail region and role of doublesex in the survival of sex-specific precursors. *Development*, *140*(8), 1830–1842.

Birkholz, O., Rickert, C., Nowak, J., Coban, I. C., & Technau, G. M. (2015). Bridging the gap between postembryonic cell lineages and identified embryonic neuroblasts in the ventral nerve cord of *Drosophila melanogaster*. *Biology Open*, *4*(4), 420–434.

Blitz, D. M., & Nusbaum, M. P. (2012). Modulation of circuit feedback specifies motor circuit output. *Journal of Neuroscience*, *32*(27), 9182–9193.

Bossing, T., & Technau, G. M. (1994). The fate of the CNS midline progenitors in Drosophila as revealed by a new method for single cell labelling. *Development*, *120*(7), 1895–1906.

Bossing, T., Udolph, G., Doe, C. Q., & Technau, G. M. (1996). The embryonic central nervous system lineages of Drosophila melanogaster. I. Neuroblast lineages derived from the ventral half of the neuroectoderm. *Developmental Biology, 179*, 41–64.

Broadus, J., Skeath, J. B., Spana, E. P., Bossing, T., Technau, G., & Doe, C. Q. (1995). New neuroblast markers and the origin of the aCC/pCC neurons in the Drosophila central nervous system. *Mechanisms of Development, 53*, 393–402.

Brody, T., & Odenwald, W. F. (2000). Programmed transformations in neuroblast gene expression during Drosophila CNS lineage development. *Developmental Biology, 226*, 34–44.

Burgos, A., Honjo, K., Ohyama, T., Qian, C. S., Shin, G. J. E., Gohl, D. M., et al. (2018). Nociceptive interneurons control modular motor pathways to promote escape behavior in Drosophila. *eLife*, 7, e26016.

Carreira-Rosario, A., Zarin, A. A., Clark, M. Q., Manning, L., Fetter, R. D., Cardona, A., et al. (2018). MDN brain descending neurons coordinately activate backward and inhibit forward locomotion. *eLife*, 7, 57.

Chang, T. N., & Keshishian, H. (1996). Laser ablation of drosophila embryonic motoneurons causes ectopic innervation of target muscle fibers. *Journal of Neuroscience, 16*(18), 5715–5726.

Choi, J. C., Park, D., & Griffith, L. C. (2004). Electrophysiological and morphological characterization of identified motor neurons in the Drosophila third instar larva central nervous system. *Journal of Neurophysiology, 91*(5), 2353–2365.

Clark, M. Q., McCumsey, S. J., Lopez-Darwin, S., Heckscher, E. S., & Doe, C. Q. (2016). Functional genetic screen to identify interneurons governing behaviorally distinct aspects of Drosophila larval motor programs. *G3: Genes, Genomes, Genetics, 6*(7), 2023–2031.

Clark, M. Q., Zarin, A. A., Carreira-Rosario, A., & Doe, C. Q. (2018). Neural circuits driving larval locomotion in drosophila. *Neural Development, 13*(1), 6. BioMed Central Ltd.

Crisp, S., Evers, J. F., Fiala, A., & Bate, M. (2008). The development of motor coordination in Drosophila embryos. *Development, 135*(22), 3707–3717.

Dana, H., Sun, Y., Mohar, B., Hulse, B. K., Kerlin, A. M., Hasseman, J. P., et al. (2019). High-performance calcium sensors for imaging activity in neuronal populations and microcompartments. *Nature Methods, 16*(7), 649–657.

de Vries, J. I. P., & Fong, B. F. (2006). Normal fetal motility: An overview. *Ultrasound in Obstetrics and Gynecology, 27*(6), 701–711.

Dixit, R., VijayRaghavan, K., & Bate, M. (2008). Hox genes and the regulation of movement in Drosophila. *Developmental Neurobiology, 68*(3), 309–316.

Doe, C. Q. (1992). Molecular markers for identified neuroblasts and ganglion mother cells in the Drosophila central nervous system. *Development, 116*(4), 855–863.

Doe, C. Q. (2017). Temporal patterning in the Drosophila CNS. *Annual Review of Cell and Developmental Biology, 33*, 219–240.

Doe, C. Q., Kuwada, J. Y., & Goodman, C. S. (1985). From epithelium to neuroblasts to neurons: The role of cell interactions and cell lineage during insect neurogenesis. *Philosophical Transactions of the Royal Society of London. Series B, Biological Sciences, 312*, 67–81.

Estacio-Gómez, A., & Díaz-Benjumea, F. J. (2014). Roles of Hox genes in the patterning of the central nervous system of Drosophila. *Fly, 8*(1), 26.

Fetcho, J. R., & McLean, D. L. (2010). Some principles of organization of spinal neurons underlying locomotion in zebrafish and their implications. *Annals of the New York Academy of Sciences, 1198*, 94–104.

Francius, C., Harris, A., Rucchin, V., Hendricks, T. J., Stam, F. J., Barber, M., et al. (2013). Identification of multiple subsets of ventral interneurons and differential distribution along the rostrocaudal axis of the developing spinal cord. *PLoS One, 8*, e70325.

Frank, C. A., James, T. D., & Müller, M. (2020). Homeostatic control of Drosophila neuromuscular junction function. *Synapse*, *74*(1), e22133. John Wiley and Sons Inc.

Friesen, W. O., Poon, M., & Stent, G. S. (1976). An oscillatory neuronal circuit generating a locomotory rhythm. *Proceedings of the National Academy of Sciences of the United States of America*, *73*(10), 3734–3738.

Fujioka, M., Emi-Sarker, Y., Yusibova, G. L., Goto, T., & Jaynes, J. B. (1999). Analysis of an even-skipped rescue transgene reveals both composite and discrete neuronal and early blastoderm enhancers, and multi-stripe positioning by gap gene repressor gradients. *Development*, *126*(11), 2527–2538.

Fushiki, A., Zwart, M. F., Kohsaka, H., Fetter, R. D., Cardona, A., & Nose, A. (2016). A circuit mechanism for the propagation of waves of muscle contraction in Drosophila. *eLife*, *5*, e13253.

Geertsen, S. S., Willerslev-Olsen, M., Lorentzen, J., & Nielsen, J. B. (2017). Development and aging of human spinal cord circuitries. *Journal of Neurophysiology*, *118*(2), 1133–1140. American Physiological Society.

Goodman, C. S., & Doe, C. Q. (1993). In M. Bate, & A. Martinez Arias (Eds.), *Embryonic development of the Drosophila central nervous system* (pp. 1131–1206).

Grosskortenhaus, R., Robinson, K. J., & Doe, C. Q. (2006). Pdm and Castor specify late-born motor neuron identity in the NB7-1 lineage. *Genes & Development*, *20*(18), 2618–2627.

Harris, R. M., Pfeiffer, B. D., Rubin, G. M., & Truman, J. W. (2015). Neuron hemilineages provide the functional ground plan for the Drosophila ventral nervous system. *eLife*, *4*.

Hartenstein, V., & Stollewerk, A. (2015). The evolution of early neurogenesis. *Developmental Cell*, *32*(4), 390–407. Cell Press.

He, L., Gulyanon, S., Mihovilovic Skanata, M., Karagyozov, D., Heckscher, E. S., Krieg, M., et al. (2019). Direction selectivity in Drosophila proprioceptors requires the mechanosensory channel Tmc. *Current Biology*, *29*(6). 945–956.e3.

Heckscher, E. S., Lockery, S. R., & Doe, C. Q. (2012). Characterization of Drosophila larval crawling at the level of organism, segment, and somatic body wall musculature. *The Journal of Neuroscience: The Official Journal of the Society for Neuroscience*, *32*(36), 12460–12471.

Heckscher, E. S., Long, F., Layden, M. J., Chuang, C. H., Manning, L., Richart, J., et al. (2014). Atlas-builder software and the eNeuro atlas: Resources for developmental biology and neuroscience. *Development*, *141*(12), 2524–2532.

Heckscher, E. S., Zarin, A. A., Faumont, S., Clark, M. Q., Manning, L., Fushiki, A., et al. (2015). Even-skipped+ interneurons are core components of a sensorimotor circuit that maintains left-right symmetric muscle contraction amplitude. *Neuron*, *88*(2), 314–329.

Hultborn, H. (2006). Spinal reflexes, mechanisms and concepts: From Eccles to Lundberg and beyond. *Progress in Neurobiology*, *78*, 215–232.

Im, S. H., & Galko, M. J. (2012). Pokes, sunburn, and hot sauce: Drosophila as an emerging model for the biology of nociception. *Developmental Dynamics*, *241*(1), 16–26.

Isshiki, T., Pearson, B., Holbrook, S., & Doe, C. Q. (2001). Drosophila neuroblasts sequentially express transcription factors which specify the temporal identity of their neuronal progeny. *Cell*, *106*(4), 511–521.

Jay, M., & McLean, D. L. (2019). Reconciling the functions of even-skipped interneurons during crawling, swimming, and walking. *Current Opinion in Physiology*, *8*, 188–192. Elsevier Ltd.

Jenett, A., Rubin, G. M., Ngo, T. T., Shepherd, D., Murphy, C., Dionne, H., et al. (2012). A GAL4-driver line resource for Drosophila neurobiology. *Cell Reports*, *2*(4), 991–1001.

Jessell, T. M. (2000). Neuronal specification in the spinal cord: Inductive signals and transcriptional codes. *Nature Reviews Genetics*, *1*(1), 20–29. European Association for Cardio-Thoracic Surgery.

Jovanic, T., Winding, M., Cardona, A., Truman, J. W., Gershow, M., & Zlatic, M. (2019). Neural substrates of drosophila larval anemotaxis. *Current Biology, 29*(4), 554–566.e4.

Jung, H., & Dasen, J. S. (2015). Evolution of patterning systems and circuit elements for locomotion. *Developmental Cell, 32*(4), 408–422.

Kambadur, R., Koizumi, K., Stivers, C., Nagle, J., Poole, S. J., & Odenwald, W. F. (1998). Regulation of POU genes by castor and hunchback establishes layered compartments in the Drosophila CNS. *Genes & Development, 12*(2), 246–260.

Knoblich, J. A. (2010). Asymmetric cell division: Recent developments and their implications for tumour biology. *Nature Reviews Molecular Cell Biology, 11*(12), 849–860.

Koch, S. C., Acton, D., & Goulding, M. (2018). Spinal circuits for touch, pain, and itch. *Annual Review of Physiology, 80*, 189–217. Annual Reviews Inc.

Kohsaka, H., Guertin, P. A., & Nose, A. (2017). Neural circuits underlying fly larval locomotion. *Current Pharmaceutical Design, 23*(12), 1722–1733.

Kohsaka, H., Zwart, M. F., Fushiki, A., Fetter, R. D., Truman, J. W., Cardona, A., et al. (2019). Regulation of forward and backward locomotion through intersegmental feedback circuits in Drosophila larvae. *Nature Communications, 10*(1), 2654.

Lacin, H., & Truman, J. W. (2016). Lineage mapping identifies molecular and architectural similarities between the larval and adult Drosophila central nervous system. *eLife, 5*.

Landgraf, M., Bossing, T., Technau, G. M., & Bate, M. (1997). The origin, location, and projections of the embryonic abdominal motorneurons of Drosophila. *Journal of Neuroscience, 17*(24), 9642–9655.

Lemon, W. C., Pulver, S. R., Höckendorf, B., McDole, K., Branson, K., Freeman, J., et al. (2015). Whole-central nervous system functional imaging in larval Drosophila. *Nature Communications, 6*, 7924.

Li, H. H., Kroll, J. R., Lennox, S. M., Ogundeyi, O., Jeter, J., Depasquale, G., et al. (2014). A GAL4 driver resource for developmental and behavioral studies on the larval CNS of drosophila. *Cell Reports, 8*(3), 897–908.

Manning, L., Heckscher, E. S., Purice, M. D., Roberts, J., Bennett, A. L., Kroll, J. R., et al. (2012). A resource for manipulating gene expression and analyzing cis-regulatory modules in the Drosophila CNS. *Cell Reports, 2*(4), 1002–1013.

Marder, E., & Bucher, D. (2001). Central pattern generators and the control of rhythmic movements. *Current Biology, 11*(23), R986–R996.

Mark, B., Lai, S. L., Zarin, A. A., Manning, L., Cardona, A., Truman, W. T., et al. (2019). The role of lineage, hemilineage and temporal identity in establishing neuronal connectivity in the Drosophila larval CNS. *bioRxiv*, 617936.

Masson, J. B., Laurent, F., Cardona, A., Barre, C., Skatchkovsky, N., Zlatic, M., et al. (2020). Identifying neural substrates of competitive interactions and sequence transitions during mechanosensory responses in Drosophila. *PLoS Genetics, 16*(2), e1008589.

McLean, D. L., & Fetcho, J. R. (2009). Spinal interneurons differentiate sequentially from those driving the fastest swimming movements in larval zebrafish to those driving the slowest ones. *Journal of Neuroscience, 29*(43), 13566–13577.

Meng, J. L., Marshall, Z. D., Lobb-Rabe, M., & Heckscher, E. S. (2019). How prolonged expression of hunchback, a temporal transcription factor, re-wires locomotor circuits. *eLife, 8*.

Meng, J. L., Wang, Y., Carrillo, R. A., & Heckscher, E. S. (2020). Temporal transcription factors determine circuit membership by permanently altering motor neuron-to-muscle synaptic partnerships. *eLife, 9*. https://doi.org/10.7554/eLife.56898.

Miroschnikow, A., Schlegel, P., Schoofs, A., Hueckesfeld, S., Li, F., Schneider-Mizell, C. M., et al. (2018). Convergence of monosynaptic and polysynaptic sensory paths onto common motor outputs in a drosophila feeding connectome. *eLife, 7*.

Namiki, S., Dickinson, M. H., Wong, A. M., Korff, W., & Card, G. M. (2018). The functional organization of descending sensory-motor pathways in Drosophila. *eLife, 7*.

Niven, J. E., Graham, C. M., & Burrows, M. (2008). Diversity and evolution of the insect ventral nerve cord. *Annual Review of Entomology*, *53*(1), 253–271.

Ohyama, T., Schneider-Mizell, C. M., Fetter, R. D., Aleman, J. V., Franconville, R., Rivera-Alba, M., et al. (2015). A multilevel multimodal circuit enhances action selection in Drosophila. *Nature*, *520*(7549), 633–639.

Osseward, P. J., & Pfaff, S. L. (2019). Cell type and circuit modules in the spinal cord. *Current Opinion in Neurobiology*, *56*, 175–184. Elsevier Ltd.

Pearson, B. J., & Doe, C. Q. (2003). Regulation of neuroblast competence in Drosophila. *Nature*, *425*(6958), 624–628.

Pearson, B. J., & Doe, C. Q. (2004). Specification of temporal identity in the developing nervous system. *Annual Review of Cell and Developmental Biology*, *20*(1), 619–647.

Pfeiffer, B. D., Jenett, A., Hammonds, A. S., Ngo, T. T. B., Misra, S., Murphy, C., et al. (2008). Tools for neuroanatomy and neurogenetics in Drosophila. *Proceedings of the National Academy of Sciences of the United States of America*, *105*(28), 9715–9720.

Philippidou, P., & Dasen, J. S. (2013). Hox genes: Choreographers in neural development, architects of circuit organization. *Neuron*, *80*(1), 12–34.

Pulver, S. R., Bayley, T. G., Taylor, A. L., Berni, J., Bate, M., & Hedwig, B. (2015). Imaging fictive locomotor patterns in larval Drosophila. *Journal of Neurophysiology*, *114*, 2564–2577.

Ravanelli, A. M., & Appel, B. (2015). Motor neurons and oligodendrocytes arise from distinct cell lineages by progenitor recruitment. *Genes & Development*, *29*, 2504–2515.

Rickert, C., Kunz, T., Harris, K.-L., Whitington, P. M., & Technau, G. M. (2011). Morphological characterization of the entire interneuron population reveals principles of neuromere organization in the ventral nerve cord of Drosophila. *The Journal of Neuroscience*, *31*, 15870–15883.

Rickert, C., Lüer, K., Vef, O., & Technau, G. M. (2018). Progressive derivation of serially homologous neuroblast lineages in the gnathal CNS of Drosophila. *PLoS One*, *13*(2), e0191453.

Rossi, A. M., Fernandes, V. M., & Desplan, C. (2017). Timing temporal transitions during brain development. *Current Opinion in Neurobiology*, *42*, 84–92.

Sagner, A., & Briscoe, J. (2019). Establishing neuronal diversity in the spinal cord: A time and a place. *Development (Cambridge)*, *146*(22). Company of Biologists Ltd.

Sanes, J. R., & Lichtman, J. W. (1999). Development of the vertebrate neuromuscular junction. *Annual Review of Neuroscience*, *22*, 389–442. https://doi.org/10.1146/annurev.neuro.22.1.389. 10202544.

Schmid, A., Chiba, A., & Doe, C. Q. (1999). Clonal analysis of Drosophila embryonic neuroblasts: Neural cell types, axon projections and muscle targets. *Development*, *126*(21), 4653–4689.

Schmidt, H., Rickert, C., Bossing, T., Vef, O., Urban, J., & Technau, G. M. (1997). The embryonic central nervous system lineages of Drosophila melanogaster II. Neuroblast lineages derived from the dorsal part of the neuroectoderm. *Developmental Biology*, *189*, 186–204.

Schneider-Mizell, C. M., Gerhard, S., Longair, M., Kazimiers, T., Li, F., Zwart, M. F., et al. (2016). Quantitative neuroanatomy for connectomics in Drosophila. *eLife*, *5*.

Sen, S. Q., Chanchani, S., Southall, T. D., & Doe, C. Q. (2019). Neuroblast-specific open chromatin allows the temporal transcription factor, Hunchback, to bind neuroblast-specific loci. *eLife*, *8*.

Seroka, A. Q., & Doe, C. Q. (2019). The Hunchback temporal transcription factor determines motor neuron axon and dendrite targeting in Drosophila. *Development*, *146*(7), dev175570.

Seroka, A., Yazejian, R. M., Lai, S. L., & Doe, C. Q. (2020). A novel temporal identity window generates alternating Eve+/Nkx6+motor neuron subtypes in a single progenitor lineage. *Neural Development*, *15*(1), 9.

Shepherd, D., & Bate, C. M. (1990). Spatial and temporal patterns of neurogenesis in the embryo of the locust (Schistocerca gregaria). *Development, 108*, 83–96.

Simpson, J. H., & Looger, L. L. (2018). Functional imaging and optogenetics in drosophila. *Genetics, 208*(4), 1291–1309.

Skeath, J. B. (1999). At the nexus between pattern formation and cell-type specification: The generation of individual neuroblast fates in the Drosophila embryonic central nervous system. *BioEssays, 21*(11), 922–931.

Skeath, J. B., & Doe, C. Q. (1998). Sanpodo and Notch act in opposition to Numb to distinguish sibling neuron fates in the Drosophila CNS. *Development, 125*, 1857–1865.

Skeath, J. B., & Thor, S. (2003). Genetic control of Drosophila nerve cord development. *Current Opinion in Neurobiology, 13*(1), 8–15. Elsevier Ltd.

Skeath, J. B., Zhang, Y., Holmgren, R., Carroll, S. B., & Doe, C. Q. (1995). Specification of neuroblast identity in the Drosophila embryonic central nervous system by gooseberry-distal. *Nature, 376*(6539), 427–430.

Snodgrass, R. E. (1935). *Principles of insect morphology*. Ithaca, NY: Cornell UP.

Stam, F. J., Hendricks, T. J., Zhang, J., Geiman, E. J., Francius, C., Labosky, P. A., et al. (2012). Renshaw cell interneuron specialization is controlled by a temporally restricted transcription factor program. *Development, 139*(1), 179–190.

Stifani, N. (2014). Motor neurons and the generation of spinal motor neuron diversity. *Frontiers in Cellular Neuroscience, 8*(Oct), 293. Frontiers Media S.A.

Stratmann, J., & Thor, S. (2017). Neuronal cell fate specification by the molecular convergence of different spatio-temporal cues on a common initiator terminal selector gene. *PLoS Genetics, 13*(4).

Sullivan, L. F., Warren, T. L., & Doe, C. Q. (2019). Temporal identity establishes columnar neuron morphology, connectivity, and function in a drosophila navigation circuit. *eLife, 8*.

Takagi, S., Cocanougher, B. T., Niki, S., Miyamoto, D., Kohsaka, H., Kazama, H., et al. (2017). Divergent connectivity of homologous command-like neurons mediates segment-specific touch responses in Drosophila. *Neuron, 96*(6), 1373–1387.e6.

Tamarelle, M., Haget, A., & Ressouches, A. (1985). Segregation, division, and early patterning of lateral thoracic neuroblasts in the embryo of Carausius morosus Br. (Phasmida: Lonchodidae). *International Journal of Insect Morphology and Embryology, 14*, 307–317.

Tastekin, I., Khandelwal, A., Tadres, D., Fessner, N. D., Truman, J. W., Zlatic, M., et al. (2018). Sensorimotor pathway controlling stopping behavior during chemotaxis in the drosophila melanogaster larva. *eLife*, 7.

Tastekin, I., & Louis, M. (2017). Manipulation of neural circuits in drosophila larvae. In *Decoding neural circuit structure and function: Cellular dissection using genetic model organisms* (pp. 171–189). Springer International Publishing.

Technau, G. M., Rogulja-Ortmann, A., Berger, C., Birkholz, O., & Rickert, C. (2014). Composition of a Neuromere and its segmental diversification under the control of Hox genes in the embryonic CNS of Drosophila. *Journal of Neurogenetics, 28*, 171–180.

Tran, K. D., & Doe, C. Q. (2008). Pdm and Castor close successive temporal identity windows in the NB3-1 lineage. *Development, 135*(21), 3491–3499.

Tripodi, M., Stepien, A. E., & Arber, S. (2011). Motor antagonism exposed by spatial segregation and timing of neurogenesis. *Nature, 479*(7371), 61–66.

Truman, J. W., & Ball, E. E. (1998). Patterns of embryonic neurogenesis in a primitive wingless insect, the silverfish, Ctenolepisma longicaudata: Comparison with those seen in flying insects. *Development Genes and Evolution, 208*(7), 357–368.

Truman, J. W., Moats, W., Altman, J., Marin, E. C., & Williams, D. W. (2010). Role of Notch signaling in establishing the hemilineages of secondary neurons in Drosophila melanogaster. *Development, 137*(1), 53–61.

Tsuji, T., Hasegawa, E., & Isshiki, T. (2008). Neuroblast entry into quiescence is regulated intrinsically by the combined action of spatial Hox proteins and temporal identity factors. *Development, 135*(23), 3859–3869.

Vaadia, R. D., Li, W., Voleti, V., Singhania, A., Hillman, E. M. C., & Grueber, W. B. (2019). Characterization of proprioceptive system dynamics in behaving Drosophila larvae using high-speed volumetric microscopy. *Current Biology*, *29*(6), 935–944.e4.
Venkatasubramanian, L., & Mann, R. S. (2019). The development and assembly of the Drosophila adult ventral nerve cord. *Current Opinion in Neurobiology*, *56*, 135–143. Elsevier Ltd.
Woodworth, M. B., Girskis, K. M., & Walsh, C. A. (2017). Building a lineage from single cells: Genetic techniques for cell lineage tracking. *Nature Reviews Genetics*, *18*(4), 230–244. Nature Publishing Group.
Wreden, C. C., Meng, J. L., Feng, W., Chi, W., Marshall, Z. D., & Heckscher, E. S. (2017). Temporal cohorts of lineage-related neurons perform analogous functions in distinct sensorimotor circuits. *Current Biology*, *27*, 1521–1528.
Zarin, A. A., Mark, B., Cardona, A., Litwin-Kumar, A., & Doe, C. Q. (2019). A multilayer circuit architecture for the generation of distinct locomotor behaviors in Drosophila. *eLife*, *8*.
Zeilig, G., Enosh, S., Rubin-Asher, D., Lehr, B., & Defrin, R. (2012). The nature and course of sensory changes following spinal cord injury: Predictive properties and implications on the mechanism of central pain. *Brain*, *135*(2), 418–430.

CHAPTER ELEVEN

The development of somatosensory neurons: Insights into pain and itch

Suna L. Cranfill and Wenqin Luo*
Department of Neuroscience, Perelman School of Medicine, University of Pennsylvania, Philadelphia, PA, United States
*Corresponding author: e-mail address: luow@pennmedicine.upenn.edu

Contents

Abstract

Primary nociceptors are a heterogeneous class of peripheral somatosensory neurons, responsible for detecting noxious, pruriceptive, and thermal stimuli. These neurons are further divided into several molecularly defined subtypes that correlate with their functional sensory modalities and morphological features. During development, all nociceptors arise from a common pool of embryonic precursors, and then segregate progressively into their mature specialized phenotypes. In this review, we summarize the intrinsic transcriptional programs and extrinsic trophic factor signaling mechanisms that interact to control nociceptor diversification. We also discuss how recent transcriptome profiling studies have significantly advanced the field of sensory neuron development.

1. Introduction

Pain and itch are sensory mechanisms essential for protecting the body from potentially harmful external agents. Although pain and itch elicit distinct sensory experiences and behavioral responses—rubbing/withdrawal

Current Topics in Developmental Biology, Volume 142
ISSN 0070-2153
https://doi.org/10.1016/bs.ctdb.2020.10.005

for pain, and scratching for itch—there is extensive overlap in the neurons and circuits mediating these sensations. Both noxious and pruriceptive (itchy) stimuli are detected by specialized sensory neurons of the dorsal root ganglia (DRG) and trigeminal ganglia (TG) called nociceptors, of which pruriceptors are considered to be a subset. Like all primary somatosensory neurons, nociceptors give rise to a single pseudounipolar axon that bifurcates into a peripheral branch, which innervates skin or other target organs, and a central branch, which terminates in the spinal cord dorsal horn or brainstem (Fig. 1). Nociceptors comprise a heterogeneous population that has historically been classified into several different subtypes based on histological and electrophysiological characteristics (Fig. 1). Thinly-myelinated Aδ fibers have intermediate-diameter cell bodies/axons and transmit fast noxious stimuli. Unmyelinated C fibers have small-diameter cell bodies/axons and transmit slow pain or itch sensations. To add complexity, combinatorial expression of molecular markers, including sensory receptors and channels, further define several molecular and functional nociceptor subclasses among both Aδ and C populations (Fig. 1).

A fundamental question in the field of sensory neuron development is how such an astonishing molecular and functional diversity of primary nociceptors is generated from a common progenitor pool of neural crest cells. In this review, we provide an overview of intrinsic and extrinsic mechanisms that drive nociceptive sensory neuron specification and give rise to specialized nociceptor populations. Most of our knowledge of sensory neuron development has been established by classical experimental techniques, such as cell lineage tracing and phenotype analysis of genetic knockout or knockin models. With recent advances in next-generation sequencing technology, several large-scale transcriptome profiling studies on DRG neurons throughout development and in adulthood have been produced in the past few years, which have provide additional insights into the molecular categorization and genesis of sensory neuron subtypes (Fig. 1). Where possible, we highlight key insights into nociceptor differentiation provided by these newly available transcriptome databases. Finally, we discuss translational research on clinical disorders that have contributed significantly to our understanding of conserved molecular mechanisms underlying nociceptor development.

2. Molecular classification of nociceptor subtypes

In the mature mouse DRG, nociceptors have been classically divided into peptidergic and nonpeptidergic subsets (Fig. 1). Peptidergic neurons are

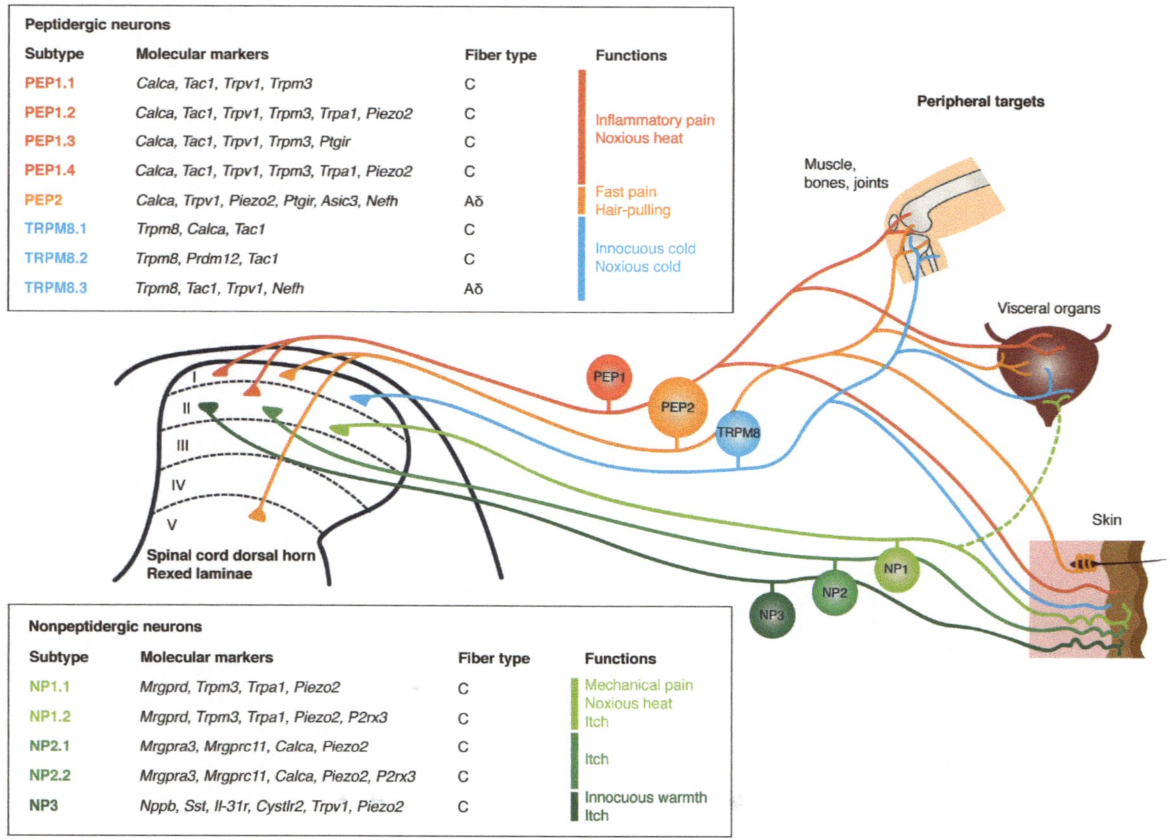

Peptidergic neurons

Subtype	Molecular markers	Fiber type	Functions
PEP1.1	*Calca, Tac1, Trpv1, Trpm3*	C	Inflammatory pain Noxious heat
PEP1.2	*Calca, Tac1, Trpv1, Trpm3, Trpa1, Piezo2*	C	
PEP1.3	*Calca, Tac1, Trpv1, Trpm3, Ptgir*	C	
PEP1.4	*Calca, Tac1, Trpv1, Trpm3, Trpa1, Piezo2*	C	
PEP2	*Calca, Trpv1, Piezo2, Ptgir, Asic3, Nefh*	Aδ	Fast pain Hair-pulling
TRPM8.1	*Trpm8, Calca, Tac1*	C	Innocuous cold Noxious cold
TRPM8.2	*Trpm8, Prdm12, Tac1*	C	
TRPM8.3	*Trpm8, Tac1, Trpv1, Nefh*	Aδ	

Nonpeptidergic neurons

Subtype	Molecular markers	Fiber type	Functions
NP1.1	*Mrgprd, Trpm3, Trpa1, Piezo2*	C	Mechanical pain Noxious heat Itch
NP1.2	*Mrgprd, Trpm3, Trpa1, Piezo2, P2rx3*	C	
NP2.1	*Mrgpra3, Mrgprc11, Calca, Piezo2*	C	Itch
NP2.2	*Mrgpra3, Mrgprc11, Calca, Piezo2, P2rx3*	C	
NP3	*Nppb, Sst, Il-31r, Cystlr2, Trpv1, Piezo2*	C	Innocuous warmth Itch

Fig. 1 Illustration of nociceptive neuron subtypes within the mouse DRGs. Traditional classifications have divided nociceptors into peptidergic and nonpeptidergic subpopulations, which display distinct patterns of innervation. Centrally, C fiber peptidergic neurons innervate laminae I and IIo, while nonpeptidergic neurons innervate lamina IIi. Aδ nociceptive fibers, most of which are peptidergic, innervate laminae I and V. Peripherally, peptidergic neurons innervate skin and deeper tissues, while the majority of nonpeptidergic neurons preferentially innervate the skin. Molecularly defined nociceptor subtypes in mouse DRG, based on recent transcriptome profiling studies, are listed in the boxes, along with their major markers and putative sensory functions.

characterized by their expression of classical neuropeptides, such as calcitonin gene-related peptide (CGRP, encoded by *Calca*) and substance P (SP, encoded by *Tac1*), and are marked by persistent expression of the nerve growth factor (NGF) receptor NTRK1 (Bennett, Averill, Clary, Priestley, & McMahon, 1996; Rosenfeld et al., 1983). Adult nonpeptidergic neurons express the glial-derived neurotrophic factor (GDNF) receptor RET, and most of this population binds isolectin B4 (IB4) (Molliver et al., 1997). It should be noted that most subtypes traditionally defined as nonpeptidergic contain neuropeptides, despite the nomenclature. In addition to their distinct molecular profiles, these two classes of nociceptors differ anatomically and functionally. Most nonpeptidergic nociceptors preferentially innervate the epidermis, while peripheral targets for peptidergic nociceptors include both skin and deeper tissues, such as dermis, muscle, bone, and visceral organs (Yang et al., 2013; Zylka, Rice, & Anderson, 2005). In the spinal cord dorsal horn, peptidergic neurons largely terminate in lamina I and the outer/dorsal part of lamina II (I/IIo), while nonpeptidergic neurons terminate in the inner part of lamina II (IIi) (Braz, Nassar, Wood, & Basbaum, 2005).

Peptidergic and nonpeptidergic nociceptor classes can be further divided into several subtypes, based on their combinatorial expression of ion channels and receptors. Within the DRG, such channels and receptors confer sensitivity to different classes of stimuli. Examples of notable markers include members of the transient receptor potential (TRP) channel family, which directly detect a variety of noxious stimuli and function downstream of sensory G-protein coupled receptors (GPCRs), and the Mas-related GPCR family (MRGPRs), which transduce a wide variety of endogenous and exogenous itch signals (Julius, 2013; Meixiong & Dong, 2017). Other important receptor types implicated in itch include histamine receptors, serotonin receptors, cytokine receptors, and toll-like receptors (Bautista, Wilson, & Hoon, 2014; Dong & Dong, 2018).

Several large-scale efforts to characterize the transcriptomes of mouse DRG neurons have further clarified the molecular categorization of nociceptors (Chiu et al., 2014; Li et al., 2016; Sharma et al., 2020; Usoskin et al., 2015; Zeisel et al., 2018; Zheng et al., 2019). For this review, we adopt a classification system derived from an unbiased single-cell RNA sequencing (scRNA-seq) experiment, which applied principal component analysis to define 11 major subtypes of sensory neurons (Emery & Ernfors, 2018; Usoskin et al., 2015). Parallel scRNA-seq analyses have confirmed the general categories delineated by Usoskin et al., with additional subclusters identified within heterogeneous thermoreceptor and nociceptor populations

(Li et al., 2016; Sharma et al., 2020; Zeisel et al., 2018). Among the peptidergic nociceptor subtypes, which express *Tac1*, PEP1 are primarily thermosensitive C nociceptors, and PEP2 are thinly-myelinated Aδ nociceptors based on expression of the A fiber-associated marker NEFH. Nonpeptidergic subtypes include NP1, which express *Mrgprd*; NP2, which express *Mrgpra3* and *Mrgprc11*; and NP3, which express the interleukin-31 receptor A (*IL-31r*) as well as neuropeptides involved in spinal cord itch transmission (natriuretic polypeptide B, *Nppb*; somatostatin, *Sst*). Based on their molecular identities, all three NP subtypes are likely itch-sensitive pruriceptors. Finally, cold-sensitive neurons expressing *Trpm8* comprise a unique, heterogeneous population of sensory neurons that includes both C fibers and Aδ fibers.

It is important to note that while these defined subtypes are derived from DRG collected at all spinal levels, sensory neurons from different segments and from TG may vary in gene expression to meet the requirements of different target tissues (Sharma et al., 2020; Smith-Anttila et al., 2020). There are also potentially major divergences between rodent and human DRG neurons that should be considered when using transcriptomic data in translational research (Haberberger, Barry, Dominguez, & Matusica, 2019; Rostock, Schrenk-Siemens, Pohle, & Siemens, 2018). For example, mice have a unique genomic expansion of the MRGPR family that is not present in other rodent species, including rats and gerbils, or in humans (Zylka, Dong, Southwell, & Anderson, 2003). In any case, comprehensive transcriptome analyses of mouse DRG neurons have contributed significantly to our knowledge of the complex molecular profiles delineating sensory neuron subtypes. In addition, recent studies have sought to characterize gene expression dynamics throughout sensory neuron development by performing scRNA-seq on neural crest cells and sensory neurons collected at critical developmental time points (Faure et al., 2020; Sharma et al., 2020). These experiments have provided a huge amount of data on sensory neuron fate specification, diversification, and the developmental trajectories of molecular marker expression in different sensory neuron subtypes. Interested readers are encouraged to explore these databases for further analysis.

3. Sensory neurogenesis and early nociceptor development

All somatosensory neurons of the DRG, including nociceptors, are derived from neural crest cells (NCCs) during early stages of embryogenesis

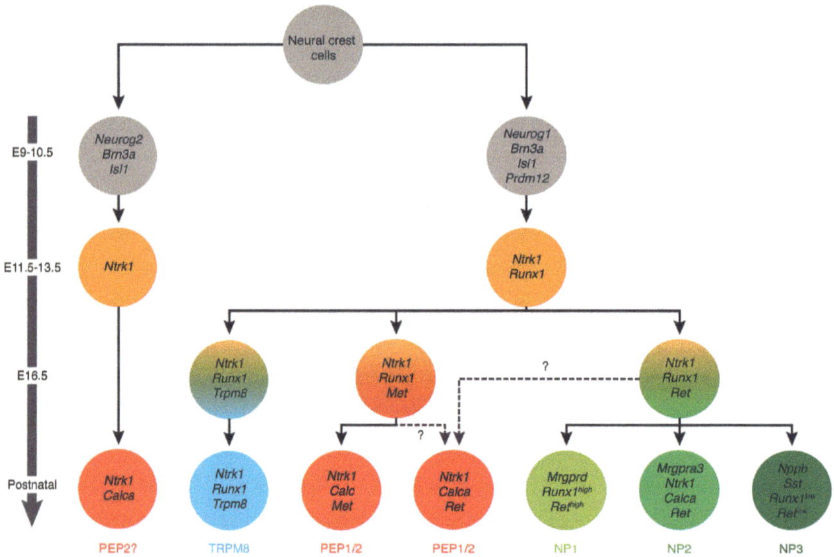

Fig. 2 Model of nociceptor diversification during DRG development in the mouse. Major known genes involved in the specification of each subtype are listed at each stage. The bottom row shows the most likely molecular subtype classification corresponding to each lineage. Dashed lines and question marks indicate unclear assignments.

(Fig. 2), while TG neurons are derived from both NCCs and cranial placodes (D'Amico-Martel & Noden, 1983; Hamburger, 1961). These NCCs are transient, multipotent stem cells that delaminate from the dorsal neural tube and migrate along a ventral pathway between embryonic day (E) 8.5 and E10 in mouse (Bronner-Fraser & Fraser, 1988; Frank & Sanes, 1991; Serbedzija, Fraser, & Bronner-Fraser, 1990). As the embryo develops from E9.5 to E13.5, these cells condense into pairs of ganglia for each spinal segment and exit the cell cycle (Kasemeier-Kulesa, Kulesa, & Lefcort, 2005; Lawson & Biscoe, 1979). The timing of NCC delamination from the neural tube is closely linked to segregation of major sensory neuron subtypes. During NCC migration, successive waves of neurogenesis are orchestrated by the proneural basic helix-loop-helix (bHLH) transcription factors neurogenin 2 (*Neurog2*) and 1 (*Neurog1*) (Frank & Sanes, 1991; Lawson & Biscoe, 1979; Ma, Fode, Guillemot, & Anderson, 1999; Marmigère & Ernfors, 2007). The first wave of neurogenesis (E9.5–E11.5) primarily contributes large-diameter myelinated neurons expressing NTRK3 (receptor for neurotrophin 3, NT-3), NTRK2 (receptor for brain-derived growth factor, BDNF), or RET (Bourane et al., 2009;

Luo, Enomoto, Rice, Milbrandt, & Ginty, 2009; Ma et al., 1999). These neurons will eventually become proprioceptors, which sense body position and movement, and low-threshold mechanoreceptors (LTMRs), which sense innocuous touch, pressure, and vibration. The second wave (E10.5–E13) mainly generates Aδ and C afferents, all of which express NTRK1. These late-born neurons are fated to become nociceptors, pruriceptors, and C-LTMRs, which may sense pleasant touch. A third wave of neurogenesis originates at E10.75 from boundary cap cells, an intermediate group of NCC derivatives positioned at the border of the spinal cord and peripheral nerve roots (Hjerling-Leffler et al., 2005; Maro et al., 2004). These progenitors form a small percentage of $NTRK1^{+}$ nociceptive neurons.

Both neurogenins are clearly required for sensory neurogenesis. No DRG neurons are generated in *Neurog1*$^{-/-}$ and *Neurog2*$^{-/-}$ double knockout mice (Ma et al., 1999; Ventéo et al., 2019). In sensory neurons, *Neurog1* and *Neurog2* are also required for expression of downstream bHLH factors *Neurod1* and *Neurod4*, which are important for neuronal differentiation (Ma, Chen, del Barco Barrantes, de la Pompa, & Anderson, 1998). The first and second NCC waves have classically been associated with *Neurog2* and *Neurog1* expression, respectively, due to the sequential timing of the onset of their expression, and the phenotypes produced in loss-of-function experiments. *Neurog1*$^{-/-}$ single knockout mice specifically lack $NTRK1^{+}$ small nociceptors in trigeminal ganglia and cervical DRG, although some $NTRK1^{+}$ neurons are present in posterior DRG (Ma et al., 1999). In *Neurog2*$^{-/-}$ knockout mice, an initial loss of $NTRK2^{+}$ and $NTRK3^{+}$ neurons is later compensated by *Neurog1*-dependent precursors in cervical DRG, but trunk DRG show a reduction in all neuronal subtypes (Ma et al., 1999; Ventéo et al., 2019). However, recent scRNA-seq studies suggest that progenitors from both waves consecutively express *Neurog2* and then *Neurog1* (Faure et al., 2020; Sharma et al., 2020). Indeed, in second-wave precursors, *Neurog2* is transiently expressed and appears to be required for an initial phase of *Neurog1* expression, during which it biases progenitors toward a somatosensory rather than melanocyte lineage (Ventéo et al., 2019). In mice that lack both *Neurog2* and second-wave progenitors, Neurog2-independent expression of *Neurog1* compensates for *Neurog2* deficiency after a delay, generating primarily $NTRK1^{+}$ nociceptors instead of $NTRK2^{+}$ and $NTRK3^{+}$ mechano/proprioceptors (Ventéo et al., 2019). Altogether, these observations suggest that neurogenins play a critical role in controlling the timing of neurogenesis of sensory neurons and specification of a sensory neuron fate.

Intriguingly, specific molecular pathways governing nociceptor specification may already exist in NCC progenitors. *Prdm12* is a member of the PRDI-BF1 and RIZ homology domain (PRDM) family of transcriptional regulators (Hohenauer & Moore, 2012). *Prdm12* is expressed by nearly all second-wave progenitors by E9.5 and is required for the maintenance, but not initiation, of *Neurog1* expression in this population (Bartesaghi et al., 2019; Chen et al., 2015; Desiderio et al., 2019). In later developmental stages and up to adulthood, *Prdm12* is restricted to all nociceptive neuron subtypes within the DRG (Bartesaghi et al., 2019; Desiderio et al., 2019; Sharma et al., 2020; Usoskin et al., 2015; Zeisel et al., 2018). Conditional deletion of *Prdm12* around E14.5 using the *Advillin-Cre* driver shows a specific loss of NTRK1 lineage neurons, likely due to a critical role for *Prdm12* in the initiation and maintenance of NTRK1 expression in embryonic nociceptors, which in turn is required for nociceptor survival and differentiation (Bartesaghi et al., 2019; Desiderio et al., 2019). On the other hand, gain-of-function experiments in both chick and *Xenopus* embryos suggest *Prdm12* is sufficient to repress a mechano/proprioceptor fate, and its expression can be induced by the transcription factor *Zic1* (Bartesaghi et al., 2019; Desiderio et al., 2019; Jaurena, Juraver-Geslin, Devotta, & Saint-Jeannet, 2015). Although *Prdm12* is not sufficient to induce nociceptor differentiation, it may modulate neurogenin activity to promote NTRK1 expression and nociceptor identity (Bartesaghi et al., 2019; Desiderio et al., 2019). These results support the existence of molecular programs committing early precursors to a nociceptive fate, although several outstanding questions remain. First, how does *Prdm12*, which lacks intrinsic histone methyltransferase activity, regulate the genetic programs specifying a nociceptor identity? Second, in addition to *Zic1*, what upstream signals converge to restrict *Prdm12* expression to nociceptive lineages during early differentiation? Finally, are other intrinsic determinants of sensory neuron subtype identity present in progenitors, before the segregation of major transcription factors controlling subtype-specific programs (e.g., nociceptor-specific Runx1 and proprioceptor-specific Runx3) (Sharma et al., 2020)?

Just prior to exiting the cell cycle, all sensory neurons express the homeodomain transcription factors *Brn3a* and *Isl1*. These factors facilitate the transition out of the neurogenic phase by repressing expression of the bHLH genes and promoting genes involved in terminal differentiation of sensory neurons (Eng, Dykes, Lanier, Fedtsova, & Turner, 2007; Lanier, Dykes, Nissen, Eng, & Turner, 2009; Ma, Lei, Eng, Turner, & Parada, 2003; Sun et al., 2008). Both factors are also required to suppress transcription factors normally associated with spinal cord (*Isl1*) and cardiac and cranial mesoderm (*Brn3a*) (Lanier et al., 2009; Sun et al., 2008), further supporting their role in biasing progenitors towards a sensory neuron fate.

After sensory neurogenesis, all embryonic nociceptors express the high-affinity NGF receptor NTRK1, and NGF/NTRK1 signaling is essential for their survival and specification (Crowley et al., 1994; Lei et al., 2005; Luo et al., 2007; Molliver & Snider, 1997; Moqrich et al., 2004; Patel, Jackman, Rice, Kucera, & Snider, 2000; Silos-Santiago et al., 1995; Smeyne et al., 1994; White et al., 1996). Mice deficient for either *NGF* or *Ntrk1* exhibit a profound loss of NTRK1 lineage neurons, and limited production of target-derived NGF at the periphery may contribute to determining nociceptor density (Crowley et al., 1994; Davies, 1996; Deppmann et al., 2008; Ginty & Segal, 2002; Lei et al., 2005; Molliver & Snider, 1997; Silos-Santiago et al., 1995; Smeyne et al., 1994; White et al., 1996). Moreover, nociceptors of $Ngf^{-/-}/Bax^{-/-}$ and $Ntrk1^{-/-}/Bax^{-/-}$ double mutant mice, in which *NGF* or *Ntrk1* is deleted in an apoptosis-deficient background, fail to establish cutaneous innervation and do not express the peptidergic markers CGRP and SP, revealing a critical role for NGF/NTRK1 signaling in nociceptor differentiation (Patel et al., 2000). Several studies have also sought to establish the specific downstream second-messenger pathways mediating the diverse functions of NGF/NTRK1 in nociceptor development. Upon NGF binding, the NTRK1 receptor dimerizes or oligomerizes, autophosphorylates, and initiates several local signaling cascades (Denk, Bennett, & McMahon, 2017; Lemmon & Schlessinger, 2010). These include phosphatidylinositol-3 kinase (PI3K), protein kinase B (AKT), and the mitogen-activated protein kinases (MAPK) ERK1, ERK2, and ERK5, which in turn activate downstream effectors to control various processes within the peripheral terminal. Substantial evidence also supports a "signaling endosome" model for retrograde NTRK1 signaling, in which NGF-NTRK1 complexes at distal terminals are internalized into endosomes that include activated members of these signal transduction pathways (Ascano, Bodmer, & Kuruvilla, 2012; Ginty & Segal, 2002; Yamashita & Kuruvilla, 2016). These signaling endosomes undergo long-distance retrograde transport along the axon to neuronal cell bodies, where they are able to further mediate the physiological functions of NGF/NTRK1 signaling by regulating expression of genes crucial for neuronal survival and differentiation.

Which downstream signaling pathways orchestrate NTRK1's diverse functions? Conditional deletion of *Erk5* in neural crest cells produced no obvious phenotype in DRG sensory neurons (Newbern et al., 2011). In addition, conditional knockout of *Erk1/2* in sensory neurons, upstream signaling components *B-raf* and *C-raf*, or the downstream effectors *Srf* and *Nfat* only perturbs NGF-mediated peripheral innervation, but is otherwise dispensable for nociceptor survival and maturation (Graef et al., 2003; Newbern et al., 2011; Wickramasinghe et al., 2008; Zhong et al., 2007). Moreover, the epidermal innervation phenotype in *Erk1/2* mutant mice is less severe than that of *Ntrk1*$^{-/-}$/*Bax*$^{-/-}$ mutants (Newbern et al., 2011; Patel et al., 2000). These observations raise the possibility that other signaling pathways downstream of NTRK1 are required for nociceptor peripheral terminal development, as well as neuronal viability and subtype specification. One potential mediator of NGF-dependent nociceptor survival is the transcription factor CREB, which is also required for establishing peripheral target innervation (Lonze, Riccio, Cohen, & Ginty, 2002). Other parallel pathways may also contribute to NGF-dependent development of NTRK1 lineage neurons, such as signaling through the promiscuous neurotrophin receptor p75^{NTR}, which binds NGF with low affinity and enhances NTRK1 activity in sensory neurons (Bergmann et al., 1997; Chen et al., 2017; Hantzopoulos, Suri, Glass, Goldfarb, & Yancopoulos, 1994; Hempstead, 2002; Lee et al., 1992; Lee, Davies, & Jaenisch, 1994; Makkerh et al., 2005; Murray, Bartlett, & Cheema, 1999). Indeed, p75^{NTR} may promote survival of NTRK1$^+$ nociceptors during a developmental window of apoptotic pruning (E12.5–14.5), although *Isl1*-mediated excision of *p75* around E12.5 appears to selectively reduce a subset of RET$^+$ nonpeptidergic nociceptors (*Barrett & Bartlett, 1994*; Chen et al., 2017; Cheng, Jin, Rose, & Deppmann, 2018). The precise contribution of different signal transduction cascades to NGF/NTRK1-mediated developmental mechanisms, as well as the role of other NGF-dependent pathways, within each nociceptor subtype requires further investigation.

4. Specification of nociceptor subtypes

During perinatal and postnatal development, embryonic NTRK1 lineage neurons proceed to undergo differentiation into peptidergic and nonpeptidergic subtypes. Broadly, peptidergic neurons (PEP1/PEP2, and possibly TRPM8) maintain NTRK1 expression into maturity, while nonpeptidergic neurons (NP1/2/3) exhibit a postnatal switch in trophic factor dependence by downregulating NTRK1 and expressing RET. The runt domain transcription factor *Runx1* is a key regulator of nociceptor segregation (Abdel Samad et al., 2010; Chen et al., 2006; Kramer et al., 2006; Marmigère et al., 2006; Qi et al., 2017; Yoshikawa et al., 2007). *Runx1* is expressed in most NTRK1$^+$ neurons shortly after NTRK1 expression begins in embryonic nociceptors (Chen et al., 2006). *Runx1* persists in the ~50% of NTRK1 lineage neurons that are fated to become nonpeptidergic, but is extinguished in prospective peptidergic neurons (Chen et al., 2006). The mechanisms by which *Runx1* orchestrates segregation of peptidergic and nonpeptidergic neurons, and further hierarchical differentiation of NP subtypes, are summarized according to each nociceptor class below.

How is dynamic expression of *Runx1* achieved in embryonic nociceptors? Although NTRK1 expression precedes the onset of *Runx1*, NGF/NTRK1 signaling appears to be required for maintenance of *Runx1* but not its initiation (Huang, O'Donovan, Turner, Zhong, & Ginty, 2015; Luo et al., 2007). *Ntrk1* and *Runx1* may share mechanisms of induction, as the transcription factors *Brn3a* and *Isl1* have both been implicated in controlling their expression (Dykes, Lanier, Eng, & Turner, 2010; Dykes, Tempest, Lee, & Turner, 2011; Eng, Lanier, Fedtsova, & Turner, 2004; Huang et al., 2015; Ma et al., 2003; Sun et al., 2008; Zou, Li, Klein, & Xiang, 2012). However, *Brn3a/Isl1* are broadly expressed by sensory neuron precursors, raising questions as to how they might confer specificity for the nociceptive lineage. Huang et al. proposed a model in which target-derived NGF signaling and *Isl1*-mediated initiation of *Runx1* expression converge to control specification of nonpeptidergic nociceptors (Huang et al., 2015). Analysis of *Ngf*$^{-/-}$/*Bax*$^{-/-}$ double mutant mice revealed that expression of core binding factor-β (CBFβ), a cofactor that forms a complex with RUNX1 to augment transcriptional activity in sensory neurons and is necessary for *Runx1*-dependent gene expression, requires NGF signaling via the ERK1/2 signaling cascade. Initiation of *Runx1* expression, however, requires *Isl1* but not NGF, suggesting *Runx1* is controlled by intrinsic transcriptional mechanisms. In this way, independent pathways regulate different components of the RUNX1/CBFβ complex, a key determinant of nonpeptidergic nociceptor subtype identity, to achieve subtype-specific differentiation. Interestingly, local induction signals within the DRG may also participate in controlling *Runx1* expression, and therefore nociceptor diversification (Hadjab et al., 2013). Members of the fibroblast growth factor (FGF) family are expressed by sensory neuron precursors at E11.5, which coincides with the onset of *Runx1* induction, and ablation of early-born *Neurog2*-dependent neurons before this time point leads to a persistent reduction in NTRK1$^+$ and RUNX1$^+$ neurons (Hadjab et al., 2013). Conditional knockout of *Fgfr1* in sensory precursors resulted in a specific but transient decrease in NTRK1$^+$/RUNX1$^+$ neurons, suggesting other mechanisms later compensate to induce NTRK1/RUNX1 expression in the absence of FGF signaling (Hadjab et al., 2013). These experiments raise the intriguing possibility that early-born neurons participate in the specification of late-born neurons through local secretion of signaling molecules. While additional transcriptional determinants of nociceptor identity remain to be identified, it will be interesting to further elucidate how intrinsic and extrinsic mechanisms might converge to regulate subtype-specific differentiation.

4.1 NTRK1$^+$ peptidergic subtypes: PEP1, PEP2, and TRPM8

In adult mice, PEP1 neurons function as important mediators of thermal and inflammatory pain. Supporting evidence includes their expression of the TRP channel family members *Trpv1*, which is activated by noxious heat (>42 °C), capsaicin, protons, and toxins, and *Trpm3*, which is activated by noxious heat (>42 °C) and pregnenolone sulfate (Julius, 2013; Vriens et al., 2011). Although PEP1 nociceptors have been traditionally grouped together by their shared expression of *Calca* and *Tac1*, further molecular analyses have revealed considerable heterogeneity (Fig. 1) (Sharma et al., 2020; Zeisel et al., 2018). For example, *Trpa1*, which responds to chemical irritants like mustard oil, is expressed in two of the four PEP1 subclusters identified in a recent resequencing study (Zeisel et al., 2018). The functional importance of such molecular diversity among the PEP1 group (PEP1.1–PEP1.4) remains to be determined. The PEP2 cluster lacks *Tac1* but expresses *Calca* and *Trpv1*, and may correspond to Aδ fibers mediating mechanical pain associated with hair pulling (Ghitani et al., 2017).

Extinction of *Runx1* expression in immature PEP1/2 neurons is required for their differentiation, as constitutive *Runx1* expression in sensory neurons abolishes NTRK1 expression and dramatically reduces CGRP expression (Abdel Samad et al., 2010; Kramer et al., 2006). For a subset of peptidergic nociceptors, this process appears to depend on *Met*, a receptor tyrosine kinase for hepatocyte growth factor (HGF) (Gascon et al., 2010). In *Met* conditional knockout mice, prospective NTRK1$^+$/MET$^+$ neurons fail to extinguish *Runx1* expression and lose markers of peptidergic neuron identity, such as CGRP (Gascon et al., 2010). In turn, *Runx1* is required for *Met* suppression, as DRG of *Runx1* conditional knockout exhibit an expansion of MET$^+$ neurons (Gascon et al., 2010). Altogether, these observations suggest *Met* and *Runx1* form a repressive genetic loop, in which *Met* persistence promotes a peptidergic fate, while *Runx1* persistence promotes a nonpeptidergic fate. It should be noted, however, that not all peptidergic nociceptors appear to extinguish *Runx1*, as a fraction of peptidergic neurons (PEP1.1 and TRPM8) exhibits detectable *Runx1* expression in adulthood (Abdel Samad et al., 2010; Usoskin et al., 2015; Zeisel et al., 2018). Moreover, some nociceptors have not been shown to require *Met* during their development. A small proportion of NTRK1$^+$ peptidergic neurons that does not express MET or RET may develop from early-born NTRK1 lineage neurons, which are generated during the *Neurog2*-dependent first wave and mature into Aδ nociceptors marked by expression of the transcription factor *Cux2* (Bachy et al., 2011; Gascon et al., 2010). These neurons may correspond with a subset of NTRK1$^+$ neurons that do not express *Runx1* at E14.5 (Bachy et al., 2011; C.-L. Chen et al., 2006), which may explain why they establish a peptidergic identity.

While *Runx1* downregulation is a critical component of PEP1/2 development, it is important to recognize that *Runx1* likely participates in the differentiation of these populations during its early broad expression. *Runx1* knockout mice show a marked loss of *Trpa1* expression and high levels of *Trpv1* expression, which are both features of PEP1/2 neuronal subtypes (Abdel Samad et al., 2010; Chen et al., 2006). This observation suggests the transient presence of *Runx1* in prospective PEP1/2 neurons is directly required for their molecular maturation. Interestingly, *Isl1*-null mice show a complete loss of *Trpv1*, supporting a role for *Isl1* in broadly coordinating *Trpv1* expression while Runx1 establishes high *Trpv1* expression in a subset of nociceptors (Sun et al., 2008).

The TRPM8 population has been proposed to form a labeled line for cold sensation, as the TRPM8 channel is activated by cold temperatures (<26 °C) and cooling agents such as menthol (Julius, 2013). TRPM8 neurons do not conform to traditional definitions of nociceptor classifications, reflecting diverse physiological functions in detecting both innocuous cool temperatures and noxious cold. All three TRPM8 subgroups express *Tac1*, which is consistent with a peptidergic categorization, but are also marked by expression of *Runx1*, which plays a key role in nonpeptidergic neuron specification (Abdel Samad et al., 2010; Sharma et al., 2020; Zeisel et al., 2018). However, as TRPM8 neurons do not appear to express $Na_v1.8$, a voltage-gated sodium channel preferentially found in nociceptors, many developmental studies using *$Na_v1.8$-Cre* to conditionally ablate genes in nociceptors may not cover TRPM8 neurons (Luiz et al., 2019). *Prdm12*, a determinant of nociceptor identity, is enriched only in the TRPM8.2 population, suggesting the other TRPM8 subtypes may not be nociceptive neurons (Bartesaghi et al., 2019; Chen et al., 2015; Desiderio et al., 2019). On the other hand, TRPM8.3 is the only subtype that coexpresses *Trpv1*. While it has been hypothesized that these $TRPV1^{+}/TRPM8^{+}$ neurons are noxious cold sensors, the lack of other nociceptive markers in the TRPM8.3 group precludes a clear functional assignment at this time (Julius, 2013). Despite this subtype's unique features, TRPM8 neurons appear to be regulated by some of the same molecular mechanisms governing classical nociceptor development. NGF/NTRK1 signaling, *Isl1*, and *Runx1* are important for the establishment of a subset of $TRPM8^{+}$ neurons. All TRPM8 subgroups maintain *Runx1* expression in maturity, and a subset of TRPM8 neurons also continues to express NTRK1 (Kobayashi et al., 2005; Sharma et al., 2020; Takashima, Ma, & McKemy, 2010; Zeisel et al., 2018). Moreover, *Runx1* conditional knockout mice exhibit a complete reduction of *Trpm8* expression, while *Islet1* conditional knockout and $Ngf^{-/-}/Bax^{-/-}$ mutant mice show a partial reduction (Abdel Samad et al., 2010; Chen et al., 2006; Lopes, Liu, Xu, & Ma, 2012; Luo et al., 2007). Curiously, one characterization of TRPM8 lineage neurons throughout development found that the NTRK1-coexpressing population increased from ~5% to ~30% from E16.5 to adulthood, providing an explanation as to why *Trpm8* expression is not completely abolished in *Ngf*-deficient mice (Takashima et al., 2010). A unique mechanism controlling NTRK1 expression dynamics in TRPM8 neurons may exist, although further studies are needed to elucidate the role of trophic factor signaling in TRPM8 neuron specification.

An essential feature of nociceptor diversity is that different subtypes display different patterns of peripheral and central innervation, which is developmentally regulated by both extrinsic and intrinsic signals. Peripherally, while the majority of nonpeptidergic nociceptor subtypes innervate the skin, peptidergic nociceptors target the skin as well as deep tissues, such as muscle, bone, and viscera (Hockley et al., 2019; Smith-Anttila et al., 2020; Song et al., 2009; Yang et al., 2013). In the spinal cord, peptidergic and TRPM8 neurons preferentially terminate in superficial laminae (lamina I and IIo), while nonpeptidergic neurons segregate to lamina II (Braz et al., 2005). Evidence suggests that NGF/NTRK1 signaling is a critical extrinsic mediator of peripheral target innervation in nociceptors. In $Ngf^{-/-}/Bax^{-/-}$ double mutant mice, nociceptive neurons fail to correctly innervate the epidermis (Crowley et al., 1994). In addition, replacing the intracellular domain of NTRK1 with that of NTRK3, which is expressed in proprioceptors, also leads to a reduction in peripheral nociceptor projections (Gorokhova et al., 2014). Intrinsic control from *Runx1* is also essential for both peripheral and central terminal establishment. Loss of *Runx1* leads to a marked reduction in cutaneous innervation, and *Runx1*-deficient IB4$^+$ nociceptors aberrantly target more superficial layers of the dorsal horn than wild-type IB4$^+$ neurons would (Abdel Samad et al., 2010; Chen et al., 2006; Hadjab et al., 2013; Yang et al., 2013).

Interestingly, *Runx1* may also play a role in determining which peripheral tissues are targeted by different nociceptor populations. Conditional deletion of *Runx1* reduces epidermal fiber density but does not affect sensory neuron innervation of visceral targets, while constitutive *Runx1* expression in DRG nociceptors markedly decreases deep tissue innervation, suggesting deep tissue-innervating NTRK1 lineage nociceptors must downregulate *Runx1* in order to project appropriately to their targets (Yang et al., 2013). Given the different physiological contexts of deep and cutaneous tissues, it is not surprising that they may require developmentally and functionally distinct nociceptor subtypes. Indeed, recent efforts to profile deep tissue-innervating sensory neurons have identified dedicated subpopulations enriched in deep tissue-specific markers, as well as divergences in molecularly defined categories between DRG from different spinal levels that are associated with different visceral regions (Hockley et al., 2019; Smith-Anttila et al., 2020; Yang et al., 2013). Further characterization of the development of deep tissue-innervating sensory neurons may yield key insights into their unique genetic and physiological properties.

4.2 RET$^+$ nonpeptidergic subtypes: NP1, NP2, and NP3

Nonpeptidergic neurons, which, despite their nomenclature, do express neuropeptides such as *Calca*, *Nmb*, and *Npbb*, are classically defined by their expression of RET and ability to bind IB4. Transcriptome profiling studies have identified three main categories of nonpeptidergic neurons, all of which are likely involved in transducing itch. MRGPRD$^+$ NP1 neurons respond to mechanical pain and noxious heat, but activation by the MRGPRD agonist β-alanine produces itch behavior and not pain, suggesting this population is polymodal (Cavanaugh et al., 2009; Liu et al., 2009). NP1 neurons are also the only nonpeptidergic subtype observed to innervate visceral tissue in addition to the epidermis (Bautzova et al., 2018; Hockley et al., 2019). MRGPRA3$^+$ NP2 neurons, which colabel with NTRK1, CGRP, and IB4, are likely itch-selective, as ablation of MRGPRA3$^+$

NP2 neurons specifically impairs itch behavior and not pain, and activation of this population triggers only itch behavior (Han et al., 2012). $SST^+/NPPB^+$ NP3 neurons, which express low levels of RET and do not bind IB4, are probably polymodal for both itch and non-noxious warmth, based on their unique expression of the warmth-sensing channel TRPM2 (Huang et al., 2018; Mishra & Hoon, 2013; Stantcheva et al., 2016; Tan & McNaughton, 2016).

Initial differentiation of NP1/2/3 neurons is defined by their developmental switch from NGF/NTRK1 dependence to GFL/RET dependence, as they begin expressing RET around E16 and progressively extinguish NTRK1 expression during the first few postnatal weeks of life (Molliver et al., 1997). $Ngf^{-/-}/Bax^{-/-}$ double mutant mice exhibit a dramatic reduction of Ret^+ neurons at P0 (Huang et al., 2015; Luo et al., 2007). *Runx1* conditional knockout mice, however, show a delayed loss of *Ret* expression during the first two postnatal weeks (Huang et al., 2015). Together, these observations suggest NGF/NTRK1 signaling is required for the initiation of *Ret* expression, and *Runx1* is required postnatally for maintenance of *Ret* expression. Just prior to the onset of *Ret* expression, prospective nociceptors begin establishing peripheral terminals and encounter target-derived cues that may instruct subtype specification. An attractive hypothesis is that peripheral endings are exposed to different levels of NGF depending on their targeted epidermal layer (Lallemend & Ernfors, 2012). Sensory neurons innervating superficial layers may be subject to stronger NGF/NTRK1 signaling, leading to increased *Runx1* and *Ret* expression and a bias towards a nonpeptidergic fate. On the other hand, neurons innervating deeper epidermal layers may encounter less NGF, which would permit peptidergic differentiation. This model highlights another key mechanism by which extrinsic target-specific signals coordinate with intrinsic determinants of identity to instruct sensory neuron specification.

During segregation of nonpeptidergic nociceptors from their peptidergic counterparts, GFL/RET signaling is essential for the extinction of *Ntrk1* expression and the establishment of nonpeptidergic identity (Luo et al., 2007). In *Ret* conditional knockout mice crossed to the *Wnt1-Cre* driver, which ablates *Ret* in all neural crest cells, postnatal *Ntrk1* extinction within the DRG is delayed until P14, at which point these mutant mice fail to survive (Luo et al., 2007). However, in *Ret* conditional knockout mice generated with $Na_v1.8$*-Cre*, which is expressed in nociceptors at E17, *Ret* appears to be dispensable for NTRK1 extinction (Golden et al., 2010). Late *Ret* conditional knockout mice also show a partial loss of nonpeptidergic neurons, highlighting a switch in trophic factor signaling dependence from NTRK1 to RET (Golden et al., 2010). Overall, these results suggest that *Ret* may be specifically required during an earlier embryonic window to initiate a program controlling postnatal suppression of NTRK1 expression. Alternatively, *Ret*-independent mechanisms may also contribute to NTRK1 extinguishment. It should be noted that both RET and NTRK1 converge on many of the same downstream signal transduction pathways, which may explain why several mouse knockout models of NTRK1 signaling components partially phenocopy features of *Ret* conditional knockout mice, including defects in peripheral innervation (Graef et al., 2003; Luo et al., 2007; Newbern et al., 2011; Wickramasinghe et al., 2008; Zhong et al., 2007).

Runx1 itself is a critical determinant of NP1/2/3 neuron identity. Loss of *Runx1* leads to a marked reduction in cutaneous innervation, and *Runx1*-deficient IB4$^+$ nociceptors aberrantly target more superficial layers of the dorsal horn than wild-type IB4$^+$ neurons would (Abdel Samad et al., 2010; Chen et al., 2006; Hadjab et al., 2013; Yang et al., 2013). *Runx1* controls the expression of a large cohort of nociceptive, pruriceptive, and thermoreceptive genes in nonpeptidergic populations, including but not limited to *Trpv1*, *Trpa1*, *Trpc3*, *Mrgrpr* family members, *Nppb*, and *Sst* (Abdel Samad et al., 2010; Chen et al., 2006; Liu et al., 2008; Lopes et al., 2012; Qi et al., 2017). A subset of these genes, such as *Trpa1* and *Mrgpra3*, are transcriptionally controlled by GFL/RET signaling downstream of *Runx1* (Luo et al., 2007). Although *Runx1* expression is broadly maintained in nonpeptidergic neurons during nociceptor segregation, it undergoes dynamic changes in expression and genetic activity to coordinate nonpeptidergic neuron diversification (Abdel Samad et al., 2010; Liu et al., 2008; Qi et al., 2017). *Runx1*, in concert with the homeodomain transcription factor *Tlx3*, is initially required to activate expression of *Mrgprd*, *Mrgprc11*, *Mrgpra3*, *Nppb*, and many other NP1/2/3 markers (Abdel Samad et al., 2010; Lopes et al., 2012). Throughout development, *Runx1* remains a constitutive activator for *Mrgprd*, but switches to a genetic repressor for *Mrgprc11*, *Mrgpra3*, and *Nppb* perinatally or postnatally. This genetic switch for a subset of nonpeptidergic genes is supported by the observation that early conditional deletion of *Runx1* using *Wnt1-Cre* leads to a complete loss of most nonpeptidergic markers, while excision of *Runx1* at a later developmental time (E18.5) using *$Na_v1.8$-Cre* disrupts only MRGPRD$^+$ neuron development, and constitutive *Runx1* expression in nociceptors results in a paradoxical loss of NP2 and NP3 markers (Abdel Samad et al., 2010; Liu et al., 2008;

Qi et al., 2017). Moreover, *Runx1* repressor activity for *Mrgprc11* and *Mrgpra3* is mediated by a C-terminal Groucho-interacting repression domain, but its mechanism for repressing *Nppb* and other NP3 markers has yet to be determined (Liu et al., 2008; Qi et al., 2017). In line with *Runx1*'s genetic switch, NP2 and NP3 neurons downregulate RUNX1 expression before establishing subtype marker expression at P0 and P7, respectively, while MRGPRD$^+$ NP1 neurons retain RUNX1 expression into adulthood (Liu et al., 2008; Qi et al., 2017). It is interesting to note that in MRGPRD$^+$ NP1 neurons, *Runx1* also maintains high levels of *Ret* expression, and *Runx1/Ret* are required for extinguishing NTRK1 expression in this population. This may explain why NP2 neurons, in which *Runx1* is only transiently active, exhibit lower levels of *Ret* expression and therefore express *Ntrk1* and *Calca* (Han et al., 2012; Sharma et al., 2020; Stantcheva et al., 2016; Usoskin et al., 2015; Zeisel et al., 2018). Additionally, the NTRK1 lineage neurons that maintain *Runx1* expression into adulthood (e.g., NP1) are all required for thermal and mechanical sensation, while *Runx1*-transient neurons (NP2 and NP3) are largely itch-selective. Through its dynamic expression and activity, *Runx1* may provide a developmental strategy for segregating nociceptors into functionally distinct sensory modalities.

5. Developmental disorders of somatosensory neurons and human disease

Animal models have been invaluable tools for determining the molecular mechanisms driving sensory neuron development. Much of the foundation laid by rodent genetic models in particular has led to the discovery of conserved themes in nociceptor development in humans. For example, the critical role of NGF/NTRK1 signaling in sensory neuron survival was first described in mouse knockout models (Crowley et al., 1994; Smeyne et al., 1994). The observation that the phenotype of these models strikingly resembled that of human individuals with congenital insensitivity to pain (CIP), inspired the identification of mutations in *NTRK1* in CIP patients just two years later (Indo et al., 1996). Translation of scientific insights has also proven bidirectional: recently, the discovery of another CIP-associated gene, *PRDM12*, spurred the development of a *Prdm12* mouse knockout line, which in turn established novel epigenetic mechanisms controlling nociceptor specification (Bartesaghi et al., 2019; Chen et al., 2015;

Desiderio et al., 2019). These examples, and others discussed below, illustrate the importance of crosstalk between research using animal models and research focusing on human diseases.

CIP is the primary clinical disorder associated with deficits in sensory neuron development. Patients with CIP lack protective responses to noxious events, and often sustain mutilating injuries from repeated trauma and burns that otherwise go unnoticed (Cox, Kurth, & Geoffrey Woods, 2018; Drissi, Woods, & Woods, 2020; Nahorski, Chen, & Woods, 2015). Most CIP cases can be attributed to Mendelian disorders, in which mutations of a single gene lead to either failed nociceptor formation or maintenance, or an inability to detect and propagate nociceptive signals (Fig. 3). Two major forms of CIP are caused by mutations in *NTRK1* and *NGF*. In both forms, patients completely lack unmyelinated C fibers and show a partial reduction of thinly-myelinated Aδ fibers. Although these patients are not always tested for sensitivity to itch stimuli, the histological absence of small-diameter nerve fibers in both sural nerve biopsies and epidermal sections would suggest they lack pruriception as well (Nolano et al., 2000; Chen et al., 2015).

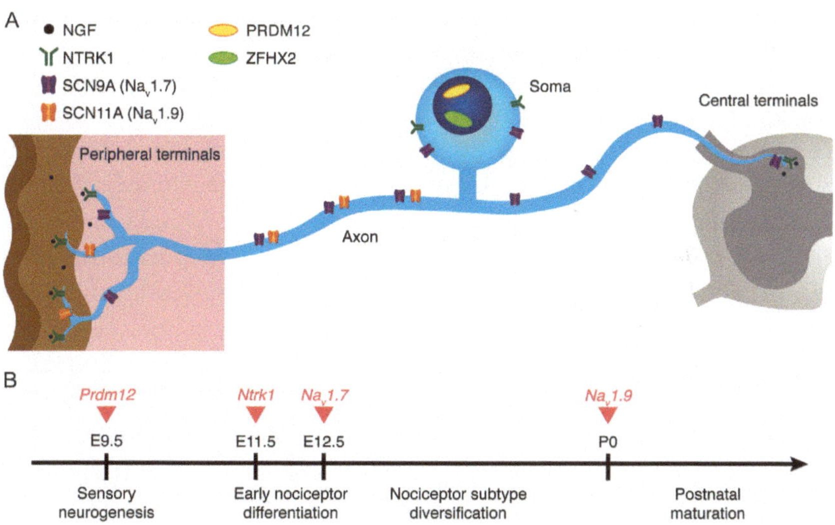

Fig. 3 Genes that cause congenital insensitivity to pain (CIP). (A) Schematic of a nociceptor showing where CIP-associated molecules function in nociceptor development and/or physiology. (B) Timeline of mouse DRG development indicating approximately when the orthologous CIP genes are expressed. *Based in part on the dataset from Sharma, N., Flaherty, K., Lezgiyeva, K., Wagner, D. E., Klein, A. M., & Ginty, D. D. (2020). The emergence of transcriptional identity in somatosensory neurons. Nature, 577(7790), 392–398.*

Clinically, the functional deficits displayed by CIP individuals with *NTRK1* or *NGF* mutations are highly consistent with the phenotypes of *Ngf* and *Ntrk1* knockout mice (Crowley et al., 1994; Patel et al., 2000; Smeyne et al., 1994). Several of the missense mutations identified in the *NTRK1* gene have been functionally characterized, revealing deficits in NTRK1 kinase activity, autophosphorylation, and activation of downstream signaling pathways that contribute to the aberrant development of nociceptors (Franco et al., 2016; Indo, 2001; Shaikh et al., 2017). A few mutations have been reported to retain some or all NTRK1 activity in a battery of functional assays, and may be associated with only partial sensory neuropathy and milder clinical phenotypes (Shaikh et al., 2017). The CIP-associated *NGF* mutations that have been identified so far disrupt cleavage of the precursor pro-protein into the biologically active product NGFβ (Carvalho et al., 2011; Einarsdottir et al., 2004; Shaikh, Nahorski, & Woods, 2018). One interesting *NGF* mutation, an amino acid substitution, also reduces processing of proNGF to the mature NGF protein but produces a nociceptor-specific phenotype (Larsson, Kuma, Norberg, Minde, & Holmberg, 2009). This specificity may be explained by the mutant NGF's decreased binding affinity for $p75^{NTR}$ but not NTRK1, and its selective inability to activate some but not all signaling pathways downstream of both NTRK1 and $p75^{NTR}$ (Capsoni et al., 2011; Covaceuszach et al., 2010; Larsson et al., 2009; Sung et al., 2018). Another possibility is that this NGF mutation does not alter trophic functions during development, but instead disrupts NGF-induced sensitization and peripheral terminal maintenance of nociceptors in adulthood (Denk et al., 2017; Gold & Gebhart, 2010). In transgenic mice modeling orthologous *Ngf* mutations, sensory neuron subtypes developed in normal proportions within the DRG, suggesting the mutated NGF is still capable of controlling nociceptor differentiation (Testa et al., 2019). It is unclear if this mutation is specific for NTRK1 signaling, therefore affecting only PEP1/2 nociceptors which maintain NTRK1 expression in adulthood, or if it contributes to a pain-insensitive phenotype through the more broadly expressed receptor $p75^{NTR}$.

Recent efforts to identify the genetic cause of CIP in patients who lack *NTRK1* and *NGF* mutations led to the discovery of another key regulator of human nociceptor development, *PRDM12* (Chen et al., 2015; Zhang et al., 2016). In vitro functional experiments show that several of the identified *PRDM12* mutations disrupt the ability of PRDM12 to promote histone methylation and induce sensory neuron markers such as *Isl1* and *Tlx3*, which likely contributes to the loss of nociceptor formation in these patients (Chen et al., 2015; Nagy et al., 2015). Additionally, *PRDM12* expression persists specifically in DRG neurons into adulthood in both mouse models and humans (Bartesaghi et al., 2019; Chen et al., 2015; Desiderio et al., 2019), raising questions as to whether PRDM12 contributes to human nociceptor maintenance and function after sensory neuron maturation. Another transcriptional regulator associated with Mendelian painlessness disorders is *ZFHX2*, a zinc finger homeobox transcription factor expressed in human nociceptors. One family with an autosomal dominant *ZFHX2* mutation has been described so far (Habib et al., 2018; Spinsanti et al., 2008). Affected individuals show hyposensitivity to thermal pain and capsaicin application, but normal intraepidermal nerve fiber density (Habib et al., 2018). Mice expressing the orthologous mutation in *Zfhx2* in a BAC transgenic model show a reduction in noxious thermal pain behaviors, but normal responses to noxious mechanical and cold stimuli (Habib et al., 2018). Although small-diameter neurons formed normally in these mice, some nociceptor-associated genes, such as *Sst, Gfra3, Ptgir*, and *Gal*, were deregulated downstream of the *Zfhx2* mutation (Habib et al., 2018). Further studies are needed to clarify the precise role *Zfhx2* plays during sensory neuron development, and which of its gene targets may be responsible for the pain-hyposensitivity phenotype in patients with *ZFHX2* mutations.

The second major group of congenital pain-related disorders are channelopathies, in which the voltage-gated sodium channels (VGSC) expressed in nociceptor populations are mutated (Bennett & Woods, 2014; Nahorski et al., 2015). These channels, which include $Na_v1.7$ (encoded by *SCN9A*), $Na_v1.8$ (*SCN10A*), and $Na_v1.9$ (*SCN11A*), play a key role in setting nociceptor excitability. Despite considerable homology among these three VGSCs, they have distinct kinetics and expression patterns (Benn, Costigan, Tate, Fitzgerald, & Woolf, 2001; Bennett, Clark, Huang, Waxman, & Dib-Hajj, 2019, Fukuoka & Noguchi, 2011; Roy & Narahashi, 1992), and appear to have unique roles in sensory neuron signal transmission. Some VGSC mutations are loss-of-function and abrogate the ability of nociceptors to detect and transmit noxious information, a phenotype well-characterized in CIP patients with bi-allelic *SCN9A* mutations (Ahmad et al., 2007; Cox et al., 2006; Emery et al., 2015; Goldberg et al., 2007; McDermott et al., 2019). *SCN9A*-associated CIP typically differs from *NGF/NTRK1*-associated CIP by its normal intraepidermal nerve fiber density, which is recapitulated in mouse knockout models of

$Na_v1.7$ (Cox et al., 2006; Gingras et al., 2014; Goldberg et al., 2007; McDermott et al., 2019). However, recent studies have reported the loss of epidermal innervation in some CIP patients, which suggests a potential function for $Na_v1.7$ in the activity-dependent development or maintenance of peripheral nociceptor terminals (Chang et al., 2018; Marchi et al., 2018; McDermott et al., 2019; Nilsen et al., 2009). In addition, conditional deletion of $Na_v1.7$ in nociceptors results in downregulation of transcription factors such as *Runx1* and *Brn3b* (Minett et al., 2015). It is possible that *SCN9A* loss-of-function mutations in CIP individuals disrupts nociceptor development by deregulating transcriptional activity, either through direct effects of intracellular sodium on transcription factor activity, or through indirect changes in intracellular calcium levels and calcium-dependent signaling pathways (Minett et al., 2015; Miyakawa, Woo, Dahl, Handler, & Kwon, 1999; Popov et al., 2012). Further studies are needed to establish the developmental time course of $Na_v1.7$ expression and its potential role in nociceptor maturation.

Another VGSC implicated in CIP is $Na_v1.9$, which is encoded by *SCN11A*. Although CIP patients with *SCN11A* mutations display insensitivity to pain, they also report severe pruritus secondary to hyperhidrosis that leads to skin ulcerations on the posterior neck, as well as intolerance to moderate hot/cold temperatures and rectal pain upon defecation. Interestingly, all *SCN11A* variants identified in CIP patients thus far have been gain-of-function mutations, which are traditionally associated with disorders of debilitating pain (Huang et al., 2017, 2014, 2019; Leipold et al., 2013, 2015; Woods, Babiker, Horrocks, Tolmie, & Kurth, 2015; Zhang et al., 2013). How does increased $Na_v1.9$ activation contribute to both painlessness and painfulness? $Na_v1.9$ channels are likely important for setting the threshold for action potential electrogenesis in nociceptors (Copel et al., 2009; Osorio, Korogod, & Delmas, 2014). In CIP patients, gain-of-function *SCN11A* mutations significantly depolarize the resting membrane potential of nociceptors, and may inactivate a sufficient number of endogenous $Na_v1.8$ channels to silence the neuron (Leipold et al., 2013; Huang et al., 2017). In patients with familial episodic pain disorder or painful neuropathy, gain-of-function *SCN11A* mutations appear to cause smaller depolarizations of resting membrane potential, which actually enhance excitability and increase spontaneous firing (Huang et al., 2014). Intriguingly, a common feature of $Na_v1.9$-associated CIP is severe pruritus, leading to self-induced skin ulcers on the back of the neck. Although mouse models of this disorder do not recapitulate all features of the clinical phenotype, such as insensitivity to noxious mechanical stimuli, mice heterozygously expressing an orthologous $Na_v1.9$ mutation also showed increased spontaneous scratching and corresponding self-inflicted lesions on the nape area (Leipold et al., 2013; Salvatierra et al., 2018). Current-clamp recordings of $MRGPRA3^+$ neurons from these mice support earlier data suggesting most nociceptors are massively depolarized due to the gain-of-function $Na_v1.9$ mutation and are subsequently unable to fire action potentials, but a small subset of $MRGPRA3^+$ neurons are instead hyperexcitable, leading to a higher percentage of chloroquine-responsive neurons in $Na_v1.9$ mutant mice (Salvatierra et al., 2018). Further studies are needed to examine how the same $Na_v1.9$ mutation can produce differential effects on excitability within nociceptor populations, perhaps through subtle variations in expression of different VGSCs or other ion channels (Han et al., 2009). Moreover, it will be interesting to determine whether this phenomenon extends to ano-rectal nociceptors, which appear to confer pain hypersensitivity upon defecation to $Na_v1.9$-associated CIP patients.

To conclude, the channelopathy-associated painlessness disorders comprise a complex range of genetic mutations and clinical phenotypes, and have provided key insights into VGSC function in nociceptors. However, most channelopathies have been characterized from the perspective of their contributions to excitability in adult DRG neurons. Given their onset of expression during crucial phases of nociceptor specification and maturation, and evidence of a potential role in regulating transcriptional activity, further analysis of $Na_v1.7$, $Na_v1.8$, and $Na_v1.9$ functions in development could be interesting (Benn et al., 2001; Minett et al., 2015). Together, CIP disorders caused by NGF/NTRK1 mutations and channelopathies highlight how human genetic studies complement in vitro and animal model systems in the advancement of basic and translational sensory disease research.

6. Conclusions

The diversification of primary nociceptor subtypes from common neural crest progenitors has provided a rich context for studying how complex intrinsic and extrinsic signals converge to specify molecularly and functionally unique neuronal populations. Traditional developmental studies primarily using molecular genetic tools in mouse models have contributed significantly to our understanding of the molecular programs orchestrating nociceptor differentiation. With the recent advancement of next-generation sequencing techniques, several recent studies have contributed an extraordinary amount of DRG transcriptomic data that will frame future investigation of sensory neuron development. Now that an unbiased classification system for sensory neurons is available, this will facilitate further studies to elucidate the assignments of classically established developmental lineages to molecularly defined subtypes, and the molecular mechanisms underlying their segregation. In addition, these transcriptome databases will be invaluable for identifying and validating novel key regulators of nociceptor differentiation, as well as unique developmental markers that can be used for lineage tracing and targeted manipulations. Altogether, knowledge of the developmental processes that establish sensory neuron subtype identity will provide an important foundation for studying sensory neuron functions in the mature animal and in pathological states in humans.

References

Abdel Samad, O., Liu, Y., Yang, F.-C., Kramer, I., Arber, S., & Ma, Q. (2010). Characterization of two Runx1-dependent nociceptor differentiation programs necessary for inflammatory versus neuropathic pain. *Molecular Pain*, *6*, 45.

Ahmad, S., Dahllund, L., Eriksson, A. B., Hellgren, D., Karlsson, U., Lund, P.-E., et al. (2007). A stop codon mutation in SCN9A causes lack of pain sensation. *Human Molecular Genetics*, *16*(17), 2114–2121.

Ascano, M., Bodmer, D., & Kuruvilla, R. (2012). Endocytic trafficking of neurotrophins in neural development. *Trends in Cell Biology*, *22*(5), 266–273.

Bachy, I., Franck, M. C. M., Li, L., Abdo, H., Pattyn, A., & Ernfors, P. (2011). The transcription factor Cux2 marks development of an A-delta sublineage of TrkA sensory neurons. *Developmental Biology*, *360*(1), 77–86.

Barrett, G. L., & Bartlett, P. F. (1994). The p75 nerve growth factor receptor mediates survival or death depending on the stage of sensory neuron development. *Proceedings of the National Academy of Sciences of the United States of America*, *91*(14), 6501–6505.

Bartesaghi, L., Wang, Y., Fontanet, P., Wanderoy, S., Berger, F., Wu, H., et al. (2019). PRDM12 is required for initiation of the nociceptive neuron lineage during neurogenesis. *Cell Reports*, *26*(13), 3484–3492.e4.

Bautista, D. M., Wilson, S. R., & Hoon, M. A. (2014). Why we scratch an itch: The molecules, cells and circuits of itch. *Nature Neuroscience*, *17*(2), 175–182.

Bautzova, T., Hockley, J. R. F., Perez-Berezo, T., Pujo, J., Tranter, M. M., Desormeaux, C., et al. (2018). 5-oxoETE triggers nociception in constipation-predominant irritable bowel syndrome through MAS-related G protein-coupled receptor D. *Science Signaling*, *11*(561), eaal2171. https://doi.org/10.1126/scisignal.aal2171.

Benn, S. C., Costigan, M., Tate, S., Fitzgerald, M., & Woolf, C. J. (2001). Developmental expression of the TTX-resistant voltage-gated sodium channels Nav1.8 (SNS) and Nav1.9 (SNS2) in primary sensory neurons. *The Journal of Neuroscience: The Official Journal of the Society for Neuroscience*, *21*(16), 6077–6085.

Bennett, D. L., Averill, S., Clary, D. O., Priestley, J. V., & McMahon, S. B. (1996). Postnatal changes in the expression of the trkA high-affinity NGF receptor in primary sensory neurons. *The European Journal of Neuroscience*, *8*(10), 2204–2208.

Bennett, D. L., Clark, A. J., Huang, J., Waxman, S. G., & Dib-Hajj, S. D. (2019). The role of voltage-gated sodium channels in pain signaling. *Physiological Reviews*, *99*(2), 1079–1151.

Bennett, D. L. H., & Woods, C. G. (2014). Painful and painless channelopathies. *Lancet Neurology, 13*(6), 587–599.

Bergmann, I., Priestley, J. V., McMahon, S. B., Bröcker, E. B., Toyka, K. V., & Koltzenburg, M. (1997). Analysis of cutaneous sensory neurons in transgenic mice lacking the low affinity neurotrophin receptor p75. *The European Journal of Neuroscience, 9*(1), 18–28.

Bourane, S., Garces, A., Venteo, S., Pattyn, A., Hubert, T., Fichard, A., et al. (2009). Low-threshold mechanoreceptor subtypes selectively express MafA and are specified by Ret signaling. *Neuron, 64*(6), 857–870.

Braz, J. M., Nassar, M. A., Wood, J. N., & Basbaum, A. I. (2005). Parallel "pain" pathways arise from subpopulations of primary afferent nociceptor. *Neuron, 47*(6), 787–793.

Bronner-Fraser, M., & Fraser, S. E. (1988). Cell lineage analysis reveals multipotency of some avian neural crest cells. *Nature, 335*(6186), 161–164.

Capsoni, S., Covaceuszach, S., Marinelli, S., Ceci, M., Bernardo, A., Minghetti, L., et al. (2011). Taking pain out of NGF: a "painless" NGF mutant, linked to hereditary sensory autonomic neuropathy type V, with full neurotrophic activity. *PLoS One, 6*(2), e17321.

Carvalho, O. P., Thornton, G. K., Hertecant, J., Houlden, H., Nicholas, A. K., Cox, J. J., et al. (2011). A novel NGF mutation clarifies the molecular mechanism and extends the phenotypic spectrum of the HSAN5 neuropathy. *Journal of Medical Genetics, 48*(2), 131–135.

Cavanaugh, D. J., Lee, H., Lo, L., Shields, S. D., Zylka, M. J., Basbaum, A. I., et al. (2009). Distinct subsets of unmyelinated primary sensory fibers mediate behavioral responses to noxious thermal and mechanical stimuli. *Proceedings of the National Academy of Sciences of the United States of America, 106*(22), 9075–9080.

Chang, W., Berta, T., Kim, Y. H., Lee, S., Lee, S.-Y., & Ji, R.-R. (2018). Expression and role of voltage-gated sodium channels in human dorsal root ganglion neurons with special focus on Nav1.7, species differences, and regulation by paclitaxel. *Neuroscience Bulletin, 34*(1), 4–12.

Chen, Y.-C., Auer-Grumbach, M., Matsukawa, S., Zitzelsberger, M., Themistocleous, A. C., Strom, T. M., et al. (2015). Transcriptional regulator PRDM12 is essential for human pain perception. *Nature Genetics, 47*(7), 803–808.

Chen, C.-L., Broom, D. C., Liu, Y., de Nooij, J. C., Li, Z., Cen, C., et al. (2006). Runx1 determines nociceptive sensory neuron phenotype and is required for thermal and neuropathic pain. *Neuron, 49*(3), 365–377.

Chen, Z., Donnelly, C. R., Dominguez, B., Harada, Y., Lin, W., Halim, A. S., et al. (2017). p75 is required for the establishment of postnatal sensory neuron diversity by potentiating ret signaling. *Cell Reports, 21*(3), 707–720.

Cheng, I., Jin, L., Rose, L. C., & Deppmann, C. D. (2018). Temporally restricted death and the role of p75NTR as a survival receptor in the developing sensory nervous system. *Developmental Neurobiology, 78*(7), 701–717.

Chiu, I. M., Barrett, L. B., Williams, E. K., Strochlic, D. E., Lee, S., Weyer, A. D., et al. (2014). Transcriptional profiling at whole population and single cell levels reveals somatosensory neuron molecular diversity. *eLife, 3*, e04660. https://doi.org/10.7554/eLife.04660.

Copel, C., Osorio, N., Crest, M., Gola, M., Delmas, P., & Clerc, N. (2009). Activation of neurokinin 3 receptor increases Nav1. 9 current in enteric neurons. *The Journal of Physiology, 587*(7), 1461–1479.

Covaceuszach, S., Capsoni, S., Marinelli, S., Pavone, F., Ceci, M., Ugolini, G., et al. (2010). In vitro receptor binding properties of a "painless" NGF mutein, linked to hereditary sensory autonomic neuropathy type V. *Biochemical and Biophysical Research Communications, 391*(1), 824–829.

Cox, J. J., Kurth, I., & Geoffrey Woods, C. (2018). Human genetics of pain. In J. N. Wood (Ed.), *The Oxford Handbook of the Neurobiology of Pain* (pp. 101–127). Oxford University Press.

Cox, J. J., Reimann, F., Nicholas, A. K., Thornton, G., Roberts, E., Springell, K., et al. (2006). An SCN9A channelopathy causes congenital inability to experience pain. *Nature, 444*(7121), 894–898.

Crowley, C., Spencer, S. D., Nishimura, M. C., Chen, K. S., Pitts-Meek, S., Armanini, M. P., et al. (1994). Mice lacking nerve growth factor display perinatal loss of sensory and sympathetic neurons yet develop basal forebrain cholinergic neurons. *Cell, 76*(6), 1001–1011.

D'Amico-Martel, A., & Noden, D. M. (1983). Contributions of placodal and neural crest cells to avian cranial peripheral ganglia. *The American Journal of Anatomy, 166*(4), 445–468.

Davies, A. M. (1996). The neurotrophic hypothesis: Where does it stand? *Philosophical Transactions of the Royal Society of London. Series B, Biological Sciences, 351*(1338), 389–394.

Denk, F., Bennett, D. L., & McMahon, S. B. (2017). Nerve growth factor and pain mechanisms. *Annual Review of Neuroscience, 40*, 307–325.

Deppmann, C. D., Mihalas, S., Sharma, N., Lonze, B. E., Niebur, E., & Ginty, D. D. (2008). A model for neuronal competition during development. *Science, 320*(5874), 369–373.

Desiderio, S., Vermeiren, S., Van Campenhout, C., Kricha, S., Malki, E., Richts, S., et al. (2019). Prdm12 directs nociceptive sensory neuron development by regulating the expression of the NGF receptor TrkA. *Cell Reports, 26*(13), 3522–3536.e5.

Dong, X., & Dong, X. (2018). Peripheral and central mechanisms of itch. *Neuron, 98*(3), 482–494.

Drissi, I., Woods, W. A., & Woods, C. G. (2020). Understanding the genetic basis of congenital insensitivity to pain. *British Medical Bulletin, 133*, 65–78. https://doi.org/10.1093/bmb/ldaa003.

Dykes, I. M., Lanier, J., Eng, S. R., & Turner, E. E. (2010). Brn3a regulates neuronal subtype specification in the trigeminal ganglion by promoting Runx expression during sensory differentiation. *Neural Development, 5*, 3.

Dykes, I. M., Tempest, L., Lee, S.-I., & Turner, E. E. (2011). Brn3a and Islet1 act epistatically to regulate the gene expression program of sensory differentiation. *The Journal of Neuroscience: The Official Journal of the Society for Neuroscience, 31*(27), 9789–9799.

Einarsdottir, E., Carlsson, A., Minde, J., Toolanen, G., Svensson, O., Solders, G., et al. (2004). A mutation in the nerve growth factor beta gene (NGFB) causes loss of pain perception. *Human Molecular Genetics, 13*(8), 799–805.

Emery, E. C., & Ernfors, P. (2018). Dorsal root ganglion neuron types and their functional specialization. In J. N. Wood (Ed.), *The Oxford Handbook of the Neurobiology of Pain* (pp. 129–155). Oxford University Press.

Emery, E. C., Habib, A. M., Cox, J. J., Nicholas, A. K., Gribble, F. M., Woods, C. G., et al. (2015). Novel SCN9A mutations underlying extreme pain phenotypes: Unexpected electrophysiological and clinical phenotype correlations. *The Journal of Neuroscience: The Official Journal of the Society for Neuroscience, 35*(20), 7674–7681.

Eng, S. R., Dykes, I. M., Lanier, J., Fedtsova, N., & Turner, E. E. (2007). POU-domain factor Brn3a regulates both distinct and common programs of gene expression in the spinal and trigeminal sensory ganglia. *Neural Development, 2*, 3.

Eng, S. R., Lanier, J., Fedtsova, N., & Turner, E. E. (2004). Coordinated regulation of gene expression by Brn3a in developing sensory ganglia. *Development, 131*(16), 3859–3870.

Faure, L., Wang, Y., Kastriti, M. E., Fontanet, P., Cheung, K. K. Y., Petitpré, C., et al. (2020). Single cell RNA sequencing identifies early diversity of sensory neurons forming via bi-potential intermediates. *Nature Communications, 11*(1), 4175.

Franco, M. L., Melero, C., Sarasola, E., Acebo, P., Luque, A., Calatayud-Baselga, I., et al. (2016). Mutations in TrkA causing congenital insensitivity to pain with anhidrosis (CIPA) induce misfolding, aggregation, and mutation-dependent neurodegeneration by dysfunction of the autophagic flux. *The Journal of Biological Chemistry, 291*(41), 21363–21374.

Frank, E., & Sanes, J. R. (1991). Lineage of neurons and glia in chick dorsal root ganglia: analysis in vivo with a recombinant retrovirus. *Development, 111*(4), 895–908.

Fukuoka, T., & Noguchi, K. (2011). Comparative study of voltage-gated sodium channel α-subunits in non-overlapping four neuronal populations in the rat dorsal root ganglion. *Neuroscience Research, 70*(2), 164–171.

Gascon, E., Gaillard, S., Malapert, P., Liu, Y., Rodat-Despoix, L., Samokhvalov, I. M., et al. (2010). Hepatocyte growth factor-Met signaling is required for Runx1 extinction and peptidergic differentiation in primary nociceptive neurons. *The Journal of Neuroscience: The Official Journal of the Society for Neuroscience, 30*(37), 12414–12423.

Ghitani, N., Barik, A., Szczot, M., Thompson, J. H., Li, C., Le Pichon, C. E., et al. (2017). Specialized mechanosensory nociceptors mediating rapid responses to hair pull. *Neuron, 95*(4), 944–954.e4.

Gingras, J., Smith, S., Matson, D. J., Johnson, D., Nye, K., Couture, L., et al. (2014). Global Nav1.7 knockout mice recapitulate the phenotype of human congenital indifference to pain. *PLoS One, 9*(9), e105895.

Ginty, D. D., & Segal, R. A. (2002). Retrograde neurotrophin signaling: Trk-ing along the axon. *Current Opinion in Neurobiology, 12*(3), 268–274.

Gold, M. S., & Gebhart, G. F. (2010). Nociceptor sensitization in pain pathogenesis. *Nature Medicine, 16*(11), 1248–1257.

Goldberg, Y. P., MacFarlane, J., MacDonald, M. L., Thompson, J., Dube, M.-P., Mattice, M., et al. (2007). Loss-of-function mutations in the Nav1.7 gene underlie congenital indifference to pain in multiple human populations. *Clinical Genetics, 71*(4), 311–319.

Golden, J. P., Hoshi, M., Nassar, M. A., Enomoto, H., Wood, J. N., Milbrandt, J., et al. (2010). RET signaling is required for survival and normal function of nonpeptidergic nociceptors. *The Journal of Neuroscience: The Official Journal of the Society for Neuroscience, 30*(11), 3983–3994.

Gorokhova, S., Gaillard, S., Urien, L., Malapert, P., Legha, W., Baronian, G., et al. (2014). Uncoupling of molecular maturation from peripheral target innervation in nociceptors expressing a chimeric TrkA/TrkC receptor. *PLoS Genetics, 10*(2), e1004081.

Graef, I. A., Wang, F., Charron, F., Chen, L., Neilson, J., Tessier-Lavigne, M., et al. (2003). Neurotrophins and netrins require calcineurin/NFAT signaling to stimulate outgrowth of embryonic axons. *Cell, 113*(5), 657–670.

Haberberger, R. V., Barry, C., Dominguez, N., & Matusica, D. (2019). Human dorsal root ganglia. *Frontiers in Cellular Neuroscience, 13*, 271.

Habib, A. M., Matsuyama, A., Okorokov, A. L., Santana-Varela, S., Bras, J. T., Aloisi, A. M., et al. (2018). A novel human pain insensitivity disorder caused by a point mutation in ZFHX2. *Brain: A Journal of Neurology, 141*(2), 365–376.

Hadjab, S., Franck, M. C. M., Wang, Y., Sterzenbach, U., Sharma, A., Ernfors, P., et al. (2013). A local source of FGF initiates development of the unmyelinated lineage of sensory neurons. *The Journal of Neuroscience: The Official Journal of the Society for Neuroscience, 33*(45), 17656–17666.

Hamburger, V. (1961). Experimental analysis of the dual origin of the trigeminal ganglion in the chick embryo. *The Journal of Experimental Zoology, 148*, 91–123.

Han, C., Dib-Hajj, S. D., Lin, Z., Li, Y., Eastman, E. M., Tyrrell, L., et al. (2009). Early- and late-onset inherited erythromelalgia: Genotype-phenotype correlation. *Brain: A Journal of Neurology, 132*(Pt 7), 1711–1722.

Han, L., Ma, C., Liu, Q., Weng, H.-J., Cui, Y., Tang, Z., et al. (2012). A subpopulation of nociceptors specifically linked to itch. *Nature Neuroscience*, *16*, 174.

Hantzopoulos, P. A., Suri, C., Glass, D. J., Goldfarb, M. P., & Yancopoulos, G. D. (1994). The low affinity NGF receptor, p75, can collaborate with each of the Trks to potentiate functional responses to the neurotrophins. *Neuron*, *13*(1), 187–201.

Hempstead, B. L. (2002). The many faces of p75NTR. *Current Opinion in Neurobiology*, *12*(3), 260–267.

Hjerling-Leffler, J., Marmigère, F., Heglind, M., Cederberg, A., Koltzenburg, M., Enerbäck, S., et al. (2005). The boundary cap: A source of neural crest stem cells that generate multiple sensory neuron subtypes. *Development*, *132*(11), 2623–2632.

Hockley, J. R. F., Taylor, T. S., Callejo, G., Wilbrey, A. L., Gutteridge, A., Bach, K., et al. (2019). Single-cell RNAseq reveals seven classes of colonic sensory neuron. *Gut*, *68*(4), 633–644.

Hohenauer, T., & Moore, A. W. (2012). The Prdm family: Expanding roles in stem cells and development. *Development*, *139*(13), 2267–2282.

Huang, J., Estacion, M., Zhao, P., Dib-Hajj, F. B., Schulman, B., Abicht, A., et al. (2019). A novel gain-of-function nav1.9 mutation in a child with episodic pain. *Frontiers in Neuroscience*, *13*, 918.

Huang, J., Han, C., Estacion, M., Vasylyev, D., Hoeijmakers, J. G. J., Gerrits, M. M., et al. (2014). Gain-of-function mutations in sodium channel Na(v)1.9 in painful neuropathy. *Brain: A Journal of Neurology*, *137*(Pt. 6), 1627–1642.

Huang, S., O'Donovan, K. J., Turner, E. E., Zhong, J., & Ginty, D. D. (2015). Extrinsic and intrinsic signals converge on the Runx1/CBFβ transcription factor for nonpeptidergic nociceptor maturation. *eLife*, *4*, e10874.

Huang, J., Polgár, E., Solinski, H. J., Mishra, S. K., Tseng, P.-Y., Iwagaki, N., et al. (2018). Circuit dissection of the role of somatostatin in itch and pain. *Nature Neuroscience*, *21*(5), 707–716.

Huang, J., Vanoye, C. G., Cutts, A., Goldberg, Y. P., Dib-Hajj, S. D., Cohen, C. J., et al. (2017). Sodium channel NaV1.9 mutations associated with insensitivity to pain dampen neuronal excitability. *The Journal of Clinical Investigation*, *127*(7), 2805–2814.

Indo, Y. (2001). Molecular basis of congenital insensitivity to pain with anhidrosis (CIPA): Mutations and polymorphisms in TRKA (NTRK1) gene encoding the receptor tyrosine kinase for nerve growth factor. *Human Mutation*, *18*(6), 462–471.

Indo, Y., Tsuruta, M., Hayashida, Y., Karim, M. A., Ohta, K., Kawano, T., et al. (1996). Mutations in the TRKA/NGF receptor gene in patients with congenital insensitivity to pain with anhidrosis. *Nature Genetics*, *13*(4), 485–488.

Jaurena, M. B., Juraver-Geslin, H., Devotta, A., & Saint-Jeannet, J.-P. (2015). Zic1 controls placode progenitor formation non-cell autonomously by regulating retinoic acid production and transport. *Nature Communications*, *6*, 7476.

Julius, D. (2013). TRP channels and pain. *Annual Review of Cell and Developmental Biology*, *29*, 355–384.

Kasemeier-Kulesa, J. C., Kulesa, P. M., & Lefcort, F. (2005). Imaging neural crest cell dynamics during formation of dorsal root ganglia and sympathetic ganglia. *Development*, *132*(2), 235–245.

Kobayashi, K., Fukuoka, T., Obata, K., Yamanaka, H., Dai, Y., Tokunaga, A., et al. (2005). Distinct expression of TRPM8, TRPA1, and TRPV1 mRNAs in rat primary afferent neurons with adelta/c-fibers and colocalization with trk receptors. *The Journal of Comparative Neurology*, *493*(4), 596–606.

Kramer, I., Sigrist, M., de Nooij, J. C., Taniuchi, I., Jessell, T. M., & Arber, S. (2006). A role for Runx transcription factor signaling in dorsal root ganglion sensory neuron diversification. *Neuron*, *49*(3), 379–393.

Lallemend, F., & Ernfors, P. (2012). Molecular interactions underlying the specification of sensory neurons. *Trends in Neurosciences*, *35*(6), 373–381.

Lanier, J., Dykes, I. M., Nissen, S., Eng, S. R., & Turner, E. E. (2009). Brn3a regulates the transition from neurogenesis to terminal differentiation and represses non-neural gene expression in the trigeminal ganglion. *Developmental Dynamics: An Official Publication of the American Association of the Anatomists*, *238*(12), 3065–3079.

Larsson, E., Kuma, R., Norberg, A., Minde, J., & Holmberg, M. (2009). Nerve growth factor R221W responsible for insensitivity to pain is defectively processed and accumulates as proNGF. *Neurobiology of Disease*, *33*(2), 221–228.

Lawson, S. N., & Biscoe, T. J. (1979). Development of mouse dorsal root ganglia: An autoradiographic and quantitative study. *Journal of Neurocytology*, *8*(3), 265–274.

Lee, K. F., Davies, A. M., & Jaenisch, R. (1994). p75-deficient embryonic dorsal root sensory and neonatal sympathetic neurons display a decreased sensitivity to NGF. *Development*, *120*(4), 1027–1033.

Lee, K. F., Li, E., Huber, L. J., Landis, S. C., Sharpe, A. H., Chao, M. V., et al. (1992). Targeted mutation of the gene encoding the low affinity NGF receptor p75 leads to deficits in the peripheral sensory nervous system. *Cell*, *69*(5), 737–749.

Lei, L., Laub, F., Lush, M., Romero, M., Zhou, J., Luikart, B., et al. (2005). The zinc finger transcription factor Klf7 is required for TrkA gene expression and development of nociceptive sensory neurons. *Genes & Development*, *19*(11), 1354–1364.

Leipold, E., Hanson-Kahn, A., Frick, M., Gong, P., Bernstein, J. A., Voigt, M., et al. (2015). Cold-aggravated pain in humans caused by a hyperactive NaV1.9 channel mutant. *Nature Communications*, *6*(1), 10049.

Leipold, E., Liebmann, L., Korenke, G. C., Heinrich, T., Giesselmann, S., Baets, J., et al. (2013). A de novo gain-of-function mutation in SCN11A causes loss of pain perception. *Nature Genetics*, *45*(11), 1399–1404.

Lemmon, M. A., & Schlessinger, J. (2010). Cell signaling by receptor tyrosine kinases. *Cell*, *141*(7), 1117–1134.

Li, C.-L., Li, K.-C., Wu, D., Chen, Y., Luo, H., Zhao, J.-R., et al. (2016). Somatosensory neuron types identified by high-coverage single-cell RNA-sequencing and functional heterogeneity. *Cell Research*, *26*(1), 83–102.

Liu, Q., Tang, Z., Surdenikova, L., Kim, S., Patel, K. N., Kim, A., et al. (2009). Sensory neuron-specific GPCR Mrgprs are itch receptors mediating chloroquine-induced pruritus. *Cell*, *139*(7), 1353–1365.

Liu, Y., Yang, F.-C., Okuda, T., Dong, X., Zylka, M. J., Chen, C.-L., et al. (2008). Mechanisms of compartmentalized expression of Mrg class G-protein-coupled sensory receptors. *The Journal of Neuroscience: The Official Journal of the Society for Neuroscience*, *28*(1), 125–132.

Lonze, B. E., Riccio, A., Cohen, S., & Ginty, D. D. (2002). Apoptosis, axonal growth defects, and degeneration of peripheral neurons in mice lacking CREB. *Neuron*, *34*(3), 371–385.

Lopes, C., Liu, Z., Xu, Y., & Ma, Q. (2012). Tlx3 and Runx1 act in combination to coordinate the development of a cohort of nociceptors, thermoceptors, and pruriceptors. *The Journal of Neuroscience: The Official Journal of the Society for Neuroscience*, *32*(28), 9706–9715.

Luiz, A. P., MacDonald, D. I., Santana-Varela, S., Millet, Q., Sikandar, S., Wood, J. N., et al. (2019). Cold sensing by NaV1.8-positive and NaV1.8-negative sensory neurons. *Proceedings of the National Academy of Sciences of the United States of America*, *116*(9), 3811–3816.

Luo, W., Enomoto, H., Rice, F. L., Milbrandt, J., & Ginty, D. D. (2009). Molecular identification of rapidly adapting mechanoreceptors and their developmental dependence on ret signaling. *Neuron*, *64*(6), 841–856.

Luo, W., Wickramasinghe, S. R., Savitt, J. M., Griffin, J. W., Dawson, T. M., & Ginty, D. D. (2007). A hierarchical NGF signaling cascade controls Ret-dependent and Ret-independent events during development of nonpeptidergic DRG neurons. *Neuron*, *54*(5), 739–754.

Ma, Q., Chen, Z., del Barco Barrantes, I., de la Pompa, J. L., & Anderson, D. J. (1998). Neurogenin1 is essential for the determination of neuronal precursors for proximal cranial sensory ganglia. *Neuron*, *20*(3), 469–482.

Ma, Q., Fode, C., Guillemot, F., & Anderson, D. J. (1999). Neurogenin1 and neurogenin2 control two distinct waves of neurogenesis in developing dorsal root ganglia. *Genes & Development*, *13*(13), 1717–1728.

Ma, L., Lei, L., Eng, S. R., Turner, E., & Parada, L. F. (2003). Brn3a regulation of TrkA/NGF receptor expression in developing sensory neurons. *Development*, *130*(15), 3525–3534.

Makkerh, J. P. S., Ceni, C., Auld, D. S., Vaillancourt, F., Dorval, G., & Barker, P. A. (2005). p75 neurotrophin receptor reduces ligand-induced Trk receptor ubiquitination and delays Trk receptor internalization and degradation. *EMBO Reports*, *6*(10), 936–941.

Marchi, M., Provitera, V., Nolano, M., Romano, M., Maccora, S., D'Amato, I., et al. (2018). A novel SCN9A splicing mutation in a compound heterozygous girl with congenital insensitivity to pain, hyposmia and hypogeusia. *Journal of the Peripheral Nervous System*, *23*(3), 202–206.

Marmigère, F., & Ernfors, P. (2007). Specification and connectivity of neuronal subtypes in the sensory lineage. *Nature Reviews Neuroscience*, *8*(2), 114–127.

Marmigère, F., Montelius, A., Wegner, M., Groner, Y., Reichardt, L. F., & Ernfors, P. (2006). The Runx1/AML1 transcription factor selectively regulates development and survival of TrkA nociceptive sensory neurons. *Nature Neuroscience*, *9*(2), 180–187.

Maro, G. S., Vermeren, M., Voiculescu, O., Melton, L., Cohen, J., Charnay, P., et al. (2004). Neural crest boundary cap cells constitute a source of neuronal and glial cells of the PNS. *Nature Neuroscience*, 7(9), 930–938.

McDermott, L. A., Weir, G. A., Themistocleous, A. C., Segerdahl, A. R., Blesneac, I., Baskozos, G., et al. (2019). Defining the functional role of NaV1.7 in human nociception. *Neuron*, *101*(5), 905–919.e8.

Meixiong, J., & Dong, X. (2017). Mas-related G protein-coupled receptors and the biology of itch sensation. *Annual Review of Genetics*, *51*, 103–121.

Minett, M. S., Pereira, V., Sikandar, S., Matsuyama, A., Lolignier, S., Kanellopoulos, A. H., et al. (2015). Endogenous opioids contribute to insensitivity to pain in humans and mice lacking sodium channel Nav1.7. *Nature Communications*, *6*, 8967.

Mishra, S. K., & Hoon, M. A. (2013). The cells and circuitry for itch responses in mice. *Science*, *340*(6135), 968–971.

Miyakawa, H., Woo, S. K., Dahl, S. C., Handler, J. S., & Kwon, H. M. (1999). Tonicity-responsive enhancer binding protein, a rel-like protein that stimulates transcription in response to hypertonicity. *Proceedings of the National Academy of Sciences of the United States of America*, *96*(5), 2538–2542.

Molliver, D. C., & Snider, W. D. (1997). Nerve growth factor receptor TrkA is down-regulated during postnatal development by a subset of dorsal root ganglion neurons. *The Journal of Comparative Neurology*, *381*(4), 428–438.

Molliver, D. C., Wright, D. E., Leitner, M. L., Parsadanian, A. S., Doster, K., Wen, D., et al. (1997). IB4-binding DRG neurons switch from NGF to GDNF dependence in early postnatal life. *Neuron*, *19*(4), 849–861.

Moqrich, A., Earley, T. J., Watson, J., Andahazy, M., Backus, C., Martin-Zanca, D., et al. (2004). Expressing TrkC from the TrkA locus causes a subset of dorsal root ganglia neurons to switch fate. *Nature Neuroscience*, 7(8), 812–818.

Murray, S. S., Bartlett, P. F., & Cheema, S. S. (1999). Differential loss of spinal sensory but not motor neurons in the p75NTR knockout mouse. *Neuroscience Letters*, *267*(1), 45–48.

Nagy, V., Cole, T., Van Campenhout, C., Khoung, T. M., Leung, C., Vermeiren, S., et al. (2015). The evolutionarily conserved transcription factor PRDM12 controls sensory neuron development and pain perception. *Cell Cycle*, *14*(12), 1799–1808.

Nahorski, M. S., Chen, Y.-C., & Woods, C. G. (2015). New mendelian disorders of painlessness. *Trends in Neurosciences*, *38*(11), 712–724.

Newbern, J. M., Li, X., Shoemaker, S. E., Zhou, J., Zhong, J., Wu, Y., et al. (2011). Specific functions for ERK/MAPK signaling during PNS development. *Neuron*, *69*(1), 91–105.

Nilsen, K. B., Nicholas, A. K., Woods, C. G., Mellgren, S. I., Nebuchennykh, M., & Aasly, J. (2009). Two novel SCN9A mutations causing insensitivity to pain. *Pain*, *143*(1–2), 155–158.

Nolano, M., Crisci, C., Santoro, L., Barbieri, F., Casale, R., Kennedy, W. R., et al. (2000). Absent innervation of skin and sweat glands in congenital insensitivity to pain with anhidrosis. *Clinical Neurophysiology: Official Journal of the International Federation of Clinical Neurophysiology*, *111*(9), 1596–1601.

Osorio, N., Korogod, S., & Delmas, P. (2014). Specialized functions of Nav1.5 and Nav1.9 channels in electrogenesis of myenteric neurons in intact mouse ganglia. *The Journal of Neuroscience: The Official Journal of the Society for Neuroscience*, *34*(15), 5233–5244.

Patel, T. D., Jackman, A., Rice, F. L., Kucera, J., & Snider, W. D. (2000). Development of sensory neurons in the absence of NGF/TrkA signaling in vivo. *Neuron*, *25*(2), 345–357.

Popov, S., Venetsanou, K., Chedrese, P. J., Pinto, V., Takemori, H., Franco-Cereceda, A., et al. (2012). Increases in intracellular sodium activate transcription and gene expression via the salt-inducible kinase 1 network in an atrial myocyte cell line. *American Journal of Physiology Heart and Circulatory Physiology*, *303*(1), H57–H65.

Qi, L., Huang, C., Wu, X., Tao, Y., Yan, J., Shi, T., et al. (2017). Hierarchical specification of pruriceptors by Runt-domain transcription factor Runx1. *The Journal of Neuroscience: The Official Journal of the Society for Neuroscience*, *37*(22), 5549–5561.

Rosenfeld, M. G., Mermod, J. J., Amara, S. G., Swanson, L. W., Sawchenko, P. E., Rivier, J., et al. (1983). Production of a novel neuropeptide encoded by the calcitonin gene via tissue-specific RNA processing. *Nature*, *304*(5922), 129–135.

Rostock, C., Schrenk-Siemens, K., Pohle, J., & Siemens, J. (2018). Human vs. mouse nociceptors-similarities and differences. *Neuroscience*, *387*, 13–27.

Roy, M. L., & Narahashi, T. (1992). Differential properties of tetrodotoxin-sensitive and tetrodotoxin-resistant sodium channels in rat dorsal root ganglion neurons. *The Journal of Neuroscience: The Official Journal of the Society for Neuroscience*, *12*(6), 2104–2111.

Salvatierra, J., Diaz-Bustamante, M., Meixiong, J., Tierney, E., Dong, X., & Bosmans, F. (2018). A disease mutation reveals a role for NaV1.9 in acute itch. *The Journal of Clinical Investigation*, *128*(12), 5434–5447.

Serbedzija, G. N., Fraser, S. E., & Bronner-Fraser, M. (1990). Pathways of trunk neural crest cell migration in the mouse embryo as revealed by vital dye labelling. *Development*, *108*(4), 605–612.

Shaikh, S. S., Chen, Y.-C., Halsall, S.-A., Nahorski, M. S., Omoto, K., Young, G. T., et al. (2017). A comprehensive functional analysis of NTRK1 missense mutations causing hereditary sensory and autonomic neuropathy type IV (HSAN IV). *Human Mutation*, *38*(1), 55–63.

Shaikh, S. S., Nahorski, M. S., & Woods, C. G. (2018). A third HSAN5 mutation disrupts the nerve growth factor furin cleavage site. *Molecular Pain*, *14*. 1744806918809223.

Sharma, N., Flaherty, K., Lezgiyeva, K., Wagner, D. E., Klein, A. M., & Ginty, D. D. (2020). The emergence of transcriptional identity in somatosensory neurons. *Nature*, *577*(7790), 392–398.

Silos-Santiago, I., Molliver, D. C., Ozaki, S., Smeyne, R. J., Fagan, A. M., Barbacid, M., et al. (1995). Non-TrkA-expressing small DRG neurons are lost in TrkA deficient mice. *The Journal of Neuroscience: The Official Journal of the Society for Neuroscience*, *15*(9), 5929–5942.

Smeyne, R. J., Klein, R., Schnapp, A., Long, L. K., Bryant, S., Lewin, A., et al. (1994). Severe sensory and sympathetic neuropathies in mice carrying a disrupted Trk/NGF receptor gene. *Nature*, *368*(6468), 246–249.

Smith-Anttila, C. J. A., Mason, E. A., Wells, C. A., Aronow, B. J., Osborne, P. B., & Keast, J. R. (2020). Identification of a sacral, visceral sensory transcriptome in embryonic and adult mice. *eNeuro*, 7(1), ENEURO.0397-19.2019 https://doi.org/10.1523/ENEURO.0397-19.2019.

Song, X., Chen, B. N., Zagorodnyuk, V. P., Lynn, P. A., Blackshaw, L. A., Grundy, D., et al. (2009). Identification of medium/high-threshold extrinsic mechanosensitive afferent nerves to the gastrointestinal tract. *Gastroenterology*, *137*(1), 274–284. 284.e1.

Spinsanti, G., Zannolli, R., Panti, C., Ceccarelli, I., Marsili, L., Bachiocco, V., et al. (2008). Quantitative Real-Time PCR detection of TRPV1-4 gene expression in human leukocytes from healthy and hyposensitive subjects. *Molecular Pain*, *4*, 51.

Stantcheva, K. K., Iovino, L., Dhandapani, R., Martinez, C., Castaldi, L., Nocchi, L., et al. (2016). A subpopulation of itch-sensing neurons marked by Ret and somatostatin expression. *EMBO Reports*, *17*(4), 585–600.

Sun, Y., Dykes, I. M., Liang, X., Eng, S. R., Evans, S. M., & Turner, E. E. (2008). A central role for Islet1 in sensory neuron development linking sensory and spinal gene regulatory programs. *Nature Neuroscience*, *11*(11), 1283–1293.

Sung, K., Ferrari, L. F., Yang, W., Chung, C., Zhao, X., Gu, Y., et al. (2018). Swedish nerve growth factor mutation (NGFR100W) defines a role for TrkA and p75NTR in nociception. *The Journal of Neuroscience: The Official Journal of the Society for Neuroscience*, *38*(14), 3394–3413.

Takashima, Y., Ma, L., & McKemy, D. D. (2010). The development of peripheral cold neural circuits based on TRPM8 expression. *Neuroscience*, *169*(2), 828–842.

Tan, C.-H., & McNaughton, P. A. (2016). The TRPM2 ion channel is required for sensitivity to warmth. *Nature*, *536*(7617), 460–463.

Testa, G., Mainardi, M., Morelli, C., Olimpico, F., Pancrazi, L., Petrella, C., et al. (2019). The NGFR100W mutation specifically impairs nociception without affecting cognitive performance in a mouse model of hereditary sensory and autonomic neuropathy type V. *The Journal of Neuroscience: The Official Journal of the Society for Neuroscience*, *39*(49), 9702–9715.

Usoskin, D., Furlan, A., Islam, S., Abdo, H., Lönnerberg, P., Lou, D., et al. (2015). Unbiased classification of sensory neuron types by large-scale single-cell RNA sequencing. *Nature Neuroscience*, *18*(1), 145–153.

Ventéo, S., Desiderio, S., Cabochette, P., Deslys, A., Carroll, P., & Pattyn, A. (2019). Neurog2 deficiency uncovers a critical period of cell fate plasticity and vulnerability among neural-crest-derived somatosensory progenitors. *Cell Reports*, *29*(10), 2953–2960.e2.

Vriens, J., Owsianik, G., Hofmann, T., Philipp, S. E., Stab, J., Chen, X., et al. (2011). TRPM3 is a nociceptor channel involved in the detection of noxious heat. *Neuron*, *70*(3), 482–494.

White, F. A., Silos-Santiago, I., Molliver, D. C., Nishimura, M., Phillips, H., Barbacid, M., et al. (1996). Synchronous onset of NGF and TrkA survival dependence in developing dorsal root ganglia. *The Journal of Neuroscience: The Official Journal of the Society for Neuroscience*, *16*(15), 4662–4672.

Wickramasinghe, S. R., Alvania, R. S., Ramanan, N., Wood, J. N., Mandai, K., & Ginty, D. D. (2008). Serum response factor mediates NGF-dependent target innervation by embryonic DRG sensory neurons. *Neuron*, *58*(4), 532–545.

Woods, C. G., Babiker, M. O. E., Horrocks, I., Tolmie, J., & Kurth, I. (2015). The phenotype of congenital insensitivity to pain due to the NaV1.9 variant p.L811P. *European Journal of Human Genetics*, *23*(5), 561–563.

Yamashita, N., & Kuruvilla, R. (2016). Neurotrophin signaling endosomes: Biogenesis, regulation, and functions. *Current Opinion in Neurobiology*, *39*, 139–145.

Yang, F.-C., Tan, T., Huang, T., Christianson, J., Samad, O. A., Liu, Y., et al. (2013). Genetic control of the segregation of pain-related sensory neurons innervating the cutaneous versus deep tissues. *Cell Reports*, *5*(5), 1353–1364.

Yoshikawa, M., Senzaki, K., Yokomizo, T., Takahashi, S., Ozaki, S., & Shiga, T. (2007). Runx1 selectively regulates cell fate specification and axonal projections of dorsal root ganglion neurons. *Developmental Biology*, *303*(2), 663–674.

Zeisel, A., Hochgerner, H., Lönnerberg, P., Johnsson, A., Memic, F., van der Zwan, J., et al. (2018). Molecular architecture of the mouse nervous system. *Cell*, *174*(4), 999–1014.e22.

Zhang, S., Malik Sharif, S., Chen, Y.-C., Valente, E.-M., Ahmed, M., Sheridan, E., et al. (2016). Clinical features for diagnosis and management of patients with PRDM12 congenital insensitivity to pain. *Journal of Medical Genetics*, *53*(8), 533–535.

Zhang, X. Y., Wen, J., Yang, W., Wang, C., Gao, L., Zheng, L. H., et al. (2013). Gain-of-function mutations in SCN11A cause familial episodic pain. *American Journal of Human Genetics*, *93*(5), 957–966.

Zheng, Y., Liu, P., Bai, L., Trimmer, J. S., Bean, B. P., & Ginty, D. D. (2019). Deep sequencing of somatosensory neurons reveals molecular determinants of intrinsic physiological properties. *Neuron*, *103*(4), 598–616.e7.

Zhong, J., Li, X., McNamee, C., Chen, A. P., Baccarini, M., & Snider, W. D. (2007). Raf kinase signaling functions in sensory neuron differentiation and axon growth in vivo. *Nature Neuroscience*, *10*(5), 598–607.

Zou, M., Li, S., Klein, W. H., & Xiang, M. (2012). Brn3a/Pou4f1 regulates dorsal root ganglion sensory neuron specification and axonal projection into the spinal cord. *Developmental Biology*, *364*(2), 114–127.

Zylka, M. J., Dong, X., Southwell, A. L., & Anderson, D. J. (2003). Atypical expansion in mice of the sensory neuron-specific Mrg G protein-coupled receptor family. *Proceedings of the National Academy of Sciences of the United States of America*, *100*(17), 10043–10048.

Zylka, M. J., Rice, F. L., & Anderson, D. J. (2005). Topographically distinct epidermal nociceptive circuits revealed by axonal tracers targeted to Mrgprd. *Neuron*, *45*(1), 17–25.

CHAPTER TWELVE

Building the brain from scratch: Engineering region-specific brain organoids from human stem cells to study neural development and disease

Fadi Jacob[a,b,c], Jordan G. Schnoll[a], Hongjun Song[a,d,e,f], and Guo-li Ming[a,d,e,g,*]

[a]Department of Neuroscience and Mahoney Institute for Neurosciences, Perelman School of Medicine, University of Pennsylvania, Philadelphia, PA, United States
[b]The Solomon H. Snyder Department of Neuroscience, Johns Hopkins University School of Medicine, Baltimore, MD, United States
[c]Medical Scientist Training Program, Johns Hopkins University School of Medicine, Baltimore, MD, United States
[d]Department of Cell and Developmental Biology, Perelman School of Medicine, University of Pennsylvania, Philadelphia, PA, United States
[e]Institute for Regenerative Medicine, University of Pennsylvania, Philadelphia, PA, United States
[f]The Epigenetics Institute, Perelman School of Medicine, University of Pennsylvania, Philadelphia, PA, United States
[g]Department of Psychiatry, Perelman School of Medicine, University of Pennsylvania, Philadelphia, PA, United States
*Corresponding author: e-mail address: gming@pennmedicine.upenn.edu

Contents

Current Topics in Developmental Biology, Volume 142
ISSN 0070-2153
https://doi.org/10.1016/bs.ctdb.2020.12.011

Abstract

Human brain development is an intricate process that involves precisely timed coordination of cell proliferation, fate specification, neuronal differentiation, migration, and integration of diverse cell types. Understanding of these fundamental processes, however, has been largely constrained by limited access to fetal brain tissue and the inability to prospectively study neurodevelopment in humans at the molecular, cellular and system levels. Although non-human model organisms have provided important insights into mechanisms underlying brain development, these systems do not fully recapitulate many human-specific features that often relate to disease. To address these challenges, human brain organoids, self-assembled three-dimensional neural aggregates, have been engineered from human pluripotent stem cells to model the architecture and cellular diversity of the developing human brain. Recent advancements in neural induction and regional patterning using small molecules and growth factors have yielded protocols for generating brain organoids that recapitulate the structure and neuronal composition of distinct brain regions. Here, we first provide an overview of early mammalian brain development with an emphasis on molecular cues that guide region specification. We then focus on recent efforts in generating human brain organoids that model the development of specific brain regions and highlight endeavors to enhance the cellular complexity to better mimic the *in vivo* developing human brain. We also provide examples of how organoid models have enhanced our understanding of human neurological diseases and conclude by discussing limitations of brain organoids with our perspectives on future advancements to maximize their potential.

1. Introduction

Brain organoids are self-organized three-dimensional (3D) neural aggregates formed from pluripotent stem cells (PSCs) that recapitulate the cytoarchitecture and cellular diversity of the developing brain (Qian, Song, & Ming, 2019). Brain organoids recapitulate key characteristic features of fetal nervous system development, including progenitor zone organization and sequential generation of neurons and glia. Studies in model organisms and *in vitro* combinatorial morphogen screening have informed protocols for differentiating PSC aggregates into organoids that generate neural progenitors and neurons of various brain regions. We first provide an overview of mammalian brain development from neural tube formation to region specification and neurogenesis, focusing on morphogens involved in specifying regional fates along the dorsal-ventral and rostral-caudal axes, which forms the basis for many region-specific brain organoid protocols. We then delve into human-specific features of brain development and compare advantages and disadvantages of monolayer and three-dimensional *in vitro* cultures. We further review different brain organoids and their applications and finally we discuss limitations and opportunities for future improvements.

1.1 Fundamentals of mammalian brain development

Mammalian brain development is a highly coordinated process that integrates diverse signals across time and space. Brain morphogenesis begins during gastrulation with neural induction of the dorsal ectoderm by signals from the mid-gastrula organizer that secretes Bone Morphogenetic Protein (BMP) inhibitors (Levine & Brivanlou, 2007). Neural fate is often described as the "default fate" as embryonic stem cells (ESCs) readily express neural markers, such as NESTIN, when deprived of any growth factors or morphogens, whereas BMP4 addition promotes epidermal differentiation (Munoz-Sanjuan & Brivanlou, 2002). After gastrulation, the dorsal ectoderm thickens to form the neural plate, which proliferates, invaginates, and separates from the surface ectoderm to form the neural tube in a process known as neurulation (Wilson & Hemmati-Brivanlou, 1997). Following neurulation, the neural tube is patterned along the rostral-caudal axis into three major brain regions, the prosencephalon (forebrain), mesencephalon (midbrain), and rhombencephalon (hindbrain) and the spinal cord. The prosencephalon further segregates into the telencephalon, which forms the cerebral cortex and basal ganglia, and the diencephalon, which forms the retina, thalamus, and hypothalamus. Likewise, the rhombencephalon further segregates into the metencephalon, which forms the pons and cerebellum, and the myelencephalon, which forms the medulla (Fig. 1A).

A widely accepted concept is that the most rostral fate constitutes the primitive identity of the neural tube (Levine & Brivanlou, 2007; Stern, 2001) and after rostral neural fate acquisition, the rostral-caudal and dorsal-ventral axes are defined by positional cues from organizing centers located near to and within the developing neural tube. Regional brain patterning relies on morphogen gradients secreted by secondary organizing centers. Generally, WNT and BMP inhibition enhances rostral neural fate, whereas retinoic acid (RA), WNTs, and FGFs enhance caudal fates in a

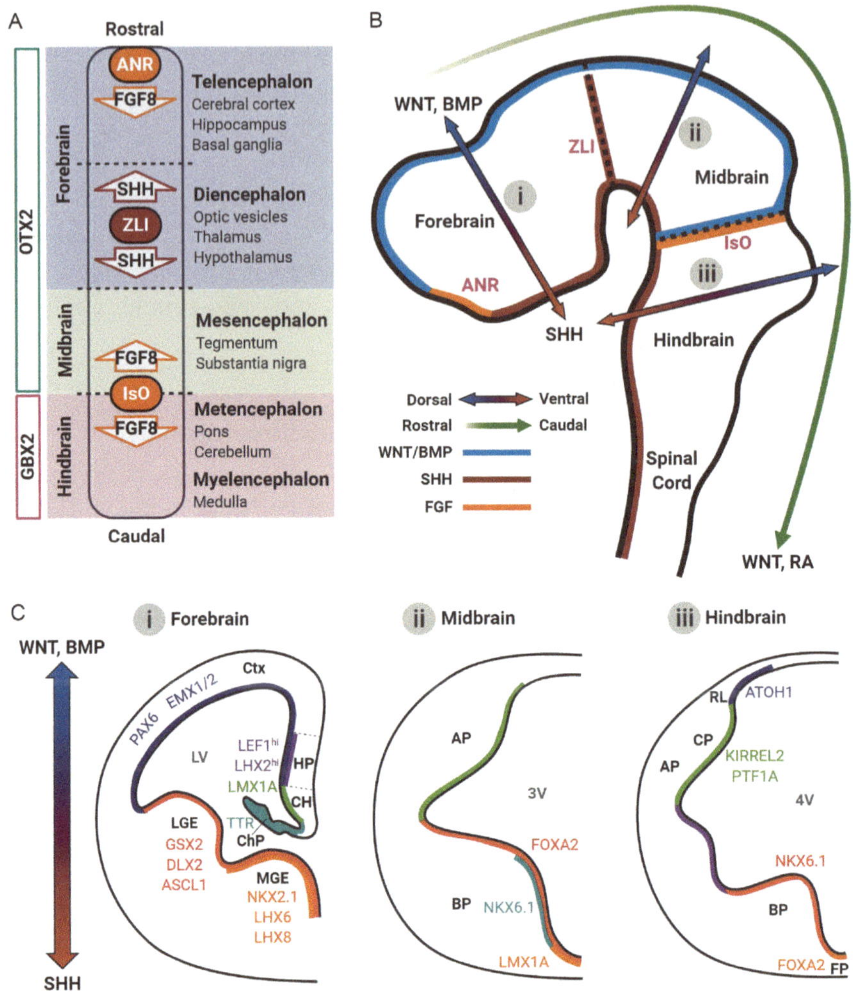

Fig. 1 Regional patterning of the developing neural tube. (A) Simplified diagram of major brain regions and their derivatives along the rostral-caudal axis of the developing neural tube with the location of important secondary organizers and their secreted morphogens. The non-overlapping expression of the transcription factors OTX2 and GBX2 distinguishes the developing forebrain and midbrain from the hindbrain. ANR, anterior neural ridge; ZLI, *zona limitans intrathalamica*; IsO, isthmic organizer. (B) Diagram of the midline sagittal plane of the embryonic mouse brain with the location of important secondary organizers and fate-inducing signals. Increasing WNT and RA specifies fates along the rostral-caudal axis and the opposing actions of WNT/BMP from the roof plate and SHH from the floor plate specifies dorsal and ventral fates. (C) Diagrams of the coronal cross sections of the embryonic mouse forebrain (i), midbrain (ii), and hindbrain (iii) outlining major progenitor zones and their defining transcription factors along the dorsal-ventral axis. Ctx, cortex; LGE, lateral ganglionic eminence; MGE, medial ganglionic eminence; HP, hippocampus; CH, cortical hem; ChP, choroid plexus; AP, alar plate; BP, basal plate; RL, rhombic lip; CP, cerebellar plate; FP, floor plate; LV, lateral ventricle; 3V, third ventricle; 4V, fourth ventricle.

dose-dependent manner (Wilson & Hemmati-Brivanlou, 1997). Secondary organizers often secrete similar molecules that exert differential downstream effects due to prepatterned cellular competencies. The rostral-caudal axis contains several secondary organizers, including the anterior neural ridge (ANR), the *zona limitans intrathalamica* (ZLI), and the isthmic organizer (IsO) (Fig. 1A). Telencephalic fate induction arises from a discrete group of cells known as the ANR located at the rostral end of the embryo. Secretion of FGF molecules, most notably FGF8, from the ANR induces the adjacent neural ectoderm to express the transcription factor FOXG1, which defines the telencephalon (Rubenstein & Beachy, 1998). The ZLI patterns the diencephalon by secreting sonic hedgehog (SHH), which diffuses locally to specify the pre-thalamus rostrally and the thalamus caudally by activation of the transcription factors DLX2 and GBX2, respectively (Scholpp, Wolf, Brand, & Lumsden, 2006). The IsO which is close to the future midbrain-hindbrain boundary secretes FGF8 and WNT1, which diffuse locally to specify the tectum rostrally and the cerebellum caudally (Chi, Martinez, Wurst, & Martin, 2003).

Neural progenitor fates along the rostral-caudal axis are further specified by signals from the dorsal and ventral poles of the neural tube. Soon after neural induction, BMP is secreted from the lateral ectoderm and diffuses to induce the formation of the roof plate, a temporary glial population that also secretes BMP and WNT to specify dorsal fates (Wilson & Maden, 2005). Conversely, ventral to the neural tube lies the notochord, a transient mesodermal structure that secretes SHH to induce the formation of the floor plate, a temporary glial population that also secretes SHH to specify ventral fates (Wilson & Maden, 2005). Dorsal-ventral patterning of the neural tube is accomplished by the opposing signals of BMP/WNT and SHH, whose combined gradients specify distinct progenitor zones (Fig. 1B). The combination of timed morphogen gradients during brain morphogenesis leads to the restricted expression of distinct transcription factors, which define the identity of early progenitor zones and their competency to generate different neural subtypes (Fig. 1C). For example, the dorsal forebrain, which produces the glutamatergic projection neurons of the cortex and hippocampus, is defined by EMX1/2 and PAX6 expression, whereas the ventral telencephalon, which produces the GABAergic interneurons of the cortex and basal ganglia, is defined by GSX2 expression (Campbell, 2003). Restricted expression of NKX2.1 toward the midline further separates the medial (MGE) from the lateral ganglionic eminence (LGE) that generate different interneuron subtypes (Campbell, 2003).

1.2 Human-specific features

The human brain has undergone a massive increase in relative size over the last few million years of evolution, with major expansion of the surface area and thickness of the neocortex, which is in part responsible for the acquisition of higher cognitive functions in humans (Hill & Walsh, 2005; Lui, Hansen, & Kriegstein, 2011). Comparative studies among the human, non-human primate, and rodent cortex reveal not only an increase in the number, but also the diversity of cortical neurons in humans. The cortical expansion has been attributed to duplications in key neural stem cell regulatory genes, such as NOTCH, and the emergence of an additional neural progenitor population termed outer radial glia cells (oRGCs) situated in the outer subventricular zone (oSVZ), which produce many upper layer neurons (Suzuki et al., 2018). This second wave of neurogenesis from oRGCs is believed to be responsible for the massive increase in cortical size and neuron abundance. For example, several genes that cause congenital microcephaly when mutated, such as abnormal spindle protein (ASPM), were shown to only have minor phenotypes in mice, which lack a distinct oSVZ with oRGCs (Pulvers et al., 2010). In contrast, brain organoids grown from ASPM-mutant human induced pluripotent stem cells (hiPSCs) derived from microcephaly patients showed robust deficits in progenitor proliferation and neuronal generation (Li, Sun, et al., 2017). In addition, human brain development occurs over a much longer timeframe than rodents and disparities in microcephaly phenotypes may relate to differential cell-cycle lengths and division properties that are intrinsic to human cells. Certain mutations may also exert greater effects on oRGC division, adding to the discrepancy. Differences in the size, cellular diversity, and synaptic properties between humans and rodents highlights the need for a cellular model system for human brain development.

1.3 Comparison of *in vitro* human cell models

Human stem cell-derived models of neural development broadly include monolayer, neurosphere, and brain organoid cultures (Fig. 2). These model systems exist along a spectrum of least to most complex and each offer their own advantages and disadvantages. Monolayer cultures are the simplest systems and consist of adherent neural progenitor cells (NPCs) that undergo neuronal differentiation upon low-density plating and/or the addition of cell-cycle inhibitors (e.g., AraC) or maturation factors, resulting in fairly uniform cultures of adherent neurons. Monolayer cultures can be easily scaled up for large studies or screens by expanding NPCs before neuronal

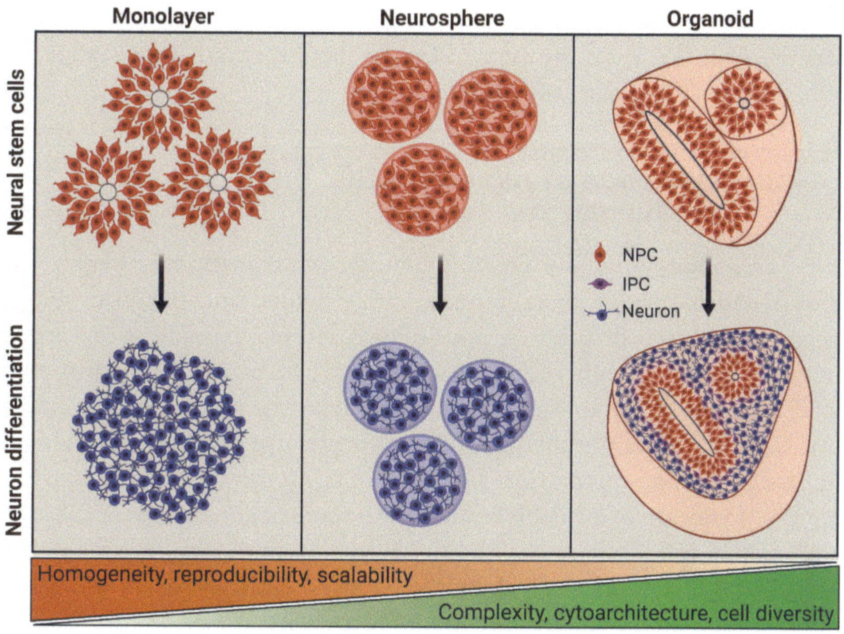

Fig. 2 Comparison of *in vitro* models of human neurodevelopment. Human pluripotent stem cell-derived neural progenitor cells can be cultured as monolayers, neurospheres, or organoids that undergo neuronal differentiation with prolonged culture. Monolayer cultures represent the most homogenous cultures that are reproducible and easily scalable by passaging. Brain organoids represent the most heterogeneous cultures that contain self-organized 3D cytoarchitecture, mixtures of progenitors and neurons, and diverse subpopulations of neural cells. Neurospheres represent an intermediate culture of neural progenitors or neurons in 3D that lack organized structures.

differentiation. Neurospheres are free-floating aggregates of neural progenitors that continuously expand with continued FGF and EGF exposure. Neurospheres permit neurons to interact with one another within a 3D environment, but do not contain progenitor structures and neuronal lamination that resemble the developing brain. Brain organoids exhibit more complex cytoarchitecture over time, consisting of both neural progenitors and developing neurons. Importantly, the structural organization of progenitors into ventricular zones followed by continuous neurogenesis for diverse neuronal subtypes and neuronal migration pattern more accurately resembles human brain development. With this increased complexity comes higher degrees of variability in structure and composition. In contrast to monolayers, 3D aggregates respond less uniformly to culture agents and undergo spontaneous self-organization. This self-organization leads to inter-organoid variability in structure size, shape, and regional identity that may complicate some analyses. This disadvantage is often outweighed by benefits of more faithfully recapitulating the structural organization and cellular diversity of the fetal brain.

2. Generation of region-specific brain organoids from human stem cells

The foundation for creating human brain organoids begins with undifferentiated cultures of human PSCs, which include both human embryonic stem cells (hESCs) and hiPSCs. Ethical considerations around the use of human embryonic tissue for research has limited the number of hESC lines available for research; however, the discovery that human somatic cells such as fibroblasts and lymphocytes can be reprogrammed into hiPSCs by forced expression of a cocktail of transcription factors has led to a massive expansion of available human pluripotent stem cell lines (Takahashi et al., 2007). hiPSCs have the additional advantage of retaining the genetic background of individuals, making them especially useful for disease modeling. The protocols for generating brain organoids start with 3D hPSC aggregates and fall into two broad categories: unguided and guided differentiation (Fig. 3). We provide an overview of unguided and guided human brain organoid differentiation protocols with an emphasis on studies that define specific morphogen formulations and exposure lengths to specify distinct regional fates (Table 1).

2.1 Unguided differentiation: Cerebral organoids

hPSC aggregates grown in suspension preferentially differentiate toward neural ectoderm in the absence of additional patterning factors. hPSC aggregates are cultured as free-floating embryoid bodies (EBs), embedded into Matrigel, and cultured in serum-free conditions in a spinning bioreactor to enhance nutrient diffusion and support long-term health and tissue growth. These aggregates, termed cerebral organoids, contain neural stem cells that self-organize to form structures with a variety of cell lineages that resemble multiple brain regions, such as the dorsal and ventral forebrain, hippocampus, retina, and choroid plexus (Lancaster & Knoblich, 2014; Lancaster et al., 2013). For example, certain regions within a cerebral organoid recapitulate structural features of the developing cortex, with well demarcated ventricular zones with apical adherens junctions and radially oriented neural progenitor cells. These radial glia cells give rise to a cortical plate with neurons of the various layers of the cortex, including neurons expressing markers such as TBR1, SATB2, CTIP2 and RELN.

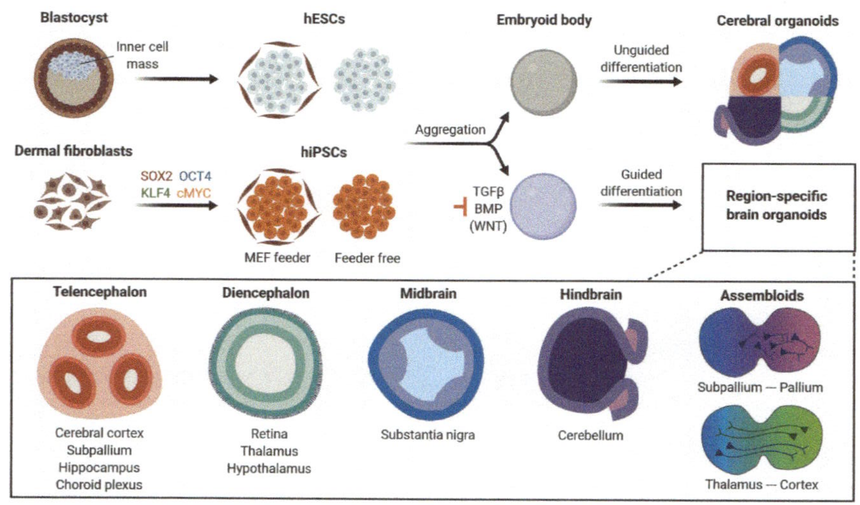

Fig. 3 Generation of brain region-specific organoids from human pluripotent stem cells. Diagram of the major steps for generating cerebral and region-specific brain organoids. Human PSCs, which include hESCs derived from the inner cell mass of human blastocysts and hiPSCs derived from cellular reprogramming of patient-derived somatic cells, such as dermal fibroblasts, are the starting material for generating brain organoids. hPSCs can be cultured on traditional mouse embryonic fibroblast feeder cells or without feeder cells using specialized media formulations. hPSCs are aggregated and cultured in suspension as embryoid bodies that are either placed in medium without morphogens for unguided differentiation toward cerebral organoids or neural induction medium with a series of patterning molecules for guided differentiation toward region-specific organoids. Organoid protocols have been developed for various brain regions of the telencephalon, diencephalon, midbrain, and hindbrain (see Table 1). Fused region-specific brain organoids, termed assembloids, have been developed to study the interactions between neurons of different brain regions.

Table 1 Region-specific brain organoid protocols.

Brain region	Starting cells	Patterning and maturation factors	ECM	Culture apparatus	Reference(s)
Brain (unguided)	ESCs or iPSCs (MEF)	NA (intrinsic)	Matrigel	Spinning bioreactor	Lancaster et al. (2013) and Lancaster and Knoblich (2014)
	iPSCs (MEF)	NA (intrinsic)	Defined hydrogel	Static culture dish	Lindborg et al. (2016)
	iPSCs (FF)	NA (intrinsic)	Matrigel	Spinning bioreactor	Quadrato et al. (2017)
Dorsal forebrain	ESCs (MEF)	SB-431542, IWR1	Matrigel	Static culture dish	Kadoshima et al. (2013)
	iPSCs (MEF or FF)	Dorsomorphin, SB-431542, (XAV-939) FGF2, EGF, BDNF, NT3	NA	Static culture plate	Pasca et al. (2015) and Yoon et al. (2018)
	iPSCs (FF)	Noggin, DKK1, bFGF, EGF, BDNF, GDNF	NA	Static culture plate	Mariani et al. (2015)
	iPSCs (MEF)	A-83, dorsomorphin; CHIR-99021, SB-431542, WNT3A	Matrigel	Miniature spinning bioreactor (Spin**Ω**)	Qian et al. (2018, 2016)
	ESCs or iPSCs (MEF or FF)	NA (intrinsic) or CycA	Matrigel	Culture plate on orbital shaker	Bagley et al. (2017)
	ESCs or iPSCs (MEF)	CHIR-99021, PLGA microfilament	Matrigel	Culture plate on orbital shaker	Lancaster et al. (2017)
	ESCs or iPSCs (FF)	SB-431542, LDN-193189, XAV-939, BDNF	NA	Culture plate on orbital shaker	Xiang et al. (2019, 2017)
	iPSCs (FF)	SB-431542, IWR1	Matrigel	Spinning bioreactor	Velasco et al. (2019)

Medial pallium	ESCs (MEF)	SB-431542, IWR1e, CHIR-99021, BMP4	NA	Permeable film culture dish	Sakaguchi et al. (2015)
Choroid plexus	ESCs or iPSCs (FF)	CHIR-99021, BMP4	Matrigel	Culture plate on orbital shaker	Pellegrini, Bonfio, et al. (2020)
	iPSCs (FF)	LDN-193189, SB-431542, IWP-2, CHIR-99021, BMP7, BDNF, GDNF	NA	Culture plate on orbital shaker	Jacob, Pather, et al. (2020)
Ventral forebrain	ESCs (MEF)	SB-431542, IWR1e, SAG	Matrigel	Static culture dish	Kadoshima et al. (2013)
	ESCs or iPSCs (MEF or FF)	IWP-2, SAG	Matrigel	Culture plate on orbital shaker	Bagley et al. (2017)
	ESCs or iPSCs (MEF)	Dorsomorphin, SB-431542, IWP-2, SAG, FGF2, EGF, BDNF, NT3	NA	Static culture plate	Birey et al. (2017) and Sloan, Andersen, Pasca, Birey, and Pasca (2018)
	ESCs or iPSCs (FF)	SB-431542, LDN-193189, XAV-939, SHH, purmorphamine, BDNF	NA	Culture plate on orbital shaker	Xiang et al. (2019, 2017)
Optic cup and retina	ESCs (MEF)	IWR1e, SAG, CHIR-99021	Matrigel	Static culture dish	Nakano et al. (2012)
	ESCs (MEF)	BMP4, CHIR-99021, SU5402	NA	Static culture dish	Kuwahara et al. (2015)
	iPSCs (FF)	NA (Intrinsic), RA (photoreceptor maturation)	Matrigel	Static culture dish	Cowan et al. (2020) and Zhong et al. (2014)
Thalamus	iPSCs (FF)	SB-431542, LDN-193189, Insulin, PD0325901, BMP7, BDNF	NA	Culture plate on orbital shaker	Xiang et al. (2019)

Continued

Table 1 Region-specific brain organoid protocols.—cont'd

Brain region	Starting cells	Patterning and maturation factors	ECM	Culture apparatus	Reference(s)
Hypothalamus	iPSCs (MEF)	LDN-193189, SB-431542, WNT3A, SHH, purmorphamine, FGF2, CNTF	NA	Miniature spinning bioreactor (SpinΩ)	Qian et al. (2018, 2016)
Midbrain	iPSCs (MEF)	LDN-193189, SB-431542, SHH, FGF8, purmorphamine, CHIR-99021, BDNF, GDNF	NA	Miniature spinning bioreactor (SpinΩ)	Qian et al, (2016, 2018)
	ESCs (FF)	SB-431542, Noggin, CHIR-99021, SHH, FGF8, BDNF, GDNF, cAMP	Matrigel	Culture plate on orbital shaker	Jo et al. (2016)
	NESCs	SB-431542, dorsomorphin, CHIR-99021, purmorphamine, BDNF, GDNF, cAMP, TGFβ	Matrigel	Culture plate on orbital shaker	Monzel et al. (2017)
Cerebellum	ESCs (MEF)	SB-431542, FGF2, Insulin, FGF19, SDF1	NA	Static culture dish	Muguruma et al. (2015)

Summary of protocols for generating region-specific brain organoids, including the starting cells, patterning and maturation factors, extracellular matrix (ECM), and special culture conditions used in each procedure. Selected references are provided for further reading. MEF, mouse embryonic fibroblast; FF, feeder-free.

While unguided differentiation of cerebral organoids generates remarkable cell diversity of the developing brain, it has two main drawbacks: (1) unguided differentiation leads to unavoidable contamination of non-neural cell lineages, such as mesoderm and endoderm; and (2) the conglomerate of many brain regions is non-uniform, with high degrees of structural and cell type variability among organoids. One effort to improve forebrain specificity and the quality of ventricular structures was to embed hPSC aggregates with engineered microfilaments, which improved the size and reproducibility of forebrain ventricular structures within cerebral organoids, although structural and cell fate variability still remains (Lancaster et al., 2017).

2.2 Guided differentiation: Region-specific brain organoids

To enhance region specificity and organoid reproducibility for developmental studies and disease modeling, several groups have devised protocols for generating regionally patterned brain organoids by influencing the fate of neuroepithelial cells. Guided differentiation protocols utilize small molecules and recombinant growth factors with precise timing of exposure to instruct hPSC aggregates to differentiate into structures and cell types of specific brain regions. The majority of protocols involves specification of major brain regions, such as the forebrain, midbrain, and hindbrain, but many protocols have been further optimized to enrich for subregions of the brain including the cerebral cortex, medial pallium, thalamus, hypothalamus, substantia nigra and cerebellum (Fig. 3). The first step toward generating regionally patterned brain organoids is to form an EB. This can be accomplished using stem cell cultures that rely on mouse embryonic feeder (MEF) cells or medium formulations that allow stem cell expansion under feeder-free conditions (Fig. 3). Stem cell colonies are enzymatically detached or dissociated and reaggregated and transferred to suspension culture to form EBs, which are guided towards neuroectoderm fate by inhibiting the BMP and TGFβ pathways, commonly referred to as "dual-SMAD inhibition." This inhibition mimics the effects of the mid-gastrula organizer during development and is generally accomplished by using a combination of small molecule inhibitors of signaling pathways of BMP (LDN-193189 or Dorsomorphin) and TGFβ (SB-431542 or A-83). Small molecules are generally preferred over recombinant growth factors due to their improved stability and lower cost. Many protocols also include simultaneous WNT inhibition with small molecules (XAV-939 and IWP-1/2), which enhance neural induction and help to specify rostral fate. Efficient neural ectoderm specification provides a blank state for further patterning to specific regions of the brain.

2.2.1 Cortical organoids

Cortical organoids are the most abundant and perhaps most well-studied group of brain organoids due to their relatively simplistic patterning procedures. Cortical fate is often observed as the default fate of neuroectodermal EBs and several protocols have emerged that apply additional small molecules, growth factors, or extracellular matrix to improve region specification, ventricular zone expansion, and neuronal layering (Qian et al., 2019). In general, cortical organoid protocols involve enrichment for neural ectoderm at the EB stage followed by a ventricular zone expansion phase using small molecules or Matrigel, and then a neurogenesis phase with basal medium containing nutrients important for neural differentiation and survival. The Sasai group first reported the generation of cortical organoids containing cortex-like structures after embedding EBs patterned to neural fate with the TGFβ and WNT antagonists SB-431542 and IWR1 into Matrigel (Kadoshima et al., 2013). The resulting organoids contain several ventricular structures, each with a defined ventricular zone enriched for $FOXG1^{+}$ forebrain progenitors. Most of the ventricular structures contain $PAX6^{+}$ dorsal forebrain radial glia cells (RGCs) that give rise to a neural plate containing neurons expressing CTIP2, TBR1, and SATB2. Interestingly, certain regions within these organoids express GSH2, a marker for ventral forebrain progenitors, and GAD65, a marker for GABAergic interneurons, contributing to increased forebrain neuronal diversity at the cost of regional purity. Ventricular RGCs within these organoids undergo stereotypical interkinetic nuclear migration as occurs in the developing mammalian cortex. The Ming group found that continued TGFβ inhibition combined with exposure to the WNT agonist CHIR99021 led to marked expansion of neural progenitors within Matrigel and the formation of large ventricular structures that are highly enriched for dorsal forebrain progenitors (Qian et al., 2018, 2016). After about 50 days in culture, these forebrain organoids contain a particularly large and well-defined oSVZ containing $HOPX^{+}$ oRGCs that resembles the developing human cortex. This protocol involves the additional use of a spinning bioreactor which improves nutrient and oxygen diffusion for long-term maintenance. In contrast, the protocol developed by the Pasca group uses FGF and EGF to expand ventricular progenitors in the absence of any added extracellular matrix (Pasca et al., 2015). Given the role of FGF and EGF in astroglial fate specification, these human cortical spheroids (hCSs) contain not only cortical neurons expressing TBR1, CTIP2 and SATB2, but are also enriched with astrocytes.

Although most protocols for cortical organoids exhibit poor lamination and layer distinction, some have shown the sequential generation of neurons expressing layer-specific markers in the correct order, such as RELN, TBR1, CTIP2, FOXP1, SATB2, BRN2, RORB, and CUX1 (Qian et al., 2016, 2020). These organoids also exhibit similar transcriptome and epitranscriptome as the developing human brain (Yoon, Ringeling, et al., 2017). A major limitation of cortical organoids, which extends to other regional organoids, is the presence of many ventricular structures in one organoid. This complicates analysis because the ventricular structures come in varying shapes and sizes and neurons produced by each can mix, blurring any potential layering. Reducing the number of ventricular structures and improving cortical layer formation and specification will be important for the next generation of cortical organoids. Another challenge in dorsal forebrain patterning is arealization—specifying the different neocortical subregions, such as the frontal, temporal, and occipital lobes; however, the factors that determine or promote subregion identity are not well understood (Cadwell, Bhaduri, Mostajo-Radji, Keefe, & Nowakowski, 2019). Accompanying this challenge is a lack of specific markers to adequately access the identity of subregions, as distinctions between these regions are not discrete and often depend on differences in layering and subcortical projections, which are not adequately modeled in organoids.

2.2.2 Choroid plexus and hippocampal organoids

The choroid plexus of the lateral ventricles arises from the dorsal telencephalon and plays a critical role in the production of cerebrospinal fluid (CSF) and formation of the blood-CSF barrier, which restricts free passage of solutes between the blood and CSF (Lun, Monuki, & Lehtinen, 2015). The most dorsal structures of the forebrain are patterned by high WNT and BMP signaling from the roof plate, with early *in vitro* studies demonstrating the sufficiency of BMP4 exposure to induce choroid plexus from neuroepithelial cells (Fig. 1C) (Watanabe et al., 2012). The Sasai group developed a protocol to produce self-organized choroid plexus epithelium from hESCs by prolonged exposure of EBs to high levels of the WNT agonist CHIR99021 and BMP4 (Sakaguchi et al., 2015). The resulting structures contained cuboidal epithelium expressing the choroid plexus-defining markers TTR and AQP1. The Lancaster group established a modified protocol which relies on exposing EBs to the same patterning molecules combined with Matrigel embedding to generate choroid plexus organoids (CPOs) that produce

CSF-containing structures (Pellegrini, Bonfio, et al., 2020). Transcriptomic analysis showed recapitulation of the diversity of human choroid plexus epithelial cells and proteomic analysis confirmed the presence of key human CSF components. These CSF-producing CPOs have an intact barrier function which recapitulated the *in vivo* CNS permeability to small molecules. More recently, the Ming group established an optimized protocol by exposing neural patterned EBs to high levels of CHIR99021 and BMP7 to generate CPOs that were highly enriched for cells expressing TTR, AQP1, OTX2 and ion channels and molecular transporters important for CSF generation (Jacob, Pather, et al., 2020). CPOs generated by this protocol form epithelial extensions resembling the choroid plexus without the addition of Matrigel, but do not form CSF-containing inclusions. While CPOs provide useful model systems to study human choroid plexus epithelial cell function, an important drawback, as in other types of brain organoids, is the lack of integration with functional vasculature and pericytes which form the complete blood-CSF barrier. Therefore, reconstructing a complete, physiological system remains a future challenge.

The hippocampus, a conserved structure among vertebrates that plays a critical role in learning and memory, arises from the medial pallium of the developing telencephalon which lies adjacent to the more midline cortical hem and choroid plexus structures (Fig. 1C) (Grove & Tole, 1999). The cortical hem, an important secondary organizer in the telencephalon, secretes WNTs, which act in combination with BMP signals from the choroid plexus to induce the medial pallium (Subramanian, Remedios, Shetty, & Tole, 2009). The neuroepithelium closest to the cortical hem receives the highest levels of WNT and becomes the dentate neuroepithelium, which produces neurons, astrocytes and adult neural stem cells of the dentate gyrus (Berg et al., 2019). The remaining portion of the medial pallium receives lower WNT and segregates into the *cornu ammonis* (CA) neuroepithelium, which produces the pyramidal neurons of CA1–3. The Sasai group discovered that medial pallium fate could be induced by exposing EBs to high levels of WNT and BMP signals, but only for a brief period of time (Sakaguchi et al., 2015). This exposure induces cortical hem tissue, which in turn induces adjacent medial pallium neural progenitor fate defined by the combined expression of FOXG1, LEF1, LHX2, and PAX6 (Fig. 1C). However, hippocampal neurons are not produced even with prolonged culturing time. Instead, dissociation of 3D medial pallium tissue and long-term maintenance as monolayer cultures produced electrically active neurons expressing the hippocampal marker ZBTB20 and more specific dentate granule neuron marker PROX1 and CA3 pyramidal neuron marker KA1. Future improvements are necessary to produce hippocampal-specific organoids not only with the full diversity of hippocampal neurons, including pyramidal neurons of CA1 and CA2, hilar mossy neurons, and GABAergic interneurons, but also to recapitulate the organized structure and typified circuitry of the mature hippocampus.

2.2.3 Subpallial organoids

The subpallium is divided into MGE, LGE, and CGE that generate different interneuron subtypes (Kepecs & Fishell, 2014). SHH from the floor plate is critical in specifying ventral progenitor fate and has been used to differentiate hPSCs into interneurons in monolayer cultures (Germain, Banda, Becker, Naegele, & Grabel, 2013; Liu et al., 2013; Maroof et al., 2013; Nicholas et al., 2013). Several protocols have emerged for generating brain organoids expressing markers for subpallium neural progenitors and the inhibitory neurons they produce (Bagley, Reumann, Bian, Levi-Strauss, & Knoblich, 2017; Birey et al., 2017; Kadoshima et al., 2013; Xiang et al., 2017). Different exposure lengths and concentrations of SHH and/or the SHH agonist purmorphamine among these protocols may be responsible for the induction of broad ventral vs MGE-specific identity and the differences in interneuron diversity. Most of the neural progenitors in these protocols express NKX2.1, a marker for the MGE, and produce inhibitory neurons characterized by the expression of GAD67 and the production of GABA. A limitation of these protocols is the production of mainly SST, CR, and/or RELN interneurons with low abundance of PV, NPY, or VIP interneurons. Refining these protocols to optimize the production of interneurons with correct subtype proportions remains a major challenge. Additionally, subpallium organoids do not contain any structures resembling the organization of the caudate or putamen nuclei, limiting studies of their development.

2.2.4 Optic cup and retinal organoids

The optic cup and neural retina arise from optic vesicles in the rostral diencephalon. The retina is a highly organized structure with distinct layers composed of defined cell types, including rod and cone photoreceptors. Generation of optic cup-like structures was first demonstrated by the Sasai group by embedding EBs into Matrigel and exposing them to WNT signaling *via* the GSK3β antagonist CHIR99021 (Nakano et al., 2012). Early retinal structures contain $CHX10^+$ progenitors that self-organize and produce a multilayered neural retina containing cells expressing the photoreceptor markers CRX and Recoverin. Many of these photoreceptors expressed the rod marker NRL or the early cone marker RXR-γ. Several protocols for retinal organoids have since emerged, each with modifications to either improve retinal cell purity, cellular organization, or function (Kuwahara et al., 2015; Zhong et al., 2014). Recently, the Roska group optimized a protocol to generate light-sensitive, multilayered human retinal organoids with functional synapses (Cowan et al., 2020). Despite these efforts, no protocol has been able to generate all layers of the retina with definitive structural and synaptic organization. A major challenge remains in reproducing the highly laminar organization of the neural retina. Single-cell transcriptomic studies comparing the mammalian retina to retinal organoids have shown recapitulation of the diverse subtypes of retinal neurons and human cell type-specific (e.g., NRL, rods; ARR3, cones) and disease-associated genes, and will provide a valuable resource to improve existing protocols (Cowan et al., 2020).

2.2.5 Thalamic and hypothalamic organoids

The thalamus and hypothalamus arise from the caudal diencephalon and play important roles in circuit integration and regulation of homeostasis. Unlike the highly laminated structure of the cortex and retina, neurons within the thalamus and hypothalamus organize into discrete nuclei. The Park group utilized caudalizing effects of early insulin exposure followed by BMP7 treatment to pattern EBs toward thalamic fate (Xiang et al., 2019). The MEK-ERK inhibitor PD0325901 was used during BMP7 treatment to prevent excess insulin-induced caudalization toward midbrain fate. This patterning combination induced robust expression of TCF7L2, a marker for the developing thalamus. By scRNA-seq they identified populations of thalamic progenitors, glutamatergic neurons, and GABAergic neurons of the thalamus, emphasizing the neuronal diversity of their system.

Specification of the hypothalamus is defined by the transcription factor RAX1, which is specific to only the hypothalamus and retina during development, which can be distinguished from each other by structural features. Hypothalamic neurons are defined by the expression of many neural peptides, including POMC, NPY and TRH that mark specific subsets of neurons. Several reports have described protocols for generating hypothalamic neurons *via* monolayer culture (Merkle et al., 2015; Rajamani et al., 2018; Wang, Egli, & Leibel, 2016), but only one study describes hypothalamic organoids specifically. The Ming group used a combination of WNT and SHH agonists, including CHIR99021, purmorphamine, and

SHH, followed by FGF2 and CNTF exposure to pattern hypothalamic organoids (Qian et al., 2018, 2016). Neural progenitors in early organoids expressed general hypothalamic marker RAX1 and subtype-specific markers NKX2.1 and FOXA2. Mature organoids contained a diverse collection of neurons expressing the neuropeptides NPY and POMC and the dopaminergic marker TH. Whether these organoids can generate neurons that represent the entire spectrum of neuropeptide secretion remains uncertain and may require protocol modifications to enrich for neuronal subtypes of interest. With both thalamic and hypothalamic organoids, the challenge remains in reconstructing organized nuclei structures and the cellular diversity seen *in vivo*.

2.2.6 Midbrain organoids

The midbrain, which includes the tegmentum and substantia nigra, arises from the developing mesencephalon and plays important functions in information relay between the forebrain and spinal cord, controlling movement and sensory processing. Several protocols for generating midbrain organoids involve the exposure to FGF8 and SHH (Fig. 1) (Jo et al., 2016; Monzel et al., 2017; Qian et al., 2018, 2016). The derivation of midbrain progenitors is defined by expression of FOXA2. Most protocols focus on generating dopaminergic neurons of the substantia nigra, which is of special interest due to their role in the pathogenesis of Parkinson's disease and other related movement disorders. Dopaminergic neurons can be distinguished by the expression of PITX3, tyrosine hydroxylase (TH), the enzyme require to make dopamine, and DAT, the dopamine transporter. Although these protocols all generate TH^+ neurons, whether they belong to the substantia nigra remains debatable. One study from the Ng group uses the presence of pigment, a defining feature of dopaminergic neurons in the substantia nigra, as a criteria to claim that their dopaminergic neurons are truly of the substantia nigra identity (Jo et al., 2016). Further studies that optimize dorsal-ventral patterning will hopefully lead to more defined protocols for generating other midbrain regions.

2.2.7 Hindbrain organoids

The hindbrain arises from the developing rhombencephalon and consists of the cerebellum, pons, and medulla, which play important roles in movement augmentation and information relay to the spinal cord (Fig. 1). Insights into the mechanisms instructing cerebellar differentiation have led to several protocols for generating cerebellar neurons from mESCs and hESCs

(Erceg et al., 2010; Muguruma et al., 2010; Salero & Hatten, 2007; Su et al., 2006; Tao et al., 2010). The Sasai group first reported a protocol to generate self-organized cerebellar tissue that includes regions of cerebellar plate neuroepithelium giving rise to Purkinje cells and interneurons, and rhombic lip neuroepithelium giving rise to granule cells and neurons of the deep cerebellar nuclei (Muguruma, Nishiyama, Kawakami, Hashimoto, & Sasai, 2015). They showed that treatment of hESC aggregates with FGF2, SB431542, and insulin suppressed forebrain fate and induced expression of caudal marker GBX2 and isthmus organizer genes FGF8 and WNT1. Continued culture resulted in neuroepithelium expressing the cerebellar plate markers KIRREL2, PTF1A, and EN2, which generated four different cerebellar neuron precursors, including Purkinje cells, Golgi cells, granule cells, and DCN projection neurons. The addition of FGF19 induced formation of neural tube-like structures that recapitulate the dorsal-ventral polarity of the developing hindbrain. On the other hand, sequential addition of FGF19 and SDF1 induced the formation of a continuous neuroepithelium that differentiates into neurons that organize into a multilayered structure seen in cerebellar development. The cerebellum organoids described by the Sasai group represent an important advancement in recapitulating the major structural organization, progenitor subtypes, and neuronal composition of the cerebellum. However, there was poor survival of long-term cultures, which limits the structural maturation beyond the first trimester of development and necessitates organoid dissociation and 2D culture to observe maturation of cerebellar neuron subtypes. Additionally, the formation of neural networks and appearance of characteristic lobular morphogenesis of the mature cerebellum is missing. Further refinement of patterning methods may lead to more specific protocols for the pons and medulla, other regions in the hindbrain.

3. Advancements in cellular complexity of brain organoids

Although improved neural ectoderm induction and regional patterning have greatly enhanced the fate specificity and reproducibility of brain organoid protocols, these improvements have the caveat of producing predominantly excitatory or inhibitory neurons, which does not recapitulate *in vivo* neuronal diversity. Efficient neural induction also restricts the production of non-neuroectodermal cells, such as microglial, endothelial, and meningeal cells that are critical to the development and function of the human brain. Additionally, despite long-term culture, brain organoids typically contain few glial cells, such as astrocytes and oligodendrocytes, and display only rudimentary synaptic circuitry and electrical activity, limiting their applicability for studying the complex physiology of the human brain and the mechanisms that go awry during disease. There are significant recent advancements in improving the cellular complexity and maturity of brain organoids to create a model system that more faithfully recapitulates the *in vivo* condition. These innovations include protocol modifications to model interactions between brain regions, enhance glial cell production, reconstitute resident immune cells and vasculature, improve healthy organoid longevity and maturity, and study long-range neuronal projections of human neurons (Fig. 4).

3.1 Fusion of region-specific brain organoids

Organoids patterned to specific brain regions contain a more homogenous population of progenitors and neurons with reduced variability among organoids, but this improvement also limits the potential for studying interactions between different brain regions. Several groups have developed methods to overcome this limitation by fusing region-specific organoids, termed "assembloids" (Fig. 3). This methodology offers the benefits of regional purity with the ability to study neuronal interactions. For example, to reproduce the mixed glutamatergic and GABAergic neuronal diversity of the cortex, the first studies fused dorsal forebrain or cortical organoids with subpallial organoids and observed the selective migration of interneurons, mimicking the process that occurs in normal forebrain development (Bagley et al., 2017; Birey et al., 2017; Xiang et al., 2017). In these assembloids, migrating interneurons undergo stepwise cell movements through the process of nucleokinesis, form synapses with glutamatergic neurons, and display increased morphological complexity (Marton & Pasca, 2020). Another study investigated the connections between the thalamus and the cortex by fusing cortical organoids with thalamic organoids and observed reciprocal axonal projections, mimicking the connections that occur in the normal brain (Xiang et al., 2019). After fusion, thalamic neurons form functional synapses with cortical neurons and exhibit higher-frequency firing, suggesting improved maturity following circuitry integration. These studies represent a step forward in improving neuronal diversity and modeling inter-regional connections in brain organoids but require more in-depth characterization of cell-type specific projections and consequences of circuitry formation on

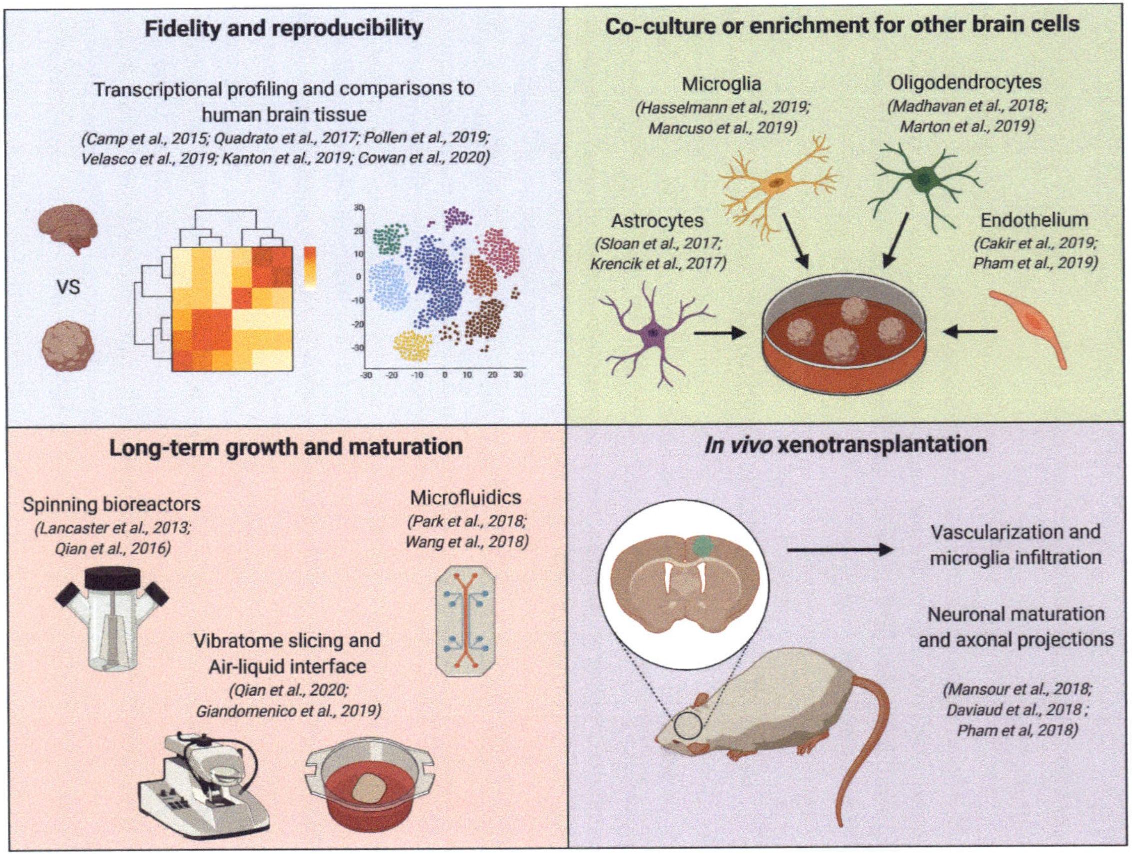

Fig. 4 Improvements and extensions of brain organoid technology. Diagram illustrating major improvements and extensions of human brain organoid technologies, including high-throughput genomics and transcriptomics to compare organoids to human brain tissue and improve reproducibility, co-culture with additional cell types to more fully reconstitute the cellular composition of the human brain, modifications to improve long-term growth and maturation, and *in vivo* xenotransplantation to improve maturation and study axonal projections.

neuronal function. The development of more region-specific protocols will allow modeling of additional regional interactions, such as cortical-striatal circuits, with the added possibility of multi-regional interactions, such as the cortex-hippocampus-amygdala fear and memory circuitry and the cortex-thalamus-spinal cord motor circuitry.

3.2 Enhancing glial cell production and maturation

In the human brain, astrocytes outnumber neurons and play critical roles in maintaining brain homeostasis, such as regulating synapse development and function, modulating blood flow and solute penetrance through the blood-brain barrier, and responding to injury (Sofroniew & Vinters, 2010). Oligodendrocytes support signal transduction by forming insulating myelin sheaths and providing metabolic support to axons (Simons & Nave, 2015). Several brain organoid protocols report the emergence of astrocytes and oligodendrocytes after long-term culture, but their appearance is variable and with low abundance. Given that gliogenesis follows neurogenesis, one approach to improving the abundance and maturation of astrocytes in organoids was to maintain cortical organoids in culture for greater than 100 days (Sloan et al., 2017). The number of astrocytes increased over time and transcriptional comparison to primary human astrocytes revealed striking similarities with the expression of mature astrocyte genes increasing with organoid age. Despite the improvement in astrocyte abundance and maturity, this approach is limited by the need to maintain cultures for very long periods of time accompanied with reduced neuronal abundance and the complete abolishment of ventricular zones and neuronal lamination. The ability of brain organoids to recapitulate the remarkable astrocyte diversity of the human brain, including fibrous and protoplasmic subtypes has just begun to be explored (Qian et al., 2020).

Several brain organoid protocols report the late emergence of oligodendrocyte precursor cells and oligodendrocytes with even some reports of myelination of neuronal axons (Pamies et al., 2017; Qian et al., 2020). In one study, ventral forebrain organoids were found to produce more oligodendrocytes at earlier timepoints than dorsal counterparts, representing the different waves of oligodendrogliogenesis in human development (Kim, Xu, et al., 2019). A recent study described a modified protocol to generate oligo-cortical spheroids that reproducibly make oligodendrocytes and myelin (Madhavan et al., 2018). After cortical patterning, treatment with platelet-derived growth factor AA (PDGFR-AA) and insulin-like growth factor 1 (IGF-1) drives the expansion of OPC populations and further treatment with thyroid hormone (T3) induces oligodendrocyte differentiation and myelination by 10 weeks in culture. Robust myelination was observed by electron microscopy. Exposure of oligo-cortical organoids to the promyelination drugs T3, clemastine, and ketoconazole increased myelination and expression of MYRF (myelin regulatory factor). This approach, however, produces oligodendrocytes at the expense of neurons and lacks obvious structural organization of neurons and axons.

3.3 Reconstitution of resident immune cells and vasculature

Because of improved neuroectoderm specification, brain region-specific organoids are relatively homogeneous for neural precursors, which consequently do not produce other cell types of the brain, including microglia, endothelial, and meningeal cells. Reconstituting these cell populations remains an important endeavor to more accurately mimic *in vivo* cellular diversity (Fig. 4). Microglia are the brain's resident macrophages that perform essential functions in maintaining homeostasis, modulating neuronal circuits, and surveying the environment (Hammond, Robinton, & Stevens, 2018; Hickman, Izzy, Sen, Morsett, & El Khoury, 2018). Microglia are also the primary responders to brain injury and infection and are implicated in the pathogenesis of many neurological disorders, such as Alzheimer's disease (Li & Barres, 2018). Cerebral organoids generated from unguided differentiation protocols were reported to innately contain IBA1$^+$ microglia-like cells, likely derived from mesodermal progenitor cells (Ormel et al., 2018; Quadrato et al., 2017). Although these cells were shown to resemble human microglia morphologically and transcriptionally, it is unknown whether these cells correspond to yolk sack-derived microglia or other tissue macrophages (Bennett & Bennett, 2020). Additionally, the variable and uncontrolled appearance of microglia-like cells in cerebral organoids limits their utility for developmental and disease modeling. The recent emergence of protocols to derive human microglia-like cells *in vitro* (Abud et al., 2017; Douvaras et al., 2017; Haenseler et al., 2017; McQuade et al., 2018; Muffat et al., 2016; Pandya et al., 2017) has created the possibility of adding pure cultures of microglia-like cells to brain organoids (Jacob & Bennett, 2020). Microglia-like cells co-cultured with cerebral organoids readily invade and integrate, expressing microglia-specific markers and displaying ramified-like morphologies (Abud et al., 2017) and microglia co-cultured with forebrain organoids show improved maturation and responses to inflammatory stimuli (Song et al., 2019). Although microglia co-culture methodologies are promising, more work is needed to improve the reproducibility and microglia function for disease modeling.

Endothelial cells lining major blood vessels and capillaries supply the brain with nutrients and play important, yet incompletely understood roles in neural development (Paredes, Himmels, & Ruiz de Almodovar, 2018). The innate presence of endothelial cells has been reported in brain organoids (Bhaduri et al., 2020), but like microglia-like cells, their presence is highly unreliable and with the caveat of a lack of organization into vessel-like structures. To vascularize brain organoids, studies have taken one of several approaches: co-culture brain organoids with endothelial cells, induce differentiation of endothelial cell precursors within organoids, or xenotransplant organoids into mouse brains and allow host vasculature to infiltrate. A pair of studies co-cultured either iPSC-derived endothelial cells or umbilical vein endothelial cells with brain organoids and observed robust engraftment with the formation of capillary-like structures (Pham et al., 2018; Shi et al., 2020). An alternative approach used hESCs engineered to ectopically express human ETV2 to produce endothelial cells within cortical organoids that formed vascular network-like structures with reported improvement to neuron maturity (Cakir et al., 2019). Transplantation of these vascularized organoids into mice showed integration with host vasculature with functional perfusion (Cakir et al., 2019). Despite the presence of endothelial cells that form structures resembling capillaries, there is a lack of concrete evidence that these cells are functional and influence the maturation or function of neurons within brain organoids. Importantly, serum-free medium formulations used commonly in brain organoid cultures may not be sufficient to promote endothelial cell maturation and function, and critically, media is unlikely to flow naturally through these vessels *in vitro*. Flow could theoretically be enhanced by integrating vascular structures with mechanical pumps, a possibility not yet explored.

3.4 Technical modifications for long-term culture

Inevitable hypoxic conditions at the center of large organoids hinders long-term brain organoid health and prevents neurons, astrocytes, and/or oligodendrocytes within brain organoids from reaching full maturity. Spinning bioreactors have been widely used to help increase nutrient and oxygen diffusion within organoids by agitating and better circulating the medium (Fig. 4). In the original cerebral organoid protocol, the Lancaster group made use of large magnetic spinning flasks to improve diffusion (Lancaster et al., 2013). These flasks have allowed organoids to maintain growth long-term but are limited by taking up extensive incubator space and are associated with a high cost. To address this constraint, the Ming group developed a miniature spinning bioreactor which operates on a standard 12-well plate (Qian et al., 2016). This innovation drastically reduces incubator space but is limited by needing to 3D print parts and maintain motors. Although spinning bioreactors have permitted organoids to grow to large sizes, the growth remains constrained by physical limits of oxygen diffusion into the organoid core. Two groups have devised methodologies to circumvent the hypoxia problem by culturing organoids as thick slices, similar to methods used for brain slice electrophysiology. Although both studies employ the same general methodology, they have distinct applications. The Lancaster group cultured their cerebral organoid slices on an air-liquid interface, which improved oxygen availability (Giandomenico et al., 2019). They witnessed improved axonal projections, with deep layer neurons having horizontal projections and upper-layer neurons projecting vertically. The Ming group maintained cortical organoid slices immersed in culture media on an orbital shaker to preserve the health of the ventricular zone that often dies due to its location at the hypoxic center of organoids (Qian et al., 2020). By preserving the progenitor zones and laminar neuronal structure within organoids, they develop more distinct layer separation and neuron identity, allowing better modeling of diseases that may have defects in layer organization or neuron identity specification. Also worth mentioning are microfluidic technologies that introduce constant medium flow through brain organoids that replenishes nutrients (Fig. 4). Using a microfluidic organoid-on-a-chip platform, the Qin group describes a system that not only increases nutrient perfusion into brain organoids, but permits the constant replacement of disease-related chemicals, such as nicotine, avoiding issues with chemical degradation during long-term culture (Wang, Wang, Zhu, & Qin, 2018). The advent of technologies to increase nutrient and oxygen diffusion into brain organoids has drastically improved cellular health and has permitted labs to culture organoids long-term to further maturation. Future improvements to bioreactor and microfluidic technologies may allow organoids to better recapitulate late gestational and early post-natal development.

3.5 *In vivo* orthotopic xenotransplantation

Neurons within the brain form immediate- and long-distance connections with neurons throughout the brain. While assembloids have been developed to investigate aspects of the synaptic connectivity between two distinct brain regions, all current organoid models fail to recapitulate the intricate network connectivity of the human brain, a major limitation that cannot be solved by current technology. As an alternative strategy, several groups have devised methods for direct orthotopic xenotransplantation of whole brain organoids into mouse brains (Daviaud, Friedel, & Zou, 2018; Dong et al., 2020; Mansour et al., 2018). In these studies, portions of the mouse brain are removed and replaced with cerebral organoids. These grafts were shown to mature and send projections throughout the mouse brain. With time, host vasculature invades organoid grafts and supplies nutrients to core regions. Additionally, host microglia migrate into grafts and disperse similarly to what is seen in the surrounding mouse brain. The xenotransplantation has led to improved organoid maturation and the integration of endothelial cells and microglia often not present in *in vitro* cultures. It has also permitted the study of human neuron synaptic targeting and electrophysiology in an *in vivo* system. Despite these advancements, xenotransplantation is limited by requiring immune-deficient animals and the different timing of neurodevelopment in humans and mouse, which results in the early depletion of progenitors and loss of human specific features.

4. Modeling neurological diseases using brain organoids

Studying and understanding the mechanistic and biological basis of brain disorders has remained a challenge due to the lack of access to human brain tissue for research, the complexity and heterogeneity of disease presentation among individuals, and the inability of current model systems to accurately recapitulate human disease phenotypes (Amin & Paşca, 2018; Wen, Christian, Song, & Ming, 2016). Although analysis of post-mortem human brain tissue has provided insight into the histological components of brain disorders, such analysis has not been fruitful toward studying disease mechanisms. Although animal models have contributed greatly in establishing a link between cellular and systems-level pathology in an intact physiological system, there remain species-specific differences that limit their ability to accurately model human disorders (Schnoll et al., 2019). Human, especially patient iPSC-based monolayer cultures have provided ways to study mechanisms of nervous system disorders in specific brain cell types with the advantage of retaining patients' genetic backgrounds, but lacking the complexity of the brain (Amin & Paşca, 2018). Brain organoids more accurately recapitulate the development, structural organization, and cellular heterogeneity of the human brain (Camp et al., 2015; Lancaster et al., 2013; ; Luo et al., 2016), therefore offering a more promising *in vitro* human model. Broadly, brain organoids permit modeling of neurological disorders of both extrinsic and intrinsic etiologies (Adams, Cugola, & Muotri, 2019; Zhang, Song, & Ming, 2020). Extrinsic etiologies include brain environmental stimuli such as chemicals, pathogens, and neoplastic cells which can be modeled by adding external agents or cells to organoid cultures (Fig. 5).

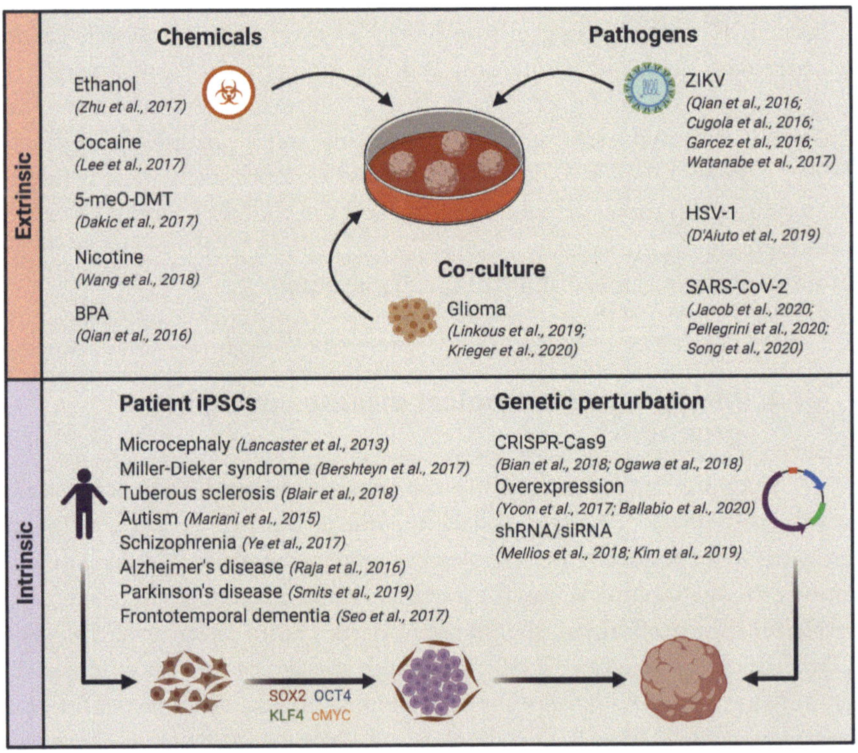

Fig. 5 Modeling neurological diseases of extrinsic and intrinsic etiology using brain organoids. Brain organoids have been used to model neurological diseases resulting from extrinsic factors (exogenous chemicals, pathogens, or neoplastic cells) and intrinsic factors (patient hiPSCs harboring disease-causing mutations or genetic perturbations that introduce aberrant gene expression into otherwise healthy PSCs or organoids). See Table 2 for a more comprehensive list of neurological diseases modeled in brain organoids.

Intrinsic etiologies include genetic abnormalities such as large chromosomal aberrations and mutations in protein coding or gene regulatory sequences, which can be modeled with organoids from patient-derived iPSCs or iPSCs modified by genomic editing or introduction of recombinant DNA (Fig. 5). We highlight examples of how brain organoids have been applied to study a wide range of both extrinsic and intrinsic nervous system pathologies, including neurodevelopmental and neuropsychiatric disorders, neurodegenerative and infectious diseases, and brain cancers (Table 2).

4.1 Neurodevelopmental and neuropsychiatric disorders

The phenotypes and mechanisms underlying many neurodevelopmental and neuropsychiatric disorders have been studied using brain organoids generated from patient iPSCs or hPSCs edited to harbor disease-associated mutations. Cephalic disorders are a group of congenital conditions characterized by gross abnormal development of the central nervous system that are often a result of genetic abnormalities, which cause severe defects in the structure and function of the brain. Brain organoids have been applied for studying a range of cephalic disorders, including microcephaly (Gabriel et al., 2016; Lancaster, 2013; Li, Sun, et al., 2017), macrocephaly (Li, Muffat, et al., 2017), and lissencephaly (Bershteyn et al., 2017; Iefremova et al., 2017). Microcephaly is characterized by a reduced head size compared to those of the same age and gender groups and may derive from several conditions or chromosomal abnormalities that lead to abnormal brain growth. Two such forms of microcephaly, autosomal recessive primary microcephaly (MCPH) (Lancaster, 2013; Li, Sun, et al., 2017) and Seckel syndrome (Gabriel et al., 2016), have been modeled in brain organoid cultures. MCPH patient-derived cerebral organoids are smaller and exhibit reduced progenitor zones, decreased neurogenesis and impaired neuronal activity compared to controls (Lancaster, 2013; Li, Sun, et al., 2017). Seckel syndrome patient-derived organoids have a similar phenotype in addition to increased length of cilia on apical neural progenitors (Gabriel et al., 2016). Conversely, macrocephaly is associated with an increased head size compared to those of the same age and gender groups, and loss-of-function mutations in *PTEN* have been shown to contribute to this phenotype (Butler et al., 2005). Organoids derived from *PTEN*-knockout hESCs exhibit increased proliferation, surface area, and organoid volume-a clinical feature of macrocephaly (Li, Muffat, et al., 2017). Lissencephaly is a cephalic disorder characterized by reduced or absent gyration in the brain, which may result from

Table 2 Neurological disease modeling using brain organoids.

Disease/Disorder	Mutation	Organoid model	Reference(s)
Neurodevelopmental			
Macrocephaly	PTEN KO iPSCs	Cerebral organoids	Li, Muffat, et al. (2017)
Microcephaly	Mutant ASPM patient iPSCs	Cortex-like organoids	Li, Sun, et al. (2017)
	WRD62 KO and mutant CDK5RAP2 patient iPSCs	Cerebral organoids	Lancaster et al. (2013)
	Mutant CPAP patient iPSCs	Cerebral organoids	Gabriel et al. (2016)
Lissencephaly	Miller-Dieker syndrome patient iPSCs	Cerebral and forebrain organoids	Bershteyn et al. (2017) and Iefremova et al. (2017)
	LIS1 heterozygous KO ESCs	Brain organoids	Karzbrun, Kshirsagar, Cohen, Hanna, and Reiner (2018)
Neuronal heterotopia	PLEKHG6 OE and knockdown iPSCs	Cerebral organoids	O'Neill et al. (2018)
	Mutant DCHS1 and FAT4 patient iPSCs	Cerebral organoids	Klaus et al. (2019)
Rett Syndrome	MECP2 mutant patient iPSCs	Cerebral organoids	Mellios et al. (2018)
Sandhoff disease	HEXB mutant patient iPSCs	Cerebral organoids	Allende et al. (2018)
Angelman syndrome	UBE3A KO and AS patient iPSCs	Cortical spheroids	Sun et al. (2019)
Leber congenital amaurosis	CEP290 mutant patient iPSCs	Retinal organoids	Parfitt et al. (2016)

Hypoxia	Control ESCs and iPSCs	Cerebral and brain organoids and cortical spheroids	Boisvert, Means, Michaud, Madri, and Katz (2019), Daviaud, Chevalier, Friedel, and Zou (2019) and Paşca et al. (2019)
Neuropsychiatric			
Autism	Idiopathic autism patient iPSCs	Cerebral organoids	Mariani et al. (2015)
	Heterozygous CHD8 KO iPSCs	Cerebral organoids	Wang et al. (2017)
	Timothy syndrome patient iPSCs	Subcortical spheroids	Birey et al. (2017)
Schizophrenia	Mutant DISC1 patient iPSCs	Forebrain and sliced cortical organoids	Qian et al. (2020) and Ye et al. (2017)
	Mutant DISC1 iPSCs	Cerebral organoids	Srikanth et al. (2018)
	16p13.11 microduplication patient iPSCs	Cerebral organoids	Johnstone et al. (2018)
	Schizophrenia patient iPSCs	Cerebral organoids	Stachowiak et al. (2017)
Neurodegenerative			
Alzheimer's disease	Mutated APP or PSEN1 AD patient iPSCs	Brain organoids	Raja et al. (2016)
Frontotemporal dementia	Mutated Tau P301L FTD patient iPSCs	Cerebral organoids	Seo et al. (2017)
Parkinson's disease	Mutant LRRK2 PD patient iPSCs	Midbrain organoids	Smits et al. (2019)
	Mutant LRRK2 iPSCs	Midbrain organoids	Kim, Park, et al. (2019)

Continued

Table 2 Neurological disease modeling using brain organoids.—cont'd

Disease/Disorder	Mutation	Organoid model	Reference(s)
Huntington's disease	Expanded CAG repeat iPSCs	Cortical organoids	Conforti et al. (2018)
Retinitis pigmentosa	RP patient iPSCs	Retinal organoids	Deng et al. (2018)
Infectious			
Zika virus	Control iPSCs	Cerebral and forebrain organoids	Cugola et al. (2016), Garcez et al. (2016), and Qian et al. (2016)
Herpes simplex virus	Control iPSCs	Brain organoids	D'Aiuto et al. (2019)
Cytomegalovirus	Control iPSCs	Brain and cortical organoids	Sison et al. (2019) and Sun et al. (2020)
Japanese encephalitis virus	Control iPSCs	Cortical organoids	Zhang et al. (2018)
SARS-CoV-2	Control iPSCs	Cerebral, cortical, hippocampus, hypothalamus, midbrain, and/or choroid plexus organoids	Jacob, Pather, et al. (2020), Pellegrini, Albecka, et al. (2020), Ramani et al. (2020), and Song et al. (2020)
Chemical exposure			
Bisphenol A	Control iPSCs	Forebrain organoids	Qian et al. (2016)
Ethanol	Control iPSCs	Brain and cerebral organoids	Arzua et al. (2020) and Zhu et al. (2017)
Cocaine	Control iPSCs	Neocortical organoids	Lee et al. (2017)

5-MeO-DMT	Control iPSCs	Cerebral organoids	Dakic et al. (2017)
Nicotine	Control iPSCs	Brain organoid-on-a-chip	Wang et al. (2018)
Cannabinoid	Control iPSCs	Microfluidic assembled cerebral organoids	Ao et al. (2020)
Neoplastic			
Tuberous Sclerosis	TSC1 and TSC2 KO iPSCs	Cortical spheroids	Blair et al. (2018)
Glioma	HRAS OE and TP53 KO	Cerebral organoids	Ogawa et al. (2018)
	MYC, EGFR, and EGFRvIII OE; CDKN2A/B, NF1, PTEN, and TP53 KO	Cerebral organoids	Bian et al. (2018)
	Control iPSCs and GSCs	Cerebral organoids	Linkous et al. (2019)
	Control iPSCs and GSCs	Cerebral organoids	Krieger et al. (2020)
Medulloblastoma	Control iPSCs	Cerebellar organoids	Ballabio et al. (2020)

Summary of neurodevelopmental and neuropsychiatric disorders, neurodegenerative and infectious diseases, chemical exposures, and neoplasms modeled using brain organoids. Selected references are provided for further reading. KO, knock-out; OE, over-expression.

defective neuronal proliferation and radial migration, leading to the appearance of a smooth cerebral cortex (Icenogle & Kaplan, 1981). Forebrain organoids generated from patients with Miller-Diekler syndrome (MDS), a form of lissencephaly, were found to be smaller with reduced proliferation in the ventricular zone-like layer (Iefremova et al., 2017). Additionally, oRGCs in organoids appeared to spend longer periods of time in mitosis, while neurons exhibited decreased ability to sustain saltatory migration, traveling less linearly at slower speeds (Bershteyn et al., 2017). Lastly, neuronal heterotopias involve the abnormal presence of neurons in areas such as the white matter, which is believed to be the consequence of abnormal neural migration during development. Modulation of PLEKHG6 primate isoform expression in cerebral organoids reproduced features of periventricular heterotopia, with defects in neural progenitor differentiation and RhoA dependent neuronal migration (O'Neill et al., 2018). Similarly, cerebral organoids derived from patient iPSCs with mutations in *DCHS1* and *FAT4* exhibit changes in the morphology of neural progenitor cells and result in defective neuronal migration dynamics (Klaus et al., 2019).

Brain organoids have also been used to model other neurodevelopmental diseases, such as Sandhoff disease, Rett syndrome, and Angelman syndrome. Sandhoff disease is an autosomal recessive lysosomal storage disorder characterized by the accumulation of lysosomal GM2 ganglioside, which results in developmental regression, neurodegeneration, and macrocephaly (Bley et al., 2011; Tavasoli et al., 2018). Patient iPSC-derived cerebral organoids showed an excess accumulation of GM2 ganglioside, along with increases in cellular proliferation and organoid size, features similar to what is observed clinically (Allende et al., 2018). Rett syndrome is a neurodevelopmental disorder caused by X-linked mutations in *MECP2*, and is characterized by developmental regression and the onset of gait abnormalities in female infants (Ip, Mellios, & Sur, 2018; Neul et al., 2010) and congenital encephalopathy and early death in male infants (Schüle, Armstrong, Vogel, Oviedo, & Francke, 2008). Rett syndrome patient-derived cerebral organoids showed an upregulation of two micro-RNAs, miR-199 and miR-214, which are involved in ERK and AKT signaling and important for neurogenesis and neural differentiation. The organoids also displayed enlarged ventricular zones, increased neural progenitor proliferation, and impaired neurogenesis (Mellios et al., 2018). Angelman syndrome is a disorder characterized by microcephaly, severe intellectual deficit, paroxysms of laughter, and hyperactivity that result from loss of function of the imprinted *UBE3A* gene (Buiting, Williams, & Horsthemke, 2016). Cortical spheroids derived from patient iPSCs or

UBE3A knock-out iPSCs showed augmented potassium channel activity, leading to increased intrinsic excitability in neurons and aberrant network synchronization (Sun et al., 2019).

Psychiatric diseases comprise a heterogeneous set of disorders with both idiopathic and known genetic causes. Schizophrenia is chronic psychiatric disorder that influences early brain development, and manifests as a combination of psychotic symptoms including hallucinations, delusions, and cognitive dysfunctions (Kahn et al., 2015). Specifically, rare mutations in *DISC1* have been associated with schizophrenia, and forebrain organoids derived from patients with a 4 base pair frameshift mutation in *DISC1* exhibited a delay in cell cycle progression related to disrupted DISC1 and NED1/NDEL1 protein interactions (Ye et al., 2017). In a separate study, forebrain organoids also displayed deficits in cortical neuron subtype and layer specification related to disrupted WNT/β-catenin signaling (Qian et al., 2020). Additionally, brain organoids derived from *DISC1*-mutated iPSCs were found to lack ventricular-like structures and had decreased progenitor proliferation (Srikanth et al., 2018).

Autism Spectrum Disorder (ASD) represents a group of heterogeneous developmental conditions that are characterized by impairments in social and emotional interactions and speech, with some patients exhibiting macrocephaly (Geschwind, 2008). Patient cortical organoids with macrocephalic features (larger size) were shown to have differential gene expression in pathways related to cell fate and proliferation, cytoskeletal regulation of dynamic cellular growth, guidance, and maintenance, synaptic assembly and channel functioning, and upregulated GABAergic enzyme synthesis (Mariani et al., 2015) . Consistent with this, cerebral organoids derived from iPSCs with heterozygous knockout of *CHD8*, a major risk factor for ASD, showed dysregulation of *DLX* genes that play a critical role in GABAergic interneuron differentiation (Wang et al., 2017). Timothy Syndrome, which is an autosomal-dominant disorder caused by a missense mutation in a gene encoding the calcium channel *CACNA1C*, is characterized by ASD. Neurons in patient-derived cortical organoids showed calcium signaling defects and delayed GABAergic interneuron migration (Birey et al., 2017).

Together, these studies demonstrate the validity of brain organoids for capturing molecular, cellular and structural phenotypes of neurodevelopmental and neuropsychiatric disorders, and provide the opportunity for investigating the underlying mechanisms. Brain organoids were first applied to model disorders with large structural deficits, as the phenotypes are usually obvious enough to detect reliably, despite variability among organoids.

In contrast, neuropsychiatric disease models are mostly limited to those with known genetic causes, which permits the use of isogenic lines with corrected mutations to confirm causality. As brain organoid technologies improve in regard to inter- and intra-organoid variability, diseases with more subtle phenotypes may be modeled with more confidence.

4.2 Neurodegenerative diseases

Neurodegenerative diseases are a diverse group of disorders that are characterized by the progressive deterioration of the nervous system resulting from neural and glial cell death or impaired neuronal function. Brain organoids generated from patient iPSCs or hPSCs edited to contain disease-causing mutations have been used to model several neurodegenerative diseases including Alzheimer's disease (AD), frontotemporal dementia (FTD), and Parkinson's disease (PD) (Grenier, Kao, & Diamandis, 2020).

AD, the most common neurodegenerative disease, is characterized by the progressive loss of memory and language with associated behavioral dysfunction. AD patients have β-amyloid plaques in the extracellular brain parenchyma and neurofibrillary tangles comprised of hyperphosphorylated tau in the intracellular neuronal spaces (Kumar, Singh, & Ekavali, 2015). Mouse models of familial AD (fAD) exhibit synaptic and memory deficits with β-amyloid pathology, but do not recapitulate other key aspects, such as neurofibrillary tangles (Duff, 2001), while AD-hiPSC-derived neurons exhibit elevated β-amyloid and phosphorylated tau, but do not develop β-amyloid plaques or neurofibrillary tangles (Israel et al., 2012; Kondo et al., 2013; Muratore et al., 2014; Yagi et al., 2011). Thick-aggregate cultures of human stem cell-derived neurons with fAD mutations provided the first evidence that 3D cultures could recapitulate both plaque and tangle pathology (Choi et al., 2014). Later, patient-derived brain organoids harboring APP duplications or PSEN1 mutations were shown to develop age-dependent extracellular β-amyloid aggregation, hyperphosphorylated tau protein and endosome abnormalities (Raja et al., 2016). Treatment with β- and γ-secretase inhibitors reduced amyloid and tau pathology.

FTD is a neurodegenerative disease characterized by atrophy in the frontal and temporal brain lobes and an increase in intracellular protein aggregates. Patient-derived cerebral organoids carrying the Tau P301L mutation showed increased levels of p25, which hyperphosphorylates tau and increases aggregation (Seo et al., 2017). Pharmacological inhibition of p25 generation reduced levels of phosphorylated tau and increased expression of synaptophysin in organoids.

PD is a neurodegenerative disease associated with abnormalities in movement, balance, and coordination (Fearnley & Lees, 1991). The most common cause of autosomal-dominant PD is a gain-of-function G2019S missense mutation in leucine-rich repeat kinase 2 (LRRK2), which has been shown to be associated with mitochondrial dysfunction (Hsieh et al., 2016; Langston, Ballard, Tetrud, & Irwin, 1983; Park, Davis, & Sue, 2018), impaired dopamine signaling leading to the degeneration and progressive loss of dopamine neurons (Fearnley & Lees, 1991), and the accumulation and aggregation of a-synuclein in the neuronal cytosol, known as Lewy bodies, causing damage to other cellular components (Kim, Park, et al., 2019; Spillantini, Crowther, Jakes, Hasegawa, & Goedert, 1998; Spillantini et al., 1997; Stefanis, 2012). Indeed, midbrain organoids generated from iPSCs with the *LRRK2* G2019S mutation exhibited similar pathological features of patient brains, including increased accumulation and aggregation of a-synuclein (Kim, Park, et al., 2019) and decreased dopaminergic neurons (Smits et al., 2019).

4.3 Chemical exposures

Brain organoids, mostly cortical organoids, have been used to study the effects of chemical exposures on fetal development and neuronal function by directly adding toxic molecules or drugs of abuse to the culture medium. Exposure to bisphenol A, a chemical found in many household plastics, reduced neural progenitor cell proliferation and ventricular zone thickness in forebrain organoids (Qian et al., 2016). Ethanol exposure induced premature neural progenitor differentiation, attenuated neuronal maturation and neurite outgrowth, and led to increased cell death in brain organoids in a concentration-dependent manner (Arzua et al., 2020; Zhu et al., 2017). Cocaine exposure led to a CYP3A5-induced generation of reactive oxygen species, reduced neural progenitor cell proliferation, and premature neuronal differentiation in cortical organoids (Lee et al., 2017). 5-MeO-DMT exposure led to altered expression of proteins associated with anti-inflammation, long-term potentiation, and dendritic spine formation (Dakic et al., 2017). Nicotine exposure induced premature neuronal differentiation, disrupted progenitor regionalization, and abnormal neurite outgrowth in a brain organoid-on-a-chip model (Wang et al., 2018). Cannabinoid exposure reduced neuronal maturation, impaired neurite outgrowth, and decreased spontaneous firing in microfluidic assembled cerebral organoids (Ao et al., 2020). Although these early studies helped in gaining understanding of the effects of many chemicals on brain development, the results must be interpreted with caution due to important limitations. Because brain organoids are isolated from other functioning organ systems, models of chemical exposure do not accurately account for metabolism and bioavailability, despite best attempts to mimic the bioactive compound and physiological concentrations. Second, brain organoids do not have an intact blood-brain barrier which functions to limit diffusion of chemicals into the brain parenchyma and cerebrospinal fluid. Third, it has been increasingly appreciated that individual genetic predispositions may affect response to certain chemicals. Most of these studies utilize a modest number of stem cell lines, limiting the broad generalizability to the general population. Improvement of organoid technology combined with high throughput screening methods that employ many iPSC lines from individuals of diverse genetic backgrounds may better inform disease mechanisms in the future.

4.4 Infectious diseases

Like chemical exposures, brain organoids have been used to study the effects of infectious diseases on neural development by directly adding pathogens to the culture medium. Following the World Health Organization's declaration of a public health emergency based on mounting epidemiological evidence associating Zika virus (ZIKV) infection in pregnant mothers with babies born with microcephaly, several groups used brain organoids to model ZIKV infection. Brain organoid cultures exposed to ZIKV exhibited selective infection of neural progenitor cells and astrocytes, resulting in cell death, reduced proliferative zones, and disrupted cortical layer organization, corroborating the microcephalic phenotype observed in babies with vertical transmission of ZIKV (Cugola et al., 2016; Dang et al., 2016; Garcez et al., 2016; Ming, Tang, & Song, 2016; Qian et al., 2016; Watanabe et al., 2017). Because of this strong and clinically relevant phenotype, brain organoids were further used to screen and test for compounds that could reduce infection (Watanabe et al., 2017; Xu et al., 2016; Zhou et al., 2017) and to better understand the mechanisms responsible for ZIKV-induced microcephaly (Dang et al., 2016; Wen, Song, & Ming, 2017; Yoon, Song, et al., 2017). 05wHerpesvirus infection, leading to a disease that infects epidermal cells and establishes latent reservoirs in neurons, has also been modeled in brain organoids and exhibited multiple changes, including viral spreading, morphological changes and cellular fusion of infected neurons (D'Aiuto et al., 2019). More recently, several studies have used various region-specific brain organoid protocols to study neurotropism of the SARS-CoV-2 virus and its potential consequences on brain cell function. Using cortical, hippocampus, hypothalamus, midbrain, and choroid plexus organoids, the Ming group revealed a specific SARS-CoV-2 tropism for choroid plexus epithelial cells that leads to productive infection, cell death, increased inflammatory cytokine transcription, and decreased transcription of genes important for cerebrospinal fluid production and barrier formation (Jacob, Pather, et al., 2020). This tropism was replicated by the Lancaster group using CSF-producing choroid organoids (Pellegrini, Albecka, et al., 2020). These studies highlight the applicability of brain organoids to model infectious disease, identify tropism and potential phenotypes, to understand pathogen-host interactions, and to serve as a platform for drug screening and developing therapeutic strategies.

4.5 Brain cancers

Malignant brain tumors are among the most devastating and lethal neoplasms affecting children and adults (Aldape et al., 2019). Although our understanding of brain tumor biology has been advanced by *in vitro* tumor stem cell lines, these model systems are limited by drifting away from human genetic signatures and poor clinical translatability (Ledur, Onzi, Zong, & Lenz, 2017; Lee et al., 2006). Brain organoids have been used to study tumorigenesis in Tuberous Sclerosis, adult glioblastoma, and pediatric medulloblastoma. Tuberous Sclerosis Complex (TSC) is a developmental disorder with heterozygous germline mutations in *TSC1* or *TSC2*, characterized by epilepsy, ASD, intellectual disabilities, and psychiatric disorders (Curatolo, Moavero, & de Vries, 2015). Patients with TSC commonly show focal regions of disorganized and dysmorphic neurons and glia, called cortical tubers, that result from haploinsufficiency or loss of heterozygosity of TSC (Martin et al., 2017). Cortical organoids derived from cells with homozygous knockout of *TSC1* and *TSC2* showed reduced and delayed expression of neuronal markers, increased glial cells, and dysmorphic neurons and glia, similar to cellular phenotypes seen in patients' tubers (Blair, Hockemeyer, & Bateup, 2018).

To model glioblastoma, one study employed a two-pronged approach to study both intrinsic tumorigenesis and extrinsic tumor cell migration using cerebral organoids (Ogawa, Pao, Shokhirev, & Verma, 2018). Combined HRasG12V overexpression and TP53 knock-out was achieved in cerebral organoids by electroporating vectors encoding Cas9 nuclease with guide RNAs. Cells with successful integration became transformed and exhibited increased proliferation and rapid invasion which destroyed surrounding organoid structures. Orthotopic xenotransplantation of dissociated cultures recapitulated key glioma phenotypes, such as extensive invasion into the surrounding brain parenchyma, migration along blood vessels, and stimulation of angiogenesis. Furthermore, when fusing glioma stem cell tumorspheres with cerebral organoids, extensive tumor cell invasion and proliferation were found. Some of these results were replicated in other studies by co-culturing glioma stem cells with cerebral organoids (Krieger et al., 2020; Linkous et al., 2019). Similar to this strategy, combined transposon-mediated oncogene insertions and CRISPR/Cas9-mediated tumor suppressor knockouts were applied to cerebral organoids to generate neoplastic cerebral organoids (Bian et al., 2018). For example, overexpression of MYC resulted in formation of tumors that resembled central nervous system primitive neuroectodermal tumors. These organoids are useful for chemotherapeutic drug screening, with EGFR inhibitors reducing the growth of tumor cells with EGFR overactivation. On the other hand, combined overexpression of OTX2 and c-MYC in cerebellar organoids induced medulloblastoma-like tissue, and EZH2 inhibition could inhibit the growth and induce apoptosis of the tumor-like tissue (Ballabio et al., 2020). These studies demonstrate the utility of brain organoids to model tumor initiation and tumor cell invasion and to test therapeutic strategies. However, the observed changes in this unique model should also be further validated because of brain organoid limitations (see below). Furthermore, modifications of a single or a few oncogenes in brain organoids do not necessarily recapitulate the entire genetic and/or epigenetic landscape of human tumors. Recent development of glioblastoma organoid models derived directly from surgically resected tumor tissue have been shown to better maintain the cellular, transcriptional, and genetic heterogeneity of patients' tumors (Hubert et al., 2016; Jacob, Ming, & Song, 2020; Jacob, Salinas, et al., 2020).

5. Limitations and perspectives

Cerebral organoids and region-specific brain organoids have expanded our toolset for studying human brain development and mechanisms underlying many neurological diseases, but they remain imperfect model systems with several important caveats, including limited protocol reproducibility among different hPSC lines, heterogeneity among organoids, lack of definitive methods and markers to accurately assess cell identity, and absence of structures that fully recapitulate the organization of the human brain. We discuss these limitations and offer some insights on minimizing their effect and improving brain organoid technology.

5.1 Variability among stem cells and heterogeneity between organoids

Inherent variability among hPSC lines limits the reproducibility of differentiation protocols and the ability to accurately compare brain organoids derived from control and disease lines. This variability arises from several sources, including starting cell type (e.g., hESC or hiPSC), reprogramming method (e.g., episomal vectors, Sendai virus), culture substrate (e.g., MEF feeder cells or Matrigel), expansion medium (e.g., knock-out serum replacement, mTeSR1, E8), and genetic heterogeneity among individuals. Pluripotency is a complex cellular state regulated by gene expression controlled by transcription factors, chromatin modifiers, and non-coding RNAs and the state of the hPSCs can have profound effects on differentiation capacity (Enver, Pera, Peterson, & Andrews, 2009; Young, 2011). Additionally, there are noticeable differences in the gene expression, epigenetic landscape, and mutational load between hESCs and hiPSCs that can impact their differentiation potential (Bilic & Izpisua Belmonte, 2012). These differences often manifest as differential responses to signaling molecules and a propensity to differentiate to certain lineages based on the identity of cells that were originally reprogrammed, the age of the individual from which they were harvested, and the reprogramming methodology (Malik & Rao, 2013). Addressing hPSC variability, thorough characterization of differentiation capacity and efficiency using multiple hESC and hiPSC lines is needed when reporting on a new brain organoid methodology. Additionally, each study should include sufficiently detailed methodology so others can reproduce culture conditions. Furthermore, when comparing brain organoids derived from both control and disease lines to report a phenotype, it is imperative that the lines are derived from the same source (e.g., dermal fibroblasts or B lymphocytes) and generated using the same reprogramming method in parallel to limit variability. Multiple control and disease lines should be used when available with more than one clone per line to rule out clone-specific effects. When possible, in-family and/or isogenic controls should be used to control for differences in genetic background. CRISPR-edited lines should be screened for off-target effects in major regulatory and protein coding regions of the genome. Comparisons between unrelated controls and disease lines should be avoided when possible and the use of lines with mutations that accurately reflect the patient condition over full gene knockouts or overexpression are preferred. Lastly, it is important to perform regular quality control on hPSCs to ensure they do not accumulate harmful mutations or microorganisms, such as mycoplasma, which can affect differentiation and disease phenotypes.

3D differentiation lends itself to much more variation in structure, size, and cell composition than 2D monolayer cultures since it relies on self-organization of stem cells and progeny as they are patterned. There is no definitive way to eliminate heterogeneity among organoids, but the effect of this heterogeneity can be limited in mechanistic and disease studies by using proper controls and devising consistent methods for quantification and analysis. Using defined medium compositions which do not rely on components with substantial batch-to-batch variations (e.g., animal serum and Matrigel) will likely improve organoid reproducibility as serum-free medium supplements and defined hydrogels have become more common. Similarly, recombinant growth factors can have large batch-to-batch variability and poor stability in culture, therefore stabilized growth factors and small molecule agonists and antagonists should be used whenever possible. Methods to control starting cell number within hPSC aggregates (e.g., 96-well self-aggregation and Aggrewell plates) should also be used when possible to reduce size variability.

5.2 Accurately accessing cell identity

Commonly used markers for characterizing neural progenitor regional identity and neuronal subtype specification in human brain organoids rely mostly on knowledge from studies in model organisms. While many of these markers are conserved, it is increasingly appreciated that there are critical differences due to human-specific features. An important advance has been the development of single cell or single nucleus RNA sequencing technologies that have enabled researchers to study the transcriptomes of individual cells of the human brain. These efforts have improved reference databases for cell type assignment, but have not solved the issue entirely. For example, most datasets are generated from careful dissections of specific brain regions, allowing researchers to classify cell identities based on preconceived knowledge of the specific brain region. Although organoids often contain ventricular structures that resemble the primitive brain, the structures alone are inadequate to assign brain region identity. Second, although there exist good marker genes for several cell populations, these markers are often only specific when looking within small subdivisions of the brain. Throughout the entire brain, different neuron types may share marker expression. Therefore, both positive and negative markers should be used when possible. For example, when assessing dorsal forebrain progenitor fate, a combination of positive markers (e.g., PAX6, FOXG1, EMX1) and negative markers (e.g.,o NKX2.1, GSX2, FOXA2) should be used. While the single cell transcriptome has been widely used for cell type classification, it should be noted that mRNA expression does not always directly translate to protein expression. Lastly, cells within brain organoids represent fetal development and are usually more immature than those found in the adult human brain, and often lack the specific signatures of human brain cells (Bhaduri et al., 2020). Because neurons in organoid culture do not reach full maturity, different neuronal subtypes may often blend in single cell analyses, with few clear distinctions for two main subtypes of glutamatergic neurons and GABAergic neurons. Therefore, it is essential to use a combination of techniques, such as immunohistochemistry, RNA sequencing, and electrophysiology for the classification of cell and regional identities. More thorough characterization of cell identity in brain organoids will provide a better framework for understanding human brain development and disorders.

5.3 Structural organization

Despite rapid advancements in patterning, most region-specific brain organoids do not fully recapitulate the intricate structural organization of neurons and cell type complexity of brain regions. Perhaps the best studied is the cerebral cortex with its highly organized laminar neuronal organization into distinct layers. Brain organoid protocols for cortical development recapitulate progenitor zone layers and rudimentary neuronal layering but have yet to reproduce the full neuronal organization of the cerebral cortex. Likewise, although medial pallium organoids generate diverse neurons of the hippocampus, they fail to recapitulate the structural organization of the CA and dentate gyrus subfields. Similarly, hypothalamic and thalamic organoids fail to recapitulate the formation of discrete nuclei. Improving structural organization will likely rely on advancements in bioengineered scaffolds or the addition of important guidance cells or molecules.

6. Summary

Recent advancements in region-specific brain organoids have contributed greatly to our understanding of the morphogens that guide neural progenitor fate specification and the processes of human neuron generation.

By utilizing precise timing and combinations of signaling molecules, researchers have guided hPSC aggregates to form primitive structures resembling the organization and neuronal diversity of specific brain regions. Although brain organoid models show great promise for developmental and disease studies, limitations in reproducibility and maturity restrict their utility. Improvements in protocol reproducibility and inter-organoid variability combined with techniques to improve cellular diversity and maturity will allow researchers to continue studying the human brain in ways that were previously impossible. Ultimately, brain organoids provide a complementary system, that in combination with other *in vitro* culture methodologies and animal models will help us better understand human brain development neurological diseases.

Acknowledgments

The research in the authors' laboratories were supported by grants from the National Institutes of Health (U19MH106434 and R35NS116843 to H.S., and U19AI131130, R35NS097370 and RF1MH123979 to G.M.). Portions of Figures were created using biorender.com.

Conflict of interest

None.

References

Abud, E. M., Ramirez, R. N., Martinez, E. S., Healy, L. M., Nguyen, C. H. H., Newman, S. A., et al. (2017). iPSC-derived human microglia-like cells to study neurological diseases. *Neuron*, *94*, 278–293. e279.

Adams, J. W., Cugola, F. R., & Muotri, A. R. (2019). Brain organoids as tools for modeling human neurodevelopmental disorders. *Physiology (Bethesda)*, *34*, 365–375.

Aldape, K., Brindle, K. M., Chesler, L., Chopra, R., Gajjar, A., Gilbert, M. R., et al. (2019). Challenges to curing primary brain tumours. *Nature Reviews. Clinical Oncology*, *16*, 509–520.

Allende, M. L., Cook, E. K., Larman, B. C., Nugent, A., Brady, J. M., Golebiowski, D., et al. (2018). Cerebral organoids derived from Sandhoff disease-induced pluripotent stem cells exhibit impaired neurodifferentiation. *Journal of Lipid Research*, *59*, 550–563.

Amin, N. D., & Paşca, S. P. (2018). Building models of brain disorders with three-dimensional organoids. *Neuron*, *100*, 389–405.

Ao, Z., Cai, H., Havert, D. J., Wu, Z., Gong, Z., Beggs, J. M., et al. (2020). One-stop microfluidic assembly of human brain organoids to model prenatal Cannabis exposure. *Analytical Chemistry*, *92*, 4630–4638.

Arzua, T., Yan, Y., Jiang, C., Logan, S., Allison, R. L., Wells, C., et al. (2020). *Modeling alcohol-induced neurotoxicity using human induced pluripotent stem cell-derived three-dimensional cerebral organoids*. *Translational Psychiatry*, *347*, 10.

Bagley, J. A., Reumann, D., Bian, S., Levi-Strauss, J., & Knoblich, J. A. (2017). Fused cerebral organoids model interactions between brain regions. *Nature Methods*, *14*, 743–751.

Ballabio, C., Anderle, M., Gianesello, M., Lago, C., Miele, E., Cardano, M., et al. (2020). Modeling medulloblastoma in vivo and with human cerebellar organoids. *Nature Communications*, *11*, 583.

Bennett, M. L., & Bennett, F. C. (2020). The influence of environment and origin on brain resident macrophages and implications for therapy. *Nature Neuroscience*, *23*, 157–166.

Berg, D. A., Su, Y., Jimenez-Cyrus, D., Patel, A., Huang, N., Morizet, D., et al. (2019). A common embryonic origin of stem cells drives developmental and adult neurogenesis. *Cell*, *177*, 654–668. e615.

Bershteyn, M., Nowakowski, T. J., Pollen, A. A., Di Lullo, E., Nene, A., Wynshaw-Boris, A., et al. (2017). Human iPSC-derived cerebral organoids model cellular features of Lissencephaly and reveal prolonged mitosis of outer radial glia. *Cell Stem Cell*, *20*, 435–449. e434.

Bhaduri, A., Andrews, M. G., Mancia Leon, W., Jung, D., Shin, D., Allen, D., et al. (2020). Cell stress in cortical organoids impairs molecular subtype specification. *Nature*, *578*, 142–148.

Bian, S., Repic, M., Guo, Z., Kavirayani, A., Burkard, T., Bagley, J. A., et al. (2018). Genetically engineered cerebral organoids model brain tumor formation. *Nature Methods*, *15*, 631–639.

Bilic, J., & Izpisua Belmonte, J. C. (2012). Concise review: Induced pluripotent stem cells versus embryonic stem cells: Close enough or yet too far apart? *Stem Cells*, *30*, 33–41.

Birey, F., Andersen, J., Makinson, C. D., Islam, S., Wei, W., Huber, N., et al. (2017). Assembly of functionally integrated human forebrain spheroids. *Nature*, *545*, 54–59.

Blair, J. D., Hockemeyer, D., & Bateup, H. S. (2018). Genetically engineered human cortical spheroid models of tuberous sclerosis. *Nature Medicine*, *24*, 1568–1578.

Bley, A. E., Giannikopoulos, O. A., Hayden, D., Kubilus, K., Tifft, C. J., & Eichler, F. S. (2011). Natural history of infantile G(M2) gangliosidosis. *Pediatrics*, *128*, e1233–e1241.

Boisvert, E. M., Means, R. E., Michaud, M., Madri, J. A., & Katz, S. G. (2019). Minocycline mitigates the effect of neonatal hypoxic insult on human brain organoids. *Cell Death & Disease*, *10*, 325.

Buiting, K., Williams, C., & Horsthemke, B. (2016). Angelman syndrome — Insights into a rare neurogenetic disorder. *Nature Reviews Neurology*, *12*, 584–593.

Butler, M. G., Dasouki, M. J., Zhou, X. P., Talebizadeh, Z., Brown, M., Takahashi, T. N., et al. (2005). Subset of individuals with autism spectrum disorders and extreme macrocephaly associated with germline PTEN tumour suppressor gene mutations. *Journal of Medical Genetics*, *42*, 318–321.

Cadwell, C. R., Bhaduri, A., Mostajo-Radji, M. A., Keefe, M. G., & Nowakowski, T. J. (2019). Development and Arealization of the cerebral cortex. *Neuron*, *103*, 980–1004.

Cakir, B., Xiang, Y., Tanaka, Y., Kural, M. H., Parent, M., Kang, Y. J., et al. (2019). Engineering of human brain organoids with a functional vascular-like system. *Nature Methods*, *16*, 1169–1175.

Camp, J. G., Badsha, F., Florio, M., Kanton, S., Gerber, T., Wilsch-Bräuninger, M., et al. (2015). Human cerebral organoids recapitulate gene expression programs of fetal neocortex development. *Proceedings of the National Academy of Sciences of the United States of America*, *112*, 15672–15677.

Campbell, K. (2003). Dorsal-ventral patterning in the mammalian telencephalon. *Current Opinion in Neurobiology*, *13*, 50–56.

Chi, C. L., Martinez, S., Wurst, W., & Martin, G. R. (2003). The isthmic organizer signal FGF8 is required for cell survival in the prospective midbrain and cerebellum. *Development*, *130*, 2633–2644.

Choi, S. H., Kim, Y. H., Hebisch, M., Sliwinski, C., Lee, S., D'Avanzo, C., et al. (2014). A three-dimensional human neural cell culture model of Alzheimer's disease. *Nature*, *515*, 274–278.

Conforti, P., Besusso, D., Bocchi, V. D., Faedo, A., Cesana, E., Rossetti, G., et al. (2018). Faulty neuronal determination and cell polarization are reverted by modulating HD early phenotypes. *Proceedings of the National Academy of Sciences of the United States of America*, *115*, E762–E771.

Cowan, C. S., Renner, M., De Gennaro, M., Gross-Scherf, B., Goldblum, D., Hou, Y., et al. (2020). Cell types of the human retina and its organoids at single-cell resolution. *Cell*, *182*, 1623–1640. e1634.

Cugola, F. R., Fernandes, I. R., Russo, F. B., Freitas, B. C., Dias, J. L., Guimaraes, K. P., et al. (2016). The Brazilian Zika virus strain causes birth defects in experimental models. *Nature*, *534*, 267–271.

Curatolo, P., Moavero, R., & de Vries, P. J. (2015). Neurological and neuropsychiatric aspects of tuberous sclerosis complex. *Lancet Neurology*, *14*, 733–745.

D'Aiuto, L., Bloom, D. C., Naciri, J. N., Smith, A., Edwards, T. G., McClain, L., et al. (2019). Modeling herpes simplex virus 1 infections in human central nervous system neuronal cells using two- and three-dimensional cultures derived from induced pluripotent stem cells. *Journal of Virology*, *93*, e000111–e000119.

Dakic, V., Minardi Nascimento, J., Costa Sartore, R., Maciel, R. M., de Araujo, D. B., Ribeiro, S., et al. (2017). Short term changes in the proteome of human cerebral organoids induced by 5-MeO-DMT. *Scientific Reports*, *7*, 12863.

Dang, J., Tiwari, S. K., Lichinchi, G., Qin, Y., Patil, V. S., Eroshkin, A. M., et al. (2016). Zika virus depletes neural progenitors in human cerebral organoids through activation of the innate immune receptor TLR3. *Cell Stem Cell*, *19*, 258–265.

Daviaud, N., Chevalier, C., Friedel, R. H., & Zou, H. (2019). Distinct vulnerability and resilience of human Neuroprogenitor subtypes in cerebral organoid model of prenatal hypoxic injury. *Frontiers in Cellular Neuroscience*, *13*, 336.

Daviaud, N., Friedel, R. H., & Zou, H. (2018). Vascularization and engraftment of transplanted human cerebral organoids in mouse cortex. *eNeuro*, *5*. ENEURO.0219-18.2018.

Deng, W. L., Gao, M. L., Lei, X. L., Lv, J. N., Zhao, H., He, K. W., et al. (2018). Gene correction reverses ciliopathy and photoreceptor loss in iPSC-derived retinal organoids from retinitis Pigmentosa patients. *Stem Cell Reports*, *10*, 1267–1281.

Dong, X., Xu, S. B., Chen, X., Tao, M., Tang, X. Y., Fang, K. H., et al. (2020). Human cerebral organoids establish subcortical projections in the mouse brain after transplantation. *Molecular Psychiatry*. https://doi.org/10.1038/s41380-020-00910-4.

Douvaras, P., Sun, B., Wang, M., Kruglikov, I., Lallos, G., Zimmer, M., et al. (2017). Directed differentiation of human pluripotent stem cells to microglia. *Stem Cell Reports*, *8*, 1516–1524.

Duff, K. (2001). Transgenic mouse models of Alzheimer's disease: Phenotype and mechanisms of pathogenesis. *Biochemical Society Symposium*, *67*, 195–202.

Enver, T., Pera, M., Peterson, C., & Andrews, P. W. (2009). Stem cell states, fates, and the rules of attraction. *Cell Stem Cell*, *4*, 387–397.

Erceg, S., Ronaghi, M., Zipancic, I., Lainez, S., Rosello, M. G., Xiong, C., et al. (2010). Efficient differentiation of human embryonic stem cells into functional cerebellar-like cells. *Stem Cells and Development*, *19*, 1745–1756.

Fearnley, J. M., & Lees, A. J. (1991). Ageing and Parkinson's disease: Substantia nigra regional selectivity. *Brain*, *114*(Pt 5), 2283–2301.

Gabriel, E., Wason, A., Ramani, A., Gooi, L. M., Keller, P., Pozniakovsky, A., et al. (2016). CPAP promotes timely cilium disassembly to maintain neural progenitor pool. *The EMBO Journal*, *35*, 803–819.

Garcez, P. P., Loiola, E. C., Madeiro da Costa, R., Higa, L. M., Trindade, P., Delvecchio, R., et al. (2016). Zika virus impairs growth in human neurospheres and brain organoids. *Science*, *352*, 816–818.

Germain, N. D., Banda, E. C., Becker, S., Naegele, J. R., & Grabel, L. B. (2013). Derivation and isolation of NKX2.1-positive basal forebrain progenitors from human embryonic stem cells. *Stem Cells and Development*, *22*, 1477–1489.

Geschwind, D. H. (2008). Autism: Many genes, common pathways? *Cell*, *135*, 391–395.

Giandomenico, S. L., Mierau, S. B., Gibbons, G. M., Wenger, L. M. D., Masullo, L., Sit, T., et al. (2019). Cerebral organoids at the air-liquid interface generate diverse nerve tracts with functional output. *Nature Neuroscience*, *22*, 669–679.

Grenier, K., Kao, J., & Diamandis, P. (2020). Three-dimensional modeling of human neurodegeneration: Brain organoids coming of age. *Molecular Psychiatry*, *25*, 254–274.

Grove, E. A., & Tole, S. (1999). Patterning events and specification signals in the developing hippocampus. *Cerebral Cortex*, *9*, 551–561.

Haenseler, W., Sansom, S. N., Buchrieser, J., Newey, S. E., Moore, C. S., Nicholls, F. J., et al. (2017). A highly efficient human pluripotent stem cell microglia model displays a neuronal-co-culture-specific expression profile and inflammatory response. *Stem Cell Reports*, *8*, 1727–1742.

Hammond, T. R., Robinton, D., & Stevens, B. (2018). Microglia and the brain: Complementary Partners in Development and Disease. *Annual Review of Cell and Developmental Biology*, *34*, 523–544.

Hickman, S., Izzy, S., Sen, P., Morsett, L., & El Khoury, J. (2018). Microglia in neurodegeneration. *Nature Neuroscience*, *21*, 1359–1369.

Hill, R. S., & Walsh, C. A. (2005). Molecular insights into human brain evolution. *Nature*, *437*, 64–67.

Hsieh, C. H., Shaltouki, A., Gonzalez, A. E., Bettencourt da Cruz, A., Burbulla, L. F., St Lawrence, E., et al. (2016). Functional impairment in Miro degradation and Mitophagy is a shared feature in familial and sporadic Parkinson's disease. *Cell Stem Cell*, *19*, 709–724.

Hubert, C. G., Rivera, M., Spangler, L. C., Wu, Q., Mack, S. C., Prager, B. C., et al. (2016). A three-dimensional organoid culture system derived from human glioblastomas recapitulates the hypoxic gradients and Cancer stem cell heterogeneity of tumors found in vivo. *Cancer Research*, *76*, 2465–2477.

Icenogle, D. A., & Kaplan, A. M. (1981). A review of congenital neurologic malformations. *Clinical Pediatrics (Phila)*, *20*, 565–576.

Iefremova, V., Manikakis, G., Krefft, O., Jabali, A., Weynans, K., Wilkens, R., et al. (2017). An organoid-based model of cortical development identifies non-cell-autonomous defects in Wnt signaling contributing to Miller-Dieker syndrome. *Cell Reports*, *19*, 50–59.

Ip, J. P. K., Mellios, N., & Sur, M. (2018). Rett syndrome: Insights into genetic, molecular and circuit mechanisms. *Nature Reviews. Neuroscience*, *19*, 368–382.

Israel, M. A., Yuan, S. H., Bardy, C., Reyna, S. M., Mu, Y., Herrera, C., et al. (2012). Probing sporadic and familial Alzheimer's disease using induced pluripotent stem cells. *Nature*, *482*, 216–220.

Jacob, F., & Bennett, M. L. (2020). Modeling neurological disease using human stem cell-derived microglia-like cells transplanted into rodent brains. *Lab Animal*, *49*, 49–51.

Jacob, F., Ming, G. L., & Song, H. (2020). Generation and biobanking of patient-derived glioblastoma organoids and their application in CAR T cell testing. *Nature Protocols*, *15*, 4000–4033.

Jacob, F., Pather, S. R., Huang, W. K., Zhang, F., Wong, S. Z. H., Zhou, H., et al. (2020). Human pluripotent stem cell-derived neural cells and brain organoids reveal SARS-CoV-2 Neurotropism predominates in choroid plexus epithelium. *Cell Stem Cell*.

Jacob, F., Salinas, R. D., Zhang, D. Y., Nguyen, P. T. T., Schnoll, J. G., Wong, S. Z. H., et al. (2020). A patient-derived glioblastoma organoid model and biobank recapitulates inter- and intra-tumoral heterogeneity. *Cell*, *180*, 188–204. e122.

Jo, J., Xiao, Y., Sun, A. X., Cukuroglu, E., Tran, H. D., Goke, J., et al. (2016). Midbrain-like organoids from human pluripotent stem cells contain functional dopaminergic and Neuromelanin-producing neurons. *Cell Stem Cell, 19*, 248–257.

Johnstone, M., Vasistha, N. A., Barbu, M. C., Dando, O., Burr, K., Christopher, E., et al. (2018). Reversal of proliferation deficits caused by chromosome 16p13.11 microduplication through targeting NFκB signaling: an integrated study of patient-derived neuronal precursor cells, cerebral organoids and in vivo brain imaging. *Molecular Psychiatry, 24*, 294–311.

Kadoshima, T., Sakaguchi, H., Nakano, T., Soen, M., Ando, S., Eiraku, M., et al. (2013). Self-organization of axial polarity, inside-out layer pattern, and species-specific progenitor dynamics in human ES cell-derived neocortex. *Proceedings of the National Academy of Sciences of the United States of America, 110*, 20284–20289.

Kahn, R. S., Sommer, I. E., Murray, R. M., Meyer-Lindenberg, A., Weinberger, D. R., Cannon, T. D., et al. (2015). Schizophrenia. *Nature reviews. Disease primers, 1*, 15067.

Karzbrun, E., Kshirsagar, A., Cohen, S. R., Hanna, J. H., & Reiner, O. (2018). Human brain organoids on a chip reveal the physics of folding. *Nature Physics, 14*, 515–522.

Kepecs, A., & Fishell, G. (2014). Interneuron cell types are fit to function. *Nature, 505*, 318–326.

Kim, H., Park, H. J., Choi, H., Chang, Y., Park, H., Shin, J., et al. (2019). Modeling G2019S-LRRK2 sporadic Parkinson's disease in 3D midbrain organoids. *Stem Cell Reports, 12*, 518–531.

Kim, H., Xu, R., Padmashri, R., Dunaevsky, A., Liu, Y., Dreyfus, C. F., et al. (2019). Pluripotent stem cell-derived cerebral organoids reveal human Oligodendrogenesis with dorsal and ventral origins. *Stem Cell Reports, 12*, 890–905.

Klaus, J., Kanton, S., Kyrousi, C., Ayo-Martin, A. C., Di Giaimo, R., Riesenberg, S., et al. (2019). Altered neuronal migratory trajectories in human cerebral organoids derived from individuals with neuronal heterotopia. *Nature Medicine, 25*, 561–568.

Kondo, T., Asai, M., Tsukita, K., Kutoku, Y., Ohsawa, Y., Sunada, Y., et al. (2013). Modeling Alzheimer's disease with iPSCs reveals stress phenotypes associated with intracellular Abeta and differential drug responsiveness. *Cell Stem Cell, 12*, 487–496.

Krieger, T. G., Tirier, S. M., Park, J., Jechow, K., Eisemann, T., Peterziel, H., et al. (2020). Modeling glioblastoma invasion using human brain organoids and single-cell transcriptomics. *Neuro-Oncology, 17*, 1138–1149.

Kumar, A., Singh, A., & Ekavali. (2015). A review on Alzheimer's disease pathophysiology and its management: An update. *Pharmacological Reports: PR, 67*, 195–203.

Kuwahara, A., Ozone, C., Nakano, T., Saito, K., Eiraku, M., & Sasai, Y. (2015). Generation of a ciliary margin-like stem cell niche from self-organizing human retinal tissue. *Nature Communications, 6*, 6286.

Lancaster, M. A., Corsini, N. S., Wolfinger, S., Gustafson, E. H., Phillips, A. W., Burkard, T. R., et al. (2017). Guided self-organization and cortical plate formation in human brain organoids. *Nature Biotechnology, 35*, 659–666.

Lancaster, M. A., & Knoblich, J. A. (2014). Generation of cerebral organoids from human pluripotent stem cells. *Nature Protocols, 9*, 2329–2340.

Lancaster, M. A., Renner, M., Martin, C. A., Wenzel, D., Bicknell, L. S., Hurles, M. E., et al. (2013). Cerebral organoids model human brain development and microcephaly. *Nature, 501*, 373–379.

Langston, J. W., Ballard, P., Tetrud, J. W., & Irwin, I. (1983). Chronic parkinsonism in humans due to a product of meperidine-analog synthesis. *Science, 219*, 979–980.

Ledur, P. F., Onzi, G. R., Zong, H., & Lenz, G. (2017). Culture conditions defining glioblastoma cells behavior: What is the impact for novel discoveries? *Oncotarget, 8*, 69185–69197.

Lee, C. T., Chen, J., Kindberg, A. A., Bendriem, R. M., Spivak, C. E., Williams, M. P., et al. (2017). CYP3A5 mediates effects of cocaine on human Neocorticogenesis: Studies using an in vitro 3D self-organized hPSC model with a single cortex-like unit. *Neuropsychopharmacology*, *42*, 774–784.

Lee, J., Kotliarova, S., Kotliarov, Y., Li, A., Su, Q., Donin, N. M., et al. (2006). Tumor stem cells derived from glioblastomas cultured in bFGF and EGF more closely mirror the phenotype and genotype of primary tumors than do serum-cultured cell lines. *Cancer Cell*, *9*, 391–403.

Levine, A. J., & Brivanlou, A. H. (2007). Proposal of a model of mammalian neural induction. *Developmental Biology*, *308*, 247–256.

Li, Q., & Barres, B. A. (2018). Microglia and macrophages in brain homeostasis and disease. *Nature Reviews. Immunology*, *18*, 225–242.

Li, Y., Muffat, J., Omer, A., Bosch, I., Lancaster, M. A., Sur, M., et al. (2017). Induction of expansion and folding in human cerebral organoids. *Cell Stem Cell*, *20*, 385–396. e383.

Li, R., Sun, L., Fang, A., Li, P., Wu, Q., & Wang, X. (2017). Recapitulating cortical development with organoid culture in vitro and modeling abnormal spindle-like (ASPM related primary) microcephaly disease. *Protein & Cell*, *8*, 823–833.

Lindborg, B. A., Brekke, J. H., Vegoe, A. L., Ulrich, C. B., Haider, K. T., Subramaniam, S., et al. (2016). Rapid induction of cerebral organoids from human induced pluripotent stem cells using a chemically defined hydrogel and defined cell culture medium. *Stem Cells Translational Medicine*, *5*, 970–979.

Linkous, A., Balamatsias, D., Snuderl, M., Edwards, L., Miyaguchi, K., Milner, T., et al. (2019). Modeling patient-derived glioblastoma with cerebral organoids. *Cell Reports*, *26*, 3203–3211. e3205.

Liu, Y., Liu, H., Sauvey, C., Yao, L., Zarnowska, E. D., & Zhang, S. C. (2013). Directed differentiation of forebrain GABA interneurons from human pluripotent stem cells. *Nature Protocols*, *8*, 1670–1679.

Lui, J. H., Hansen, D. V., & Kriegstein, A. R. (2011). Development and evolution of the human neocortex. *Cell*, *146*, 18–36.

Lun, M. P., Monuki, E. S., & Lehtinen, M. K. (2015). Development and functions of the choroid plexus-cerebrospinal fluid system. *Nature Reviews. Neuroscience*, *16*, 445–457.

Luo, C., Lancaster, M. A., Castanon, R., Nery, J. R., Knoblich, J. A., & Ecker, J. R. (2016). Cerebral organoids recapitulate Epigenomic signatures of the human fetal brain. *Cell Reports*, *17*, 3369–3384.

Madhavan, M., Nevin, Z. S., Shick, H. E., Garrison, E., Clarkson-Paredes, C., Karl, M., et al. (2018). Induction of myelinating oligodendrocytes in human cortical spheroids. *Nature Methods*, *15*, 700–706.

Malik, N., & Rao, M. S. (2013). A review of the methods for human iPSC derivation. *Methods in Molecular Biology*, *997*, 23–33.

Mansour, A. A., Goncalves, J. T., Bloyd, C. W., Li, H., Fernandes, S., Quang, D., et al. (2018). An in vivo model of functional and vascularized human brain organoids. *Nature Biotechnology*, *36*, 432–441.

Mariani, J., Coppola, G., Zhang, P., Abyzov, A., Provini, L., Tomasini, L., et al. (2015). FOXG1-dependent dysregulation of GABA/glutamate neuron differentiation in autism Spectrum disorders. *Cell*, *162*, 375–390.

Maroof, A. M., Keros, S., Tyson, J. A., Ying, S. W., Ganat, Y. M., Merkle, F. T., et al. (2013). Directed differentiation and functional maturation of cortical interneurons from human embryonic stem cells. *Cell Stem Cell*, *12*, 559–572.

Martin, K. R., Zhou, W., Bowman, M. J., Shih, J., Au, K. S., Dittenhafer-Reed, K. E., et al. (2017). The genomic landscape of tuberous sclerosis complex. *Nature Communications*, *8*, 15816.

Marton, R. M., & Pasca, S. P. (2020). Organoid and Assembloid Technologies for Investigating Cellular Crosstalk in human brain development and disease. *Trends in Cell Biology*, *30*, 133–143.

McQuade, A., Coburn, M., Tu, C. H., Hasselmann, J., Davtyan, H., & Blurton-Jones, M. (2018). Development and validation of a simplified method to generate human microglia from pluripotent stem cells. *Molecular Neurodegeneration*, *13*, 67.

Mellios, N., Feldman, D. A., Sheridan, S. D., Ip, J. P. K., Kwok, S., Amoah, S. K., et al. (2018). MeCP2-regulated miRNAs control early human neurogenesis through differential effects on ERK and AKT signaling. *Molecular Psychiatry*, *23*, 1051–1065.

Merkle, F. T., Maroof, A., Wataya, T., Sasai, Y., Studer, L., Eggan, K., et al. (2015). Generation of neuropeptidergic hypothalamic neurons from human pluripotent stem cells. *Development*, *142*, 633–643.

Ming, G. L., Tang, H., & Song, H. (2016). Advances in Zika virus research: Stem cell models, challenges, and opportunities. *Cell Stem Cell*, *19*, 690–702.

Monzel, A. S., Smits, L. M., Hemmer, K., Hachi, S., Moreno, E. L., van Wuellen, T., et al. (2017). Derivation of human midbrain-specific organoids from Neuroepithelial stem cells. *Stem Cell Reports*, *8*, 1144–1154.

Muffat, J., Li, Y., Yuan, B., Mitalipova, M., Omer, A., Corcoran, S., et al. (2016). Efficient derivation of microglia-like cells from human pluripotent stem cells. *Nature Medicine*, *22*, 1358–1367.

Muguruma, K., Nishiyama, A., Kawakami, H., Hashimoto, K., & Sasai, Y. (2015). Self-organization of polarized cerebellar tissue in 3D culture of human pluripotent stem cells. *Cell Reports*, *10*, 537–550.

Muguruma, K., Nishiyama, A., Ono, Y., Miyawaki, H., Mizuhara, E., Hori, S., et al. (2010). Ontogeny-recapitulating generation and tissue integration of ES cell-derived Purkinje cells. *Nature Neuroscience*, *13*, 1171–1180.

Munoz-Sanjuan, I., & Brivanlou, A. H. (2002). Neural induction, the default model and embryonic stem cells. *Nature Reviews. Neuroscience*, *3*, 271–280.

Muratore, C. R., Rice, H. C., Srikanth, P., Callahan, D. G., Shin, T., Benjamin, L. N., et al. (2014). The familial Alzheimer's disease APPV717I mutation alters APP processing and tau expression in iPSC-derived neurons. *Human Molecular Genetics*, *23*, 3523–3536.

Nakano, T., Ando, S., Takata, N., Kawada, M., Muguruma, K., Sekiguchi, K., et al. (2012). Self-formation of optic cups and storable stratified neural retina from human ESCs. *Cell Stem Cell*, *10*, 771–785.

Neul, J. L., Kaufmann, W. E., Glaze, D. G., Christodoulou, J., Clarke, A. J., Bahi-Buisson, N., et al. (2010). Rett syndrome: Revised diagnostic criteria and nomenclature. *Annals of Neurology*, *68*, 944–950.

Nicholas, C. R., Chen, J., Tang, Y., Southwell, D. G., Chalmers, N., Vogt, D., et al. (2013). Functional maturation of hPSC-derived forebrain interneurons requires an extended timeline and mimics human neural development. *Cell Stem Cell*, *12*, 573–586.

Ogawa, J., Pao, G. M., Shokhirev, M. N., & Verma, I. M. (2018). Glioblastoma Model Using Human Cerebral Organoids. *Cell Reports*, *23*, 1220–1229.

O'Neill, A. C., Kyrousi, C., Klaus, J., Leventer, R. J., Kirk, E. P., Fry, A., et al. (2018). A primate-specific isoform of PLEKHG6 regulates neurogenesis and neuronal migration. *Cell Reports*, *25*, 2729–2741. e2726.

Ormel, P. R., Vieira de Sa, R., van Bodegraven, E. J., Karst, H., Harschnitz, O., Sneeboer, M. A. M., et al. (2018). Microglia innately develop within cerebral organoids. *Nature Communications*, *9*, 4167.

Pamies, D., Barreras, P., Block, K., Makri, G., Kumar, A., Wiersma, D., et al. (2017). A human brain microphysiological system derived from induced pluripotent stem cells to study neurological diseases and toxicity. *ALTEX*, *34*, 362–376.

Pandya, H., Shen, M. J., Ichikawa, D. M., Sedlock, A. B., Choi, Y., Johnson, K. R., et al. (2017). Differentiation of human and murine induced pluripotent stem cells to microglia-like cells. *Nature Neuroscience*, *20*, 753–759.

Paredes, I., Himmels, P., & Ruiz de Almodovar, C. (2018). Neurovascular communication during CNS development. *Developmental Cell*, *45*, 10–32.

Parfitt, D. A., Lane, A., Ramsden, C. M., Carr, A. J., Munro, P. M., Jovanovic, K., et al. (2016). Identification and correction of mechanisms underlying inherited blindness in human iPSC-derived optic cups. *Cell Stem Cell*, *18*, 769–781.

Park, J. S., Davis, R. L., & Sue, C. M. (2018). Mitochondrial dysfunction in Parkinson's disease: New mechanistic insights and therapeutic perspectives. *Current Neurology and Neuroscience Reports*, *18*, 21.

Paşca, A. M., Park, J.-Y., Shin, H.-W., Qi, Q., Revah, O., Krasnoff, R., et al. (2019). Human 3D cellular model of hypoxic brain injury of prematurity. *Nature Medicine*, *25*, 784–791.

Pasca, A. M., Sloan, S. A., Clarke, L. E., Tian, Y., Makinson, C. D., Huber, N., et al. (2015). Functional cortical neurons and astrocytes from human pluripotent stem cells in 3D culture. *Nature Methods*, *12*, 671–678.

Pellegrini, L., Albecka, A., Mallery, D. L., Kellner, M. J., Paul, D., Carter, A. P., et al. (2020). SARS-CoV-2 infects the brain choroid plexus and disrupts the blood-CSF-barrier in human brain organoids. *Cell Stem Cell*, *27*, 951–961.

Pellegrini, L., Bonfio, C., Chadwick, J., Begum, F., Skehel, M., & Lancaster, M. A. (2020). Human CNS barrier-forming organoids with cerebrospinal fluid production. *Science*, *369*, eaaz5626.

Pham, M. T., Pollock, K. M., Rose, M. D., Cary, W. A., Stewart, H. R., Zhou, P., et al. (2018). Generation of human vascularized brain organoids. *Neuroreport*, *29*, 588–593.

Pulvers, J. N., Bryk, J., Fish, J. L., Wilsch-Brauninger, M., Arai, Y., Schreier, D., et al. (2010). Mutations in mouse Aspm (abnormal spindle-like microcephaly associated) cause not only microcephaly but also major defects in the germline. *Proceedings of the National Academy of Sciences of the United States of America*, *107*, 16595–16600.

Qian, X., Jacob, F., Song, M. M., Nguyen, H. N., Song, H., & Ming, G. L. (2018). Generation of human brain region-specific organoids using a miniaturized spinning bioreactor. *Nature Protocols*, *13*, 565–580.

Qian, X., Nguyen, H. N., Song, M. M., Hadiono, C., Ogden, S. C., Hammack, C., et al. (2016). Brain-region-specific organoids using mini-bioreactors for modeling ZIKV exposure. *Cell*, *165*, 1238–1254.

Qian, X., Song, H., & Ming, G. L. (2019). Brain organoids: Advances, applications and challenges. *Development*, *146*, dev166074.

Qian, X., Su, Y., Adam, C. D., Deutschmann, A. U., Pather, S. R., Goldberg, E. M., et al. (2020). Sliced human cortical organoids for modeling distinct cortical layer formation. *Cell Stem Cell*, *26*(766–781). e9.

Quadrato, G., Nguyen, T., Macosko, E. Z., Sherwood, J. L., Min Yang, S., Berger, D. R., et al. (2017). Cell diversity and network dynamics in photosensitive human brain organoids. *Nature*, *545*, 48–53.

Raja, W. K., Mungenast, A. E., Lin, Y. T., Ko, T., Abdurrob, F., Seo, J., et al. (2016). Self-organizing 3D human neural tissue derived from induced pluripotent stem cells recapitulate Alzheimer's disease phenotypes. *PLoS One*, *11*, e0161969.

Rajamani, U., Gross, A. R., Hjelm, B. E., Sequeira, A., Vawter, M. P., Tang, J., et al. (2018). Super-obese patient-derived iPSC hypothalamic neurons exhibit obesogenic signatures and hormone responses. *Cell Stem Cell*, *22*, 698–712. e699.

Ramani, A., Muller, L., Ostermann, P. N., Gabriel, E., Abida-Islam, P., Muller-Schiffmann, A., et al. (2020). SARS-CoV-2 targets neurons of 3D human brain organoids. *The EMBO Journal*, *39*, e106230.

Rubenstein, J. L., & Beachy, P. A. (1998). Patterning of the embryonic forebrain. *Current Opinion in Neurobiology*, *8*, 18–26.

Sakaguchi, H., Kadoshima, T., Soen, M., Narii, N., Ishida, Y., Ohgushi, M., et al. (2015). Generation of functional hippocampal neurons from self-organizing human embryonic stem cell-derived dorsomedial telencephalic tissue. *Nature Communications*, *6*, 8896.

Salero, E., & Hatten, M. E. (2007). Differentiation of ES cells into cerebellar neurons. *Proceedings of the National Academy of Sciences of the United States of America*, *104*, 2997–3002.

Schnoll, J. G., Temsamrit, B., Zhang, D., Song, H., Ming, G. L., & Christian, K. M. (2019). Evaluating neurodevelopmental consequences of perinatal exposure to antiretroviral drugs: Current challenges and new approaches. *Journal of Neuroimmune Pharmacology*. https://doi.org/10.1007/s11481-019-09880-z.

Scholpp, S., Wolf, O., Brand, M., & Lumsden, A. (2006). Hedgehog signalling from the zona limitans intrathalamica orchestrates patterning of the zebrafish diencephalon. *Development*, *133*, 855–864.

Schüle, B., Armstrong, D. D., Vogel, H., Oviedo, A., & Francke, U. (2008). Severe congenital encephalopathy caused by MECP2 null mutations in males: Central hypoxia and reduced neuronal dendritic structure. *Clinical Genetics*, *74*, 116–126.

Seo, J., Kritskiy, O., Watson, L. A., Barker, S. J., Dey, D., Raja, W. K., et al. (2017). Inhibition of p25/Cdk5 attenuates Tauopathy in mouse and iPSC models of frontotemporal dementia. *The Journal of Neuroscience*, *37*, 9917–9924.

Shi, Y., Sun, L., Wang, M., Liu, J., Zhong, S., Li, R., et al. (2020). Vascularized human cortical organoids (vOrganoids) model cortical development in vivo. *PLoS Biology*, *18*, e3000705.

Simons, M., & Nave, K. A. (2015). Oligodendrocytes: Myelination and axonal support. *Cold Spring Harbor Perspectives in Biology*, *8*, a020479.

Sison, S. L., O'Brien, B. S., Johnson, A. J., Seminary, E. R., Terhune, S. S., Ebert, A. D., et al. (2019). Human cytomegalovirus disruption of calcium signaling in neural progenitor cells and organoids. *Journal of Virology*, *93*. e00954-19.

Sloan, S. A., Andersen, J., Pasca, A. M., Birey, F., & Pasca, S. P. (2018). Generation and assembly of human brain region-specific three-dimensional cultures. *Nature Protocols*, *13*, 2062–2085.

Sloan, S. A., Darmanis, S., Huber, N., Khan, T. A., Birey, F., Caneda, C., et al. (2017). Human astrocyte maturation captured in 3D cerebral cortical spheroids derived from pluripotent stem cells. *Neuron*, *95*, 779–790. e776.

Smits, L. M., Reinhardt, L., Reinhardt, P., Glatza, M., Monzel, A. S., Stanslowsky, N., et al. (2019). Modeling Parkinson's disease in midbrain-like organoids. *NPJ Parkinson's Disease*, *5*, 5.

Sofroniew, M. V., & Vinters, H. V. (2010). Astrocytes: Biology and pathology. *Acta Neuropathologica*, *119*, 7–35.

Song, L., Yuan, X., Jones, Z., Vied, C., Miao, Y., Marzano, M., et al. (2019). Functionalization of brain region-specific spheroids with isogenic microglia-like cells. *Scientific Reports*, *9*, 11055.

Song, E., Zhang, C., Israelow, B., Lu-Culligan, A., Prado, A. V., Skriabine, S., et al. (2020). Neuroinvasion of SARS-CoV-2 in human and mouse brain. *bioRxiv*.

Spillantini, M. G., Crowther, R. A., Jakes, R., Hasegawa, M., & Goedert, M. (1998). Alpha-Synuclein in filamentous inclusions of Lewy bodies from Parkinson's disease and dementia with lewy bodies. *Proceedings of the National Academy of Sciences of the United States of America*, *95*, 6469–6473.

Spillantini, M. G., Schmidt, M. L., Lee, V. M., Trojanowski, J. Q., Jakes, R., & Goedert, M. (1997). Alpha-synuclein in Lewy bodies. *Nature*, *388*, 839–840.

Srikanth, P., Lagomarsino, V. N., Muratore, C. R., Ryu, S. C., He, A., Taylor, W. M., et al. (2018). Shared effects of DISC1 disruption and elevated WNT signaling in human cerebral organoids. *Translational Psychiatry, 8*, 77.

Stachowiak, E. K., Benson, C. A., Narla, S. T., Dimitri, A., Chuye, L. E. B., Dhiman, S., et al. (2017). Cerebral organoids reveal early cortical maldevelopment in schizophrenia—Computational anatomy and genomics, role of FGFR1. *Translational Psychiatry, 7*, 6.

Stefanis, L. (2012). α-Synuclein in Parkinson's disease. *Cold Spring Harbor Perspectives in Medicine, 2*, a009399.

Stern, C. D. (2001). Initial patterning of the central nervous system: How many organizers? *Nature Reviews. Neuroscience, 2*, 92–98.

Su, H. L., Muguruma, K., Matsuo-Takasaki, M., Kengaku, M., Watanabe, K., & Sasai, Y. (2006). Generation of cerebellar neuron precursors from embryonic stem cells. *Developmental Biology, 290*, 287–296.

Subramanian, L., Remedios, R., Shetty, A., & Tole, S. (2009). Signals from the edges: The cortical hem and antihem in telencephalic development. *Seminars in Cell & Developmental Biology, 20*, 712–718.

Sun, G., Chiuppesi, F., Chen, X., Wang, C., Tian, E., Nguyen, J., et al. (2020). Modeling human cytomegalovirus-induced microcephaly in human iPSC-derived brain organoids. *Cell Reports Medicine, 1*, 100002.

Sun, A. X., Yuan, Q., Fukuda, M., Yu, W., Yan, H., Lim, G. G. Y., et al. (2019). Potassium channel dysfunction in human neuronal models of Angelman syndrome. *Science, 366*, 1486–1492.

Suzuki, I. K., Gacquer, D., Van Heurck, R., Kumar, D., Wojno, M., Bilheu, A., et al. (2018). Human-specific NOTCH2NL genes expand cortical neurogenesis through Delta/Notch regulation. *Cell, 173*, 1370–1384. e1316.

Takahashi, K., Tanabe, K., Ohnuki, M., Narita, M., Ichisaka, T., Tomoda, K., et al. (2007). Induction of pluripotent stem cells from adult human fibroblasts by defined factors. *Cell, 131*, 861–872.

Tao, O., Shimazaki, T., Okada, Y., Naka, H., Kohda, K., Yuzaki, M., et al. (2010). Efficient generation of mature cerebellar Purkinje cells from mouse embryonic stem cells. *Journal of Neuroscience Research, 88*, 234–247.

Tavasoli, A. R., Parvaneh, N., Ashrafi, M. R., Rezaei, Z., Zschocke, J., & Rostami, P. (2018). Clinical presentation and outcome in infantile Sandhoff disease: A case series of 25 patients from Iranian neurometabolic bioregistry with five novel mutations. *Orphanet Journal of Rare Diseases, 13*, 130.

Velasco, S., Kedaigle, A. J., Simmons, S. K., Nash, A., Rocha, M., Quadrato, G., et al. (2019). Individual brain organoids reproducibly form cell diversity of the human cerebral cortex. *Nature, 570*, 523–527.

Wang, L., Egli, D., & Leibel, R. L. (2016). Efficient generation of hypothalamic neurons from human pluripotent stem cells. *Current Protocols in Human Genetics, 90*. 21 25 21-21 25 14.

Wang, P., Mokhtari, R., Pedrosa, E., Kirschenbaum, M., Bayrak, C., Zheng, D., et al. (2017). CRISPR/Cas9-mediated heterozygous knockout of the autism gene CHD8 and characterization of its transcriptional networks in cerebral organoids derived from iPS cells. *Molecular Autism, 8*, 11.

Wang, Y., Wang, L., Zhu, Y., & Qin, J. (2018). Human brain organoid-on-a-chip to model prenatal nicotine exposure. *Lab on a Chip, 18*, 851–860.

Watanabe, M., Buth, J. E., Vishlaghi, N., de la Torre-Ubieta, L., Taxidis, J., Khakh, B. S., et al. (2017). Self-organized cerebral organoids with human-specific features predict effective drugs to combat Zika virus infection. *Cell Reports, 21*, 517–532.

Watanabe, M., Kang, Y. J., Davies, L. M., Meghpara, S., Lau, K., Chung, C. Y., et al. (2012). BMP4 sufficiency to induce choroid plexus epithelial fate from embryonic stem cell-derived neuroepithelial progenitors. *The Journal of Neuroscience, 32*, 15934–15945.

Wen, Z., Christian, K. M., Song, H., & Ming, G. L. (2016). Modeling psychiatric disorders with patient-derived iPSCs. *Current Opinion in Neurobiology*, *36*, 118–127.

Wen, Z., Song, H., & Ming, G. L. (2017). How does Zika virus cause microcephaly? *Genes & Development*, *31*, 849–861.

Wilson, P. A., & Hemmati-Brivanlou, A. (1997). Vertebrate neural induction: Inducers, inhibitors, and a new synthesis. *Neuron*, *18*, 699–710.

Wilson, L., & Maden, M. (2005). The mechanisms of dorsoventral patterning in the vertebrate neural tube. *Developmental Biology*, *282*, 1–13.

Xiang, Y., Tanaka, Y., Cakir, B., Patterson, B., Kim, K. Y., Sun, P., et al. (2019). hESC-derived thalamic organoids form reciprocal projections when fused with cortical organoids. *Cell Stem Cell*, *24*, 487–497. e487.

Xiang, Y., Tanaka, Y., Patterson, B., Kang, Y. J., Govindaiah, G., Roselaar, N., et al. (2017). Fusion of regionally specified hPSC-derived organoids models human brain development and interneuron migration. *Cell Stem Cell*, *21*, 383–398. e387.

Xu, M., Lee, E. M., Wen, Z., Cheng, Y., Huang, W. K., Qian, X., et al. (2016). Identification of small-molecule inhibitors of Zika virus infection and induced neural cell death via a drug repurposing screen. *Nature Medicine*, *22*, 1101–1107.

Yagi, T., Ito, D., Okada, Y., Akamatsu, W., Nihei, Y., Yoshizaki, T., et al. (2011). Modeling familial Alzheimer's disease with induced pluripotent stem cells. *Human Molecular Genetics*, *20*, 4530–4539.

Ye, F., Kang, E., Yu, C., Qian, X., Jacob, F., Mao, M., et al. (2017). DISC1 regulates neurogenesis via modulating kinetochore attachment of Ndel1/Nde1 during mitosis. *Neuron*, *96*, 1041–1054.

Yoon, S.-J., Elahi, L. S., Paşca, A. M., Marton, R. M., Gordon, A., Revah, O., et al. (2018). Reliability of human cortical organoid generation. *Nature Methods*, *16*, 75–78.

Yoon, K. J., Ringeling, F. R., Vissers, C., Jacob, F., Pokrass, M., Jimenez-Cyrus, D., et al. (2017). Temporal control of mammalian cortical neurogenesis by m(6)A methylation. *Cell*, *171*, 877–889. e817.

Yoon, K. J., Song, G., Qian, X., Pan, J., Xu, D., Rho, H. S., et al. (2017). Zika-virus-encoded NS2A disrupts mammalian cortical neurogenesis by degrading Adherens junction proteins. *Cell Stem Cell*, *21*, 349–358. e346.

Young, R. A. (2011). Control of the embryonic stem cell state. *Cell*, *144*, 940–954.

Zhang, B., He, Y., Xu, Y., Mo, F., Mi, T., Shen, Q. S., et al. (2018). Differential antiviral immunity to Japanese encephalitis virus in developing cortical organoids. *Cell Death & Disease*, *9*, 719.

Zhang, D. Y., Song, H., & Ming, G. L. (2020). Modeling neurological disorders using brain organoids. *Seminars in Cell & Developmental Biology*.

Zhong, X., Gutierrez, C., Xue, T., Hampton, C., Vergara, M. N., Cao, L.-H., et al. (2014). Generation of three-dimensional retinal tissue with functional photoreceptors from human iPSCs. *Nature Communications*, *5*, 4047.

Zhou, T., Tan, L., Cederquist, G. Y., Fan, Y., Hartley, B. J., Mukherjee, S., et al. (2017). High-content screening in hPSC-neural progenitors identifies drug candidates that inhibit Zika virus infection in fetal-like organoids and adult brain. *Cell Stem Cell*, *21*, 274–283. e275.

Zhu, Y., Wang, L., Yin, F., Yu, Y., Wang, Y., Shepard, M. J., et al. (2017). Probing impaired neurogenesis in human brain organoids exposed to alcohol. *Integrative Biology (Camb)*, *9*, 968–978.

Kolibri 360 GmbH
Ferdinand-Jühlke-Straße 7
99095 Erfurt, Deutschland
produktsicherheit@kolibri360.de